Günter Mayer

Interval Analysis

De Gruyter Studies in Mathematics

Volume 65

Günter Mayer

Interval Analysis

and Automatic Result Verification

DE GRUYTER

Mathematics Subject Classification 2010
Primary: 65-01, 65Fxx, 65Gxx; Secondary: 65Y04, 68N15

Author
Prof. Dr. Günter Mayer
Universität Rostock
Institut für Mathematik
Ulmenstr. 69
Haus 3
18051 Rostock
guenter.mayer@uni-rostock.de

ISBN 978-3-11-050063-9
e-ISBN (PDF) 978-3-11-049946-9
e-ISBN (EPUB) 978-3-11-049805-9
Set-ISBN 978-3-11-049947-6
ISSN 0179-0986

Library of Congress Cataloging-in-Publication Data
A CIP catalog record for this book has been applied for at the Library of Congress.

Bibliographic information published by the Deutsche Nationalbibliothek
The Deutsche Nationalbibliothek lists this publication in the Deutsche Nationalbibliografie; detailed bibliographic data are available on the Internet at http://dnb.dnb.de.

Typesetting: PTP-Berlin, Protago-TEX-Production GmbH, Berlin
Printing and binding: CPI books GmbH, Leck
♾ Printed on acid-free paper
Printed in Germany

www.degruyter.com

To my wife Judith

Preface

The aim of this book is to give an introduction into the interesting field of interval computation. To this end we tried to keep the material self-contained, presenting also some basic facts of mathematics which normally are taught in elementary mathematics courses. Emphasis was laid on understanding and thorough proofs.

One of the basic duties of mathematicians is to provide solutions to a given mathematical problem. Sometimes they want to know whether such a solution exists, whether it is unique, and how it behaves when the input data vary. They may find and prove properties on it, and sometimes they have to construct it, or at least to approximate it. In the latter case bounds for the approximation error are interesting, and when using a computer in order to solve the problem, rounding errors can occur and should be estimated. All these topics form the starting point of interval computation. It deals with (primarily) compact intervals, and verifies in many cases existence and uniqueness of a solution even if the input data of the mathematical problem vary to a certain extent. The verification process should be realizable by means of a computer so that at the end a statement with mathematical rigor can be made. In order to illustrate the meaning of these sentences assume that one wants to know whether the nonlinear system

$$\begin{aligned} x &= 1 + 2\sin(x+y) + \cos(x-y) \\ y &= -1 + \sin(x-y) - 3\cos(x+y) \end{aligned}$$

has a solution. It is no problem to prove the existence of such a solution by traditional mathematical tools. (How would you do it?) For instance, define the vector function f by the right-hand side of this system and show by simple estimates that the continuous function f maps the rectangle $D = [-2, 4] \times [-5, 3]$ into itself so that Brouwer's fixed point theorem applies. It is the self-mapping property of Brouwer's fixed point theorem that can be verified very easily by means of interval analysis and a computer since interval arithmetic provides a simple tool to enclose the range of a function. Ranges themselves play a subordinate role in classical computation, but become important in interval computation. This means a change of paradigm when working with intervals.

Up to now we did not ask for an actual solution of our toy problem. Banach's fixed point theorem cannot be applied since f is not a contraction on D – at least with respect to any monotone norm (cf. p. 9), as the points $(\pi/4, \pi/4)$ and $(0, 0)$ show. Does Newton's method work? Who can trust a direct calculation on a computer when even a software like MATLAB returns a result for the expression $0.3 - (0.1 + 0.1 + 0.1)$ which

DOI 10.1515/9783110499469-001

differs from zero or even computes four different results for the equivalent expressions

$$\begin{aligned}&10^{18}+9-10^{18}-17+23,\\&10^{18}+23-10^{18}-17+9,\\&10^{18}+23-17+9-10^{18},\\&10^{18}-10^{18}+23-17+9.\end{aligned}$$

The verification of solutions combined with safe enclosures of the result although calculated by a computer is the philosophy of interval analysis and interval machine arithmetic. And if an interval algorithm cannot verify a solution, it should inform the user about this failure. That interval methods sometimes can also verify the nonexistence of a solution is an agreeable add-on. Miracles, however, cannot be expected. If a problem is ill-conditioned, interval analysis can perhaps verify a solution, but within bounds which may not be tight. Solve a linear system with the Hilbert matrix $H = (1/(i+j-1))$ of dimension 15 and a right-hand side $b = H \cdot (1, 1, \dots, 1)^T$ on a computer with a – traditional or interval – method of your choice and compare the result with the solution $(1, 1, \dots, 1)^T$. Unless you use an algebra system or increase the precision, you will see what is meant.

Nevertheless famous conjectures could be proved, assisted by a computer and interval analysis: Kepler conjectured that the faced-centered cubic packing is the densest packing of equally sized balls (cf. also Hilbert's 18th problem). In Hales [126] and Hales, Ferguson [127] the authors showed that Kepler was right. The double bubble conjecture says that the double bubble is the surface of the smallest area enclosing two equal, given volumes. Hass and Schlafly proved it in [137]. Smale's 14th problem asks if the structure of the solution to the Lorenz equations is that of a strange attractor. Tucker answered this question in [352]. More such problems, their solutions and the role of interval analysis in this connection are described in Frommer [104].

In Chapter 1 we start with our notation which is used throughout the book. We continue with basic facts from classical mathematics relevant for specific subjects later on. Some of the topics are metric spaces, norms, the mapping degree, fixed point theorems, normal forms of matrices, results on eigenvalues and eigenvectors, and on nonnegative matrices.

Chapter 2 introduces the basics of interval analysis. Here we define the interval arithmetic and present various auxiliary functions. In addition, we introduce elementary interval functions and comment on particularities on a machine interval arithmetic.

Chapter 3 generalizes the previous chapter to interval vectors and interval matrices. We also add some characterizations of special interval matrices like M-, H- and inverse stable matrices.

In Chapter 4 we define expressions and their interval arithmetic evaluation. We consider more general interval functions, among them centered forms and, in particular, the mean value form. We discuss the quality of approximation between range and interval expression, and show how higher order approximations can be achieved. As a basic property in interval analysis we mention P-contractions, which allow the application of Banach's fixed point theorem. As an important tool in verification algorithms we introduce the so-called ε-inflation and prove properties on it.

In Chapter 5 we consider interval linear systems. First we motivate such systems and present characterizations of the corresponding solution set S and particular subsets of it such as the symmetric or skew-symmetric solution set. In a section of its own we deal with the interval hull of S. Direct methods like the interval Gaussian algorithm and the interval Cholesky method are discussed extensively, followed by a section on iterative interval methods for interval linear systems.

Nonlinear equations and nonlinear systems are studied in Chapter 6. Here we start with the one-dimensional case and show how interval analysis can verify the existence and uniqueness of a solution and how it can also prove the nonexistence of it. To this end we use the one-dimensional interval Newton method and some modifications. Afterwards we present several methods for the multidimensional case, among them the multidimensional interval Newton method, the often used Krawczyk method, several modifications of it, the Hansen–Sengupta method, and some methods by Frommer, Lang et al. based on Miranda's theorem, on the mapping degree, and on Borsuk's theorem.

Chapter 7 is devoted to the algebraic eigenvalue problem and to related ones. Many of the methods to be presented can be considered as a quadratic system. Therefore, we verify solutions of such systems first. As a method for the verification of a simple eigenvalue of a matrix and a corresponding normalized eigenvector we consider a Krawczyk-like method. For symmetric matrices we remind of Lohner's method, which is a combination of the classical Jacobi method, the application of Gershgorin's theorem and an error estimate for eigenvectors found by Wilkinson. For double or nearly double eigenvalues we apply a method due to Alefeld and Spreuer. For the generalized eigenvalue problem we study a method by Alefeld similar to the Krawczyk-like method for simple eigenvalues, and a method originating from Behnke which is based on a theorem of Lehmann. We continue with ideas for verifying singular values, and we finish the chapter with some inverse eigenvalue problem.

In Chapter 8 we present automatic differentiation in the forward and the backward mode with which one can compute the Taylor coefficients at a given point for a function which is defined by a programmable expression. In particular, one can use these modes in order to compute the functional matrix needed for the Newton method in Chapter 6.

Our final Chapter 9 deals with complex intervals represented as rectangles or circular discs. For both kinds of complex intervals we introduce an arithmetic and prove some properties.

In an appendix we list some longer proofs which we skipped in our main text. So we recall Filippov's proof of the Jordan normal form and present two additional proofs of Brouwer's fixed point theorem which turns out to be very important in verification numerics. We add Ortega's and Rheinboldt's proof on the famous Kantorovich theorem, Forsythe's and Henrici's convergence proof on the row cyclic Jacobi method, and our own proof on Hladík's characterization of the symmetric solution set. For the interested reader we indicate a way to program a variety of elementary functions using the CORDIC algorithm. Finally, we give a rough overview on INTLAB, a specific software originating from Siegfried M. Rump. It is based on interval arithmetic and embedded in MATLAB. We used Version 9 of it in our book unless we refer to some older results in the literature which are obtained by means of the software package PASCAL-XSC written by Ulrich Kulisch and his staff.

We mention that a lot of our material is contained in the excellent text books of Alefeld, Herzberger [25, 26], Golub, van Loan [121], Heuser [146], Neumaier [257], Ortega, Rheinboldt [267], and Varga [356]. We do not always cite these sources when using results or proofs from there.

I want to thank everybody – students as well as colleagues – who helped to improve this book. Among them are Dipl-Math. Willi Gerbig, MSc Henning Schröder, Dr Ming Zhou (all three from the University of Rostock) who read several parts of the manuscript and made various suggestions. I am deeply indebted to my scientific teacher, Prof. Dr Götz Alefeld, Karlsruhe Institute of Technology (KIT), who undertook the very time-consuming and hard task of reading all the pages. He pointed out to me many inconsistencies, gaps, and errors, and made a variety of helpful suggestions. Over the decades he has not only accompanied and inspired my scientific life but he also never stopped gently reminding me to write this book. I am also grateful to the staff of de Gruyter for a painstaking job of copy editing.

Finally, I want to thank my wife to whom this book is dedicated. She gave the impulse to start my PhD thesis after having finished my studies in mathematics and physics, she always supported me selflessly in my profession, and from time to time she brought to my mind that sometimes there is a life beyond the university. A great big 'Thanks!' to you, Judith!

Rostock and Speyer, February 2017
Günter Mayer

Contents

Preface — vii

1 Preliminaries — 1
1.1 Notations and basic definitions — **1**
1.2 Metric spaces — **2**
1.3 Normed linear spaces — **7**
1.4 Polynomials — **12**
1.5 Zeros and fixed points of functions — **20**
1.6 Mean value theorems — **28**
1.7 Normal forms of matrices — **29**
1.8 Eigenvalues — **36**
1.9 Nonnegative matrices — **56**
1.10 Particular matrices — **65**

2 Real intervals — 75
2.1 Intervals, partial ordering — **75**
2.2 Interval arithmetic — **77**
2.3 Algebraic properties, χ-function — **80**
2.4 Auxiliary functions — **88**
2.5 Distance and topology — **93**
2.6 Elementary interval functions — **97**
2.7 Machine interval arithmetic — **99**

3 Interval vectors, interval matrices — 109
3.1 Basics — **109**
3.2 Powers of interval matrices — **114**
3.3 Particular interval matrices — **121**

4 Expressions, P-contraction, ε-inflation — 125
4.1 Expressions, range — **125**
4.2 P-contraction — **139**
4.3 ε-inflation — **150**

5 Linear systems of equations — 159
5.1 Motivation — **159**
5.2 Solution sets — **160**
5.3 Interval hull — **177**
5.4 Direct methods — **191**
5.5 Iterative methods — **239**

6 **Nonlinear systems of equations — 289**
6.1 Newton method – one-dimensional case — **289**
6.2 Newton method – multidimensional case — **297**
6.3 Krawczyk method — **303**
6.4 Hansen–Sengupta method — **320**
6.5 Further existence tests — **325**
6.6 Bisection method — **342**

7 **Eigenvalue problems — 345**
7.1 Quadratic systems — **346**
7.2 A Krawczyk-like method — **349**
7.3 Lohner method — **363**
7.4 Double or nearly double eigenvalues — **371**
7.5 The generalized eigenvalue problem — **379**
7.6 A method due to Behnke — **382**
7.7 Verification of singular values — **393**
7.8 An inverse eigenvalue problem — **398**

8 **Automatic differentiation — 407**
8.1 Forward mode — **407**
8.2 Backward mode — **413**

9 **Complex intervals — 417**
9.1 Rectangular complex intervals — **418**
9.2 Circular complex intervals — **423**
9.3 Applications of complex intervals — **429**

Final Remarks — 433

Appendix

A **Jordan normal form — 437**

B **Brouwer's fixed point theorem — 439**

C **Theorem of Newton–Kantorovich — 445**

D **The row cyclic Jacobi method — 451**

E **The CORDIC Algorithm — 457**

F The symmetric solution set — 463

G INTLAB — 469

Bibliography — 483

Symbol Index — 499

Author Index — 505

Subject Index — 509

1 Preliminaries

In this chapter we provide some definitions and theorems which we will frequently use throughout this book – often without mentioning them again.

1.1 Notations and basic definitions

We use the standard notation $\mathbb{Z}$ for the integer numbers, $\mathbb{N}$, for the positive integers, $\mathbb{N}_0$ for the nonnegative integers, $\mathbb{R}$ for the reals, $\mathbb{R}^+$ for the positive reals, $\mathbb{R}_0^+$ for the nonnegative reals and $\mathbb{C}$ for the complex numbers. If we mean $\mathbb{R}$ *or* $\mathbb{C}$ we use the symbol $\mathbb{K}$. We write $\mathbb{R}^n$ for the set of column vectors $x = (x_1, \dots, x_n)^T = (x_i)$ with n real components x_i, and $\mathbb{R}^{m\times n}$ for the set of $m \times n$ matrices $A = (a_{ij})$ with real entries a_{ij}, where the row index i ranges from 1 to m and the column index j ranges from 1 to n. We denote by $A_{*,j}$ the j-th column of A and by $A_{i,*}$ the i-th row. Sometimes we tacitly identify $\mathbb{R}^n$ with $\mathbb{R}^{n\times 1}$ and $\mathbb{R}$ with $\mathbb{R}^1$ or $\mathbb{R}^{1\times 1}$. Analogously we define $\mathbb{C}^n$, $\mathbb{C}^{m\times n}$, $\mathbb{K}^n$ and $\mathbb{K}^{m\times n}$. We denote the zero matrix by O and the identity matrix by I, sometimes also by I_n in order to indicate the dimension of I. In addition, we use the i-th column $e^{(i)}$ of I and the vector $e = (1, 1, \dots, 1)^T \in \mathbb{R}^n$. By $A^T = (c_{ij})$ we denote the transpose of a matrix A, i.e., $c_{ij} = a_{ji}$ for all indices i, j, and by $A^H = (c_{ij})$ we denote the conjugate transpose of A, i.e., $c_{ij} = \bar{a}_{ji}$. We call A symmetric if $A = A^T$, Hermitian if $A = A^H$, skew-symmetric if $A = -A^T$, orthogonal if A is regular and $A^{-1} = A^T$, and unitarian if A is regular and $A^{-1} = A^H$. Moreover, we use the symbol A^{-T} for $(A^{-1})^T$ and A^k, $k = 0, 1, \dots$, for the k-th power of $A \in \mathbb{K}^{n\times n}$, where $A^0 = I$. Since a diagonal matrix $D \in \mathbb{C}^{n\times n}$ is determined by its diagonal entries $d_{ii} = z_i$, $i = 1, \dots, n$, we often write $D = D_z = \operatorname{diag}(z_1, \dots, z_n)$.

A signature matrix D is a real diagonal matrix with diagonal entries $d_{ii} \in \{-1, 1\}$. Trivially, a signature matrix D satisfies $D = D^{-1}$ and $D^2 = I$. The determinant of a matrix $A \in \mathbb{K}^{n\times n}$ is denoted by $\det A$.

A square matrix $(a_{ij})_{i,j=i_1,\dots,i_k}$ with $1 \le i_1 < i_2 < \dots < i_k \le n$ is called a principal submatrix of $A \in \mathbb{K}^{n\times n}$ and its determinant a principal minor of order k. If, in addition, $i_j = j$, $j = 1, \dots, k$, then the principal submatrix is called a leading principal submatrix.

Generalizing this concept the determinant

$$\det(a_{ij})_{\substack{i=i_1,\dots,i_k\\ j=j_1,\dots,j_k}} \quad \text{with } 1 \le i_1 < i_2 < \dots < i_k \le n,\ 1 \le j_1 < j_2 < \dots < j_k \le n$$

is called a minor of order k.

Following good tradition we use $f(x)$ and $f(x_1, \dots, x_n)$ simultaneously for the values of functions $f\colon \mathbb{R}^n \to \mathbb{R}^m$ although somebody might claim that the vector argument in the second case should be transposed. We thus recall that $\mathbb{R}^n$ originally abbreviates the Cartesian product $X_{i=1}^n \mathbb{R} = \mathbb{R} \times \mathbb{R} \times \dots \times \mathbb{R}$ (n factors) with the n-tuples $(x_1, \dots, x_n)$ as elements, although we introduced it here as a set of column vectors.

DOI 10.1515/9783110499469-002

By interpreting $(x_1, \dots, x_n)$ as a finite sequence we generalize this notation by writing infinite sequences as (x_k) or (x^k). It will always be clear from the context whether (x^k) denotes the geometric sequence (with powers of x as members) or a general one.

For real intervals, brackets are used if the corresponding endpoint does not belong to the set and square brackets are used otherwise. An example is $(\underline{a}, \overline{a}] = \{ a \mid a \in \mathbb{R}, \underline{a} < a \leq \overline{a} \}$.

By δ_{ij} we denote the Kronecker symbol, i.e., $\delta_{ij} = 1$ if $i = j$ and $\delta_{ij} = 0$ if $i \neq j$. The symbols $o(\cdot)$, $O(\cdot)$ are the usual Landau symbols. The symbols := and =: indicate definition. They are not used consistently in this book. Empty sums like $\sum_{i>n}^{n} \dots$ are defined to be zero, empty products like $\prod_{i>n}^{n} \dots$ or x^0 are defined as one.

We often identify functions $f\colon D \to Y$ with their values $f(x)$ if the sets D and Y are clear or not of primary interest for the momentary problem. In the whole book we assume that the domain D of a function f is not empty. By $R_f(D)$ we denote the range of values of f, by $f^{-1}(T)$ the set of all elements of D whose image is contained in $T \subseteq Y$.

For $D \subseteq \mathbb{R}^m$, $Y \subseteq \mathbb{R}^n$ we define the set $C(D, Y) = C^0(D, Y)$ of continuous functions $f\colon D \to Y$, and the set $C^k(D, Y)$ of functions $f \in C(D, Y)$ which have continuous partial derivatives up to the order k. If in the latter case D is not open we assume that there is an open superset of D on which f is defined or to which f can be extended such that the required smoothness holds. If $Y = D$ or if Y is obvious we omit it and simply write $C(D)$, $C^0(D)$, $C^k(D)$.

1.2 Metric spaces

Metric spaces form one of the bases for reasonable working in the field of applied analysis. Aside from properties like continuity, convergence and compactness, which can be defined in a more general setting, they provide the definition of a distance which allows to consider errors without having any algebraic structure on the underlying set.

Definition 1.2.1. Let S be any nonempty set and let $m\colon S \times S \to \mathbb{R}$ be a function with the following properties for all $x, y, z \in S$.

$$m(x, y) \geq 0 \quad \text{with } m(x, y) = 0 \text{ if and only if } x = y \tag{1.2.1}$$

$$m(x, y) = m(y, x) \tag{1.2.2}$$

$$m(x, y) \leq m(x, z) + m(z, y) \tag{1.2.3}$$

Then m is called a metric on S and (S, m) is called a metric space. If the metric is obvious and of no particular importance, we also call S a metric space.

Property (1.2.1) is called the definiteness of a metric, (1.2.2) the symmetry and (1.2.3) the triangular inequality.

Simple examples are the metric spaces (S, m) with $S = \mathbb{R}$ and $m(x, y) = |x - y|$ or $S = \mathbb{R}^n$ and $m(x, y) = \max\{ |x_i - y_i| \mid i = 1, \ldots, n\}$; cf. Exercise 1.2.2. By reasons which will be explained in the next section, we write $(\mathbb{R}, |\cdot|)$, and $(\mathbb{R}^n, \|\cdot\|_\infty)$, respectively, for these metric spaces. The discrete metric m defined by $m(x, y) = 0$ if $x = y$, and $m(x, y) = 1$ otherwise, shows that each nonempty set S can be equipped with a metric.

Each metric space (S, m) has a topological structure in a canonical way. We can see this from the following definition.

Definition 1.2.2. Let (S, m) be a metric space. The set

$$B(x, r) = \{z \mid z \in S,\ m(z, x) < r\}, \quad x \in S,\ 0 < r \in \mathbb{R},$$

is called an open ball with center (or midpoint) x and radius r.

The set

$$\overline{B}(x, r) = \{z \mid z \in S,\ m(z, x) \le r\}, \quad x \in S,\ 0 \le r \in \mathbb{R},$$

is called a closed ball with center (or midpoint) x and radius r.

A subset T of S is defined to be open if T contains an open ball $B(x, r_x)$ for each element $x \in T$, where $r_x > 0$ may depend on x. The set T is named closed if $S \setminus T$ is open. The boundary ∂T of T consists of those elements x of S for which each open ball $B(x, r)$ contains at least one element of T and at least one of $S \setminus T$. The set T is called bounded if it is contained in some open ball. The closure $\overline{T}$ of T is the intersection of all closed supersets of T.

Using the triangular inequality of a metric it can immediately be seen that an open ball is an open set in the above-mentioned sense and a closed ball is closed. A subset T of S is closed if and only if it contains its boundary ∂T. The sets $\emptyset$ and S are open and closed at the same time.

According to Definition 1.2.2 the set of open sets forms a topology $\mathcal{T}$ on S. This is a family of subsets of S (i.e., a subset of the power set of S) with the following properties: The empty set $\emptyset$ and the set S itself are elements of $\mathcal{T}$, the intersection of two elements of $\mathcal{T}$ and the union of arbitrary many elements of $\mathcal{T}$ are elements of $\mathcal{T}$. Notice that for a topology no metric is needed. The topology defined by the open sets in Definition 1.2.2 is called a metric topology.

The closure $\overline{T}$ of a subset T of S is a closed set. Using De Morgan's rule this can be seen by considering the complement of $\overline{T}$ which is the union of open sets. Notice that the closure $\overline{B(x, r)}$ of an open ball $B(x, r)$ is contained in the corresponding closed ball $\overline{B}(x, r)$. But it does not necessarily coincide with it; cf. Exercise 1.2.3.

Now we introduce convergence in metric spaces.

Definition 1.2.3. Let (S, m) be a metric space and let (x_k) be a sequence in S.

We define (x_k) to be convergent to some limit $x \in S$ if $\lim_{k\to\infty} m(x_k, x) = 0$. In this case we write $\lim_{k\to\infty} x_k = x$.

The sequence (x_k) is a Cauchy sequence if for each positive real number ε there is an index $k_0 = k_0(\varepsilon)$ such that $m(x_k, x_l) < \varepsilon$ for all indices $k, l \ge k_0$.

If each Cauchy sequence in S is convergent we call (S, m) a complete metric space. A sequence is bounded if the set of its members is bounded.

We leave it to the reader to show that each subsequence of a convergent sequence is convergent and has the same limit. Moreover, each convergent sequence is a Cauchy sequence and bounded. The example (S, m) with $S = (0, 1] \subseteq \mathbb{R}$, $m(x, y) = |x - y|$, $x_k = 1/k$ shows however, that metric spaces are not necessarily complete ones. They can be completed as is shown, e.g., in Heuser [146].

For complete metric spaces Banach's famous fixed point theorem is valid.

Theorem 1.2.4 (Banach). *Let (S, m) be a complete metric space and let T be a non-empty closed subset of S. Let $f : T \to T$ be a contraction on T, i.e., there is a constant $\alpha \in [0, 1)$, the so-called contraction constant, such that*

$$m(f(x), f(y)) \le \alpha\, m(x, y) \quad \text{for all } x, y \in T. \tag{1.2.4}$$

Then for each sequence $(x_k)_{k=0}^{\infty}$ with

$$x_{k+1} = f(x_k), \quad k = 0, 1, \ldots, \quad x_0 \in T, \tag{1.2.5}$$

the following properties hold.

(a) *The sequence (x_k) is well-defined (i.e., $x_k \in T$) and converges to a fixed point x^* of f which is unique in T.*

(b) *There is an a priori error estimate*

$$m(x^k, x^*) \le \frac{\alpha^k}{1-\alpha} m(x^1, x^0), \quad k = 1, 2, \ldots, \tag{1.2.6}$$

and an a posteriori error estimate

$$m(x^k, x^*) \le \frac{\alpha}{1-\alpha} m(x^k, x^{k-1}), \quad k = 1, 2, \ldots .$$

Proof. (a) Since by assumption $R_f(T) \subseteq T$ the sequence (x_k) is well-defined. From (1.2.4) we get the estimate

$$m(x_{k+1}, x_k) = m(f(x_k), f(x_{k-1})) \le \alpha m(x_k, x_{k-1}) \le \cdots \le \alpha^k m(x_1, x_0).$$

Together with the triangular inequality this implies

$$\begin{aligned} m(x_{k+d}, x_k) &\le m(x_{k+d}, x_{k+d-1}) + m(x_{k+d-1}, x_{k+d-2}) + \cdots + m(x_{k+1}, x_k) \\ &= \sum_{i=0}^{d-1} m(x_{k+i+1}, x_{k+i}) \le \sum_{i=0}^{d-1} \alpha^{k+i} m(x_1, x_0) \\ &= \alpha^k \frac{1-\alpha^d}{1-\alpha} m(x_1, x_0) \le \frac{\alpha^k}{1-\alpha} m(x_1, x_0). \end{aligned} \tag{1.2.7}$$

This estimate proves (x_k) to be a Cauchy sequence. By completeness it converges to some limit $x^* \in S$. Since T is closed, x^* lies in T. From (1.2.4) with $x = x_k$ and $y = x^*$ we get $\lim_{k\to\infty} f(x_k) = f(x^*)$ and $k \to \infty$ in (1.2.5) proves $x^* = f(x^*)$. If there is another fixed point $y^* \in T$, $y^* \neq x^*$, we obtain the contradiction

$$m(x^*, y^*) = m(f(x^*), f(y^*)) \le \alpha m(x^*, y^*) < m(x^*, y^*).$$

Therefore, the fixed point x^* is unique and so is the limit of the sequences (x_k).

(b) Let d tend to infinity in (1.2.7) in order to get the a priori estimate. For the a posteriori estimate choose $k = 1$ in (1.2.6), then replace x_0, x_1 by x_{k-1}, x_k. □

Next we define compactness in metric spaces.

Definition 1.2.5. Let (S, m) be a metric space and $T \subseteq S$. If each sequence out of T contains a convergent subsequence with limit in T, then we call T compact.

In metric spaces the following equivalence can be shown.

Theorem 1.2.6 (Heine–Borel property). *Let (S, m) be a metric space and $T \subseteq S$. Then T is compact if and only if out of each set of open sets whose union contains T one can find finitely many such sets whose union already contains T. ('Each open covering of T contains a finite subcovering of T.')*

The proof of this theorem can be found in Ahlfors [4]. We leave it to the reader to show that a compact set is closed and bounded; cf. Exercise 1.2.6. The converse is, however, not true in general. Notice that we defined compactness only in metric spaces. In a more general setting one introduces compactness via the Heine–Borel property and calls that of Definition 1.2.5 sequentially compact.

Now we address continuity.

Definition 1.2.7. Let (S, m), (S', m') be metric spaces and $f : S \to S'$. Then f is called continuous in $x^* \in S$ if $\lim_{k\to\infty} f(x_k) = f(x^*)$ holds for all sequences (x_k) with $\lim_{k\to\infty} x_k = x^*$. If f is continuous in all elements of S, then we shortly say that f is continuous (in S).

It is easy to see that contractive functions are continuous. We leave it to the reader to show (mostly by contradiction) that the following theorem holds.

Theorem 1.2.8. *Let (S, m), (S', m') be metric spaces and $f : S \to S'$. Then the following statements are equivalent.*

(a) *The function f is continuous in S.*
(b) *For each element $x^* \in S$ and each positive real number ε there is a positive real number δ which may depend on x^* and ε such that $m(x, x^*) < \delta$ implies $m'(f(x), f(x^*)) < \varepsilon$.*
(c) *For each open subset U' of S' the set $f^{-1}(U')$ is open.*
(d) *For each closed subset U' of S' the set $f^{-1}(U')$ is closed.*

If δ in Theorem 1.2.8 does not depend on x^* but only on ε, then we call f *uniformly* continuous on S. Trivially, each uniformly continuous function is continuous but not vice versa as the example $(\mathbb{R} \setminus \{0\}, |\cdot|)$, $f(x) = 1/x$ shows.

Our next theorem combines continuity and compactness.

Theorem 1.2.9. *Let (S, m), (S', m') be metric spaces, $T \subseteq S$ compact and $f: S \to S'$ continuous.*

(a) *The image $R_f(T)$ is compact.*
(b) *The function f is uniformly continuous on T.*
(c) *If $(S', m') = (\mathbb{R}, |\cdot|)$, then there are elements $\underline{x}, \overline{x} \in T$ such that $f(\underline{x}) = \min_{x \in T} f(x)$ and $f(\overline{x}) = \max_{x \in T} f(x)$.*

The simple proof is left to the reader.

We conclude this section by the definition of Lipschitz continuity.

Definition 1.2.10. Let (S, m), (S', m') be metric spaces and let $f: S \to S'$. If there is a constant $\alpha \geq 0$ such that

$$m(f(x), f(y)) \leq \alpha\, m(x, y)$$

holds for all $x, y \in S$, then f is called Lipschitz continuous with the Lipschitz constant α.

Trivially, contractions are Lipschitz continuous but not vice versa. Lipschitz continuous functions are uniformly continuous; cf. Exercise 1.2.8.

Exercises

Ex. 1.2.1. Let (S, m) be a metric space. Show that $(S, \alpha m)$ with $\alpha \in \mathbb{R}^+$ is a metric space, too.

Ex. 1.2.2. Show that $m(x, y) = |x - y|$ is a metric on $\mathbb{R}$ and $m(x, y) = \max_{1 \leq i \leq n} |x_i - y_i|$ is a metric on $\mathbb{R}^n$.

Ex. 1.2.3. Show that $m(x, y) = 0$ if $x = y$ and $m(x, y) = 1$ otherwise is a metric (the so-called discrete metric) on any nonempty set S. What do the open balls look like, what do the closed ones look like? Which are the open resp. closed sets?

If S contains at least two elements show that $B(x, 1) = \overline{B(x, 1)} \subset \overline{B}(x, 1) = S$ holds, where B is defined in Definition 1.2.2.

Ex. 1.2.4. Show that $q([x], [y]) = \max\{|\underline{x} - \underline{y}|, |\overline{x} - \overline{y}|\}$ is a metric on the set $\mathbb{IR}$ of all nonempty, closed, bounded intervals $[a] = [\underline{a}, \overline{a}]$ of $\mathbb{R}$.

Ex. 1.2.5. Let S be the set of all nonempty compact subsets of $\mathbb{R}$ and let $X, Y \in S$. Show that

$$q(X, Y) = \max\{\sup_{y\in Y}\inf_{x\in X}|x-y|,\ \sup_{x\in X}\inf_{y\in Y}|x-y|\}$$

is a metric on S, the so-called Hausdorff metric.

Ex. 1.2.6. Show that in a metric space a compact set is closed and bounded.

Let ℓ^2 be the set of infinite sequences (a_i), $a_i \in \mathbb{R}$, which satisfy $(\sum_{i=1}^{\infty} a_i^2)^{1/2} < \infty$. Show that $m(x, y) = (\sum_{i=1}^{\infty}(x_i - y_i)^2)^{1/2}$ is a metric on ℓ^2.

Fix $a = (a_i) \in \ell^2$ and define A to be the set with the elements

$$x^{(k)} = (a_1, \dots, a_{k-1}, a_k + 1, a_{k+1}, \dots), \quad k = 1, 2, \dots.$$

Show that A is a closed and bounded subset of ℓ^2 but not a compact one.

Ex. 1.2.7. Show that the function $y = f(x) = \sqrt{x}$ is not Lipschitz continuous on $[0, 1]$.

Ex. 1.2.8. Show that Lipschitz continuous functions are uniformly continuous.

1.3 Normed linear spaces

Whereas for metric spaces no algebraic structure is assumed, we start now with linear spaces V over the scalar field $\mathbb{R}$ or $\mathbb{C}$ with the usual symbols for the operations. Recall the symbol $\mathbb{K} \in \{\mathbb{R}, \mathbb{C}\}$ from Section 1.1. We first define a mapping $\|\cdot\|$ similar to a metric which in fact induces a metric by using one of the algebraic operations.

Definition 1.3.1. Let V be a linear space over $\mathbb{K}$ and let $\|\cdot\|: V \to \mathbb{R}$ be a mapping with the following properties for all $x, y \in V$ and all $\alpha \in \mathbb{K}$.

$$\|x\| \ge 0 \quad \text{with } \|x\| = 0 \text{ if and only if } x = 0 \tag{1.3.1}$$

$$\|\alpha x\| = |\alpha|\,\|x\| \tag{1.3.2}$$

$$\|x + y\| \le \|x\| + \|y\| \tag{1.3.3}$$

Then $\|\cdot\|$ is called a (vector) norm on V and $(V, \|\cdot\|)$ is called a normed linear space. If the norm is obvious and of no particular importance, we also call V a normed linear space.

Property (1.3.1) is called definiteness, (1.3.2) homogeneity and (1.3.3) triangular inequality.

Replacing x in (1.3.3) by $x - y$ yields $\|x\| - \|y\| \le \|x - y\|$; interchanging the roles of x and y and combining the result with the preceding inequality leads to

$$\big|\,\|x\| - \|y\|\,\big| \le \|x - y\|,$$

which shows that the norm is a continuous function.

An example of a normed linear space is $(\mathbb{R}, |\cdot|)$ with the usual absolute value $|\cdot|$. The linear space $\mathbb{R}^n$ as well as $\mathbb{C}^n$ can be normed by the maximum norm $\|x\|_\infty = \max\{|x_i| \mid i = 1, \dots, n\}$ and by the l_p norms $\|x\|_p = \left(\sum_{i=1}^n |x_i|^p\right)^{1/p}$, $p \ge 1$, among them the l_1-norm $\|x\|_1 = \sum_{i=1}^n |x_i|$ and the Euclidean norm $\|x\|_2 = \left(\sum_{i=1}^n |x_i|^2\right)^{1/2}$.

It is easy to see that $m(x, y) = \|x - y\|$ is a metric. In this way each normed linear space is a metric space in a canonical way. If we speak of convergence or continuity in a normed linear space, we always use this terminology with respect to the canonical metric. Normed linear spaces which are complete with respect to this metric are called Banach spaces.

A particular class of normed linear spaces are the finite dimensional ones since for them each norm generates the same topology, i.e., the same open sets, as we are going to see. To this end we need the following lemma which also shows that in finite dimensional spaces convergence is equivalent with componentwise convergence.

Lemma 1.3.2. *Let $b^1, \dots, b^n$ be linearly independent elements of a vector space $(V, \|\cdot\|)$ over $\mathbb{K}$. Then there is a positive constant μ such that*

$$|\alpha_1| + \cdots + |\alpha_n| \le \mu \, \|\alpha_1 b^1 + \cdots + \alpha_n b^n\| \tag{1.3.4}$$

for all numbers $\alpha_1, \dots, \alpha_n \in \mathbb{K}$.

Proof. The equality holds with any $\mu > 0$ if all α_i are zero. Therefore, we assume that at least one of them differs from zero. Let

$$\gamma = \inf_{\sum_{i=1}^n |\alpha_i| = 1} \|\alpha_1 b^1 + \cdots + \alpha_n b^n\|.$$

Then $\gamma > 0$, and there is a vector sequence $(y^{(k)})$ with

$$y^{(k)} = \sum_{i=1}^n \alpha_i^{(k)} b^i, \quad \sum_{i=1}^n |\alpha_i^{(k)}| = 1, \quad \text{and} \quad \lim_{k\to\infty} \|y^{(k)}\| = \gamma.$$

Since the numbers $\alpha_i^{(k)}$, $k = 1, 2, \dots$ are contained in the compact set $\overline{B}(0, 1)$ (with respect to $(\mathbb{K}, |\cdot|)$) there is a subsequence $(\alpha_i^{(k_l)})_{l=1}^\infty$ which converges to some β_i. This holds for each $i = 1, \dots, n$. W.l.o.g. we can assume that k_l is independent of i. Then $\sum_{i=1}^n |\beta_i| = 1$, hence $x = \sum_{i=1}^n \beta_i b^i \ne 0$ and $\|x\| > 0$. With $\lim_{k\to\infty} \|y^{(k_l)}\| = \|x\| = \gamma$ we get

$$1 \le \frac{1}{\gamma} \|\alpha_1 b^1 + \cdots + \alpha_n b^n\| \quad \text{for} \ \sum_{i=1}^n |\alpha_i| = 1.$$

Replace α_i by $\alpha_i / \sum_{i=1}^n |\alpha_i|$ in order to end up with (1.3.4) choosing $\mu = 1/\gamma > 0$. □

Definition 1.3.3. Two norms $\|\cdot\|$ and $\|\cdot\|'$ on a linear space V are called equivalent if there are positive constants α, β such that

$$\alpha\|x\| \le \|x\|' \le \beta\|x\|$$

holds for all $x \in V$.

Equivalent norms have the advantage that they generate the same open sets and therefore the same topology. Sequences which are convergent with respect to one norm are automatically convergent with respect to any other equivalent norm. For finite dimensional linear spaces, and, therefore, for $\mathbb{R}^n$, $\mathbb{C}^n$, we are going to show that all norms are equivalent. This proves our remark preceding Lemma 1.3.2.

Theorem 1.3.4. *All norms on a finite dimensional vector space V are equivalent.*

Proof. Let $\{b^1, \ldots, b^n\}$ be a basis of V and

$$x = \sum_{i=1}^{n} \alpha_i b^i, \quad \gamma = \max_{1\le i\le n} \|b^i\|, \quad \gamma' = \max_{1\le i\le n} \|b^i\|',$$

where $\|\cdot\|$, $\|\cdot\|'$ denote any two norms on V.

Using Lemma 1.3.2 with the constants μ and μ', respectively, we get

$$\begin{aligned}\|x\| &\le \sum_{i=1}^{n} |\alpha_i|\,\|b^i\| \le \gamma \sum_{i=1}^{n} |\alpha_i| \le \gamma\mu' \|x\|' \\ &\le \gamma\mu' \sum_{i=1}^{n} |\alpha_i|\,\|b^i\|' \le \gamma\gamma'\mu' \sum_{i=1}^{n} |\alpha_i| \le \gamma\gamma'\mu\mu' \|x\|.\end{aligned}$$

The equivalence follows with $\alpha = 1/(\gamma\mu')$, $\beta = \gamma'\mu$. □

Now we introduce a partial ordering on $\mathbb{R}^n$.

Definition 1.3.5.
(a) For $x \in \mathbb{K}^n$ we define the absolute value $|x| = (|x_i|) \in \mathbb{R}^n$.
(b) For $x, y \in \mathbb{R}^n$ we define the partial ordering $x \le y$ entrywise by $x_i \le y_i$, $i = 1, \ldots, n$. Similarly, we define $x \ge y$. If strict inequality holds for all components we write $x < y$, and $x > y$, respectively. If $x \ge 0$, we call x nonnegative, and positive if $x > 0$.

The relation '$\le$' is a partial ordering in $\mathbb{R}^n$ but not a total one as can be seen by the vectors $(1, 2)^T, (2, 1)^T \in \mathbb{R}^2$, which are not comparable with respect to '$\le$'.

We introduce this new terminology into norms.

Definition 1.3.6. A vector norm $\|\cdot\|$ on $\mathbb{K}^n$ is called
(a) Absolute if $\||x|\| = \|x\|$ holds for all $x \in \mathbb{K}^n$.
(b) Monotone if $|x| \le |y|$ implies $\|x\| \le \|y\|$ for all $x, y \in \mathbb{K}^n$.

The maximum norm and the l_p-norms are absolute and monotone norms. The norm $\|x\| = \|Ax\|_\infty$ with $A = \left(\begin{smallmatrix} 2 & -1 \\ 1 & 0 \end{smallmatrix}\right)$ is neither absolute nor monotone as can be demonstrated by the vectors $x = (1, -2)^T$, and $x = (0, 3)^T$, $y = (2, 4)^T$, respectively.

The next theorem shows that both properties are equivalent.

Theorem 1.3.7. *A vector norm $\|\cdot\|$ on $\mathbb{K}^n$ is absolute if and only if it is monotone.*

Proof. Let $\|\cdot\|$ be monotone and apply $\|\cdot\|$ to $-|x| \le x$ and to $x \le |x|$ in order to prove $\|\cdot\|$ to be absolute.

For the converse let $\|\cdot\|$ be absolute and $|x| \le |y|$, $x, y \in \mathbb{K}^n$. Consider $x^{(1)}(s) = (s, x_2, \ldots, x_n)^T$, $s \in \mathbb{K}$. For $|s| \le |t|$ there is a real number $\alpha \in [0, 1]$ such that $x^{(1)}(|s|) = \alpha x^{(1)}(-|t|) + (1-\alpha)x^{(1)}(|t|)$, whence $\|x^{(1)}(s)\| = \|x^{(1)}(|s|)\| \le \alpha\|x^{(1)}(-|t|)\| + (1-\alpha)\,\|x^{(1)}(|t|)\| = \|x^{(1)}(|t|)\| = \|x^{(1)}(t)\|$. Choose $s = x_1$ and $t = y_1$ which implies $x \le x^{(1)}(y_1) = (y_1, x_2, \ldots, x_n)^T$. Repeat the steps with $x^{(2)}(s) = (y_1, s, x_3, \ldots, x_n)^T$, $s \in \mathbb{K}$. Choose $s = x_2$, $t = y_2$, whence $x^{(1)}(y_1) \le x^{(2)}(y_2)$. An inductive argument finally generates the components of y successively and concludes the proof. □

The linear space of $n \times n$ matrices can easily be equipped with a vector norm by considering $\mathbb{K}^{n^2}$. It is usual, however, to add a fourth condition on matrix norms $\|\cdot\|$, namely the submultiplicativity

$$\|A \cdot B\| \le \|A\| \cdot \|B\|. \tag{1.3.5}$$

This property is required for matrix norms in the whole book. If $\|\cdot\|$ is a vector norm of $\mathbb{K}^n$ one obtains immediately a matrix norm for $A \in \mathbb{K}^{n\times n}$ by defining

$$\|A\| = \sup_{x\in\mathbb{K}^n\setminus\{0\}} \frac{\|Ax\|}{\|x\|}. \tag{1.3.6}$$

In this case the matrix norm is called the operator norm generated by the vector norm $\|\cdot\|$. Notice that we use the same symbol both for the vector norm and the corresponding operator norm. Since in finite dimensional linear spaces 'compact' and 'closed and bounded' are equivalent and since $x/\|x\|$ is a unit vector, one can replace the supremum on the right-hand side of (1.3.6) by the maximum. Moreover, it is sufficient to consider the right-hand side only for elements of the unit sphere.

The reader should show that the maximum norm $\|\cdot\|_\infty$ for vectors generates the row sum norm $\|A\|_\infty = \max\{\sum_{j=1}^n |a_{ij}| \mid i = 1, \ldots, n\}$, the l_1-norm induces the column sum norm $\|A\|_1 = \max\{\sum_{i=1}^n |a_{ij}| \mid j = 1, \ldots, n\}$, and the Euclidean norm $\|\cdot\|_2$ yields the spectral norm $\|A\|_2 = \sqrt{\rho(A^H A)}$, cf. Exercise 1.3.2, where here and in the sequel

$$\rho(A) = \max\{|\lambda| \mid \lambda \text{ eigenvalue of } A\}$$

denotes the spectral radius of a square matrix $A \in \mathbb{K}^{n\times n}$. Obviously, $\|A\|_2 = \sqrt{\rho(A^T A)}$ for $A \in \mathbb{R}^{n\times n}$. If S is a regular matrix then $\|x\|_S = \|S^{-1}x\|_\infty$ is a vector norm which generates the operator norm $\|A\|_S = \|S^{-1}AS\|_\infty$; cf. Exercise 1.3.3. Since each operator norm fulfills $\|I\| = 1$, it is clear that the Frobenius norm $\|A\|_F = \left(\sum_{i=1}^n \sum_{j=1}^n |a_{ij}|^2\right)^{1/2}$ is a matrix norm, but not an operator norm. Later on we will frequently use a vector norm $\|\cdot\|_S$ with a particular matrix S which is generated by means of a positive vector.

Definition 1.3.8. With $0 < z \in \mathbb{R}^n$ define the diagonal matrix $D_z = \operatorname{diag}(z_1, \ldots, z_n) \in \mathbb{R}^{n\times n}$. Then

$$\|x\|_z = \|D_z^{-1}x\|_\infty \tag{1.3.7}$$

and

$$\|A\|_z = \|D_z^{-1}AD_z\|_\infty. \tag{1.3.8}$$

Clearly, the vector norm $\|\cdot\|_z$ is absolute and monotone with $\|z\|_z = 1$, and the matrix norm $\|\cdot\|_z$ is the associated operator norm.

It is easy to see that a general operator norm satisfies the inequality

$$\|A \cdot x\| \leq \|A\| \cdot \|x\|. \tag{1.3.9}$$

This leads to the following definition.

Definition 1.3.9. Let $\|\cdot\|'$ be a vector norm on $\mathbb{K}^n$ and $\|\cdot\|$ a matrix norm on $\mathbb{K}^{n\times n}$. If

$$\|A \cdot x\|' \leq \|A\| \cdot \|x\|'$$

holds we call the matrix norm compatible with the vector norm.

According to (1.3.7) each operator norm is compatible with its generating vector norm. The Frobenius norm is compatible with the Euclidean norm although it is not an operator norm; cf. Exercise 1.3.4. In particular, it is not generated by $\|\cdot\|_2$.

For any matrix norm $\|\cdot\|$ one can find a vector norm $\|\cdot\|'$ with which $\|\cdot\|$ is compatible. Simply define $\|x\|' = \|(x, 0, \dots, 0)\|$ for vectors x, where the zeros denote zero columns. The submultiplicativity (1.3.5) of matrix norms ensures the compatibility.

We will now prove a standard estimate for the spectral radius $\rho(A)$ of a matrix $A \in \mathbb{K}^{n\times n}$.

Theorem 1.3.10. *Let $A \in \mathbb{K}^{n\times n}$ and $\|\cdot\|$ be a matrix norm on $\mathbb{K}^{n\times n}$. Then the spectral radius $\rho(A)$ of A satisfies*

$$\rho(A) \leq \|A\|. \tag{1.3.10}$$

Proof. Let $\|\cdot\|'$ be a vector norm on $\mathbb{K}^n$ with which the given matrix norm on $\mathbb{K}^{n\times n}$ is compatible.

Case $\mathbb{K} = \mathbb{C}$: Apply the vector norm to the equation $Az = \lambda z$, $z \in \mathbb{C}$, $\|z\|' = 1$ in order to get $|\lambda| = |\lambda|\,\|z\|' = \|Az\|' \leq \|A\|\,\|z\|' = \|A\|$. The assertion follows immediately.

Case $\mathbb{K} = \mathbb{R}$: First notice that now the matrix norm of the theorem and the abovementioned construction $\|\cdot\|'$ of a compatible vector norm refer to *real* quantities while the eigenvalue/eigenvector equation involves *complex* entries. Therefore, define $\|z\|' = \|\mathrm{Re}\, z\|' + \|\mathrm{Im}\, z\|'$ for $z \in \mathbb{C}^n$ and consider $\mathbb{C}^n$ as vector space *over* $\mathbb{R}$ instead of $\mathbb{C}$. (This is necessary in order to prove the homogeneity of $\|\cdot\|'$ in the new space.) Then $\|\cdot\|'$ is a norm which is compatible with the real matrix norm $\|\cdot\|$, and the ideas above with $\|z\|' = 1$ and $\theta = -\arg(\lambda)$ for $\lambda \neq 0$ yield

$$\begin{aligned} |\lambda|^k &= |\lambda|^k\,\|z\|' = \|\,|\lambda|^k z\,\|' = \|e^{ik\theta}\lambda^k z\|' = \|e^{ik\theta}A^k z\|' \leq \|A\|^k\,\|e^{ik\theta}z\|' \\ &= \|A\|^k\,(\|\cos(k\theta)\,\mathrm{Re}\, z - \sin(k\theta)\,\mathrm{Im}\, z\|' + \|\sin(k\theta)\,\mathrm{Re}\, z + \cos(k\theta)\,\mathrm{Im}\, z\|') \\ &\leq 2\|A\|^k, \quad k = 1, 2, \dots . \end{aligned} \tag{1.3.11}$$

Assume that $|\lambda| > \|A\|$ holds. Divide (1.3.11) by $|\lambda|^k$. Then the left-hand side is one while the right-hand side tends to zero if k tends to infinity. This contradiction concludes the proof. □

The example $A = (1, 0)^T(0, 1)$ shows that there are matrices for which equality can never hold in (1.3.10). In Section 1.7, however, we will see that there are matrix norms $\|\cdot\|$ such that $\|A\|$ approximates $\rho(A)$ arbitrarily well.

Exercises

Ex. 1.3.1. Let $\|\cdot\|$ be any vector norm on $\mathbb{K}^n$. Show that the operator norm defined in (1.3.6) is a matrix norm on $\mathbb{K}^{n\times n}$.

Hint: Use results of Section 1.8.

Ex. 1.3.2. Show that the Euclidean norm on $\mathbb{K}^n$ generates the spectral norm $\|A\|_2 = \sqrt{\rho(A^HA)}$ on $\mathbb{R}^{n\times n}$. In addition, prove $\|A^H\|_2 = \|A\|_2$ and $\|UA\|_2 = \|AU\|_2$ if $U \in \mathbb{K}^{n\times n}$ is a Hermitian matrix.

Hint: Use basic results of linear algebra; see for instance Section 1.8.

Ex. 1.3.3. Let S be a regular matrix. Show that $\|x\|_S = \|S^{-1}x\|_\infty$ is a vector norm on $\mathbb{R}^n$ which generates the operator norm $\|A\|_S = \|S^{-1}AS\|_\infty$ on $\mathbb{R}^{n\times n}$.

Ex. 1.3.4. Prove the following properties of the Frobenius norm $\|\cdot\|_F$ for $A \in \mathbb{R}^{n\times n}$.

(a) It is compatible with the Euclidean norm $\|\cdot\|_2$ on $\mathbb{R}^n$.

(b) It can be represented as

$$\|A\|_F = \left\| \begin{pmatrix} a^1 \\ \vdots \\ a^n \end{pmatrix} \right\|_2 ,$$

where $a^i = A_{*,i}$, $i = 1, \dots, n$, and where $\|\cdot\|_2$ denotes the Euclidean norm on $\mathbb{R}^{n^2}$.

(c) $\|A\|_F = \|A^T\|_F$.

(d) $\|UA\|_F = \|AU\|_F = \|A\|_F$ if $U \in \mathbb{R}^{n\times n}$ is orthogonal.

Hint: Use (b) and (c).

1.4 Polynomials

In this section we consider polynomials p and some of their properties. Our first result shows that the zeros of a polynomial p depend continuously on the coefficients of p.

Theorem 1.4.1. *Let the polynomial p be defined by*

$$p(x) = p(x; a_0, \dots, a_n) = \sum_{j=0}^{n} a_j x^j, \quad a_n \neq 0,\ n \geq 1,$$

and denote its zeros by $x_i = x_i(a_0, \dots, a_n)$, $i = 1, \dots, n$ *(counted according to their multiplicity).*

(a) *The zeros of p depend continuously on the coefficients* a_k.

(b) *A simple zero* $x_{i_0} = x_{i_0}(a_0, \dots, a_n)$ *of p is infinitely often continuously differentiable with respect to the coefficients* a_k. *One obtains*

$$\frac{\partial x_{i_0}(a_0, \dots, a_n)}{\partial a_k} = \frac{-x_{i_0}^k}{a_n \prod_{\substack{i=1\\ i\neq i_0}}^n (x_{i_0} - x_i)} = \frac{-x_{i_0}^k}{p'(x_{i_0})}, \qquad k = 0, 1, \dots, n. \tag{1.4.1}$$

Proof. (a) Assume that the assertion is wrong. Then there is a positive real number ε, an index i_0, coefficients $a_0^*, \dots, a_n^*$, $a_n^* \neq 0$, and a sequence of coefficients $(a_0^{(k)}, \dots, a_n^{(k)})_{k=1}^\infty$ such that for all $k \in \mathbb{N}$ the inequalities

$$|a_j^{(k)} - a_j^*| \le \frac{1}{k}, \quad j = 0, 1, \dots, n,$$

and

$$|x_i^{(k)} - x_{i_0}^*| \ge \varepsilon, \quad i = 1, \dots, n,$$

hold with $x_i^{(k)} = x_i(a_0^{(k)}, \dots, a_n^{(k)})$ and $x_{i_0}^* = x_{i_0}(a_0^*, \dots, a_n^*)$. Let k tend to infinity in the inequality

$$|p(x_{i_0}^*; a_0^{(k)}, \dots, a_n^{(k)})| = |a_n^{(k)}| \prod_{i=1}^n |x_{i_0}^* - x_i^{(k)}| \ge |a_n^{(k)}|\, \varepsilon^n.$$

Then we obtain the contradiction

$$0 = |p(x_{i_0}^*; a_0^*, \dots, a_n^*)| \ge |a_n^*|\, \varepsilon^n > 0.$$

(b) W.l.o.g. let $i_0 = 1$. Define the polynomial $p(x, h) = p(x; a_0, \dots, a_{k-1}, a_k + h, a_{k+1}, \dots, a_n)$ and denote its zeros by $x_i(h)$, $i = 1, \dots, n$, with the indices such that – with (a) – $\lim_{h\to\infty} x_i(h) = x_i$, $i = 1, \dots, n$, holds. Then

$$a_n \prod_{i=1}^n (x_1 - x_i(h)) - 0 = p(x_1, h) - p(x_1) = [(a_k + h) - a_k] x_1^k,$$

whence for $h \neq 0$, $|h| \ll 1$, we get

$$\frac{x_1(h) - x_1}{h} = \frac{-x_1^k}{a_n \prod_{i=2}^n (x_1 - x_i(h))}.$$

Let h tend to zero. One sees at once that $\frac{\partial x_1(a_0, \dots, a_n)}{\partial a_k}$ exists, is continuous and fulfills the first equality in (b). In order to prove the second one, differentiate the equation

$$p(x_1(a_0, \dots, a_n);\ a_0, \dots, a_n) = 0$$

with respect to a_k. Then

$$0 = x_1^k + p'(x_1) \frac{\partial x_1(a_0, \dots, a_n)}{\partial a_k}.$$

An inductive argument applied to (1.4.1) proves the higher smoothness. □

Next we consider Weierstrass' famous approximation theorem. To prove it we need the following Bernstein polynomials.

Definition 1.4.2. The k-th Bernstein polynomial b_{nk} of degree n (with respect to the interval $[0, 1]$) is defined by

$$b_{nk}(x) = \binom{n}{k} x^k (1-x)^{n-k}, \quad x \in \mathbb{R},\ n, k \in \mathbb{N}_0.$$

This definition shows that $b_{nk}(x)$ is nonnegative when x is restricted to the interval $[0, 1]$. Among the many additional nice results of Bernstein polynomials we use the following ones.

Lemma 1.4.3. *The Bernstein polynomials b_{nk} satisfy*

$$\sum_{k=0}^{n} b_{nk}(x) = 1, \tag{1.4.2}$$

$$\sum_{k=0}^{n} k b_{nk}(x) = nx, \tag{1.4.3}$$

$$\sum_{k=0}^{n} k^2 b_{nk}(x) = nx(1-x) + (nx)^2. \tag{1.4.4}$$

Proof. The first equality follows from

$$(x+\alpha)^n = \sum_{k=0}^{n} \binom{n}{k} x^k \alpha^{n-k} \tag{1.4.5}$$

with $\alpha = 1 - x$. Now differentiate (1.4.5) by x, then multiply by x and substitute $\alpha = 1 - x$ again in order to obtain (1.4.3). Finally, differentiate (1.4.5) twice with respect to x, multiply by x^2 and substitute $\alpha = 1 - x$ once more in order to get

$$n(n-1)x^2 = \sum_{k=0}^{n} k(k-1)b_{nk}(x) = \sum_{k=0}^{n} k^2 b_{nk}(x) - nx,$$

whence the third equality follows immediately. □

Bernstein polynomials are often applied in computer aided geometric design (CAGD). For the equations in Lemma 1.4.3 there is also a nice probabilistic interpretation: Let A denote some event which occurs with probability $x \in [0, 1]$. Then $b_{nk}(x)$ can be viewed as the probability with which A occurs exactly k times in a Bernoulli chain of length n. Thus for $x \in [0, 1]$ the vector $(b_{n0}(x), b_{n1}(x), \ldots, b_{nn}(x))$ can be interpreted as the distribution vector of the binomial distribution with the parameters n and x. Hence the first equation of Lemma 1.4.3 says that the total probability of the binomial distribution is one (which is trivial). The second equation shows that the expected value E of this distribution is nx, and if one subtracts $E^2 = (nx)^2$ from both sides of (1.4.4) one ends up with the variance $V = nx(1-x)$. We will no longer pursue this probabilistic interpretation but prove the following simplest version of the Weierstrass approximation theorem.

Theorem 1.4.4 (Weierstrass – one-dimensional). *Let $f : [0,1] \to \mathbb{R}$ be continuous. Then for any positive real number ε there is an integer n such that the polynomial p_n defined by*

$$p_n(x) = \sum_{j=0}^{n} f\left(\frac{j}{n}\right) b_{nj}(x) \tag{1.4.6}$$

satisfies

$$\|f - p_n\|_\infty = \max_{x\in[0,1]} |f(x) - p_n(x)| < \varepsilon.$$

If f is defined on the more general interval $[a,b]$, $a < b$, the assertion holds with the summands in (1.4.6) *being replaced by $f(a + \frac{j}{n}(b-a))\, b_{nj}(\frac{x-a}{b-a})$.*

Proof. Since f is uniformly continuous on the compact set $[0,1]$ we can associate with any $\varepsilon > 0$ a real number $\delta > 0$ such that for any real numbers x, x' with $|x - x'| < \delta$ we have $|f(x) - f(x')| < \varepsilon/2$. Fix $x \in [0,1]$ and define the index sets

$$N' = \left\{ j \in \{0, 1, \ldots, n\} \;\middle|\; \left| x - \frac{j}{n} \right| < \delta \right\},$$

$$N'' = \left\{ j \in \{0, 1, \ldots, n\} \;\middle|\; \left| x - \frac{j}{n} \right| \ge \delta \right\}.$$

With $M = \max_{0\le x\le 1} |f(x)|$ and repeated application of Lemma 1.4.3 we get

$$\begin{aligned}
|f(x) - p_n(x)| &\le \sum_{j=0}^{n} \left| f(x) - f\left(\frac{j}{n}\right) \right| b_{nj}(x) \\
&\le \sum_{j\in N'} \left| f(x) - f\left(\frac{j}{n}\right) \right| b_{nj}(x) + \sum_{j\in N''} \left(|f(x)| + \left| f\left(\frac{j}{n}\right) \right| \right) b_{nj}(x) \\
&\le \frac{\varepsilon}{2} \sum_{j=0}^{n} b_{nj}(x) + 2M \sum_{j\in N''} b_{nj}(x) \cdot \underbrace{\frac{\left(x - \frac{j}{n}\right)^2}{\delta^2}}_{\ge 1} \\
&\le \frac{\varepsilon}{2} + \frac{2M}{\delta^2} \sum_{j=0}^{n} b_{nj}(x) \left[x^2 - 2x\frac{j}{n} + \left(\frac{j}{n}\right)^2 \right] \\
&= \frac{\varepsilon}{2} + \frac{2M}{\delta^2} \left[x^2 - 2x^2 + x^2 + \frac{x}{n}(1-x) \right] \\
&\le \frac{\varepsilon}{2} + \frac{2M}{\delta^2} \cdot \frac{1}{4n} \le \frac{\varepsilon}{2} + \frac{\varepsilon}{2} = \varepsilon
\end{aligned}$$

independently of $x \in [0,1]$ and N', N'', provided that $n \ge \frac{M}{\delta^2 \varepsilon}$. This proves the theorem for the interval $[a,b] = [0,1]$.

For general intervals $[a,b]$ set $x = a + t(b-a)$, $t \in [0,1]$, and apply the preceding results to the function $\tilde{f}(t) = f(a + t(b-a))$ with $t \in [0,1]$. □

Now we generalize our results to the multidimensional case.

Theorem 1.4.5 (Weierstrass – multidimensional). *Let $f: D \to \mathbb{R}$ be continuous on the m-dimensional unit cube $D = [0,1]^m$. Then for any positive real number ε there is an integer n such that the multivariate polynomial p_n defined by*

$$p_n(x) = p_n(x_1,\dots,x_m) = \sum_{j_1=0}^{n}\cdots\sum_{j_m=0}^{n} f\left(\frac{j_1}{n},\dots,\frac{j_m}{n}\right) b_{nj_1}(x_1)\cdots b_{nj_m}(x_m) \tag{1.4.7}$$

satisfies

$$\|f - p_n\|_\infty = \max_{x\in D}|f(x) - p_n(x)| < \varepsilon.$$

If f is defined on the more general m-dimensional rectangle $D = [a_1,b_1]\times[a_2,b_2]\times\cdots\times[a_m,b_m]$, $a_i < b_i$, the assertion holds with the summands in (1.4.7) *being modified analogously to Theorem* 1.4.4.

Proof. Let $D = [0,1]^m$. By virtue of Lemma 1.4.3 the equality

$$\sum_{j_1=0}^{n}\cdots\sum_{j_m=0}^{n} c\cdot b_{nj_1}(x_1)\cdots b_{nj_m}(x_m) = r$$

holds for the following choices of (c,r): $(c,r) = (1,1)$, $(c,r) = (j_l, nx_l)$, $(c,r) = (j_l^2, nx_l(1-x_l) + (nx_l)^2)$. We proceed as in Theorem 1.4.4 using

$$N' = \left\{(j_1,\dots,j_m)\in\{0,1,\dots,n\}^m \;\middle|\; \left\|x - \left(\frac{j_1}{n},\dots,\frac{j_m}{n}\right)^T\right\|_2 < \delta\right\},$$

$$N'' = \left\{(j_1,\dots,j_m)\in\{0,1,\dots,n\}^m \;\middle|\; \left\|x - \left(\frac{j_1}{n},\dots,\frac{j_m}{n}\right)^T\right\|_2 \ge \delta\right\}$$

and replacing

$$\frac{\left(x-\frac{j}{n}\right)^2}{\delta^2} \quad\text{by}\quad \frac{1}{\delta^2}\left\|x - \left(\frac{j_1}{n},\dots,\frac{j_m}{n}\right)^T\right\|_2^2 = \frac{1}{\delta^2}\sum_{l=1}^{m}\left(x_l - \frac{j_l}{n}\right)^2.$$

We finally obtain

$$\begin{aligned}|f(x) - p_n(x)| &\le \frac{\varepsilon}{2} + \frac{2M}{\delta^2}\sum_{l=1}^{m}\left[x_l^2 - 2x_l^2 + x_l^2 + \frac{x_l}{n}(1-x_l)\right]\\ &\le \frac{\varepsilon}{2} + \frac{2M}{\delta^2}\cdot\sum_{l=1}^{m}\frac{1}{4n} = \frac{\varepsilon}{2} + \frac{Mm}{2\delta^2}\cdot\frac{1}{n} < \varepsilon,\end{aligned}$$

provided that $n > \frac{Mm}{\delta^2\varepsilon}$. □

Our final version of Weierstrass' approximation theorem reads as follows.

Theorem 1.4.6 (Weierstrass – general). *Let $D \subseteq \mathbb{R}^m$ be compact and $f\colon D \to \mathbb{R}$ continuous. Then for each $\varepsilon > 0$ there is a polynomial p of the form*

$$p(x) = p(x_1, \dots, x_m) = \sum_{j_1=0}^{n_1} \cdots \sum_{j_m=0}^{n_m} a_{j_1 \dots j_m} \cdot x_1^{j_1} \cdot x_2^{j_2} \cdots x_m^{j_m}$$

such that

$$\|f - p\|_\infty = \max_{x \in D} |f(x) - p(x)| < \varepsilon.$$

Proof. Since D is assumed to be compact there exists an m-dimensional rectangle $I = [a_1, b_1] \times \cdots \times [a_m, b_m]$ which encloses D. We will construct a continuous function $h\colon I \to \mathbb{R}$ with

$$\max_{x \in D} |f(x) - h(x)| < \frac{\varepsilon}{2}.$$

On I we will approximate h by a polynomial p according to Theorem 1.4.5 with

$$\max_{x \in I} |h(x) - p(x)| < \frac{\varepsilon}{2}.$$

Then Theorem 1.4.6 is proved by virtue of

$$\begin{aligned}\max_{x \in D} |f(x) - p(x)| &\le \max_{x \in D} |f(x) - h(x)| + \max_{x \in I} |h(x) - p(x)| \\ &< \frac{\varepsilon}{2} + \frac{\varepsilon}{2} = \varepsilon.\end{aligned}$$

In view of h we temporarily fix some $y \in D$, choose $z \in D$ and define the function $g_z\colon I \to \mathbb{R}$ by

$$g_z(x) = \begin{cases} f(y), & \text{if } f(y) = f(z), \\ f(y) + \dfrac{\|x - y\|_\infty}{\|z - y\|_\infty} [f(z) - f(y)], & \text{if } f(y) \ne f(z). \end{cases}$$

(Here $\|x\|_\infty = \max_{1 \le i \le m} |x_i|$.)

In each of the two cases g_z is continuous in I and has the values

$$g_z(y) = f(y), \quad g_z(z) = f(z).$$

Hence there is an open ball $B(z, \delta(z)) \subseteq \mathbb{R}^m$ (with respect to $\|\cdot\|_2$, e.g.) such that

$$g_z(x) < f(x) + \frac{\varepsilon}{2} \quad \text{for all } x \in B(z; \delta(z)) \cap D.$$

Trivially, $D \subseteq \bigcup_{z \in D} B(z, \delta(z))$. Since D is compact there are finitely many elements $z_1, \dots, z_k$ of D with $D \subseteq \bigcup_{i=1}^k B(z_i, \delta(z_i))$. The function

$$h_y(x) = \min\{g_{z_1}(x), \dots, g_{z_k}(x)\}$$

is continuous on I and satisfies

$$h_y(x) < f(x) + \frac{\varepsilon}{2} \quad \text{for all } x \in D.$$

By virtue of $h_y(y) = f(y)$ there is an open ball $\tilde{B}(y, \tilde{\delta}(y))$ such that

$$h_y(x) > f(x) - \frac{\varepsilon}{2} \quad \text{for all } x \in \tilde{B}(y, \tilde{\delta}(y)) \cap D.$$

By a similar argument as above there are finitely many elements $y_1, \ldots, y_l \in D$ with $D \subseteq \bigcup_{i=1}^{l} \tilde{B}(y_i, \tilde{\delta}(y_i))$. Define the continuous function h by

$$h(x) = \max\{h_{y_1}(x), \ldots, h_{y_l}(x)\}, \quad x \in I.$$

Then for all $x \in D$ the double inequality

$$f(x) - \frac{\varepsilon}{2} < h(x) < f(x) + \frac{\varepsilon}{2}$$

holds, which proves the approximation property of h. □

The polynomial in the Weierstrass theorem approximates $f(x)$ but does not necessarily coincide with $f(x)$ somewhere in D. Things change when considering the so-called Hermite interpolation which contains the classical polynomial interpolation as a particular case.

Theorem 1.4.7 (Hermite interpolation). *Let $f \in C^{m+1}(D)$, where D is an open subset of $\mathbb{R}$ and m is some nonnegative integer. Let $x_0, \ldots, x_n \in [x] \subset D$ be pairwise different and let $m_0, \ldots, m_n$ be positive integers which satisfy*

$$m + 1 = \sum_{i=0}^{n} m_i.$$

Then there is a unique polynomial p_m of degree at most m such that

$$p_m^{(j)}(x_i) = f^{(j)}(x_i), \quad j = 0, \ldots, m_i - 1, \ i = 0, \ldots, n. \tag{1.4.8}$$

For each element $\tilde{x} \in [x]$ there is a number $\tilde{x}_\xi \in [x]$ such that

$$f(\tilde{x}) = p_m(\tilde{x}) + \frac{1}{(m+1)!} f^{(m+1)}(\tilde{x}_\xi) \prod_{i=0}^{n} (\tilde{x} - x_i)^{m_i}. \tag{1.4.9}$$

Proof. First we prove the uniqueness. To this end assume that there are two polynomials $p_m, \tilde{p}_m$ of degree at most m with $p_m \neq \tilde{p}_m$. Then the degree of the polynomial $q = p_m - \tilde{p}_m$ is at most m, but q has at least $m + 1$ zeros $x_0, \ldots, x_n$, if these are counted according to their multiplicity. Hence $q = 0$ which contradicts $p_m \neq \tilde{p}_m$.

In order to show existence we start with $p_m = \sum_{i=0}^{m} a_i x^i$. Then the conditions (1.4.8) lead to a linear system $Az = b$, where $z = (a_0, \ldots, a_m)^T$, where A is a square matrix which is independent of f, and where the components of the vector b consist of the values of the right-hand side in (1.4.8) in some order. We first assume $f = 0$ which implies $b = 0$. Then $Az = b$ has the solution $z = 0$ which is unique as we have already proved. Therefore A is regular, hence $Az = b$ is (uniquely) solvable for $b \neq 0$, too.

In order to prove (1.4.9) we first remark that this equality certainly holds for $\tilde{x} = x_i$. Therefore, we may assume $\tilde{x} \in [x] \setminus \{x_0, \ldots, x_n\}$. Consider

$$r(x) = f(x) - p_m(x) - c\prod_{i=0}^{n}(x - x_i)^{m_i},$$

where the constant c is chosen such that $r(\tilde{x}) = 0$. Then $x_0, \ldots, x_n, \tilde{x}$ are $m + 2$ zeros of r if one takes into account multiplicities. Hence the generalized Rolle theorem guarantees a number $x_\xi \in [x]$ such that $r^{(m+1)}(x_\xi) = 0$, whence $0 = f^{(m+1)}(x_\xi) - c(m + 1)!$ follows. Solving for c proves the assertion. □

The Hermite interpolation polynomial in Theorem 1.4.7 can be constructed explicitly – see Stoer, Bulirsch [348], pp. 52–57. If not all successive derivatives of p_m are prescribed by (1.4.8) (while decreasing m correspondingly) one speaks of Birkhoff interpolation. For this generalization the assertion of Theorem 1.4.7 may be false; cf. Exercise 1.4.2.

Exercises

Ex. 1.4.1. Show that the polynomial in (1.4.8) can be represented as

$$p_m(x) = \sum_{i=0}^{n}\sum_{j=0}^{m_i} f^{(j)}(x_i)L_{ij},$$

where the generalized Lagrange polynomials L_{ij} are defined by means of

$$l_{ij}(x) = \frac{(x - x_i)^j}{j!}\prod_{\substack{k=0\\k\neq i}}^{n}\left(\frac{x - x_k}{x_i - x_k}\right)^{m_k}, \qquad 0 \le i \le n,\ 0 \le j < m_i,$$

as

$$L_{i,m_i-1}(x) = l_{i,m_i-1}(x), \quad i = 0, \ldots, n$$

and recursively for $j = m_i - 2, m_i - 3, \ldots, 0$ by

$$L_{ij}(x) = l_{ij}(x) - \sum_{k=j+1}^{m_i-1} l_{ij}^{(k)}(x_i)L_{ik}(x).$$

Hint: Show by induction $L_{ij}^{(k)}(x_\ell) = \delta_{i\ell}\delta_{jk}$.

Ex. 1.4.2. Show that there is no unique polynomial p of degree at most two which satisfies $p(-1) = p(1) = 1$ and either $p'(0) = 0$ or $p'(0) = 1$.

1.5 Zeros and fixed points of functions

Many problems in applications, but also proper mathematical problems such as discretizations of partial differential equations finally lead to systems of equations which always can be written as a zero problem

$$f(x) = 0 \tag{1.5.1}$$

or – in the case $f\colon D \subseteq \mathbb{R}^n \to \mathbb{R}^n$ sometimes more appropriate – as an equivalent fixed point problem

$$g(x) = x. \tag{1.5.2}$$

Therefore, it is interesting to know when a function has a zero, and a fixed point, respectively, and then to compute it or at least to enclose it by some simple kind of set. The current section primarily studies the existence of zeros and fixed points. We will present a variety of theorems in this respect which are not always standard in introductory courses of analysis. In later chapters these theorems form the outset of tests for verifying zeros and fixed points within boxes, i.e., n-dimensional rectangles. The proofs of these theorems will be relatively simple if one introduces the concept of mapping degree. Since such an introduction is very lengthy, we restrict ourselves to the definition of the degree following the way in Ortega, Rheinboldt [267]. Then we list without proof the essential properties necessary to derive the crucial facts to follow.

In the whole section we assume that D is a Jordan measurable subset of $\mathbb{R}^n$, i.e., the Riemann integral $\int_D 1\,dx$ exists.

Definition 1.5.1.
(a) Let $f\colon D \subseteq \mathbb{R}^n \to \mathbb{R}^m$. Then the set

$$\operatorname{supp}(f) = \overline{\{x \mid f(x) \neq 0\} \cap D}$$

is called the support of f.
(b) Given $\alpha > 0$ the set W_α consists of all functions φ with the following properties.
   (i) $\varphi \in C([0,\infty), \mathbb{R})$,
   (ii) $\operatorname{supp}(\varphi) \subseteq (0, \alpha]$,
   (iii) $\int_{\mathbb{R}^n} \varphi(\|x\|_2)\,dx = 1$.

Notice that in Definition 1.5.1 (b) for each element $\varphi \in W_\alpha$ there is some $\delta \in (0, \alpha]$ such that $\varphi(t) = 0$ if $t \notin [\delta, \alpha]$. Therefore, the range $\mathbb{R}^n$ of the integral in (iii) can be replaced by the ball $\overline{B}(0, \alpha)$.

Next we introduce the degree integral by means of which we can finally define the mapping degree.

Definition 1.5.2. Let D, E be open and bounded subsets of $\mathbb{R}^n$ with $\overline{D} \subseteq E$. Let $f \in C(\overline{D}, \mathbb{R}^n)$, $y \in \mathbb{R}^n$, $f(x) \neq y$ for all $x \in \partial D$, $0 < \alpha < \min\{\|f(x) - y\|_2 \mid x \in \partial D\}$, $\varphi \in W_\alpha$.

(a) For $\tilde{f} \in C^1(E, \mathbb{R}^n)$ we define the mapping $\phi\colon \mathbb{R}^n \to \mathbb{R}$ by

$$\phi(x) = \begin{cases} \varphi(\|\tilde{f}(x) - y\|_2) \det \tilde{f}'(x), & x \in D \\ 0, & \text{otherwise} \end{cases}$$

and the degree integral

$$d_\varphi(\tilde{f}, D, y) = \int_{\mathbb{R}^n} \phi(x)\, dx.$$

(b) The (topological) mapping degree of f at y with respect to D is defined by

$$\deg(f, D, y) = \lim_{k\to\infty} d_\varphi(f_k, D, y),$$

where the functions f_k, $k = 1, 2, \ldots$, satisfy $f_k \in C^1(E, \mathbb{R}^n)$ and $\lim_{k\to\infty}\|f_k - f\|_D = 0$ with $\|g\|_D := \max_{x\in\overline{D}}\|g(x)\|_2$.

With some effort it can be shown that the definition of the mapping degree does not depend on the choice of α nor on the particular choice of $\varphi \in W_\alpha$, nor on the particular choice of the approximating sequence (f_k). So it is well-defined and apparently coincides with the value of the degree integral if $f \in C^1(E, \mathbb{R}^n)$. Moreover, it can be proved that it is an integer. Out of its many additional nice properties we only mention the following without proof, where we no longer mention the set E of Definition 1.5.2 which comes into play for the functions f_k in (b) only. It guarantees the existence of $d_\varphi(f_k, D, y)$ since then $\det(f_k'(x))$ is bounded in D.

Theorem 1.5.3 (Properties of the mapping degree). *Let $D \subseteq \mathbb{R}^n$ be open and bounded and $f \in C(\overline{D}, \mathbb{R}^n)$. Moreover, let $y \in \mathbb{R}^n$ be such that $f(x) \neq y$ for all $x \in \partial D$.*

(a) *If $f \in C^1(D, \mathbb{R}^n)$ has a nonsingular Jacobian $f'(x)$ at all points $x \in D$ where $f(x) = y$, then*

$$\deg(f, D, y) = \sum_{\substack{x\in D,\\ f(x)=y}} \operatorname{sign}(\det f'(x)),$$

where the sum turns out to have only finitely many summands.

(b) *If $h\colon [0, 1] \times \overline{D} \to \mathbb{R}^n$ is continuous and $h(t, x) \neq y$ for all $t \in [0, 1]$ and for all $x \in \partial D$, then $\deg(h(t, \cdot), D, y)$ does not depend on t.*

(c) *If $\deg(f, D, y) \neq 0$, then there exists an $x \in D$ such that $f(x) = y$.*

(d) *Let D be symmetric with respect to 0 (i.e., $x \in D \Leftrightarrow -x \in D$) and let $0 \in D$. If f is odd, i.e., $f(-x) = -f(x)$, and if $0 \notin f(\partial D)$, then $\deg(f, D, 0)$ is odd.*

Now we are ready to address the main results of this section.

Theorem 1.5.4 (Borsuk). *Let $D \subseteq \mathbb{R}^n$ be open, bounded, convex and symmetric with respect to some $x_0 \in D$ (i.e., $x_0 + x \in D$ implies $x_0 - x \in D$). Let $f \in C(\overline{D}, \mathbb{R}^n)$ and assume that*

$$f(x_0 + x) \neq \lambda f(x_0 - x) \quad \text{for all } \lambda > 0 \text{ and for all } x_0 + x \in \partial D. \tag{1.5.3}$$

Then f has a zero in $\overline{D}$.

Proof. W.l.o.g we can assume $x_0 = 0$. (Otherwise consider the function $g(x) = f(x_0 + x)$.) If $f(x) = 0$ for some $x \in \partial D$ we are done. Therefore, we assume from now on that $0 \notin R_f(\partial D)$. We define the function h by $h(t, x) = f(x) - tf(-x)$, $0 \le t \le 1$, $x \in \overline{D}$. Then $h(0, x) = f(x)$ and $h(1, x) = f(x) - f(-x)$. In particular this latter function is odd. For $x \in \partial D$ we have $h(t, x) \ne 0$. For $t = 0$ this follows from the assumption $0 \notin R_f(\partial D)$ and for $t \in (0, 1]$ the assumption $h(t, x) = 0$, $x \in \partial D$ leads to $f(x) = tf(-x)$ contradicting (1.5.3). Therefore, Theorem 1.5.3 (b)–(d) guarantees that $\deg(f, D, 0) = \deg(h(0, \cdot), D, 0) = \deg(h(1, \cdot), D, 0) \ne 0$ holds and that f has a zero in D. □

Theorem 1.5.5 (Leray–Schauder). *Let $D \subseteq \mathbb{R}^n$ be open and bounded with $0 \in D$ and let $f \in C(\overline{D}, \mathbb{R}^n)$ satisfy*

$$f(x) \ne \lambda x \quad \textit{for all } \lambda > 1 \textit{ and for all } x \in \partial D. \tag{1.5.4}$$

Then f has a fixed point in $\overline{D}$.

Proof. With $h(t, x) = x - tf(x)$, $0 \le t \le 1$, $x \in \overline{D}$ we obtain $h(0, x) = x \ne 0$ on ∂D since $0 \in D$ by assumption and since D is open. Moreover, assume that $h(1, x) = x - f(x) \ne 0$ holds on ∂D since otherwise we are done. For $t \in (0, 1)$, the assumption $h(t, x) = 0$, $x \in \partial D$ leads to $f(x) = (1/t)x$ contradicting (1.5.4). Hence Theorem 1.5.3 (a)–(c) yields $\deg(h(1, \cdot), D, 0) = \deg(h(0, \cdot), D, 0) = 1 \ne 0$ and the existence of a fixed point. □

Notice the following equivalent formulation of the preceding fixed point theorem; cf. Exercise 1.5.2.

Theorem 1.5.6 (Leray–Schauder). *Let $D \subseteq \mathbb{R}^n$ be open and bounded with $0 \in D$ and let $f \in C(\overline{D}, \mathbb{R}^n)$ satisfy*

$$f(x) \ne \lambda x \quad \textit{for all } \lambda > 0 \textit{ and for all } x \in \partial D. \tag{1.5.5}$$

Then f has a zero in $\overline{D}$.

Now we present Brouwer's fixed point theorem for continuous self-mappings of the n-dimensional closed zero-centered unit ball $\overline{B}(0, 1)$.

Theorem 1.5.7 (Brouwer – unit ball). *Each function $f \in C(\overline{B}(0, 1), \overline{B}(0, 1))$ with $B(0, 1) \subseteq \mathbb{R}^n$ has a fixed point.*

Proof. Let $D = B(0, 1)$ and define $h(t, x) = x - tf(x)$, $0 \le t \le 1$, $x \in \overline{D}$. For $\|x\|_2 = 1$ and $t \in [0, 1)$, we obtain $\|h(t, x)\|_2 \ge \|x\|_2 - t\|f(x)\|_2 \ge 1 - t > 0$. For $t = 1$, we can also assume $h(1, x) \ne 0$ on ∂D since otherwise the theorem is proved. The proof now terminates by virtue of $\deg(h(1, \cdot), D, 0) = \deg(h(0, \cdot), D, 0) = 1 \ne 0$. □

Two elementary proofs of Brouwer's fixed point theorem can be found in Appendix B. One is based on the multidimensional transformation rule and does not use the mapping degree. The second one applies to the Sperner lemma.

Notice that the Leray–Schauder fixed point theorem, too, implies Brouwer's fixed point theorem for the unit ball. To see this choose $D = B(0, 1)$ and let $f \in$

$C(\overline{B}(0,1), \overline{B}(0,1))$. If $f(x) = \lambda x$ for some $\lambda > 1$ and some x with $\|x\|_2 = 1$, then we get the contradiction $1 \geq \|f(x)\|_2 = |\lambda| \, \|x\|_2 = |\lambda|$.

Conversely, if one requires the domain D of the Leray–Schauder fixed point theorem to be itself a ball, say $D = B(0, \rho)$, this theorem can be easily deduced from Brouwer's fixed point theorem: Assume that $f \in C(\overline{D}, \mathbb{R}^n)$ does not have a fixed point in $\overline{D}$. Then the function $\hat{f}$ with

$$\hat{f}(x) = \frac{f(\rho x) - \rho x}{\|f(\rho x) - \rho x\|_2}, \qquad x \in \overline{B}(0,1),$$

is well-defined and maps $\overline{B}(0,1)$ continuously into itself. Hence it has a fixed point x^*. Clearly, $\|x^*\|_2 = \|\hat{f}(x^*)\|_2 = 1$ hence $\rho x^* \in \partial D$. From $x^* = \hat{f}(x^*)$ we finally obtain $f(\rho x^*) = (1 + \|f(\rho x^*) - \rho x^*\|_2/\rho)\rho x^*$ which contradicts (1.5.4).

Now we generalize Brouwer's fixed point theorem to general compact convex sets.

Theorem 1.5.8 (Brouwer – general). *If $D \subseteq \mathbb{R}^n$ is a nonempty, convex and compact set then each function $f \in C(D, D)$ has a fixed point.*

Proof. We first enclose the set D by a closed ball $\overline{B}(0, \rho)$. Since D is compact the nonnegative number $\gamma_x = \min_{z \in D} \|z - x\|_2$ is defined for every $x \in \overline{B}(0, \rho)$. Moreover, by the same reason there is a point $z_x \in D$ such that $\|z_x - x\|_2 = \gamma_x$. Assume that there is another point $\hat{z}_x$ with the same property. Then

$$\begin{aligned} \|z_x - \hat{z}_x\|_2^2 &= 2\,\|z_x - x\|_2^2 + 2\,\|\hat{z}_x - x\|_2^2 - 4\left\|\frac{z_x + \hat{z}_x}{2} - x\right\|_2^2 \\ &= 4\gamma_x^2 - 4\left\|\frac{z_x + \hat{z}_x}{2} - x\right\|_2^2 \leq 4\gamma_x^2 - 4\gamma_x^2 = 0, \end{aligned} \tag{1.5.6}$$

whence $z_x = \hat{z}_x$. Notice that for the inequality in (1.5.6) we used that the midpoint of $(z_x + \hat{z}_x)/2$ is an element of the convex set D.

We consider the function $g\colon \overline{B}(0,\rho) \to D$ with $g(x) = z_x$. For $x, \hat{x} \in \overline{B}(0, \rho)$ we get $\gamma_x \leq \|z_{\hat{x}} - x\|_2 \leq \gamma_{\hat{x}} + \|\hat{x} - x\|_2$. Interchanging the roles of x and $\hat{x}$ and combining both double inequalities finally leads to

$$|\gamma_x - \gamma_{\hat{x}}| \leq \|x - \hat{x}\|_2. \tag{1.5.7}$$

Analogously to (1.5.6) we find

$$\|z_x - z_{\hat{x}}\|_2^2 \leq 2\gamma_x^2 + 2\,\|z_{\hat{x}} - x\|_2^2 - 4\gamma_x^2 \leq 2(\gamma_{\hat{x}} + \|\hat{x} - x\|_2)^2 - 2\gamma_x^2.$$

Together with (1.5.7) this shows that g is continuous and so is the function $h\colon \overline{B}(0,1) \to \overline{B}(0,1)$ with $h(x) = f(g(\rho x))/\rho$. According to Theorem 1.5.7 the function h has a fixed point $x^* \in \overline{B}(0,1)$, whence $\rho x^* = f(g(\rho x^*)) \in D$. Since g restricted to D is the identity, the point ρx^* is a fixed point for f in D. □

Next we will consider Miranda's theorem for zeros of continuous functions in boxes. It turns out that this theorem is equivalent to Brouwer's fixed point theorem.

Theorem 1.5.9 (Miranda). *Let* $[a] = [\underline{a}_1, \overline{a}_1] \times \ldots \times [\underline{a}_n, \overline{a}_n] \subset \mathbb{R}^n$ *and let* $f = (f_i) \in C([a], \mathbb{R}^n)$ *satisfy*

$$f_i(x_1, \ldots, x_{i-1}, \underline{a}_i, x_{i+1}, \ldots, x_n) \le 0, \quad f_i(x_1, \ldots, x_{i-1}, \overline{a}_i, x_{i+1}, \ldots, x_n) \ge 0 \quad (1.5.8)$$

for all $x_j \in [\underline{a}_j, \overline{a}_j]$, $j = 1, \ldots, n$, $j \ne i$, *and for all* $i = 1, \ldots, n$. *Then* f *has a zero in* $[a]$.

This assertion remains true if for arbitrary coordinate functions f_i *the two inequality signs in* (1.5.8) *are reversed simultaneously.*

Before we prove this theorem we illustrate the assumptions in the two-dimensional case (Figure 1.5.1).

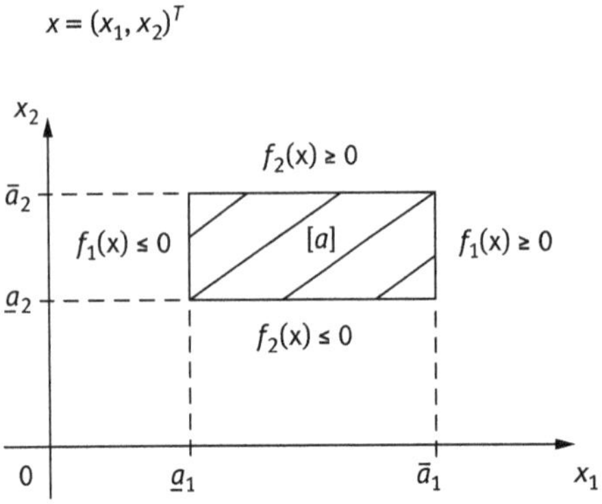

Fig. 1.5.1: Miranda's theorem in $\mathbb{R}^2$.

Proof. First we assume that $\underline{a}_i < \overline{a}_i$ holds for all i and that (1.5.8) holds with strict inequalities.

Define $S_i^+ = \{x \in [a] \mid f_i(x) > 0\}$ and $S_i^- = \{x \in [a] \mid f_i(x) < 0\}$. Since f is continuous, $f_i(x)$ remains negative near the facet $\{x \mid x \in [a], x_i = \underline{a}_i\}$ of $[a]$. Hence $\delta_i^+ = \inf\{|x_i - \underline{a}_i| \mid x \in S_i^+\}$ is positive, and analogously $\delta_i^- = \inf\{|x_i - \overline{a}_i| \mid x \in S_i^-\} > 0$ (cf. Figure 1.5.2).

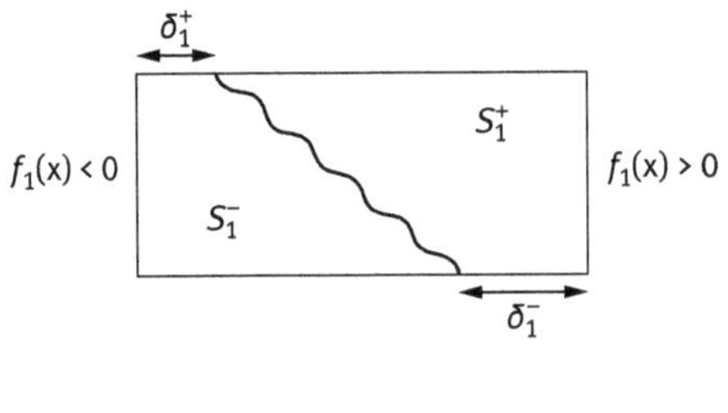

Fig. 1.5.2: Illustration of $S_1^\pm$ in $\mathbb{R}^2$.

Let $m_i = \min_{x \in [a]} f_i(x)$ (< 0), $M_i = \max_{x \in [a]} f_i(x)$ (> 0) and choose ε_i such that

$$0 < \varepsilon_i < \min\left\{-\frac{\delta_i^-}{m_i}, \frac{\delta_i^+}{M_i}\right\}.$$

The function $g\colon [a] \to \mathbb{R}^n$ defined by $g_i(x) = x_i - \varepsilon_i f_i(x)$, $i = 1, \dots, n$ has the following properties for $x \in [a]$. If $\overline{a}_i - \delta_i^- < x_i \le \overline{a}_i$ the values of $f_i(x)$ are nonnegative, hence $g_i(x) \le x_i \le \overline{a}_i$. If $\underline{a}_i \le x_i \le \overline{a}_i - \delta_i^-$ we have $g_i(x) \le \overline{a}_i - \delta_i^- - \varepsilon_i \cdot m_i < \overline{a}_i - \delta_i^- + \frac{\delta_i^-}{m_i} \cdot m_i = \overline{a}_i$. By considering the cases $\underline{a}_i \le x_i < \underline{a}_i + \delta_i^+$ and $\underline{a}_i + \delta_i^+ \le x_i \le \overline{a}_i$ we similarly get $g_i(x) \ge \underline{a}_i$. Hence $g \in C([a], [a])$ and Brouwer's fixed point theorem ensures a fixed point x^* of g which turns out to be a zero of f.

Next we keep the restriction $\underline{a}_i < \overline{a}_i$, $i = 1, \dots n$, and consider the general assumption (1.5.8).

With $\check{a} = (\underline{a} + \overline{a})/2$ define $h_i^k(x) = f_i(x) + (x_i - \check{a}_i)/k$, $i = 1, \dots, n$, $k \in \mathbb{N}$. Then $h_i^k(x) < 0$ for $x_i = \underline{a}_i$ and $h_i^k(x) > 0$ for $x_i = \overline{a}_i$. Hence each function h^k has a zero $x^k \in [a]$. Since $[a]$ is compact there is a subsequence of (x^k) which converges to some limit $x^* \in [a]$. It is easy to see that this limit is a zero of f.

Now we consider the degenerate case $\underline{a}_{i_0} = \overline{a}_{i_0}$ for some index i_0.

The assumption (1.5.8) implies $f_{i_0}(x) = 0$ for all $x \in [a]$. One can delete the i_0-th component in $[a]$ and f and obtain the associated function

$$\tilde{f}\colon ([\underline{a}_1, \overline{a}_1] \dots, [\underline{a}_{i_0-1}, \overline{a}_{i_0-1}], [\underline{a}_{i_0+1}, \overline{a}_{i_0+1}] \dots, [\underline{a}_n, \overline{a}_n])^T \to \mathbb{R}^{n-1}$$

for which one can repeat the considerations above until one ends up with a function $\hat{f}$, which either satisfies $\hat{f} = 0$ or satisfies the assumptions (1.5.8) with $\underline{a}_i < \overline{a}_i$ for the remaining indices i.

If the inequality signs are pairwise interchanged in (1.5.8), then apply the theorem to the function g defined by

$$g_i = \begin{cases} f_i, & \text{if (1.5.8) holds for } i, \\ -f_i, & \text{otherwise.} \end{cases}$$ □

We used Brouwer's fixed point theorem in order to prove Miranda's theorem. Now we show that the latter theorem can be used in order to prove the first one. To this end we first prove Brouwer's theorem for the normalized n-cube $[a] = [-1, 1]^n \subset \mathbb{R}^n$. We start with $f \in C([a], [a])$ and define $g = (g_i)$ on $[a]$ by $g_i(x) = x_i - f_i(x)$, $i = 1, \dots, n$. Apparently, the function g fulfills the assumptions of Miranda's theorem, hence it has a zero x^* which is a fixed point of f.

Next we prove Brouwer's theorem for the unit ball. We will extend $f \in C(\overline{B}(0, 1), \overline{B}(0, 1))$ continuously to the n-cube $[a]$ above. Let y_z be the unique element of the intersection of $\partial[a]$ with the ray from 0 through $z \in \mathbb{R}^n \setminus \{0\}$. Define the continuous function $\varphi\colon \overline{B}(0, 1) \to [a]$ by

$$\varphi(z) = \begin{cases} \|z\|_2\, y_z, & z \ne 0 \\ 0, & z = 0. \end{cases}$$

This function has the inverse function φ^{-1} which can be represented as

$$\varphi^{-1}(x) = \begin{cases} \dfrac{x}{\|y_x\|_2}, & x \neq 0 \\ 0, & x = 0 \end{cases}$$

and which is continuous. Hence the continuous function $\hat{f} = \varphi \circ f \circ \varphi^{-1}$ maps $[a]$ into itself and therefore has a fixed point $x^* \in [a]$. Then $\varphi^{-1}(x^*)$ is a fixed point of f in $\in \overline{B}(0, 1)$.

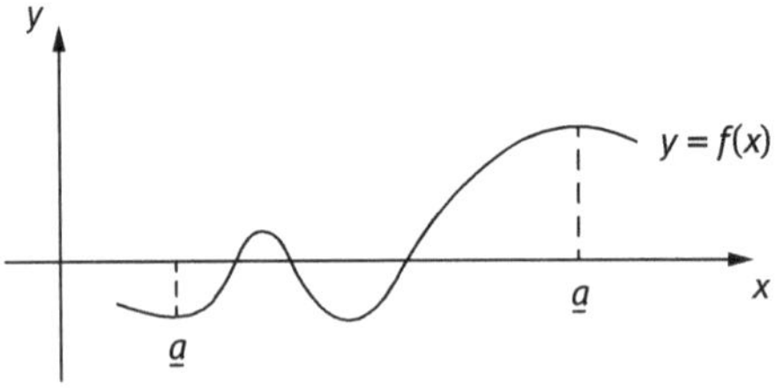

Fig. 1.5.3: Miranda's theorem in $\mathbb{R}^1$.

Notice that the one-dimensional case in the preceding theorems on zeros of f leads to a famous theorem of elementary analysis known as Bolzano's theorem: By virtue of convexity, $\overline{D}$ is a closed interval $[a] = [\underline{a}, \overline{a}]$. The assumptions (1.5.3) of the Borsuk theorem, (1.5.5) of the Leray–Schauder theorem and (1.5.8) of Miranda's theorem (cf. Figure 1.5.3) all mean that $f(\underline{a})$ and $f(\overline{a})$ cannot have the same positive, respectively negative, sign and imply that then the continuous function $f\colon [a] \to \mathbb{R}$ has a zero in $[a]$. This is exactly what is assumed and stated in Bolzano's theorem.

We close this section with the famous Newton–Kantorovich Theorem which guarantees convergence of the multidimensional Newton iteration

$$x^{k+1} = x^k - f'(x^k)^{-1} f(x^k), \quad k = 0, 1, \ldots. \tag{1.5.9}$$

Theorem 1.5.10 (Newton–Kantorovich). *Assume that $f\colon D \to \mathbb{R}^n$ is continuously differentiable on an open convex set $D \subseteq \mathbb{R}^n$. Denote by $\|\cdot\|$ any vector norm of $\mathbb{R}^n$ and the corresponding operator norm, respectively. Let $\beta > 0$, $\gamma > 0$, $\eta \geq 0$ be some constants and $\alpha = \beta\gamma\eta$. Define*

$$\underline{t} = \frac{1 - \sqrt{1 - 2\alpha}}{\beta\gamma}, \quad \overline{t} = \frac{1 + \sqrt{1 - 2\alpha}}{\beta\gamma} \tag{1.5.10}$$

and let the following conditions hold.

(i) $\|f'(x) - f'(y)\| \leq \gamma \|x - y\|$ *for all* $x, y \in D$,
(ii) $\|f'(x^0)^{-1}\| \leq \beta$,
(iii) $\|f'(x^0)^{-1} f(x^0)\| \leq \eta$,
(iv) $\alpha \leq \frac{1}{2}$,
(v) $\overline{B}(x^0, \underline{t}) \subseteq D$.

Then the Newton iterates x^k from (1.5.9) are well-defined (i.e., $f'(x^k)$ is nonsingular), remain in $\overline{B}(x^0, \underline{t})$ and converge to a zero $x^* \in \overline{B}(x^0, \underline{t})$ of f. If $0 \le \alpha < 1/2$ this zero is unique in $B(x^0, \bar{t}) \cap D$, if $\alpha = 1/2$ it is unique in $\overline{B}(x^0, \underline{t}) = \overline{B}(x^0, \bar{t})$. In addition, the error estimate

$$\|x^k - x^*\| \le \frac{1}{\beta\gamma 2^k}(2\alpha)^{2^k}, \quad k = 0, 1, \ldots \tag{1.5.11}$$

holds.

The lengthy proof of this theorem can be found in Appendix C. It is worth noticing that there are some relations between variants of the Theorems of Newton–Kantorovich, Borsuk and Miranda; for details see Alefeld, Frommer, Heindl, Mayer [21].

For the example $f(x) = x^2$, $x^0 = 1$ Theorem 1.5.10 can be applied with $\beta = \eta = 1/2$, $\gamma = 2$ whence $\alpha = 1/2$ follows. This implies $\underline{t} = \bar{t} = 1$, $B(x^0, \underline{t}) = B(x^0, \bar{t}) = (0, 2)$ and $x^* = 0 \in \partial B(x^0, \underline{t}) = \partial B(x^0, \bar{t})$. In particular, $B(x^0, \bar{t})$ does not contain any zero of f. This is the reason why we have two cases for α in the uniqueness statement of Theorem 1.5.10. For $0 \le \alpha < 1/2$, we always have $\underline{t} < \bar{t}$. Therefore, $\overline{B}(x^0, \underline{t}) \subset B(x^0, \bar{t})$ and $x^* \in B(x^0, \bar{t})$ follow trivially in this case.

By virtue of $\beta > 0$, $\gamma > 0$, the particular case $\alpha = 0$ is equivalent to $\eta = 0$ and to $\underline{t} = 0$. From (iii) one gets $f(x^0) = 0$, and the Newton iterates x^k satisfy $x^k = x^0 = x^* \in \overline{B}(x^0, \underline{t}) = [x^0, x^0]$. Therefore, the existence statement of Theorem 1.5.10 and the error estimate (1.5.11) are trivially true if $\alpha = 0$.

Exercises

Ex. 1.5.1. Define $f = (f_1, f_2)^T : \mathbb{R}^2 \to \mathbb{R}^2$ by $f_1(x) = 8x_1$, $f_2(x) = 8x_2$ for $x = (x_1, x_2)^T \in \mathbb{R}^2$ and choose $y = (0, 0)^T$. Let D be the open unit ball $B(0, 1)$ centered at $(0, 0)^T$ and define $\varphi : [0, \infty) \to \mathbb{R}$ by

$$\varphi(t) = \begin{cases} \frac{3}{16\pi}(t - 1)(3 - t), & 1 \le t \le 3, \\ 0, & \text{otherwise}. \end{cases}$$

(a) Compute the support of f when f is restricted to D.
(b) Show that φ has the property of Definition 1.5.1 (b) for $\alpha = 4$.
(c) Compute $\deg(f, D, y)$ according to Definition 1.5.2 (b) using $f_k = f$, $k = 1, 2, \ldots$.
(d) Compute $\deg(f, D, y)$ using Theorem 1.5.3 (a).
(e) Check whether Theorem 1.5.3 (c), (d) is applicable with the data of the example.

Ex. 1.5.2. Prove the equivalence of the Theorems 1.5.5 and 1.5.6.

1.6 Mean value theorems

In elementary courses on analysis one learns that functions $f\colon (a,b) \to \mathbb{R}$ that are at least $n+1$ times continuously differentiable can be represented as

$$f(x) = \sum_{k=0}^{n} \frac{f^{(k)}(x_0)}{k!}(x-x_0)^k + \frac{f^{(n+1)}(\xi)}{(n+1)!}(x-x_0)^{n+1}, \tag{1.6.1}$$

where x and x_0 can be chosen arbitrarily from (a,b) and ξ lies between x and x_0. The first sum is called the Taylor polynomial, the last summand is the remainder term of this so-called Taylor expansion. With $n = 0$ one obtains the mean value theorem

$$f(x) = f(x_0) + f'(\xi)(x-x_0), \tag{1.6.2}$$

which still holds for functions f, which are defined on open, convex subsets D of $\mathbb{R}^n$ as long as their values lie in $\mathbb{R}$. Here, the intermediate point ξ lies on the line segment connecting x with x_0. If f maps into $\mathbb{R}^m$, $m > 1$, formula (1.6.2) no longer holds with one and the same intermediate point ξ as argument of the Jacobian f'. This can be seen by the example $f(x) = (\cos x, \sin x)^T$, $x \in \mathbb{R}$. For $x = 2\pi$, $x_0 = 0$, one obtains

$$f(2\pi) - f(0) = (0,0)^T \neq \begin{pmatrix} -\sin\xi \\ \cos\xi \end{pmatrix} \cdot 2\pi = f'(\xi)(x-x_0),$$

for any $\xi \in [0, 2\pi]$ since $\|f'(\xi)\|_2 = 1$. Therefore, one either must use n separate intermediate points ξ_i, one for each component function f_i of f in the general case, or the analogue of (1.6.2) must look different. In fact, one can start with the equivalent integral version of (1.6.2) which finally leads to the following mean value theorem.

Theorem 1.6.1 (Mean value theorem). *Let $f\colon D \to \mathbb{R}^m$ be continuously differentiable (i.e., $f \in C^1(D)$) on an open convex set $D \subseteq \mathbb{R}^n$. Defining*

$$J(x,y) = \int_0^1 f'(x + t(y-x))\,dt \in \mathbb{R}^{m\times n}, \quad x, y \in D, \tag{1.6.3}$$

by entrywise integration one gets $J(x,y) = J(y,x)$ and

$$f(x) - f(x_0) = J(x, x_0)(x - x_0). \tag{1.6.4}$$

The matrix $J(x,y)$ depends continuously on x and y.

Proof. Define $\varphi_i(t) = f_i(x + t(y-x))$, $i, \dots, n$. Then

$$f_i(y) - f_i(x) = \int_0^1 \varphi_i'(t)\,dt = \int_0^1 f_i'(x + t(y-x)) \cdot (y-x)\,dt$$

which implies the assertion (1.6.4). Substituting $s = 1 - t$ results in $J(x,y) = J(y,x)$. Since f' is continuous it is uniformly continuous on each common compact convex subset $S \subseteq D$ of $x, y \in D$. Therefore the order of limit and integration can be interchanged which proves the continuity of J. □

Next we show that continuously differentiable functions are Lipschitz continuous on compact sets.

Theorem 1.6.2. *Let $f : D \to \mathbb{R}^n$ be continuously differentiable on an open convex set $D \subseteq \mathbb{R}^n$ and let $D_0 \subseteq D$ be compact. If $\|\cdot\|$ is any norm on $\mathbb{R}^n$, then there is a positive constant c such that*

$$\|f(x) - f(y)\| \le c\|x - y\|$$

holds for all $x, y \in D_0$.

Proof. From (1.6.4) we get $\|f(x) - f(y)\| \le \|J(x, y)\| \cdot \|x - y\|$, where we used the corresponding operator norm for the matrix $J(x, y)$. By means of Riemann sums for instance, one can see that $\|J(x, y)\| \le \int_0^1 \|f'(x + t(y - x))\| \, dt$ holds. Since the last integrand is continuous on the compact set D_0 we can bound it by some constant $c > 0$ independently of $x, y \in D_0$. This concludes the proof. □

1.7 Normal forms of matrices

Normal forms of matrices are representatives of larger classes of matrices. They show particularities of these classes which otherwise are hidden. Examples of normal forms are named after Jordan, Schur, Smith or bear names like reducible normal form or cyclic normal form. Sometimes they are also referred to as canonical forms. Some of these forms are used in this book and will be recalled in this section and in Section 1.9.

We start with the well-known Jordan form which is used, e.g., in connection with eigenvalues and powers of matrices.

Theorem 1.7.1 (with definition; Jordan normal form). *For each matrix $A \in \mathbb{K}^{n\times n}$ there are integers $m, n_1, \dots, n_m \in \mathbb{N}$ and a regular matrix $S \in \mathbb{C}^{n\times n}$ such that*

$$S^{-1}AS = J = \begin{pmatrix} J_1 & & & & O \\ & J_2 & & & \\ & & \ddots & & \\ & & & J_{m-1} & \\ O & & & & J_m \end{pmatrix} \in \mathbb{C}^{n\times n},$$

where

$$J_l = \begin{pmatrix} \lambda_l & 1 & & O \\ & \ddots & \ddots & \\ & & \lambda_l & 1 \\ O & & & \lambda_l \end{pmatrix} \in \mathbb{C}^{n_l\times n_l}, \quad l = 1, \dots, m.$$

The matrix J is called the Jordan normal form of A, J_l is called the Jordan block associated with λ_l. The characteristic polynomials $\det(\mu I - J_l) = (\mu - \lambda_l)^{n_l}$, $l = 1, \dots, m$ are called the elementary divisors of A. If $n_l = 1$, the corresponding elementary divisor is called linear.

The Jordan normal form is unique up to the order of its Jordan blocks. The complex numbers λ_l are the eigenvalues of A. For each eigenvalue λ of A there are $n - \operatorname{rank}(\lambda I - A)$ Jordan blocks and equally many linearly independent eigenvectors. This number is called the geometric multiplicity of λ. The total number of columns of those Jordan blocks that are associated with the same λ is called the algebraic multiplicity of λ or simply multiplicity of λ. It coincides with the multiplicity of λ as a zero of the characteristic polynomial $\det(\mu I - A) = (-1)^n \det(A - \mu I)$. Eigenvalues of algebraic multiplicity one are called simple eigenvalues. They are at the same time geometrically simple but not vice versa.

A proof of this important theorem is extensive. Usually it is presented in a course of linear algebra. Therefore, we omit it here but defer it to Appendix A.

The matrix S in Theorem 1.7.1 describes a change of basis in $\mathbb{C}^n$ (or in the vector space for which $\mathbb{C}^n$ is the corresponding coordinate space), where the new basis vectors are just the columns of S when represented in the old basis. In this way S represents a mapping from $\mathbb{C}^n$ with the new basis into $\mathbb{C}^n$ with the old one. From the Jordan normal form it can immediately be seen that there is a basis $\{b^1, \dots, b^n\}$ of $\mathbb{C}^n$ such that Ab^i either satisfies

$$Ab^i = \lambda b^i \tag{1.7.1}$$

or

$$Ab^i = \lambda b^i + b^{i-1}. \tag{1.7.2}$$

Here λ is an eigenvalue of A and b^i from (1.7.1) is a corresponding eigenvector. Together they form an eigenpair (b^i, λ).

Definition 1.7.2. Let $\lambda \in \mathbb{C}$ be an eigenvalue of $A \in \mathbb{K}^{n\times n}$.

(a) If $x \in \mathbb{C}^n$ satisfies

$$(A - \lambda I)^m x = 0, \quad (A - \lambda I)^{m-1} x \neq 0$$

for some $m \in \mathbb{N}$, then x is called a principal vector of degree m (of A associated with the eigenvalue λ).

(b) A sequence $x^1, x^2, \dots, x^m$ of principal vectors x^i of degree i forms a Jordan chain of length m with leading vector x^1 (= head vector) and end vector x^m if

$$\begin{aligned}(A - \lambda I)x^1 &= 0,\\ (A - \lambda I)x^i &= x^{i-1}, \quad i = 2, \dots, m,\\ (A - \lambda I)z &= x^m \text{ is unsolvable.}\end{aligned}$$

According to this definition eigenvectors are principal vectors of degree 1. Each of them initializes some Jordan chain. It is easily seen that for a principal vector x of degree m (associated with an eigenvalue λ of $A \in \mathbb{K}^{n\times n}$) the vectors

$$b^i = (A - \lambda I)^{m-i} x, \quad i = 1, \dots, m$$

are principal vectors of degree i which satisfy (1.7.1) for $i = m$ and (1.7.2) for $i = 1, \dots, m-1$.

As a first application of the Jordan normal form we show that the spectral radius of a matrix can be arbitrarily well approximated by a matrix norm.

Theorem 1.7.3. *Let $A \in \mathbb{K}^{n\times n}$, $J = S^{-1}AS$ its Jordan normal form, and $D_\varepsilon = \operatorname{diag}(1, \varepsilon, \varepsilon^2, \dots, \varepsilon^{n-1}) \in \mathbb{R}^{n\times n}$ with a real parameter $\varepsilon > 0$. Then*

$$\|A\|_{SD_\varepsilon} \le \rho(A) + \varepsilon,$$

where the norm was introduced in Section 1.3.

Proof. The inequality follows at once from

$$(SD_\varepsilon)^{-1}ASD_\varepsilon = D_\varepsilon^{-1}JD_\varepsilon = \operatorname{diag}(\hat{J}_1, \dots, \hat{J}_m)$$

with the modified Jordan blocks $\hat{J}_l = \begin{pmatrix} \lambda_l & \varepsilon & & O \\ & \ddots & \ddots & \\ & & \lambda_l & \varepsilon \\ O & & & \lambda_l \end{pmatrix}$. □

Next we consider the powers of a Jordan block.

Theorem 1.7.4. *The powers of a Jordan block J_l of Theorem* 1.7.1 *can be represented as $J_l^k = (c_{ij}^{(k,l)})$ with*

$$c_{ij}^{(k,l)} = \begin{cases} \binom{k}{j-i} \lambda_l^{k-(j-i)}, & \text{if } i \le j \le \min\{n_l,\ k+i\} \\ 0, & \text{otherwise.} \end{cases}$$

Proof. By induction. □

Definition 1.7.5. A matrix $A \in \mathbb{K}^{n\times n}$ is called semiconvergent if $A^\infty = \lim_{k\to\infty} A^k$ exists. It is called convergent if it is semiconvergent with $A^\infty = O$.

From $A^k = SJ^kS^{-1} = \operatorname{diag}(J_1^k, \dots, J_m^k)$ and from Theorem 1.7.4 we immediately get the following result.

Theorem 1.7.6. *The matrix $A \in \mathbb{K}^{n\times n}$ is convergent if and only if $\rho(A) < 1$. It is semiconvergent if and only if $\rho(A) \le 1$, where in the case $\rho(A) = 1$ the only eigenvalue λ with $|\lambda| = 1$ is $\lambda = 1$, which, in addition, must have only linear elementary divisors.*

Definition 1.7.7. The series $\sum_{k=0}^{\infty} A^k$, $A \in \mathbb{K}^{n\times n}$ is called the Neumann series.

Theorem 1.7.8. *For $A \in \mathbb{K}^{n\times n}$ the Neumann series converges if and only if $\rho(A) < 1$. In this case the inverse $(I - A)^{-1}$ exists and is equal to this series.*

Proof. Let $\rho(A) < 1$ and choose $\varepsilon > 0$ such that $\rho(A) + \varepsilon < 1$. Use the norm $\|\cdot\| = \|\cdot\|_{SD_\varepsilon}$ from Theorem 1.7.3 in order to prove

$$\left\| \sum_{k=k_0}^{\infty} A^k \right\| \le \sum_{k=k_0}^{\infty} \|A^k\| \le \{\rho(A) + \varepsilon\}^{k_0} / (1 - \rho(A) - \varepsilon).$$

Thus the remainder of the series tends to zero if k_0 tends to infinity. This proves the convergence of the series to some matrix L. Let k_0 tend to infinity in $(I - A) \sum_{k=0}^{k_0 - 1} A^k = I - A^{k_0}$, whence $(I - A)L = I$. Therefore, $(I - A)^{-1}$ exists and equals L.

Conversely, let the series be convergent to some matrix L. Then A^k tends to zero as the difference of two successive partial sums which both tend to L. Hence $\rho(A) < 1$ by Theorem 1.7.6. □

Now we present a second normal form of a matrix A. It grows out of restricting the class of transformation matrices S in the Jordan normal form to the class of unitarian matrices.

Theorem 1.7.9 (with definition; Schur normal form). *Let $A \in \mathbb{K}^{n\times n}$. Then there is a unitarian matrix $U \in \mathbb{C}^{n\times n}$ and an upper triangular matrix $R \in \mathbb{C}^{n\times n}$ such that $U^H AU = R$. The matrix R is called the Schur normal form of A, the representation $A = URU^H$ Schur decomposition.*

Proof. We proceed by induction on the dimension n of A.

If $n = 1$, then one can choose $U = I$. Assume now that the assertion holds for all matrices of dimension at most $n - 1$. Choose an eigenpair (x^1, λ^1) of A with $\|x^1\|_2 = 1$. Extend x^1 by $x^2, \ldots, x^n$ to an orthonormal basis of $\mathbb{C}^n$, i.e., $(x^i)^H x^j = \delta_{ij}$. Define the matrix $M = (x^2, \ldots, x^n) \in \mathbb{C}^{n\times(n-1)}$. Then

$$(x^1, M)^H A(x^1, M) = \begin{pmatrix} \lambda^1 & (x^1)^H AM \\ 0 & M^H AM \end{pmatrix}.$$

By virtue of the induction hypothesis there are a unitarian matrix $\tilde{U}$ and an upper triangular matrix $\tilde{R}$ such that $\tilde{U}^H (M^H AM)\tilde{U} = \tilde{R}$. Let

$$U = (x^1, M) \begin{pmatrix} 1 & 0 \\ 0 & \tilde{U} \end{pmatrix}.$$

Then U is unitarian and

$$U^H AU = \begin{pmatrix} \lambda^1 & (x^1)^H AM\tilde{U} \\ 0 & \tilde{R} \end{pmatrix} = R.$$ □

Notice that for $A \in \mathbb{R}^{n\times n}$ the matrices U and R cannot necessarily be chosen to be real: since $AU = UR$ the first column of U is an eigenvector of A associated with the eigenvalue r_{11}. The matrix

$$A = \begin{pmatrix} 0 & 1 \\ -1 & 0 \end{pmatrix},$$

however, has the eigenvalues $\pm i$ and eigenvectors which all are multiples of $(1, \pm i)^T$.

The Schur decomposition is not unique: one can always replace U, R by UD, DRD, where D is any signature matrix. If A is singular one can modify U even further as the extreme example $A = O$ shows.

Our next normal form concerns rectangular matrices.

Theorem 1.7.10 (with definition; singular value decomposition). *Let $A \in \mathbb{K}^{m\times n}$. Then there are unitarian matrices $U \in \mathbb{K}^{n\times n}$, $V \in \mathbb{K}^{m\times m}$ and a rectangular matrix $\Sigma \in \mathbb{R}^{m\times n}$ such that*

$$A = V\Sigma U^H \tag{1.7.3}$$

holds where Σ has the form

$$\Sigma = \begin{pmatrix} \sigma_1 & 0 & 0 & \dots & 0 \\ 0 & \sigma_2 & 0 & \dots & 0 \\ \vdots & \ddots & \ddots & \ddots & \vdots \\ 0 & \dots & 0 & \sigma_{n-1} & 0 \\ 0 & \dots & \dots & 0 & \sigma_n \\ & & O & & \end{pmatrix}$$

in the case $m \geq n$, and

$$\Sigma = \begin{pmatrix} \sigma_1 & 0 & 0 & \dots & \dots & \dots & 0 \\ 0 & \sigma_2 & 0 & \dots & \dots & \dots & 0 \\ \vdots & \ddots & \ddots & \ddots & & & \vdots \\ 0 & \dots & 0 & \sigma_m & 0 & \dots & 0 \end{pmatrix}$$

in the case $m \leq n$ with $\sigma_1 \geq \sigma_2 \geq \sigma_r > \sigma_{r+1} = 0 = \dots = \sigma_{\min\{m,n\}}$. The matrix Σ is unique. The decomposition (1.7.3) is called the singular value decomposition of A, the values σ_i are denoted as singular values, the columns of U as right singular vectors, those of V as left singular vectors; r is the rank of A.

Proof. Let (x, λ) be an eigenpair of A^HA with $\|x\|_2 = 1$. Then $\|Ax\|_2^2 = x^HA^HAx = \lambda x^Hx = \lambda$, i.e., $\lambda \in \mathbb{R}_0^+$. Now choose (x, λ) such that $\lambda = \rho(A^HA)$, $x^Hx = 1$, and define $\sigma := \sqrt{\lambda} \geq 0$. If $\sigma \neq 0$ let $y := \frac{1}{\sigma}Ax$ which implies $\|y\|_2^2 = y^Hy = 1$. Otherwise choose y arbitrarily such that $\|y\|_2 = 1$ and notice that $\|Ax\|_2^2 = x^HA^HAx = 0$ holds which implies $Ax = 0$. Let $U = (x, U_1) \in \mathbb{C}^{n\times n}$, $V = (y, V_1) \in \mathbb{C}^{m\times m}$ be unitarian matrices. (Such matrices exist since it is always possible to extend an orthonormal set of vectors in $\mathbb{K}^n$ to an orthonormal basis of $\mathbb{K}^n$.) Then

$$\hat{A} := V^HAU = \begin{pmatrix} \sigma & w^H \\ 0 & A' \end{pmatrix}, \quad \hat{A}^T\hat{A} = U^HA^HAU$$

holds, and we get

$$\left\| \hat{A} \begin{pmatrix} \sigma \\ w \end{pmatrix} \right\|_2^2 \geq (\sigma^2 + w^Hw)^2 = \left\| \begin{pmatrix} \sigma \\ w \end{pmatrix} \right\|_2^4 .$$

Hence

$$\|\hat{A}\|_2^2 \geq \frac{\left\|\hat{A}\binom{\sigma}{w}\right\|_2^2}{\left\|\binom{\sigma}{w}\right\|_2^2} \geq \left\|\binom{\sigma}{w}\right\|_2^2 = \sigma^2 + w^H w$$
$$= \|A\|_2^2 + w^H w = \|\hat{A}\|_2^2 + w^H w,$$

and $\|w\|_2^2 = w^H w = 0$ follows. Here, we used the fact that any two matrices A and $B = S^{-1}AS$ have the same eigenvalues; cf. Exercise 1.8.2. An easy induction proves the existence of the decomposition $A = V^H \Sigma U$, the nonnegativity of σ_i and its monotony with respect to i. The meaning of r can be seen from $\mathrm{rank}(A) = \mathrm{rank}(\Sigma)$ since U and V are regular. Results of Section 1.8 show that Σ is unique while U, V are not. Replace for instance U, V by $-U$, $-V$ or consider the example $A = O \in \mathbb{R}^{n\times n}$, where $\Sigma = O$ and U, V can be chosen to be any orthogonal matrices. □

From (1.7.3) we get $A^H A U = U\Sigma^H \Sigma$ and $V^H A A^H = \Sigma\Sigma^H V^H$ i.e., the columns of U are eigenvectors of $A^H A$ with respect to the eigenvalues σ_i^2, and, similarly, the columns of V are left eigenvectors of AA^H (i.e., eigenvectors of $(AA^H)^H$) with respect to the same eigenvalues.

In order to define another normal form for $A \in \mathbb{K}^{n\times n}$ some preparations are necessary.

Definition 1.7.11. The (directed) graph $G(A) = (N, E)$ of $A \in \mathbb{K}^{n\times n}$ consists of the set $N = \{1, \dots, n\}$ of nodes and of the set $E = \{(i, j) \mid a_{ij} \neq 0\}$ of (directed) edges.

The sequence $(i_0, i_1, \dots, i_k)$ of nodes with $(i_{l-1}, i_l) \in E$ for $l = 1, \dots, k$ is called a (directed) path of length k from i_0 to i_k. We sometimes write $i_0 \to i_1 \to \cdots \to i_k$ for this path and say 'i_0 is connected to i_k'. If $i_0 = i_k$ we call this path a cycle through i_0. Cycles of length 1 will be denoted as loops, abbreviated by '$\curvearrowright$'.

Since we will only deal with directed graphs in this book we will always omit the specification 'directed'. Loops do not matter in our applications.

Example 1.7.12.

$$A = \begin{pmatrix} 10 & 0 & 12 & 13 \\ 0 & 0 & 14 & 0 \\ 0 & 15 & 0 & 0 \\ 16 & 0 & 0 & 17 \end{pmatrix}, \qquad G(A): \begin{array}{ccc} \curvearrowright & & \curvearrowright \\ 1 & \longleftrightarrow & 4 \\ \downarrow & & \\ 3 & \longleftrightarrow & 2 \end{array}$$

In the next definition we use permutation matrices. These are matrices from $\mathbb{R}^{n\times n}$ which in each column and in each row have exactly one entry equal to one and zeros otherwise.

Definition 1.7.13. The matrix $A \in \mathbb{K}^{n\times n}$, $n > 1$, is called reducible if there is a permutation matrix $P \in \mathbb{R}^{n\times n}$ such that the matrix PAP^T has the block form

$$PAP^T = \begin{pmatrix} A_{11} & O \\ A_{21} & A_{22} \end{pmatrix}, \tag{1.7.4}$$

where A_{11}, A_{22} are square matrices. In the case $n = 1$ the matrix is called reducible if $A = O$.

The matrix A is called irreducible if it is not reducible.

With the exception of the case $n = 1$, the diagonal entries of a matrix are irrelevant for the property reducible/irreducible. It can be seen at once that an irreducible matrix can neither have a zero row nor a zero column.

Theorem 1.7.14. *For $A \in \mathbb{K}^{n\times n}$, $n > 1$ the following statements are equivalent.*
(a) *A is reducible.*
(b) *There are two nodes i, j of $G(A) = (N, E)$ which are not connected by a path from i to j.*
(c) *There are two nonempty disjoint sets N_1, N_2 of nodes with $N_1 \cup N_2 = N$ and $a_{ij} = 0$ for all $(i, j) \in N_1 \times N_2$.*

Proof. '(a) ⇒ (b)': Let $\tilde{A} = PAP^T$ be the matrix in (1.7.4) and denote by $\pi: N \to N$ the permutation which describes the row and column permutation effected by P, i.e., $a_{ij} = \tilde{a}_{\pi(i),\pi(j)}$. According to the representation (1.7.4) there is certainly no path in $G(\tilde{A})$ from 1 to n. Therefore, there is no path in $G(A)$ from $\pi^{-1}(1)$ to $\pi^{-1}(n)$.

'(b) ⇒ (c)': Let i, j be as in (b) and define

$$N_1 = \{k \mid k \in N, i \text{ is connected with } k\} \cup \{i\} \quad \text{and} \quad N_2 = N \setminus N_1,$$

and notice that $j \in N_2$.

'(c) ⇒ (a)': Choose P such that the first rows in PAP^T are just those rows of A with the indices from N_1. □

Clearly, for irreducible matrices the negation of the preceding theorem holds. We restate a part of it for easy reference.

Theorem 1.7.15. *A matrix $A \in \mathbb{K}^{n\times n}$ is irreducible if and only if any two nodes i, j of $G(A) = (N, E)$ are connected by a path from i to j.*

The matrix in Example 1.7.12 is reducible since there is no path from 3 to 1. According to the proof of the preceding theorem one can choose $N_1 = \{2, 3\}$ and $N_2 = \{1, 4\}$. With

$$P = \begin{pmatrix} 0 & 0 & 1 & 0 \\ 0 & 1 & 0 & 0 \\ 1 & 0 & 0 & 0 \\ 0 & 0 & 0 & 1 \end{pmatrix}$$

we get the reducible form

$$PAP^T = \begin{pmatrix} 0 & 15 & 0 & 0 \\ 14 & 0 & 0 & 0 \\ 12 & 0 & 10 & 13 \\ 0 & 0 & 16 & 17 \end{pmatrix}.$$

The idea of permuting rows and columns simultaneously can be repeated for the whole matrix in order to aim at a reducible structure for the diagonal blocks A_{11} and A_{22}. Repeating the process until no further decomposition can be achieved finally results in the so-called reducible normal form introduced in the next theorem.

Theorem 1.7.16 (Reducible normal form; definition). *For each matrix $A \in \mathbb{K}^{n\times n}$ there are positive integers r, s and a permutation matrix $P \in \mathbb{R}^{n\times n}$ such that PAP^T has the triangular block form*

$$\begin{pmatrix} A_{11} & & & & & O \\ & \ddots & & & & \\ O & & A_{rr} & & & \\ A_{r+1,1} & \dots & \dots & A_{r+1,r+1} & & \\ \vdots & & & & \ddots & \\ A_{s1} & \dots & \dots & \dots & \dots & A_{ss} \end{pmatrix} =: R_A \tag{1.7.5}$$

with block matrices A_{ij} that are square in the diagonal and there they are either irreducible or 1×1 zero matrices.

The matrix R_A is called the reducible normal form of A.

It is unique with the exception of permutations of block rows and corresponding block columns and of permutations of rows/columns within fixed block rows and corresponding block columns.

In practice it is not so easy to generate the reducible normal form of a matrix. Search algorithms are necessary in the corresponding graph. See Pissanetzky [272] for details. If, however, a block structure like that in (1.7.4) or (1.7.5) is known, the amount of work for solving linear systems or computing eigenvalues can be diminished drastically by decomposing the task into several smaller ones.

In theory the reducible normal form is interesting because it often allows to study a problem for irreducible matrices first and then to generalize to matrices of the form (1.7.5).

Exercises

Ex. 1.7.1. Show that principal vectors $x^1, \dots, x^m$ of pairwise different degrees are linearly independent. In particular, the elements of a Jordan chain have this property.

1.8 Eigenvalues

The eigenvalues of a matrix $A \in \mathbb{K}^{n\times n}$ are the zeros of the characteristic polynomial $\det(\lambda I - A)$. Therefore, they can be complex even if A is real. Since one can prove that generally the zeros of polynomials of degree greater than 4 cannot be computed explicitly in finitely many steps, the same holds for the computation of the eigenvalues, the Jordan normal form, and the Schur normal form of general square matrices.

In the previous section we already learnt some basics of the classification of eigenvalues. Now, we will recall some further results on them.

Theorem 1.8.1. *Let $\lambda_1, \dots, \lambda_n$ be the eigenvalues of $A \in \mathbb{C}^{n\times n}$ (counted according to their multiplicity). Then*

$$\operatorname{trace}(A) = a_{11} + \cdots + a_{nn} = \sum_{i=1}^{n} \lambda_i, \quad \det A = \prod_{i=1}^{n} \lambda_i. \tag{1.8.1}$$

If A is a real matrix with a nonreal eigenvalue λ, then the conjugate complex value $\bar{\lambda}$ is also an eigenvalue of A.

Proof. The formulae (1.8.1) follow immediately from the factorization $\det(\lambda I - A) = \prod_{i=1}^{n}(\lambda - \lambda_i)$ of the characteristic polynomial when expanding the expressions on both sides and comparing the coefficients of λ^{n-1} and λ^0. The remaining part of the theorem can be seen from $0 = \bar{0} = \overline{\det(\lambda I - A)} = \det(\bar{\lambda} I - A)$ which holds since A is real. □

Theorem 1.8.2. *Let $A \in \mathbb{C}^{n\times n}$.*

(a) *The eigenvalues of A depend continuously on the entries a_{ij}.*
(b) *All algebraically simple eigenvalues of A and the corresponding eigenvectors normalized by $x_{i_0} = \alpha \neq 0$ for some fixed index i_0 are functions of the entries a_{ij} which are infinitely often continuously differentiable.*

Proof. The assertion in (a) follows immediately from the characteristic polynomial and Theorem 1.4.1. By the same reasons, the statement in (b) holds for eigenvalues. Therefore, it remains to prove the smoothness for the eigenvectors.

Let λ^* be a simple eigenvalue of A. Then $\operatorname{rank}(\lambda^* I - A) = n - 1$. Let x^* be a corresponding eigenvector normalized by $x_n^* = \alpha$. (W.l.o.g. we choose $i_0 = n$.) According to the rank condition there is an index k such that the components x_i^*, $i = 1, \dots, n-1$, form the unique solution of the linear system

$$\sum_{j=1}^{n-1} (\delta_{ij}\lambda^* - a_{ij})x_j = -(\delta_{in}\lambda^* - a_{in})\alpha, \quad i = 1, \dots, n, \ i \neq k. \tag{1.8.2}$$

For simplicity let $k = n$. Consider any matrix $E \in \mathbb{K}^{n\times n}$ with $\|E\|_\infty$ being sufficiently small. Denote by $\lambda(E)$ an eigenvalue of $A + E$ such that $\lim_{E\to O} \lambda(E) = \lambda^*$. Such an eigenvalue exists according to (a). By the same reason we may assume that $\lambda(E)$ is simple. The linear system

$$\sum_{j=1}^{n-1} (\delta_{ij}\lambda(E) - a_{ij} - e_{ij})x_j = -(\delta_{in}\lambda(E) - a_{in} - e_{in})\alpha, \quad i = 1, \dots, n-1, \tag{1.8.3}$$

has a coefficient matrix $B(E)$ which we assume to be regular. This is possible because of $\lim_{E\to O} \det B(E) = \det B(O) \neq 0$ where $B(O)$ is the coefficient matrix of the regular system (1.8.2) for $k = n$. Let $x(E)$ be an eigenvector associated with $\lambda(E)$. Then $x_n(E) \neq 0$ since otherwise $B(E)(x_1(E), \dots, x_{n-1}(E))^T = 0$ implies $x(E) = 0$, hence $x(E)$ cannot be an eigenvector. Therefore, we can assume $x_n(E) = \alpha$, and $x_i = x_i(E)$, $i = 1, \dots, n-1$, is the unique solution of (1.8.3). Represent $x_i(E)$ by means of Cramer's rule. This shows that $x_i(E)$ depends continuously on a_{ij} and $\lambda(E)$, and can be differentiated infinitely often with continuous derivatives. □

For multiple eigenvalues part (b) of the theorem may become false as the example $A = I \in \mathbb{R}^{2\times 2}$, $A_\varepsilon = I + (1,0)^T(0,\varepsilon)$ shows. Choose $x^* = e$ as eigenvector of A. Each eigenvector of A_ε can be written as $x_\varepsilon = \beta(1,0)^T$, $\beta \neq 0$, so that $\|x_\varepsilon - x^*\|_\infty \geq 1$ for arbitrary β. Hence $\lim_{\varepsilon\to 0} x_\varepsilon = x^*$ is not possible.

Our next theorem characterizes algebraically simple eigenvalues.

Theorem 1.8.3. *Let (x^*, λ^*) be an eigenpair of $A \in \mathbb{C}^{n\times n}$ with $x_n^* \neq 0$. Then the following assertions are equivalent.*
(i) *λ^* is an algebraically simple eigenvalue.*
(ii) $B^* = \begin{pmatrix} A - \lambda^* I & -x^* \\ (e^{(n)})^T & 0 \end{pmatrix}$ *is nonsingular.*
(iii) $(B^*)' = ((A - \lambda^* I)_{*,1}, \dots, (A - \lambda^* I)_{*,n-1}, -x^*)$ *is nonsingular.*

Proof. '(i) ⇒ (ii)': Assume B^* to be singular with

$$B^* z = 0 \tag{1.8.4}$$

for some $z \in \mathbb{C}^{n+1} \setminus \{0\}$. With z we construct the vector $\hat{z} = (z_1, \dots, z_n)^T \in \mathbb{C}^n$. From (1.8.4) we get

$$(A - \lambda^* I)\hat{z} = z_{n+1}\, x^* \tag{1.8.5}$$

and

$$z_n = 0. \tag{1.8.6}$$

By (1.8.5) we obtain $\hat{z} \neq 0$ since otherwise we get $z_{n+1} = 0$ from (1.8.5) which implies the contradiction $z = 0$. If $z_{n+1} = 0$, then (1.8.5), (1.8.6) and $x_n^* \neq 0$ show that $\hat{z}$ and x^* are linearly independent eigenvectors of A. If $z_{n+1} \neq 0$, then (1.8.5) yields

$$(A - \lambda^* I)^2 \frac{\hat{z}}{z_{n+1}} = 0,$$

hence $\frac{\hat{z}}{z_{n+1}}$ is a principal vector of A. Thus in both cases λ^* is an algebraically multiple eigenvalue. This contradicts statement (i).

'(ii) ⇒ (iii)': follows by evaluating $\det(B^*)$ along the $(n+1)$-th row which gives $-\det(B^*)'$.

'(iii) ⇒ (i)': Assume that λ^* is an algebraically multiple eigenvalue. Then A has a left eigenvector y^* which is associated with this eigenvalue and which satisfies $(y^*)^H x^* = 0$. This can be easily seen from the Jordan normal form of A. Thus $(y^*)^H (B^*)' = 0$ so that the matrix $(B^*)'$ is singular, contradicting statement (iii). □

Next we show that the order of multiplication of two matrices is irrelevant for the spectral radius of their product.

Theorem 1.8.4. *Let* $A \in \mathbb{C}^{m\times n}$, $B \in \mathbb{C}^{n\times m}$. *Then* $\rho(AB) = \rho(BA)$.

Proof. Assume $\rho(AB) < \rho(BA)$ and let λ be an eigenvalue of BA with $|\lambda| = \rho(BA)$. Let x be a corresponding eigenvector of BA. From $\rho(BA) > 0$ we get $BAx = \lambda x \neq 0$, hence $x' = Ax$ differs from zero and fulfills $ABx' = \lambda x'$. Hence λ is an eigenvalue of AB, which leads to the contradiction $\rho(AB) \geq \rho(BA)$. Interchanging the roles of A and B shows that $\rho(AB) > \rho(BA)$ cannot hold either. □

Now we consider symmetric matrices. We start with a result which is well-known from elementary courses on linear algebra.

Theorem 1.8.5. *Let* $A = A^T \in \mathbb{R}^{n\times n}$. *Then all eigenvalues* λ_i *of* A *are real, there is an orthonormal basis of real eigenvectors* x^i, *and* $X^TAX = \operatorname{diag}(\lambda_1, \dots, \lambda_n)$, *where* $X = (x^1, \dots, x^n)$.

Proof. The Schur decomposition $A = URU^H = A^T = A^H = UR^HU^H$ of A implies $R = R^H$, hence R is a real diagonal matrix, identical with the Jordan normal form, all eigenvalues of A are real, and there is a basis of (at the moment still complex) orthonormal eigenvectors, namely the columns of U. Since the matrix $A - \lambda_1 I$ is real, there is a real eigenvector x^1 with $\|x^1\|_2 = 1$, and the induction in the proof of Theorem 1.7.9 (Schur normal form) can be done completely with real vectors and real matrices resulting in $U \in \mathbb{R}^{n\times n}$. This proves the assertion with $X = U$. □

Definition 1.8.6. The inertia $\operatorname{ind}(A)$ (= index of inertia) of a symmetric matrix $A \in \mathbb{R}^{n\times n}$ is the triplet (i, j, k) of nonnegative integers which count the number of negative, zero, and positive eigenvalues of A.

Theorem 1.8.7 (Sylvester's law of inertia). *Let* $A = A^T \in \mathbb{R}^{n\times n}$ *and let* $S \in \mathbb{R}^{n\times n}$ *be regular. Then* $\operatorname{ind}(A) = \operatorname{ind}(S^TAS)$.

Proof. Define $B = S^TAS$. According to Theorem 1.8.5 there are orthogonal matrices X, X and diagonal matrices $D, \tilde{D}$ such that $X^TAX = D$, $\tilde{X}^TB\tilde{X} = \tilde{D}$. Hence $\tilde{D} = \tilde{S}^TD\tilde{S}$ with the regular matrix $\tilde{S} = X^TS\tilde{X}$. Assume that the number of positive eigenvalues of B and therefore of $\tilde{D}$, is less than that of A and therefore of D. Then the homogeneous linear system

$$\begin{aligned} x_i &= 0 \quad \text{for } i \text{ with } \tilde{d}_{ii} > 0 \\ (\tilde{S}x)_i &= 0 \quad \text{for } i \text{ with } d_{ii} \leq 0 \end{aligned}$$

consists of less than n equations. Therefore, it has a solution $x \neq 0$. This implies the contradiction

$$0 \geq x^T\tilde{D}x = (\tilde{S}x)^TD(\tilde{S}x) > 0.$$

Notice that at least one component of $\tilde{S}x$ differs from zero by virtue of the regularity of $\tilde{S}$. Interchanging the roles of A and B shows that the numbers of the positive eigenvalues are equal. Since S is regular, both matrices have the same rank and thus the same number of zero and negative eigenvalues. □

Sylvester's law of inertia considers transformations of the form S^TAS which are called congruence transformations. In contrast to the similarity transformations $S^{-1}AS$, the eigenvalues of A may change with congruence transformations, but for *symmetric* matrices A, at least their signs remain stable. This will become important in one of our verification algorithms for eigenvalues in Chapter 7. There, we will use a congruence transformation with an appropriate lower triangular matrix L. With some permutation matrix P and a block diagonal matrix D with at most 2×2 diagonal blocks we will obtain the representation $PAP^T = LDL^T$. The eigenvalues of $A = A^T$ coincide with those of PAP^T. The eigenvalues of D and hence $\operatorname{ind}(D)$ can be computed (theoretically) without any problems. Sylvester's law of inertia finally guarantees $\operatorname{ind}(A) = \operatorname{ind}(D)$. The matrices P, D, L are constructed by the subsequent algorithm of Bunch, Kaufman, and Parlett [74] which proceeds recursively and is based on the representation

$$A' = \tilde{P}A\tilde{P}^T = \begin{pmatrix} E & C^T \\ C & B \end{pmatrix} = (A')^T \tag{1.8.7}$$

with an appropriate permutation matrix $\tilde{P}$, and $E \in \mathbb{R}^{s\times s}$, $C \in \mathbb{R}^{(n-s)\times s}$, $B \in \mathbb{R}^{(n-s)\times(n-s)}$ with $s = 1$ or $s = 2$. The matrix E is symmetric and can be chosen to be regular if A is regular or if $C \neq O$. It is the starting block of D. If $C \neq O$, then

$$\begin{pmatrix} E & C^T \\ C & B \end{pmatrix} = \begin{pmatrix} I_s & O \\ CE^{-1} & I_{n-s} \end{pmatrix} \begin{pmatrix} E & O \\ O & B' \end{pmatrix} \begin{pmatrix} I_s & (CE^{-1})^T \\ O & I_{n-s} \end{pmatrix} \tag{1.8.8}$$

with $B' = B - CE^{-1}C^T$; otherwise

$$\begin{pmatrix} E & C^T \\ C & B \end{pmatrix} = I \begin{pmatrix} E & O \\ O & B' \end{pmatrix} I^T \tag{1.8.9}$$

with $B' = B$. More precisely, the algorithm runs as follows.

Algorithm 1.8.8 (Bunch, Kaufman, Parlett). Let $A = A^T \in \mathbb{R}^{n\times n}$.
Choose $\alpha \in (0, 1)$ and let $m = n$;
if $m = 1$, then $s = 1$, $\tilde{P} = I$, $e_{11} = a_{11}$;
if $m > 1$,
  choose $r \in \{2, \dots, m\}$ such that $|a_{r1}| = \max_{2\le i\le m}|a_{i1}|$;
  if

$$|a_{11}| \geq \alpha|a_{r1}|, \tag{1.8.10}$$

  then $s = 1$, $\tilde{P} = I$, $e_{11} = a_{11}$,
  else
    choose $p \in \{1, \dots, m\} \setminus \{r\}$ such that $|a_{pr}| = \max_{1\le i\le m,\ i\neq r}|a_{ir}|$;

if

$$|a_{11}| \cdot |a_{pr}| \geq \alpha |a_{r1}|^2, \tag{1.8.11}$$

then $s = 1$, $\tilde{P} = I$, $e_{11} = a_{11}$,
else
  if

$$|a_{rr}| \geq \alpha |a_{pr}|, \tag{1.8.12}$$

  then $s = 1$ and $\tilde{P}$ such that $e_{11} = a_{rr}$,
  else
    choose $s = 2$ and $\tilde{P}$ such that

$$E = \begin{pmatrix} a_{11} & a_{r1} \\ a_{r1} & a_{rr} \end{pmatrix}; \tag{1.8.13}$$

(notice $a_{r1} = a_{1r}$)
redefine m by $m - s$;
if $m > 0$, then repeat the steps for

$$B' = \begin{cases} B, & \text{if } C = O \\ B - CE^{-1}C^T, & \text{if } C \neq O, \end{cases} \tag{1.8.14}$$

else terminate the algorithm; the diagonal block matrix D consists of the individual 1×1 or 2×2 blocks E.

When defining B' in the case $C \neq O$ one needs the regularity of E. This property is shown in the proof of the subsequent theorem which is based on the foregoing algorithm.

Theorem 1.8.9. *Let $A = A^T \in \mathbb{R}^{n \times n}$. Then there exist matrices $P, L, D \in \mathbb{R}^{n \times n}$ with the following properties. The matrix P is a permutation matrix, L is a lower triangular matrix with $l_{ii} = 1$ for $i = 1, \ldots, n$, and D is a symmetric block diagonal matrix with some combination of 1×1 and 2×2 diagonal blocks such that*

$$PAP^T = LDL^T. \tag{1.8.15}$$

Each 2×2 matrix associated with one of the 2×2 blocks is symmetric, regular, and nondiagonal with two nonzero real eigenvalues of opposite signs.

The matrices P, L, D can be constructed recursively by means of the Algorithm 1.8.8 *of Bunch, Kaufman, and Parlett.*

Proof. We will use the notation of (1.8.7) and of Algorithm 1.8.8. We show first that E is regular if $C \neq O$.

Case 1, $a_{r1} = 0$: In this case (1.8.10) holds and (1.8.7) follows with $s = 1$, $e_{11} = a_{11}$, $\tilde{P} = I$, $C = 0$. The latter contradicts our assumption.

Case 2, $a_{r1} \neq 0$: If (1.8.10) or (1.8.11) holds, we choose $s = 1$, $e_{11} = a_{11} \neq 0$, $\tilde{P} = I$. If (1.8.10) and (1.8.11) are false, but (1.8.12) is true, then $|a_{rr}| \geq \alpha |a_{pr}| \geq \alpha |a_{1r}| = \alpha |a_{r1}| > 0$.

Choose $\tilde{P}$ such that $(\tilde{P}A\tilde{P}^T)_{11} = a_{rr}$. Then (1.8.7) holds with $s = 1$, $e_{11} = a_{rr} \neq 0$. Finally, if (1.8.10)–(1.8.12) are false, then we choose

$$s = 2, \quad E = \begin{pmatrix} a_{11} & a_{r1} \\ a_{r1} & a_{rr} \end{pmatrix},$$

and adapt $\tilde{P}$ correspondingly. Then we get

$$|a_{11}| \cdot |a_{rr}| \leq \alpha |a_{11}| \cdot |a_{pr}| < \alpha^2 |a_{r1}|^2 < |a_{r1}|^2. \tag{1.8.16}$$

The first inequality follows from the failure of (1.8.12) and the possibility $a_{11} = 0$, the second results from the failure of (1.8.11). Now (1.8.16) implies

$$0 < |a_{r1}|^2 - |a_{11}| \cdot |a_{rr}| \leq a_{r1}^2 - a_{11}a_{rr} = -\det(E), \tag{1.8.17}$$

whence E is regular, nondiagonal, with two real eigenvalues λ_1, λ_2 which – by virtue of Theorem 1.8.1 – satisfy $\lambda_1\lambda_2 = \det(E) < 0$.

Theorem 1.8.9 is now proved by induction on the dimension n. If $n = 1$, then things are trivial. If the theorem is true for symmetric matrices up to dimension $n - 1$, then B' in Algorithm 1.8.8 can be written as

$$P'B'P'^T = L'D'L'^T$$

with an analogous meaning of P', D', L' as in (1.8.15). Define

$$\tilde{P}' = \begin{pmatrix} I_s & O \\ O & P' \end{pmatrix}, \quad \tilde{L}' = \begin{pmatrix} I_s & O \\ O & L' \end{pmatrix}, \quad D = \begin{pmatrix} E & O \\ O & D' \end{pmatrix}.$$

If $C \neq O$ use (1.8.8) in order to end up with

$$\begin{aligned}
\tilde{P}'\tilde{P}A(\tilde{P}'\tilde{P})^T &= \tilde{P}' \begin{pmatrix} E & C^T \\ C & B \end{pmatrix} \tilde{P}'^T \\
&= \tilde{P}' \begin{pmatrix} I_s & O \\ CE^{-1} & I_{n-s} \end{pmatrix} \tilde{P}'^T \tilde{P}' \begin{pmatrix} E & O \\ O & B' \end{pmatrix} \tilde{P}'^T \tilde{P}' \begin{pmatrix} I_s & (CE^{-1})^T \\ O & I_{n-s} \end{pmatrix} \tilde{P}'^T \\
&= \begin{pmatrix} I_s & O \\ P'CE^{-1} & I_{n-s} \end{pmatrix} \begin{pmatrix} E & O \\ O & P'B'P'^T \end{pmatrix} \begin{pmatrix} I_s & (P'CE^{-1})^T \\ O & I_{n-s} \end{pmatrix} \\
&= \begin{pmatrix} I_s & O \\ P'CE^{-1} & I_{n-s} \end{pmatrix} \tilde{L}'D\tilde{L}'^T \begin{pmatrix} I_s & (P'CE^{-1})^T \\ O & I_{n-s} \end{pmatrix} \\
&= \begin{pmatrix} I_s & O \\ P'CE^{-1} & L' \end{pmatrix} D \begin{pmatrix} I_s & (P'CE^{-1})^T \\ O & L'^T \end{pmatrix}
\end{aligned}$$

which proves (1.8.15). In the case $C = O$ the proof is based on (1.8.9) and proceeds analogously. □

Now we address the choice of α in Algorithm 1.8.8: The number of arithmetic operations there is $n^3/6 + O(n^2)$; cf. Exercise 1.8.7. A bound for the number of comparisons is $n^2 + O(n)$. In view of element growth, the choice of α should be such that after two steps with $s = 1$ the modified entries a'_{ij} of A (cf. (1.8.7)) should have approximately the same magnitude as those after one step with $s = 2$. With $B' = (b'_{ij})_{i,j=s+1,\ldots,n}$ from (1.8.14) and

$$m = \max_{i,j=1,\ldots,n} |a_{ij}| \geq \max\{|a_{pr}|, |a_{r1}|\},$$

$$m' = \max_{i,j} |b'_{ij}|,$$

one can show by distinguishing the cases (1.8.10)–(1.8.12) that

$$|b'_{ij}| = \left| b_{ij} - \frac{a'_{i1}a'_{1j}}{a'_{11}} \right| \leq \left(1 + \frac{1}{\alpha}\right) m \tag{1.8.18}$$

holds if $s = 1$ and $C \neq O$. Hence

$$m' \leq \left(1 + \frac{1}{\alpha}\right) m$$

is true in this case, and trivially also if $C = O$; cf. Exercise 1.8.7. Thus two steps of this kind result in the bound $(1 + \frac{1}{\alpha})^2 m$. If $s = 2$ one similarly gets with some estimations (cf. Exercise 1.8.7)

$$\begin{aligned} |b'_{ij}| &= \left| b_{ij} - \frac{1}{a'_{11}a'_{22} - (a'_{12})^2} \left\{ a'_{i1}(a'_{22}a'_{1j} - a'_{21}a'_{2j}) + a'_{i2}(-a'_{21}a'_{1j} + a'_{11}a'_{2j}) \right\} \right| \\ &\leq \left(1 + \frac{2}{1-\alpha}\right) m. \end{aligned} \tag{1.8.19}$$

Equating the bounds according to our requirement yields

$$\left(1 + \frac{1}{\alpha}\right)^2 = 1 + \frac{2}{1-\alpha},$$

whence $0 < \alpha = (1 + \sqrt{17})/8 \approx 0.6404$.

The decomposition (1.8.15) can be used to solve linear systems of equations with a symmetric matrix. Moreover, it can be applied in order to test whether there are k eigenvalues of $A = A^T \in \mathbb{R}^{n\times n}$ which are greater than some number $c \in \mathbb{R}$. To this end apply Theorem 1.8.9 to $A - cI$ in order to get the representation

$$P(A - cI)P^T = LDL^T, \tag{1.8.20}$$

and use Sylvester's law of inertia: Since each 2×2 block E of D has exactly one positive eigenvalue, the number of these 2×2 blocks plus the number of the positive 1×1 blocks E of D is the number of eigenvalues of A which are greater than c.

The arithmetic amount of work for (1.8.20) is $n^3/6 + O(n^2)$ if no $\pm$-operations are counted, while a reduction of A to a tridiagonal matrix T using Householder matrices, and exploiting the Sturm chain of leading principal minors of T costs $2n^3/3 + O(n^2)$ arithmetic operations of the same kind.

Definition 1.8.10. Let $A \in \mathbb{R}^{n\times n}$. If $x^TAx > 0$ for all vectors $x \in \mathbb{R} \setminus \{0\}$, then A is called positive definite.

Notice that positive definite matrices have positive diagonal entries. This follows from $0 < (e^{(i)})^TAe^{(i)} = a_{ii}$.

Theorem 1.8.11. *Let $A = A^T \in \mathbb{R}^{n\times n}$. Then A is positive definite if and only if all its eigenvalues are positive.*

Proof. Let A be positive definite. Since A is symmetric, every eigenvalue λ is real. Therefore the eigenvectors can be chosen to be real. Let x be an eigenvector associated with λ and normalized by $\|x\|_2 = 1$. Then $Ax = \lambda x$ implies $0 < x^TAx = \lambda x^Tx = \lambda$.

Conversely, let all eigenvalues be positive. Since A is symmetric, there is a basis of orthonormal eigenvectors $x^1, \dots, x^n$ associated with the eigenvalues $\lambda_1, \dots, \lambda_n$ of A. Let $x = \sum_{j=1}^n \alpha_j x^j \neq 0$. By virtue of the orthonormality we obtain $x^TAx = \sum_{j=1}^n \lambda_j (x^j)^T x^j = \sum_{j=1}^n \lambda_j > 0$. □

It is an immediate consequence of Theorem 1.8.11 and Exercise 1.8.1 that A is symmetric and positive definite if and only if A^{-1} has this property.

Theorem 1.8.12. *Let $A \in \mathbb{R}^{n\times n}$ be symmetric and positive definite. Then there is a unique matrix $A^{1/2}$, the so-called square root of A, which is symmetric positive definite and satisfies $A^{1/2} \cdot A^{1/2} = A$.*

Proof. According to Theorem 1.8.5 there is an orthogonal matrix $X = (x^1, \dots, x^n) \in \mathbb{R}^{n\times n}$ which satisfies $X^TAX = \operatorname{diag}(\lambda_1, \dots, \lambda_n)$, whence $Ax^i = \lambda_i x^i$, $i = 1, \dots, n$. Define

$$A^{1/2} = X \operatorname{diag}(\sqrt{\lambda_1}, \dots, \sqrt{\lambda_n}) X^T.$$

Then $(A^{1/2})^2 = A$. This proves the existence of the square root. In order to prove its uniqueness, assume that the matrix B has the same properties as $A^{1/2}$. Let (v, μ) be an eigenpair of B and let $\lambda_1, \dots, \lambda_n$ be the eigenvalues of A which are all positive by Theorem 1.8.11. Then $Av = B^2v = \mu^2 v$, hence there is an eigenvalue λ_i of A which satisfies $\mu = \sqrt{\lambda_i}$. Both statements together imply

$$\operatorname{kernel}(B - \sqrt{\lambda_i} I) \subseteq \operatorname{kernel}(A - \lambda_i I) \tag{1.8.21}$$

from which $\dim \operatorname{kernel}(B - \sqrt{\lambda_i} I) \le \dim \operatorname{kernel}(A - \lambda_i I)$ follows. Consider the maximal set of eigenvalues λ_i of A such that they are pairwise different and denote by Λ_A the set of their indices. Denote by Λ_B that subset of Λ_A which consists only of indices i with $\sqrt{\lambda_i}$ being an eigenvalue of B. Then

$$n = \sum_{i\in\Lambda_B} \dim \operatorname{kernel}(B - \sqrt{\lambda_i} I) \le \sum_{i\in\Lambda_A} \dim \operatorname{kernel}(A - \lambda_i I) = n.$$

Therefore, equality must hold in (1.8.21), whence $Bx^i = \sqrt{\lambda_i} x^i$ for $i = 1, \dots, n$. This implies $X^TBX = \operatorname{diag}(\sqrt{\lambda_1}, \dots, \sqrt{\lambda_n})$, hence $B = A^{1/2}$. □

Theorem 1.8.13 (Cholesky decomposition). *The matrix $A \in \mathbb{R}^{n\times n}$ is symmetric and positive definite if and only if there is a lower triangular matrix $L^C \in \mathbb{R}^{n\times n}$ with positive diagonal entries such that $A = L^C(L^C)^T$. The matrix L^C is unique.*

Proof. Let A be symmetric and positive definite. Then $a_{11} > 0$. Choose α in Algorithm 1.8.8 so small that (1.8.10) holds, whence $s = 1$, $\tilde{P} = I$, $E = (a_{11}) \in \mathbb{R}^{1\times 1}$, $C^T \equiv c^T = (a_{12}, \dots, a_{1n})$ in the notation there. This implies that B' in (1.8.8) and (1.8.9) is symmetric and positive definite. For (1.8.8) this can be seen from

$$0 < y^T Ay = x^T B' x = x^T(B - cc^T/a_{11})x \tag{1.8.22}$$

$$\text{with } A = \begin{pmatrix} a_{11} & c^T \\ c & B \end{pmatrix}, \quad x \in \mathbb{R}^{n-1} \setminus \{0\}, \quad y^T = (-c^T x/a_{11}, x^T).$$

For (1.8.9) use $B' = B$ and $y^T = (0, x^T)$ in (1.8.22). Therefore, one can choose $\alpha \in (0, 1)$ in Algorithm 1.8.8 so small that (1.8.10) holds in each step. This leads to $P = I$ and $A = LDL^T = L^C(L^C)^T$, where D is a diagonal matrix with positive diagonal entries and $L^C = LD^{1/2}$.

The converse follows from $A = L^C(L^C)^T$ and $x^T Ax = \|(L^C)^T x\|_2^2 > 0$ for $x \neq 0$.

In order to prove the uniqueness of L^C assume that there is another lower triangular matrix L with positive diagonal entries such that

$$A = LL^T = L^C(L^C)^T. \tag{1.8.23}$$

Theorem 1.8.11 guarantees that A is regular, and so are all matrices in (1.8.23). Hence

$$(L^C)^{-1}L = (L^C)^T L^{-T}. \tag{1.8.24}$$

Since the inverse of a regular triangular matrix is a triangular matrix of the same kind, and since the same holds for products of triangular matrices of the same kind, both sides in (1.8.24) are diagonal, say $(L^C)^{-1}L = D$, whence $L = L^C D$ with $d_{ii} > 0$, $i = 1, \dots, n$, and $A = LL^T = L^C D^2 (L^C)^T$. From (1.8.23) and (1.8.24) we get $(L^C)^{-1}A(L^C)^{-T} = I = D^2$ which implies $D = I$ and $L = L^C$. □

An even more elementary proof of Theorem 1.8.13 is given in Section 5.4.

Lemma 1.8.14. *Let $A = A^T$ and let p be a polynomial. If $x^T p(A)x \le 0$ for a nonzero vector $x \in \mathbb{R}^n$, then there is at least one eigenvalue λ of A such that $p(\lambda) \le 0$.*

Proof. Let $J = S^{-1}AS$ be the Jordan normal form of A. Then by virtue of $S^{-1}A^k S = (S^{-1}AS)^k$ we get

$$S^{-1}p(A)S = p(S^{-1}AS) = p(J) = \begin{pmatrix} p(\lambda_1) & * & \dots & * \\ & \ddots & \ddots & \vdots \\ & & \ddots & * \\ O & & & p(\lambda_n) \end{pmatrix},$$

where the stars denote real numbers. If λ_i, $i = 1, \dots, n$, denote the eigenvalues of A, we see immediately that exactly $p(\lambda_i)$, $i = 1, \dots, n$, are the eigenvalues of $p(A) = p(A)^T$. By the assumption, $p(A)$ is not positive definite. Hence, by Theorem 1.8.11 there is at least one nonpositive eigenvalue of $p(A)$. □

We now address localization theorems for eigenvalues and eigenvectors. These are theorems which provide error bounds for these items. A first one was already given in Theorem 1.3.10. In order to prove additional ones we need some preparation.

Definition 1.8.15. Let $A = A^T \in \mathbb{R}^{n\times n}$, $x \in \mathbb{R}^n \setminus \{0\}$, $\alpha \in \mathbb{R}$. Then we define

$$m_i = x^T A^i x, \quad i = 0, 1, \dots \qquad \text{moments}$$

$$R_x = \frac{x^T A x}{x^T x} = \frac{m_1}{m_0} \qquad \text{Rayleigh quotient}$$

$$T_x(\alpha) = \frac{m_2 - \alpha m_1}{m_1 - \alpha m_0}, \quad \alpha \neq R_x, \qquad \text{Temple quotient}$$

$$\varepsilon_x^2 = \frac{m_2}{m_0} - R_x^2.$$

Theorem 1.8.16. *Let $A = A^T \in \mathbb{R}^{n\times n}$, $x \in \mathbb{R} \setminus \{0\}$, and $\alpha \in \mathbb{R} \setminus \{R_x\}$.*
(a) *There is an eigenvalue λ of A such that $|\lambda - R_x| \le \sqrt{\varepsilon_x^2}$.*
(b) *If $\lambda_1 \le \lambda_2 \le \dots \le \lambda_n$ denote the eigenvalues of A, then*

$$\lambda_1 \le R_x \le \lambda_n.$$

(c) *There is an eigenvalue of A which lies between α and $T_x(\alpha)$.*

Proof. (a) Define $p(t) = (t - R_x)^2 - \varepsilon_x^2$. Then $x^T p(A) x = m_2 - 2m_1 R_x + R_x^2 m_0 - m_2 + R_x^2 m_0 = 0$, and Lemma 1.8.14 proves $0 \le (\lambda - R_x)^2 \le \varepsilon_x^2$ for some eigenvalue λ of A which implies the assertion.

(b) Let $x = \sum_{i=1}^n \alpha_i x^i \neq 0$, $\alpha_i \in \mathbb{R}$, where $\{x^1, \dots, x^n\}$ is an orthonormal basis of eigenvectors of A associated with the eigenvalues $\lambda_1, \dots, \lambda_n$. Then

$$x^T A x = \sum_{i=1}^n \alpha_i^2 \lambda_i \begin{cases} \le \lambda_n x^T x \\ \ge \lambda_1 x^T x, \end{cases}$$

which proves the assertion.

(c) Define $p(t) = (t - T_x(\alpha))(t - \alpha)$. Then $x^T p(A) x = m_2 - \alpha m_1 - T_x(\alpha)(m_1 - \alpha m_0) = 0$. Again by Lemma 1.8.14 there is an eigenvalue λ of A such that $(\lambda - T_x(\alpha))(\lambda - \alpha) \le 0$, which concludes the proof. □

Notice that part (a) of Theorem 1.8.16 is due to Krylov, Bogolyubov and Weinstein while part (c) originates from Temple and Wielandt. Both parts remain valid if '$A \in \mathbb{R}^{n\times n}$ symmetric' is replaced by '$A \in \mathbb{C}^{n\times n}$ normal', i.e., $AA^H = A^H A$, and if one modifies the definition of ε_x^2 and $T_x(\alpha)$ slightly. Lemma 1.8.14 then has to be changed appropriately. For details see Mayer [208] or Neumaier [253].

Our next theorem refines part (b) of Theorem 1.8.16 on the Rayleigh quotient.

Theorem 1.8.17. *Let $\lambda_1 \le \lambda_2 \le \cdots \le \lambda_n$ denote the eigenvalues of $A = A^T \in \mathbb{R}^{n\times n}$, and let $\{x^1, \dots, x^n\}$ be an orthonormal basis in $\mathbb{R}^n$ of corresponding eigenvectors.*

(a) (min–max principle):
For each subspace U of $\mathbb{R}^n$ with $\dim U = j$ we have

$$\lambda_j \le \max_{x\in U\setminus\{0\}} R_x \le \lambda_n$$

and for varying subspaces $U \subseteq \mathbb{R}^n$ we get

$$\lambda_j = \min\{\max_{x\in U\setminus\{0\}} R_x \mid \dim U = j\}, \quad j = 1, \dots n.$$

The minimum is attained for U spanned by $x^1, x^2, \dots, x^j$.

(b) (max–min principle):
For each subspace U of $\mathbb{R}^n$ with $\dim U = n + 1 - j$ we have

$$\lambda_j \ge \min_{x\in U\setminus\{0\}} R_x \ge \lambda_1$$

and for varying subspaces $U \subseteq \mathbb{R}^n$ we get

$$\lambda_j = \max\{\min_{x\in U\setminus\{0\}} R_x \mid \dim U = n + 1 - j\}, \quad j = 1, \dots n.$$

The maximum is attained for U spanned by $x^j, x^{j+1}, \dots, x^n$.

Proof. (a) If $j = 1$, the assertion follows from Theorem 1.8.16 (b) with equality on the left for U spanned by x^1. If $j > 1$, the right inequality follows from Theorem 1.8.16 (b). For the left inequality let $\{w^1, \dots, w^j\}$ be an orthonormal basis of U, where the vectors w^i are not necessarily eigenvectors. Let $\{w^{j+1}, \dots, w^n\}$ be an orthonormal basis of the orthogonal complement $U^\perp$ of U. Then the underdetermined homogeneous linear system

$$\sum_{l=j}^{n} \alpha_l (x^l)^T w^k = 0, \quad k = j + 1, \dots n, \tag{1.8.25}$$

has a nontrivial solution $\alpha_j^*, \dots, \alpha_n^*$ with which we form the vector $z = \sum_{l=j}^n \alpha_l^* x^l$. From (1.8.25) we get $z^T w^k = 0$, $k = j + 1, \dots n$, whence $z \in (U^\perp)^\perp = U$ and

$$z^T A z = \sum_{l=j}^{n} (\alpha_l^*)^2 \lambda_l \ge \lambda_j \sum_{l=j}^{n} (\alpha_l^*)^2 = \lambda_j z^T z$$

follows. This leads immediately to $\lambda_j \le R_z$ which implies the left inequality in (a). The last statement of (a) is obvious.

(b) is proved analogously; cf. Exercise 1.8.13. □

Theorem 1.8.17 is often named after R. Courant, E. Fischer, and H. Weyl, part (b) is sometimes associated with H. Poincaré and W. Ritz.

By means of Theorem 1.8.17 we can prove the following Theorem of Weyl which we need in Appendix D.

Theorem 1.8.18 (Weyl). *Assume that the eigenvalues λ_i, $i = 1, \ldots, n$, of $A = A^T \in \mathbb{R}^{n\times n}$ are ordered increasingly and that the same holds for the eigenvalues μ_i of $B = B^T$ and for the eigenvalues δ_i of $B - A$. Then for all $i = 1, \ldots, n$ and each matrix norm $\|\cdot\|$ the inequalities*

$$\lambda_i + \delta_1 \le \mu_i \le \lambda_i + \delta_n \tag{1.8.26}$$

and

$$|\mu_i - \lambda_i| \le \|B - A\| \tag{1.8.27}$$

hold.

Proof. Let $\{x^1, \ldots, x^n\}$ be an orthonormal basis of eigenvectors of A and let the subspace U be spanned by $x^1, \ldots, x^i$. Denote the Rayleigh quotient of $A, B, B - A$ by $R_x(A)$, $R_x(B)$, and $R_x(B - A)$, respectively. Then Theorem 1.8.17 (a) implies

$$\begin{aligned}\mu_i &\le \max_{x\in U\setminus\{0\}} R_x(B) \le \max_{x\in U\setminus\{0\}} R_x(A) + \max_{x\in U\setminus\{0\}} R_x(B-A)\\ &= \lambda_i + \max_{x\in U\setminus\{0\}} R_x(B-A) \le \lambda_i + \delta_n .\end{aligned}$$

Interchanging the roles of A and B yields $\lambda_i \le \mu_i - \delta_1$. Thus we obtain finally $\delta_1 \le \mu_i - \lambda_i \le \delta_n$ and herewith $|\mu_i - \lambda_i| \le \max\{|\delta_1|, |\delta_n|\} = \rho(B - A) \le \|B - A\|$. □

Theorem 1.8.19 (Gershgorin). *Let $A \in \mathbb{C}^{n\times n}$. Define the Gershgorin discs*

$$B_i(A) = \overline{B}(a_{ii}, r_i) \subseteq \mathbb{C}, \quad i = 1, \ldots, n, \quad \textit{where } r_i = \sum_{j=1,\, j\ne i}^{n} |a_{ij}|,$$

and let $B_G(A) = \bigcup_{i=1}^n B_i(A)$.

(a) *All eigenvalues of A lie in $B_G(A)$.*
(b) *Let $K \cup L = \{1, \ldots, n\}$, where K has exactly $k \in (0, n)$ elements. Then $U_1 = \bigcup_{k\in K} B_k(A)$ contains exactly k eigenvalues of A while $U_2 = \bigcup_{l\in L} B_l(A)$ contains the remaining $n - k$ ones provided that $U_1 \cap U_2 = \emptyset$.*
(c) *If A is irreducible and has an eigenvalue at the boundary of $B_G(A)$, then this eigenvalue lies at the boundary of each individual Gershgorin disc and there is a corresponding eigenvector x of A with $|x| = e$.*
(d) *If A is real and symmetric, then for each $i \in \{1, \ldots, n\}$ there is an eigenvalue λ_i of A such that*

$$|\lambda_i - a_{ii}| \le \sqrt{\sum_{j=1,\, j\ne i}^{n} |a_{ij}|^2} \le r_i .$$

In particular, for symmetric matrices A each Gershgorin disc contains at least one eigenvalue of A.

Proof. (a) Let $Ax = \lambda x$, $\max\{|x_i| \mid i = 1, \ldots, n\} = 1 = |x_s|$ for some index s. Then

$$|a_{ss} - \lambda| = |a_{ss} - \lambda|\,|x_s| = |a_{ss}x_s - \lambda x_s| = \left|\sum_{j=1,\, j\ne s}^{n} a_{sj}x_j\right| \le r_s, \tag{1.8.28}$$

whence $\lambda \in B_s(A) \subseteq B_G(A)$.

(b) Let $D = \operatorname{diag}(a_{11}, \dots, a_{nn})$, $C = A - D$, $A_\varepsilon = D + \varepsilon C$, $0 \le \varepsilon \le 1$. Then $B_i(A_\varepsilon) \subseteq B_i(A_1) = B_i(A)$ for all $\varepsilon \in [0, 1]$ and $i = 1, \dots, n$. From $U_1 \cap U_2 = \emptyset$ we see that U_1 contains the k eigenvalues a_{ii}, $i \in K$, of A_0 while U_2 contains the remaining $n - k$ ones, namely a_{ii}, $i \in L$. By the same reason and by virtue of Theorem 1.8.2 (a) the same holds for $\varepsilon \in (0, 1]$. Choosing $\varepsilon = 1$ proves the assertion.

(c) Let $\lambda \in \partial B_G(A)$. Then $\lambda \in \partial B_s(A)$ with s as in (1.8.28). Hence equality holds there and $a_{sj} \ne 0$ implies $|x_j| = 1$. Choose any $k \in \{1, \dots, n\}$. According to Theorem 1.7.14 there is a path $s \to i_1 \to i_2 \to \cdots \to i_l \to k$, whence $a_{si_1} a_{i_1 i_2} \cdots a_{i_l k} \ne 0$. In particular, $a_{si_1} \ne 0$ which implies $|x_{i_1}| = 1$. Now we can repeat the steps in (1.8.28) with i_1 instead of s in order to obtain $|x_{i_2}| = 1$ and, finally, $|x_k| = 1$. Hence $\lambda \in \partial B_k(A)$.

(d) The first inequality is an immediate consequence of Theorem 1.8.16 (a) with $x = e^{(i)}$. The second is trivial when considered in squared form. □

Theorem 1.8.19 (d) also holds for complex Hermitian matrices. It remains true if A is normal, see for instance Mayer [208].

In our next theorem we consider the generalized eigenvalue problem

$$Ax = \lambda Bx, \quad x \ne 0, \tag{1.8.29}$$

where $A, B \in \mathbb{R}^{n\times n}$ are symmetric and B is positive definite. For such matrices we first list some properties.

Theorem 1.8.20. *Let $A, B \in \mathbb{R}^{n\times n}$ be symmetric, B positive definite with the Cholesky decomposition $B = L^C (L^C)^T$.*

(a) *The generalized eigenvalue problem (1.8.29) is equivalent to the ordinary eigenvalue problems*

$$B^{-1}Ax = \lambda x$$

and

$$(L^C)^{-1} A (L^C)^{-T} y = \lambda y \quad \text{with } y = (L^C)^T x. \tag{1.8.30}$$

The matrix in the left equality of (1.8.30) is symmetric; in particular, (x, λ) is an eigenpair of (1.8.29) if and only if $((L^C)^T x, \lambda)$ is an eigenpair of (1.8.30); λ is real and x can be chosen to be real. The eigenvalue problems (1.8.29) and (1.8.30) have exactly n eigenvalues if these are counted according to their multiplicity, i.e., the multiplicity of the zeros of $\det(A - \lambda B) = (\det B) \det(B^{-1}A - \lambda I)$.

(b) *The product $(x, y)_B = x^T B y$ is a scalar product on $\mathbb{R}^n$.*
If $(x, y)_B = 0$, we call x, y B-orthogonal. If $(x, x)_B = (y, y)_B = 1$ and $(x, y)_B = 0$, we call x, y B-orthonormal.

(c) *There are eigenpairs $(x^1, \lambda_1), \dots, (x^n, \lambda_n)$ of (1.8.29) such that the eigenvectors x^i form a B-orthonormal basis of $\mathbb{R}^n$.*

(d) *If the eigenvalues $\lambda_1, \dots, \lambda_n$ of (1.8.29) are ordered increasingly then*

$$\lambda_1 \le \frac{x^T A x}{x^T B x} \le \lambda_n \tag{1.8.31}$$

for any vector $x \in \mathbb{R}^n \setminus \{0\}$.

Proof. (a)–(c) are obvious; (d) follows from Theorem 1.8.16 (b) applied to (1.8.30); (d) reduces to this theorem if $B = I$. □

In our next result on (1.8.29) we use (1.8.30) together with Sylvester's law of inertia.

Theorem 1.8.21. *Let $c \in \mathbb{R}$ and assume that $A, B \in \mathbb{R}^{n\times n}$ are symmetric and B is positive definite. Then the number of eigenvalues λ of $Ax = \lambda Bx$ that are greater than c, equal to c, and less than c, respectively, coincides with that of the eigenvalues μ of $A - cB$ which satisfy $\mu > 0$, $\mu = 0$, $\mu < 0$, respectively.*

Proof. Let $B = L^C(L^C)^T$ and $y = (L^C)^T x$. From $Ax = \lambda Bx$ we get equivalently $My = (\lambda - c)y$ with $M = (L^C)^{-1}(A - cB)(L^C)^{-T}$. Hence Sylvester's law of inertia guarantees $\mathrm{ind}(A - cB) = \mathrm{ind}(M)$ which proves the assertion. □

Now we will enclose eigenvalues of a larger generalized eigenvalue problem (1.8.29) by using those of a smaller one.

Theorem 1.8.22 (Lehmann). *Let A, $B \in \mathbb{R}^{n\times n}$ be symmetric and B, in addition, positive definite. Choose $k \le n$ and $U \in \mathbb{R}^{n\times k}$ with full rank. Assume that $\alpha \in \mathbb{R}$ is not an eigenvalue of* (1.8.29). *Define the symmetric $k \times k$ matrices*

$$\begin{aligned} A_\alpha &= U^T(A - \alpha B)U, \\ B_\alpha &= U^T(AB^{-1}A - 2\alpha A + \alpha^2 B)U = U^T(A - \alpha B)B^{-1}(A - \alpha B)U, \end{aligned}$$

and denote by $\mu_1 \le \cdots \le \mu_k$ the eigenvalues of the $k \times k$ generalized eigenvalue problem

$$A_\alpha x = \mu B_\alpha x. \tag{1.8.32}$$

(a) *If $\mu_\ell < 0$, then the interval $\left[\alpha + \frac{1}{\mu_\ell}, \alpha\right)$ contains at least ℓ eigenvalues λ_i of* (1.8.29).

(b) *If $\mu_\ell > 0$, then the interval $\left(\alpha, \alpha + \frac{1}{\mu_\ell}\right]$ contains at least $k + 1 - \ell$ eigenvalues λ_i of* (1.8.29).

Proof. Choose $z \in \mathbb{R}^k \setminus \{0\}$ and define $v = Uz$, $w = (A - \alpha B)v$. By virtue of the assumptions, v differs from zero and $A - \alpha B$ is regular, whence $w \ne 0$. From

$$z^T B_\alpha z = v^T(A - \alpha B)B^{-1}(A - \alpha B)v = w^T B^{-1} w$$

and the assumptions on B we deduce that B_α is symmetric and positive definite.

(a) Assume that the assertion is false for some ℓ, i.e., $[\eta, \alpha)$ contains less than ℓ eigenvalues λ_i of (1.8.29), where $\eta = \alpha + 1/\mu_\ell$. Choose a B-orthonormal basis $\{x^1, \dots, x^n\}$ of eigenvectors of (1.8.29), and a B_α-orthonormal basis $\{y^1, \dots, y^k\}$ of eigenvectors of (1.8.32), respectively, corresponding to the eigenvalues $\lambda_1, \dots, \lambda_n$, and $\mu_1, \dots, \mu_k$. Let $Y = (y^1, \dots, y^\ell) \in \mathbb{R}^{k\times\ell}$. Then $Y^T B_\alpha Y = I \in \mathbb{R}^{\ell\times\ell}$. Choose $b \in \mathbb{R}^\ell \setminus \{0\}$ such that the vector $u = UYb \in \mathbb{R}^n \setminus \{0\}$ satisfies

$$u^T Bx^i = 0 \quad \text{for all } i \in J = \{\, j \mid \lambda_j \in [\eta, \alpha)\,\}. \tag{1.8.33}$$

Such a vector $b \neq 0$ exists since (1.8.33) is equivalent to the homogeneous linear system $(x^i)^T BUYb = 0$, $i \in J$, which consists of less than ℓ equations by our assumption at the beginning of the proof of (a). Taking into account $\mu_i \le \mu_\ell < 0$ for $i = 1, \dots, \ell$ we get

$$\begin{aligned} &u^T(\alpha B - A)B^{-1}(\eta B - A)u \\ &= b^T Y^T\{B_\alpha + (\alpha - \eta)A_\alpha\}Yb = b^T\{I + (\alpha - \eta)Y^T A_\alpha Y\}b \\ &= b^T\left\{I - \frac{1}{\mu_\ell}\operatorname{diag}(\mu_1, \dots, \mu_\ell)\right\}b = \sum_{i=1}^{\ell}\left(1 - \frac{\mu_i}{\mu_\ell}\right)b_i^2 \le 0. \end{aligned} \tag{1.8.34}$$

Represent u as $u = \sum_{i=1}^n y_i x^i$. Then $(x^j)^T Bu = y_j$, and

$$\begin{aligned} &u^T(\alpha B - A)B^{-1}(\eta B - A)u \\ &= \sum_{i=1}^n\sum_{j=1}^n (x^i)^T(\alpha\eta B - \eta A - \alpha A + AB^{-1}A)x^j y_i y_j \\ &= \sum_{i=1}^n\sum_{j=1}^n (x^i)^T(\alpha\eta - \eta\lambda_j - \alpha\lambda_j + \lambda_j^2)Bx^j y_i y_j \\ &= \sum_{i=1}^n (\alpha - \lambda_i)(\eta - \lambda_i)((x^i)^T Bu)^2 \ge 0 \end{aligned} \tag{1.8.35}$$

since $(x^i)^T Bu = 0$ for $\lambda_i \in [\eta, \alpha)$ and $(\alpha - \lambda_i)(\eta - \lambda_i) \ge 0$ for $\lambda_i \notin [\eta, \alpha)$. Together with (1.8.34) this implies $u^T(\alpha B - A)B^{-1}(\eta B - A)u = 0$, and, by virtue of (1.8.35) and $\lambda_i \neq \alpha$, finally

$$(x^i)^T Bu = 0, \quad i = 1, \dots, n. \tag{1.8.36}$$

From (1.8.36) we get $Bu = 0$ and the contradiction $u = 0$.

(b) Let $\tilde{A} = -A$, $\tilde{\lambda}_i = -\lambda_i$, $\tilde{\alpha} = -\alpha$, and consider the generalized eigenvalue problem

$$\tilde{A}x = \tilde{\lambda}Bx \tag{1.8.37}$$

which is equivalent to (1.8.29). Define $\tilde{A}_{\tilde{\alpha}}$, $B_{\tilde{\alpha}}$, $\tilde{\mu}_i$ analogously to A_α, B_α, μ_i. Then $\tilde{A}_{\tilde{\alpha}} = -A_\alpha$, $B_{\tilde{\alpha}} = B_\alpha$, and

$$\tilde{A}_{\tilde{\alpha}}x = \tilde{\mu}B_{\tilde{\alpha}}x \tag{1.8.38}$$

is equivalent to $A_\alpha x = (-\tilde{\mu})B_\alpha x$. Hence $\tilde{\mu}_i = -\mu_{k+1-i}$. Now apply (a) to (1.8.37), (1.8.38): If $\tilde{\mu}_{k+1-\ell} < 0$, then $[\tilde{\alpha} + \frac{1}{\tilde{\mu}_{k+1-\ell}}, \tilde{\alpha})$ contains at least $k + 1 - \ell$ eigenvalues $\tilde{\lambda}_i$ of (1.8.37), which implies the assertion. □

Remark 1.8.1.

(a) If in Theorem 1.8.22 the eigenvalues μ_ℓ are ordered monotonously decreasing instead of increasing, then part (b) of this theorem reads as follows:
If $\mu_\ell > 0$, then the interval $(\alpha, \alpha + \frac{1}{\mu_\ell}]$ contains at least ℓ eigenvalues λ_i of (1.8.29).

(b) In Theorem 1.8.22 the parameter α can be chosen as an approximate eigenvalue of (1.8.29). The columns of U can be chosen as approximately B-orthonormal eigenvectors. Then the matrices U^TBU, U^TAU, $U^TAB^{-1}AU$ are approximately diagonal matrices.

Corollary 1.8.23. *Let the assumptions of Theorem 1.8.22 hold and denote by $\tau_1 \le \cdots \le \tau_k$ the eigenvalues of the generalized eigenvalue problem*

$$\{U^T(AB^{-1}A - \alpha A)U\}x = \tau\{U^T(A - \alpha B)U\}x. \tag{1.8.39}$$

(a) *If p is the number of eigenvalues $\tau_j < \alpha$ and if $1 \le i \le p$, then the interval $[\tau_i, \alpha)$ contains at least $p + 1 - i$ eigenvalues of* (1.8.29).
(b) *If q is the number of eigenvalues $\tau_j > \alpha$ and if $k + 1 - q \le i \le k$, then the interval $(\alpha, \tau_i]$ contains at least $q + i - k$ eigenvalues of* (1.8.29).

Proof. Let $\mu \ne 0$ and define A_α, B_α as in Theorem 1.8.22. Then $A_\alpha x = \mu B_\alpha x$ is equivalent to

$$U^T(AB^{-1}A - \alpha A)Ux = (B_\alpha + \alpha A_\alpha)x = \left(\frac{1}{\mu} + \alpha\right)A_\alpha x = \left(\frac{1}{\mu} + \alpha\right)U^T(A - \alpha B)Ux,$$

i.e., $\mu \ne 0$ is an eigenvalue of (1.8.32) if and only if $\tau = \frac{1}{\mu} + \alpha$ is an eigenvalue of (1.8.39). Hence $\mu < 0$ if and only if $\tau < \alpha$. Therefore, the assumption of (a) implies

$$\tau_i = \frac{1}{\mu_{p+1-i}} + \alpha, \quad i = 1, \dots, p,$$

whence

$$[\tau_i, \alpha) = \left[\alpha + \frac{1}{\mu_{p+1-i}}, \alpha\right),$$

and (a) follows by virtue of Theorem 1.8.22 (a).

From

$$\tau_i = \frac{1}{\mu_{2k+1-q-i}} + \alpha, \quad i = (k+1) - q, \dots, k,$$

we get

$$(\alpha, \tau_i] = \left(\alpha, \alpha + \frac{1}{\mu_{2k+1-q-i}}\right],$$

and (b) follows from Theorem 1.8.22 (b), since $\ell = 2k + 1 - q - i$ implies $k + 1 - \ell = k + 1 - (2k + 1 - q - i) = q + i - k$. □

Remark 1.8.2. In the case $k = 1$, $B = I$ we identify the corresponding matrix $U \in \mathbb{R}^{n\times 1}$ with an approximate eigenvector $\tilde{x} \ne 0$ of (1.8.29). Then (1.8.39) reads

$$\tilde{x}^T(A^2 - \alpha A)\tilde{x} = \tau\tilde{x}^T(A - \alpha I)\tilde{x}, \quad \tilde{x} \ne 0$$

which is equivalent to

$$m_2 - \alpha m_1 = \tau(m_1 - \alpha m_0)$$

with m_i as in Definition 1.8.15 with $\tilde{x}$ instead of x. Assuming $m_1 - \alpha m_0 \neq 0$, the parameter τ is just the Temple quotient $T_{\tilde{x}}(\alpha)$. Thus Corollary 1.8.23 is a generalization of Theorem 1.8.16 (c).

Notice that $m_1 - \alpha m_0 = 0$ is possible as the example

$$A = \begin{pmatrix} 1 & 0 \\ 0 & -1 \end{pmatrix}, \quad \alpha = 0, \quad x = e$$

shows.

Corollary 1.8.24. *Let the assumptions of Theorem 1.8.22 hold and denote by $\tau_1 \leq \cdots \leq \tau_k$ the eigenvalues of the generalized eigenvalue problem*

$$U^T A U x = \tau U^T B U x. \tag{1.8.40}$$

Then there are at least $k + 1 - i$ eigenvalues of (1.8.29) *which are greater than or equal to τ_i and at least i eigenvalues which are less than or equal to τ_i.*

Proof. Choose $\alpha > 0$ greater than all eigenvalues of (1.8.29) and denote by $\tau_1^{(\alpha)} \leq \cdots \leq \tau_k^{(\alpha)}$ the eigenvalues of (1.8.39) ordered by magnitude as in Corollary 1.8.23.

First we show that α is not an eigenvalue of (1.8.39). Otherwise

$$\{U^T(AB^{-1}A - 2\alpha A + \alpha^2 B)U\}x = 0$$

for some $x \in \mathbb{R}^k \setminus \{0\}$, whence

$$x^T U^T (A - \alpha B) B^{-1} (A - \alpha B) U x = 0. \tag{1.8.41}$$

By the choice of α, the matrix $A - \alpha B$ is nonsingular, and the assumption on U implies $y = (A - \alpha B)Ux \neq 0$. Since B is symmetric positive definite, we get $y^T B^{-1} y > 0$ which contradicts (1.8.41).

Corollary 1.8.23 guarantees

$$\tau_k^{(\alpha)} < \alpha, \tag{1.8.42}$$

since otherwise $(\alpha, \tau_k^{(\alpha)}]$ contains at least one eigenvalue of (1.8.29) contradicting the choice of α. Together with part (a) of this corollary the inequality (1.8.42) implies that there are at least $k + 1 - i$ eigenvalues of (1.8.29) which are not less than $\tau_i^{(\alpha)}$. Divide (1.8.39) by $-\alpha$. This does not change the eigenvalues $\tau_i^{(\alpha)}$. Now let α tend to infinity. Then the transformed equality (1.8.39) tends to (1.8.40). Since the eigenvalues $\tau_i^{(\alpha)}$ depend continuously on α they tend to the eigenvalues $\tau_1 \leq \cdots \leq \tau_k$ of (1.8.40), and there remain at least $k + 1 - i$ eigenvalues of (1.8.29) which are not less than τ_i.

If $\alpha < 0$ is a lower bound of the eigenvalues of (1.8.29), the limit $\alpha \to -\infty$ (after division of (1.8.39) by $-\alpha$) implies again the equality (1.8.40), and Corollary 1.8.23 (b) with $q = k$ guarantees at least i eigenvalues of (1.8.29) which are not greater than τ_i. □

Notice that Corollary 1.8.24 does not imply that there are exactly i eigenvalues which are less than or equal to τ_i. This can be seen from the example $A = B = I_3 \in \mathbb{R}^{3\times 3}$, $U \equiv e^{(1)} \in \mathbb{R}^{3\times 1}$, $\lambda_1 = \lambda_2 = \lambda_3 = 1 \le \tau_1 = 1$.

We conclude this section with a localization theorem for eigenvectors.

Theorem 1.8.25 (Wilkinson). *Let $A = A^T \in \mathbb{R}^{n\times n}$, $\tilde{x} \in \mathbb{R}^n$, $\|\tilde{x}\|_2 = 1$, $\tilde{\lambda} \in \mathbb{R}$, $\delta = \|A\tilde{x} - \tilde{\lambda}\tilde{x}\|_2 > 0$. If the interval $[\tilde{\lambda} - \delta, \tilde{\lambda} + \delta]$ contains a simple eigenvalue λ of A and if the remaining $n-1$ eigenvalues lie outside of the interval $[\tilde{\lambda} - a, \tilde{\lambda} + a]$ with some $a \ge \delta$, then there is an eigenvector x associated with λ such that $\|x\|_2 = 1$ and*

$$\|x - \tilde{x}\|_2^2 \le \frac{\delta^2}{a^2} + \left(1 - \left(1 - \frac{\delta^2}{a^2}\right)^{\frac{1}{2}}\right)^2 . \tag{1.8.43}$$

Proof. Let $\{x^1, \dots, x^n\}$ be an orthonormal basis of eigenvectors of A with $Ax^i = \lambda_i x^i$, $\tilde{x} = \sum_{i=1}^n \alpha_i x^i$, where the eigenvalues λ_i are not assumed to be ordered. W.l.o.g. let $\alpha_1 \ge 0$ and $\lambda_1 \in [\tilde{\lambda} - \delta, \tilde{\lambda} + \delta]$. Then

$$\delta^2 = (A\tilde{x} - \tilde{\lambda}\tilde{x})^T (A\tilde{x} - \tilde{\lambda}\tilde{x}) = \sum_{i=1}^n \alpha_i^2 (\lambda_i - \tilde{\lambda})^2 \ge a^2 \sum_{i=2}^n \alpha_i^2 ,$$

whence $\sum_{i=2}^n \alpha_i^2 \le \delta^2/a^2 \le 1$. Since $\|\tilde{x}\|_2^2 = \sum_{i=1}^n \alpha_i^2 = 1$ we get

$$\alpha_1 = \left(1 - \sum_{i=2}^n \alpha_i^2\right)^{\frac{1}{2}} \ge \left(1 - \frac{\delta^2}{a^2}\right)^{\frac{1}{2}}$$

and finally

$$\|x^1 - \tilde{x}\|_2^2 = (1 - \alpha_1)^2 + \sum_{i=2}^n \alpha_i^2 \le \left(1 - \left(1 - \frac{\delta^2}{a^2}\right)^{\frac{1}{2}}\right)^2 + \frac{\delta^2}{a^2} .$$ □

Exercises

Ex. 1.8.1. Show that λ is an eigenvalue of a regular matrix $A \in \mathbb{R}^{n\times n}$ if and only if $1/\lambda$ is an eigenvalue of A^{-1}. What can be said about the corresponding algebraic and geometric multiplicities and the corresponding eigenvectors?

Ex. 1.8.2. Show that for an arbitrary regular matrix $S \in \mathbb{C}^{n\times n}$ the matrices $A \in \mathbb{C}^{n\times n}$ and $S^{-1}AS$ have the same eigenvalues with the same algebraic and geometric multiplicity. How do the eigenvectors change?

Ex. 1.8.3.

(a) Show that the sequence of normalized eigenvectors x^k, $x_1^k = 1$, of the matrix sequence

$$A_k = \begin{pmatrix} 1 & 1/k \\ 0 & 1 \end{pmatrix}, \quad k = 1, 2, \dots ,$$

does not tend to the eigenvector $x^* = e$ of the limit matrix $I = \lim_{k\to\infty} A_k$.

(b) (J. W. Givens, Example E 3.1.5 in Ortega [266])
Show that $\lambda_1(\varepsilon) = 1 - \varepsilon$ and $\lambda_2(\varepsilon) = 1 + \varepsilon$ are the eigenvalues of the matrix

$$A(\varepsilon) = \begin{pmatrix} 1 + \varepsilon\cos(2/\varepsilon) & \varepsilon\sin(2/\varepsilon) \\ \varepsilon\sin(2/\varepsilon) & 1 - \varepsilon\cos(2/\varepsilon) \end{pmatrix}, \quad \varepsilon \neq 0,$$

and that $x^1(\varepsilon) = (\sin(1/\varepsilon), -\cos(1/\varepsilon))^T$, $x^2(\varepsilon) = (\cos(1/\varepsilon), \sin(1/\varepsilon))^T$ are corresponding eigenvectors normalized by one with respect to the Euclidean norm. Show that these eigenvectors do not tend to a limit as $\varepsilon \to 0$ even though $\lim_{\varepsilon\to 0} A(\varepsilon)$ exists.

(c) Use one of the previous examples to show that Theorem 1.8.2 (b) can become false for multiple eigenvalues.

Ex. 1.8.4. Let $A \in \mathbb{C}^{m\times n}$, $B \in \mathbb{C}^{n\times m}$. Show that the matrices AB and BA have the same nonzero eigenvalues with the same algebraic and geometric multiplicity. Find two matrices A, B such that zero is an eigenvalue of AB but not of BA.

Ex. 1.8.5. Let $R \in \mathbb{C}^{n\times n}$ be an upper triangular matrix. Show that $e^{(1)}$ is an eigenvector of R. Which are the eigenvalues of R?

Ex. 1.8.6. Show that the eigenvalues of a matrix $A \in \mathbb{R}^{n\times n}$ are the diagonal entries of its Schur normal form R.

Ex. 1.8.7. Show that the number of arithmetic operations in Algorithm 1.8.8 is $n^3/6 + O(n^2)$ and that a bound for the number of comparisons is $n^2 + O(n)$. Prove (1.8.18) and (1.8.19).

Ex. 1.8.8. Show that any matrix $A \in \mathbb{R}^{n\times n}$ is symmetric positive definite if and only if A is regular and its inverse is symmetric and positive definite, too.

Ex. 1.8.9. Let $A \in \mathbb{R}^{n\times n}$ be a skew-symmetric matrix. Show that $a_{ii} = 0$ holds for $i = 1, \dots, n$, and $x^TAx = 0$ for all $x \in \mathbb{R}^n$. Show also that λ and $-\lambda$ are eigenvalues of A simultaneously. If n is odd, prove that $\det(A) = 0$, A is singular, and zero is one of its eigenvalues. Does this also hold if n is even?

Ex. 1.8.10. Use the splitting $A = A_{\text{sym}} + A_{\text{skew}}$ with the symmetric part $A_{\text{sym}} = (A + A^T)/2$ and the skew-symmetric part $A_{\text{skew}} = (A - A^T)/2$ of $A \in \mathbb{R}^{n\times n}$ in order to show that $x^TAx > 0$ is equivalent to $x^TA_{\text{sym}}x > 0$.

Ex. 1.8.11. Show that each of the following algebraic structures is a group:
(i) the set of all lower triangular matrices from $\mathbb{R}^{n\times n}$ together with the addition of matrices;
(ii) the set of all regular lower triangular matrices from $\mathbb{R}^{n\times n}$ together with the multiplication of matrices;
(iii) the set of all regular lower triangular matrices from $\mathbb{R}^{n\times n}$ with all diagonal entries equal to one, together with the multiplication of matrices.

Find additional classes of matrices which form a group with respect to addition or multiplication of matrices.

Ex. 1.8.12. Let x, y be real eigenvectors of $A \in \mathbb{R}^{n\times n}$ associated with a *simple* real eigenvalue λ. Let $i_0 \in \{1,\dots,n\}$ be some fixed index and assume $x_{i_0} = \alpha > 0$, $y^T y = 1$, $y_{i_0} > 0$. Express y by x and show that Theorem 1.8.2 (b) remains true if the normalization $x_{i_0} = \alpha \neq 0$ there is replaced by $x^T x = 1$ with $x_{i_0} > 0$.

Ex. 1.8.13. Prove part (b) of Theorem 1.8.17.

Ex. 1.8.14. Show that the number of eigenvalues of the generalized eigenvalue problem (1.8.29) can differ from n if the positive definiteness of the symmetric matrix B is dropped. To this end consider the three cases (i) $A = I$, $B = O$, (ii) $A = O$, $B = e^{(1)}(e^{(1)})^T$, and (iii) $A = I$, $B = e^{(1)}(e^{(1)})^T$. (Compute the multiplicity of the eigenvalue in the last case!) What is the result if one requires B to be symmetric and regular?

Ex. 1.8.15. Show that each two eigenvectors of the generalized eigenvalue problem (1.8.29) associated with two different eigenvalues are B-orthogonal.

1.9 Nonnegative matrices

We first generalize the absolute value and the partial ordering for vectors introduced in Definition 1.3.5.

Definition 1.9.1.
(a) For $A \in \mathbb{K}^{m\times n}$ we define the absolute value $|A| = (|a_{ij}|) \in \mathbb{R}^{m\times n}$.
(b) For A, $B \in \mathbb{R}^{m\times n}$ we define the partial ordering $A \leq B$ entrywise by $a_{ij} \leq b_{ij}$, $i = 1,\dots,m$, $j = 1,\dots,n$. Similarly we define $A \geq B$. If strict inequality holds for all entries, we write $A < B$, and $A > B$, respectively. If $A \geq O$, we call A nonnegative, and positive if $A > O$.

Trivially, the absolute value $|A|$ of $A \in \mathbb{C}^{m\times n}$ is a particular nonnegative matrix which satisfies

$$-|A| \leq \operatorname{Re} A \leq |A|, \quad -|A| \leq \operatorname{Im} A \leq |A|$$

and

$$|A \cdot B| \leq |A| \cdot |B| \quad \text{for } A \in \mathbb{K}^{m\times n},\ B \in \mathbb{K}^{n\times p}.$$

Now we address properties of nonnegative matrices. To this end we first restrict the class of such matrices to nonnegative irreducible ones. It will turn out that this subclass has slightly stronger properties than the general one.

Irreducible nonnegative matrices

Theorem 1.9.2. *Let $O \le A \in \mathbb{R}^{n\times n}$. Then A is irreducible if and only if for each index pair (i_0, j_0) there is a power $A^k = (a_{ij}^{(k)})$ such that $a_{i_0j_0}^{(k)}$ is positive. The exponent k can be chosen to be less than n in the case $i_0 \ne j_0$ and less than $n+1$ otherwise.*

Proof. Let A be irreducible. Then there is a path $i_0 \to i_1 \to \dots i_{k-1} \to j_0$, $k \le n-1$, if $i_0 \ne j_0$ and $k \le n$ otherwise (if necessary shorten the path by excising cycles). Hence $a_{i_0i_1}a_{i_1i_2}\cdots a_{i_{k-1}j_0} > 0$ and

$$\begin{aligned} a_{i_0j_0}^{(k)} &= \sum_{l_1=1}^{n} \cdots \sum_{l_{k-1}=1}^{n} a_{i_0l_1}a_{l_1l_2}\cdots a_{l_{k-1}j_0} \\ &\ge a_{i_0i_1}a_{i_1i_2}\cdots a_{i_{k-1}j_0} > 0. \end{aligned} \tag{1.9.1}$$

Conversely, choose i_0, j_0 arbitrarily. The assumption guarantees a positive entry $a_{i_0j_0}^{(k)}$ for some power A^k. Therefore, at least one summand of the representation (1.9.1) must be positive. The index pairs of this summand form a path which connects i_0 with j_0 and which has length k. This length must eventually be shortened as above in order to see that k can be restricted as required. □

Theorem 1.9.3. *Let $A \in \mathbb{R}^{n\times n}$ be nonnegative and irreducible. Then $(I+A)^{n-1} > O$.*

Proof. The theorem follows directly from the representation

$$(I+A)^{n-1} = \sum_{k=0}^{n-1} \binom{n-1}{k} A^k$$

and Theorem 1.9.2. □

We will use Theorem 1.9.3 in order to transform a problem with a nonnegative irreducible matrix into one with a positive matrix while the eigenvectors do not change and while the eigenvalues transform in a definite way.

First we consider nonnegative eigenvectors of A.

Theorem 1.9.4. *Let $A \in \mathbb{R}^{n\times n}$ be nonnegative and irreducible. Then A has at most one linear independent nonnegative eigenvector x which then is positive together with its eigenvalue.*

Proof. Let $x \ge 0$ be an eigenvector of A associated with some eigenvalue λ. Since x has at least one positive component, say $x_{i_0} > 0$, we get $0 \le (Ax)_{i_0} = \lambda x_{i_0}$, hence $\lambda \ge 0$. We apply Theorem 1.9.3 ending up with $0 < (I+A)^{n-1}x = (1+\lambda)^{n-1}x$, whence $x > 0$. By virtue of the irreducibility of A, we get $\lambda x = Ax > 0$, hence $\lambda > 0$.

Let $y \ge 0$ be a second eigenvector of A associated with some eigenvalue μ and linearly independent of x. As above, we have $\mu \ge 0$ and $y > 0$. Assume $\lambda \le \mu$. Choose $\alpha > 0$ such that $\alpha x \ge y$ with equality in at least one component. Then

$$0 < (I+A)^{n-1}(\alpha x - y) = (1+\lambda)^{n-1}\alpha x - (1+\mu)^{n-1} y \le (1+\mu)^{n-1}(\alpha x - y)$$

which is a contradiction for the zero component of $\alpha x - y$. If $\lambda > \mu$ interchange the roles of x and y in order to get a contradiction once more. □

We are now ready to prove the famous Theorem of Perron and Frobenius.

Theorem 1.9.5 (Perron–Frobenius, irreducible matrices). *Let $A \in \mathbb{R}^{n\times n}$ be nonnegative and irreducible. Then the spectral radius $\rho(A)$ of A is a positive eigenvalue of A and there is a corresponding positive eigenvector u.*

Proof. Let $S = \{x \mid x \in \mathbb{R}^n,\ x \geq 0,\ \sum_{i=1}^n x_i = 1\}$. This set is nonempty, convex and compact. Define the function $f: S \to \mathbb{R}^n$ by

$$f(x) = \frac{Ax}{\sum_{i=1}^n \sum_{j=1}^n a_{ij}x_j}.$$

Then f is well-defined (i.e., the denominator differs from zero), continuous, and satisfies $f(S) \subseteq S$. By Brouwer's fixed point theorem f has a fixed point $u \in S$. With $\lambda = \sum_{i=1}^n \sum_{j=1}^n a_{ij}u_j$ the fixed point equation transforms to $Au = \lambda u$ and from the definition of S we see that u is a nonnegative eigenvector. Theorem 1.9.4 implies $\lambda > 0$ and $u > 0$ and the inequality $\rho(A) \leq \|A\|_u = \lambda \leq \rho(A)$ concludes the proof. □

Each *positive* eigenvector u of a nonnegative matrix A is called a *Perron vector*. According to Theorem 1.9.5 each *irreducible* nonnegative matrix has a Perron vector which, by Theorem 1.9.4, is unique up to a multiplicative scalar $\alpha > 0$. Notice that Perron vectors are often normalized by $\|u\|_2 = 1$ in order to make them unique. This is not done here.

Theorem 1.9.6. *Let $A \in \mathbb{R}^{n\times n}$ be nonnegative and irreducible. Then the spectral radius $\rho(A)$ is an algebraically simple eigenvalue of A.*

Proof. We already know that $\rho(A)$ is a positive eigenvalue of A associated with a Perron vector u. Assume that $\rho(A)$ is not simple.

Case 1: The Jordan normal form has at least two blocks associated with $\rho(A)$. In addition to u there is another eigenvector v associated with $\rho(A)$ such that u, v are linearly independent. Let v be real; otherwise replace v by its imaginary part. W.l.o.g. we can assume that $u - v$ is positive. Since $A(u - v) = \rho(A)(u - v)$ Theorem 1.9.4 implies $u - v = \alpha u$, whence $v = (1 - \alpha)u$. This contradicts the linear independence of u and v.

Case 2: The Jordan normal form J has exactly one block associated with $\rho(A)$ and this block is at least of size 2×2. There is a principal vector v such that $(A - \rho(A)I)v = u$. From $(A - \rho(A)I)\,\mathrm{Im}\, v = 0$ we can conclude that v is real. Let $t_0 = \min\{t \mid t \in \mathbb{R},\ tu - v \geq 0\}$. Then

$$0 \leq A(t_0u - v) = \rho(A)\left[\left(t_0 - \frac{1}{\rho(A)}\right)u - v\right],$$

whence $(t_0 - \frac{1}{\rho(A)})u - v \geq 0$. Thus t_0 was not minimal, which is a contradiction. □

As the example

$$A_\alpha = \begin{pmatrix} 1 & \alpha \\ 1 & 1 \end{pmatrix} \quad \text{with } \alpha = -4,\ \alpha = 1,\ \alpha = 4$$

shows the spectral radius of a matrix A is not monotone with respect to the entries of A even if this matrix is irreducible. The situation changes for nonnegative matrices.

Theorem 1.9.7 (Comparison theorem, irreducible case). *Let $A \in \mathbb{R}^{n\times n}$ be nonnegative and irreducible and let $B \in \mathbb{C}^{n\times n}$ such that $|B| \le A$. Then $\rho(B) \le \rho(A)$ with equality if and only if $B = e^{i\varphi}DAD^{-1}$, where $\lambda = e^{i\varphi}\rho(A)$ is an eigenvalue of B and where $D \in \mathbb{C}^{n\times n}$, $|D| = I$. In particular, $\rho(A)$ is a strictly monotonously increasing function of the entries of A.*

Proof. The inequality follows from

$$\rho(B) \le \|B\|_u = \|\,|B|\,\|_u \le \|A\|_u = \rho(A),$$

where u is a Perron vector of A.

In order to prove the equivalence let $\rho = \rho(A) = \rho(B)$, $\lambda = e^{i\varphi}\rho$, $Bz = \lambda z$, $z \ne 0$. Then

$$\rho|z| = |\lambda|\,|z| = |Bz| \le |B|\,|z| \le A|z|. \tag{1.9.2}$$

Define $y = (I + A)^{n-1}|z|$ which is positive by virtue of Theorem 1.9.3. If strict inequality holds in (1.9.2) for at least one component we get

$$\rho y < (I + A)^{n-1}A|z| = Ay,$$

which yields the contradiction $\rho\|y\|_u < \|A\|_u\|y\|_u = \rho\|y\|_u$. Therefore, equality holds in (1.9.2), whence $|z|$ is a Perron vector of A. In particular, it is positive. From (1.9.2) we obtain $(A - |B|)|z| = 0$, hence $A = |B|$.

Let $D = \operatorname{diag}(\frac{z_1}{|z_1|}, \dots, \frac{z_n}{|z_n|})$. Then $z = D|z|$ and $BD|z| = Bz = e^{i\varphi}\rho z = \rho e^{i\varphi}D|z|$. With $C = e^{-i\varphi}D^{-1}BD$ we deduce $\rho|z| = C|z| = A|z|$, where we also used (1.9.2) with equality. This implies $(A - C)|z| = 0$. Since $|C| = |B| = A$, we can replace A by $|C|$ in order to end up with $(|C| - C)|z| = 0$. Therefore, $(|C| - \operatorname{Re} C)|z| = 0$, whence $|C| = \operatorname{Re} C$ and $C = |C| = A$. The definition of C finally yields $B = e^{i\varphi}DAD^{-1}$.

The converse is trivial.

Let $O \le A \le A'$, $A \ne A'$. Then the preceding results (with A, A' in place of B, A) imply $\rho(A) < \rho(A')$, which proves the monotony. □

We continue our results with an interesting alternative.

Theorem 1.9.8 (Quotient theorem, irreducible case). *Let $A \in \mathbb{R}^{n\times n}$ be nonnegative and irreducible and let $x \in \mathbb{R}^n$ be positive. Then either*

$$\frac{(Ax)_i}{x_i} = \rho(A) \quad \textit{for } i = 1, \dots, n$$

or

$$\min\left\{\frac{(Ax)_i}{x_i}\ \middle|\ i = 1, \dots, n\right\} < \rho(A) < \max\left\{\frac{(Ax)_i}{x_i}\ \middle|\ i = 1, \dots, n\right\}.$$

Proof. Let v be a Perron vector of A^T. Then $A^T v = \rho(A^T)v = \rho(A)v$. The assertion can be deduced from

$$0 = \rho(A)v^T x - v^T Ax = \sum_{i=1}^{n} \left(\rho(A) - \frac{(Ax)_i}{x_i} \right) v_i x_i$$

taking into account that $v_i x_i$ is positive for all i. □

General nonnegative matrices

Now we consider nonnegative matrices which are not necessarily irreducible. Most of the preceding results can be shown when replacing the strict inequality sign by '≤'. The proof is then based on a continuity argument.

Theorem 1.9.9 (Perron–Frobenius, general case). *Let $A \in \mathbb{R}^{n\times n}$ be nonnegative. Then the spectral radius $\rho(A)$ of A is a nonnegative eigenvalue of A and there is a corresponding nonnegative eigenvector u. Moreover, the spectral radius is zero if and only if A is reducible with a strict triangular matrix as reducible normal form.*

Proof. Let $k \in \mathbb{N}$ and $A_k = A + ee^T/k > O$. According to Theorem 1.9.5 the matrix A_k has a Perron vector u^k which we choose such that $\|u^k\|_2 = 1$. The sequence (u^k) is bounded and thus has a convergent subsequence. For simplicity we assume that the whole sequence is convergent to some limit u which certainly is nonnegative and satisfies $\|u\|_2 = 1$. Let k tend to infinity in $A_k u^k = \rho(A_k)u^k$. Then $Au = \rho(A)u$ which proves the first part of the theorem.

The equivalence follows for instance from the reducible normal form $R_A = (A_{ij})_{i,j=1,\dots,s}$ of A and the subsequent Theorem 1.9.10 applied to the block matrix $B = (B_{ij})_{i,j=1,\dots,s}$ with the same block structure as R_A, $B_{ii} = A_{ii}$, and $B_{ij} = O$ otherwise. It implies $\rho(A_{ii}) = \rho(B_{ii}) \le \rho(B) \le \rho(R_A) = \rho(A) = 0$ whence $A_{ii} = O$ follows by virtue of Theorem 1.9.5 and the definition of R_A in Theorem 1.7.16. □

The example $A = \operatorname{diag}(1, 1/2)$ shows that not every nonnegative matrix has a Perron vector. In addition, the 2×2 matrices I and $I + (1, 0)^T(0, 1)$ illustrate that the spectral radius does not strictly increase with the entries. However, the following slightly weaker result can be deduced immediately from Theorem 1.9.7 using the continuity argument in the proof of Theorem 1.9.9.

Theorem 1.9.10 (Comparison theorem, general case). *Let $A \in \mathbb{R}^{n\times n}$ be nonnegative and let $B \in \mathbb{C}^{n\times n}$ such that $|B| \le A$. Then $\rho(B) \le \rho(A)$. In particular, $\rho(A)$ is a (weakly) monotonously increasing function of the entries of A.*

By a similar continuity argument as above we obtain a generalization of the quotient theorem.

Theorem 1.9.11 (Quotient theorem, general case). *Let $A \in \mathbb{R}^{n\times n}$ be nonnegative and let $x \in \mathbb{R}^n$ be positive. Then*

$$\min\left\{\frac{(Ax)_i}{x_i}\;\middle|\; i=1,\dots,n\right\} \le \rho(A) \le \max\left\{\frac{(Ax)_i}{x_i}\;\middle|\; i=1,\dots,n\right\}.$$

Again, the result cannot be improved. Consider the 2×2 matrix $A = I + (1,0)^T(0,1)$ and the vector $x = e$.

In case of nonnegative matrices, we can sharpen Theorem 1.7.3 to operator norms which are generated by *monotone* vector norms. To this end we notice that the norm $\|\cdot\|_v$ defined in (1.3.8) satisfies $\|A\|_v = \rho(A)$ if $v > 0$ is a Perron vector of the nonnegative irreducible matrix A.

Theorem 1.9.12. *For each nonnegative matrix $A \in \mathbb{R}^{n\times n}$ and any positive real number ε there is a monotone vector norm $\|\cdot\|$ such that the generated operator norm satisfies $\|A\| \le \rho(A) + \varepsilon$. If, in addition, A is irreducible, then $\|\cdot\|$ can be chosen such that $\|A\| = \rho(A)$ holds.*

Proof. Let $A_\delta = A + \delta ee^T$, $\delta > 0$. Then A_δ is positive, hence irreducible. Choose δ such that $\rho(A_\delta) < \rho(A) + \varepsilon$. Denote by $v > 0$ a Perron vector of A_δ. Then $\|A\|_v \le \|A_\delta\|_v = \rho(A_\delta) < \rho(A) + \varepsilon$.

If $A \ge O$ is irreducible, choose v as a Perron vector of A instead of A_δ. □

Notice that the norm in Theorem 1.9.12 depends on $A \ge O$ and (in the general case) on ε.

Next we prove localization theorems for $\rho(A)$ for which no norm is needed.

Theorem 1.9.13. *Let $A \in \mathbb{R}^{n\times n}$ be nonnegative, $x \in \mathbb{R}^n$ be positive and $s \in \mathbb{R}_0^+$.*

(a) *If $Ax < sx$, then $\rho(A) < s$.*
(b) *If $Ax \le sx$, then $\rho(A) \le s$.*
(c) *If $Ax \le sx$, $Ax \ne sx$ and if A is irreducible, then $\rho(A) < s$.*

Proof. The parts (a) and (b) are direct consequences of Theorem 1.9.11. Part (c) follows from Theorem 1.9.8. □

Theorem 1.9.14. *Let $A \in \mathbb{R}^{n\times n}$ be nonnegative and $s \in \mathbb{R}$.*

(a) *The inverse $(sI - A)^{-1}$ exists and is nonnegative if and only if $\rho(A) < s$.*
(b) *The inverse $(sI - A)^{-1}$ exists and is positive if and only if $\rho(A) < s$ and A is irreducible.*

Proof. (a) Let $(sI - A)^{-1} \ge O$ and $Ax = \lambda x$, $x \ne 0$. Then $|\lambda||x| \le A|x|$ and $(s - |\lambda|)|x| \ge (sI - A)|x|$, whence $(s - |\lambda|)(sI - A)^{-1}|x| \ge |x| \ge 0$. Since $x \ne 0$ we have $s > |\lambda|$.

Conversely, let $\rho(A) < s$. Then $s > 0$ and $\rho(\frac{1}{s}A) = \frac{1}{s}\rho(A) < 1$. Hence Theorem 1.7.6 guarantees

$$(sI - A)^{-1} = \frac{1}{s}\left(I - \frac{1}{s}A\right)^{-1} = \frac{1}{s}\sum_{k=0}^{\infty}\left(\frac{1}{s}A\right)^k \ge O. \tag{1.9.3}$$

(b) Let $(sI - A)^{-1} > O$. Then $\rho(A) < s$ by virtue of (a), whence $\rho(\frac{1}{s}A) < 1$. The irreducibility of A follows from the positivity of the sum in (1.9.3) and from Theorem 1.9.2. □

Structure of irreducible nonnegative matrices

Now we reconsider irreducible nonnegative matrices and divide them in two large classes: those for which $\lambda = \rho(A)$ is the only eigenvalue with $|\lambda| = \rho(A)$, and the remaining ones. More precisely, we use the following terminology.

Definition 1.9.15. Let $A \in \mathbb{R}^{n\times n}$ be nonnegative and irreducible. In addition, let A have exactly h eigenvalues λ_j, $j = 0, 1, \ldots, h-1$, with $|\lambda_j| = \rho(A)$. (All eigenvalues are counted according to their multiplicity.) If $h = 1$, then A is called primitive, if $h > 1$, it is called h-cyclic.

Frobenius showed that cyclic matrices have very interesting properties. Some of them are listed in our next theorem.

Theorem 1.9.16 (and Definition). *Let $A \in \mathbb{R}^{n\times n}$ be nonnegative, irreducible and h-cyclic with the eigenvalues $\lambda_j = e^{i\theta_j}\rho(A)$, $j = 0, 1, \ldots, h-1$, where $0 = \theta_0 \le \theta_1 \le \cdots \le \theta_{h-1} < 2\pi$. With $\varphi = 2\pi/h$ the following properties hold.*

(a) *The exponents θ_j are equally spaced and satisfy $\theta_j = j\varphi$, $j = 0, 1, \ldots, h-1$. In particular, λ_j is a simple eigenvalue of A.*
(b) *The spectrum of A is invariant when being rotated around zero in the complex plane by an angle of φ.*
(c) *There is a permutation matrix $P \in \mathbb{R}^{n\times n}$ such that A transforms to the block form*

$$PAP^T = (A_{ij}) = \begin{pmatrix} O & A_{12} & O & \ldots & O \\ \vdots & O & A_{23} & & \vdots \\ \vdots & & \ddots & \ddots & O \\ O & & & \ddots & A_{h-1,h} \\ A_{h1} & O & \ldots & \ldots & O \end{pmatrix} \tag{1.9.4}$$

with quadratic diagonal blocks $A_{ii} = O$ and $A_{h1} \ne O$, $A_{i,i+1} \ne O$, $i = 1, \ldots, h-1$. The block form (1.9.4) is called the cyclic normal form of A.

Proof. (a) According to Theorem 1.9.7 applied to $B = A$ and $\lambda_m = e^{i\theta_m}\rho(A)$, there is a diagonal matrix $D_m \in \mathbb{C}^{n\times n}$ such that $|D_m| = I$ and $A = e^{i\theta_m}D_mAD_m^{-1}$. Then $\det(A - \lambda I) = \det(e^{i\theta_m}D_mAD_m^{-1} - \lambda I)$, whence (with the eigenvalues λ_j, $j = 0, \ldots, n-1$ of A) we get

$$\prod_{j=0}^{n-1}(\lambda_j - \lambda) = \prod_{k=0}^{n-1}(e^{i\theta_m}\lambda_k - \lambda). \tag{1.9.5}$$

Therefore, $\{\lambda_j \mid j = 0, 1, \ldots, h-1\} = \{e^{i\theta_m}\lambda_k \mid k = 0, 1, \ldots, h-1\}$. Since λ_0 is simple by Theorem 1.9.6 and since m was arbitrary, all the eigenvalues λ_j, $j = 0, 1, \ldots, h-1$ must be simple. Hence $0 = \theta_0 < \theta_1 < \cdots < \theta_{h-1} < 2\pi$, and from (1.9.5) with $\lambda = 0$ we get $\prod_{j=0}^{h-1} \lambda_j = \prod_{k=0}^{h-1} e^{i\theta_m}\lambda_k = e^{ih\theta_m}\prod_{k=0}^{h-1} \lambda_k$. By virtue of $|\lambda_j| = \rho(A) > 0$ we can divide by the product and obtain $e^{ih\theta_m} = 1$. Thus θ_m is an h-th unit root. Since m was arbitrary and since the individual values $\theta_m \in [0, 2\pi)$ differ from each other we must have $\theta_m = \frac{2\pi}{h}m$.

(b) follows immediately from (1.9.5).

(c) Let the diagonal matrix $D = D_1$ from the proof of (a) (with $m = 1$) have s different diagonal entries $e^{i\varphi_k}$, $0 \le \varphi_1 < \varphi_2 < \cdots < \varphi_s < 2\pi$. W.l.o.g. let $\varphi_1 = 0$, otherwise choose $D = e^{-i\varphi_1}D_1$. From the definition of D_1 we get

$$\begin{aligned} A &= e^{i\theta_1}DAD^{-1} = e^{i\theta_1}D(e^{i\theta_1}DAD^{-1})D^{-1} \\ &= e^{i2\theta_1}D^2AD^{-2} = \cdots = e^{ih\theta_1}D^hAD^{-h} = D^hAD^{-h} \end{aligned} \tag{1.9.6}$$

and with a Perron vector u of A we end up with $AD^hu = D^hAD^{-h}D^hu = \lambda_0D^hu$. Since λ_0 is a simple eigenvalue, the eigenvector D^hu must satisfy $D^hu = \alpha u$. From $D = \operatorname{diag}(\ldots, e^{i\varphi_1}, \ldots) = \operatorname{diag}(\ldots, 1, \ldots)$ we deduce $\alpha = 1$, hence $D^h = I$. This results in $\varphi_k = \frac{2\pi}{h}n_k$ with some integers $0 = n_1 < n_2 < \cdots < n_s < h$, in particular, $s \le h$. There is a permutation matrix $P \in \mathbb{R}^{n\times n}$ with $PDP^T = \operatorname{diag}(e^{i\varphi_1}I^{(1)}, \ldots, e^{i\varphi_s}I^{(s)})$, where $I^{(k)}$ are appropriate unit matrices. Let $PAP^T = (A_{kj})$ be divided into blocks according to PDP^T. Then the diagonal blocks A_{kk} are quadratic and from (1.9.6) we get $PAP^T = e^{i\theta_1}(PDP^T)(PAP^T)(PD^{-1}P^T)$. Hence $A_{kj} = e^{i\theta_1}e^{i\varphi_k}A_{kj}e^{-i\varphi_j}$. This implies $e^{i\theta_1(1+n_k-n_j)} = 1$ if $A_{kj} \ne O$, or, equivalently, $\theta_1(1 + n_k - n_j) \equiv 0 \mod 2\pi$ and

$$n_k + 1 \equiv n_j \mod h, \quad \text{if } A_{kj} \ne O. \tag{1.9.7}$$

Choose n_j, $n_k \in \{0, \ldots, h-1\}$ and notice that PAP^T is irreducible. For each block index $k \in \{0, \ldots, s\}$ we can find an index j in the same index set such that $A_{kj} \ne O$, whence, by virtue of (1.9.7), $n_k + 1 \equiv n_j \mod h$. Start with $k = 1$. From $n_1 = 0$ we get $n_j = 1$ and $\varphi_j = 2\pi/h$. Since this is the smallest possible positive angle we must have $\varphi_2 = \varphi_j$, in particular, $j = 2$, $n_2 = 1$, $A_{12} \ne O$. Step by step one shows that $n_{k+1} = k$, $A_{k,k+1} \ne O$ for $k = 2, \ldots, s-1$. The irreducibility of A together with $A_{k1} = O$ for $k = 1, \ldots, s-1$ implies $A_{s1} \ne O$, whence, by (1.9.7) and $n_s = s-1$, $n_1 = 0$, $0 < s \le h$ we finally obtain $s = h$. □

Corollary 1.9.17. *If $A \in \mathbb{R}^{n\times n}$ is nonnegative and irreducible with $a_{ii} > 0$ for at least one diagonal entry then A is primitive.*

Theorem 1.9.18. *Let $A \in \mathbb{R}^{n\times n}$ be nonnegative and irreducible. If A is h-cyclic then there is a permutation matrix P and matrices B_j, $j = 1, \ldots, h$ such that*

$$PA^{hk}P^T = \operatorname{diag}(B_1^k, \ldots B_h^k), \quad k = 1, 2, \ldots, \tag{1.9.8}$$

where the diagonal blocks B_j are primitive and satisfy $\rho(B_j) = \rho(A)^h$.

Proof. Choose P as for the cyclic normal form (1.9.4). Multiply PAP^T with itself several times to see that the upper nonzero diagonal moves to the right upper corner while the lower one tends to the diagonal. With

$$B_i = A_{i,i+1}A_{i+1,i+2}\cdots A_{h1}A_{12}\cdots A_{i-1,i}$$

one ends up with (1.9.8). With Theorem 1.8.4 we get $\rho(B_i) = \rho(B_j)$ for $i, j = 1, \ldots, h$. Choose $k = 1$ in (1.9.8). This proves $\rho(B_i) = \rho(A^h) = [\rho(A)]^h$. If at least one matrix B_i is not primitive there would be at least $h + 1$ eigenvalues λ of A^h with $|\lambda| = [\rho(A)]^h$. Then A would have at least $h + 1$ eigenvalues μ with $|\mu| = \rho(A)$ contradicting h-cyclicity. □

Example 1.9.19. Let

$$A = \begin{pmatrix} O & A_1 & O & O \\ O & O & A_2 & O \\ O & O & O & A_3 \\ A_4 & O & O & O \end{pmatrix}$$

be 4-cyclic and denote by $A_{i_1i_2\ldots i_s}$ the product $A_{i_1}A_{i_2}\cdots A_{i_s}$.

Then

$$A^2 = \begin{pmatrix} O & O & A_{12} & O \\ O & O & O & A_{23} \\ A_{34} & O & O & O \\ O & A_{41} & O & O \end{pmatrix},$$

$$A^3 = \begin{pmatrix} O & O & O & A_{123} \\ A_{234} & O & O & O \\ O & A_{341} & O & O \\ O & O & A_{412} & O \end{pmatrix},$$

$$A^4 = \begin{pmatrix} A_{1234} & O & O & O \\ O & A_{2341} & O & O \\ O & O & A_{3412} & O \\ O & O & O & A_{4123} \end{pmatrix},$$

$$A^5 = \begin{pmatrix} O & A_{12341} & O & O \\ O & O & A_{23412} & O \\ O & O & O & A_{34123} \\ A_{41234} & O & O & O \end{pmatrix},$$

etc.

The powers A^k of an arbitrary h-cyclic matrix (in cyclic normal form) show an analogous behavior.

1.10 Particular matrices

In this section we introduce some more classes of matrices which are connected with nonnegativity. We will consider M-matrices, H-matrices, and inverse positive matrices: M-matrices are real matrices with a particular sign structure and a nonnegative inverse. H-matrices are M-matrices if one changes the signs of their entries to those of an M-matrix. Inverse positive matrices have a nonnegative inverse. All matrices will be of importance in interval analysis.

Definition 1.10.1. The set of all matrices $A \in \mathbb{R}^{n\times n}$ with $a_{ij} \le 0$ for $i \ne j$ is denoted by $Z^{n\times n}$.

Definition 1.10.2. A regular matrix $A \in Z^{n\times n}$ is called an M-matrix if $A^{-1} \ge O$. A symmetric M-matrix is called a Stieltjes matrix.

Example 1.10.3. The matrix

$$A = \begin{pmatrix} 2 & -1 \\ -3 & 2 \end{pmatrix}$$

has the nonnegative inverse

$$A^{-1} = \begin{pmatrix} 2 & 1 \\ 3 & 2 \end{pmatrix},$$

hence it is an M-matrix.

Theorem 1.10.4. *For $A \in Z^{n\times n}$ the following statements are equivalent.*
(a) *The matrix A is an M-matrix.*
(b) *There is a vector $u > 0$ such that $Au > 0$.*
(c) *Each principal submatrix of A is an M-matrix.*
(d) *There is a matrix $B \ge O$ and a real number $s > \rho(B)$ such that $A = sI - B$.*
(e) *All principal minors of A are positive.*

Proof. '(a) ⇒ (b)': Choose $u = A^{-1}e$.

'(b) ⇒ (d)': Choose $s = \max\{|a_{ii}| \mid i = 1, \dots, n\}$. Then $B = sI - A$ is nonnegative and $Bu = su - Au < su$. Hence Theorem 1.9.13 guarantees $\rho(B) < s$.

'(d) ⇒ (a)': $A^{-1} = (sI - B)^{-1}$ exists and is nonnegative by Theorem 1.9.14. The sign pattern of A follows from $A = sI - B$.

'(b) ⇒ (c)': Let A' by any principal submatrix of A and let $u > 0$ be a vector for A as guaranteed by (b). Shorten this vector by those components whose indices do not belong to A' and denote the result by u'. Then A' has the sign pattern of an M-matrix and fulfills $A'u' > 0$. Hence it is an M-matrix.

'(c) ⇒ (a)': Is trivial since A is a principal submatrix itself.

'(c), (d) ⇒ (e)': If λ is a real eigenvalue of A with $A = sI - B$ as in (d), then $\lambda \ge s - \rho(B)$ which is positive by (d). Since the nonreal eigenvalues of A occur in conjugate complex pairs $\lambda, \bar{\lambda}$ we have $\det A = \prod_{i=1}^{n} \lambda_i > 0$. This argumentation can be repeated for any principal minors different from $\det A$ by virtue of (c).

'(e) ⇒ (d)': Let all principal minors of A be positive. Choose $s \in \mathbb{R}^+$, $O \le B \in \mathbb{R}^{n\times n}$ such that $A = sI - B$, and define the matrices $A(\lambda) = \lambda I - B$ and the function $f(\lambda) = \det A(\lambda)$.

We first show by induction on their order k that all principal minors of $A(\lambda)$ are positive for $\lambda \ge s$. If $k = 1$ this is trivial. Assume that our subsidiary statement holds for the dimension $k - 1 < n$. Since the principal submatrices of A, and $A(\lambda)$, respectively, are matrices of the form $A(\lambda)$ with smaller dimension for which (d), and the induction hypothesis, respectively, hold we may assume w.l.o.g. that $k = n - 1$. Then all principal minors of $A(\lambda)$, $\lambda \ge s$, are positive provided their order is less than n. Since (proved by induction with respect to n)

$$f'(\lambda) = \sum_{j=1}^{n} \det(A(\lambda)_{*,1}, \ldots, A(\lambda)_{*,j-1}, e^{(j)}, A(\lambda)_{*,j+1}, \ldots, A(\lambda)_{*,n})$$

is a sum of principal minors of the order $n - 1$ the derivative is positive. Hence $f(\lambda)$ is monotonously increasing and therefore positive for $\lambda \ge s$ since $f(s) > 0$ by (e). This proves our subsidiary statement.

From $f(\lambda) = \det(\lambda I - A) > 0$ for $\lambda \ge s$ we see at once that the eigenvalue $\rho(B)$ of B must be less than s, since $\det(\rho(B)I - B) = 0$. Thus we get (d). □

Criterion (b) is the famous criterion by Fan [92] which is often used in order to verify an M-matrix. From (b) and the sign pattern of an M-matrix we obtain immediately the following corollary.

Corollary 1.10.5. *Let $A \in Z^{n\times n}$ be an M-matrix. Then the diagonal entries of A are positive and each matrix B with $A \le B \in Z^{n\times n}$ is an M-matrix.*

For irreducible matrices the criteria in Theorem 1.10.4 can be slightly modified.

Theorem 1.10.6. *Let $A \in Z^{n\times n}$ be irreducible. Then the following statements are equivalent.*
(a) *The matrix A is an M-matrix.*
(b) *The matrix A is regular with $A^{-1} > O$.*
(c) *There is a vector $u > 0$ such that $Au \ge 0$ with $(Au)_i > 0$ for at least one component.*

Proof. '(a) ⇔ (b)': Let $A = sI - B$ as in Theorem 1.10.4. Then $B \ge O$ is irreducible and $A^{-1} = \frac{1}{s}(I - \frac{1}{s}B)^{-1} > O$ using the Neumann series and Theorem 1.9.2. The converse follows directly from the Definition 1.10.2.

'(a) ⇔ (c)': The implication '⇒' follows directly from Theorem 1.10.4. The converse direction is proved nearly literally as '(b) ⇒ (d)' of the same theorem, this time applying Theorem 1.9.13 (c). □

Theorem 1.10.7 (with definition). *Let $A \in Z^{n\times n}$ have only positive diagonal entries. Then each of the following matrices is an M-matrix.*

(a) *The matrix A is strictly diagonally dominant, i.e.,*

$$|a_{ii}| > \sum_{\substack{j=1 \\ j \neq i}}^{n} |a_{ij}|, \quad i = 1, \ldots, n.$$

(b) *The matrix A is irreducibly diagonally dominant, i.e., A is irreducible and*

$$|a_{ii}| \geq \sum_{\substack{j=1 \\ j \neq i}}^{n} |a_{ij}|, \quad i = 1, \ldots, n,$$

with strict inequality for at least one index i.

(c) *The matrix A is regular and diagonally dominant, i.e.,*

$$|a_{ii}| \geq \sum_{\substack{j=1 \\ j \neq i}}^{n} |a_{ij}|, \quad i = 1, \ldots, n.$$

(d) *The matrix A is triangular with positive diagonal entries.*

Proof. (a), (b) follow immediately from Theorem 1.10.4 and Theorem 1.10.6 with $u = e$.

(c) The matrix $A + \varepsilon I$, $\varepsilon > 0$, fulfills the assumptions of (a). Hence $(A + \varepsilon I)^{-1} \geq O$ which proves the assertion if $\varepsilon \to 0$.

(d) Here, a positive vector u can be constructed componentwise such that $Au > 0$. □

Theorem 1.10.8. *A matrix $A \in Z^{n\times n}$ is a Stieltjes matrix if and only if A is symmetric and positive definite.*

Proof. Let A be a Stieltjes matrix and assume that A has an eigenvalue $\lambda \leq 0$. Then $A - \lambda I$ is singular. Moreover, $A \leq A - \lambda I \in Z^{n\times n}$, hence $A - \lambda I$ is an M-matrix by Corollary 1.10.5. Thus $A - \lambda I$ is regular which is a contradiction. Therefore, A has only positive eigenvalues and must be positive definite.

Conversely, let A be symmetric and positive definite. Choose

$$s > \max\{|a_{ii}| \mid i = 1, \ldots, n\}, \quad B = sI - A.$$

Then B is nonnegative and $s - \rho(B)$ is an eigenvalue of A which must be positive. Hence $\rho(B) < s$, and A is an M-matrix. □

Example 1.10.9. If one discretizes the boundary value problem

$$\begin{gathered} -u'' + q(x)u = f(x), \quad q(x) > 0, \quad 0 < x < 1, \\ u(0) = \alpha, \quad u(1) = \beta \end{gathered}$$

on an equidistant grid $\{x_i \mid x_i = ih,\ h = 1/(n+1),\ i = 0, \ldots, n+1\}$ using the approximation $u(x_i) \approx \eta_i$, $u''(x_i) \approx (\eta_{i-1} - 2\eta_i + \eta_{i+1})/h^2$ and $\eta_0 = \alpha$, $\eta_{n+1} = \beta$, then one ends up with a linear system $A\eta = b$ with

$$A = \begin{pmatrix} 2 & -1 & & & O \\ -1 & 2 & -1 & & \\ & \ddots & \ddots & \ddots & \\ & & -1 & 2 & -1 \\ O & & & -1 & 2 \end{pmatrix} + h^2 \operatorname{diag}(q(x_1), \dots, q(x_n)) \in Z^{n\times n},$$

$$b = h^2 (f(x_1) + \alpha/h^2, f(x_2), \dots, f(x_{n-1}), f(x_n) + \beta/h^2)^T \in \mathbb{R}^n.$$

By virtue of Theorem 1.10.7 (b) the matrix A is an M-matrix. Moreover, by Definition 1.10.2 it is a Stieltjes matrix.

Example 1.10.10. If one discretizes the elliptic boundary value problem

$$-\Delta u + q(x,y)u = f(x,y), \quad q(x,y) \ge 0, \quad (x,y) \in Q = (0,1) \times (0,1),$$
$$u(x,y) = g(x,y), \quad (x,y) \in \partial Q$$

on an equidistant grid $\{(x_i, y_j) \mid x_i = ih,\ y_j = jh,\ h = 1/(n+1),\ i,j = 0,\dots,n+1\}$ using the approximation $u(x_i,y_j) \approx \eta_{ij}$, the five point star approximation $\Delta u(x_i, y_j) := u_{xx}(x_i,y_j) + u_{yy}(x_i,y_j) \approx (\eta_{i-1,j} + \eta_{i,j-1} - 4\eta_{ij} + \eta_{i+1,j} + \eta_{i,j+1})/h^2$ and $\eta_{ij} = g(x_i,y_j)$, if $i \in \{0, n+1\}$ or $j \in \{0, n+1\}$, then one ends up with a linear system $A\eta = b$, where $\eta \in \mathbb{R}^{n^2}$ contains the components η_{ij} in a row-wise ordering, i.e., $\eta_k = \eta_{ij}$ with $k = (j-1)n + i$, $i,j = 1,\dots,n$. With $q_{ij} = q(x_i, y_j)$ the matrix A has the form

$$A = \begin{pmatrix} T & -I & & & O \\ -I & T & -I & & \\ & \ddots & \ddots & \ddots & \\ & & -I & T & -I \\ O & & & -I & T \end{pmatrix} + h^2 \operatorname{diag}(q_{11}, q_{21}, \dots, q_{nn}) \in \mathbb{R}^{n^2\times n^2},$$

where

$$T = \begin{pmatrix} 4 & -1 & & & O \\ -1 & 4 & -1 & & \\ & \ddots & \ddots & \ddots & \\ & & -1 & 4 & -1 \\ O & & & -1 & 4 \end{pmatrix} \in \mathbb{R}^{n\times n}.$$

With $f_{ij} = f(x_i, y_j)$ and $g_{ij} = g(x_i, y_j)$ the vector b reads

$$b = h^2 \Big(f_{11} + \frac{g_{10} + g_{01}}{h^2}, f_{21} + \frac{g_{20}}{h^2}, \dots, f_{n1} + \frac{g_{n0} + g_{n+1,1}}{h^2},$$
$$f_{12} + \frac{g_{02}}{h^2}, \dots, f_{nn} + \frac{g_{n+1,n} + g_{n,n+1}}{h^2} \Big)^T \in \mathbb{R}^{n^2}.$$

By the same reasons as in Example 1.10.9, the matrix A is a Stieltjes matrix.

Singular M-matrices occur in applications like the closed input–output model of Leontief. Their definition is similar to the condition (d) in Theorem 1.10.4, which is equivalent to a regular M-matrix.

Definition 1.10.11. A matrix $A \in Z^{n\times n}$ of the form $A = \rho(B)I - B$, $B \geq O$, is called a singular M-matrix.

Notice that M-matrices without the supplement 'singular' are always assumed to be regular in this book.

Theorem 1.10.12. *Let $0 < u \in \mathbb{R}^n$, $A \in Z^{n\times n}$ irreducible, $Au = 0$. Then A is a singular M-matrix.*

Proof. Let $s = \max\{a_{ii} \mid i = 1, \dots, n\}$. Then the assumptions of the theorem imply $s > 0$, $B := sI - A \geq O$, B irreducible, $Bu = su$. Theorem 1.9.8 applied to B and u shows $\rho(B) = s$ and finishes the proof because of $A = sI - B$. □

Theorem 1.10.13. *Let $A \in Z^{n\times n}$ be a singular irreducible M-matrix. Then*
(i) *A has rank $n - 1$.*
(ii) *There is a positive vector u such that $Au = 0$.*
(iii) *Each principal submatrix of A other than A itself is a regular M-matrix.*

Proof. Let $A = \rho(B)I - B$, $B \geq O$. Then B is irreducible.

(i) Since $\rho(B)$ is a simple eigenvalue of B, zero is a simple eigenvalue of A.

(ii) Any Perron vector u of B satisfies $u > 0$, $Au = 0$.

(iii) Let $A_k \in \mathbb{R}^{k\times k}$, $k < n$, be any principal submatrix of A. It grows out from A by simultaneously deleting certain rows and columns. Replace them in A by rows and columns of the identity matrix and multiply the arising entry one in the diagonal by $\rho(B)$. Denote the new matrix by A'. Then $A' = \rho(B)I - B'$ with $O \leq B' \leq B$. Since B was irreducible, each column contains at least one positive nondiagonal entry. By construction, B differs from B' just by zeroing out those rows and columns which had to be deleted for A_k. Therefore, $B' \neq B$, whence $\rho(B') < \rho(B)$ by Theorem 1.9.7. By virtue of Theorem 1.10.4 (d), A' is a regular M-matrix and so is A_k as a principal submatrix of it. □

Next we associate with $A \in \mathbb{R}^{n\times n}$ a matrix which has the sign pattern of an M-matrix.

Definition 1.10.14. Let $A = D - B \in \mathbb{R}^{n\times n}$ with $D = \mathrm{diag}(a_{11}, \dots, a_{nn})$. Then the matrix $\langle A\rangle = |D| - |B|$ is called a comparison (or Ostrowski) matrix of A.

Definition 1.10.15. A matrix $A \in \mathbb{R}^{n\times n}$ is called an H-matrix if $\langle A\rangle$ is an M-matrix.

Theorem 1.10.16. *Let A, D, B be as in Definition 1.10.14. Then the following statements are equivalent.*
(a) *A is an H-matrix.*
(b) *There is a positive vector u such that $\langle A\rangle u > 0$.*

(c) $\rho(|D^{-1}B|) = \rho(|D|^{-1}|B|) < 1$.
(d) *If* $0 \le u \in \mathbb{R}^n$ *and* $\langle A\rangle u \le 0$, *then* $u = 0$.

Proof. The equivalence of (a) and (b) follows immediately from Definition 1.10.15 and Theorem 1.10.4.

'(a) ⇒ (c)': Let A be an H-matrix. By virtue of Corollary 1.10.5 the diagonal matrix $|D|^{-1}$ exists and is nonnegative. Since $O \le \langle A\rangle^{-1}|D| = (I - |D|^{-1}|B|)^{-1}$ Theorem 1.9.14 proves the assertion.

Conversely, let $\rho(|D|^{-1}|B|) < 1$. By virtue of Theorem 1.9.14 we get $(I - |D|^{-1}|B|)^{-1} \ge O$, whence $\langle A\rangle^{-1} \ge O$. Therefore, $\langle A\rangle$ is an M-matrix.

'(a) ⇒ (d)': The implication follows from $\langle A\rangle^{-1} \ge O$ and the assumptions of (d) which imply $0 \le u = \langle A\rangle^{-1}\langle A\rangle u \le 0$, i.e., $u = 0$.

'(d) ⇒ (c)': Choose $u = e^{(j)}$. Since $u \ne 0$ and $(\langle A\rangle u)_i = -|a_{ij}| \le 0$ for $i \ne j$ we must have $|a_{jj}| = |d_{jj}| > 0$. Otherwise $\langle A\rangle u \le 0$ and (d) implies the contradiction $u = e^{(j)} = 0$. Therefore, $|D^{-1}B| = |D^{-1}| \cdot |B|$ exists and has an eigenvector $v \ge 0$ associated with the eigenvalue $\rho := \rho(|D^{-1}B|)$. If $\rho \ge 1$, then $\langle A\rangle v = (|D| - |B|)v = |D|(1-\rho)v \le 0$ implies $v = 0$, which is again a contradiction. □

Corollary 1.10.17. *Each H-matrix is regular.*

Proof. Let A, D, B be as in Definition 1.10.14 and assume that there is a vector $x \ne 0$ with $Ax = 0$. Then $(I - D^{-1}B)x = 0$, whence $1 \le \rho(D^{-1}B) \le \rho(|D^{-1}B|)$. This contradicts Theorem 1.10.16. □

Theorem 1.10.18. *Each of the following conditions guarantees that $A \in \mathbb{R}^{n\times n}$ is an H-matrix.*

(a) *The matrix A is an M-matrix.*
(b) *The matrix A is diagonally dominant and $\langle A\rangle$ is regular.*
(c) *The matrix A is strictly diagonally dominant.*
(d) *The matrix A is irreducibly diagonally dominant.*
(e) *The matrix $\langle A\rangle$ is symmetric and positive definite.*
(f) *The matrix A is symmetric, positive definite and tridiagonal.*
(g) *The matrix A is regular and triangular.*

Proof. (a) is trivial.

(b) The assumption shows that $\langle A\rangle$ is diagonally dominant and regular. Therefore, Theorem 1.10.7 can be applied.

(c), (d) follow analogously to (b).

(e) follows from Theorem 1.10.8.

(f) With the notation of Definition 1.10.14 we notice first that $|D| = D$. Choose $\sigma_1 = 1$ and $\sigma_i \in \{-1, 1\}$ such that $-a_{i,i+1} = \sigma_i|a_{i,i+1}|\sigma_{i+1}$, $i = 1, \ldots, n-1$. Let $D_\sigma = \operatorname{diag}(\sigma_1, \ldots, \sigma_n)$. For any vector $x \in \mathbb{R}^n \setminus \{0\}$ define $\tilde{x} = D_\sigma^{-1}x \ne 0$. Then

$$x^T\langle A\rangle x = \tilde{x}^T(D_\sigma|D|D_\sigma - D_\sigma|B|D_\sigma)\tilde{x} = \tilde{x}^T A\tilde{x} > 0,$$

where we exploited for B that A is tridiagonal. Hence $\langle A\rangle$ is symmetric and positive definite, and part (e) concludes the proof.

(g) Here, the matrix $\langle A\rangle$ has positive diagonal entries and is triangular. Hence Theorem 1.10.7 can be applied again. □

Next we introduce matrices that do not necessarily have the sign pattern of M-matrices but, like them, do have a nonnegative inverse.

Definition 1.10.19. A matrix $A \in \mathbb{R}^{n\times n}$ is called inverse positive if A is regular and $A^{-1} \geq O$.

Notice that in Definition 1.10.19 we followed the traditional terminology although the terminology ‘inverse nonnegative’ would be more appropriate. As an example, consider the permutation matrix $P = \left(\begin{smallmatrix} 0 & 1 \\ 1 & 0 \end{smallmatrix}\right)$ which is obviously not an M-matrix but an inverse positive one since $P^{-1} = P^T = P \geq O$.

Definition 1.10.20. The pair (M, N) is called a regular splitting of $A \in \mathbb{R}^{n\times n}$ if $A = M - N$, M regular, and $M^{-1} \geq O$, $N \geq O$.

Theorem 1.10.21. *Let (M, N) be a regular splitting of $A \in \mathbb{R}^{n\times n}$. Then A is inverse positive if and only if $\rho(M^{-1}N) < 1$.*

Proof. Let $P = M^{-1}N$ and $S_k = \sum_{i=0}^{k} P^i$. By the assumption, P is nonnegative.

Let A be inverse positive. From $O \leq M^{-1} = (I - P)A^{-1}$ we get

$$O \leq S_k M^{-1} = S_k(I - P)A^{-1} = (I - P^{k+1})A^{-1} \leq A^{-1}.$$

Since each row of $M^{-1} \geq O$ contains at least one positive entry, the sum S_k is bounded from above for all k. Therefore, since $P \geq O$, $\lim_{k\to\infty} S_k$ exists, whence $\rho(P) < 1$ by virtue of Theorem 1.7.8.

If, conversely, $\rho(P) < 1$, then

$$O \leq \sum_{i=0}^{\infty} P^i M^{-1} = (I - P)^{-1} M^{-1} = A^{-1}.$$ □

Theorem 1.10.22. *Let (M, N) and $(\hat{M}, \hat{N})$ be two regular splittings of the inverse positive matrix $A \in \mathbb{R}^{n\times n}$ which satisfy $N \leq \hat{N}$ or, equivalently $M \leq \hat{M}$. Then*

$$\rho(M^{-1}N) \leq \rho(\hat{M}^{-1}\hat{N}) < 1. \tag{1.10.1}$$

Proof. The equivalence follows from $M - N = A = \hat{M} - \hat{N}$, which implies $M - \hat{M} = N - \hat{N} \leq O$. Next we show that the eigenvalues μ of

$$M^{-1}N = (I + A^{-1}N)^{-1}A^{-1}N \tag{1.10.2}$$

are coupled with the eigenvalues τ of $A^{-1}N$ via

$$\mu = \frac{\tau}{1 + \tau}. \tag{1.10.3}$$

To this end we start with an eigenpair (x, τ) of $A^{-1}N$. The regularity of $I + A^{-1}N$ implies $1 + \tau \neq 0$, and from (1.10.2) we get $M^{-1}Nx = \frac{\tau}{1+\tau}x$, i.e., $(x, \frac{\tau}{1+\tau})$ is an eigenpair μ of $M^{-1}N$. Conversely, let (x, μ) be an eigenpair of $M^{-1}N$. Then $\mu x = M^{-1}Nx = (I + A^{-1}N)^{-1}A^{-1}Nx$, whence

$$\mu(I + A^{-1}N)x = A^{-1}Nx \tag{1.10.4}$$

and $\frac{\mu}{1-\mu}x = A^{-1}Nx$. Thus $\tau = \frac{\mu}{1-\mu}$ is an eigenvalue of $A^{-1}N$ which satisfies (1.10.3) again. Notice that $\mu \neq 1$ holds by virtue of (1.10.4). The Theorem of Perron and Frobenius shows that $\rho(M^{-1}N)$ is an eigenvalue of $M^{-1}N \geq O$, and the same holds for $\rho(A^{-1}N)$ and $A^{-1}N \geq O$. Since $\frac{\tau}{1+\tau}$ is monotone increasing for $\tau \geq 0$, the nonnegative eigenvalues of $M^{-1}N$ are maximized by choosing $\tau = \rho(A^{-1}N)$ in (1.10.3). This leads to

$$\rho(M^{-1}N) = \frac{\rho(A^{-1}N)}{1 + \rho(A^{-1}N)}. \tag{1.10.5}$$

By the hypothesis, we have $N \leq \hat{N}$ which implies $\rho(A^{-1}N) \leq \rho(A^{-1}\hat{N})$. The monotone behavior of the right-hand side in (1.10.5) together with Theorem 1.10.21 finally proves (1.10.1). □

Notes to Chapter 1

To 1.3: Most of the material can be found in any textbook on multidimensional calculus or functional analysis; cf. for instance Heuser [146]. Absolute and monotone norms are used for instance in Horn, Johnson [150] or Ortega, Rheinboldt [267]. The proof of Theorem 1.3.10 in the case $\mathbb{K} = \mathbb{R}$ is due to Markus Neher [250].

To 1.4: The proof of Theorem 1.4.1 (a) was communicated to me by Martin Koeber [167]. The proof of Weierstrass' approximation theorem follows traditional lines; cf. for instance Hämmerlin, Hoffmann [125]. Hermite interpolation can be found there, too; see also Stoer, Bulirsch [348].

To 1.5: Most of the results are contained in the book of Ortega, Rheinboldt [267]. See also Franklin [100], Neumaier [257], and Istrăţescu [152]. Borsuk's Theorem 1.5.4 can be found in Borsuk [71] or Deimling [83], Brouwer's fixed point theorem 1.5.7 is contained in Brouwer [73]. The proof of Miranda's theorem 1.5.9 is the original one in Miranda [228]. For an alternative proof and a generalization of Miranda's theorem see Vrahatis [358]. Some of the convergence results on Newton's method in the Newton–Kantorovich theorem 1.5.10 are contained in a more general setting in Kantorovich [158].

To 1.6: The mean value theorem and the counter-example above Theorem 1.6.1 are from Heuser [145], but certainly not presented there for the first time.

To 1.7: This section owes much Stoer, Bulirsch [348] and Varga [356].

To 1.8: For Section 1.8 we refer to Golub, van Loan [121], Ortega [266], and Stoer, Bulirsch [348]. Algorithm 1.8.8 originates from Bunch, Kaufman, Parlett [74]; see also Golub, van Loan [121]. The proof of Theorem 1.8.16 (a) and (c) is from Neumaier [253], Lehmann's Theorem 1.8.22 can be found in Lehmann [188, 189], Wilkinson's Theorem 1.8.25 is taken from Wilkinson [361].

To 1.9: Here we essentially follow Varga [356].

To 1.10: Many of the results can be found in Berman, Plemmons [68], Ortega [266], or Varga [356].

2 Real intervals

2.1 Intervals, partial ordering

Real compact intervals are particular subsets of $\mathbb{R}$ which form the main objects of this book. They are denoted by square brackets and are defined by

$$[a] = [\underline{a}, \overline{a}] = \{\, \tilde{a} \mid \underline{a} \le \tilde{a} \le \overline{a} \,\}. \tag{2.1.1}$$

Here, the lower bound $\inf([a]) = \min([a]) = \underline{a}$ and the upper bound $\sup([a]) = \max((a]) = \overline{a}$ are real numbers which satisfy $\underline{a} \le \overline{a}$. In brief, we call $[a]$ an interval, nearly always dropping the specifications real and compact. The set of all intervals $[a]$ is denoted by $\mathbb{IR}$, that of all intervals $[a]$ contained in some given subset S of $\mathbb{R}$ by $\mathbb{I}(S)$. Functions which map $\mathbb{I}(S)$ into $\mathbb{IR}$ are called interval functions. They are denoted, for instance, by $[f]\colon \mathbb{I}(S) \to \mathbb{IR}$ or, in short, by $[f]$.

If the bounds $\underline{a}, \overline{a}$ coincide, then the interval $[a]$ contains exactly one element a. In this case we call $[a]$ degenerate or a point interval. Otherwise we call $[a]$ nondegenerate. We identify a point interval with its element, i.e., $a \equiv [a] = [a, a]$. In this way $\mathbb{R}$ is embedded in $\mathbb{IR}$, and we will not distinguish between a and $[a, a]$ in the sequel. If the bounds of an interval coincide in the leading digits, we sometimes present these common digits and add the differing digits as subscript, and superscript, respectively, in order to indicate the digits of the lower and the upper bound, respectively. Thus $[1.234_{56}^{78}]$ means $[1.23456, 1.23478]$.

The interior of $[a]$ is the set $\{\, \tilde{a} \mid \underline{a} < \tilde{a} < \overline{a} \,\}$. It is denoted by $(\underline{a}, \overline{a})$ or $\operatorname{int}([a])$.

The midpoint $\check{a} = \operatorname{mid}([a])$ of $[a]$ is given by

$$\check{a} = \operatorname{mid}([a]) = (\underline{a} + \overline{a})/2, \tag{2.1.2}$$

the radius $r_a = \operatorname{rad}([a])$ by

$$r_a = \operatorname{rad}([a]) = (\overline{a} - \underline{a})/2, \tag{2.1.3}$$

and the diameter $d([a])$ (sometimes also called width of $[a]$) by

$$d([a]) = \overline{a} - \underline{a}. \tag{2.1.4}$$

We use the notations $\operatorname{mid}([a])$ and $\operatorname{rad}([a])$ if the argument $[a]$ is a more complex expression. Obviously, $d([a]) = 2r_a$ and

$$\tilde{a} \in [a] \Leftrightarrow |\tilde{a} - \check{a}| \le r_a.$$

We call $[a]$ zero-symmetric if $\check{a} = 0$. Trivially, zero-symmetric intervals have the form $[a] = [-r_a, r_a]$ and can be characterized by $\underline{a} = -\overline{a}$.

Equality $=$, intersection $\cap$, union $\cup$ and the subset signs $\subseteq, \supseteq$ are applied in the usual set theoretical sense. In particular,

$$[a] = [b] \Leftrightarrow \underline{a} = \underline{b} \text{ and } \overline{a} = \overline{b},$$

DOI 10.1515/9783110499469-003

and

$$[a] \cup [b] \in \mathbb{IR} \Leftrightarrow [a] \cap [b] \neq \emptyset.$$

The tightest interval containing $[a]$ and $[b]$ is called the convex union of $[a]$ and $[b]$ and is denoted by $[a] \underline{\cup} [b]$. Obviously,

$$[a] \underline{\cup} [b] = [\min\{\underline{a}, \underline{b}\},\ \max\{\overline{a}, \overline{b}\}].$$

The interval hull $\square S$ of a bounded set $S \subseteq \mathbb{R}$ is defined by

$$\square S = [\inf(S), \sup(S)],$$

where the infimum $\inf(S)$ is the greatest lower bound of S and the supremum $\sup(S)$ is the smallest upper one. We leave it to the reader to show

$$\square S = \bigcap_{S \subseteq [a]} [a], \quad \square\{\underline{a}, \overline{a}\} = [a], \quad \text{and} \quad \square([a] \cup [b]) = [a] \underline{\cup} [b].$$

In $\mathbb{IR}$ we define

$$\begin{aligned} [a] &\le [b], \quad \text{if } \underline{a} \le \underline{b} \text{ and } \overline{a} \le \overline{b}, \\ [a] &< [b], \quad \text{if } \underline{a} < \underline{b} \text{ and } \overline{a} < \overline{b}. \end{aligned}$$

Notice that the relation $\le$ satisfies

$$\begin{aligned} &[a] \le [a] && \text{(reflexivity)} \\ &[a] \le [b] \text{ and } [b] \le [a] \Rightarrow [a] = [b] && \text{(antisymmetry)} \\ &[a] \le [b] \text{ and } [b] \le [c] \Rightarrow [a] \le [c] && \text{(transitivity)} \end{aligned}$$

Therefore, the relation $\le$ is a partial ordering which extends the classical ordering on $\mathbb{R}$. But it is not a complete ordering, since not any two intervals are comparable as the example $[a] = [2, 3]$, $[1, 4]$ shows. As usual, we define $\ge$ by

$$[a] \ge [b] \quad \text{if } [b] \le [a],$$

and we proceed similarly for $>$. We mostly use these relations for expressing $0 \le [a]$, $[a] \le 0$, etc. Notice that $[a] < [b]$ is not equivalent to $([a] \le [b]) \wedge ([a] \neq [b])$.

Another partial ordering on $\mathbb{IR}$ is given by the subset relation $\subseteq$.

Exercises

Ex. 2.1.1. Why is the definition $[a] \le [b] \Leftrightarrow \overline{a} \le \underline{b}$ not a partial ordering in $\mathbb{IR}$?

2.2 Interval arithmetic

We equip $\mathbb{IR}$ with an arithmetic which extends that of $\mathbb{R}$.

Definition 2.2.1. For $[a], [b] \in \mathbb{IR}$ the binary operation $\circ \in \{+, -, \cdot, /\}$ is defined on $\mathbb{IR}$ by

$$[a] \circ [b] = \{\tilde{a} \circ \tilde{b} \mid \tilde{a} \in [a],\ \tilde{b} \in [b]\}, \tag{2.2.1}$$

where we assume $0 \notin [b]$ in the case of division.

The unary operations $+[a]$, $-[a]$ are defined by $0 + [a]$ and $0 - [a]$, respectively. Instead of $1/[a]$ we also write $[a]^{-1}$.

As usual we often suppress the multiplication sign and the unary plus. From (2.2.1) we immediately get the following explicit representation of $[a] \circ [b]$ by means of the interval bounds $\underline{a}, \overline{a}, \underline{b}, \overline{b}$.

Theorem 2.2.2. *For $[a], [b] \in \mathbb{IR}$ and $\circ \in \{+, -, \cdot, /\}$ we have*

$$\begin{aligned}
[a] + [b] &= [\underline{a} + \underline{b}, \overline{a} + \overline{b}],\\
[a] - [b] &= [\underline{a} - \overline{b}, \overline{a} - \underline{b}],\\
[a] \cdot [b] &= [\min S, \max S], \text{ where } S = \{\underline{a}\underline{b}, \underline{a}\overline{b}, \overline{a}\underline{b}, \overline{a}\overline{b}\},\\
1/[b] &= \left[\frac{1}{\overline{b}}, \frac{1}{\underline{b}}\right], \quad \text{if } 0 \notin [b],\\
[a]/[b] &= [a] \cdot \frac{1}{[b]}, \quad \text{if } 0 \notin [b].
\end{aligned}$$

In particular, $[a] \circ [b]$ is again an interval.

Proof. We restrict ourselves to the addition. Then the set R on the right-hand side of (2.2.1) is obviously contained in $[c] = [a] + [b]$. Since the function $f(x, y) = x + y$ is continuous on the compact set $[a] \times [b]$ it assumes each value between its minimum $\underline{c} = f(\underline{a}, \underline{b})$ and its maximum $\overline{c} = f(\overline{a}, \overline{b})$. This shows $[c] \subseteq R$ and finally proves $[c] = R$.

For the remaining operations the proof proceeds analogously. □

The multiplication and division require case distinctions as the Tables 2.2.1–2.2.3 show.

Tab. 2.2.1: Multiplication $[a] \cdot [b]$.

	$\overline{b} \le 0$	$\underline{b} < 0 < \overline{b}$	$0 \le \underline{b}$
$\overline{a} \le 0$	$[\overline{a}\overline{b}, \underline{a}\underline{b}]$	$[\underline{a}\overline{b}, \underline{a}\underline{b}]$	$[\underline{a}\overline{b}, \overline{a}\underline{b}]$
$\underline{a} < 0 < \overline{a}$	$[\overline{a}\underline{b}, \underline{a}\underline{b}]$	$[\min\{\overline{a}\underline{b}, \underline{a}\overline{b}\}, \max\{\underline{a}\underline{b}, \overline{a}\overline{b}\}]$	$[\underline{a}\overline{b}, \overline{a}\overline{b}]$
$0 \le \underline{a}$	$[\overline{a}\underline{b}, \underline{a}\overline{b}]$	$[\overline{a}\underline{b}, \overline{a}\overline{b}]$	$[\underline{a}\underline{b}, \overline{a}\overline{b}]$

The case

$$\underline{a} < 0 < \overline{a}, \quad \underline{b} < 0 < \overline{b} \tag{2.2.2}$$

in Table 2.2.1 can be further resolved if one uses the midpoint/radius representation

$$[a] = \check{a} + [-r_a, r_a] = \check{a} + r_a[-1, 1] = [\check{a} - r_a, \check{a} + r_a] \tag{2.2.3}$$

of intervals. Since in (2.2.2) we have $r_a > 0$, $r_b > 0$, the inequalities there are equivalent to

$$-1 < \frac{\check{a}}{r_a} < 1, \quad -1 < \frac{\check{b}}{r_b} < 1,$$

and we get Table 2.2.2 as a result.

For the division we obtain Table 2.2.3.

Tab. 2.2.2: Multiplication $[a] \cdot [b]$ in the case $\underline{a} < 0 < \overline{a}$, $\underline{b} < 0 < \overline{b}$.

	$-1 < \frac{\check{a}}{r_a} \le -\frac{\check{b}}{r_b} < 1$	$-1 < -\frac{\check{b}}{r_b} < \frac{\check{a}}{r_a} < 1$
$-1 < \frac{\check{a}}{r_a} \le \frac{\check{b}}{r_b} < 1$	$[\underline{a}\overline{b}, \underline{a}\underline{b}]$	$[\underline{a}\overline{b}, \overline{a}\overline{b}]$
$-1 < \frac{\check{b}}{r_b} < \frac{\check{a}}{r_a} < 1$	$[\overline{a}\underline{b}, \underline{a}\underline{b}]$	$[\overline{a}\underline{b}, \overline{a}\overline{b}]$

Tab. 2.2.3: Division $[a]/[b]$.

	$\overline{b} < 0$	$0 < \underline{b}$
$\overline{a} \le 0$	$[\overline{a}/\underline{b}, \underline{a}/\overline{b}]$	$[\underline{a}/\underline{b}, \overline{a}/\overline{b}]$
$\underline{a} < 0 < \overline{a}$	$[\overline{a}/\overline{b}, \underline{a}/\overline{b}]$	$[\underline{a}/\underline{b}, \overline{a}/\underline{b}]$
$0 \le \underline{a}$	$[\overline{a}/\overline{b}, \underline{a}/\underline{b}]$	$[\underline{a}/\overline{b}, \overline{a}/\underline{b}]$

Example 2.2.3.

$$[0, 1] + [-2, 3] = [-2, 4], \quad [0, 1] - [-2, 3] = [-3, 3],$$
$$[0, 1] \cdot [-2, 3] = [-2, 3], \quad [-2, 4] \,/\, [1, 2] = [-2, 4].$$

By the Definition 2.2.1 the product $[a] \cdot [b]$ of two intervals is the range of the function $z = f(x, y) = x \cdot y$, $x \in [a]$, $y \in [b]$. In three-dimensional space, the graph of this function describes a surface which is part of the hyperboloid $z = x \cdot y$, $(x, y) \in \mathbb{R}^2$. The range of f is itself an interval $[c]$ whose bounds are attained at two (at least!) of the four corners of the rectangle $R = [a] \times [b]$. If one moves such a rectangle over the plane $\mathbb{R}^2$, one sees how the positions of the extremal points change with respect to the position and

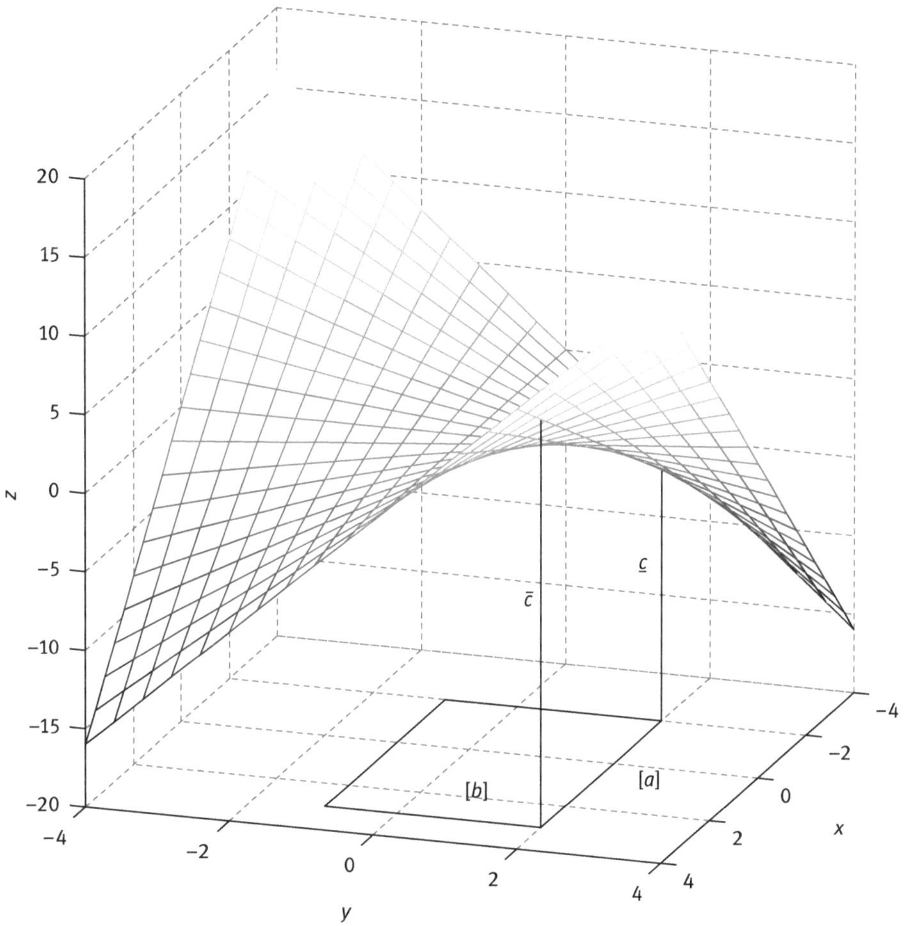

Fig. 2.2.1: $[c] = [a] \cdot [b]$ for $[a] = [-2, 3]$, $[b] = [-1, 2]$.

the radius of the operands $[a], [b]$. Here, particularly the case $(0, 0) \in \operatorname{int}([a] \times [b])$ is interesting; see Figure 2.2.1, where we plotted the hyperboloid and R. In order to increase the visibility, we moved R from the plane $z = 0$ into the plane $z = -20$ so that $\underline{c}$ and $\overline{c}$ in Figure 2.2.1 must be interpreted correspondingly.

Obviously, $(\mathbb{R}, +, -, \cdot, /)$ with the standard operations is isomorphic to the set of point intervals equipped with the corresponding operations $+, -, \cdot, /$. In addition, the following two crucial properties of interval arithmetic are immediate consequences of Definition 2.2.1.

Theorem 2.2.4 (Inclusion property, inclusion isotony).
(a) *For* $[a], [b] \in \mathbb{IR}$ *and* $\circ \in \{+, -, \cdot, /\}$ *we have the inclusion property*

$$\tilde{a} \circ \tilde{b} \in [a] \circ [b] \quad \textit{for all elements } \tilde{a} \in [a],\ \tilde{b} \in [b]. \tag{2.2.4}$$

(b) *If* $[a], [a]', [b], [b]' \in \mathbb{IR}$ *satisfy* $[a] \subseteq [a]'$, $[b] \subseteq [b]'$, *then*

$$[a] \circ [b] \subseteq [a]' \circ [b]' \tag{2.2.5}$$

for each operation $\circ \in \{+, -, \cdot, /\}$.
Property (2.2.5) *is called inclusion isotony or, less specifically, inclusion monotony.*

In view of (2.2.4) the definition of $[a] \circ [b]$ is optimal, i.e., the resulting interval is – trivially – the smallest one such that the inclusion property (2.2.4) holds.

Exercises

Ex. 2.2.1. Verify the Tables 2.2.1–2.2.3.

Ex. 2.2.2. Compute the following expressions for $[a] = [-1, 3]$, $[b] = [3, 4]$, $[c] = [2, 3]$.

$$[a] + [b], \quad [b] + [a], \quad [a] - [b], \quad [b] - [a],$$
$$[a] \cdot [b], \quad [b] \cdot [a], \quad [a]/[b],$$
$$[a] \cdot ([b] + [c]), \quad [a] \cdot [b] + [a] \cdot [c], \quad [a] \cdot ([b] \cdot [c]), \quad ([a] \cdot [b]) \cdot [c],$$
$$\check{a}, \ r_a, \ d([a]).$$

Ex. 2.2.3. For $\circ \in \{+, -, \cdot, /\}$ define the binary operation $\diamondsuit$ on $\mathbb{IR}$ by

$$[a] \diamondsuit [b] := [a] \circ [b] + [-1, 1].$$

Show that Theorem 2.2.4 holds also for $\diamondsuit$, but $\diamondsuit$ is not optimal in view of (2.2.4).

2.3 Algebraic properties, χ-function

In this section we study algebraic properties of the interval arithmetic defined in Section 2.2. We start with an example.

Example 2.3.1.

$$[1, 2] \cdot (1 - 1) = 0 \neq [1, 2] - [1, 2] = [-1, 1],$$
$$[1, 2]/[1, 2] = [1/2, 2] \neq 1.$$

Apparently, the distributive law does not hold and inverses for the addition and multiplication do not exist or can at least not be expressed as expected. Despite these discouraging features, the interval arithmetic has a variety of 'good' properties which are listed in our next theorem.

Theorem 2.3.2. *For intervals* $[a], [b], [c] \in \mathbb{IR}$ *the following properties hold.*

(a) $\left.\begin{aligned}[a]+[b]&=[b]+[a]\\ [a]\cdot[b]&=[b]\cdot[a]\end{aligned}\right\}$ *commutative laws*

(b) $\left.\begin{aligned}[a]+([b]+[c])&=([a]+[b])+[c]\\ [a]\cdot([b]\cdot[c])&=([a]\cdot[b])\cdot[c]\end{aligned}\right\}$ *associative laws*

(c) $[a]([b]+[c]) \subseteq [a][b]+[a][c]$ *subdistributive law*

(d) $\left.\begin{aligned}[a]+0&=[a]\\ [a]\cdot 1&=[a]\end{aligned}\right\}$ *neutral elements*

(e) $[a]\cdot[b]=0 \Leftrightarrow [a]=0$ *or* $[b]=0$ *zero divisor free*

(f) $0\in[a]\cdot[b] \Leftrightarrow 0\in[a]$ *or* $0\in[b]$

(g) *Nondegenerate intervals have neither an inverse with respect to addition nor with respect to multiplication.*
But $0\in[a]-[a]$ *and* $1\in[a]/[a]$ *holds.*

(h) $\left.\begin{aligned}[a]\circ[c]&=[b]\circ[c]\\ [c]\,/\,[a]&=[c]\,/\,[b]\end{aligned}\right\}\Rightarrow [a]=[b]$ *reduction rules*
where $\circ\in\{+,-,\cdot,/\}$ *and* $0\notin[c]$ *in case of multiplication and division.*

(i) $[a]+[c]\subseteq[b]+[c]\Rightarrow[a]\subseteq[b]$.

Proof. (a) and (f) follow directly from Theorem 2.2.2.

(b) The first equality follows by a simple calculation. In order to prove the second one let $[l]=[a]\cdot([b]\cdot[c])$. There are elements $\tilde a\in[a]$, $\tilde d\in[b]\cdot[c]$ such that $\tilde a\cdot\tilde d=\underline{l}$, and by Theorem 2.2.2 there are elements $\tilde b\in[b]$, $\tilde c\in[c]$ with $\tilde b\tilde c=\tilde d$. Hence $\underline{l}\in[r]=([a]\cdot[b])\cdot[c]$. Similarly, $\overline{l}\in[r]$, whence $[l]\subseteq[r]$. An analogous argumentation yields the converse subset property which proves equality.

(c) Is proved similarly as the subset property in the proof for (b).

(d) Is trivial.

(e) If $[a][b]=0$, then by the definition of the multiplication we have $\max S=\min S=0$, where $S=\{\underline{a}\underline{b},\underline{a}\overline{b},\overline{a}\underline{b},\overline{a}\overline{b}\}$. Assume that $\overline{b}\neq 0$. Then necessarily $\underline{a}=0=\overline{a}$, which implies $[a]=0$. Similarly, the remaining three cases $\underline{b}\neq 0$, $\overline{a}\neq 0$ and $\underline{a}\neq 0$ yield an analogous implication. The converse part of the statement is trivial.

(g) Let $[a]+[x]=0$, $r_a>0$. Then $\underline{x}=-\underline{a}>-\overline{a}=\overline{x}$, which contradicts $\underline{x}\le\overline{x}$. Hence $[a]$ cannot have an inverse with respect to addition.

Assume $[a]\cdot[x]=1$, $r_a>0$. Then (f) implies $0\notin[a]$ and $0\notin[x]$. W.l.o.g. let $\underline{a}>0$. Then necessarily $\underline{x}>0$, whence $\underline{a}\underline{x}=1=\overline{a}\overline{x}$. From this equality and $r_a>0$ we get the contradiction $\underline{a}=1/\underline{x}\ge 1/\overline{x}=\overline{a}>\underline{a}$.

The remaining part of the statement is trivial.

(h) The statements for addition and subtraction follow from a simple calculation. In order to prove the statement for multiplication we assume w.l.o.g. that $\overline{a}\ge 0$ and $\underline{c}>0$ holds.

If $\underline{a} \geq 0$, then $[a][c] = [\underline{a}\underline{c}, \overline{a}\overline{c}] \geq 0$, whence $\underline{b} \geq 0$. This implies $[b][c] = [\underline{b}\underline{c}, \overline{b}\overline{c}]$ and a simple comparison of the bounds yields $[a] = [b]$.

If $\underline{a} < 0$, then $0 \in [a][c] = [b][c]$, whence $0 \in [b]$. We obtain $[a][c] = [\underline{a}\overline{c}, \overline{a}\overline{c}] = [b][c] = [\underline{b}\overline{c}, \overline{b}\overline{c}]$ and finally $[a] = [b]$.

Both statements for division are reduced to that of multiplication by virtue of the definition $[a]/[c] = [a] \cdot (1/[c])$ and $[c]/[a] = [c] \cdot (1/[a])$. In the last case one first obtains $1/[a] = 1/[b]$ from which the statement follows by comparing and inverting the bounds.

(i) The premiss implies $\underline{a} + \underline{c} \geq \underline{b} + \underline{c}$ and $\overline{a} + \overline{c} \leq \overline{b} + \overline{c}$, whence $\underline{a} \geq \underline{b}$ and $\overline{a} \leq \overline{b}$. This is equivalent to $[a] \subseteq [b]$. □

Theorem 2.3.2 shows that $(\mathbb{IR}, +)$ and $(\mathbb{IR} \setminus \{0\}, \cdot)$ are commutative semigroups with neutral elements, but neither of them is a group. Inverses are missing for nondegenerate intervals.

For zero-symmetric intervals we get particularly nice results which can immediately be verified by simple computations.

Theorem 2.3.3 (Zero-symmetric intervals). *If $[a], [b] \in \mathbb{IR}$ are zero-symmetric and $[c]$ is any interval, then $[a] \pm [b]$ and $[a] \cdot [c]$, $[a]/[c]$ are zero-symmetric provided that $0 \notin [c]$ in the last case.*

Now we take a closer look to the subdistributivity

$$[a]([b] + [c]) \subseteq [a][b] + [a][c] \tag{2.3.1}$$

of the interval arithmetic. We want to find out under which conditions equality holds in (2.3.1). To this purpose we introduce Ratschek's χ-function which measures the relations between the bounds of a nonzero interval. It goes back to Ljapin [190] and was used by Ratschek in [281, 282, 283].

Definition 2.3.4. The function $\chi\colon \mathbb{IR} \setminus \{0\} \to [-1, 1]$ is defined by

$$\chi([a]) = \begin{cases} \underline{a}/\overline{a}, & \text{if } |\underline{a}| \leq |\overline{a}| \quad (\Leftrightarrow \check{a} \geq 0 \text{ or } \underline{a} = \overline{a} = \check{a}) \\ \overline{a}/\underline{a}, & \text{if } |\underline{a}| > |\overline{a}| \quad (\Leftrightarrow \check{a} < 0 \text{ and } \underline{a} \neq \overline{a}). \end{cases}$$

Example 2.3.5.

$$[a] = [-2, 6], \quad \check{a} = 2, \quad \chi([a]) = -2/6 = -1/3,$$
$$[a] = [-6, 2], \quad \check{a} = -2, \quad \chi([a]) = 2/(-6) = -1/3.$$

Representing an interval $[\underline{a}, \overline{a}]$ by the point $(\underline{a}, \overline{a})$ in the half-plane $y \geq x$ yields Figure 2.3.1 in which the rays indicate the level curves of χ. The function decreases if one follows the dashed curved arrows.

Theorem 2.3.6 (Properties of the χ-function). *Let $[a] \in \mathbb{IR} \setminus \{0\}$. Then*

(a) $\chi([a]) = 1$ *if and only if* $[a]$ *is degenerate.*
$\chi([a]) = -1$ *if and only if* $[a]$ *is symmetric.*
$\chi([a]) = 0$ *if and only if* $[a] = [\underline{a}, 0]$ *or* $[a] = [0, \overline{a}]$.

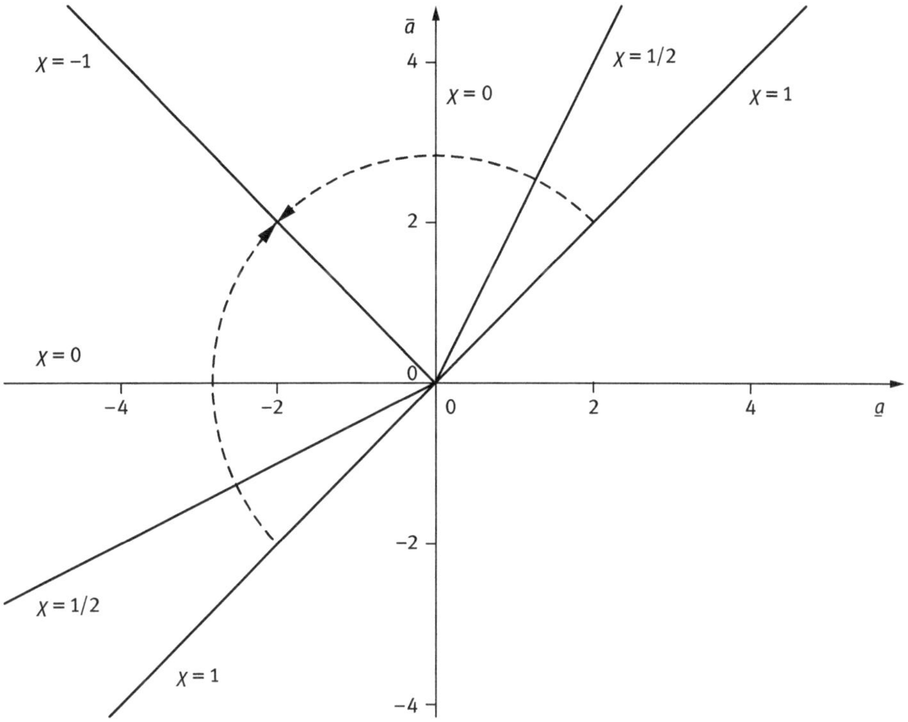

Fig. 2.3.1: Level curves of the χ-function.

(b) $\chi([a]) < 1$ *if and only if* $[a]$ *is not degenerate.*
$\chi([a]) \le 0$ *if and only if* $0 \in [a]$.
$\chi([a]) < 0$ *if and only if* $\underline{a} < 0 < \overline{a}$.
$\chi([a]) \ge 0$ *if and only if* $\overline{a} \le 0$ *or* $0 \le \underline{a}$.
$\chi([a]) > 0$ *if and only if* $\overline{a} < 0$ *or* $0 < \underline{a}$.

(c) $\chi(t[a]) = \chi([a])$, $t \in \mathbb{R} \setminus \{0\}$. *In particular,* $\chi(-[a]) = \chi([a])$.

(d) $\chi(1/[a]) = \chi([a])$.

(e) $$\chi([a][b]) = \begin{cases} \chi([a]) \cdot \chi([b]), & \text{if } 0 \notin [a][b], \\ \min\{\chi([a]), \chi([b])\} & \text{otherwise.} \end{cases}$$

(f) $$\chi([a]/[b]) = \begin{cases} \chi([a]) \cdot \chi([b]), & \text{if } 0 \notin [a], \\ \min\{\chi([a]), \chi([b])\} & \text{otherwise.} \end{cases}$$

Proof. (a), (b) and (c) follow directly from Definition 2.3.4.

(d) By virtue of (c) we may assume $\underline{a} > 0$. Then $\chi(1/[a])) = (1/\overline{a})/(1/\underline{a}) = \underline{a}/\overline{a} = \chi([a])$.

(e) Again by virtue of (c) we assume $\check{a} \geq 0$, $\check{b} \geq 0$, whence $\chi([a]) = \underline{a}/\overline{a}$, $\chi([b]) = \underline{b}/\overline{b}$. In addition, the definition of χ requires $[a] \neq 0$, $[b] \neq 0$ so that our assumption implies $\overline{a} > 0$, $\overline{b} > 0$.

If $0 \notin [a][b]$, then $0 < \underline{a}$, $0 < \underline{b}$, hence $[a][b] = [\underline{a}\underline{b}, \overline{a}\overline{b}] > 0$. This yields $\chi([a][b]) = (\underline{a}\underline{b})/(\overline{a}\overline{b}) = \chi([a])\chi([b])$.

If $0 \in [a][b]$ we may assume $0 \in [a]$. Then $\underline{a} \leq 0 < \overline{a}$, whence $[a][b] = [\min\{\underline{a}\overline{b}, \overline{a}\underline{b}\}, \overline{a}\overline{b}]$ with $\chi([a][b]) = \min\{\underline{a}\overline{b}, \overline{a}\underline{b}\}/(\overline{a}\overline{b}) = \min\{\chi([a]), \chi([b])\}$.

(f) follows from $[a]/[b] = [a] \cdot \frac{1}{[b]}$, $0 \notin [b]$, and (d), (e). □

The χ-function is appropriate to fix bounds in the product $[a][b]$. For example, if $\chi([a]) \leq \chi([b]) < 0$ and $\check{a} \geq 0$, $\check{b} \leq 0$ is known, then $\underline{a} < 0 < \overline{a}$, $\underline{b} < 0 < \overline{b}$, and $\overline{a} \geq |\underline{a}|$, $|\underline{b}| \geq \overline{b}$ holds. Therefore, the lower bound of $[a][b]$ is $\overline{a}\underline{b}$. For the upper bound we must choose the maximum of the two values $\underline{a}\underline{b}$ and $\overline{a}\overline{b}$, cf. Table 2.2.1. From the assumptions we obtain

$$\chi([a]) = \underline{a}/\overline{a} \leq \chi([b]) = \overline{b}/\underline{b}.$$

Since $\underline{b} < 0$ we get $\underline{a}\underline{b} \geq \overline{a}\overline{b}$, whence $[a][b] = [\overline{a}\underline{b}, \underline{a}\underline{b}]$.

This observation and similar ones are the basis of our next theorem which completely clarifies equality in (2.3.1). We even can generalize the situation there by allowing more than two summands.

Theorem 2.3.7 (Distributivity). *Let $[a], [b]_i \in \mathbb{IR} \setminus \{0\}$, $i = 1, \ldots, n$. Then equality*

$$[l] = [a] \sum_{i=1}^{n} [b]_i = \sum_{i=1}^{n} [a][b]_i = [r] \tag{2.3.2}$$

holds if and only if one of the following conditions is fulfilled.

(D1) $\chi([a]) = 1$ *(i.e., $[a]$ is degenerate).*

(D2) $0 \leq \chi([a]) < 1$ *and $[b]_i[b]_j \geq 0$ for all $i, j = 1, \ldots, n$.*

(D3) $0 \leq \chi([a]) < 1$ *and $\chi([b]_i) \leq 0$ for all $i = 1, \ldots, n$.*

(D4) $\chi([a]) < 0$ *and $\chi([a]) \leq \chi([b]_i)$ and $\check{b}_i\check{b}_j \geq 0$ for all $i, j = 1, \ldots, n$.*

(D5) $\chi([a]) < 0$ *and $\chi([a]) \geq \chi([b]_i)$ for all $i = 1, \ldots, n$.*

Proof. We prove only the case $n = 2$ and set $[b] = [b]_1$, $[c] = [b]_2$. The general case can be found in Ratschek [282].

First we notice that if we have proved (D1) we get $-[r] = -[a][b] - [a][c]$, hence $[l] = [r]$ holds if and only if $-[a]([b] + [c]) = -[a][b] - [a][c]$. Therefore, with the exception of (D1) we may assume w.l.o.g.

$$\check{a} \geq 0, \quad \text{and} \quad \check{b} \geq |\check{c}| \geq 0,$$

whence

$$\begin{cases} \overline{a} \geq |\underline{a}|, & \overline{b} \geq |\underline{b}|, \quad \overline{a}, \overline{b} > 0, \\ \chi([a]) = \underline{a}/\overline{a}, & \chi([b]) = \underline{b}/\overline{b}. \end{cases} \tag{2.3.3}$$

(Notice that $\chi([a]) = \chi(-[a])$ holds by Theorem 2.3.6 (c), and that $\overline{a} > 0$, $\overline{b} > 0$ in (2.3.3) is mandatory because $\overline{a} = 0$ or $\overline{b} = 0$ yields the trivial cases $[a] = 0$ or $[b] = 0$ which we excluded in our theorem.)

'$\Leftarrow$':

(D1) The assertion follows by virtue of

$$[r] = \{\underline{a}\tilde{b} + \underline{a}\tilde{c} \mid \tilde{b} \in [b],\ \tilde{c} \in [c]\} = [l].$$

(D2) The assumptions imply $[a],\ [b],\ [c] \geq 0$, hence

$$[r] = [\underline{a}\underline{b}, \overline{a}\overline{b}] + [\underline{a}\underline{c}, \overline{a}\overline{c}] = [l].$$

(D3) Here, the assumptions imply $[a] \geq 0$, $0 \in [b]$, $0 \in [c]$, whence

$$[r] = [\overline{a}\underline{b}, \overline{a}\overline{b}] + [\overline{a}\underline{c}, \overline{a}\overline{c}] = [l].$$

(D4) From $\check{b} \geq |\check{c}|$ and $\check{b}\check{c} \geq 0$ we get $\check{c} \geq 0$ and therefore $\chi([c]) = \underline{c}/\overline{c}$. In addition, the assumptions of (D4) imply $\underline{a} < 0 < \overline{a}$ and $\underline{a}/\overline{a} \leq \underline{b}/\overline{b}$, $\underline{a}/\overline{a} \leq \underline{c}/\overline{c}$. Thus, $\underline{a}\overline{b} \leq \overline{a}\underline{b}$ and $\underline{a}\overline{c} \leq \overline{a}\underline{c}$ holds with nonpositive left-hand sides, which results in

$$[r] = [\underline{a}\overline{b}, \overline{a}\overline{b}] + [\underline{a}\overline{c}, \overline{a}\overline{c}] = [\underline{a}(\overline{b} + \overline{c}), \overline{a}(\overline{b} + \overline{c})] = [a](\overline{b} + \overline{c}) \subseteq [l].$$

The converse subset relation follows from the subdistributivity of the interval arithmetic.

(D5) The assumptions imply $\underline{a} < 0 < \overline{a}$ and $\chi([b]) < 0$, $\chi([c]) < 0$, whence $\underline{b} < 0 < \overline{b}$ and $\underline{c} < 0 < \overline{c}$. Together with (2.3.3) one obtains $0 \leq \underline{a}\underline{b} \leq \overline{a}\overline{b}$, $\underline{b}/\overline{b} \leq \underline{a}/\overline{a}$, and finally $\overline{a}\underline{b} \leq \underline{a}\overline{b} < 0$. If $\check{c} < 0$, then $\underline{c} < -|\overline{c}|$ and $\chi([c]) = \overline{c}/\underline{c}$. Hence the assumptions – including (2.3.3) – imply $\overline{a}\underline{c} \leq |\underline{a}|\underline{c} \leq |\underline{a}|(-|\overline{c}|) = \underline{a}\overline{c} \leq 0$ and $\overline{c}/\underline{c} = \chi([c]) \leq \chi([a]) = \underline{a}/\overline{a}$, whence $\overline{a}\overline{c} \geq \underline{a}\underline{c} \geq 0$. Similarly, if $\check{c} > 0$, then $|\underline{c}| \leq \overline{c}$ and $\chi([c]) = \underline{c}/\overline{c}$ holds. Thus we get $0 \leq \underline{a}\underline{c} = |\underline{a}|\,|\underline{c}| \leq \overline{a}\overline{c}$ and $\underline{c}/\overline{c} = \chi([c]) \leq \chi([a]) = \underline{a}/\overline{a}$, whence $\overline{a}\underline{c} \leq \underline{a}\overline{c} \leq 0$. Using (D1) both cases result in

$$[r] = [\overline{a}\underline{b}, \overline{a}\overline{b}] + [\overline{a}\underline{c}, \overline{a}\overline{c}] = \overline{a}([b] + [c]) \subseteq [l],$$

and the proof finishes as with (D4).

'$\Rightarrow$': We show $[l] \subset [r]$ if none of the conditions (D1)–(D5) are fulfilled. To this end let again $\check{a} \geq 0$, $\check{b} \geq 0$.

Case 1: $0 \leq \chi([a]) < 1$, whence $0 \leq \underline{a} < \overline{a}$.

If neither (D2) nor (D3) holds, then $[b][c] \ngeq 0$, and $\chi([b]) > 0$ or $\chi([c]) > 0$. In the case $\chi([b]) > 0$ we get $0 < \underline{b} \leq \overline{b}$ and $\underline{c} < 0$ which yields

$$\underline{r} = \underline{a}\underline{b} + \overline{a}\underline{c} < \min\{\underline{a}(\underline{b} + \underline{c}), \overline{a}(\underline{b} + \underline{c})\} = \underline{l}.$$

Thus $\underline{r} \notin [l]$, hence $[l] \subset [r]$ must hold by virtue of the subdistributivity. The case $\chi([c]) > 0$ can be reduced to the previous one by interchanging the roles of $[b]$ and $[c]$.

Case 2: $\chi([a]) < 0$, whence $\underline{a} < 0 < \overline{a}$.

If neither (D4) nor (D5) holds, then there are four subcases.

(i) $\chi([a]) < \chi([b]), \chi([a]) > \chi([c])$.
Here, $\underline{a}/\overline{a} < \underline{b}/\overline{b}$, i.e., $\underline{a}\overline{b} < \overline{a}\underline{b}$. Moreover, $\chi([c]) < 0$ holds such that $\underline{c} < 0 < \overline{c}$. If $\check{c} \geq 0$, then $\chi([c]) = \underline{c}/\overline{c} < \chi([a]) = \underline{a}/\overline{a}$ by assumption, and $\overline{a}\underline{c} < \underline{a}\overline{c} < 0$ follows. This is also true if $\check{c} < 0$, since $\check{a} \geq 0$ by (2.3.3). Therefore,

$$\underline{r} = \underline{a}\overline{b} + \overline{a}\underline{c} < \min\{\, \underline{a}(\overline{b} + \overline{c}),\ \overline{a}(\underline{b} + \underline{c})\,\} = \underline{l},$$

and the proof finishes as above.

(ii) $\chi([a]) < \chi([b]), \chi([a]) \leq \chi([c]),\ \check{b}\check{c} < 0$.
Together with (2.3.3) we get $\check{c} < 0$, which implies $\overline{a}\underline{c} < \underline{a}\overline{c}$. As in (i) we have $\underline{a}\overline{b} < \overline{a}\underline{b}$, and the proof terminates as there.

The remaining cases (iii) $\chi([a]) > \chi([b]), \chi([a]) < \chi([c])$ and (iv) $\chi([a]) \leq \chi([b])$, $\chi([a]) < \chi([c]),\ \check{b}\check{c} < 0$ can be reduced to (i) and (ii), respectively. □

Example 2.3.8. For $[a] = [-6, 2]$, $[b] = [-2, 6]$, $[c] = [0, 2]$ we have $\chi([a]) = -1/3 = \chi([b]) < 0 = \chi([c])$, $\check{b} = 2$, $\check{c} = 1$. Hence $[a]([b] + [c]) = [a][b] + [a][c]$ by (D4) of Theorem 2.3.7.

Although one could assume by Theorem 2.3.7 that distributivity rarely holds, there are some important simple cases which frequently occur and for which it can be deduced. These cases are listed in our next corollary which is an immediate consequence of Theorem 2.3.7.

Corollary 2.3.9. *Let $a \in \mathbb{R}$, $[a], [b], [c] \in \mathbb{IR}$. Then the following assertions hold.*
(a) $a([b] + [c]) = a[b] + a[c]$.
(b) $[a]([b] + [c]) = [a][b] + [a][c]$, *if* $0 \leq [b], 0 \leq [c]$ *or if* $0 \geq [b], 0 \geq [c]$.

As an application of the diameter d and the function χ we address the question whether there is an interval $[x]$ such that the equation

$$[a] \circ [x] = [b] \tag{2.3.4}$$

holds for given intervals $[a]$ and $[b]$ and $\circ \in \{+, -, \cdot\}$. If so we will call $[x]$ an algebraic solution of (2.3.4). We already know that nondegenerate intervals do not have inverses with respect to addition and multiplication. So, solving (2.3.4) is not quite trivial.

Theorem 2.3.10. *Let $[a], [b] \in \mathbb{IR}$.*
(a) *The interval equations*

$$[a] + [x] = [b], \quad [a] - [y] = [b] \tag{2.3.5}$$

have an algebraic solution if and only if $d([a]) \leq d([b])$; in this case both solutions are unique and given by

$$[x] = [\underline{b} - \underline{a},\ \overline{b} - \overline{a}].$$

(b) *If* $[a] \neq 0$, $[b] \neq 0$, *then the interval equation*

$$[a][x] = [b] \tag{2.3.6}$$

has an algebraic solution if and only if $\chi([a]) \geq \chi([b])$. *This solution is not unique if and only if* $\chi([a]) = \chi([b]) \leq 0$.

Proof. (a) The equality $[a] + [x] = [b]$ holds if and only if the two equations $\underline{a} + \underline{x} = \underline{b}$ and $\overline{a} + \overline{x} = \overline{b}$ are solvable with $\underline{x} \leq \overline{x}$, i.e., $\underline{x} = \underline{b} - \underline{a} \leq \overline{x} = \overline{b} - \overline{a}$, or, equivalently, $d([a]) = \overline{a} - \underline{a} \leq \overline{b} - \underline{b} = d([b])$ must be true.

The second equation can be reduced to the first one via $[a] + (-[y]) = [b]$.

(b) W.l.o.g. we may assume $\check{a} \geq 0$ and $\check{b} \geq 0$. (Otherwise multiply $[x]$ or the equation by -1.) Then

$$\overline{a} > 0, \quad \overline{b} > 0, \quad \chi([a]) = \underline{a}/\overline{a}, \quad \chi([b]) = \underline{b}/\overline{b}.$$

'$\Rightarrow$': Let $[x]$ be a solution of (2.3.6). Then

$$\chi([b]) = \chi([a][x]) = \begin{cases} \chi([a])\chi([x]) \leq \chi([a]) \cdot 1, & \text{if } 0 \notin [a][x], \\ \min\{\chi([a]), \chi([x])\} \leq \chi([a]) & \text{otherwise.} \end{cases}$$

Here, $0 \notin [a][x]$ implies $0 \notin [a]$, $0 \notin [x]$, whence $\chi([a]) \geq 0$, $\chi([x]) \geq 0$.

'$\Leftarrow$': Let $\chi([a]) \geq \chi([b])$ hold. Then we consider four cases.

(i) $\chi([a]) = \chi([b]) \leq 0$.
Here,

$$\underline{a} \leq 0 < \overline{a}, \quad \underline{b} \leq 0 < \overline{b}, \quad \underline{a}/\overline{a} = \underline{b}/\overline{b}. \tag{2.3.7}$$

Define $\overline{x} = \overline{b}/\overline{a}$ and choose any $\underline{x} \in [0, \overline{x}]$. Then $[a][x] = [\underline{a}\overline{x}, \overline{a}\overline{x}] = [b]$ holds by virtue of (2.3.7).

(ii) $0 \leq \chi([b]) \leq \chi([a])$, $\chi([a]) \neq 0$.
The assumptions imply $0 \leq \underline{b}$, $0 < \underline{a}$, hence $0 \leq \underline{x}$, and we get $[a][x] = [\underline{a}\underline{x}, \overline{a}\overline{x}]$. This equals $[b]$ if $\underline{x} = \underline{b}/\underline{a}$, $\overline{x} = \overline{b}/\overline{a}$, and $\underline{x} \leq \overline{x}$. The latter is true because of the assumption of the theorem.

(iii) $\chi([b]) < 0 \leq \chi([a])$.
Here, we get $\underline{b} < 0 < \overline{b}$ and $0 \leq \underline{a}$, whence $\underline{x} < 0 < \overline{x}$ and $[a][x] = [\overline{a}\underline{x}, \overline{a}\overline{x}] = \overline{a}[x]$. Therefore, we must have $[x] = [b]/\overline{a}$ in order to fulfill (2.3.6).

(iv) $\chi([b]) < \chi([a]) < 0$.
Now $\underline{b} < 0 < \overline{b}$, $\underline{a} < 0 < \overline{a}$, hence $\chi([b]) = \chi([a][x]) = \min\{\chi([a]), \chi([x])\} = \chi([x])$ by the assumption of (iv). This implies $\underline{x} < 0 < \overline{x}$.
If $\check{x} \geq 0$, then $\chi([x]) = \underline{x}/\overline{x} < \underline{a}/\overline{a}$ and furthermore $\overline{a}\underline{x} < \underline{a}\overline{x}$. Therefore, $[a][x] = [\overline{a}\underline{x}, \overline{a}\overline{x}] = \overline{a}[x]$ as in the previous case.
If $\check{x} < 0$, then we similarly get $\overline{a}\overline{x} > \underline{a}\underline{x}$, whence $[a][x] = [\overline{a}\underline{x}, \overline{a}\overline{x}] = \overline{a}[x] = [b]$. This implies the contradiction $0 \leq \check{b} = \overline{a}\check{x} < 0$.

Since only (i) allows multiple solutions, the proof is complete. □

Remark.

(a) If $0 \notin [a]$ and if (2.3.6) is solvable, then certainly $[x] \subseteq [b]/[a]$, but if $0 \in [a]$, then $[b]/[a]$ is not defined although (2.3.6) may have a solution. This is illustrated by the examples $[3,4][x] = [1,2]$ and $[-2,6][x] = [-1,1]$ with the algebraic solutions $[x] = [1/3, 1/2] \subset [1,2]/[3,4] = [1/4, 2/3]$, and $[x] = [-1,1]/6$, respectively.

(b) It will turn out in later sections that algebraic solutions do not play such an important role when computing with intervals as they do when computing with real numbers. We will see that – due to the lack of inverses – enclosures of the solution set

$$S = \{\, x \mid a \circ x = b,\ a \in [a],\ b \in [b] \,\}$$

will be much more interesting where $\circ \in \{+, -, \cdot\}$.

Exercises

Ex. 2.3.1. Let $[a] = [1,2]$, $[b] = 2$, $[c] = [-1,1]$, $[d] = [1,3]$. Compute $[a] \cdot [c]$, $[b] \cdot [c]$ and $[c]/[d]$, $[c]/[d]$ and show that the two corresponding reduction rules of Theorem 2.3.2 (h) do not hold here. Why?

Ex. 2.3.2. Prove Theorem 2.3.3 on the arithmetic for symmetric intervals.

2.4 Auxiliary functions

In this section we introduce the auxiliary functions mignitude and magnitude (better known as absolute value of an interval), which play an important role in interval arithmetic. Moreover, we derive rules for them and for the midpoint and the radius/diameter of an interval.

Definition 2.4.1 (Mignitude, magnitude, and absolute value). Let $[a] \in \mathbb{IR}$. Then the mignitude $\langle [a] \rangle$ of $[a]$ is defined by

$$\langle [a] \rangle = \min\{\, |\tilde{a}| \mid \tilde{a} \in [a] \,\}$$

and the magnitude or absolute value $|[a]|$ by

$$|[a]| = \max\{\, |\tilde{a}| \mid \tilde{a} \in [a] \,\}.$$

It is obvious that the mignitude denotes the smallest distance from zero which an element of $[a]$ can have while the magnitude is the largest one. In particular, both functions are invariant against a multiplication of $[a]$ by -1. They reduce to the usual absolute value of reals if $[a]$ is degenerate.

Our first theorem gathers some properties of mignitude and magnitude.

Theorem 2.4.2 (Mignitude and magnitude). *Let* $[a],\ [b] \in \mathbb{IR}$. *Then we have*

(a) $|[a]| = \max\{|\underline{a}|, |\overline{a}|\} \geq \langle[a]\rangle = \begin{cases} 0, & \text{if } 0 \in [a], \\ \min\{|\underline{a}|, |\overline{a}|\}, & \text{otherwise}, \end{cases}$

(b) $\langle[a]\rangle \geq 0, \quad \langle[a]\rangle = 0 \Leftrightarrow 0 \in [a],$

$\quad |[a]| \geq 0, \quad |[a]| = 0 \Leftrightarrow [a] = 0,$

(c) $\langle t[a]\rangle = |t| \cdot \langle[a]\rangle$ *and* $|t[a]| = |t| \cdot |[a]|$ *for* $t \in \mathbb{R}$,

(d) $\langle[a]\rangle - |[b]| \leq \langle[a] \pm [b]\rangle \leq \langle[a]\rangle + \langle[b]\rangle,$

$\quad |[a]| - |[b]| \leq |[a]| - \langle[b]\rangle \leq |[a] \pm [b]| \leq |[a]| + |[b]|,$

(e) $\langle[a][b]\rangle = \langle[a]\rangle \cdot \langle[b]\rangle, \quad |[a][b]| = |[a]| \cdot |[b]|,$

(f) $\langle[a]^{-1}\rangle = (|[a]|)^{-1}, \quad |[a]^{-1}| = (\langle[a]\rangle)^{-1},$

(g) $\langle[a]/[b]\rangle = \langle[a]\rangle/|[b]|, \quad |[a]/[b]| = |[a]|/\langle[b]\rangle,$

(h) $[a] \subseteq [b] \Rightarrow \begin{cases} \langle[a]\rangle \geq \langle[b]\rangle, \\ |[a]| \leq |[b]|, \end{cases}$

(i) *If* $[a]$ *is symmetric, then* $[a] \cdot [b] = [a] \cdot |[b]|$.

Proof. We only prove (d) and (e); the remaining items are obvious or follow directly from the Definition 2.4.1.

(d) Choose any $\tilde{a} \in [a],\ \tilde{b} \in [b]$. Then $\langle[a]\rangle - |[b]| \leq |\tilde{a}| - |\tilde{b}| \leq |\tilde{a} \pm \tilde{b}|$. Since $\tilde{a}, \tilde{b}$ are arbitrary within the corresponding intervals, the first inequality is proved. Next choose $\tilde{a} \in [a],\ \tilde{b} \in [b]$ such that $|\tilde{a}| = \langle[a]\rangle,\ |\tilde{b}| = \langle[b]\rangle$. Then

$$\langle[a] \pm [b]\rangle \leq |\tilde{a} \pm \tilde{b}| \leq |\tilde{a}| + |\tilde{b}| = \langle[a]\rangle + \langle[b]\rangle,$$

which terminates the proof of the first line of (d).

The first inequality of the second line is obvious. In order to see the second one choose $\tilde{a} \in [a],\ \tilde{b} \in [b]$ such that $|\tilde{a}| = |[a]|,\ |\tilde{b}| = \langle[b]\rangle$. Then we get

$$|[a]| = |\tilde{a}| = |\tilde{a} \pm \tilde{b} \mp \tilde{b}| \leq |\tilde{a} \pm \tilde{b}| + |\tilde{b}| \leq |[a] \pm [b]| + \langle[b]\rangle.$$

Subtracting the last summand yields the second inequality. For the third one choose any $\tilde{a} \in [a],\ \tilde{b} \in [b]$. Then

$$|\tilde{a} \pm \tilde{b}| \leq |\tilde{a}| + |\tilde{b}| \leq |[a]| + |[b]|.$$

Since $\tilde{a}, \tilde{b}$ were arbitrary we get the assertion.

(e) $\langle[a] \cdot [b]\rangle = \min\{\, |\tilde{a}\tilde{b}| \mid \tilde{a} \in [a],\ \tilde{b} \in [b]\,\} = \langle[a]\rangle \cdot \langle[b]\rangle$.

The second equality is proved analogously. □

Both functions frequently occur in connection with the midpoint $\text{mid}([a]) = \check{a}$ and the radius $\text{rad}([a]) = r_a$ of an interval $[a]$ as can be seen from the subsequent theorems which we prove simultaneously after the second one. In order to formulate them we need the following definition.

Definition 2.4.3. Let $[a] \in \mathbb{IR}$. Then we define

$$r_a^- = \min\{\,|\check{a}|,\ r_a\,\} = \big||\overline{a}| - |\underline{a}|\big|/2, \qquad r_a^+ = \max\{\,|\check{a}|,\ r_a\,\} = (|\overline{a}| + |\underline{a}|)/2.$$

Theorem 2.4.4 (Midpoint). *Let* $[a], [b] \in \mathbb{IR}$. *Then we get*

(a) $[a] = \check{a} + r_a[-1, 1], \quad \underline{a} = \check{a} - r_a, \quad \overline{a} = \check{a} + r_a, \quad |[a]| = |\check{a}| + r_a \ge r_a,$

(b) $\mathrm{mid}(t[a]) = t\check{a}$ *for* $t \in \mathbb{R}$; *in particular,* $\mathrm{mid}(-[a]) = -\check{a}$,

(c) $\mathrm{mid}([a] \pm [b]) = \check{a} \pm \check{b}$,

(d) $$\mathrm{mid}([a][b]) = \check{a}\check{b} + \mathrm{sign}(\check{a}\check{b}) \min\{\,r_a|\check{b}|,\ |\check{a}|r_b,\ r_a r_b\,\}$$
$$= \check{a}\check{b} + \mathrm{sign}(\check{a})\,\mathrm{sign}(\check{b}) \cdot \begin{cases} r_a^- r_b^-, & \text{if } 0 \notin \mathrm{int}([a] \cap [b]) \\ \min\{\,r_a|\check{b}|,\ |\check{a}|r_b\,\}, & \text{if } 0 \in \mathrm{int}([a] \cap [b]) \end{cases},$$

(e) $$\mathrm{mid}([a]^{-1}) = \frac{\check{a}}{|[a]|\langle[a]\rangle},$$

(f) $$\mathrm{mid}([a]/[b]) = \frac{\mathrm{mid}([a][b])}{|[b]|\langle[b]\rangle}.$$

Theorem 2.4.5 (Radius). *Let* $[a], [b] \in \mathbb{IR}$. *Then we get*

(a) $r_a = |[a] - \check{a}| \ge 0, \quad r_a = 0 \Leftrightarrow \underline{a} = \overline{a}$,

(b) $\mathrm{rad}(t[a]) = |t|r_a$ *for* $t \in \mathbb{R}$; *in particular,* $\mathrm{rad}(-[a]) = r_a$,

(c) $\mathrm{rad}([a] \pm [b]) = r_a + r_b$,

(d) $$\mathrm{rad}([a][b]) = \max\{\,r_a|[b]|,\ |[a]|r_b,\ r_a|\check{b}| + |\check{a}|r_b\,\}$$
$$= \begin{cases} r_a^+ r_b + r_a r_b^+, & \text{if } 0 \notin \mathrm{int}([a] \cap [b]) \\ \max\{r_a|[b]|,\ |[a]|r_b\}, & \text{if } 0 \in \mathrm{int}([a] \cap [b]), \end{cases}$$

(e) $$\mathrm{rad}([a]^{-1}) = \frac{r_a}{|[a]|\langle[a]\rangle},$$

(f) $$\mathrm{rad}([a]/[b]) = \frac{\mathrm{rad}([a][b])}{|[b]|\langle[b]\rangle}.$$

Proof of the Theorems 2.4.4 *and* 2.4.5. The assertions (a) and (b) of both theorems are obvious from the definition.

(c) We only prove the minus case. To this end let $[c] = [a] - [b] = [\underline{a} - \overline{b}, \overline{a} - \underline{b}]$. Then $\check{c} = \{(\underline{a} - \overline{b}) + (\overline{a} - \underline{b})\}/2 = (\underline{a} + \overline{a})/2 - (\underline{b} + \overline{b})/2 = \check{a} - \check{b}$; $r_c = r_a + r_b$ follows analogously.

(d) Let $[c] = [a][b]$. By virtue of (b) we may assume w.l.o.g. $\check{a} \ge 0$ and $\check{b} \ge 0$, hence $\overline{a} \ge 0$, $\overline{b} \ge 0$. Then

$$\overline{c} = (\check{a} + r_a)(\check{b} + r_b) = \check{a}\check{b} + r_a\check{b} + \check{a}r_b + r_a r_b,$$
$$\underline{c} = \min\{\,(\check{a} + r_a)(\check{b} - r_b),\ (\check{a} - r_a)(\check{b} + r_b),\ (\check{a} - r_a)(\check{b} - r_b)\,\}$$
$$= \check{a}\check{b} - r_a\check{b} - \check{a}r_b - r_a r_b + 2\min\{\,r_a\check{b},\ \check{a}r_b,\ r_a r_b\,\}.$$

Taking $\check{a} \ge 0$, $\check{b} \ge 0$ into account we get

$$\check{c} = (\underline{c} + \overline{c})/2 = \check{a}\check{b} + \mathrm{sign}(\check{a}\check{b}) \min\{\,r_a|\check{b}|,\ |\check{a}|r_b,\ r_a r_b\,\} \qquad (2.4.1)$$

and

$$
\begin{aligned}
r_c = (\overline{c} - \underline{c})/2 &= r_a\check{b} + \check{a}r_b + r_a r_b - \min\{\, r_a\check{b},\ \check{a}r_b,\ r_a r_b \,\}\\
&= \max\{\check{a}r_b + r_a r_b,\ r_a\check{b} + r_a r_b,\ r_a\check{b} + \check{a}r_b\,\}\\
&= \max\{\, |[a]| r_b,\ r_a|[b]|,\ r_a|\check{b}| + |\check{a}|r_b \,\}.
\end{aligned}
\tag{2.4.2}
$$

If $0 \notin \operatorname{int}([a] \cap [b])$, then w.l.o.g. let $0 \notin \operatorname{int}([a])$. This implies $|\check{a}| \geq r_a = r_a^-$, and from (2.4.1), (2.4.2) we obtain

$$\check{c} = \check{a}\check{b} + \operatorname{sign}(\check{a}\check{b}) \min\{\, r_a|\check{b}|,\ r_a r_b \,\} = \check{a}\check{b} + \operatorname{sign}(\check{a}) \operatorname{sign}(\check{b}) r_a \min\{\, |\check{b}|,\ r_b \,\}$$

and

$$r_c = \max\{\, |[a]|r_b,\ r_a|\check{b}| + |\check{a}|r_b \,\} = |\check{a}|r_b + r_a \max\{\, r_b,\ |\check{b}| \,\},$$

whence the remaining assertions follow.

If $0 \in \operatorname{int}([a] \cap [b])$, then $|\check{a}| < r_a$, $|\check{b}| < r_b$. Hence the remaining assertions follow directly from (2.4.1) and (2.4.2), respectively.

(e) $\operatorname{mid}([a]^{-1}) = (1/\overline{a} + 1/\underline{a})/2 = \check{a}/(\overline{a}\underline{a}) = \check{a}/(|[a]|\langle[a]\rangle)$,

$\operatorname{rad}([a]^{-1}) = (1/\underline{a} - 1/\overline{a})/2 = r_a/(\underline{a}\overline{a}) = r_a/(|[a]|\langle[a]\rangle)$.

(f) By means of (d) and (e) we get

$$
\begin{aligned}
\operatorname{mid}([a]/[b]) &= \operatorname{mid}([a] \cdot [b]^{-1})\\
&= \check{a}\check{b}/(|[b]|\langle[b]\rangle) + \operatorname{sign}(\check{a}\check{b}) \min\{\, r_a|\check{b}|,\ \check{a}r_b,\ r_a r_b \,\}/(|[b]|\langle[b]\rangle)\\
&= \operatorname{mid}([a][b])/(|[b]|\langle[b]\rangle)
\end{aligned}
$$

and

$$
\begin{aligned}
\operatorname{rad}([a]/[b]) &= \operatorname{rad}([a] \cdot [b]^{-1})\\
&= \max\{\, |[a]|r_b,\ r_a|[b]|,\ r_a|\check{b}| + |\check{a}|r_b \,\}/(|[b]|\langle[b]\rangle)\\
&= \operatorname{rad}([a][b])/(|[b]|\langle[b]\rangle).
\end{aligned}
$$

□

We continue with additional properties of the radius also using the diameter $d([a])$.

Theorem 2.4.6 (Inequalities related to the radius). *Let* $[a], [b] \in \mathbb{IR}$. *Then we get*

(a) $r_a \leq |[a]| - \langle[a]\rangle \leq d([a])$,

(b) $r_a \leq |[a] - \tilde{a}| \leq d([a])$ *for all* $\tilde{a} \in [a]$,

(c) $[a] \subseteq [b] \Rightarrow r_a \leq r_b$,

(d) $\operatorname{rad}([a][b]) \leq r_a^+ r_b + r_a r_b^+ \leq |[a]|r_b + r_a|[b]|$ *with '=' in the first inequality if and only if* $0 \notin \operatorname{int}([a] \cap [b])$,

(e) $|[a]|r_b \leq \operatorname{rad}([a][b]) \leq |[a]|r_b + r_a|\check{b}| \leq |[a]|r_b + r_a|[b]|$,

(f) $r_a|[b]| \leq \operatorname{rad}([a][b]) \leq r_a|[b]| + |\check{a}|r_b < |[a]|r_b + r_a|[b]|$,

(g) $r_a/\langle[b]\rangle \leq \operatorname{rad}([a]/[b]) \leq (r_a|[b]| + |\check{a}|r_b)/(|[b]|\langle[b]\rangle)$.

Proof. (a)–(c) are obvious.

(d) follows from Theorem 2.4.5 (d) and $|[a]| = |\check{a}| + r_a \leq 2r_a$ for intervals $[a]$ with $0 \in [a]$.

(e) follows from Theorem 2.4.5 (d) and the inequality

$$r_a|[b]| = r_a(|\check{b}| + r_b) \le r_a|\check{b}| + |[a]|r_b.$$

(f) is an immediate consequence of (e), and (g) follows from (f) and Theorem 2.4.5 (e). □

Notice that the inequality in Theorem 2.4.6 (d) is certainly not worse than the corresponding ones in (e) and (f) of this theorem provided that $0 \notin \mathrm{int}([a] \cap [b])$. If $0 \in \mathrm{int}([a] \cap [b])$, there are cases in which it is better than the second to last one in (e), (f), and others in which it is worse, as the examples $[a] = [b] = [-2, 8]$ and $[a] = [b] = [-1, 1]$ show.

Theorem 2.4.7 (Radius for particular cases). *Let $[a], [b] \in \mathbb{IR}$ with $0 \in [a]$. Then we have*

(a) $\mathrm{rad}([a][b]) \le 2r_a r_b = r_a^+ r_b + r_a r_b^+$ *if* $0 \in [b]$,
(b) $\mathrm{rad}([a][b]) = r_a|[b]|$ *if* $[a]$ *is symmetric or if* $0 \notin \mathrm{int}([b])$,
(c) $\mathrm{rad}([a]/[b]) = r_a/\langle[b]\rangle$.

Proof. (a) is an immediate consequence of Theorem 2.4.6 (d).

(b) If $\check{a} = 0$, then the equality follows from Theorem 2.4.6 (f). If $0 \notin \mathrm{int}([b])$ it follows from Theorem 2.4.5 (d), since the assumptions imply $|\check{a}| \le r_a$, $|\check{b}| \ge r_b$, hence $r_a^+ = r_a$, $r_b^+ = |\check{b}|$, $r_b + |\check{b}| = |[b]|$.

(c) follows from (b). □

It is obvious that all results on the radius can be reformulated for the diameter $d([a]) = 2r_a$ by taking into account the factor two. Thus Theorem 2.4.5 remains true if one replaces the radii by diameters, Theorem 2.4.7 (a) simplifies to

$$d([a][b]) \le d([a])d([b]) \quad \text{if } 0 \in [a] \text{ and } 0 \in [b],$$

and as an additional property we mention

$$d([a]) = |[a] - [a]| = \max\{\, \tilde{a} - \tilde{a}' \mid \tilde{a}, \tilde{a}' \in [a] \,\}. \tag{2.4.3}$$

Our final result deals with equivalences of some set theoretic operations.

Theorem 2.4.8. *Let $[a], [b] \in \mathbb{IR}$. Then we have*

(a) $[a] \subseteq [b] \Leftrightarrow |[a] - \check{b}| \le r_b \Leftrightarrow |\check{b} - \check{a}| \le r_b - r_a$,
(b) $[a] \subseteq \mathrm{int}([b]) \Leftrightarrow |[a] - \check{b}| < r_b \Leftrightarrow |\check{b} - \check{a}| < r_b - r_a$,
(c) $[a] \cap [b] \ne \emptyset \Leftrightarrow |\check{a} - \check{b}| \le r_a + r_b \Leftrightarrow \underline{a} \le \overline{b}$ *and* $\underline{b} \le \overline{a}$.

Proof. (a) From $[a] \subseteq [b] \Leftrightarrow [a] - \check{b} \subseteq [b] - \check{b} = r_b[-1, 1]$ we get equivalently $|[a] - \check{b}| \le r_b$, i.e., $|\check{a} - \check{b} + r_a[-1, 1]| \le r_b$. Since the left-hand side equals $|\check{a} - \check{b}| + r_a$, the assertion follows.

(b) is proved analogously to (a), and (c) is obvious by geometric inspection. □

Exercises

Ex. 2.4.1. Prove the following formulae for the diameter.
(a) $d([a][b]) = d([a])d([b]) + \langle[a]\rangle d(b]) + d([a])\langle[b]\rangle$ if $0 \notin \operatorname{int}([a] \cap [b])$.
(b) $d(\frac{[a]}{[b]}) = \frac{d([a])}{\langle[b]\rangle} + \frac{\langle[a]\rangle}{\langle[b]\rangle|[b]|}d([b])$ if $0 \notin [b]$.
What are the corresponding formulae for the radius?

2.5 Distance and topology

Up to now we mainly considered algebraic aspects of interval computation. In order to study iterative processes we need topological tools. To this end we introduce the following function q.

Definition 2.5.1. Let $[a], [b] \in \mathbb{IR}$. Then the mapping $q: \mathbb{IR} \times \mathbb{IR} \to \mathbb{R}$ is defined by

$$q([a], [b]) = \max\{\, |\underline{a} - \underline{b}|,\ |\overline{a} - \overline{b}|\, \}.$$

As we are going to see, the function q is a metric which is called the Hausdorff metric or (Hausdorff) distance. In fact it is a particular case of the well-known Hausdorff metric for compact sets $U, V \subseteq \mathbb{R}$ which is defined as

$$q(U, V) = \max\{\sup_{u \in U} \inf_{v \in V} |u - v|,\ \sup_{v \in V} \inf_{u \in U} |u - v|\,\}. \tag{2.5.1}$$

Theorem 2.5.2. *The function q of Definition 2.5.1 is a metric. It reduces to the standard metric in $\mathbb{R}$ if its operands are degenerate intervals, i.e., $q([a], [b]) = |a - b|$ for $[a] \equiv a$, $[b] \equiv b$.*

Proof. The definiteness (1.2.1) and the symmetry (1.2.2) are seen directly from the definition of q. The triangular inequality (1.2.3) is an immediate consequence of the triangular inequality for the absolute value $|\cdot|$ in $\mathbb{R}$. The reduction to the standard metric in $\mathbb{R}$ for degenerate intervals is obvious from Definition 2.5.1. □

As every metric does, q induces a metric topology on $\mathbb{IR}$ with which convergence and continuity are defined as usual – cf. Section 1.2.

Definition 2.5.3.
(a) We call an interval sequence $([a]_k)$ convergent to an interval $[a] \in \mathbb{IR}$ and write $\lim_{k\to\infty}[a]_k = [a]$ if $\lim_{k\to\infty} q([a]_k, [a]) = 0$.
(b) Let $D \subseteq \mathbb{R}$. We call an interval function $[f]: \mathbb{I}(D) \to \mathbb{IR}$ continuous in $[a] \in \mathbb{I}(D)$ if $\lim_{k\to\infty}[f]([a]_k) = [f]([a])$ for all sequences $([a]_k)$ from $\mathbb{I}(D)$ which converge to $[a]$.

We call $[f]$ continuous in $\mathbb{I}(D)$ if it is continuous for any $[a] \in \mathbb{I}(D)$.

Remark. Whenever we speak of distance, convergence or continuity in $\mathbb{IR}$ we always tacitly assume that this terminology is used with respect to q.

Now we show how convergence in $\mathbb{IR}$ can be expressed by convergence in $\mathbb{R}$. This is the key for proving that the metric space $(\mathbb{IR}, q)$ is complete – an important remark since then we can apply Banach's fixed point theorem.

Theorem 2.5.4.
(a) *Let* $[a] \in \mathbb{IR}$ *and* $[a]_k = [\underline{a}_k, \overline{a}_k] \in \mathbb{IR}$ *for* $k \in \mathbb{N}$. *Then we have*

$$\lim_{k\to\infty} [a]_k = [a] \Leftrightarrow \lim_{k\to\infty} \underline{a}_k = \underline{a} \text{ and } \lim_{k\to\infty} \overline{a}_k = \overline{a}$$
$$\Leftrightarrow \lim_{k\to\infty} \check{a}_k = \check{a} \text{ and } \lim_{k\to\infty} \operatorname{rad}([a]_k) = r_a.$$

(b) *The space* $(\mathbb{IR}, q)$ *is a complete metric space.*

Proof. (a) follows immediately from the Definition 2.5.1 of the metric and the definition of midpoint and radius.

(b) Let $([a]_k)$ be a Cauchy sequence. Then by Definition 2.5.1 the same holds for the two real sequences $(\underline{a}_k)$ and $(\overline{a}_k)$. Since $(\mathbb{R}, |\cdot|)$ is a complete metric space and since $\underline{a}_k \le \overline{a}_k$ for all k, these two sequences converge to some limits $\underline{a}$, $\overline{a}$ with $\underline{a} \le \overline{a}$, and it is easy to see that $\lim_{k\to\infty}[a]_k = [a]$ holds. The converse can be seen similarly. □

In applications we are often faced with sequences of intervals which are nested with respect to '⊆'. We shall see that such a property is already sufficient to guarantee convergence.

Theorem 2.5.5. *Let* $([a]_k)$ *be a sequence in* $\mathbb{IR}$ *which satisfies*

$$[a]_0 \supseteq [a]_1 \supseteq [a]_2 \supseteq \cdots .$$

Then this sequence is convergent to some limit $[a]$ *which satisfies*

$$[a] = \bigcap_{k=0}^{\infty} [a]_k \in \mathbb{IR}. \tag{2.5.2}$$

The lower bounds $\underline{a}_k$ *converge monotonically increasing to* $\underline{a}$, *and the upper bounds* $\overline{a}_k$ *converge monotonically decreasing to* $\overline{a}$.

Proof. The nested sequence $([a]_k)$ satisfies

$$\underline{a}_0 \le \underline{a}_1 \le \cdots \le \underline{a}_k \le \cdots \le \overline{a}_k \le \cdots \le \overline{a}_1 \le \overline{a}_0.$$

Since monotone and bounded sequences in $\mathbb{R}$ are convergent there are real numbers $\underline{a}, \overline{a}$ such that

$$\lim_{k\to\infty} \underline{a}_k = \underline{a} \qquad \lim_{k\to\infty} \overline{a}_k = \overline{a}.$$

Therefore, by Theorem 2.5.4 we get $\lim_{k\to\infty}[a]_k = [a]$. Moreover, by the monotonicity of the bounding sequences the representation (2.5.2) is immediate. □

Next we shortly address continuity, where this time we also consider functions from $\mathbb{IR}$ to $\mathbb{R}$ or $\mathbb{IR} \times \mathbb{IR}$ to $\mathbb{R}$. Therefore, we have to deal with two metrics as indicated in Definition 1.2.7, or we should interpret the image as a degenerate interval and keep q as metric for the range, too.

Theorem 2.5.6. *The functions* $\inf(\cdot)$, $\sup(\cdot)$, $\operatorname{mid}(\cdot)$, $\operatorname{rad}(\cdot)$, $d(\cdot)$, $|\cdot|$, $\langle\cdot\rangle$ *are continuous, and the same holds for the binary operations* $\circ \in \{+, -, \cdot/\}$.

Proof. The proof of Theorem 2.5.6 is immediate by virtue of Theorem 2.5.4 and the continuity of the functions max, min and the corresponding operations $\circ$ in $\mathbb{R}$. □

Now we state several properties of the Hausdorff distance q.

Theorem 2.5.7. *Let* $[a], [b], [c], [d] \in \mathbb{IR}$. *Then the Hausdorff metric* q *satisfies the following properties.*

(a) $q([a], [b]) = \min\{\, t \in \mathbb{R} \mid t \geq 0,\ [a] \subseteq [b] + [-t, t],\ [b] \subseteq [a] + [-t, t]\,\}$,
(b) $q([a], [b]) = |\check{a} - \check{b}| + |r_a - r_b|$,
(c) $q([a], t) = |[a] - t|$ *for* $t \in \mathbb{R}$; *in particular,* $q([a], 0) = |[a]|$,
(d) $q(t[a], t[b]) = |t| q([a], [b])$ *for* $t \in \mathbb{R}$,
(e) $q([a] + [c], [b] + [d]) \leq q([a], [b]) + q([c], [d])$,
(f) $q([a] \pm [c], [b] \pm [c]) = q([a], [b])$,
(g) $q([a][c], [b][c]) \leq |[c]| q([a], [b])$,
(h) $q([a]/[c], [b]/[c]) \leq \dfrac{1}{\langle [c] \rangle} q([a], [b])$,
(i) $q([c]/[a], [c]/[b]) \leq \dfrac{|[c]|}{\langle [a] \rangle \langle [b] \rangle} q([a], [b])$,
(j) $r_b \leq r_a + q([a], [b])$,
(k) $|[b]| \leq |[a]| + q([a], [b])$,
(l) $[b] \subseteq [a] + q([a], [b])[-1, 1]$,
(m) $[a] \subseteq [b] \Rightarrow q([a], [b]) = |[b] - \check{a}| - r_a$,
(n) $[a] \subseteq [b] \Rightarrow r_b - r_a \leq q([a], [b]) \leq d([b]) - d([a])$,
(o) $[a] \subseteq [b] \subseteq [c] \Rightarrow \max\{\, q([a], [b]), q([b], [c])\,\} \leq q([a], [c])$,
(p) $[a] \subseteq [b] \subseteq [d],\ [a] \subseteq [c] \subseteq [d] \Rightarrow q([b], [c]) \leq \max\{\, q([a], [b]), q([b], [d])\,\}$,
(q) $b \in [d],\ [c] \subseteq [d] \Rightarrow q(b, [c]) \leq q(b, [d])$.

Proof. Let $t \geq 0$ and $q = q([a], [b])$.

(a) The equivalences

$$[a] \subseteq [b] + [-t, t] \quad\Leftrightarrow\quad \underline{b} - t \leq \underline{a} \text{ and } \overline{a} \leq \overline{b} + t \quad\Leftrightarrow\quad \underline{b} - \underline{a} \leq t \text{ and } \overline{a} - \overline{b} \leq t, \tag{2.5.3}$$

and, analogously,

$$[b] \subseteq [a] + [-t, t] \quad\Leftrightarrow\quad \underline{a} - \underline{b} \leq t \text{ and } \overline{b} - \overline{a} \leq t, \tag{2.5.4}$$

imply

$$q = \max\{\, |\underline{a} - \underline{b}|,\ |\overline{a} - \overline{b}|\,\} \leq t.$$

Hence

$$q \le m = \min\{\, t \in \mathbb{R} \mid t \ge 0,\ [a] \subseteq [b] + [-t, t],\ [b] \subseteq [a] + [-t, t]\,\}.$$

Since (2.5.3) and (2.5.4) hold for $t = q$, we get $m \le q$, which proves the assertion.

(b) Choose $t \ge 0$ such that (2.5.3) and (2.5.4) hold. By virtue of Theorem 2.4.8 (a) this is equivalent to

$$|\check{b} - \check{a}| \le \operatorname{rad}([b] + [-t, t]) - r_a = r_b + t - r_a \quad \text{and} \quad |\check{a} - \check{b}| \le r_a + t - r_b,$$

and, furthermore, to $t \ge |\check{a} - \check{b}| + |r_a - r_b| =: s$. Hence $q \ge s$. For the converse inequality set $t = s$.

(c)–(f) follow directly from Definition 2.5.1 and the triangular inequality for (e).

(g) According to (a) we have

$$\begin{cases} [a] \subseteq [b] + q[-1, 1] & \Rightarrow \quad [a][c] \subseteq [b][c] + q[-1, 1]\,|[c]|, \\ [b] \subseteq [a] + q[-1, 1] & \Rightarrow \quad [b][c] \subseteq [a][c] + q[-1, 1]\,|[c]|. \end{cases}$$

Applying (a) once more finishes the proof.

(h) follows from (g).

(i) From

$$\left|\frac{1}{\overline{a}} - \frac{1}{\overline{b}}\right| \le \frac{|\overline{b} - \overline{a}|}{\langle [a] \rangle \langle [b] \rangle}$$

and a similar relation for the lower bounds we get

$$q([a]^{-1}, [b]^{-1}) \le \frac{1}{\langle [a] \rangle \langle [b] \rangle}\, q.$$

Together with (g) this implies the assertion.

(j) and (k) follow from (l) which is true in turn by (a).

(m) Since the assumption implies $r_a \le r_b$, we get $q = |\check{a} - \check{b}| + r_b - r_a$ by virtue of (b). Thus, $q = |\check{b} - \check{a}| + \operatorname{rad}([b] - \check{a}) - r_a = |[b] - \check{a}| - r_a$.

(n) The first inequality is (j), the second follows from (b) and Theorem 2.4.8 (a) via

$$q = |\check{a} - \check{b}| + r_b - r_a \le 2(r_b - r_a).$$

(o) With (m) we get

$$q = |[b] - \check{a}| - r_a \le |[c] - \check{a}| - r_a = q([a], [c]).$$

Again by means of (m) and Theorem 2.4.8 (a) we similarly obtain

$$q([b], [c]) = |[c] - \check{b}| - r_b \le |[c] - \check{a}| + |\check{a} - \check{b}| - r_b \le |[c] - \check{a}| + r_b - r_a - r_b = q([a], [c]).$$

(p) If $\underline{b} \le \underline{c}$, then the assumption implies $|\underline{c} - \underline{b}| = \underline{c} - \underline{b} \le \underline{a} - \underline{b} \le q([a], [b])$. Otherwise we obtain $|\underline{c} - \underline{b}| = \underline{b} - \underline{c} \le \underline{b} - \underline{d} \le q([b], [d])$. Similarly we get $|\overline{c} - \overline{b}| \le \max\{q([b], [d]),\ q([a], [b])\}$ which proves the assertion.

(q) is proved similarly as (p). □

The example $[c] \subset [d] = [b]$ shows that (q) is no longer true if the assumption '$b \in [d]$' is replaced by '$[b] \subseteq [d]$'.

We close this section by introducing Neumaier's β-function from Neumaier [257], which is used in Chapter 5.

Definition 2.5.8. Let $[a], [b] \in \mathbb{IR}$. Then the mapping $\beta\colon \mathbb{IR} \times \mathbb{IR} \to \mathbb{R}$ is defined by

$$\beta([a],[b]) = |[a]| + q([a],[b]).$$

Theorem 2.5.9. *The β-function has the following properties.*
(a) *If $[a], [b], [c], [d] \in \mathbb{IR}$, $[a] \subseteq [b]$, $[c] \subseteq [d]$, then*

$$\beta([a]\cdot[c],[b]\cdot[d]) \le \beta([a],[b])\cdot\beta([c],[d]). \tag{2.5.5}$$

(b) *If $[a], [b] \in \mathbb{IR}$, $[a] \subseteq [b]$, $\langle [a]\rangle > q([a],[b])$, then*

$$\beta([a]^{-1},[b]^{-1}) \le \left(\langle [a]\rangle - q([a],[b])\right)^{-1}. \tag{2.5.6}$$

Proof. (a) Let $q = q([a],[b])$, $q' = q([c],[d])$. Theorem 2.5.7 implies $[b] \subseteq [a] + q[-1,1]$, $[d] \subseteq [c] + q'[-1,1]$, hence $[b][d] \subseteq [a][c] + (|[a]|q' + q|[c]| + qq')[-1,1]$. From $[a][c] \subseteq [b][d]$ we get $q([a][c],[b][d]) \le |[a]|q' + q|[c]| + qq'$, whence $\beta([a]\cdot[c],[b]\cdot[d]) \le |[a]|\cdot|[c]| + |[a]|q' + q|[c]| + qq' = \beta([a],[b])\beta([c],[d])$.

(b) With q as in (a) and $[b]' = [a] + [-q,q]$ we get $\langle [b]'\rangle \ge \langle [a]\rangle - q > 0$ by virtue of the assumption. Therefore, $0 \notin [b]'$, and $[a] \subseteq [b] \subseteq [b]'$ implies $[a]^{-1} \subseteq [b]^{-1} \subseteq ([b]')^{-1}$. By virtue of Theorem 2.5.7 we obtain

$$\begin{aligned} q([a]^{-1},[b]^{-1}) &\le q([a]^{-1},([b]')^{-1}) = \max(|(\underline{b}')^{-1} - \underline{a}^{-1}|, |(\overline{b}')^{-1} - \overline{a}^{-1}|) \\ &= \max(q/|\underline{a}(\underline{a}-q)|, q/|\overline{a}(\overline{a}+q)|) \\ &\le q/(\langle [a]\rangle(\langle [a]\rangle - q)) = (\langle [a]\rangle - q)^{-1} - \langle [a]\rangle^{-1}, \end{aligned}$$

from which (2.5.6) follows immediately. □

2.6 Elementary interval functions

In order to define elementary interval functions similar to those in $\mathbb{R}$ we are guided by the inclusion property (2.2.4) and the optimality of the interval arithmetic with respect to this property. Bearing these properties in mind we are automatically led to the following definition.

Definition 2.6.1. By $\mathbb{F}$ we define the set of all real elementary functions φ whose function term $\varphi(x)$ is given by $\mathrm{abs}(x) = |x|$, $\mathrm{pow}_\alpha(x) = x^\alpha$ ($\alpha \in \mathbb{R}$), $\exp(x) = e^x$, $\ln(x)$, $\sin(x)$, $\cos(x)$, or $\arctan(x)$, and whose domain of definition D_φ is maximal. In accordance with Section 1.1 we define $R_\varphi([x])$ as the range of φ restricted to $[x] \in \mathbb{IR}$.

We extend $\varphi \in \mathbb{F}$ to an interval function on the set $\{ [x] \in \mathbb{IR} \mid [x] \subseteq D_\varphi \}$ by defining

$$\varphi([x]) = R_\varphi([x]). \tag{2.6.1}$$

We call this extension an elementary interval function.

Remark.

(a) We write $\varphi([a])$ instead of $[\varphi]([a])$ in accordance with Definition 4.1.2 below.

(b) We are going to show that $R_\varphi([x])$ is an interval so that we completely stay in $\mathbb{IR}$ as we did when defining the interval arithmetic in Section 2.2. For degenerate intervals, the elementary interval functions reduce to the corresponding elementary functions in $\mathbb{R}$. Whenever $\varphi([x])$ occurs we tacitly assume that $[x] \subseteq D_\varphi$.

(c) We restricted $\mathbb{F}$ to a subset of what is usually called an elementary function in $\mathbb{R}$, including, for instance, all trigonometrical functions and their inverses as well as the hyperbolic functions and their inverses. We made this restriction only in order to keep certain succeeding proofs within a limited frame. The reader is faced with a similar restriction in many programming languages. He is requested to adapt $\mathbb{F}$ according to his needs.

(d) Notice that for nondegenerate intervals $[x]$, the value $\operatorname{abs}([x])$ differs from $|[x]|$.

(e) As usual we often replace $\operatorname{pow}_{1/n}([x])$ by $\sqrt[n]{[x]}$ or $[x]^{\frac{1}{n}}$ for $n \in \mathbb{N}$, and we proceed similarly for pow_{-n}, $n \in \mathbb{N}$.

Simple examples are $\operatorname{abs}([-1, 2]) = [0, 2]$, $\sin([-10, 10]) = [-1, 1]$, and $\operatorname{pow}_2([-1, 2]) = [-1, 2]^2 = [0, 4] \neq [-1, 2] \cdot [-1, 2] = [-2, 4]$.

Theorem 2.6.2 (Representation). *Let $\varphi \in \mathbb{F}$, $[x] \subseteq D_\varphi$. Then the following properties hold.*

(a) $\varphi([x]) = [\min_{\tilde{x} \in [x]} \varphi(\tilde{x}), \max_{\tilde{x} \in [x]} \varphi(\tilde{x})]$; *in particular, $\varphi([x])$ is an interval.*

(b) $\varphi([x]) = [\varphi(\underline{x}), \varphi(\overline{x})]$ *for* $\varphi \in \{ \operatorname{pow}_n$ *for* $n \in \mathbb{N}, \exp, \ln, \arctan \}$.

(c) $\operatorname{abs}([x]) = [\langle [x] \rangle, |[x]|]$.

Proof. (a) is based on the fact that elementary functions are continuous and therefore have an absolute maximum and minimum on compact intervals.

(b) and (c) are obvious. □

Our next result is a copy of Theorem 2.2.4 applied to elementary interval functions. It follows immediately from Definition 2.6.1.

Theorem 2.6.3 (Inclusion property, inclusion isotony). *Let $\varphi \in \mathbb{F}$ and $[x], [y] \in D_\varphi$. Then φ fulfills the inclusion property*

$$\varphi(\tilde{x}) \in \varphi([x]) \quad \text{for all } \tilde{x} \in [x]$$

and is inclusion isotone, or – less specifically – inclusion monotone, i.e.,

$$[x] \subseteq [y] \quad \Rightarrow \quad \varphi([x]) \subseteq \varphi([y]).$$

More general interval functions and specific properties are considered in Chapter 4 after we have introduced interval vectors and interval matrices.

2.7 Machine interval arithmetic

Because of its straightforward implementation in digital electronic circuitry, the binary number system is used internally by almost all modern computers. Decimal numbers $r \neq 0$ have the binary representation

$$r = \pm \sum_{i=-\infty}^{n} b_i 2^i, \; b_i \in \{0, 1\}, \; b_n = 1, \; b_i \neq 1 \text{ for infinitely many indices } i < n, \quad (2.7.1)$$

or in the short form

$$r = \pm b_n b_{n-1} \dots b_1 b_0 \, . \, b_{-1} b_{-2} \dots = (\pm b_n b_{n-1} \dots b_1 b_0 \, . \, b_{-1} b_{-2} \dots)_2 . \qquad (2.7.2)$$

Here, n is an appropriate integer which depends on r. If n is negative, we insert zeros for $b_0, b_{-1}, \dots, b_{n+1}$ in the representation (2.7.2). If $b_i = 0$ for all $i \leq i_0$ and some $i_0 < 0$, then (2.7.2) can be shortened as in the decimal system. In this case we call the representation (2.7.2) finite. Notice that a decimal number r with a finite decimal representation can have a periodic infinite binary representation (2.7.2) as the example $0.1 = 1/10 = 0.000\overline{1100}$ shows. Here, periodicity results from the fact that the integer 10 in the denominator of the decimal number $1/10$ has the integral divisor 5 which is not a power of 2.

By virtue of (2.7.1), positive integers w can be represented as the evaluation of the polynomial

$$p(x) = \sum_{i=0}^{n} b_i x^i, \quad b_i \in \{0, 1\}, \; b_n = 1,$$

at $x = 2$. Therefore the conversion of positive decimal integers to binary ones can be realized by representing $p(2)$ in a Horner-like manner and by computing $b_0, \dots, b_n$ successively by integral division '$\div$' from

$$\begin{aligned} w &= 2w_0 + b_0, & &\text{where } w_0 = w \div 2, \\ w_k &= 2w_{k+1} + b_{k+1}, & &\text{where } w_{k+1} = w_k \div 2, \quad k = 0, 1, \dots, n-1. \end{aligned}$$

The well-known subtraction method with powers of 2 can also be applied to convert decimal to binary – in particular, if w is not an integer.

Because of the limited storage space for a binary integer z – say n bits – only a finite subset $\mathbb{Z}_M$ (= set of machine integers) of $\mathbb{Z}$ can be represented on a computer. Since one bit is needed to store the sign of z, the maximal machine integer is $z_{\max} = \sum_{i=0}^{n-2} 2^i = 2^{n-1} - 1$. Negative integers are usually stored as two's complement of $|z|$ with respect to n bits, which facilitates subtraction as we are going to see below. Here, the two's complement z'' of an integer $z \in \mathbb{Z}_M$ is defined as $z'' = 2^n - z$. It can easily be implemented as one's complement z' of z plus one, where this latter complement with respect to n bits is the binary representation of z with n bits (including leading zeros if necessary), in which all zeros are replaced by ones and vice versa. The relation

$z'' = z' + 1$ can be seen at once from $z + z' = (1, 1, \ldots, 1)_2 = \sum_{i=0}^{n-1} 2^i = 2^n - 1$. Thus if a positive integer z is coded with a leading zero and $n - 1$ subsequent bits, then the corresponding negative integer $-z$ necessarily has a leading 1 in front. Notice that $z = 0$ implies $z'' = 2^n$, which cannot be stored with n bits, and the equation $z'' = 2^n - z = z$ implies $z = 2^{n-1}$, which has a one in front and, therefore, encodes the negative integer $-2^{n-1} = -z_{\max} - 1$. This means that $\mathbb{Z}_M \neq -\mathbb{Z}_M$, which formerly led to the unexpected result $(z_{\max} + 1) + z_{\max} = -1$ in the programming environment Turbo Pascal if warnings were suppressed.

There is a second kind of machine numbers, the so-called floating point numbers, which form a subset $\mathbb{R}_M$ of $\mathbb{R}$. These numbers are represented with ℓ bits as

$$\tilde{r} = \pm m \cdot 2^{\pm e}, \tag{2.7.3}$$

where $m \geq 0$ is called the mantissa and $e \in \mathbb{N}_M$ is called the exponent with some finite subset $\mathbb{N}_M$ of $\mathbb{N}_0$. For $\tilde{r} \neq 0$ the mantissa often has the form $1. * \ldots *$, $* \in \{0, 1\}$, which defines the normalized floating point numbers in binary representation. These numbers form the basis for most calculations on a computer. The leading 1 of m is usually not stored; this bit is then called a hidden bit. The exponent $-e < 0$ is often made positive by adding a constant integer called bias. Usually, this bias equals $2^{\ell''} - 1 = (1, 1, \ldots, 1)_2$ with ℓ'' ones, where ℓ'' denotes the number of bits provided for the storage of $e > 0$ (without sign bit). Thus the signed exponent $\pm e$ is coded as binary integer by $\ell'' + 1$ bits. By virtue of the bias, $e > 0$ has a leading 1 while the biased binary representation of $-e \leq 0$ starts with a zero. In particular, the largest positive exponent $e_{\max}$ is given in unbiased form with ℓ'' bits as

$$e_{\max} = (1, 1, \ldots, 1)_2 = 2^{\ell''} - 1$$

and in biased form with $\ell'' + 1$ bits as $(1, 1, \ldots, 1, 0)_2$ so that the biased exponent $(1, 1, \ldots, 1)_2 = 2^{\ell''+1} - 1$ is missing and can be used in combination with the mantissa to store additional information, for instance particularities such as NaN (not a number; mantissa $m \neq 0$) or $\pm$ infinity (mantissa $m = 0$).

In the commonly accepted IEEE Standard 754 [50], Bohlender, Ullrich [70] the smallest negative exponent $e_{\min}$ for the format 'double' is $e_{\min} = -e_{\max} + 1$ so that the biased exponent zero would not occur. Together with the bits of the signed mantissa, this biased exponent zero can be used to code denormalized floating point numbers (including signed zero) which are defined to have a mantissa of the form $0. * \ldots *$, $* \in \{0, 1\}$ and the unbiased exponent $2^{e_{\min}-1} = 2^{-e_{\max}}$. We assume that these particular denormalized numbers also belong to $\mathbb{R}_M$.

If ℓ' denotes the number of bits provided to store m without counting the hidden bit, then $(1 + \ell') + (1 + \ell'') = \ell$, where the two ones on the left-hand side count the storage of the signs for the mantissa and the exponent (hidden behind the addition of the bias). In the IEEE Standard 754 the number format 'double' has the specifications $\ell = 64$, $\ell' = 52$, $\ell'' = 10$, bias $= 1023$, where the highest bit stores the sign of the mantissa (0 for positive and 1 for negative floating point numbers) while the next 11 bits store the bit sequence of the biased exponent. Therefore, both pieces of information are stored in the three highest places of the corresponding hexadecimal code.

By virtue of (2.7.3) the largest representable normalized floating point number $\tilde{r}_{\max}$ is given with binary representation of the mantissa as

$$\tilde{r}_{\max} = 1.1\ldots1\cdot 2^{+e_{\max}} = 2^{e_{\max}+1} - 2^{e_{\max}-\ell'} = 2^{2^{\ell''}} - 2^{2^{\ell''}-\ell'-1}.$$

Thus for the IEEE Standard 'double' we obtain $e_{\max} = 2^{10} - 1 = 1023$ and $\tilde{r}_{\max} = 2^{1024} - 2^{971} = 2^{971}\cdot(2^{53}-1) \doteq 1.797693134862316\cdot 10^{308}$, where $\doteq$ means 'equal to the computed result'. This result can be obtained when invoking for instance the MATLAB command `realmax`.

Similarly, we can compute the smallest positive normalized floating point number $\tilde{r}_{\min}$, which is given by

$$\tilde{r}_{\min} = 1.0\cdot 2^{-(e_{\max}-1)} = 2^{-1022} \doteq 2.225073858507201\cdot 10^{-308}$$

if the IEEE Standard is fulfilled. This number is the output of the MATLAB command `realmin`, for example. The interval $(-\tilde{r}_{\min}, \tilde{r}_{\min})$ does not contain any normalized floating point number. It can be diminished by allowing denormalized floating point numbers. The smallest positive denormalized floating point number is

$$\tilde{r}^{\mathrm{ext}}_{\min} = \tilde{r}_{\min}\cdot 2^{-\ell'} = 2^{-1074} \doteq 4.940656458412465\cdot 10^{-324}.$$

The set $(-\tilde{r}^{\mathrm{ext}}_{\min}, \tilde{r}^{\mathrm{ext}}_{\min})\setminus\{0\}$ is called underflow, the set $\mathbb{R}\setminus\mathbb{R}_B$ is called overflow, where $\mathbb{R}_B := [-\tilde{r}_{\max}, \tilde{r}_{\max}]$.

Working with real numbers, $r \in \mathbb{R}_B$ means approximating r by some floating point number $\tilde{r} \in \mathbb{R}_M$. The corresponding mapping $fl : \mathbb{R}_B \to \mathbb{R}_M$ with $fl(r) = \tilde{r}$ is called rounding if $fl(\tilde{r}) = \tilde{r}$ holds for all $\tilde{r} \in \mathbb{R}_M$. Such a rounding is called monotone if $r \le s$ implies $fl(r) \le fl(s)$ for all $r, s, \in \mathbb{R}_B$. It is called downward directed if $fl(r) \le r$ holds for all $r \in \mathbb{R}_B$. Similarly, we use the terminology upward directed if $fl(r) \ge r$ is fulfilled for all $r \in \mathbb{R}_B$. If a rounding fl satisfies $fl(-r) = -fl(r)$, it is called antisymmetric.

Notice that any upward directed rounding $fl_\uparrow$ defines a downward directed $fl_\downarrow$ one via $fl_\downarrow(r) := -fl_\uparrow(-r)$.

There are several particular kinds of roundings which we describe for simplicity only for real numbers

$$r = \pm\sum_{i=-\infty}^{n} b_i 2^i = \pm\left(\sum_{i=-\infty}^{0} b_{i+n}2^i\right)2^n$$

with $b_n = 1$, $b_i \neq 1$ for infinitely many indices $i < n$, $\tilde{r}_{\min} \le |r| \le \tilde{r}_{\max}$, so that

$$r = m\cdot 2^n \quad \text{with } m = 1.b_{n-1}b_{n-2}\ldots b_{n-\ell'} \mid b_{n-\ell'-1}\ldots \tag{2.7.4}$$

and $n \in [e_{\min}, e_{\max}] \cap \mathbb{Z}$. The subsequent definitions can be extended to $r \in (-\tilde{r}_{\min}, \tilde{r}_{\min})$ without any problems, i.e., to the underflow and the denormalized domain.

(i) Rounding by chopping

Here the trailing end of m in (2.7.4) is simply discarded so that

$$\tilde{r} = fl_c(r) := \pm\left(\sum_{i=-\ell'}^{0} b_{i+n}2^i\right)2^n, \quad b_n = 1. \tag{2.7.5}$$

The rounding is monotone and antisymmetric but not directed, as the example $\ell' = \ell'' = 1$, $fl_c(-1.01 \cdot 2^0) = -1.0 \geq -1.01$, $fl_c(1.01 \cdot 2^0) = 1.0 \not\geq 1.01$ shows.

(ii) Rounding to nearest

Nomen est omen! The rounding depends on the value b_i of the largest index i of (2.7.4) which was not taken into account in (2.7.5):

$$\tilde{r} = fl_n(r) := \begin{cases} fl_c(r), & \text{if } b_{n-\ell'-1} = 0, \\ fl_c(r) + \operatorname{sign}(r)2^{n-\ell'}, & \text{if } b_{n-\ell'-1} = 1. \end{cases} \tag{2.7.6}$$

The rounding is monotone and antisymmetric but not directed, as the example $\ell' = \ell'' = 1$, $fl_n(1.010 \cdot 2^0) = 1.1 \geq 1.010$, $fl_c(1.001 \cdot 2^0) = 1.0 \not\geq 1.001$ shows.

(iii) Upward rounding $fl_\triangle$

Here $\tilde{r} = fl_\triangle(r)$ is the nearest machine number which is not smaller than r. This rounding is monotone but not antisymmetric, as the example $\ell' = \ell'' = 1$, $fl_\triangle(-1.01 \cdot 2^0) = -1.0 \neq -1.1 = -fl_\triangle(1.01 \cdot 2^0)$ shows. Clearly, $fl_\triangle$ is upward directed.

(iv) Downward rounding $fl_\triangledown$

Here $\tilde{r} = fl_\triangledown$ is the nearest machine number which is not greater than r. This rounding satisfies $fl_\triangledown(r) = -fl_\triangle(-r)$. Hence it is monotone, downward directed, but not antisymmetric.

By virtue of the finite format of machine numbers $\tilde{r}$, real numbers r are usually represented on a computer with a so-called conversion error. It is one kind of a rounding error. For the decimal $r = 0.1$, rounding to nearest and the above-mentioned IEEE Standard format 'double' results in the normalized floating point representation

$$\tilde{r} = fl_n(r) = fl_n(1.\overline{1001} \cdot 2^{-4}) = 1.1001\ldots1001\,1010 \cdot 2^{-4}$$

with 52 bits behind the comma and a relative error $|\frac{\tilde{r}-r}{r}|$ which is bounded by the machine precision (or machine epsilon) eps $= 2^{-\ell'-1}$, cf. Anderson et al. [49], for example.

Notice that the definition of machine epsilon is not unique in the literature. Some authors like Chaitin-Chatelin [76] or Higham [147] use the terminology 'unit rounding' respectively 'unit roundoff' for our eps and define the machine epsilon as in MATLAB by $2 \cdot \text{eps}$ in order to measure the distance between 1.0 and the next larger floating point number.

A second kind of rounding error may result from the four arithmetic operations $+, -, *, /$. These operations can be realized on a computer simply by an adder with the possibility to

- add bitwise two binaries,
- shift the bit sequence in the register one position to the left or right,
- compare two binaries in order to decide which one is the largest,
- interchange ones with zeros and vice versa bitwise in order to compute the two complements of a binary integer.

The last property is interesting since the subtraction $z - w$ of two integers z and w can be done by addition via

$$z - w = z - 2^n + (2^n - w) = (z + w'') - 2^n = -(2^n - (z + w'')) = -(z + w'')''. \qquad (2.7.7)$$

One can show (Exercise 2.7.2) that

$$2^n < z + w'' < 2^{n+1} \quad \text{holds if } z > w \qquad (2.7.8)$$

so that the subtraction of 2^n in (2.7.7) can be done simply by deleting the leading bit (cf. the second equality in (2.7.7)). In addition,

$$2^{n-1} < z + w'' < 2^n \quad \text{holds if } z < w \qquad (2.7.9)$$

(Exercise 2.7.2), i.e., $z + w''$ encodes the negative integer $z - w$ as its two's complement (cf. the last equality in (2.7.7)).

Adding two normalized floating point numbers $z_1 = \pm m_1 2^{e_1}$, $z_2 = \pm m_2 2^{e_2}$ is realized in essentially four steps:

- check, whether e_1, e_2 differ, and determine the larger one in this case,
- if $e_1 \neq e_2$, adapt the smaller exponent to the larger one by shifting the comma in the mantissa appropriately to the left,
- add the modified mantissas bitwise,
- renormalize the sum.

It is obvious that the last step in this procedure can cause errors even if the first three steps can be done accurately, for instance by using a sufficiently long accumulator. So, adding the representable floating point numbers $\tilde{r}_1 = 1.0$ and $\tilde{r}_2 = 1.0 \cdot 2^{-\ell'-2}$ and normalizing either by chopping or by rounding to nearest results in $\tilde{r}_1 + \tilde{r}_2 \doteq \tilde{r}_1$.

Such rounding errors combined with conversion errors yield the unexpected result

$$((0.1 + 0.1) + 0.1) - 0.3 \doteq 5.551115123125783 \cdot 10^{-17} \qquad (2.7.10)$$

in MATLAB which computes in the format 'double' by default. The result in (2.7.10) is represented by the hexadecimal code $(3C90000000000000)_{16}$ which encodes $2^{-54} \doteq 5.551115123125783 \cdot 10^{-17}$; cf. Exercise 2.7.1.

The division $1/r$ can be realized by comparison and subtraction although the result can be received faster by Newton's method applied to the function $f(x) = 1/x - r$. This leads to the iteration

$$x_{k+1} = (2 - rx_k)x_k, \quad k = 0, 1, \ldots, \tag{2.7.11}$$

which contains no division and which converges very fast if the initial approximation x_0 is sufficiently good; cf. Exercise 2.7.3.

A third kind of rounding error results from the computation of elementary function values. If for instance $\sin(r)$, $r \notin [-\pi/2, \pi/2]$, is required, the system first reduces the argument r by a multiple of π in order to end up with an argument $r - k\pi \in I = [-\pi/2, \pi/2]$. Then it computes $\sin(r - k\pi)$, mostly in a way unknown to the user. Even if r is a machine number and if the latter action could be done with an error of 1 ulp (:= unit of the last place), the reduction to I causes rounding errors since π is not a machine number. For details see Muller [240], for example. Safety bit, guard bit, and a tricky algorithm are necessary in order to get a sufficiently good approximation of $\sin r$. Volder's CORDIC algorithm in Volder [357] and in Appendix E may represent an example of how to compute elementary functions iteratively with arguments in a reference interval like I.

Computing with intervals on a computer must satisfy a simple *paradigm*:

All roundings must be such that the resulting theoretical interval is contained in the corresponding machine interval.

This already starts with the conversion of real numbers and implies, for instance, that the decimal 0.1 must be enclosed by a machine interval, i.e., by an interval whose bounds are machine numbers. Generally, this can be achieved by using any directed roundings, for instance by $[fl_\nabla(\underline{a}), fl_\Delta(\overline{a})]$ for $\underline{a}, \overline{a} \in \mathbb{R}_B$, $\underline{a} \le \overline{a}$, fl_∇, fl_Δ being the downward resp. upward rounding in (iv), resp. (iii) above. Define $\mathbb{IR}_B$ as set of all intervals $[a] = [\underline{a}, \overline{a}]$ with $\underline{a}, \overline{a} \in \mathbb{R}_B$ and let $\mathbb{IR}_M$ be the set of all machine intervals. Furthermore, let $fl_\uparrow$ denote any monotone upward directed rounding on $\mathbb{R}_B$ and define $fl_\downarrow$ on $\mathbb{R}_B$ by $fl_\downarrow(r) := -fl_\uparrow(-r) \in \mathbb{R}_M$ and $fl_\diamond$ on $\mathbb{IR}_B$ by

$$fl_\diamond([a]) := [fl_\downarrow(\underline{a}), fl_\uparrow(\overline{a})] \in \mathbb{IR}_M. \tag{2.7.12}$$

Then $fl_\diamond$ fulfills the following properties for $[a], [b] \in \mathbb{IR}_B$:

(i) $fl_\diamond([a]) = [a]$ for all $[a] \in \mathbb{IR}_M$,

(ii) $[a] \subseteq [b]$ implies $fl_\diamond([a]) \subseteq fl_\diamond([b])$,

(iii) $[a] \subseteq fl_\diamond([a])$,

(iv) $fl_\diamond(-[a]) = -fl_\diamond([a])$.

Therefore, $fl_\diamond$ is a monotone, directed and antisymmetric rounding on $\mathbb{IR}_B$, where the partial ordering is '$\subseteq$'. Because of (iii) it is called an outward rounding.

If $\uparrow = \triangle$, then $fl_\diamond$ represents that machine interval which encloses $[a]$ the closest. In this case, $fl_\diamond$ is called an optimal outward rounding. Optimality may be lost if arbitrary monotone directed roundings are chosen to define $fl_\diamond$.

For $\circ \in \{+, -, \cdot, \backslash\}$, $[a], [b] \in \mathbb{R}_M$, $\varphi \in \mathbb{F}$, $fl_\diamond$ as in (2.7.12) we define a machine interval arithmetic by

$$[a] \circledcirc [b] := fl_\diamond([a] \circ [b]), \quad \text{if } [a] \circ [b] \in \mathbb{IR}_B, \tag{2.7.13}$$

$$\varphi_\diamond([a]) := fl_\diamond(\varphi([a]), \quad \text{if } \varphi([a]) \in \mathbb{IR}_B. \tag{2.7.14}$$

The following obvious properties hold.

Theorem 2.7.1. *Let* $[a], [a]', [b], [b]' \in \mathbb{IR}_B$. *Then the arithmetic in* (2.7.13), (2.7.14) *satisfies*

(a) $[a] \circ [b] \subseteq [a] \circledcirc [b], \quad \varphi([a]) \subseteq \varphi_\diamond([a])$.

(b) $[a] \subseteq [b]$, $[a]' \subseteq [b]'$ *implies* $[a] \circledcirc [a]' \subseteq [b] \circledcirc [b]'$.

Directed roundings are part of the IEEE Standard 754. They are used in Rump's MATLAB tool INTLAB (cf. Rump [320] and Appendix G) in order to provide an efficient interval arithmetic which follows the lines above and in this way realizes the paradigm on page 104.

In some of our computations we need a storage of floating point numbers z in double length or even longer. Such multiple precision numbers z can be represented as a sum

$$z = z_1 + \cdots + z_k. \tag{2.7.15}$$

Here z_i are floating point numbers of the MATLAB format `double`. Usually the very long mantissa of z is decomposed into appropriate pieces which form the mantissas of the numbers z_i. Such a decomposition of z is often nonoverlapping, but this is not a must. The individual z_i can be stored as cells `z{i}` of a MATLAB cell array. In the literature the decomposition and storage can be found in Stetter [346] and is denoted there as staggered correction format; see also Auzinger, Stetter [58] and Lohner [194]. In order to benefit from (2.7.15) a processor with a long accumulator is necessary, or at least an appropriate software handling for (2.7.15); cf. Kulisch [182], Kulisch, Miranker [183, 184], and Rump [320].

Exercises

Ex. 2.7.1.

(a) Convert 203474 to binary using two different ways.

(b) Convert the decimals 0.1 and 0.3 to binary and verify (2.7.10); use rounding to nearest already for the conversion and compute $fl_n(fl_n(fl_n(fl_n(0.1) + fl_n(0.1)) + fl_n(0.1)) - fl_n(0.3))$.

Ex. 2.7.2. Prove (2.7.8) and (2.7.9).

Ex. 2.7.3. Derive the iteration (2.7.11). Prove that this iteration converges if and only if $0 < rx_0 < 2$ holds, i.e., if the starting point x_0 satisfies $|\frac{1}{r} - x_0| < \frac{1}{|r|}$. This is certainly fulfilled if x_0 is chosen between zero and $1/r$. Show that for $r = m \cdot 2^e$ with normalized mantissa m and nonnegative exponent e the inequality

$$0 < 2^{-(e+1)} < \frac{1}{r} < 2^{-e}$$

holds so that $x_0 = 2^{-(e+1)}$ is a starting point (not necessarily a good one) which implies convergence of the Newton sequence (2.7.11).

Hint: Prove first $x_k - \frac{1}{r} = -r(x_{k-1} - \frac{1}{r})^2 = -\frac{1}{r}(rx_{k-1} - 1)^2$.

Ex. 2.7.4. Compute $\sin(2^k \cdot \pi)$ in MATLAB for several integer values of k, for instance for $k = 1, 10, 50, 100$, and compare the computed result with the exact one which is zero.

Ex. 2.7.5. Evaluate the expression $9x^4 - y^4 + 2y^2$ for $x = 10864$ and $y = 18817$ using your pocket calculator, MATLAB, and MAPLE. Repeat your trial in MAPLE with the decimal numbers $x = 10864.0$ and $y = 18817.0$ which should yield the same result as the corresponding integer values.

Repeat your calculations with the same numbers substituted in the equivalent expression $(3x^2 - y^2 + 1)(3x^2 + y^2 - 1) + 1$. (The exact result is 1.)

Ex. 2.7.6. Show by the example $\underline{a} = 7.16$, $\overline{a} = 7.18$ that in a *normalized* decimal floating point system with two decimals after the decimal point and two decimals for the exponent (plus storage for its sign) the midpoint $\check{a}$ of $[a] = [\underline{a}, \overline{a}]$ is *not* contained in $[a]$ if computed via $\check{a} = (\underline{a} + \overline{a})/2$. What happens if $\check{a}$ is computed via $\check{a} = \underline{a} + (\overline{a} - \underline{a})/2$? What can be said on this midpoint problem if one works in a normalized *binary* system with rounding by chopping or rounding to nearest?

Notes to Chapter 2

To 2.2: Real interval arithmetic and basic properties can be derived from an arithmetic for general sets of numbers in Young [366]. Intervals in the context of error control in numerical analysis were also considered in Dwyer [88]. Interval arithmetic, properties, and midpoint-radius form of intervals can be found in Sunaga [351]. Moore used intervals and interval arithmetic in applications; see Moore [230, 231, 232].

To 2.3: The cases in which distributivity holds in interval arithmetic were completely discussed in Ratschek [282] and Spaniol [345]. For extensions see Popova [275]. For a generalization of (2.3.6) to a 'linear' interval equation with several variables see Ratschek, Sauer [288].

To 2.4: The mignitude and its rules originate from Neumaier [254], the definition of r_a^-, r_a^+ and properties with them are contained in Mayer [218]. For the width of interval products see also Ris [294, 295], and Ratschek, Rokne [286]. The formula in Exercise 2.4.1 (b) on the diameter is from Kreß [176], that from part (a) seems to be new. Most of the remaining facts – also of the following sections – can be found in Kulisch [180], Alefeld, Herzberger [26], and Neumaier [257].

To 2.5: Neumaier's β-function as well as its properties in Theorem 2.5.9 are repeated from Neumaier [257].

To 2.7: A theoretical and practical concept of rounding is presented in Kulisch [181], Kulisch, Miranker [183], Kulisch [182], where the latter also contains a huge bibliography on interval analysis (more than 660 publications). Details of the IEEE Standard 754 are explained in Bohlender, Ullrich [70]. Properties and applications can also be found in Rump [322].

3 Interval vectors, interval matrices

3.1 Basics

Interval vectors and interval matrices are defined as vectors and matrices with interval entries. Thus

$$[x] = ([x]_i) = \begin{pmatrix} [x]_1 \\ \vdots \\ [x]_n \end{pmatrix}$$

is a real interval (column) vector with n components $[x]_i \in \mathbb{IR}$, $i = 1, \ldots, n$, and

$$[A] = ([a]_{ij}) = \begin{pmatrix} [a]_{11} & \ldots & [a]_{1n} \\ \vdots & & \vdots \\ [a]_{m1} & \ldots & [a]_{mn} \end{pmatrix}$$

is a real $m \times n$ interval matrix with entries $[a]_{ij} \in \mathbb{IR}$, $i = 1, \ldots, m$, $j = 1, \ldots, n$. The set of all real interval vectors with n components is denoted by $\mathbb{IR}^n$, that of all real $m \times n$ interval matrices by $\mathbb{IR}^{m\times n}$. Analogously to Chapter 2 we use the notation $\mathbb{I}(S)$ for $S \subseteq \mathbb{R}^n$ or $S \subseteq \mathbb{R}^{m\times n}$.

We call n and $m \times n$ the dimension of $[x]$ and $[A]$, respectively, noting that this terminology only refers to the form of $[x]$ and $[A]$ and not to the dimension of some vector space. As usual we drop the specification 'real', since in nearly all chapters of this book we consider only real intervals and, correspondingly, real interval vectors and real interval matrices. We also identify $\mathbb{IR}^1$ and $\mathbb{IR}^{1\times 1}$ with $\mathbb{IR}$, and $\mathbb{IR}^{m\times 1}$ with $\mathbb{IR}^m$. Therefore, we no longer mention definitions or properties for interval vectors if they are stated and proved for general interval matrices.

We define $\tilde{A} = (\tilde{a}_{ij}) \in [A] = ([a]_{ij}) \in \mathbb{IR}^{m\times n}$ if $\tilde{a}_{ij} \in [\tilde{a}]_{ij}$ holds for $i = 1, \ldots, m$, $j = 1, \ldots, n$. Thus with $[a]_{ij} = [\underline{a}_{ij}, \overline{a}_{ij}]$, with $\underline{A} = (\underline{a}_{ij})$, $\overline{A} = (\overline{a}_{ij}) \in \mathbb{R}^{m\times n}$, and with

$$[\underline{A}, \overline{A}] = \{\tilde{A} \in \mathbb{R}^{m\times n} \mid \underline{A} \le \tilde{A} \le \overline{A}\}$$

we have $[A] = [\underline{A}, \overline{A}]$, i.e., an interval matrix is at the same time a matrix interval and vice versa.

For matrices $[A], [B] \in \mathbb{IR}^{m\times n}$ the relation $[A] \circ [B]$ for $\circ \in \{=, \subseteq, \supseteq, \cap, \cup, \le, <, \ge, >\}$ and the function value $h([A])$ for $h \in \{|\cdot|, \text{mid}, \text{rad}, d(\cdot)\}$ as well as $q([A], [B])$ are defined entrywise. Although these function values all are real $m \times n$ matrices, we keep the terminology of Chapter 2. For instance, we speak of the absolute value $|[A]|$ of $[A]$ instead of an absolute value *matrix*, and we denote the matrix $q([A], [B])$ as distance, keeping in mind that one can easily get a metric in the usual sense via $\|q([A], [B])\|$ with any monotone matrix norm $\|\cdot\|$ (i.e., $|A| \le |B|$, $A, B \in \mathbb{R}^{m\times n}$ implies $\|A\| \le \|B\|$). For the midpoint of $[A]$ we often write $\check{A} = (\check{a}_{ij})$ instead of mid$([A])$, and for the radius we sometimes use the symbol r_A instead of rad$([A])$. In addition, for $[A] \in \mathbb{IR}^{m\times n}$

DOI 10.1515/9783110499469-004

we define the real $m \times n$ matrices $r_A^- = (r_{a_{ij}}^-)$ and $r_A^+ = (r_{a_{ij}}^+)$, and as in MATLAB we use the symbol $.*$ for the Hadamard product, i.e., the entrywise product $[A] .* [B] = ([a]_{ij} \cdot [b]_{ij})$ for matrices $[A]$, $[B]$ of the same dimension.

Notice that the entrywise definition of $=, \subseteq, \supseteq, \cap, \cup$ for interval matrices is equivalent to the usual set-theoretic meaning of these symbols. Only in this latter sense, i.e., not by an entrywise definition, we will use the symbols $\neq, \subset, \not\subseteq, \supset, \not\supseteq$ for interval matrices. With the partial ordering for real matrices as introduced in Definition 1.9.1 (b) the interval hull $\square S$ of a bounded set $S \subseteq \mathbb{R}^{m\times n}$ is defined analogously to that of $S \subseteq \mathbb{R}$.

As with intervals we call an interval matrix $[A]$, written as $[A] = [\underline{A}, \overline{A}]$, degenerate or a point matrix if $\underline{A} = \overline{A} = \tilde{A}$ holds. Otherwise we call it nondegenerate. We identify point interval matrices with their element matrix, i.e., $\tilde{A} \equiv [A] = [\tilde{A}, \tilde{A}]$. In this way $\mathbb{R}^{m\times n}$ is embedded in the set $\mathbb{IR}^{m\times n}$ of interval matrices $[A]$. Again we will not distinguish between $\tilde{A}$ and $[\tilde{A}, \tilde{A}]$ in the sequel.

As indicated above, $[A] \leq [B]$ means $\underline{A} \leq \underline{B}$ and $\overline{A} \leq \overline{B}$. In particular, $O \leq [A]$ is equivalent to $O \leq \underline{A}$. For point matrices this partial ordering reduces to the natural partial ordering in $\mathbb{R}^{m\times n}$ as introduced in Definition 1.9.1 (b).

For matrices $[A] \in \mathbb{IR}^{n\times n}$ the so-called comparison matrix or Ostrowski matrix $\langle [A] \rangle = (c_{ij}) \in \mathbb{R}^{n\times n}$ is defined by

$$c_{ij} = \begin{cases} \langle [a]_{ii} \rangle, & \text{if } i = j, \\ -|[a]_{ij}|, & \text{if } i \neq j, \end{cases}$$

where we used the mignitude of Section 2.4 for the symbol and the definition of the diagonal entries.

Example 3.1.1. Let

$$[A] = \begin{pmatrix} [0,1] & [-1,5] \\ [3,7] & [-4,-2] \end{pmatrix}, \quad [B] = \begin{pmatrix} [3,6] & [1,2] \\ [-4,-3] & -9 \end{pmatrix}.$$

Then

$$|[A]| = \begin{pmatrix} 1 & 5 \\ 7 & 4 \end{pmatrix}, \quad \langle [A] \rangle = \begin{pmatrix} 0 & -5 \\ -7 & 2 \end{pmatrix}, \quad \check{A} = \begin{pmatrix} 1/2 & 2 \\ 5 & -3 \end{pmatrix},$$

$$d([A]) = \begin{pmatrix} 1 & 6 \\ 4 & 2 \end{pmatrix}, \quad q([A],[B]) = \begin{pmatrix} 5 & 3 \\ 10 & 7 \end{pmatrix}.$$

The transpose $[A]^T$ of $[A] = ([a]_{ij}) \in \mathbb{IR}^{m\times n}$ is defined as usual by $[A]^T = ([a]_{ji}) \in \mathbb{IR}^{n\times m}$. If $m = n$ and $[A] = [A]^T$ we call $[A]$ symmetric while we use the terminology zero-symmetric for $[A] \in \mathbb{IR}^{m\times n}$ if $\check{A} = O$.

The definition of the arithmetic for interval matrices follows exactly the lines of traditional arithmetic, i.e.,

$$[A] + [B] = ([a]_{ij} + [b]_{ij}), \quad [A] - [B] = ([a]_{ij} - [b]_{ij}), \quad [c] \cdot [A] = ([c][a]_{ij}) \tag{3.1.1}$$

for matrices $[A], [B] \in \mathbb{IR}^{m\times n}$ and any constant $[c] \in \mathbb{IR}$. Moreover,

$$[A]\cdot[B] = \left(\sum_{k=1}^{\ell} [a]_{ik}[b]_{kj}\right) \in \mathbb{IR}^{m\times n}$$

for matrices $[A] \in \mathbb{IR}^{m\times \ell}$, $[B] \in \mathbb{IR}^{\ell\times n}$.

We start with some properties which can easily be proved entrywise by means of the corresponding ones in Chapter 2.

Theorem 3.1.2. *For $\circ \in \{+,-,\cdot\}$ and matrices of the appropriate dimensions the following properties hold.*

(a) $\{A \circ B \mid A \in [A],\ B \in [B]\} \subseteq [A] \circ [B]$ *inclusion property*
with equality in the case of $\circ \in \{+,-\}$

(b) $[A]' \subseteq [A]$, $[B]' \subseteq [B]$ *implies* $[A]' \circ [B]' \subseteq [A] \circ [B]$ *inclusion monotony*

(c) $[A] + [B] = [B] + [A]$ *commutative law*

(d) $[A] + ([B] + [C]) = ([A] + [B]) + [C]$ *associative law*

(e) $[A] + O = [A]$, $[A] \cdot I = [A]$ *neutral elements*

(f) $\left.\begin{array}{l} [A]([B]C) \subseteq ([A][B])C \\ (A[B])[C] \subseteq A([B][C]) \\ (A[B])C = A([B]C) \end{array}\right\}$ *for point matrices A, C*

(g) $\left.\begin{array}{l} [A]([B]+[C]) \subseteq [A][B] + [A][C] \\ ([B]+[C])[A] \subseteq [B][A] + [C][A] \end{array}\right\}$ *subdistributive law*
with equality in the following cases
(i) $[A] \equiv A$
(ii) $O \le \underline{B}$ *and* $O \le \underline{C}$
(iii) $O \ge \overline{B}$ *and* $O \ge \overline{C}$.

Equality in Theorem 3.1.2 (a) can be easily seen for $\circ \in \{+,-\}$ by constructing matrices $\tilde{A} \in [A]$, $\tilde{B} \in [B]$ entrywise such that $\tilde{A} + \tilde{B} = \tilde{C}$ for any given matrix $\tilde{C} \in [A] + [B]$. Alternatively, use Theorem 4.1.5.

Our next example shows that strict enclosure can hold in Theorem 3.1.2 (a) for multiplication, and that the multiplication of matrices is not even potential associative.

Example 3.1.3. Let

$$[A] = \begin{pmatrix} 1 & [0,1] \\ 1 & -1 \end{pmatrix}.$$

Then

$$[A]\cdot[A] = \begin{pmatrix} [1,2] & [-1,1] \\ 0 & [1,2] \end{pmatrix},$$

hence

$$[A]\cdot([A]\cdot[A]) = \begin{pmatrix} [1,2] & [-1,3] \\ [1,2] & [-3,0] \end{pmatrix} \left\{\begin{matrix} \neq \\ \not\subseteq \\ \not\supseteq \end{matrix}\right\} ([A]\cdot[A])\cdot[A] = \begin{pmatrix} [0,3] & [-1,3] \\ [1,2] & [-2,-1] \end{pmatrix}.$$

Moreover, $\tilde{A} \in [A]$ implies

$$\tilde{A} = \begin{pmatrix} 1 & \alpha \\ 1 & -1 \end{pmatrix} \quad \text{for some } \alpha \in [0,1].$$

Therefore, $\tilde{A}\cdot\tilde{A} = (1+\alpha)I$, and

$$R = \{\tilde{A}\cdot\tilde{A} \mid \tilde{A} \in [A]\} = \{\beta I \mid \beta \in [1,2]\} \subset [A]\cdot[A]. \tag{3.1.2}$$

Notice that by the last definition in (3.1.1) the matrix $[1,2]I$ means the interval matrix

$$\begin{pmatrix} [1,2] & 0 \\ 0 & [1,2] \end{pmatrix},$$

which represents all diagonal matrices in which the entry $a_{11} \in [1,2]$ may vary *independently* of $a_{22} \in [1,2]$. Obviously, the range R in (3.1.2) is not representable here by an interval matrix.

We call an interval matrix $[A] \in \mathbb{IR}^{n\times n}$ singular if it contains at least one singular matrix $\tilde{A} \in \mathbb{R}^{n\times n}$. Otherwise we call it regular. We use the terminology strongly regular if $\check{A}^{-1}[A]$ is regular.

As in the one-dimensional case, the inverse $[A]^{-1}$ of a regular interval matrix $[A] \in \mathbb{IR}^{n\times n}$ cannot exist if $[A]$ has at least one nondegenerate entry. Therefore, we define $[A]^{-1}$ by means of the interval hull

$$[A]^{-1} = \square\{\tilde{A}^{-1} \mid \tilde{A} \in [A]\} \tag{3.1.3}$$

for regular matrices $[A] \in \mathbb{IR}^{n\times n}$ keeping in mind that $[A]\cdot[A]^{-1} = I$ does not hold if $d([A]) \neq O$.

It is generally not possible to represent $[A]^{-1}$ by means of the inverses of two particular matrices $\tilde{A} \in [A]$. This can be seen from the following example.

Example 3.1.4. Let

$$[A] = \begin{pmatrix} [1,2] & 0 \\ [-2,2] & [-2,-1] \end{pmatrix} \quad \text{and} \quad \tilde{A} = \begin{pmatrix} a & 0 \\ b & c \end{pmatrix} \in [A].$$

Then

$$\tilde{A}^{-1} = \begin{pmatrix} 1/a & 0 \\ -b/(ac) & 1/c \end{pmatrix},$$

whence, by virtue of (3.1.3),

$$[A]^{-1} = \begin{pmatrix} 1/[a]_{11} & 0 \\ -[a]_{21}/([a]_{11}[a]_{22}) & 1/[a]_{22} \end{pmatrix} = \begin{pmatrix} [1/2, 1] & 0 \\ [-2, 2] & [-1, -1/2] \end{pmatrix}.$$

The lower bound $\underline{C}$ of $[A]^{-1}$ is

$$\underline{C} = \begin{pmatrix} 1/2 & 0 \\ -2 & -1 \end{pmatrix} \quad \text{with } \underline{C}^{-1} = \begin{pmatrix} 2 & 0 \\ -4 & -1 \end{pmatrix} \notin [A].$$

Therefore, there is no matrix $\tilde{A} \in [A]$ such that $\tilde{A}^{-1} = \underline{C}$.

For some classes of matrices one can find $[A]^{-1}$ very easily. For instance, if $[A]$ is a regular diagonal matrix, then clearly

$$[A]^{-1} = \operatorname{diag}(1/[a]_{11}, \ldots, 1/[a]_{nn}).$$

Other classes can be found in Section 3.3.

We continue with some frequently used properties for the auxiliary functions $|\cdot|$, $\langle\cdot\rangle$, $\operatorname{rad}(\cdot)$, etc. with matrix arguments.

Theorem 3.1.5. *Let $[A], [B], [C]$ be interval matrices of appropriate dimensions and define the matrices $S_{\check{A}} = (\operatorname{sign}(\check{a}_{ij}))$, $S_{\check{B}} = (\operatorname{sign}(\check{b}_{ij}))$. Then the following properties hold.*

(a) $|[A] + [B]| \le |[A]| + |[B]|$,

(b) $|[A] \cdot [B]| \le |[A]| \cdot |[B]|$,

(c) $\operatorname{mid}([A][B]) = \check{A}\check{B} + (S_{\check{A}} .* r_A^-)(S_{\check{B}} .* r_B^-)$ *if* $0 \notin \operatorname{int}([a]_{ik} \cap [b]_{kj})$ *for all indices* i, j, k,

(d) $\operatorname{rad}([A] \pm [B]) = r_A + r_B$,

(e) $\operatorname{rad}(A[B]) = |A| r_B$, $\operatorname{rad}([A]B) = r_A |B|$,

(f) $\operatorname{rad}([A][B]) \le r_A^+ r_B + r_A r_B^+ \le |[A]| r_B + r_A |[B]|$
with '=' in the first inequality if and only if $0 \notin \operatorname{int}([a]_{ik} \cap [b]_{kj})$ *for all indices* i, j, k, *and with strict inequality for those entries* $(\operatorname{rad}([A][B]))_{ij}$ *for which* $0 \in \operatorname{int}([a]_{ik} \cap [b]_{kj})$ *occurs for at least one index* k,

(g) $|[A]| r_B \le \operatorname{rad}([A][B]) \le |[A]| r_B + r_A |\check{B}| \le |[A]| r_B + r_A |[B]|$,

(h) $r_A |[B]| \le \operatorname{rad}([A][B]) \le r_A |[B]| + |\check{A}| r_B \le |[A]| r_B + r_A |[B]|$,

(i) $O \in [A]$ *and* $O \in [B]$ *implies* $\operatorname{rad}([A][B]) \le 2 r_A \cdot r_B$,

(j) $q([C][A], [C][B]) \le |[C]| q([A], [B])$, $\quad q([A][C], [B][C]) \le q([A], [B]) |[C]|$.

Proof. The proof can be easily achieved by means of the Theorems 2.4.2–2.4.7 and 2.5.7. As an example we prove here only (e) and (j) explicitly.

$$\text{(e)}\ (\operatorname{rad}(A[B]))_{ij} = \operatorname{rad}\left(\sum_{k=1}^{n} a_{ik}[b]_{kj} \right) = \sum_{k=1}^{n} \operatorname{rad}(a_{ik}[b]_{kj})$$

$$= \sum_{k=1}^{n} |a_{ik}| \operatorname{rad}([b]_{kj}) = (|A| \operatorname{rad}([B]))_{ij}.$$

(j)
$$\begin{aligned}(q([C][A],[C][B]))_{ij} &= q\left(\sum_{k=1}^{m}[c]_{ik}[a]_{kj}, \sum_{k=1}^{m}[c]_{ik}[b]_{kj}\right)\\ &\le \sum_{k=1}^{m} q([c]_{ik}[a]_{kj},[c]_{ik}[b]_{kj}) \le \sum_{k=1}^{m}|[c]_{ik}|\,q([a]_{kj},[b]_{kj})\\ &= (|[C]|q([A],[B]))_{ij}.\end{aligned}$$
□

Theorem 3.1.6. *Let $[b], [c] \in \mathbb{IR}^n$, $[A], [B] \in \mathbb{IR}^{n\times n}$ and denote the spectral radius by $\rho(\cdot)$.*

(a) *If* $\operatorname{rad}([b] + [A][c]) < \operatorname{rad}([c])$ *holds, then* $\rho(\tilde{A}) \le \rho(|\tilde{A}|) \le \rho(|[A]|) < 1$ *follows for all matrices* $\tilde{A} \in [A]$.
(b) *If* $\operatorname{rad}([b] + (I - [A][B])[c]) < \operatorname{rad}([c])$ *holds, then* $[A], [B]$ *are regular.*

Proof. (a) is an immediate consequence of the inequality

$$|\tilde{A}|\operatorname{rad}([c]) \le |[A]|\operatorname{rad}([c]) \le \operatorname{rad}([A][c]) + \operatorname{rad}([b]) = \operatorname{rad}([b] + [A][c]) < \operatorname{rad}([c])$$

and the Theorems 1.9.10 and 1.9.13.

In order to prove (b), choose $\tilde{A} \in [A]$, $\tilde{B} \in [B]$. Then (a) guarantees $\rho(I - \tilde{A}\tilde{B}) < 1$. Therefore, the Neumann series for $(I - \tilde{A}\tilde{B})$ converges and represents $(I - (I - \tilde{A}\tilde{B}))^{-1} = (\tilde{A}\tilde{B})^{-1}$. Hence $\tilde{A}, \tilde{B}$ are regular, and so are $[A]$ and $[B]$. □

3.2 Powers of interval matrices

Powers A^k of square matrices $A \in \mathbb{R}^{n\times n}$ occur, for instance, in the Neumann series $\sum_{k=0}^{\infty} A^k$ or when discussing stability of initial value problems

$$y'(t) = \{B + C(t)\}\,y(t) + \{b + c(t)\}, \quad t > t_0, \tag{3.2.1}$$

$$y(t_0) = y^0, \tag{3.2.2}$$

where $t, t_0 \in \mathbb{R}$, $b, c(t), y^0, y(t) \in \mathbb{R}^n$, $B, C(t) \in \mathbb{R}^{n\times n}$, C and c sufficiently smooth periodic functions with the same period $T > 0$. In the latter case the fundamental matrix $\Phi(t) \in \mathbb{R}^{n\times n}$ of the associated homogeneous system

$$y'(t) = \{B + C(t)\}\,y(t), \quad t \in \mathbb{R}, \tag{3.2.3}$$

with the initial value

$$\Phi(t_0) = I \tag{3.2.4}$$

satisfies

$$\Phi(t + T) = \Phi(t) \cdot \Phi(t_0 + T)$$

since both sides are (matrix) solutions of (3.2.3) and have the same initial value $\Phi(t_0 + T)$. An inductive argument leads to

$$\Phi(t + kT) = \Phi(t) \cdot \{\Phi(t_0 + T)\}^k, \quad t_0 \le t \le t_0 + T,\ k \in \mathbb{N}_0. \tag{3.2.5}$$

By virtue of (3.2.4) the solution y_h of (3.2.3) with initial value $y_h(t_0) = y^0$ can be written as $y_h(t) = \Phi(t)y^0$. Therefore, if y_h^ε denotes the solution of (3.2.3) with the initial value $y_h^\varepsilon(t_0) = y^0 + \varepsilon$, ($\varepsilon \in \mathbb{R}^n$) we get

$$y_h^\varepsilon(t) - y_h(t) = \Phi(t)\varepsilon$$

and finally

$$y_h^\varepsilon(t + kT) - y_h(t + kT) = \Phi(t) \cdot \{\Phi(t_0 + T)\}^k \varepsilon, \quad t_0 \le t \le t_0 + T, \; k \in \mathbb{N}_0. \tag{3.2.6}$$

This also holds if y_h^ε, y_h are replaced by the solutions of the corresponding inhomogeneous system (3.2.1) with the same initial values (Jordan, Smith [157], pp. 226–227). Therefore, the powers A^k of $A = \Phi(t_0 + T)$ determine whether the difference in (3.2.6) is bounded (i.e., the solution of (3.2.1), (3.2.2) is stable) and, in addition, tends to zero if t tends to infinity (which means that the solution of (3.2.1), (3.2.2) is asymptotically stable). The first case certainly occurs if A is semiconvergent, the second case is equivalent to A being convergent (cf. Definition 1.7.5). With an appropriate software like AWA (see Lohner [192], Rihm [293]) it is possible to enclose $\Phi(t_0 + T)$. Thus the powers of such enclosures can be used in order to derive sufficient criteria for stability, respectively asymptotic stability, of (3.2.1), (3.2.2). This was done in Cordes [77] and leads us to the study of powers $[A]^k$ of interval matrices $[A] \in \mathbb{IR}^{n\times n}$. By virtue of the missing power associativity of interval matrices – cf. Example 3.1.3 – we must first define what we mean by a power of $[A]$.

Definition 3.2.1. Let $[A] \in \mathbb{IR}^{n\times n}$.

(a) We define

$$[A]^0 = ([a]_{ij}^{(0)}) = I, \quad [A]^k = ([a]_{ij}^{(k)}) = [A]^{k-1} \cdot [A], \quad k \in \mathbb{N}. \tag{3.2.7}$$

(b) Analogously to Definition 1.7.4 we call $[A]$ semiconvergent if $[A]^\infty = \lim_{k\to\infty} [A]^k$ exists, and convergent if, in addition, $[A]^\infty = O$ holds.

One might ask why we did not define the powers of $[A]$ by

$$^0[A] = I, \quad ^k[A] = [A] \cdot {^{k-1}[A]}, \quad k \in \mathbb{N}, \tag{3.2.8}$$

i.e., by multiplication with $[A]$ from the left. We oriented on (3.2.5), (3.2.6) where one can think of multiplying $\Phi(t_0 + T)$ from the right in order to get the next section $t_0 + (k + 1)T \le t + (k + 1)T \le t_0 + (k + 2)T$, $t \in [t_0, t_0 + T]$. But this is not a must, of course. We leave it to the reader as Exercise 3.2.2 to show that $(^k[A])$ is convergent if and only if $(([A]^T)^k)$ has this property, and that $(^k[A])$ and $([A]^k)$ do not need to converge, respectively diverge, simultaneously.

Next we introduce (ir)reducibility for interval matrices similarly to singularity/regularity in Section 3.1. An interval matrix $[A] \in \mathbb{IR}^{n\times n}$ is denoted to be irreducible if it contains at least one irreducible matrix $\tilde{A} \in \mathbb{R}^{n\times n}$, and reducible otherwise. It is obvious that $[A]$ is (ir)reducible if and only if the real matrix $|[A]|$ has the same property.

If P is a permutation matrix such that $R_{|[A]|} = P|[A]|P^T$ is 'the' reducible normal form of $|[A]|$ according to (1.7.5), then we call $R_{[A]} = P[A]P^T$ the reducible normal form of the interval matrix $[A]$. Here we dropped brackets because of Theorem 3.1.2 (f).

Our first result handles convergence of nondegenerate irreducible interval matrices $[A]$.

Theorem 3.2.2. *Let $[A] \in \mathbb{IR}^{n\times n}$ be nondegenerate and irreducible. Then $[A]$ is convergent if and only if $\rho(|[A]|) < 1$.*

Proof. '⇒': Let $\rho = \rho(|[A]|)$, $u > 0$ be a Perron vector of $|[A]|$, and $d([a]_{i_0 j_0}) > 0$. Then the inequality

$$d([A]^{k+1})u \geq d([A])|[A]|^k u = \rho^k d([A])u$$

implies

$$\left(d([A]^{k+1})u\right)_{i_0} \geq \rho^k d([a]_{i_0 j_0})u_{j_0} > 0.$$

Since $[A]$ is convergent so is (ρ^k), which proves $\rho < 1$.

'⇐': follows immediately from $|[A]^k| \leq |[A]|^k$ and Theorem 1.7.6 applied to $|[A]|$. □

If one of the assumptions of Theorem 3.2.2 is dropped, only the if-part of the theorem remains true as can be seen from its proof and from the convergent matrix

$$[A] \equiv A = \begin{pmatrix} 1 & -1 \\ 1 & -1 \end{pmatrix}$$

with $[A]^2 = O$, $\rho(A) = 0$, $\rho(|[A]|) = 2 > 1$. Apparently, the missing nondegenerate intervals in $[A]$ cause the failure of Theorem 3.2.2. Therefore, we will now characterize those columns j of $[A]$ for which at least some power of $[A]$ exists which contains a nondegenerate entry somewhere in this same column j.

Definition 3.2.3. Let $[A] \in \mathbb{IR}^{n\times n}$. Then we define the column index set $C \subseteq \{1, \ldots, n\}$ by

$$C = \{\, j \mid \exists k_0, i_0 : \, d([a]_{i_0 j}^{(k_0)}) > 0 \,\}. \tag{3.2.9}$$

Since C is difficult to determine from this definition we present an equivalent formulation of C.

Lemma 3.2.4. *Let $[A] \in \mathbb{IR}^{n\times n}$, define C as in Definition 3.2.3, and denote by $G(|[A]|)$ the directed graph of $|[A]|$. Then $j \in C$ if and only if there is a path $(i_0, i_1, \ldots, i_s = j)$ in $G(|[A]|)$ with at least one edge (i_α, i_β) such that $d([a]_{i_\alpha i_\beta}) > 0$.*

Proof. '⇒': Assume that there are some columns $j \in C$ for which no path exists as described in the lemma. By definition there are positive integers i, k with $d([a]_{ij}^{(k)}) > 0$. Choose i, j in such a way that k is minimal. Then $d([a]_{lj}) = d([a]_{lj}^{(1)}) = 0$, $l = 1, \ldots, n$, since in the case $d([a]_{l_0 j}) > 0$ the path (l_0, j) contradicts the assumption. Therefore, we get $k > 1$ and

$$0 < d([a]_{ij}^{(k)}) \le \sum_{l=1}^{n} d([a]_{il}^{(k-1)})\,|[a]_{lj}| + \sum_{l=1}^{n} |[a]_{il}^{(k-1)}|\,d([a]_{lj})$$
$$= \sum_{l=1}^{n} d([a]_{il}^{(k-1)})\,|[a]_{lj}|.$$

There must be an s with $|[a]_{sj}| \ne 0$, $d([a]_{is}^{(k-1)}) > 0$, which proves $s \in C$. Since k was minimal there is a path as described in the lemma which ends at s. Taking into account $|[a]_{sj}| \ne 0$ this path can be lengthened by the edge (s, j) which contradicts our assumption.

'$\Leftarrow$': Let $(\alpha, \beta, i_2, \dots, i_s = j)$ be a path as described in the lemma. Without loss of generality assume $d([a]_{\alpha,\beta}) > 0$. Then we obtain

$$d([a]_{\alpha j}^{(s)}) = \sum_{l=1}^{n} d([a]_{\alpha l}^{(s-1)}[a]_{lj}) \ge d([a]_{\alpha,i_{s-1}}^{(s-1)}[a]_{i_{s-1},j})$$
$$\ge d([a]_{\alpha,i_{s-1}}^{(s-1)})|[a]_{i_{s-1},j}| \ge d([a]_{\alpha,i_{s-2}}^{(s-2)}) \cdot |[a]_{i_{s-2},i_{s-1}}| \cdot |[a]_{i_{s-1},j}|$$
$$\ge \cdots \ge d([a]_{\alpha\beta}) \cdot |[a]_{\beta,i_2}| \cdot |[a]_{i_2,i_3}| \cdots |[a]_{i_{s-1},j}| > 0$$

which completes the proof. □

By virtue of Lemma 3.2.4 the column index set C of a nondegenerate interval matrix $[A]$ with irreducible absolute value is the full index set, i.e., $C = \{1, \dots, n\}$.

Now we are ready to formulate a necessary and sufficient criterion for the convergence of general interval matrices $[A] \in \mathbb{IR}^{n\times n}$.

Theorem 3.2.5. *Let $[A] \in \mathbb{IR}^{n\times n}$ and $C \subseteq \{1, \dots, n\}$ as in Definition 3.2.3. Construct the matrix $B \in \mathbb{R}^{n\times n}$ by*

$$b_{ij} = \begin{cases} |[a]_{ij}|, & \text{if } j \in C, \\ \lfloor a \rfloor_{ij} \equiv a_{ij}, & \text{if } j \notin C. \end{cases} \tag{3.2.10}$$

Then $[A]$ is convergent if and only if $\rho(B) < 1$.

Proof. If $[A]$ is irreducible and $d([A]) \ne O$, then Theorem 3.2.5 reduces completely to Theorem 3.2.2, hence it is proved in this case. If $[A]$ is a point matrix, then $[A] \equiv B$, and the assertion follows from Theorem 1.7.6. Now let $[A]$ be reducible. By virtue of Exercise 3.2.3 (b) we can assume that $[A]$ has reducible normal form, i.e., $[A]$ has the block form

$$[A] = \begin{pmatrix} [A]_{11} & O & \dots & O \\ [A]_{21} & [A]_{22} & \ddots & \vdots \\ \vdots & \ddots & \ddots & O \\ [A]_{s1} & \dots & [A]_{s,s-1} & [A]_{ss} \end{pmatrix} \tag{3.2.11}$$

where each diagonal submatrix $[A]_{ii}$ is either a 1×1 zero matrix or irreducible. Let $[A]_{ij}^{(k)}, B_{ij}, B_{ij}^{(k)}, I_{ij}$ be the corresponding blocks of $[A]^k, B, B^k$, and I, respectively.

'$\Rightarrow$': Let $[A]$ be convergent. Then the same holds for all diagonal blocks $[A]_{ii}$. Using Lemma 3.2.4 one sees that either all columns of $[A]_{ii}$ belong to C (which always refers to the whole matrix $[A]$) or none. In the first case for any column j of $[A]_{ii}$ there are integers m, p with $d([a]_{pj}^{(m)}) > 0$. From

$$d([a]_{pl}^{(m+k)}) \geq \{d([A]^m)\,|[A]|^k\}_{pl} \geq d([a]_{pj}^{(m)})\,(|[A]|^k)_{jl} \geq 0 \tag{3.2.12}$$

for all $k \in \mathbb{N}$ and all column indices l of $[A]_{ii}$ we get $\lim_{k\to\infty} B_{ii}^k = \lim_{k\to\infty} |[A]_{ii}|^k = O$, whence $\rho(B_{ii}) < 1$. In the second case we have $B_{ii} \equiv [A]_{ii}$. Hence $\rho(B_{ii}) < 1$ holds again because of $B_{ii}^k \equiv [A]_{ii}^k$. Since $\det(B - \lambda I) = \prod_{i=1}^s \det(B_{ii} - \lambda I_{ii})$ we obtain $\rho(B) < 1$.

'$\Leftarrow$': Let B be convergent. Then $B_{ii} = |[A]_{ii}|$ or $B_{ii} \equiv [A]_{ii}$ implies the convergence of $[A]_{ii}$, $i = 1, \ldots, s$. Assume that $[A]$ is not convergent. Then there is a block $[A]_{ij}$ with $i > j$ and maximal $j < s$ such that the matrix sequence $([A]_{ij}^{(k)})$ does not converge to the zero matrix. Let

$$[R] = \begin{pmatrix} [A]_{j+1,j} \\ \vdots \\ [A]_{ij} \\ \vdots \\ [A]_{sj} \end{pmatrix}, \qquad [Q] = \begin{pmatrix} [A]_{j+1,j+1} & O & \ldots & O \\ \vdots & \ddots & \ddots & \vdots \\ \vdots & & \ddots & O \\ [A]_{s,j+1} & \ldots & \ldots & [A]_{ss} \end{pmatrix}.$$

Let $[Q]^{(k)}, [R]^{(k)}$ be the submatrices of $[A]^k$ which correspond to $[Q]$ and $[R]$, respectively. Because of the extremal choice of j, the sequence $([Q]^{(k)}) = ([Q]^k)$ converges to the zero matrix. From $[R]^{(k+1)} = [R]^{(k)}[A]_{jj} + [Q]^k[R]$ and by induction we get

$$|[R]^{(k+1)}| \leq \sum_{l=0}^{k} |[Q]^l|\,|[R]|\,|[A]|_{jj}^{k-l}.$$

If $[A]_{jj}$ is a 1×1 zero matrix we obtain $|[R]^{(k+1)}| \leq |[Q]^k|\,|[R]|$ and the convergence of $([R]^{(k)})$ and $([A]_{ij}^{(k)})$ to zero matrices follows, contradicting the assumption. If $|[A]_{jj}|$ is irreducible and $j' \notin C$ for all columns j' of $[A]_{jj}$, then we will prove by induction

$$[A]_{ij}^{(k)} = B_{ij}^{(k)}, \quad k = 1, 2, \ldots, \tag{3.2.13}$$

which implies the contradiction $\lim_{k\to\infty} [A]_{ij}^{(k)} = O$.

The definition of B guarantees (3.2.13) for $k = 1$. Let (3.2.13) hold for some $k \in \mathbb{N}$ and let (i', j') be an arbitrary index pair which belongs to B_{ij}. Then we obtain

$$b_{i'j'}^{(k+1)} = \sum_{l \in C} b_{i'l}^{(k)} a_{lj'} + \sum_{l \notin C} b_{i'l}^{(k)} a_{lj'}. \tag{3.2.14}$$

All the entries $a_{lj'}$ of the first sum are zero because otherwise the contradiction $j' \in C$ can be easily shown using Lemma 3.2.4. Therefore, the summands $b_{i'l}^{(k)} a_{lj'}$ can be replaced by $[a]_{i'l}^{(k)} a_{lj'}$ in both sums without changing the value in (3.2.14). This terminates the induction for (3.2.13).

In order to finish the proof it remains to consider the case of $|[A]_{jj}|$ being irreducible with all columns in C. Let $u > 0$ be a Perron vector of $|[A]_{jj}|$. From the beginning of the proof we already know $\rho = \rho(|[A]_{jj}|) < 1$. Taking into account the convergence of $[Q]$, there is a vector $v > 0$ with $0 \le |[Q]^l|\,|[R]|u \le v$ for all $l = 0, 1, \ldots$. If $\varepsilon > 0$ is given, we can choose a positive integer m such that

$$\sum_{l=0}^{k-m} |[Q]^l|\,|[R]|\,|[A]_{jj}|^{k-l} u = \sum_{l=0}^{k-m} \rho^{k-l} |[Q]^l|\,|[R]|u \le \frac{\rho^m}{1-\rho} v < \varepsilon v$$

holds for all $k > m$. Choosing k so large that $|[Q]^l| \le \frac{\varepsilon}{m} ee^T$ holds for all $l > k - m$, we get

$$\begin{aligned} 0 \le |[R]^{(k+1)}|u &\le \sum_{l=0}^{k} |[Q]^l|\,|[R]|\,|[A]_{jj}|^{k-l} u \\ &\le \sum_{l=k-m+1}^{k} |[Q]^l|\,|[R]|u + \varepsilon v \le \varepsilon(ee^T|[R]|u + v). \end{aligned}$$

Since ε was arbitrary, $([R]^{(k)})$ converges to zero and the same is true for $([A]_{ij}^{(k)})$. □

Example 3.2.6. Let

$$[A] = \begin{pmatrix} 1 & -1 & 0 \\ 1 & -1 & 0 \\ 0 & 0 & [0, \frac{1}{2}] \end{pmatrix}.$$

Then

$$B = \begin{pmatrix} 1 & -1 & 0 \\ 1 & -1 & 0 \\ 0 & 0 & \frac{1}{2} \end{pmatrix} \quad \text{with } \rho(B) = \frac{1}{2} < 1.$$

Hence $[A]$ is convergent. In fact,

$$[A]^k = \frac{1}{2^k} \begin{pmatrix} 0 & 0 & 0 \\ 0 & 0 & 0 \\ 0 & 0 & [0, 1] \end{pmatrix} \qquad k = 2, 3, \ldots,$$

and $\rho(|[A]|) = 2$.

Without proof we cite two results from Mayer [199] on the semiconvergence of interval matrices with irreducible absolute value.

Theorem 3.2.7. *Let $[A] \in \mathbb{IR}^{n\times n}$ be nondegenerate and irreducible. Then $[A]$ is semiconvergent with limit $[A]^\infty \ne O$ if and only if the following two conditions hold.*

(i) *$|[A]|$ is semiconvergent with limit $|[A]|^\infty \ne O$ (which may differ from $|[A]^\infty|$.*

(ii) *If $[A]$ contains exactly one real matrix $\tilde{A}$ with $|\tilde{A}| = |[A]|$, then $\tilde{A} \ne -D|[A]|D$ where D is a signature matrix.*

In many examples the limit $[A]^\infty$ in Theorem 3.2.7 is zero-symmetric in the sense of Section 3.1. Therefore, it is interesting to know when this property does not occur.

Theorem 3.2.8. *Let $[A] \in \mathbb{IR}^{n\times n}$ be nondegenerate and irreducible. Assume that $[A]$ is semiconvergent with limit $[A]^\infty \neq O$. Then $[A]^\infty$ is not zero-symmetric if and only if $[A]$ contains exactly one matrix $\tilde{A}$ with $|\tilde{A}| = |[A]|$ and this matrix $\tilde{A}$ has, in addition, the representation $\tilde{A} = D|[A]|D$ with a signature matrix D.*

Exercises

Ex. 3.2.1. Show that Mathieu's differential equation

$$y''(t) + cy'(t) + \{a + b\cos(2t)\}\, y(t) = 0, \quad c > 0,\ a, b \in \mathbb{R}$$

is an example for (3.2.3).

Ex. 3.2.2. Let $[A] \in \mathbb{IR}^{n\times n}$.

(a) Show that ${}^\infty[A] = \lim_{k\to\infty} {}^k[A]$ exists if and only if $([A]^T)^\infty = \lim_{k\to\infty}([A]^T)^k$ exists and that ${}^\infty[A] = O$ and $([A]^T)^\infty = O$ hold simultaneously.

(b) Show that each matrix

$$\tilde{A} \in [A] = \begin{pmatrix} 1 & -1 & 0 \\ 1 & -1 & 0 \\ 0 & [0, \frac{1}{2}] & 0 \end{pmatrix}$$

has the spectral radius $\rho(\tilde{A}) = 0$.

(c) Show that $\lim_{k\to\infty}[A]^k$ does not exist while $\lim_{k\to\infty} {}^k[A]$ does.

Ex. 3.2.3. Let $[A], [B] \in \mathbb{IR}^{n\times n}$ and let $P \in \mathbb{R}^{n\times n}$ contain in each column and each row exactly one entry not equal to zero. (For instance, let P be a permutation matrix.) Prove the following equalities.

(a) $P([A][B]) = (P[A])[B]$; $[A]([B]P) = ([A][B])P$; $[A](P[B]) = ([A]P)[B]$.

(b) $P[A]^kP^{-1} = (P[A]P^{-1})^k$, where $P[A][B] := (P[A])[B]$.

Ex. 3.2.4. Check which of the following matrices are semiconvergent. If so, check whether their limit is zero-symmetric. If possible, compute this limit.

$$[A]_1 = \frac{1}{2}\begin{pmatrix} [0,1] & [0,1] \\ [0,1] & [0,1] \end{pmatrix}, \qquad [A]_2 = \frac{1}{2}\begin{pmatrix} [-1,0] & [0,1] \\ [0,1] & [-1,0] \end{pmatrix},$$

$$[A]_3 = \frac{1}{2}\begin{pmatrix} [0,1] & [-1,10] \\ [-1,0] & [0,1] \end{pmatrix}, \quad [A]_4 = \frac{1}{2}\begin{pmatrix} [0,1] & [0,1] \\ [0,1] & [-1,0] \end{pmatrix},$$

$$[A]_5 = \frac{1}{2}\begin{pmatrix} [-1,1] & [1,1] \\ [1,1] & [-1,-1] \end{pmatrix}.$$

Ex. 3.2.5. Show be means of the example

$$[A] \equiv \frac{1}{2}\begin{pmatrix} 1 & -1 \\ 1 & 1 \end{pmatrix}$$

that $\rho(|[A]^m|) < 1$ for some $m \in \mathbb{N}$ does not imply $\rho(|[A]|) < 1$.

3.3 Particular interval matrices

In this subsection we will extend the concept of particular matrices from Section 1.10 to interval matrices.

Definition 3.3.1. An interval matrix $[A] \in \mathbb{IR}^{n\times n}$ is called a diagonal matrix, an upper or lower triangular matrix, an M-matrix, an H-matrix, and an inverse positive matrix, respectively, if all element matrices $\tilde{A} \in [A]$ have the corresponding property.

By virtue of their definition, M-, H- and inverse positive matrices $[A] \in \mathbb{IR}^{n\times n}$ are regular interval matrices. For each of these classes of matrices we now list some crucial properties. We start with a criterion on inverse positive matrices which is due to Kuttler [185].

Theorem 3.3.2 (Inverse positive matrices; Kuttler's theorem). *Let $[A] \in \mathbb{IR}^{n\times n}$. Then $[A] \in \mathbb{IR}^{n\times n}$ is inverse positive if and only if $\underline{A}, \overline{A}$ are inverse positive. In this case we have $[A]^{-1} = [\overline{A}^{-1}, \underline{A}^{-1}]$, i.e., $\overline{A}^{-1} \le \tilde{A}^{-1} \le \underline{A}^{-1}$ for all matrices $\tilde{A} \in [A]$.*

Proof. If $[A]$ is inverse positive, then so are $\underline{A}$ and $\overline{A}$ by definition. For the converse assume that $\underline{A}^{-1} \ge O$, $\overline{A}^{-1} \ge O$ and let $0 < u \in \mathbb{R}^n$. Then $v = \underline{A}^{-1}u > 0$. Choose $\tilde{A} \in [A]$. Then $\underline{A} \le \tilde{A} \le \overline{A}$ implies

$$\overline{A}^{-1}\tilde{A} \le I \le \underline{A}^{-1}\tilde{A}. \tag{3.3.1}$$

From the first inequality we get $B = \overline{A}^{-1}\tilde{A} \in Z^{n\times n}$ and from $Bv = \overline{A}^{-1}\tilde{A}v \ge \overline{A}^{-1}\underline{A}v = \overline{A}^{-1}u > 0$ we see that B is an M-matrix. Therefore, B and $\tilde{A} = \overline{A}B$ are regular and $\tilde{A}^{-1} = B^{-1}\overline{A}^{-1} \ge O$. The final double inequality of the theorem follows directly from (3.3.1) multiplied by $\tilde{A}^{-1} \ge O$. □

Sometimes not $[A]$ but $D[A]D'$ must be tested for inverse positivity where D, D' are two signature matrices defined in Section 1.1. The following corollary is an immediate consequence of Kuttler's theorem.

Corollary 3.3.3. *Let D, D' be two signature matrices and $[A] \in \mathbb{IR}^{n\times n}$. Then $D[A]D'$ is inverse positive if and only if at least one of the two subsequent conditions* (i), (ii) *hold.*

(i) *$[A]$ is regular and $D'\tilde{A}^{-1}D \ge O$ for each $\tilde{A} \in [A]$.*
(ii) *$\check{A} - D\,\mathrm{rad}([A])D'$ and $\check{A} + D\,\mathrm{rad}([A])D'$ are regular and satisfy*

$$D'(\check{A} - D\,\mathrm{rad}([A])D')^{-1}D \ge O \quad \text{and} \quad D'(\check{A} + D\,\mathrm{rad}([A])D')^{-1}D \ge O.$$

Proof. The equivalence with (i) follows from $D[A]D' = \{D\tilde{A}D' \mid \tilde{A} \in [A]\}$ and $D^{-1} = D$, $D'^{-1} = D'$. The equivalence with (ii) follows with Kuttler's theorem from $\min(D[A]D') = \min\{D(\check{A} + [-\operatorname{rad}[A], \operatorname{rad}([A])])D'\} = D\check{A}D' - \operatorname{rad}([A]) = D(\check{A} - D\operatorname{rad}([A])D')D'$ and an analogous expression for $\max(D[A]D')$. □

We continue with properties of M- and H-matrices.

Theorem 3.3.4 (M-matrices). *Let* $[A] \in \mathbb{IR}^{n\times n}$.

(a) *The matrix* $[A] \in \mathbb{IR}^{n\times n}$ *is an* M*-matrix if and only if* $\underline{A}$ *is an* M*-matrix and* $\overline{A} \in Z^{n\times n}$. *In this case we have* $[A]^{-1} = [\overline{A}^{-1}, \underline{A}^{-1}]$, *i.e.,* $\overline{A}^{-1} \le \tilde{A}^{-1} \le \underline{A}^{-1}$ *for all matrices* $\tilde{A} \in [A]$.

(b) *A triangular matrix with* $\underline{a}_{ii} > 0$ *for* $i = 1, \dots, n$ *and* $\overline{a}_{ij} \le 0$ *for* $i \ne j$ *is an* M*-matrix.*

Proof. (a) If $[A]$ is an M-matrix, so are $\underline{A}$ and $\overline{A}$.

Assume conversely that $\underline{A}$ is an M-matrix and $\overline{A} \in Z^{n\times n}$. Then $\tilde{A} \in Z^{n\times n}$ for all matrices $\tilde{A} \in [A]$. Moreover, by virtue of Corollary 1.10.5 they are M-matrices, and so is $[A]$.

The representation of $[A]^{-1}$ follows from $\overline{A}^{-1} \le \tilde{A}^{-1} \le \underline{A}^{-1}$ for $\underline{A} \le \tilde{A} \le \overline{A}$ or from Theorem 3.3.2.

(b) follows immediately from Theorem 1.10.7 (d). □

Theorem 3.3.5 (H-matrices). *Let* $[A] \in \mathbb{IR}^{n\times n}$.

(a) *The matrix* $[A] \in \mathbb{IR}^{n\times n}$ *is an* H*-matrix if and only if* $\langle [A] \rangle$ *is an* M*-matrix.*

(b) *If* $[A]$ *is an* M*-matrix, then it is an* H*-matrix.*

(c) *If* $\langle [A] \rangle$ *is strictly diagonally dominant, then* $[A]$ *is an* H*-matrix.*

(d) *If* $\langle [A] \rangle$ *is irreducibly diagonally dominant, then* $[A]$ *is an* H*-matrix.*

(e) *If* $\langle [A] \rangle$ *is regular and diagonally dominant, then* $[A]$ *is an* H*-matrix.*

(f) *If* $[A]$ *is a triangular matrix with* $0 \notin [a]_{ii}$ *for* $i = 1, \dots, n$, *then it is an* H*-matrix.*

Proof. (a) Let $[A]$ be an H-matrix and choose $\tilde{A} \in [A]$ such that $\langle \tilde{A} \rangle = \langle [A] \rangle$. Since $\tilde{A}$ is an H-matrix, $\langle \tilde{A} \rangle$ is an M-matrix and so is $\langle [A] \rangle$.

Conversely, let $\langle [A] \rangle$ be an M-matrix and let $\tilde{A} \in [A]$. Then $\langle [A] \rangle \le \langle \tilde{A} \rangle \in Z^{n\times n}$, hence $\langle \tilde{A} \rangle$ is an M-matrix by virtue of Corollary 1.10.5. Therefore, $\tilde{A}$ is an H-matrix, and so is $[A]$.

(b)–(f) follow immediately from the Theorems 1.10.7 and 1.10.18. □

We next consider regular interval matrices $[A]$ for which the entry $(\tilde{A}^{-1})_{ij}$ of the inverse does not change its sign when $\tilde{A}$ varies in $[A]$ and i, j are chosen arbitrary but held fixed.

Definition 3.3.6. A regular interval matrix $[A] \in \mathbb{IR}^{n\times n}$ is called inverse stable if for each $i, j \in \{1, \dots, n\}$ either $(\tilde{A}^{-1})_{ij} \ge 0$ for each $\tilde{A} \in [A]$ or $(\tilde{A}^{-1})_{ij} \le 0$ for each $\tilde{A} \in [A]$.

Obviously, M-matrices and inverse positive matrices $[A] \in \mathbb{IR}^{n\times n}$ belong to the class of inverse stable matrices. A general sufficient criterion for inverse stability is given in the following theorem.

Theorem 3.3.7. *Let* $[A] \in \mathbb{IR}^{n\times n}$ *have a regular midpoint matrix* $\check{A}$ *which satisfies*

$$\rho(R) < 1, \quad \text{where } R = |\check{A}^{-1}|\,\mathrm{rad}([A]). \tag{3.3.2}$$

Then $[A]$ *is regular. Moreover, if*

$$Q|\check{A}^{-1}| \le |\check{A}^{-1}| \quad \text{with } Q = R(I-R)^{-1}, \tag{3.3.3}$$

then $[A]$ *is inverse stable.*

Proof. Choose $\tilde{A} = \check{A} - \Delta \in [A]$. Then $|\check{A}^{-1}\Delta| \le R$, whence $\rho(\check{A}^{-1}\Delta) < 1$. Therefore, the corresponding Neumann series converges, and one obtains $\sum_{k=0}^{\infty}(\check{A}^{-1}\Delta)^k\check{A}^{-1} = (I - \check{A}^{-1}\Delta)^{-1}\check{A}^{-1} = \tilde{A}^{-1}$. Hence $[A]$ is regular. Moreover,

$$\begin{aligned} |\tilde{A}^{-1} - \check{A}^{-1}| &= |((\check{A}-\Delta)^{-1}\check{A} - I)\check{A}^{-1}| \le |(I - \check{A}^{-1}\Delta)^{-1} - I| \cdot |\check{A}^{-1}| \\ &= \left|\sum_{k=1}^{\infty}(\check{A}^{-1}\Delta)^k\right| |\check{A}^{-1}| \le \sum_{k=1}^{\infty} R^k \cdot |\check{A}^{-1}| = Q|\check{A}^{-1}| \le |\check{A}^{-1}| \end{aligned}$$

from which one can immediately conclude inverse stability by inspecting the entries. □

Notes to Chapter 3

To 3.1: Interval vectors and basic operations with them are already contained in Sunaga [351]. For interval matrices see Moore [232] and Apostolatos, Kulisch [51].

To 3.2: Powers of interval matrices and convergence are first considered in Mayer [198]. The results in Section 3.2 are taken from there. Based on these results the convergence of the Neumann series with $[A] \in \mathbb{IR}^{n\times n}$ was completely discussed in Mayer [200]. Semiconvergence for interval matrices with irreducible absolute value was introduced in Mayer [199], necessary and sufficient results for the convergence as well as results on the shape of the limit are given there. The results were generalized to arbitrary matrices in Arndt, Mayer [55, 56], but a necessary and sufficient criterion for the semiconvergence of a general interval matrix $[A]$ is still missing. An alternative proof for Theorem 3.2.5 can be found in Pang, Lur, Guu [268]. Results on the stability of linear homogeneous systems of differential equations with periodic coefficients can be found in the literature as Floquet theory; cf. Yakubovich, Starzhinskii [365], for example. Applications of stability of such systems using interval analysis are given in Cordes [77].

To 3.3: Interval M-matrices are considered in Barth, Nuding [59], interval H-matrices and their properties are discussed in Neumaier [254]. They are already used implicitly in Alefeld [7].

4 Expressions, P-contraction, ε-inflation

4.1 Expressions, range

In Section 2.6 we defined the elementary interval functions by their range. This definition could also be applied for general continuous functions $f\colon D \subseteq \mathbb{R}^n \to \mathbb{R}$ since such functions assume their maximum and minimum on each compact set, particularly on each interval vector $[x] \subseteq D$. Unfortunately, in most cases neither these extrema nor their position can be easily determined. Therefore, one replaces these optimal definitions by a coarser one such that the resulting image for an input $[x]$ encloses at least the range $R_f([x])$ of f restricted to $[x] \subseteq D$. To this end we define a certain set $\mathbb{E}$ of admissible expressions and the interval arithmetic evaluation of their induced functions which all belong to the class of so-called factorable functions.

Definition 4.1.1. Let $\mathbb{F}_u$ denote the set of unary operations $+, -$, $\mathbb{F}_b$ the set of binary operations $+, -, \cdot, /$, and $\mathbb{F}$ the set of elementary functions as in Definition 2.6.1 (which can be extended if necessary). A function $f\colon D \subseteq \mathbb{R}^n \to \mathbb{R}$ with $x = (x_1, \dots, x_n)^T \mapsto f(x)$ is called factorable with respect to $\mathbb{F}_u \cup \mathbb{F}_b \cup \mathbb{F}$ if there exists a sequence $f^1, \dots, f^s$ of functions $f^j : D \to \mathbb{R}$, $j = 1, \dots, s$, such that the following properties hold.
(1) $f^k(x) = x_k$, $k = 1, \dots, n$,
(2) $f^k(x) = c_{k-n}$, $k = n+1, \dots, m$, where c_j denotes a constant,
(3) $f^k(x) = f^\ell(x) \circ f^r(x)$, $\circ \in \mathbb{F}_b$ or $f^k(x) = \pm f^\ell(x)$, or $f^k(x) = \varphi(f^\ell(x))$ with $\varphi \in \mathbb{F}$, and $\ell, r < k$, $k = m+1, \dots s$,
(4) $f(x) = f^s(x)$.
We call (f^k) a factor sequence for f.

Definition 4.1.2. We define $\mathbb{E}$ as the set of all scalar mathematical expressions which are built in finitely may steps by the vector variable $x = (x_i) \in \mathbb{R}^m$, real constants, the unary operations $+$, $-$, the binary operations $+, -, \cdot, /$, and the elementary functions $\varphi \in \mathbb{F}$. Let $f\colon D \subseteq \mathbb{R}^m \to \mathbb{R}^n$ be a function with $f(x) = (f_i(x))$ such that $f_i(x) \in \mathbb{E}$, $i = 1, \dots, n$. Replace each variable x_i by an interval $[x]_i$ with $[x] = ([x]_i) \subseteq D$ using the corresponding interval operations and elementary interval functions. If all interval operations are defined, we end up with an interval vector which we denote by $f([x])$ and which we call the interval arithmetic evaluation of f at $[x]$. If one varies $[x] \in \mathbb{IR}^m$, $[x] \subseteq D$, one obtains an interval function $[f]$ with expression $[f]([x]) = f([x])$, which we also denote as interval arithmetic evaluation of f. In the sequel we will mostly write $f([x])$ if we mean this particular interval function $[f]$.

Apparently $[f]$ from Definition 4.1.2 may differ if the form of the original expression $f(x)$ is changed equivalently. For instance the equivalent expressions $f(x) = 0$ and $f(x) = x - x$ yield different interval functions. Therefore, when speaking of the interval arithmetic evaluation of $f = (f_i)$ one always has a particular expression for

DOI 10.1515/9783110499469-005

$f_i(x) \in \mathbb{E}$ in mind which represents f_i. We assume this tacitly for f when we write $f([x])$. In addition, we assume that $f([x])$ exists whenever $f([x])$ is used in the sequel. We emphasize that $f([x])$ does not denote the range $R_f([x])$ of f over $[x]$ as is sometimes done in the literature. In fact we will see that only $R_f([x]) \subseteq f([x])$ can be guaranteed; cf. Theorem 4.1.4 and Example 4.1.3.

It is obvious that one can identify $\mathbb{E}$ with the set of all factor sequences $f^1, \ldots, f^s$ defined in Definition 4.1.1, but not with the set of all factorable functions f since f can be represented by more than one such sequence.

Composition of functions is contained in Definition 4.1.2 by using an appropriate formula tree.

Example 4.1.3.

(a) Let $f(x) = x - x$, hence $f([x]) = [x] - [x]$. Then $f([-1, 2]) = [-1, 2] - [-1, 2] = [-3, 3] \neq 0$. In contrast with this, the equivalent expression $f(x) \equiv 0$ yields $f([x]) \equiv 0 \equiv [0, 0] = R_f([x])$.

(b) The expression

$$f(x) = \frac{1}{x - x + 1}$$

is defined for all $x \in \mathbb{R}$, while

$$f([-1, 2]) = \frac{1}{[-1, 2] - [-1, 2] + 1} = \frac{1}{[-2, 4]}$$

is not defined because of $0 \in [-2, 4]$.

(c) Let

$$f(x) = \frac{x + x^3}{2 + x^2 - \sin x}.$$

Then

$$f([x]) = \frac{[x] + [x]^3}{2 + [x]^2 - \sin([x])},$$

whence

$$f([-2, 2]) = \frac{[-2, 2] + [-8, 8]}{2 + [0, 4] - [-1, 1]} = \frac{[-10, 10]}{[1, 7]} = [-10, 10].$$

By virtue of the definition of $+, -, \cdot, /$ and $\varphi \in \mathbb{F}$ for intervals, one immediately gets the following property.

Theorem 4.1.4 (Inclusion property, inclusion monotony). *Let $f : D \subseteq \mathbb{R}^m \to \mathbb{R}^n$ and $[x], [y] \subseteq D$. Then*

$$R_f([x]) \subseteq f([x]) \quad \textit{(inclusion property)} \tag{4.1.1}$$

holds. In addition,

$$[x] \subseteq [y] \textit{ implies } f([x]) \subseteq f([y]) \quad \textit{(inclusion isotony, or inclusion monotony).}$$

In particular, $f([x])$ exists in this case if $f([y])$ does.

Notice that a meaningful machine interval arithmetic must always take care that rounding is such that (4.1.1) remains true. Thus $f([x])$ should be computed with outward rounding for each operation.

Before defining interval functions with property (4.1.1) in a way different from interval arithmetic evaluations, we state and prove a simple but important result by Moore [232] which guarantees equality in (4.1.1).

Theorem 4.1.5. *Let $f: D \subseteq \mathbb{R}^m \to \mathbb{R}$ be defined by $f(x) \in \mathbb{E}$ such that each variable x_i of $x = (x_i)$ appears at most once in $f(x)$. Then $R_f([x]) = f([x])$ for all $[x] \in \mathbb{IR}^n$ for which $f([x])$ is defined.*

Proof. Due to (4.1.1) it remains to prove

$$R_f([x]) \supseteq f([x]). \tag{4.1.2}$$

To this end we will show by induction on the number of operations in $f(x)$ that there exist vectors $x^1, x^2 \in [x]$ such that

$$f([x]) = [f(x^1), f(x^2)]. \tag{4.1.3}$$

Since, by virtue of $f(x) \in \mathbb{E}$, the function f is continuous, it attains all values between $f(x^1)$ and $f(x^2)$ so that (4.1.2) then holds.

Let us now start the induction with $k = 0$. Then the following possibilities for $f(x)$ can occur:

$$f(x) \equiv c \in \mathbb{R}, \quad \text{whence } f([x]) = [c, c] = [f(\underline{x}), f(\overline{x})],$$

and

$$f(x) = x_i, \quad \text{whence } f([x]) = [x]_i = [\underline{x}_i, \overline{x}_i] = [f(\underline{x}), f(\overline{x})].$$

Assume now that (4.1.3) holds for all expressions in $\mathbb{E}$ which have at most k operations, where we count the application of an elementary function as an operation, too. Assume that the expression $f(x)$ has $k + 1$ operations.

Case 1: $f = \pm g$ with (4.1.3) for g. By the induction hypothesis there are vectors v^1, v^2 such that $g([x]) = [g(v^1), g(v^2)]$ whence $f([x]) = [f(v^1), f(v^2)]$ if $f = +g$ and $f([x]) = [-g(v^2), -g(v^1)] = [f(v^2), f(v^1)]$ if $f = -g$.

Case 2: $f = g \circ h$ with $\circ \in \{+, -, \cdot, /\}$ and with (4.1.3) for g, h. By the induction hypothesis we get

$$f([x]) = g([x]) \circ h([x]) = [g(v^1), g(v^2)] \circ [h(w^1), h(w^2)]$$

with appropriate vectors $v^i, w^j \in [x]$, $i, j = 1, 2$. Theorem 2.2.2 guarantees

$$\min f([x]) = g(v^{i_0}) \circ h(w^{j_0}) \quad \text{for appropriate indices } i_0, j_0 \in \{1, 2\}.$$

According to our assumptions, for each fixed $l \in \{1, \ldots, n\}$ at least one of the two functions g, h does not depend on x_l. Without loss of generality assume that the function

h does not depend on $x_1, \dots, x_\sigma$ ($\sigma \le n$) and the function g does not depend on $x_{\sigma+1}, \dots, x_n$.

Define $x^1 \in \mathbb{R}^n$ by

$$x_k^1 = \begin{cases} v_k^{i_0}, & k = 1, \dots, \sigma \\ w_k^{j_0}, & k = \sigma + 1, \dots, n. \end{cases}$$

Then

$$x^1 \in [x], \quad g(x^1) = g(v^{i_0}), \quad h(x^1) = h(w^{j_0}),$$

whence

$$\min f([x]) = g(x^1) \circ h(x^1) = f(x^1).$$

Analogously there is a vector $x^2 \in [x]$ such that $\max f([x]) = f(x^2)$ holds.

This implies (4.1.3).

Case 3: $f = \varphi(g)$ with $\varphi \in \mathbb{F}$ and with (4.1.3) for g. By virtue of the continuity of φ we get

$$\min f([x]) = \min \varphi(g([x])) = \varphi(\tilde{g})$$

with $\tilde{g} \in g([x]) = [g(v^1), g(v^2)]$ and appropriate vectors $v^1, v^2 \in [x]$, guaranteed by the induction hypothesis. Since g is continuous, there exists a vector $x^1 \in [x]$ such that $g(x^1) = \tilde{g}$, whence min $f([x]) = \varphi(g(x^1)) = f(x^1)$. Analogously there is a vector $x^2 \in [x]$ with $\max f([x]) = f(x^2)$. □

It is obvious that one can also prove Theorem 4.1.5 using the factor sequence of f and an induction on the length of such sequences. If in an expression $f(x) \in \mathbb{E}$ some variable x_i of $x = (x_i)$ appears more than once, then $R_f([x]) = f([x])$ need no longer hold because the interval arithmetic does not take into account the linkage of x_i in $f(x)$. In fact, for the interval arithmetic evaluation $f([x])$, each occurrence of x_i in the expression $f(x)$ is treated as if a new variable substitutes x_i, which varies in the same interval $[x]_i$ as x_i. This may lead to an overestimation of $f([x])$ over the range $R_f([x])$, which is sometimes complained of. In the case of such multiple occurrences of x_i, one often speaks of dependent variables or dependency in the variables.

Example 4.1.6. Let $f(x_1, x_2) = \frac{5+x_1}{6-x_2}$ for $(x_1, x_2) \in [1, 2] \times [-2, 2]$. Then

$$f([1,2],[-2,2]) = \frac{5 + [1,2]}{6 - [-2,2]} = \frac{[6,7]}{[4,8]} = \left[\frac{3}{4}, \frac{7}{4}\right] = R_f([1,2] \times [-2,2]),$$

where we applied Theorem 4.1.5 for the last equality.

Another possibility to enclose the range of a function f can be constructed by means of the following centered forms.

Definition 4.1.7 (Centered form). Let $f: D \subseteq \mathbb{R}^n \to \mathbb{R}$, $[x] \subseteq D$, $z \in D$. Assume that there is a vector $[s] \in \mathbb{IR}^n$ such that for all $x \in [x]$ there exists a value $s(x, z) \in [s]$, the so-called slope (between x and z), with

$$f(x) = f(z) + s(x,z)^T(x-z). \tag{4.1.4}$$

Then the right-hand side of (4.1.4) is called the centered form of $f(x)$ with center z. It induces an interval

$$[f]([x]) = f(z) + [s]^T \cdot ([x]-z). \tag{4.1.5}$$

For fixed z, variable $[x] \subseteq D$, and $[s] = [s]([x], z)$ with the property above we thus get an interval function $[f]\colon \mathbb{I}(D) \to \mathbb{IR}$.

Notice that we do not require $z \in [x]$ for general centered forms.

From (4.1.4) we can immediately deduce the analogue of property (4.1.1).

Theorem 4.1.8. *For the interval function $[f]$ in Definition 4.1.7 we get $R_f([x]) \subseteq [f]([x])$.*

Example 4.1.9. If $n = 1$ and $x \neq z$, then (4.1.4) necessarily leads to the slope

$$s(x,z) = \frac{f(x)-f(z)}{x-z}.$$

If $n = 2$, $x = (x_1, x_2)^T$, $f(x) = x_1 \cdot x_2$, and $z = (z_1, z_2)^T$ we have at the same time

$$f(z) + (x_2, z_1)(x-z) = z_1z_2 + x_2(x_1-z_1) + z_1(x_2-z_2) = x_1x_2 = f(x)$$

$$\text{with} \quad s(x,z) = \begin{pmatrix} x_2 \\ z_1 \end{pmatrix}$$

and

$$f(z) + (z_2, x_1)(x-z) = z_1z_2 + z_2(x_1-z_1) + x_1(x_2-z_2) = x_1x_2 = f(x)$$

$$\text{with} \quad s(x,z) = \begin{pmatrix} z_2 \\ x_1 \end{pmatrix}.$$

In particular, $s(x,z)$ is not unique.

A special case of a centered form is the following mean value form which is an immediate consequence of the mean value theorem.

Theorem 4.1.10 (Mean value form). *Let $D \subseteq \mathbb{R}^n$ be a region with $z \in [x] \subseteq D$, and let $f\colon D \to \mathbb{R}$ be continuously differentiable. Moreover, let $f'([x]) = (\operatorname{grad} f([x]))^T$ exist. Then the mean value form*

$$f(x) = f(z) + f'(\xi)(x-z), \quad \xi = z + \theta(x-z),\ \theta = \theta(x,z) \in (0,1) \text{ appropriately},$$

is a centered form which induces the interval

$$[f]_M([x]) = f(z) + f'([x])([x]-z).$$

Notice that we require $z \in [x]$ for the mean value form.

Our next example illustrates this centered form and shows that it delivers an enclosure of the range $R_f([x])$ which at least here is better than the interval arithmetic evaluation $f([x])$.

Example 4.1.11. Let $f(x) = x - x$ and $z \in [x] \in \mathbb{IR}$. Then $f'(x) \equiv 0$ and $[f]_M([x]) = f(z) + f'([x]) \cdot ([x] - z) = [0, 0] = R_f([x])$, while $f([x]) = [x] - [x] \supset R_f([x])$ for nondegenerate intervals $[x]$.

In order to assess the quality of enclosure of $R_f([x])$, we introduce the concept of the order of approximation.

Definition 4.1.12 (Order of approximation). Let $[x]^0 \subseteq D \subseteq \mathbb{R}^n$. If $f: D \to \mathbb{R}$ is continuous and if $[f]: \mathbb{I}([x]^0) \to \mathbb{IR}$ is an interval function which fulfills the conditions

$$R_f([x]) \subseteq [f]([x]), \tag{4.1.6}$$

$$q([f]([x]), R_f([x])) \leq c \, \|d([x])\|_\infty^m \tag{4.1.7}$$

for all $[x] \in \mathbb{I}([x]^0)$ with a constant c independent of $[x]$, then we say that $[f]([x])$ approximates the range $R_f([x])$ (on $\mathbb{I}([x]^0)$) with order m.

By virtue of the norm equivalence on $\mathbb{R}^n$ the maximum norm in Definition 4.1.12 can be replaced by any other norm on $\mathbb{R}^n$.

Our next theorem motivates this definition.

Theorem 4.1.13. *Let $f: D \subseteq \mathbb{R}^n \to \mathbb{R}$ be continuous, $[x]^0 \subseteq D$, $[f]: \mathbb{I}([x]^0) \to \mathbb{IR}$. Let $[f]([x])$ approximate $R_f([x])$ on $\mathbb{I}([x]^0)$ of order m. Decompose $[x]^0$ in k^n boxes such that every i-th edge of these boxes has length $\frac{d([x]_i^0)}{k}$. Denote by M_k the set of all these boxes and define*

$$[f]_k([x]^0) = \bigcup_{[x] \in M_k} [f]([x]).$$

Then the following properties hold.

(a) $R_f([x]^0) \subseteq [f]_k([x]^0) \in \mathbb{IR}$.

(b) $q([f]_k([x]^0), R_f([x]^0)) \leq \frac{\tilde{c}}{k^m}$, *($\tilde{c} \in \mathbb{R}$ independent of k).*

(c) $\lim_{k \to \infty} [f]_k([x]^0) = R_f([x]^0)$, *i.e., the finer the grid the better the approximation of* $R_f([x]^0)$.

Notice the appearance of the order m in part (b) of the theorem. Summarizing the theorem says that the finer the grid and the higher the order, the better the enclosure of $R_f([x]^0)$ by $[f]_k([x]^0)$.

Proof. (a) By virtue of $\bigcup_{[x] \in M_k} [x] = [x]^0$ and (4.1.6) we obtain $R_f([x]^0) \subseteq [f]_k([x]^0)$. If $[x]$, $[x]' \in M_k$ are neighbors, then $[x] \cap [x]' \neq \emptyset$, hence $[f]([x]) \cap [f]([x]') \neq \emptyset$ and $[f]([x]) \cup [f]([x]') \in \mathbb{IR}$. Now an inductive argument implies $[f]_k([x]^0) \in \mathbb{IR}$.

(b) Let $[f]_k([x]^0) = [\underline{f}_k, \overline{f}_k]$ and $R_f([x]^0) = [\underline{w}, \overline{w}]$. There is a vector $[x]^1 \in M_k$ with $\underline{f}_k = \min [f]([x]^1)$. Let $\underline{w}^1 = \min R_f([x]^1)$. Then by (a) we have $|\underline{w} - \underline{f}_k| = \underline{w} - \underline{f}_k \leq \underline{w}^1 - \underline{f}_k = |\underline{w}^1 - \underline{f}_k|$.

Analogously there exists a vector $[x]^2 \in M_k$ with $\overline{f}_k = \max [f]([x]^2)$ and $|\overline{w} - \overline{f}_k| \leq |\overline{w}^2 - \overline{f}_k|$, where $\overline{w}^2 = \max R_f([x]^2)$.

By means of (4.1.7) and of the assumption we finally get

$$\begin{aligned} q([f]_k([x]^0), R_f([x]^0)) &= \max\{\, |\underline{f}_k - \underline{w}|, |\overline{f}_k - \overline{w}|\,\} \\ &\le \max\{\, q([f]([x]^1), R_f([x]^1)),\; q([f]([x]^2), R_f([x]^2))\,\} \\ &\le \max\{\, c\,\|d([x]^1)\|_\infty^m,\; c\,\|d([x]^2)\|_\infty^m\,\} \le \frac{\tilde{c}}{k^m} \end{aligned}$$

with $\tilde{c} = c\,\|d([x]^0)\|_\infty^m$.

(c) follows from (b). □

Example 4.1.14. Let $f(x) = x - x$ and divide $[x]^0 = [-1, 1]$ into k equally spaced intervals $[x]^j = [-1 + \frac{2}{k} \cdot (j - 1), -1 + \frac{2}{k} \cdot j]$, $j = 1, \ldots, k$. Define $[f]$ on $\mathbb{I}([x]^0)$ by $[f]([x]) = f([x])$. Then $d([x]^j) = \frac{2}{k}$, $M_k = \{[x]^j \mid j = 1, \ldots, k\}$, $f([x]^j) = [-\frac{2}{k}, \frac{2}{k}]$, $[f]_k([x]^0) = \bigcup_{j=1}^k f([x]^j) = [-\frac{2}{k}, \frac{2}{k}]$, and $q([f]_k([x]^0), R_f([x]^0)) = q([f]_k([x]^0), [0, 0]) = \frac{2}{k}$ which tends to zero if k tends to infinity.

In order to determine the order of approximation of the interval arithmetic evaluation of a function f and of its mean value form we need a smoothness property which is fulfilled by nearly all expressions in $\mathbb{E}$.

Definition 4.1.15 (Lipschitz continuity). Let $[x]^0 \in \mathbb{IR}^m$. An interval function

$$[f] : \mathbb{I}([x]^0) \to \mathbb{IR}^n$$

is called Lipschitz continuous if there is a positive constant $l_f \in \mathbb{R}$ such that

$$\|q([f]([x]), [f]([y]))\|_\infty \le l_f \cdot \|q([x], [y])\|_\infty$$

holds for all $[x], [y] \in \mathbb{I}([x]^0)$. The constant l_f is called Lipschitz constant; the set of all functions which are Lipschitz continuous in $\mathbb{I}([x]^0)$ is denoted by $L([x]^0)$.

As in Definition 4.1.12 the maximum norm in Definition 4.1.15 can be replaced by any other vector norm on $\mathbb{R}^m$ and $\mathbb{R}^n$, respectively. Definition 4.1.15 generalizes the definition of Lipschitz continuity for real functions f. In particular, if the interval arithmetic evaluation of f is Lipschitz continuous, then the real function f itself is Lipschitz continuous on $[x]^0$ in the traditional way.

If $[f] \in L([x]^0)$ with $[f]([x]) = f([x])$ we often shorten this by $f \in L([x]^0)$ or $f([x]) \in L([x]^0)$. If $m > 1$, $n = 1$ in Definition 4.1.15, we write $f'([x]) \in L([x]^0)$ or $f' \in L([x]^0)$, if we mean $\operatorname{grad} f = (f')^T \in L([x]^0)$. For $n > 1$ and $f = (f_i)$ we use $f' \in L([x]^0)$ if $f_i' \in L([x]^0)$ for $i = 1, \ldots, n$.

Theorem 4.1.16. *Let $f : D \subseteq \mathbb{R}^m \to \mathbb{R}$ be given by an expression $f(x) \in \mathbb{E}$. Assume that $f([x]^0)$ exists for some interval vector $[x]^0 \subseteq D$ and define $[f]$ on $\mathbb{I}([x]^0)$ by $[f]([x]) = f([x])$. If $f([x]^0)$ does not contain a root whose radicand contains zero, then $[f]$ is Lipschitz continuous.*

Proof. We prove the theorem by induction with respect to the number of operations, similar to the proof of Theorem 4.1.5. To this end let $[x], [y] \in \mathbb{I}([x]^0)$.

Case 1: $f = c$ (= constant) or $f(x) = x_i$. Here, $f \in L([x]^0)$ is immediate with $l_f = 1$ for instance.

Case 2: $f = \pm g$, where $g \in L([x]^0)$ with Lipschitz constant l_g.

The subcase $f = g$ is trivial, the subcase $f = -g$ follows from $q(f([x]), f([y])) = q(-g([x]), -g([y])) = |(-1)|q(g([x]), g([y])) \le l_g \|q([x], [y])\|_\infty$.

Case 3: $f = g \circ h$, where $\circ \in \{+, -, *, /\}$ and $g, h \in L([x]^0)$ with Lipschitz constants l_g and l_h, respectively.

The subcases $f = g \pm h$ follow from

$$\begin{aligned} q(f([x]), f([y])) &= q(g([x]) \pm h([x]), g([y]) \pm h([y])) \\ &\le q(g([x]), g([y])) + q(h([x]), h([y])) \le (l_g + l_h)\|q([x], [y])\|_\infty . \end{aligned}$$

The subcase $f = g \cdot h$ is shown by

$$\begin{aligned} q(f([x]), f([y])) &= q(g([x])h([x]), g([y])h([y])) \\ &\le q(g([x])h([x]), g([x])h([y])) + q(g([x])h([y]), g([y])h([y])) \\ &\le |g([x])|q(h([x]), h([y])) + |h([y])|q(g([x]), g([y])) \\ &\le |g([x]^0)|l_h\|q([x], [y])\|_\infty + |h([x]^0)|l_g\|q([x], [y])\|_\infty \\ &= l_f \|q([x], [y])\|_\infty \end{aligned}$$

with $l_f = |g([x]^0)|l_h + |h([x]^0)|l_g$.

The proof of the final subcase $f = g/h$ is based on Theorem 2.5.7 (h), (i):

$$\begin{aligned} q(f([x]), f([y])) &= q\left(\frac{g([x])}{h([x])}, \frac{g([y])}{h([y])}\right) \\ &\le q\left(\frac{g([x])}{h([x])}, \frac{g([y])}{h([x])}\right) + q\left(\frac{g([y])}{h([x])}, \frac{g([y])}{h([y])}\right) \\ &\le \frac{1}{\langle h([x])\rangle} q(g([x]), g([y])) + \frac{|g([y])|}{\langle h([x])\rangle\langle h([y])\rangle} q(h([x]), h([y])) \\ &\le l_f\|q([x], [y])\|_\infty \end{aligned}$$

with

$$l_f = \frac{1}{\langle h([x]^0)\rangle} l_g + \frac{|g([x]^0)|}{\langle h([x]^0)\rangle^2} l_h .$$

Case 4: $f = \varphi(g)$ with $\varphi \in \mathbb{F}$ and $g \in L([x]^0)$ with Lipschitz constant l_g. If φ is a root we assume that $\min(g([x]^0)) > 0$.

Let $[u] = g([x])$, $[v] = g([y])$, and $q = q([u], [v])$. By virtue of Theorem 2.5.7 (a) we have $[u] \subseteq [v] + [-q, q]$.

Choose $\tilde{u} \in [u]$ arbitrarily. Then there exist elements $\tilde{v} \in [v]$, $\tilde{q} \in [-q, q]$ such that $\tilde{u} = \tilde{v} + \tilde{q}$ holds. Apply the mean value theorem in order to get $\varphi(\tilde{u}) = \varphi(\tilde{v}) + \varphi'(\xi)\tilde{q} \in \varphi([v]) + \varphi'(g([x]^0))[-q, q]$.

Since $\varphi([u]) = R_\varphi([u])$ we obtain $\varphi([u]) \subseteq \varphi([v]) + [-\hat{q}, \hat{q}]$ with $\hat{q} = |\varphi'(g([x]^0))| \cdot q$.

Analogously we can show $\varphi([v]) \subseteq \varphi([u]) + [-\hat{q}, \hat{q}]$, and Theorem 2.5.7 (a) implies

$$\begin{aligned} q(\varphi(g([x])), \varphi(g([y]))) &= q(\varphi([u]), \varphi([v])) \le \hat{q} \\ &= |\varphi'(g([x]^0))| \cdot q(g([x]), g([y])) \\ &\le l_f \|q([x], [y])\|_\infty \quad \text{with } l_f = |\varphi'(g([x]^0))| \cdot l_g. \end{aligned}$$

□

Theorem 4.1.17. *If $[x]^0 \subseteq D \subseteq \mathbb{R}^n$, $f: D \to \mathbb{R}$, $f([x]) \in L([x]^0)$, then the following properties hold.*

(a) $f([x]) \subseteq f(\check{x}) + l_f e^T([x] - \check{x})$.

(b) $d(f([x])) \le l_f e^T d([x])$.

Proof. (a) follows from

$$\begin{aligned} |f([x]) - f(\check{x})| &= q(f([x]) - f(\check{x}), 0) = q(f([x]), f(\check{x})) \\ &\le l_f \|q([x], \check{x})\|_\infty = l_f \||[x] - \check{x}|\|_\infty \le l_f e^T |[x] - \check{x}|, \end{aligned}$$

whence

$$f([x]) - f(\check{x}) \subseteq \sum_{i=1}^{n} l_f |[x]_i - \check{x}_i| [-1, 1] = l_f e^T([x] - \check{x}).$$

(b) is an immediate consequence of (a). □

We will extend the result in Theorem 4.1.17 (b) a little bit, allowing now $f(x) \in \mathbb{R}^n$ instead of merely $f(x) \in \mathbb{R}$.

Theorem 4.1.18. *Let $[x]^0 \subseteq D \subseteq \mathbb{R}^n$, $f = (f_i): D \to \mathbb{R}^n$.*

(a) *If $f([x]) \in L([x]^0)$, then $\|d(f([x]))\|_\infty \le \gamma \|d([x])\|_\infty$ follows for arbitrary interval vectors $[x] \subseteq [x]^0$ with some independent constant $\gamma > 0$.*

(b) *If, conversely, $\|d(f([x]))\|_\infty \le \gamma \|d([x])\|_\infty$ holds for all $[x] \subseteq [x]^0$, then the real function $f: D \to \mathbb{R}^n$ is Lipschitz continuous on $[x]^0$ in the usual sense.*

Proof. (a) follows immediately from Theorem 4.1.17 (b): One sees at once that with $f([x])$ each of its components $f_i([x])$ is also Lipschitz continuous with the same Lipschitz constant l_f. Therefore, $d(f_i([x])) \le l_f e^T d([x])$ holds and implies $\|d(f([x]))\|_\infty \le l_f e^T d([x]) \le n l_f \|d([x])\|_\infty$.

In order to prove (b), choose $x, y \in [x]^0$ arbitrarily and define

$$[z] = ([\min\{x_i, y_i\}, \ \max\{x_i, y_i\}]) \subseteq [x]^0.$$

Obviously, $d([z]) = |x - y|$ holds and from $|f(x) - f(y)| \le d(f([z]))$ we get the classical Lipschitz condition

$$\|f(x) - f(y)\|_\infty \le \|d(f([z]))\|_\infty \le \gamma \|d([z])\|_\infty = \gamma \|x - y\|_\infty$$

as required. □

It is unknown to the author whether in (b) the Lipschitz continuity follows for the interval arithmetic evaluation $f([x])$, too.

After these smoothness results we return to the order of approximation.

Theorem 4.1.19. *If $f: D \subseteq \mathbb{R}^n \to \mathbb{R}$, $[x]^0 \subseteq D$, then the following properties hold.*
(a) *If $f([x]) \in L([x]^0)$, then $q(f([x]), R_f([x])) \le c\,\|d([x])\|_\infty$*
(b) *If $f'([x]) \in L([x]^0)$ and $z \in [x]$, then $q([f]_M([x]), R_f([x])) \le \tilde{c}\,\|d([x])\|_\infty^2$*
with constants c, $\tilde{c}$, which do not depend on $[x] \subseteq [x]^0$.

Proof. (a) We first remark that $\check{x} \in [x] \subseteq [y]$ implies $q([x],[y]) \le q(\check{x},[y])$; cf. Figure 4.1.1 for the one-dimensional case. With $f(\check{x}) \in R_f([x]) \in \mathbb{IR}$ we therefore obtain

$$q(f([x]), R_f([x])) \le q(f([x]), f(\check{x})) \le l_f \|q([x], \check{x})\|_\infty = l_f \cdot \frac{1}{2}\|d([x])\|_\infty$$

which proves the assertion.

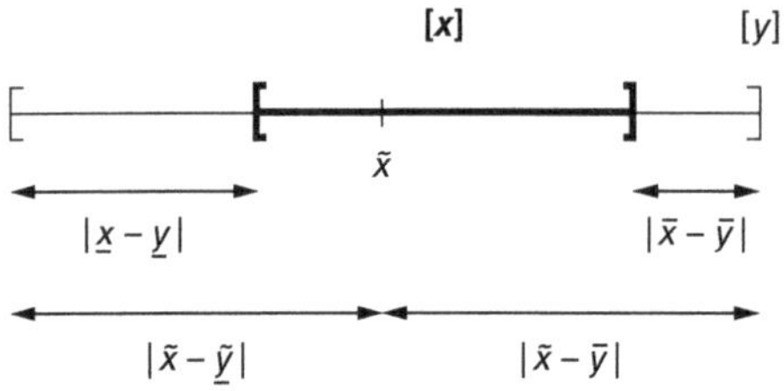

Fig. 4.1.1: $q([x],[y]) \le q(\check{x},[y])$.

(b) Let $[f]_M([x]) = [\underline{f}_M, \overline{f}_M]$ and $R_f([x]) = [f_{\min}, f_{\max}]$. There are elements $\tilde{x} \in [x]$, $\tilde{f}' \in f'([x])$ such that $\min\left(f'([x]) \cdot ([x] - z)\right) = \tilde{f}' \cdot (\tilde{x} - z)$, where we use $\tilde{f}'$ only symbolically. (It is not necessarily a derivative.) W.l.o.g. let $q([f]_M([x]), R_f([x])) = f_{\min} - \underline{f}_M$. Then by means of the mean value theorem we obtain

$$\begin{aligned}
q([f]_M([x]), R_f([x])) &= f_{\min} - \{f(z) + \tilde{f}' \cdot (\tilde{x} - z)\} \\
&\le f(\tilde{x}) - f(z) - \tilde{f}' \cdot (\tilde{x} - z) \\
&\le |f'(\xi) - \tilde{f}'| \cdot |\tilde{x} - z| \le q(f'(\xi), f'([x]))\, d([x]) \\
&\le n \cdot \|(q(f'(\xi), f'([x])))^T\|_\infty \,\|d([x])\|_\infty \\
&\le n \cdot l_{f'} \|q(\xi, [x])\|_\infty \|d([x])\|_\infty \\
&\le n \cdot l_{f'} \|d([x])\|_\infty^2,
\end{aligned}$$

where the final inequality can be seen from Figure 4.1.2 in the one-dimensional case. □

We illustrate the theorem by the following example.

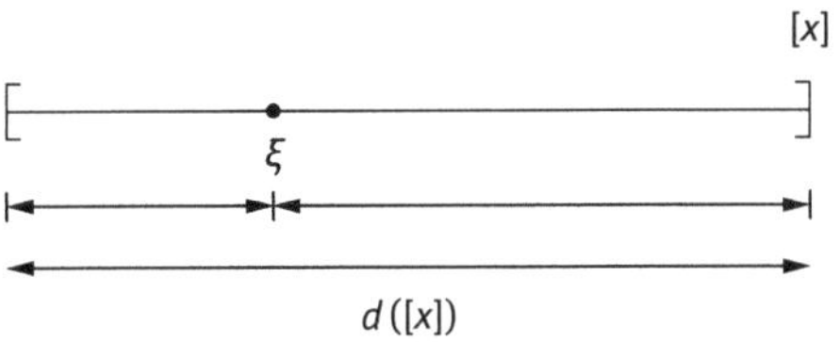

Fig. 4.1.2: $q(\xi,[x]) \le d([x])$.

Example 4.1.20. Let $f(x) = \frac{x+1}{x-1} = 1 + \frac{2}{x-1}$, $x \in [x]_\varepsilon = 2 + [-\varepsilon, \varepsilon]$ for $0 < \varepsilon \le \frac{1}{2}$, $[x]^0 = [\frac{3}{2}, \frac{5}{2}]$.

Use $f'(x)$ in the form

$$f'(x) = -\frac{2}{(x-1)^2}.$$

Then $f([x]) \in L([x]^0)$, $f'([x]) \in L([x]^0)$, and a simple computation yields the following results:

– Range:

$$R_f([x]_\varepsilon) = 1 + \frac{2}{[x]_\varepsilon - 1} = \left[3 - \frac{2\varepsilon}{1+\varepsilon}, 3 + \frac{2\varepsilon}{1-\varepsilon}\right].$$

– Interval arithmetic evaluation of the first expression of $f(x)$:

$$f([x]_\varepsilon) = \left[3 - \frac{4\varepsilon}{1+\varepsilon}, 3 + \frac{4\varepsilon}{1-\varepsilon}\right]$$

with

$$q_A = q(f([x]_\varepsilon), R_f([x]_\varepsilon)) = \frac{2\varepsilon}{1-\varepsilon} = \frac{d([x]_\varepsilon)}{1-\varepsilon} = O(\varepsilon), \quad \varepsilon \to +0.$$

– Mean value form with $z = 2$:

$$[f]_M([x]_\varepsilon) = 3 - \left[-\frac{2\varepsilon}{(1-\varepsilon)^2}, \frac{2\varepsilon}{(1-\varepsilon)^2}\right]$$

with

$$\begin{aligned} q_M = q([f]_M([x]_\varepsilon), R_f([x]_\varepsilon)) &= \frac{2\varepsilon^2}{(1-\varepsilon)^2}\,\frac{3-\varepsilon}{1+\varepsilon} \\ &= \frac{3-\varepsilon}{2(1-\varepsilon)^2(1+\varepsilon)}\big(d([x]_\varepsilon)\big)^2 = O(\varepsilon^2), \quad \varepsilon \to +0. \end{aligned}$$

For $\varepsilon = 10^{-k}$ and various integers k we represent the overestimation in Table 4.1.1, where we rounded the mantissas to one decimal place after the decimal point. The table should highlight the difference between the interval arithmetic evaluation and the mean value form when assessing the quality of enclosure of $R_f([x])$.

Example 4.1.20 shows that the order of approximation in Theorem 4.1.19 cannot generally be increased for $f([x])$ and $[f]_M([x])$, respectively. Under certain conditions Hertling showed in [143] that the order two cannot be improved in $\mathbb{IR}^1$ by rearranging terms in the given expression. Nevertheless, there are methods of order higher than two even in $\mathbb{IR}^1$. A result towards this direction is contained in the following lemma of Cornelius and Lohner [80].

Tab. 4.1.1: Results of Example 4.1.20.

k	q_A	q_M
1	$2.2 \cdot 10^{-1}$	$6.5 \cdot 10^{-2}$
2	$2.0 \cdot 10^{-2}$	$6.0 \cdot 10^{-4}$
3	$2.0 \cdot 10^{-3}$	$6.0 \cdot 10^{-6}$
4	$2.0 \cdot 10^{-4}$	$6.0 \cdot 10^{-8}$
5	$2.0 \cdot 10^{-5}$	$6.0 \cdot 10^{-10}$

Lemma 4.1.21. *Let* $[h]$, $[x] \in \mathbb{IR}$, $g, h \in C^0([x], \mathbb{R})$, $R_h([x]) \subseteq [h]$, $f = g + h$, $F([x]) = R_g([x]) + [h]$. *Then* $R_f([x]) \subseteq F([x])$ *and*

$$q(R_f([x]), F([x])) \le d([h]) \le 2|[h]|. \tag{4.1.8}$$

Proof. The first assertion is trivial. In order to prove (4.1.8) we choose $x_1, x_2, y_1, y_2 \in [x]$ such that $R_f([x]) = [f(x_1), f(x_2)]$, $R_g([x]) = [g(y_1), g(y_2)]$. Then

$$q(R_f([x]), F([x])) = \max\{\, |f(x_1) - g(y_1) - \underline{h}|,\ |f(x_2) - g(y_2) - \overline{h}|\,\}.$$

By virtue of our first assertion we get

$$\begin{aligned} |f(x_1) - g(y_1) - \underline{h}| &= f(x_1) - g(y_1) - \underline{h} \le f(y_1) - g(y_1) - \underline{h} \\ &= h(y_1) - \underline{h} \le \overline{h} - \underline{h} = d([h]) \end{aligned}$$

and, similarly,

$$|f(x_2) - g(y_2) - \overline{h}| = g(y_2) + \overline{h} - f(x_2) \le \overline{h} - h(y_2) \le d([h]).$$

This proves the first inequality of (4.1.8); the second is obvious. □

Based on this lemma and on Hermite interpolation (see Theorem 1.4.7 and the notation there) we immediately get the following result of Cornelius and Lohner [80].

Theorem 4.1.22. *Let D be an open subset of $\mathbb{R}$, and let $f \in C^{m+1}(D)$, $[x]^0 \subset D$, $f^{(m+1)} \in L([x]^0)$. For fixed $n \in \mathbb{N}_0$ and each $[x] \subseteq [x]^0$ let $x_0, \ldots, x_n \in [x]$, $m_0, \ldots, m_n \in \mathbb{N}_0$ be given numbers with $\sum_{i=0}^{n} m_i = m + 1$. Denote by p_m the corresponding Hermite interpolation polynomial according to Theorem 1.4.7 (which depends on x_i and m_i). Choose a number $\tilde{f}^{(m+1)} \in f^{(m+1)}([x])$ and define the functions q_{m+1}, F_{mn}, G_{mn} by*

$$q_{m+1}(x) = p_m(x) + \frac{\tilde{f}^{(m+1)}}{(m+1)!} \prod_{i=0}^{n} (x - x_i)^{m_i}$$

$$F_{mn}([x]) = R_{p_m}([x]) + \frac{1}{(m+1)!} f^{(m+1)}([x]) \prod_{i=0}^{n} ([x] - x_i)^{m_i}$$

$$G_{mn}([x]) = R_{q_{m+1}}([x]) + \frac{1}{(m+1)!} \{ f^{(m+1)}([x]) - \tilde{f}^{(m+1)} \} \prod_{i=0}^{n} ([x] - x_i)^{m_i}.$$

With appropriate constants c_1, c_2, which depend on $[x]^0, m, f$ only, we obtain
(a) $R_f([x]) \subseteq F_{mn}([x])$, $R_f([x]) \subseteq G_{mn}([x])$,
(b) $q(R_f([x]), F_{mn}([x])) \le c_1(d([x]))^{m+1}$,
(c) $q(R_f([x]), G_{mn}([x])) \le c_2(d([x]))^{m+2}$.

Proof. (a) follows from Theorem 1.4.7 and Lemma 4.1.21.
(b) Lemma 4.1.21 implies

$$\begin{aligned} q(R_f([x]), F_{mn}([x])) &\le \frac{2}{(m+1)!}\,|f^{(m+1)}([x])| \prod_{i=0}^{n} |[x]-x_i|^{m_i} \\ &\le \frac{2}{(m+1)!}\,|f^{(m+1)}([x]^0)| \prod_{i=0}^{n} (d([x]))^{m_i} = c_1(d([x]))^{m+1}. \end{aligned}$$

(c) Using Lemma 4.1.21 and Theorem 4.1.17 (b) we obtain

$$\begin{aligned} q(R_f([x]), G_{mn}([x])) &\le \frac{2}{(m+1)!}\,|f^{(m+1)}([x]) - \tilde{f}^{(m+1)}| \prod_{i=0}^{n} |[x]-x_i|^{m_i} \\ &\le \frac{2}{(m+1)!}\, d(f^{(m+1)}([x])) \prod_{i=0}^{n} |[x]-x_i|^{m_i} \\ &\le \frac{2}{(m+1)!}\, l_{f^{(m+1)}}\, d([x])\,(d([x]))^{m+1} = c_2(d([x]))^{m+2}. \qquad \square \end{aligned}$$

Notice that the exponents $m+1$ and $m+2$ in Theorem 4.1.22 are only lower bounds of the order of approximation as the mean value form shows – cf. Theorem 4.1.19 (b) and the subsequent Corollary 4.1.23 (a).

Now we specialize m in Theorem 4.1.22 in order to end up with various range enclosures of different approximation orders.

Corollary 4.1.23. *With the notation and assumptions of Theorem 4.1.22 we get the following enclosures of $R_f([x])$:*
(a) $m = 0,\ n = 0,\ m_0 = 1$:

$$\begin{aligned} F_{00}([x]) &= f(x_0) + f'([x])([x]-x_0) = f_M([x]) && (\textit{Order} \ge 1) \\ G_{00}([x]) &= \{f(x_0) + \tilde{f}' \cdot ([x]-x_0)\} + \{f'([x]) - \tilde{f}'\}([x]-x_0) && (\textit{Order} \ge 2) \end{aligned}$$

(b) $m = 1,\ n = 0,\ m_0 = 2,\ \tilde{f}'' \ne 0$:

$$F_{10}([x]) = \{f(x_0) + f'(x_0)([x]-x_0)\} + \frac{1}{2} f''([x])([x]-x_0)^2 \qquad (\textit{Order} \ge 2)$$

$$\begin{aligned} G_{10}([x]) = &\left\{ f(x_0) + \frac{\tilde{f}''}{2}\left(\left(([x]-x_0) - \frac{f'(x_0)}{\tilde{f}''} \right)^2 - \left(\frac{f'(x_0)}{\tilde{f}''} \right)^2 \right) \right\} \\ &+ \frac{1}{2}(f''([x]) - \tilde{f}'')([x]-x_0)^2 \qquad (\textit{Order} \ge 3) \end{aligned}$$

(c) $m = 1,\ n = 1,\ m_0 = 1,\ m_1 = 1,\ x_0 = \underline{x} \neq x_1 = \overline{x},\ \tilde{f}'' \neq 0$:

$$F_{11}([x]) = \{f(\underline{x}) + \Delta([x] - \underline{x})\} + \frac{1}{2} f''([x])[-(\mathrm{rad}([x]))^2, 0] \qquad (\textit{Order} \geq 2)$$

$$G_{11}([x]) = \left\{\frac{f(\underline{x}) + f(\overline{x})}{2} + \frac{\tilde{f}''}{2}\left(\left([x] - \check{x} + \frac{\Delta}{\tilde{f}''}\right)^2 - \left(\frac{\Delta}{\tilde{f}''}\right)^2 - (\mathrm{rad}([x]))^2\right)\right\}$$
$$+ \frac{1}{2}(f''([x]) - \tilde{f}'')\left[-(\mathrm{rad}([x]))^2, 0\right], \qquad (\textit{Order} \geq 3)$$

where $\Delta = \dfrac{f(\overline{x}) - f(\underline{x})}{\overline{x} - \underline{x}}$.

Proof. (a) is obvious.

(b) uses the interval square function and a representation which allows the application of Theorem 4.1.5.

(c) proceeds similarly to (b). In addition it modifies the part $([x] - \underline{x})([x] - \overline{x})$ of the remainder term in Theorem 4.1.22 in order to get the range $[h(\check{x}), 0] = [-(\mathrm{rad}([x]))^2, 0]$ of the function $h(x) = (x - \underline{x})(x - \overline{x})$ for $x \in [x]$.

The details of (b) and (c) are left to the reader as Exercise 4.1.1. □

We will apply Corollary 4.1.23 to the function f of Example 4.1.20.

Example 4.1.24. Let $f(x) = \frac{x+1}{x-1} = 1 + \frac{2}{x-1}$, $x \in [x]_\varepsilon = 2 + [-\varepsilon, \varepsilon]$ for $0 < \varepsilon \leq \frac{1}{2}$, $[x]^0 = [\frac{3}{2}, \frac{5}{2}]$. Use $f'(x)$ in the form $f'(x) = -\frac{2}{(x-1)^2}$ and $f''(x)$ in the form $f''(x) = \frac{4}{(x-1)^3}$, and choose $x_0 = 2$, $\tilde{f}' = f'(2) = -2$, $\tilde{f}'' = f''(2) = 4$. Then analogously to Example 4.1.20 we get the following results:

- Range:

$$R_f([x]_\varepsilon) = \left[3 - \frac{2\varepsilon}{1+\varepsilon}, 3 + \frac{2\varepsilon}{1-\varepsilon}\right] = \left[3 - 2\varepsilon + \frac{2\varepsilon^2}{1+\varepsilon}, 3 + 2\varepsilon + \frac{2\varepsilon^2}{1-\varepsilon}\right].$$

- $m = 0,\ n = 0,\ m_0 = 1$:

$$F_{00}([x]_\varepsilon) = f_M([x]_\varepsilon) = 3 + \frac{2\varepsilon}{(1-\varepsilon)^2}[-1, 1]$$

$$q_{F_{00}} = q(R_f([x]_\varepsilon), F_{00}) = \frac{2\varepsilon^2}{(1-\varepsilon)^2} \cdot \frac{3-\varepsilon}{1+\varepsilon} = O(\varepsilon^2), \quad \varepsilon \to +0.$$

$$G_{00}([x]_\varepsilon) = [3 - 2\varepsilon, 3 + 2\varepsilon] + \frac{2\varepsilon^2(2-\varepsilon)}{(1-\varepsilon)^2}[-1, 1]$$

$$q_{G_{00}} = q(R_f([x]_\varepsilon), G_{00}) = \frac{2\varepsilon^2}{(1-\varepsilon)^2} \cdot \frac{3-\varepsilon}{1+\varepsilon} = q_{F_{00}} = O(\varepsilon^2), \quad \varepsilon \to +0.$$

- $m = 1,\ n = 0,\ m_0 = 2$:

$$F_{10}([x]) = [3 - 2\varepsilon, 3 + 2\varepsilon] + \left[0, \frac{2\varepsilon^2}{(1-\varepsilon)^3}\right]$$

$$q_{F_{10}} = q(R_f([x]_\varepsilon), F_{10}) = \frac{2\varepsilon^2}{1+\varepsilon} = O(\varepsilon^2), \quad \varepsilon \to +0.$$

$$G_{10}([x]) = [3 - 2\varepsilon, 3 + 2\varepsilon] + 2\varepsilon^2 \left[\frac{1}{(1+\varepsilon)^3}, \frac{1}{(1-\varepsilon)^3}\right]$$

$$q_{G_{10}} = q(R_f([x]_\varepsilon), G_{10}) = 2\varepsilon^3 \frac{2-\varepsilon}{(1-\varepsilon)^3} = O(\varepsilon^3), \quad \varepsilon \to +0.$$

For $\varepsilon = 10^{-k}$, $k = 0, \ldots, 5$, we list our results in Table 4.1.2, where we rounded again to one decimal place after the decimal point.

Tab. 4.1.2: Results of Example 4.1.24.

k	$q_{F_{00}} = q_{G_{00}}$	$q_{F_{10}}$	$q_{G_{10}}$
1	$6.5 \cdot 10^{-2}$	$1.8 \cdot 10^{-2}$	$5.2 \cdot 10^{-3}$
2	$6.0 \cdot 10^{-4}$	$2.0 \cdot 10^{-4}$	$4.1 \cdot 10^{-6}$
3	$6.0 \cdot 10^{-6}$	$2.0 \cdot 10^{-6}$	$4.0 \cdot 10^{-9}$
4	$6.0 \cdot 10^{-8}$	$2.0 \cdot 10^{-8}$	$4.0 \cdot 10^{-12}$
5	$6.0 \cdot 10^{-10}$	$2.0 \cdot 10^{-10}$	$4.0 \cdot 10^{-15}$

Exercises

Ex. 4.1.1. Fill in the details in the proof of Corollary 4.1.23.

Ex. 4.1.2. Verify the results of Example 4.1.24.

Ex. 4.1.3 (Cornelius, Lohner [80]). Compute similar tables to Tables 4.1.1 and 4.1.2 for the function $f(x) = \ln(x)/x$, $x_0 = 1.5$, and the intervals $[x]_i = 1.5 + 10^{-i}[-1, 1]$, $i = 0, 1, \ldots, 7$.

Do the same for the function $f(x) = e^{x - \sin x} - 1$, $x_0 = -1.5$, $[x]_i = -1.5 + 10^{-i}[-1, 1]$, $i = 0, 1, \ldots, 7$.

4.2 *P*-contraction

As we remarked in Section 3.1, we will apply the Hausdorff distance q for vectors $[x] \in \mathbb{IR}^n$ componentwise. Then by virtue of any monotone vector norm $\|\cdot\|$ on $\mathbb{R}^n$, we will get a metric on $\mathbb{IR}^n$ via $\|q([x], [y])\|$ with which $\mathbb{IR}^n$ becomes a complete metric space. With this metric one can introduce contractive mappings in order to apply Banach's fixed point theorem. Often it is simpler to work with q on $\mathbb{IR}^n$ directly instead of composing it with an additional norm. This leads to the concept of P-contraction introduced in Schröder [333].

Definition 4.2.1. Let $[f]: \mathbb{IR}^n \to \mathbb{IR}^n$, $O \le P \in \mathbb{R}^{n\times n}$ with $\rho(P) < 1$. If

$$q([f]([x])), [f]([y])) \le Pq([x],[y]) \tag{4.2.1}$$

holds for all $[x], [y] \in \mathbb{IR}^n$, then $[f]$ is called a P-contraction.

If $[f]: \mathbb{I}([z]) \to \mathbb{IR}^n$ for some given $[z] \in \mathbb{IR}^n$ and if (4.2.1) holds for $[x], [y] \in \mathbb{I}([z])$, then we call $[f]$ a P-contraction on $[z]$.

Our first result relates P-contractions to contractions in the usual sense.

Theorem 4.2.2. *P-contractions are contractions in the usual sense with respect to some particular metric, but not vice versa.*

Proof. Let $[f]$ be a P-contraction as in Definition 4.2.1. Choose $\varepsilon > 0$ such that $\alpha = \rho(P) + \varepsilon < 1$. By virtue of Theorem 1.9.12 there is a monotone vector norm $\|\cdot\|$ whose operator norm satisfies $\|P\| \le \alpha$. Then (4.2.1) implies the contraction condition

$$\|q([f]([x]), [f]([y]))\| \le \|P\|\,\|q([x],[y])\| \le \alpha\|q([x],[y])\|.$$

In order to prove the missing converse implication we consider the function

$$[f]: \begin{cases} \mathbb{IR}^2 \to \mathbb{IR}^2 \\ [x] \mapsto \frac{1}{2}\|\,|[x]|\,\|_\infty \binom{1}{1} \end{cases}$$

and recall that the norm $\|\cdot\|_\infty$ is monotone on $\mathbb{R}^2$. Using results from Theorem 2.5.7 we obtain

$$\begin{aligned}
\|q([f]([x]), [f]([y]))\|_\infty &= \|[f]([x]) - [f]([y])\|_\infty \\
&= \tfrac{1}{2}\big|\,\|\,|[x]|\,\|_\infty - \|\,|[y]|\,\|_\infty\,\big| \le \tfrac{1}{2}\|\,|[x]| - |[y]|\,\|_\infty \\
&= \tfrac{1}{2}\|\,|\check{x}| + r_x - |\check{y}| - r_y\,\|_\infty \\
&\le \tfrac{1}{2}\|\,\big||\check{x}| - |\check{y}|\big| + |r_x - r_y|\,\|_\infty \\
&\le \tfrac{1}{2}\|\,|\check{x} - \check{y}| + |r_x - r_y|\,\|_\infty = \tfrac{1}{2}\|q([x],[y])\|_\infty
\end{aligned}$$

for all $[x], [y] \in \mathbb{IR}^n$, hence $[f]$ is a contraction with respect to $\|q(\cdot,\cdot)\|_\infty$.

Assume now that f is a P-contraction. Then

$$|[f](x)| = |[f](x) - [f](0)| = q([f](x), [f](0)) \le Pq(x,0) = P|x|$$

for all $x \in \mathbb{R}^2$. Choosing $x = (1,0)^T$ and $x = (0,1)^T$, respectively, implies

$$\frac{1}{2}\begin{pmatrix}1\\1\end{pmatrix} \le \begin{pmatrix}p_{11}\\p_{21}\end{pmatrix} \quad \text{and} \quad \frac{1}{2}\begin{pmatrix}1\\1\end{pmatrix} \le \begin{pmatrix}p_{12}\\p_{22}\end{pmatrix},$$

respectively.

Thus

$$P \ge \frac{1}{2}\begin{pmatrix}1 & 1\\1 & 1\end{pmatrix},$$

hence

$$\rho(P) \geq \rho\left(\frac{1}{2}\begin{pmatrix} 1 & 1 \\ 1 & 1 \end{pmatrix}\right) = 1$$

by virtue of Theorem 1.9.10, which contradicts the definition of a P-contraction. □

The example in the proof of Theorem 4.2.2 is from Martin Koeber [167]; see Mayer [211]. Our next example presents one of the most important P-contractions.

Example 4.2.3. Let $[f]: \mathbb{IR}^n \to \mathbb{IR}^n$ be defined by

$$[f]([x]) = [A][x] + [b], \tag{4.2.2}$$

where $[A] \in \mathbb{IR}^{n\times n}$ satisfies $\rho(|[A]|) < 1$. Then $[f]$ is a P-contraction with $P = |[A]|$.

The proof is immediate since

$$q([f]([x]), [f]([y])) = q([A][x], [A][y]) \leq |[A]|q([x], [y]). \qquad \square$$

In order to facilitate the verification of P-contractions different from (4.2.2) we need some preparations.

Lemma 4.2.4. *Let $\underline{f}, \overline{f}: \mathbb{R} \to \mathbb{R}$ satisfy*

$$|\underline{f}(x) - \underline{f}(y)| \leq \alpha|x - y|, \quad |\overline{f}(x) - \overline{f}(y)| \leq \alpha|x - y| \tag{4.2.3}$$

for some fixed $\alpha > 0$ and for all $x, y \in \mathbb{R}$. Assume that $[f]: \mathbb{IR} \to \mathbb{IR}$ is defined by

$$[f]([x]) = [\underline{f}(x_1), \overline{f}(x_2)], \tag{4.2.4}$$

where x_1, x_2 are some elements from $[x]$ with $\underline{f}(x_1) = \min\{\underline{f}(x) \mid x \in [x]\} \leq \overline{f}(x_2) = \max\{\overline{f}(x) \mid x \in [x]\}$. Then

$$q([f]([x]), [f]([y])) \leq \alpha q([x], [y]) \tag{4.2.5}$$

holds.

Proof. Let $[x], [y] \in \mathbb{IR}$ and let x_i, y_i, $i = 1, 2$ be the elements from $[x]$, $[y]$ which define $[f]([x])$ and $[f]([y])$, respectively. W.l.o.g. we assume $\underline{f}(y_1) \geq \underline{f}(x_1)$. Then by (4.2.3) we obtain

$$\begin{aligned} |\underline{f}(y_1) - \underline{f}(x_1)| &= \underline{f}(y_1) - \underline{f}(x_1) \\ &\leq \begin{cases} \underline{f}(\overline{y}) - \underline{f}(x_1) \leq \alpha|\overline{y} - x_1| \\ \underline{f}(\underline{y}) - \underline{f}(x_1) \leq \alpha|\underline{y} - x_1| \end{cases}. \end{aligned}$$

If $x_1 > \overline{y}$, the upper inequality yields

$$|\underline{f}(y_1) - \underline{f}(x_1)| \leq \alpha|\overline{y} - \overline{x}| \leq \alpha q([x], [y]).$$

If $x_1 < \underline{y}$, then the lower inequality yields

$$|\underline{f}(y_1) - \underline{f}(x_1)| \le \alpha|\underline{y} - \underline{x}| \le \alpha q([x],[y]).$$

If $\underline{y} \le x_1 \le \overline{y}$, then $\underline{f}(y_1) \le \underline{f}(x_1)$, hence $\underline{f}(y_1) = \underline{f}(x_1)$, which trivially implies

$$|\underline{f}(y_1) - \underline{f}(x_1)| \le \alpha q([x],[y]).$$

Similarly, one shows

$$|\overline{f}(y_2) - \overline{f}(x_2)| \le \alpha q([x],[y]),$$

which proves (4.2.5). □

We apply Lemma 4.2.4 in order to verify P-contractivity for the four interval functions in Example 4.2.5.

Example 4.2.5. The following functions are P-contractions on $\mathbb{IR}$.
(i) $[f]([x]) = \frac{1}{2}\arctan([x])$.
(ii) $[f]([x]) = e^{-[x]^2}$.
(iii) $[f]([x]) = \beta\sin([x])$, where $|\beta| < 1$.
(iv) $[f]([x]) = \gamma\cos([x])$, where $|\gamma| < 1$.
In order to verify the property for (i) we apply Lemma 4.2.4 with $\underline{f}(x) = \overline{f}(x) = \frac{1}{2}\arctan x$, $x_1 = \underline{x}$, $x_2 = \overline{x}$, and $\alpha = \max\{\,(\frac{1}{2}\arctan x)' \mid x \in \mathbb{R}\,\} = \max\{\,\frac{1}{2(1+x^2)} \mid x \in \mathbb{R}\,\} = \frac{1}{2} < 1$.
For (ii) we apply Lemma 4.2.4 with $\underline{f}(x) = \overline{f}(x) = e^{-x^2}$,

$$x_1 = \begin{cases} \overline{x}, & \text{if } |\overline{x}| \ge |\underline{x}| \\ \underline{x}, & \text{if } |\overline{x}| < |\underline{x}| \end{cases}, \qquad x_2 = \begin{cases} 0, & \text{if } 0 \in [x] \\ \underline{x}, & \text{if } 0 < \underline{x} \\ \overline{x}, & \text{if } 0 > \overline{x} \end{cases},$$

and

$$\alpha = \max\{\,2xe^{-x^2} \mid x \in \mathbb{R}^n\,\} = \frac{2}{\sqrt{2}}e^{-1/2} = \sqrt{\frac{2}{e}} < 1.$$

In order to prove the P-contractivity in (iii) we use Lemma 4.2.4 with $\underline{f}(x) = \overline{f}(x) = \beta\sin x$. As x_1 we choose the smallest value in $[x]$, where $\beta\sin x$ assumes its minimum on $[x]$, as x_2 we choose the largest value in $[x]$ where $\beta\sin x$ assumes its maximum on $[x]$. With $\alpha = \max\{\beta\cos x \mid x \in \mathbb{R}\} = |\beta| < 1$ we obtain the assertion for (iii).
The P-contractivity for the function in (iv) is proved analogously to (iii). □

Our next theorem is basic for many results in the subsequent chapters.

Theorem 4.2.6. *Let $[f]: \mathbb{IR}^n \to \mathbb{IR}^n$ be a P-contraction. Then $[f]$ is Lipschitz continuous with Lipschitz constant $\|P\|_\infty$. Each sequence of iterates*

$$[x]^{k+1} = [f]([x]^k), \quad k = 0, 1, \ldots \tag{4.2.6}$$

converges to the same limit $[x]^*$ which is the unique fixed point of $[f]$.

If, in addition,

$$[f](x) \in \mathbb{R}^n \tag{4.2.7}$$

holds for all $x \in \mathbb{R}^n$*, then* $[x]^*$ *is a point vector.*

If a function $f : \mathbb{R}^n \to \mathbb{R}^n$ *satisfies the inclusion property*

$$f(x) \in [f]([x]) \tag{4.2.8}$$

for all $x \in [x]$ *and arbitrary* $[x] \in \mathbb{IR}^n$*, then* $[x]^*$ *contains all fixed points of* f*. If* f *is continuous, it has at least one fixed point in* $[x]^*$*.*

If $[f]([x])$ *is the interval arithmetic evaluation* $f([x])$ *of some function* $f : \mathbb{R}^n \to \mathbb{R}^n$*, then* f *is a* P*-contraction on* $\mathbb{R}^n$*. It has a unique fixed point* x^* *to which the interval sequence (4.2.6) contracts independently of* $[x]^0 \in \mathbb{IR}^n$*, and which can be identified with* $[x]^*$*.*

Proof. Applying the maximum norm to the defining inequality of a P-contraction results in the Lipschitz continuity of $[f]$ and the corresponding Lipschitz constant. The convergence of the sequence in (4.2.6) can be seen immediately from Theorem 4.2.2 and Banach's fixed point theorem.

If (4.2.7) holds, then start (4.2.6) with any point vector $[x]^0$. In this case all iterates $[x]^k$ are point vectors and so is $[x]^*$.

If (4.2.8) holds and if x^* is a fixed point of f, then start (4.2.6) with $[x]^0 \equiv x^*$. This implies

$$x^* = f(x^*) \in [f](x^*) = [x]^1,$$

and a simple induction shows $x^* \in [x]^k$, $k = 0, 1, \dots$, whence $x^* \in [x]^*$.

If (4.2.8) holds and if f is continuous, then Brouwer's fixed point theorem applies to $[x]^*$ since $f(x) \in [f]([x]^*) = [x]^*$ for all $x \in [x]^*$, i.e., f maps $[x]^*$ into itself. This guarantees the existence of at least one fixed point of f in $[x]^*$.

The last part of the theorem follows from the previous one taking into account that now (4.2.7) is fulfilled. □

Our next result extends Theorem 4.2.6.

Theorem 4.2.7. *Let* $[f] : \mathbb{IR}^n \to \mathbb{IR}^n$ *be a* P*-contraction. With the notation of Theorem 4.2.6 we obtain the following properties.*

(a) $[x]^* \subseteq [x]^0 + (I - P)^{-1} q([x]^1, [x]^0)[-1, 1]$.

(b) *If* $\alpha = \|P\|_\infty < 1$*, then*

$$[x]^* \subseteq [x]^0 + \frac{1}{1-\alpha} \|q([x]^1, [x]^0)\|_\infty e[-1, 1].$$

(c) *If* $[x]^1 \subseteq [x]^0$ *and if* $[f]$ *is inclusion monotone, then*

$$[x]^* \subseteq [x]^k \subseteq [x]^{k-1} \subseteq \cdots \subseteq [x]^0 \tag{4.2.9}$$

for all $k \in \mathbb{N}_0$*.*

(d) *If* $[x]^1 \subseteq [x]^0$ *and* $[f]([x]) = f([x])$ *for some function* $f\colon \mathbb{R} \to \mathbb{R}$, *then* f *has a unique fixed point* x^*, *and* (4.2.9) *holds with* $[x]^* \equiv x^*$.

Proof. (a) Analogously to the proof of Banach's fixed point theorem we have

$$\begin{aligned} q([x]^{k+1},[x]^0) &\le \sum_{i=0}^{k} q([x]^{i+1},[x]^i) \\ &= \sum_{i=1}^{k} q([f]([x]^i),[f]([x]^{i-1})) + q([x]^1,[x]^0) \\ &\le \sum_{i=1}^{k} Pq([x]^i,[x]^{i-1}) + q([x]^1,[x]^0) \le \cdots \\ &\le \left(\sum_{i=0}^{k} P^i\right) q([x]^1,[x]^0) \le \left(\sum_{i=0}^{\infty} P^i\right) q([x]^1,[x]^0) \\ &= (I-P)^{-1} q([x]^1,[x]^0), \end{aligned} \tag{4.2.10}$$

and in the limit $k \to \infty$ we get

$$q([x]^*,[x]^0) \le (I-P)^{-1} q([x]^1,[x]^0).$$

This proves (a) by virtue of Theorem 2.5.7 (a).

(b) follows from (a),

$$(I-P)^{-1} q([x]^1,[x]^0) \le \|(I-P)^{-1} q([x]^1,[x]^0)\|_\infty \cdot e,$$

and

$$\|(I-P)^{-1}\|_\infty = \left\|\sum_{k=0}^{\infty} P^k\right\|_\infty \le \sum_{k=0}^{\infty} \|P\|_\infty^k = \frac{1}{1-\alpha}.$$

(c) The hypothesis implies $[x]^2 \subseteq [f]([x]^1) \subseteq [f]([x]^0) = [x]^1$, and by induction we get $[x]^k \subseteq [x]^{k-1} \subseteq \cdots \subseteq [x]^0$. The rest is immediate since $\lim_{k\to\infty}[x]^k = [x]^*$ by Theorem 4.2.6.

(d) follows at once from (c) and Theorem 4.2.6. □

The remaining part of this section is devoted to local P-contractions, i.e., to P-contractions on some fixed vector $[z]$. Our first result can be proved analogously to Theorem 4.2.6.

Theorem 4.2.8. *Let* $[f]\colon \mathbb{I}([z]) \to \mathbb{I}([z])$ *be a* P*-contraction on* $[z] \in \mathbb{IR}^n$. *Then the assertions of Theorem 4.2.6 hold analogously with* $[z]$ *replacing* $\mathbb{R}^n$ *and* $\mathbb{I}([z])$ *replacing* $\mathbb{IR}^n$.

Notice that the definition of $[f]$ in Theorem 4.2.8 implies $[f]([z]) \subseteq [z]$. Therefore, each continuous function $f\colon [z] \to \mathbb{R}^n$ has at least one fixed point in $[z]$, provided

$$f(x) \in [f]([x]) \text{ holds for all } x \in [x] \text{ and arbitrary } [x] \in \mathbb{I}([z]).$$

No P-contraction property is needed for this observation which, again, is based on Brouwer's fixed point theorem. If in this case $[f]$ is inclusion monotone, then the fixed points of $[f]$ are contained in each of the iterates

$$[x]^{k+1} = [f]([x]^k), \quad k = 0, 1, \ldots$$

when starting this process with $[x]^0 = [z]$.

Theorem 4.2.9. *For a fixed vector $[z] \in \mathbb{IR}^n$ and an $n \times n$ matrix $P \geq O$ with $\rho(P) < 1$, let $[z]_P \in \mathbb{IR}^n$ be a vector which satisfies*

$$[z]_P \supseteq [z] + (I - P)^{-1} d([z])[-1, 1]. \tag{4.2.11}$$

Let $[f]: \mathbb{I}([z]_P) \to \mathbb{IR}^n$ be a P-contraction on $[z]_P$ with the contraction matrix P from (4.2.11). Choose

$$[x]^0 \subseteq [z] \tag{4.2.12}$$

and assume

$$[x]^1 = [f]([x]^0) \subseteq [z]. \tag{4.2.13}$$

Then the iterates

$$[x]^{k+1} = [f]([x]^k) \tag{4.2.14}$$

are defined for $k = 0, 1, \ldots$ and converge to some vector $[x]^* \subseteq [z]_P$ which is independent of $[x]^0$ as long as (4.2.12) and (4.2.13) are fulfilled. This limit is a point vector if $[f]([x])$ is the interval arithmetic evaluation $f([x])$ of some function $f: [z]_P \to \mathbb{R}^n$.

Proof. Assume that $[x]^i \subseteq [z]_P$ holds for $i = 0, 1, 2, \ldots, k$. This is certainly true for $k = 0$ and $k = 1$. Then by virtue of (4.2.10) we get

$$q([x]^{k+1}, [x]^0) \leq (I - P)^{-1} q([x]^1, [x]^0) \leq (I - P)^{-1} d([z]),$$

where the last inequality follows from (4.2.12), (4.2.13) and from the definition of $q(\cdot, \cdot)$. Hence

$$[x]^{k+1} \subseteq [x]^0 + (I - P)^{-1} d([z])[-1, 1] \subseteq [z]_P, \tag{4.2.15}$$

and $[x]^k$ exists for all nonnegative integers k. Since

$$\begin{aligned} q([x]^{k+m}, [x]^k) &= q([f]([x]^{k-1+m}), [f]([x]^{k-1})) \\ &\leq P q([x]^{k-1+m}, [x]^{k-1}) \leq \cdots \leq P^k q([x]^m, [x]^0) \\ &\leq P^k (I - P)^{-1} d([z]) \quad \text{for all } m = 0, 1, \ldots, \end{aligned}$$

and since $\rho(P) < 1$ by assumption, the sequences $(\underline{x}^k)$, $(\overline{x}^k)$ are Cauchy sequences in $\mathbb{R}^n$, hence they converge to limits $\underline{x}^*$ and $\overline{x}^*$, respectively, with $\underline{x}^* \leq \overline{x}^*$. Therefore, $\lim_{k\to\infty} [x]^k = [\underline{x}^*, \overline{x}^*] =: [x]^*$ with $[x]^* \subseteq [z]_P$ by (4.2.15). To show the uniqueness, let $[y]^k$ be constructed analogously to (4.2.14) and assume that (4.2.12) and (4.2.13) hold for $[y]^0$ and $[y]^1$. Let $[y]^*$ be the limit of $([y]^k)$. Since P-contractions are continuous functions, we get

$$q([x]^*, [y]^*) = q([f]([x]^*), [f]([y]^*)) \le Pq([x]^*, [y]^*)$$
$$\le \cdots \le P^k q([x]^*, [y]^*), \quad k = 0, 1, \ldots$$

which implies $q([x]^*, [y]^*) = 0$ and $[x]^* = [y]^*$. □

We point out that Theorem 4.2.9 suffers from the dependency of $[f]$ and $[z]_P$ on the same matrix P. How it can be applied is shown by the following steps for which we assume $[f]([x]) = f([x])$ with $f\colon [z]_P \to \mathbb{R}^n$ being some function for which a fixed point is looked for.

(1) Choose a 'sufficiently good' approximation $\tilde{x}$ of x^*, say, by some noninterval method.
(2) Let $[x]^0 = [\tilde{x}, \tilde{x}]$.
(3) Define $[z] = \tilde{x} + \delta[-1, 1]e$, where $\delta = \|\tilde{x} - f(\tilde{x})\|_\infty$, or let $[z] = \tilde{x} + \delta[-1, 1]$, where $\delta = |\tilde{x} - f(\tilde{x})|$. (Both possibilities imply $[x]^1 = [f]([x]^0) = [f(\tilde{x}), f(\tilde{x})] = \tilde{x} + [f(\tilde{x}) - \tilde{x}, f(\tilde{x}) - \tilde{x}] \subseteq [z]$.)
(4) Check whether $[f]$ is an (αP)-contraction on $[z]$, where $\alpha \in (0, 1)$ is sufficiently small, say $\alpha = 0.1$, and where $O \le P \in \mathbb{IR}^{n\times n}$ with $\rho(P) < 1$.
(5) Choose $[z]_P = [z] + (I - P)^{-1} d([z])\,[-1, 1] \supseteq [z]$, and check whether $[f]$ is a P-contraction on $[z]_P$.

If $\tilde{x}$ approximates x^* sufficiently well, then δ is small, hence $d([z])$ will be small. Assuming $[f]$ to be an (αP)-contraction as required in (4) lets expect $[f]$ to be a P-contraction on $[z]_P$ since $[z]_P$ differs 'only slightly' from $[z]$. This is heuristics, of course. That it can work is illustrated by the following example which deals with enclosures of eigenpairs as we shall see in Section 7.2 and which modifies the Steps (1)–(5) slightly.

Example 4.2.10. Let $(\tilde{x}, \tilde{\lambda})$ be an approximation of some eigenpair (x^*, λ^*) of a given matrix $A \in \mathbb{R}^{n\times n}$ with λ^* being an algebraic simple eigenvalue. Assume x^*, $\tilde{x}$ to be normalized by $x_n^* = \tilde{x}_n = 1$ (if possible). Define the functions $f, g\colon \mathbb{R}^{n+1} \to \mathbb{R}^{n+1}$ by

$$f(x, \lambda) = \begin{pmatrix} Ax - \lambda x \\ (e^{(n)})^T x - 1 \end{pmatrix},$$

$$g(x, \lambda) = \begin{pmatrix} \tilde{x} \\ \tilde{\lambda} \end{pmatrix} - Cf(\tilde{x}, \tilde{\lambda}) + \left\{ I - C \begin{pmatrix} A - \tilde{\lambda} I & -x \\ (e^{(s)})^T & 0 \end{pmatrix} \right\} \begin{pmatrix} x - \tilde{x} \\ \lambda - \tilde{\lambda} \end{pmatrix},$$

with some matrix $C \in \mathbb{R}^{(n+1)\times(n+1)}$, and choose

$$[x]^0 = \tilde{z} = \begin{pmatrix} \tilde{x} \\ \tilde{\lambda} \end{pmatrix}, \quad \delta = \|Cf(\tilde{x}, \tilde{\lambda})\|_\infty, \quad [z] = \tilde{z} + \delta[-1, 1]e \in \mathbb{R}^{n+1}.$$

Let P be a nonnegative $(n + 1) \times (n + 1)$ matrix to be determined later such that

$$\|P\|_\infty \le \frac{1}{2} \tag{4.2.16}$$

holds, the bound $\frac{1}{2}$ being arbitrary within the interval $(0, 1)$. Then

$$[z]_P = [z] + 2d([z])[-1, 1] = \tilde{z} + 5\delta[-1, 1]e$$

fulfills the assumption (4.2.11) of Theorem 4.2.9 since

$$\|(I - P)^{-1}\|_\infty = \left\| \sum_{k=0}^{\infty} P^k \right\|_\infty \le \sum_{k=0}^{\infty} \|P\|_\infty^k = (1 - \|P\|_\infty)^{-1} \le 2.$$

The assumptions (4.2.12) and (4.2.13) of this theorem can also be seen at once if $[f]$ is defined by $[f]([x], [\lambda]) = g([x], [\lambda])$. We will show that $[f]$ is a P-contraction on $[z]_P$ with the same P as above, provided that δ is small enough, i.e. $(\tilde{x}, \tilde{\lambda})$ approximates (x^*, λ^*) sufficiently well. To this end let

$$[v] = \begin{pmatrix} [v]' \\ [v]_{n+1} \end{pmatrix} \subseteq [z]_P - \tilde{z}, \quad [w] = \begin{pmatrix} [w]' \\ [w]_{n+1} \end{pmatrix} \subseteq [z]_P - \tilde{z},$$

and choose

$$C = \begin{pmatrix} A - \tilde{\lambda} I & -\tilde{x} \\ (e^{(n)})^T & 0 \end{pmatrix}^{-1}.$$

By virtue of Theorem 1.8.3 the inverse of

$$\begin{pmatrix} A - \lambda I & -x \\ (e^{(n)})^T & 0 \end{pmatrix}$$

exists at least in some neighborhood of $(x, \lambda) = (x^*, \lambda^*)$, whence

$$\begin{pmatrix} A - \tilde{\lambda} I & -x \\ (e^{(n)})^T & 0 \end{pmatrix} = C^{-1} + \begin{pmatrix} O & -x + \tilde{x} \\ O & 0 \end{pmatrix}.$$

Therefore,

$$\begin{aligned}
&q(g(\tilde{z} + [v]), g(\tilde{z} + [w])) \\
&\quad = q\left(\left\{ C \begin{pmatrix} O & [v]' \\ O & 0 \end{pmatrix} \right\} [v], \left\{ C \begin{pmatrix} O & -[w]' \\ O & 0 \end{pmatrix} \right\} [w] \right) \\
&\quad \le q\left(\left\{ C \begin{pmatrix} O & -[v]' \\ O & 0 \end{pmatrix} \right\} [v], \left\{ C \begin{pmatrix} O & -[v]' \\ O & 0 \end{pmatrix} \right\} [w] \right) \\
&\qquad + q\left(\left\{ C \begin{pmatrix} O & -[v]' \\ O & 0 \end{pmatrix} \right\} [w], \left\{ C \begin{pmatrix} O & -[w]' \\ O & 0 \end{pmatrix} \right\} [w] \right) \\
&\quad \le \left| C \begin{pmatrix} O & -[v]' \\ O & 0 \end{pmatrix} \right| q([v], [w]) + |C| q\left(\begin{pmatrix} O & -[v]' \\ O & 0 \end{pmatrix}, \begin{pmatrix} O & -[w]' \\ O & 0 \end{pmatrix} \right) |[w]| \\
&\quad \le |C| \left\{ \left| \begin{pmatrix} O & -[v]' \\ O & 0 \end{pmatrix} \right| q([v], [w]) + q\left(\begin{pmatrix} -[v]' \\ 0 \end{pmatrix}, \begin{pmatrix} -[w]' \\ 0 \end{pmatrix} \right) |[w]_{n+1}| \right\} \\
&\quad \le |C| \left\{ \left| \begin{pmatrix} O & -[v]' \\ O & 0 \end{pmatrix} \right| + |[w]_{n+1}| I \right\} q([v], [w]).
\end{aligned}$$

With

$$P = 5\delta|C| \begin{pmatrix} 1 & 0 & \dots & 0 & 1 \\ & 1 & \ddots & \vdots & 1 \\ & & \ddots & 0 & \vdots \\ & O & & 1 & 1 \\ & & & & 1 \end{pmatrix} = 5\delta|C|(I + (e - e^{(n+1)})(e^{(n+1)})^T)$$

we thus get

$$q(g(\tilde{z} + [v]), g(\tilde{z} + [w])) \le Pq([v], [w]) = Pq(\tilde{z} + [v], \tilde{z} + [w]).$$

For sufficiently small δ, i.e., for sufficiently good approximations $(\tilde{x}, \tilde{\lambda})$, the condition $\|P\|_\infty \le \frac{1}{2}$ is certainly fulfilled, so that Theorem 4.2.9 can be applied. Notice that C depends on $(\tilde{x}, \tilde{\lambda})$, but it is bounded if $(\tilde{x}, \tilde{\lambda})$ lies in some fixed neighborhood of (x^*, λ^*). □

If it is known that $[f]$ is a P-contraction on some interval vector $[z]_P$, one can often *construct* a vector $[z]$ such that

$$[z]_P = [z] + (I - P)^{-1} d([z])\,[-1, 1]$$

holds. How this can be done is stated in our next theorem.

Theorem 4.2.11. *Let $[f]$ be a P-contraction on $[z]_P \in \mathbb{IR}^n$ and assume*

$$\underline{z}_P + (3I - P)^{-1} d([z]_P) \le \overline{z}_P - (3I - P)^{-1} d([z]_P). \tag{4.2.17}$$

Then

$$[z]_P = [z] + (I - P)^{-1} d([z])[-1, 1] \tag{4.2.18}$$

holds if and only if

$$[z] = [\underline{z}_P + (3I - P)^{-1} d([z]_P),\ \overline{z}_P - (3I - P)^{-1} d([z]_P)]. \tag{4.2.19}$$

The assumption (4.2.17) is certainly satisfied if

$$Pd([z]_P) \le d([z]_P). \tag{4.2.20}$$

Proof. Since (4.2.18) implies $\underline{z}_P = \underline{z} - (I - P)^{-1}(\overline{z} - \underline{z})$ we get

$$\overline{z} = \underline{z} + (I - P)(\underline{z} - \underline{z}_P). \tag{4.2.21}$$

By (4.2.18) and (4.2.21), it follows

$$\overline{z}_P = \overline{z} + (I - P)^{-1}(\overline{z} - \underline{z}) = \underline{z} + (I - P)(\underline{z} - \underline{z}_P) + \underline{z} - \underline{z}_P,$$

hence $d([z]_P) = (3I - P)(\underline{z} - \underline{z}_P)$, and

$$\underline{z} = \underline{z}_P + (3I - P)^{-1} d([z]_P). \tag{4.2.22}$$

Substituting (4.2.22) in (4.2.21) yields

$$\begin{aligned} \overline{z} &= \underline{z}_P + (3I - P)^{-1} d([z]_P) + (I - P)(3I - P)^{-1} d([z]_P) \\ &= \overline{z}_P - d([z]_P) + (3I - P)^{-1}(2I - P) d([z]_P) \\ &= \overline{z}_P - (3I - P)^{-1}\{3I - P - (2I - P)\} d([z]_P) \\ &= \overline{z}_P - (3I - P)^{-1} d([z]_P). \end{aligned}$$

Notice that $(3I - P)^{-1} = \frac{1}{3}(I - \frac{1}{3}P)^{-1}$ exists by Neumann's series, since $\rho(\frac{1}{3}P) = \frac{1}{3}\rho(P) < \frac{1}{3} < 1$.

If (4.2.20) holds, then $2d([z]_P) \le (3I - P) d([z]_P)$ which implies $2(3I - P)^{-1} d([z]_P) \le d([z]_P)$. This inequality is equivalent to (4.2.17). □

In the case $n = 1$, the condition (4.2.20) and hence (4.2.17) always holds if $\rho(P) < 1$.

For $n > 1$, the inequality (4.2.17) can fail, as is shown by the example

$$P = \begin{pmatrix} 1/2 & 1/4 \\ 1/4 & 1/2 \end{pmatrix}, \quad d([z]_P) = \begin{pmatrix} 0.001 \\ 1 \end{pmatrix},$$

which yields $\rho(P) = 3/4 < 1$ and

$$\begin{aligned} 2(3I - P)^{-1} d([z]_P) &= \frac{2}{3} \begin{pmatrix} 5/6 & -1/12 \\ -1/12 & 5/6 \end{pmatrix}^{-1} \begin{pmatrix} 0.001 \\ 1 \end{pmatrix} \\ &= \frac{8}{99} \begin{pmatrix} 10 & 1 \\ 1 & 10 \end{pmatrix} \begin{pmatrix} 0.001 \\ 1 \end{pmatrix} \\ &= \frac{8}{99} \begin{pmatrix} 1.01 \\ 10.001 \end{pmatrix} \not\le d([z]_P). \end{aligned}$$

Exercises

Ex. 4.2.1. Show that the real function $f: \mathbb{R} \to \mathbb{R}$ with $f(x) = x - x$ is a contraction while its interval arithmetic evaluation is not.

Ex. 4.2.2. Let $[f]: \mathbb{IR}^n \to \mathbb{IR}^n$ be a P-contraction and let $[g]: \mathbb{IR}^n \to \mathbb{IR}^n$ be a Q-contraction. Show that the following assertions hold.

(a) $[f] \pm [g]$ is a $(P + Q)$-contraction, provided that $\rho(P + Q) < 1$.

(b) $[f]([g])$ is a PQ-contraction, provided that $\rho(PQ) < 1$.

(c) Let $n = 1$. Then $[f] \cdot [g]$ is an $(\alpha Q + \beta P)$-contraction, provided that $|[f]([x])| \le \alpha$, $|[g]([x])| \le \beta$ for all $[x] \in \mathbb{IR}$, and $|\alpha Q + \beta P| < 1$.

Ex. 4.2.3. Show that the function $[f]\colon \mathbb{IR}^2 \to \mathbb{IR}^2$, defined by

$$[f]([x]_1, [x]_2) = \begin{pmatrix} \frac{1}{2} \arctan\left(\frac{1}{2}\sin([x]_1)\right) + \frac{1}{2} e^{-[x]_2^2} \\ \frac{1}{4}\sin([x]_1) - \frac{1}{4}\sin([x]_2) \end{pmatrix},$$

is a P-contraction. Hint: Use Exercise 4.2.2.

Ex. 4.2.4. Show that the function $[f]\colon \mathbb{IR} \to \mathbb{IR}$ with $[f]([x]) = [x] \cdot [x]$ is not a P-contraction on $\mathbb{IR}$. Find a nondegenerate interval $[z]$ such that $[f]$ is a P-contraction on $[z]$ and maps $\mathbb{I}([z])$ into itself.

Ex. 4.2.5. Reformulate and prove the results for P-contractions for contractive interval mappings as well.

4.3 ε-inflation

Verifying and enclosing solutions of mathematical problems and improving the enclosures are basic aims of interval computations. If $[f]\colon \mathbb{IR}^n \to \mathbb{IR}^n$ is a P-contraction with $[f]([x]) = f([x])$ for some function $f\colon \mathbb{R} \to \mathbb{R}$, then Theorem 4.2.6 guarantees that f has a unique fixed point x^* to which the iterates $[x]^{k+1} = f([x]^k)$ converge independently of the starting vector $[x]^0$. Unfortunately, no iterate needs to contain x^* as can be seen from the simple example $f(x) = x/2$ when starting with $[x]^0 \equiv \tilde{x} = 1$. For P-contractions, Theorem 4.2.7 (a) provides a vector $[z]$ which encloses x^*. If one starts the iteration with this vector, then x^* remains enclosed in each iterate $[x]^k$. But often the contraction property or even the existence of a fixed point are unknown. In this case the additional assumptions of Theorem 4.2.7 (d) become important since they will guarantee a fixed point of f and an iterative behavior as in (4.2.9), even if $[f]$ is not a P-contraction.

Theorem 4.3.1. *Let $D \subseteq \mathbb{R}^n$ and let $f\colon D \to \mathbb{R}^n$ be continuous. If $[f]\colon \mathbb{I}(D) \to \mathbb{IR}^n$ is inclusion monotone and satisfies*

$$f(x) \in [f]([x]) \quad \text{for all } x \in [x] \text{ and arbitrary } [x] \subseteq D \tag{4.3.1}$$

and

$$[f]([z]) \subseteq [z] \quad \text{for some } [z] \subseteq D, \tag{4.3.2}$$

then f has at least one fixed point $x^* \in [z]$ and the iteration

$$[x]^0 = [z], \quad [x]^{k+1} = [f]([x]^k),\ k = 0, 1, \ldots$$

converges to some limit $[x]^*$ such that

$$x^* \in [x]^* \subseteq [x]^k \subseteq [x]^{k-1} \subseteq \cdots \subseteq [x]^0 = [z] \tag{4.3.3}$$

holds for all $k \in \mathbb{N}_0$ and all fixed points $x^* \in [z]$ of f.

In particular, these statements are true if (4.3.2) *and* $[f]([x]) = f([x]) \in L([z])$ *hold. In this case* $[x]^*$ *is a fixed point of* $[f]$.

Proof. The assumptions of the theorem imply $f(x) \in [f]([z]) \subseteq [z]$ for all vectors $x \in [z]$. Therefore, f maps $[z]$ into itself, and Brouwer's fixed point theorem 1.5.8 guarantees the existence of a fixed point x^* in $[z] = [x]^0$. The inclusion property of $[f]$ and (4.3.2) yield

$$x^* = f(x^*) \in [f]([x]^0) = [x]^1 \subseteq [x]^0,$$

and an inductive argument proves (4.3.3), where we used Theorem 2.5.5 for the existence of the limit $[x]^*$.

The rest of the theorem follows from the inclusion monotony, enclosure property and continuity of the interval arithmetic evaluation $f([x])$. □

Theorem 4.3.1 does not imply that $[x]^*$ is a point vector even if $[f]([x]) = f([x])$ as can be seen from the simple example $f(x) = x$, $[x]^0 \neq 0$.

There are essentially two ways to fulfill (4.3.2). Both start with an approximation $\tilde{x} \in \mathbb{R}^n$ of x^* which can be thought to be computed by some traditional (i.e., noninterval) method.

The first way uses an ansatz

$$[z] = \tilde{x} + \delta[-1, 1]e \tag{4.3.4}$$

and tries to compute $0 \leq \delta \in \mathbb{R}$ from (4.3.2). For instance if we choose $\tilde{x} = 1$ in the simple example $f(x) = x/2$, $[f]([x]) = f([x])$, we require

$$(1 + \delta[-1, 1])/2 = f(1 + \delta[-1, 1]) \subseteq 1 + \delta[-1, 1],$$

which leads to $1/2 + \delta/2 \leq 1 + \delta$ and $1/2 - \delta/2 \geq 1 - \delta$. The second inequality implies $\delta \geq 1$, whence the first holds trivially for these values. Thus $[z] = [0, 2]$ will fulfill (4.3.2) and contains the (only) fixed point $x^* = 0$ of f. Unfortunately, for nonlinear functions f an ansatz like (4.3.4) does not always work since it leads to nonlinear inequalities for δ. Therefore, we must look for a second way.

This second way iterates according to

$$[x]^{k+1} = f([x]^k_\varepsilon), \quad k = 0, 1, \ldots, \tag{4.3.5}$$

where frequently $[x]^0 = \tilde{x}$, and where $[x]^k_\varepsilon$ is any interval vector which contains $[x]^k$ in its interior. This means that in each step the method enlarges the momentary iterate $[x]^k$ slightly to a superset and iterates with it in place of $[x]^k$. The construction of $[x]^k_\varepsilon$ from $[x]^k$ is called ε-inflation. It usually depends on a small real parameter $\varepsilon > 0$ which provides the name. It goes back to Caprani and Madson [75] and particularly to Rump [311, 312]. One iterates according to (4.3.5) until

$$[x]^{k_0+1} \subseteq [x]^{k_0}_\varepsilon \tag{4.3.6}$$

holds for some index k_0, which is (4.3.2) with $[z] = [x]_\varepsilon^{k_0}$, or until some index $k_{\max}$ is reached. In the latter case one can continue by inflating with a different value of ε or stop with a remark concerning the failure of verification.

There are various possibilities for ε-inflation which, for simplicity, we demonstrate for $[x] \in \mathbb{IR}$. By means of the vector e there should be no problem to generalize them to $[x] \in \mathbb{IR}^n$. One often chooses $\varepsilon = 0.1$ and sometimes uses a second parameter η which is much smaller than ε and which takes care that $[x]$ is inflated even in particular cases. Frequently, η is chosen to be the smallest positive machine number.

Possibilities for ε-inflation

(i) $[x]_\varepsilon = [x] + [-\varepsilon, \varepsilon]$ (absolute inflation)

(ii) $$[x]_\varepsilon = \begin{cases} [x] + \varepsilon[-|\underline{x}|, |\overline{x}|], & \text{if } \underline{x} \cdot \overline{x} \neq 0 \\ [x] + [-\eta, \eta], & \text{if } \underline{x} \cdot \overline{x} = 0 \end{cases}$$
(inflation relatively to the bounds $\underline{x}, \overline{x}$)

(iii) $[x]_\varepsilon = [x] + \varepsilon[-|\underline{x}|, |\overline{x}|] + [-\eta, \eta]$

(iv) $$[x]_\varepsilon = \begin{cases} [x] + |[x]|[-\varepsilon, \varepsilon], & \text{if } [x] \neq 0 \\ [x] + [-\eta, \eta], & \text{if } [x] = 0 \end{cases}$$
(inflation relatively to the absolute value $|[x]|$)

(v) $[x]_\varepsilon = [x] + |[x]|[-\varepsilon, \varepsilon] + [-\eta, \eta]$

(vi) $$[x]_\varepsilon = \begin{cases} [x] + d([x])[-\varepsilon, \varepsilon] = (1+\varepsilon)[x] - \varepsilon[x], & \text{if } d([x]) = \overline{x} - \underline{x} \neq 0 \\ [x] + [-\eta, \eta], & \text{if } d[x] = 0 \end{cases}$$
(inflation relatively to the diameter $d([x])$)

(vii) $[x]_\varepsilon = [x] + d([x])[-\varepsilon, \varepsilon] + [-\eta, \eta] = (1+\varepsilon)[x] - \varepsilon[x] + [-\eta, \eta]$

(viii) $[x]_\varepsilon = [x] \cdot [1-\varepsilon, 1+\varepsilon] + [-\eta, \eta]$

(ix) $$\begin{cases} [y] = (1+\varepsilon)[x] - \varepsilon[x] \\ [x]_\varepsilon = [\underline{y}', \overline{y}'] \end{cases}$$

where $\underline{y}'$, $\overline{y}'$ denote the next preceding and the next succeeding machine number of $\underline{y}$ and $\overline{y}$, respectively.

This list from Mayer [211] could certainly be continued. The definition in (vii) was used in Lohner's software package AWA for verifying and enclosing solutions of initial value problems and boundary value problems; cf. Lohner [192]. The last definition for $[x]_\varepsilon$ was implemented in the scientific language PASCAL-XSC (cf. Klatte et al. [164]).

For our simple example above we choose $[x]_\varepsilon = [x] + \varepsilon[-1, 1]$ with $\varepsilon = 0.1$ and start the iteration with $[x]^0 \equiv \tilde{x} = 1$. Then $[x]^4 \subseteq [x]_\varepsilon^3$, i.e., the iteration with ε-inflation succeeds in (4.3.6) within four steps.

It is an open question under which conditions ε-inflation results in success. The interval arithmetic evaluation of $f(x) = 2x$ will certainly never imply (4.3.2) unless $[z] = 0$. For P-contractions $[f]$, however, success is guaranteed as we shall see, if the inflated vectors $[x]_\varepsilon^k$ can be written in the form

$$[x]_\varepsilon^{k+1} = [f]([x]_\varepsilon^k) + [\delta]^k$$

with $[\delta]^k$ tending to some limit $[\delta]$ which contains 0 in its interior. For inflations like (vii) these assumptions seem to be reasonable at least for sequences $([x]^k)$ which converge to some limit $[x]^*$ since then, for $k \to \infty$ and (4.3.5), the intervals $[\delta]^k = d([x]^{k+1})[-\varepsilon, \varepsilon] + [-\eta, \eta]$ tend to $[\delta] = d([x]^*)[-\varepsilon, \varepsilon] + [-\eta, \eta]$ which apparently fulfills $0 \in \mathrm{int}([\delta])$.

Notice that the enclosure of rounding errors can also be incorporated in $[f]$: think of $[f]([x])$ to be the outward rounded interval arithmetic evaluation $f([x])$, so that

$$f([x]) \subseteq [f]([x])$$

holds. The crucial condition, however, is the P-contraction property for $[f]$.

Theorem 4.3.2. *Let $[f]: D = \mathbb{IR}^n \to \mathbb{IR}^n$ be a P-contraction satisfying* (4.3.1). *Iterate by ε-inflation according to*

$$[x]_\varepsilon^{k+1} = [f]([x]_\varepsilon^k) + [\delta]^k\,, \tag{4.3.7}$$

where $[\delta]^k \in \mathbb{IR}^n$ are vectors which converge to some limit $[\delta]$. If $[\delta]$ contains 0 in its interior, then there is an integer $k_0 = k_0([x]_\varepsilon^0)$ such that

$$[f]([x]_\varepsilon^{k_0}) \subseteq \mathrm{int}([x]_\varepsilon^{k_0}) \tag{4.3.8}$$

holds.

Proof. Let $[h]([x]) = [f]([x]) + [\delta]$. Then by Theorem 2.5.7 we get

$$q([h]([x]), [h]([y])) = q([f]([x]), [f]([y])) \le P\, q([x], [y]), \tag{4.3.9}$$

hence $[h]$ is a P-contraction. By Theorem 4.2.6 it has a unique fixed point $[x]^*$ which satisfies

$$[x]^* = [f]([x]^*) + [\delta]. \tag{4.3.10}$$

Assume for the moment that

$$\lim_{k\to\infty} [x]_\varepsilon^k = [x]^* \tag{4.3.11}$$

holds for the sequence in (4.3.7). By the continuity of $[f]$ we have

$$\lim_{k\to\infty} [f]([x]_\varepsilon^k) = [f]([x]^*). \tag{4.3.12}$$

Since $0 \in \operatorname{int}([\delta])$, equation (4.3.10) implies

$$[f]([x]^*) \subseteq \operatorname{int}([x]^*).$$

Together with (4.3.11) and (4.3.12) this yields (4.3.8) for all sufficiently large integers k_0.

We will prove now the assumption (4.3.11). With Theorem 2.5.7 we obtain

$$\begin{aligned} q([x]_\varepsilon^{k+1}, [x]^*) &= q([f]([x]_\varepsilon^k) + [\delta]^k, [f]([x]^*) + [\delta]) \\ &\le Pq([x]_\varepsilon^k, [x]^*) + q([\delta]^k, [\delta]) \\ &\le P^2 q([x]_\varepsilon^{k-1}, [x]^*) + Pq([\delta]^{k-1}, [\delta]) + q([\delta]^k, [\delta]) \\ &\le \cdots \le P^{k+1} q([x]_\varepsilon^0, [x]^*) + \sum_{i=0}^{k} P^i q([\delta]^{k-i}, [\delta]). \end{aligned} \tag{4.3.13}$$

Fix $\theta > 0$. Since $\rho(P) < 1$, there is an integer m such that

$$P^i \le \theta ee^T \quad \text{for all } i \ge m.$$

Since $\lim_{k\to\infty} [\delta]^k = [\delta]$, there are a constant $c > 0$ and an integer $k' > m$ with

$$q([\delta]^i, [\delta]) \le ce \quad i = 0, 1, \ldots,$$

and

$$q([\delta]^{k-i}, [\delta]) \le \theta e, \quad k \ge k',\ i = 0, 1, \ldots, m.$$

Thus with (4.3.13) and $k \ge k'$ we get

$$\begin{aligned} q([x]_\varepsilon^{k+1}, [x]^*) &\le \theta ee^T\, q([x]_\varepsilon^0, [x]^*) + \sum_{i=0}^{m} P^i \theta e + P^{m+1} \sum_{i=m+1}^{k} P^{i-(m+1)} ce \\ &\le \theta \{ ee^T q([x]_\varepsilon^0, [x]^*) + (I - P)^{-1} e + ee^T (I - P)^{-1} ce \}. \end{aligned}$$

Since the expression in braces is independent of θ, m and k, and since θ can be chosen arbitrarily small, (4.3.11) holds. □

Example 4.3.3. Let $[b] \in \mathbb{IR}^n$, $[A] \in \mathbb{IR}^{n\times n}$, and let $[D]_\varepsilon \in \mathbb{IR}^{n\times n}$ be diagonal with $I \in [D]_\varepsilon$ and $\rho(|[D]_\varepsilon [A]|) < 1$. Define $[f]$ by $[f]([x]) = [D]_\varepsilon([b] + [A][x])$. Then Theorem 4.3.2 applies. □

In many cases, $[f]$ is only a local P-contraction with respect to some vector $[z]$. Then based on Theorem 4.2.6 one can formulate a local variant of Theorem 4.3.2 which we first formulate and prove for $[\delta]^k = [\delta]$, $k = 0, 1, \ldots$.

Theorem 4.3.4. *Let $[f] : \mathbb{IR}^n \to \mathbb{IR}^n$ satisfy (4.3.1) and let $[\delta]$, $[z] \in \mathbb{IR}^n$ fulfill the following properties:*

(i) $0 \in \operatorname{int}([\delta])$,

(ii) *$[f]$ is a local P-contraction on*

$$[z]_P \supseteq [z] + (I - P)^{-1} d([z])[-1, 1]. \tag{4.3.14}$$

If $[x]^0_\varepsilon \subseteq [z]$ and $[x]^1_\varepsilon \subseteq [z]$ hold for the iterates from (4.3.7) with $[\delta]^k = [\delta]$, then there is an integer $k_0 = k_0([x]^0_\varepsilon)$ such that (4.3.8) is true. In particular, f from (4.3.1) has a fixed point in $[x]^{k_0}_\varepsilon$.

Proof. Since $[h]([x]) = [f]([x]) + [\delta]$ fulfills (4.3.9) for all $[x], [y] \subseteq [z]_P$, the function $[h]$ is a local P-contraction on $[z]_P$. By Theorem 4.2.9 there is a vector $[x]^* \subseteq [z]_P$ which satisfies

$$\lim_{k\to\infty} [x]^k_\varepsilon = [x]^* = h([x]^*) = [f]([x]^*) + [\delta]. \tag{4.3.15}$$

Since $0 \in \operatorname{int}([\delta])$, this yields

$$[f]([x]^*) \subseteq \operatorname{int}([x]^*), \tag{4.3.16}$$

and the assertion follows from (4.3.12), (4.3.16) and from the first equality in (4.3.15). □

In order to get the full analogy of Theorem 4.3.2, condition (4.3.14) has to be replaced by a slightly more complicated one.

Theorem 4.3.5. *Let $[f]\colon \mathbb{IR}^n \to \mathbb{IR}^n$ satisfy (4.3.1), and let $[\delta]^k, [\delta], [z] \in \mathbb{IR}^n$ fulfill the following properties:*
(i) $\lim_{k\to\infty}[\delta]^k = [\delta]$,
(ii) $0 \in \operatorname{int}([\delta])$,
(iii) *$[f]$ is a local P-contraction on*

$$[z]_P \supseteq [z] + (I - P)^{-1}(d([z]) + v)\,[-1, 1]$$

with

$$v = \max\{\, w, d([z]) + (I - P)^{-1} d([z])\,\},$$

where

$$w = (\max_k\{\, q([\delta]^k_i, [\delta]_i)\,\}) \in \mathbb{R}^n.$$

If $[x]^0_\varepsilon \subseteq [z]$ and $[x]^1_\varepsilon \subseteq [z]$ hold for the iterates from (4.3.7), then there is an integer $k_0 = k_0([x]^0_\varepsilon)$ such that (4.3.8) is true. In particular, f from (4.3.1) has a fixed point in $[x]^{k_0}_\varepsilon$.

Proof. As in the proof of Theorem 4.3.4 there is a vector $[x]^* \subseteq [z]_P$ which satisfies

$$[x]^* = [f]\,([x]^*) + [\delta].$$

It remains to show

$$\lim_{k\to\infty} [x]^k_\varepsilon = [x]^*,$$

then the proof can be finished as in Theorem 4.3.2. To this end, we prove by induction

$$[x]^k_\varepsilon \subseteq [z]_P, \quad k = 0, 1, \ldots. \tag{4.3.17}$$

For $k = 0$ and $k = 1$ this holds by assumption.

Let $[x]_\varepsilon^0, [x]_\varepsilon^1, \ldots, [x]_\varepsilon^k \subseteq [z]_P$. Then as in (4.3.13) we have

$$q([x]_\varepsilon^{k+1}, [x]^*) \le P^{k+1} q([x]_\varepsilon^0, [x]^*) + \sum_{i=0}^{k} P^i q([\delta]^{k-i}, [\delta]). \tag{4.3.18}$$

Since $[x]_\varepsilon^0 \subseteq [z]$ by assumption, and since $[x]^* \subseteq [z] + (I - P)^{-1} d([z])\,[-1, 1]$ by Theorem 4.2.9, we get

$$q([x]_\varepsilon^0, [x]^*) \le |\underline{z} - \overline{z}| + (I - P)^{-1} d([z]) = d([z]) + (I - P)^{-1} d([z]),$$

hence (4.3.18) implies

$$\begin{aligned} q([x]_\varepsilon^{k+1}, [x]^*) &\le P^{k+1}\{\, d([z]) + (I - P)^{-1} d([z]) \,\} + \sum_{i=0}^{k} P^i w \\ &\le P^{k+1} v + \sum_{i=0}^{k} P^i v \le \sum_{i=0}^{\infty} P^i v = (I - P)^{-1} v. \end{aligned}$$

Therefore,

$$\begin{aligned} [x]_\varepsilon^{k+1} &\subseteq [x]^* + (I - P)^{-1} v[-1, 1] \\ &\subseteq [z] + (I - P)^{-1} d([z])\,[-1, 1] + (I - P)^{-1} v[-1, 1] \\ &\subseteq [z]_P, \end{aligned}$$

and (4.3.17) is true.

The proof finishes now following the lines of Theorem 4.3.2. The inclusion (4.3.17) is needed to justify the first estimate in (4.3.13) which is now based on the P-contraction property on $[z]_P$. □

We illustrate our theoretical results by a simple example which fulfills the conditions of Theorem 4.3.4.

Example 4.3.6. Let $f(x) = x \cdot x$, $[f]([x]) = [x] \cdot [x]$, $[z] = [-2, 2] \cdot 10^{-2}$, $P = \frac{1}{2}$, $[x]_\varepsilon = [x] + [\delta]$ with $[\delta] = [-1, 1] \cdot 10^{-4}$. Then $0 \in \operatorname{int}([\delta])$ and

$$\begin{aligned} [z]_P &= [z] + (I - P)^{-1} d([z])\,[-1, 1] = [-0.1, 0.1], \\ q([x] \cdot [x], [y] \cdot [y]) &\le q([x] \cdot [x], [x] \cdot [y]) + q([x] \cdot [y], [y] \cdot [y]) \\ &\le (|[x]| + |[y]|)\, q([x], [y]) \\ &\le 0.2\, q([x], [y]) \quad \text{for all } [x], [y] \subseteq [z]_P, \end{aligned}$$

hence $[f]$ is a local P-contraction on $[z]_P$.

With $[x]^0 = 10^{-2}$ we get

$$\begin{aligned} [x]_\varepsilon^0 &= [0.99, 1.01] \cdot 10^{-2} \subseteq [z], \\ [x]^1 &= [f]([x]_\varepsilon^0) = [0.9801, 1.0201] \cdot 10^{-4}, \\ [x]_\varepsilon^1 &= [-0.0199, 2.0201] \cdot 10^{-4} \subseteq [z]. \end{aligned}$$

Thus the assumptions of Theorem 4.3.4 are fulfilled, whence (4.3.8) holds for some integer k_0. Indeed,

$$[x]^2 = [f]([x]_\varepsilon^1) = [-0.04019999, 4.08080401] \cdot 10^{-8}$$
$$\subseteq \operatorname{int}([x]_\varepsilon^1). \qquad \square$$

How to choose ε generally in verification algorithms remains an open question. Some remarks on this problem – primarily based on experience – can be found in König [170] and Rump [311].

For an effective estimate of $(I - P)^{-1}$ which occurs in the Theorems 4.2.9, 4.3.4 and 4.3.5, we refer to Hansen [131].

Exercises

Ex. 4.3.1. Prove Example 4.3.3.

Notes to Chapter 4

To 4.1: Factorable functions are used in Fischer [97]. Higher order methods are considered in Cornelius, Lohner [80] where also Lemma 4.1.21, Theorem 4.1.22, and Corollary 4.1.23 can be found. Extensions are contained in Alefeld, Lohner [34]. The remaining part of Section 4.1 is standard knowledge in interval analysis. See also Ratschek, Rokne [284]. The quality of enclosure is discussed in Alefeld [9, 16], and Nickel [261].

To 4.2: The concept of P-contraction for *intervals* was already used in Alefeld, Herzberger [26]. Theorem 4.2.6 is an extension of Schröder's theorem in Schröder [333] and Neumaier [257], Example 4.2.3 goes back to Otto Mayer [227]. Most of the remaining material of Section 4.2 was published in Mayer [211]; see also Mayer [212].

To 4.3: The idea of ε-inflation can be found in Alefeld, Apostolatos [19] and Caprani, Madsen [75] although it was not so named there. Independently, the tool was introduced for applications by Siegfried M. Rump in [311, 312]. In Rump's paper [315] a first theoretical result on the success of ε-inflation was proved for the affine case which we repeated in this book as Example 4.3.3. Most of the remaining results in Section 4.3 on ε-inflation were published in Mayer [211, 212].

5 Linear systems of equations

5.1 Motivation

Linear systems of equations are basic in mathematical modeling and in numerical analysis. Even if a model is nonlinear or nondiscrete, linearization and discretization finally can lead to a linear system. The reason why one tries to end up with such systems lies in their simplicity and their 'easy' numerical treatment. A typical example is Newton's method for the computation of a zero x^* of a nonlinear (sufficiently smooth) function $f: D \subseteq \mathbb{R}^n \to \mathbb{R}^n$. It results from a Taylor expansion of $f(x^*)$ at some vector $x^k \neq x^*$ cut off behind the linear term. Thus the nonlinear system $f(x^*) = 0$ transforms to the linear approximation $0 \approx f(x^k) + f'(x^k)(x^* - x^k)$. Changing this approximation to an equality by replacing x^* by some vector x^{k+1} leads to the linear system

$$f'(x^k)(x^{k+1} - x^k) = -f(x^k).$$

This is the Newton iteration in the form which is used in practice.

In some cases a linearization is hidden behind some assumptions at the beginning. This holds for instance if linear proportionality is assumed. We will illustrate this phenomenon by Leontief's input–output model, which will serve us at the same time as motivation for *interval* linear systems.

The static open input–output model was invented in order to describe the relations between different sectors of a complex economic system on which future developments are based. To be more precise, let a national economy be divided into n sectors each of which is assumed to produce a single kind of goods. The goods are consumed by the n sectors in order to maintain their production and by an additional sector called the open sector or final demand. In order to be comparable, the gross outputs x_i are measured in a common monetary unit. That part of x_i which is consumed by the j-th sector is assumed to be proportional to the production x_j of this sector itself. Introducing so-called input coefficients c_{ij} for this proportionality and denoting by b_i the final demand or net output of the i-th sector yields the equilibrium equations

$$x_i = \sum_{i=1}^{n} c_{ij}x_j + b_i, \quad i = 1, \ldots, n, \tag{5.1.1}$$

which describe the distribution of the total production x_i of the i-th sector. Introducing vectors, we can write the system (5.1.1) equivalently in fixed point form $x = Cx + b$ with $C \geq O$, $b \geq 0$, $x \geq 0$. In order to satisfy each demand of the open sector it is necessary that the matrix $A = I - C \in Z^{n\times n}$ has a nonnegative inverse, hence it is an M-matrix.

Although originally the model was assumed to be static, the input coefficients are often subject to uncertainties at the start or they do not remain constant over time.

DOI 10.1515/9783110499469-006

Mostly lower and upper bounds for c_{ij} and b_i are available which then form enclosing intervals $[c]_{ij}$ and $[b]_i$. Therefore, a whole set of linear systems $Ax = b$ must be studied with $A \in [A] = I - [C]$ and $b \in [b]$. The set S of the corresponding individual solutions will be considered in the next section. There we are going to show that S can be represented in a simple but costly manner. In fact, the description of S is NP-hard, since in the general case one has to study the solution of all the 2^{n^2+n} linear systems $Ax = b$ with $A \in \partial[A]$ and $b \in \partial[b]$. (Cf. for instance Garey, Johnson [113] or van Leeuwen [355] for a definition of NP-hardness.) Therefore, one looks for simpler types of sets which enclose S rather than representing it. For our purposes this means that we look for corresponding interval vectors. But we point out that not all of our algorithms to construct such vectors will be polynomial-time ones. For details in this respect see Rohn's contribution in Herzberger [144] or in Fiedler et al. [95], or consider Kreinovich et al. [175], Rohn [303], Rohn, Kreinovich [309], for example.

5.2 Solution sets

We start with a basic definition suggested by the preceding section.

Definition 5.2.1. Let $[A] \in \mathbb{IR}^{n\times n}$, $[b] \in \mathbb{IR}^n$.

(a) The symbolic equation
$$[A]x = [b] \tag{5.2.1}$$
is called an interval linear system. It abbreviates the set of all linear systems $Ax = b$ with $A \in [A]$ and $b \in [b]$.

(b) The set
$$S = \{x \mid Ax = b,\ A \in [A],\ b \in [b]\} \tag{5.2.2}$$
is called the solution set of (5.2.1).

(c) Any interval vector $[x]$ which contains at least one solution of each linear system $Ax = b$ with $A \in [A]$, $b \in [b]$ is called a solution of (5.2.1).

Notice that $[A][x]$ does not represent a linear mapping from $\mathbb{IR}^n$ to $\mathbb{IR}^n$ although the word 'linear' appears in the definition of (5.2.1). Here we followed tradition. Moreover, a solution $[x]$ of (5.2.1) usually does not satisfy $[A][x] = [b]$ algebraically.

If $[A]$ is regular, a solution of (5.2.1) is a superset of S. This follows directly from Definition 5.2.1. If $[A]$ is singular, S contains a nontrivial affine subspace which clearly cannot be enclosed by a compact interval vector. This justifies our definition in (c), which at a first glance seems to be somehow strange.

Solutions of (5.2.1) will be considered in the next sections. In the present one we will take a closer look to the solution set itself. Our first theorem contains criteria which guarantee that a fixed vector $x \in \mathbb{R}^n$ belongs to S.

Theorem 5.2.2. *Denote by S the solution set (5.2.2), where $[A] \in \mathbb{IR}^{n\times n}$ and $[b] \in \mathbb{IR}^n$. Then the following properties are equivalent, where $[A]x$ denotes the product of $[A] \in \mathbb{IR}^{n\times n}$ and $x \in \mathbb{R}^n$.*

(a) $x \in S$,

(b) $[A]x \cap [b] \neq \emptyset$,

(c) $0 \in [A]x - [b]$,

(d) $|\check{A}x - \check{b}| \le \mathrm{rad}([A])|x| + \mathrm{rad}([b])$, *(Oettli–Prager inequality)*

(e) $\exists D \in \mathbb{R}^{n\times n}\colon |D| \le I \wedge \check{A}x - \check{b} = D\,(\mathrm{rad}([A])|x| + \mathrm{rad}([b]))$,

(f) $\left\{\begin{matrix} \underline{b}_i - \sum_{j=1}^n \hat{a}_{ij}^+ x_j \le 0 \\ -\overline{b}_i + \sum_{j=1}^n \hat{a}_{ij}^- x_j \le 0 \end{matrix}\right\}\quad i = 1, \dots, n,$

where $\hat{a}_{ij}^-$ and $\hat{a}_{ij}^+$ are defined by the equality

$$[a]_{ij} = \begin{cases} [\hat{a}_{ij}^-, \hat{a}_{ij}^+] & \text{if } x_j \ge 0 \\ [\hat{a}_{ij}^+, \hat{a}_{ij}^-] & \text{if } x_j < 0 \end{cases}.$$

Proof. The implications '(a) $\Rightarrow$ (b) $\Leftrightarrow$ (c)' and '(d) $\Leftrightarrow$ (e)' can be seen immediately. Notice that the property $|D| \le I$ in (e) implies that D is a diagonal matrix. The equivalence of (b) and (d) follows from the Theorems 2.4.4, 2.4.5 and 2.4.8 (c). Property (f) follows from (c) and the definition of $a_{ij}^\pm$, whence

$$\min([A]x - [b])_i = -\overline{b}_i + \sum_{j=1}^n \hat{a}_{ij}^- x_j \le 0 \le -\underline{b}_i + \sum_{j=1}^n \hat{a}_{ij}^+ x_j = \max([A]x - [b])_i,$$
$$i = 1, \dots, n.$$

The implication '(f) $\Rightarrow$ (a)' can be seen as follows. From (f) we obtain

$$\underline{([A]x)_i} = \sum_{j=1}^n \hat{a}_{ij}^- x_j \le \overline{b}_i, \quad \overline{([A]x)_i} = \sum_{j=1}^n \hat{a}_{ij}^+ x_j \ge \underline{b}_i.$$

Define $\alpha_{ij}(t) = t\hat{a}_{ij}^+ + (1-t)\hat{a}_{ij}^-$, $\beta_i(t) = \sum_{j=1}^n \alpha_{ij}(t)x_j$, $t \in [0,1]$, $i, j = 1, \dots, n$. Since $\beta_i(t)$ is continuous, it assumes for $0 \le t \le 1$ all values between $\beta_i(0)$ and $\beta_i(1)$. From $\beta_i(0) \le \overline{b}_i$, $\underline{b}_i \le \beta_i(1)$ and $\underline{b}_i \le \overline{b}_i$ there is some value $t = t_i \in [0,1]$ such that $\beta_i(t_i) \in [b]_i$. With $\hat{A} = A(t_1, \dots, t_n) = (\alpha_{ij}(t_i))$, $\hat{b} = (\beta_i(t_i))$ we get $\hat{A}x = \hat{b}$, $\hat{A} \in [A]$, $\hat{b} \in [b]$, hence $x \in S$. □

The equivalence '(a) $\Leftrightarrow$ (d)' is the famous *Oettli–Prager theorem* in Oettli and Prager [264] whose interval formulation '(a) $\Leftrightarrow$ (b) $\Leftrightarrow$ (c)' is due to Beeck [62, 63]. Property (e) goes back to Rohn [308] while (f) can be found in Hartfiel [136]. Since in (f) the coefficients of the inequalities remain fixed as long as one stays in a fixed orthant O, we immediately obtain the first part of (a) of the subsequent theorem.

Theorem 5.2.3. *Denote by O any fixed orthant in $\mathbb{R}^n$ and let S be the solution set (5.2.2). Then the following statements hold.*

(a) *The intersection $S \cap O$ can be represented as the intersection of finitely many half-spaces. It is convex.*
(b) *If $[A]$ is regular, then S is connected and compact, but not necessarily convex. If $[A]$ is singular and $S \neq \emptyset$, then S is not compact and need not be connected.*

Proof. Since half-spaces are convex, the same holds for intersections among them. Together with (f) of Theorem 5.2.2 this proves (a). If $[A]$ is regular, then the mapping $f(A, b) = A^{-1}b$ exists and is continuous on the compact, connected set $[A] \times [b]$. Therefore, its range S has the same properties. If $[A]$ is singular, then clearly S is unbounded if there is a singular matrix $\tilde{A} \in [A]$ and a right-hand side $\tilde{b} \in [b]$ such that $\tilde{A}x = \tilde{b}$ is solvable. But even if no such singular linear system has a solution, Cramer's rule shows (how?) that S is unbounded provided that it is not empty. Connectivity may be lost as the example $[A] = [-1, 1]$, $[b] = 1$, $S = (-\infty, -1] \cup [1, \infty)$ in Jansson [156] shows. □

We illustrate the last two theorems by an example.

Example 5.2.4. Let

$$[A] = \begin{pmatrix} 1 & [0, 1] \\ [0, 1] & [-4, -1] \end{pmatrix}, \quad [b] = \begin{pmatrix} [0, 2] \\ [0, 2] \end{pmatrix}.$$

For

$$A = \begin{pmatrix} 1 & \alpha \\ \beta & -\gamma \end{pmatrix} \in [A]$$

we get

$$A^{-1} = \frac{1}{\gamma + \alpha\beta} \begin{pmatrix} \gamma & \alpha \\ \beta & -1 \end{pmatrix}$$

with $\alpha, \beta \in [0, 1]$, $\gamma \in [1, 4]$. Since $\underline{b} \geq 0$ the first component of $A^{-1}b$ is nonnegative for all $b \in [b]$. Therefore, S is completely contained in the union $O_1 \cup O_4$ of the first and the fourth quadrant. According to Theorem 5.2.2 (f) the intersection $S \cap O_1$ is characterized by

$$0 \leq x_1 \leq 2, \quad 0 \leq x_2 \leq x_1,$$

while $S \cap O_4$ can be described by

$$-2 \leq x_2 \leq 0, \quad 0 \leq x_1 \leq 2 - x_2.$$

Figure 5.2.1 shows S together with some subsets which will be explained later on.

Asking for S or an enclosure of S is not the only possible task in connection with linear systems. When considering Leontief's input–output model of Section 5.1 one might be interested in the question of which gross outputs x lead to a final demand $b \in [b]$ for *each* $A \in [A]$. The so-called *tolerance solution set*

$$S_{\text{tol}} = \{x \mid (\forall A \in [A])\,(\exists b \in [b])\,(Ax = b)\}$$

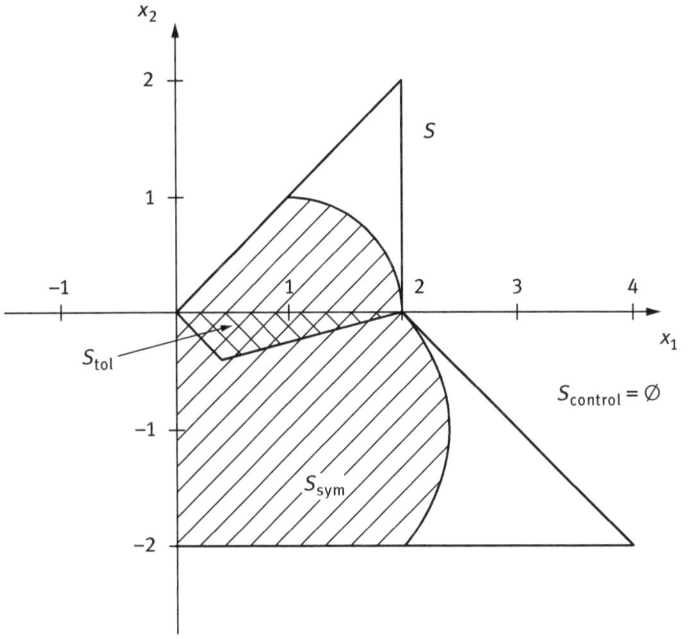

Fig. 5.2.1: Solution sets of Example 5.2.4.

gathers these particular solutions. Moreover, one might ask which gross outputs x can satisfy *each* final demand b if $A \in [A]$ is chosen appropriately (dependent on b). This leads to the control solution set

$$S_{\text{control}} = \{\, x \mid (\forall b \in [b])\, (\exists A \in [A])\, (Ax = b)\,\}.$$

Both new solution sets can be expressed by means of set theoretic symbols, partly combined with interval arithmetic, or by a midpoint-radius formulation.

Theorem 5.2.5. *Let* $[A] \in \mathbb{IR}^{n\times n}$, $[b] \in \mathbb{IR}^n$. *Then the following equivalences hold.*

(a) $x \in S_{\text{tol}} \Leftrightarrow [A]x \subseteq [b] \Leftrightarrow |\check{b} - \check{A}x| \le \text{rad}([b]) - \text{rad}([A])|x|$. (Rohn [299])

(b) $x \in S_{\text{control}} \Leftrightarrow [A]x \supseteq [b] \Leftrightarrow |\check{b} - \check{A}x| \le \text{rad}([A])|x| - \text{rad}([b])$. (Lakeyev, Noskov [186])

Proof. The proof follows immediately from the Theorems 2.4.4, 2.4.5, and 2.4.8 (a). □

Theorem 5.2.5 shows that the solution sets S_{tol} and S_{control} can also be expressed by means of appropriate linear inequalities. Using the subset property there we can see that for the tolerance solution set S_{tol} in Example 5.2.4 they read

$$0 \le x_1, \quad 0 \le -4x_2, \quad x_1 + x_2 \le 2, \quad x_1 - x_2 \le 2$$

in the first quadrant O_1 and

$$0 \le x_1 + x_2, \quad 0 \le -x_2, \quad x_1 \le 2, \quad x_1 - 4x_2 \le 2$$

in the forth quadrant O_4. Here we have already used that $x_1 \geq 0$ holds for the elements of S_{tol} as a subset of S. An easy calculation yields $S_{\text{tol}} \cap O_1 = [0, 2] \times [0, 0]$ while $S_{\text{tol}} \cap O_4$ is characterized by the two inequalities $x_2 \geq -x_1$, $x_2 \geq -1/2 + x_1/4$ for $(x_1, x_2) \in O_4$. The tolerance solution set is illustrated in Figure 5.2.1.

The corresponding control solution set S_{control} is empty. This can be seen from the inequalities

$$x_1 \leq 0, \quad -4x_2 \leq 0, \quad 2 \leq x_1 + x_2, \quad 2 \leq x_1 - x_2$$

in O_1 and

$$x_1 + x_2 \leq 0, \quad -x_2 \leq 0, \quad 2 \leq x_1, \quad 2 \leq x_1 - 4x_2$$

in O_4.

From Theorem 5.2.5 we can also deduce that for a regular matrix $[A] \in \mathbb{IR}^{n\times n}$ each of the solution sets S_{tol}, S_{control} is either empty or a closed polyhedron. Moreover, the first one is convex. Additional properties for these sets as well as further references can be found, e.g., in Fiedler et al. [95], Chapter 2, and Shary [342].

While the three solution sets described above do not restrict $[A]$ it is sometimes clear from the origin of the problem that the matrices $A \in [A]$ of the underlying point systems share some property such as symmetry or Toeplitz form. Therefore, it is useful to modify S in this respect. In Neumaier [257] the *symmetric* solution set

$$S_{\text{sym}} = \{\, x \mid Ax = b,\ A = A^T,\ A \in [A],\ b \in [b] \,\} \subseteq S \tag{5.2.3}$$

was considered. It can be described by means of linear and quadratic inequalities. In order to show this we proceed in two ways: The first one describes S_{sym} by means of a particular set of inequalities which extend the Oettli–Prager criterion in Theorem 5.2.2. The second one indicates a way to describe even more general solution sets than (5.2.3). Both are based on some kind of Fourier–Motzkin elimination known from linear programming; cf. for instance Schrijver [332]. How it works can be seen by proving the equivalence '(a) ⇔ (f)' of Theorem 5.2.2 once more, but with different means. For simplicity we restrict ourselves to $S_1 = S \cap D$, where $D = O_1$ denotes the first orthant of $\mathbb{R}^n$. Trivially, S_1 is characterized by

$$S_1 = \{\, x \in D \mid \exists a_{ij}, b_i \in \mathbb{R}\colon \text{(5.2.4)–(5.2.6) hold} \,\}$$

where

$$\sum_{j=1}^{n} a_{ij}x_j \leq b_i \leq \sum_{j=1}^{n} a_{ij}x_j, \quad i = 1, \ldots, n, \quad (\text{i.e., } (Ax)_i = b_i) \tag{5.2.4}$$

$$\underline{a}_{ij} \leq a_{ij} \leq \overline{a}_{ij}, \qquad i, j = 1, \ldots, n, \tag{5.2.5}$$

$$\underline{b}_i \leq b_i \leq \overline{b}_i, \qquad i = 1, \ldots, n. \tag{5.2.6}$$

Those inequalities in (5.2.4)–(5.2.5) which contain a_{11} can be rewritten as

$$b_1 - \sum_{j=2}^{n} a_{1j}x_j \le a_{11}x_1, \tag{5.2.7}$$

$$\underline{a}_{11} \le a_{11}, \tag{5.2.8}$$

$$a_{11}x_1 \le b_1 - \sum_{j=2}^{n} a_{1j}x_j, \tag{5.2.9}$$

$$a_{11} \le \overline{a}_{11}. \tag{5.2.10}$$

Multiply (5.2.8) and (5.2.10) by x_1, combine each left-hand side of (5.2.7), (5.2.8) with each right-hand side of (5.2.9), (5.2.10) and drop the two trivial inequalities. Then this action results in the two nontrivial inequalities

$$b_1 - \sum_{j=2}^{n} a_{1j}x_j \le \overline{a}_{11}x_1, \tag{5.2.11}$$

$$\underline{a}_{11}x_1 \le b_1 - \sum_{j=2}^{n} a_{1j}x_j, \tag{5.2.12}$$

which are supplemented by

$$\text{the original } a_{11}\text{-free inequalities.} \tag{5.2.13}$$

Hence

$$S_1 \subseteq S_2 = \{\, x \in D \mid \exists a_{ij}\ (i \ne 1 \text{ if } i = j),\ b_i \in \mathbb{R}\colon \text{(5.2.11)–(5.2.13) holds} \,\}.$$

Since the converse $S_2 \subseteq S_1$ is also true (see the proof of the subsequent Theorem 5.2.7), one ends up with $S_1 = S_2$, where in S_2 the entry a_{11} is replaced by the bounds $\underline{a}_{11}, \overline{a}_{11}$ of the given interval $[a]_{11}$. It is obvious that this process can be repeated for the remaining entries a_{ij} and b_i. One finally gets the inequalities of Theorem 5.2.2 (f).

We will summarize the crucial steps of this procedure.

(1) Fix an orthant and consider only those inequalities which contain a parameter to be replaced.
(2) Isolate this parameter in each of these inequalities.
(3) Multiply each of the newly arranged inequalities by appropriate functions in $x_1, \dots, x_n$ such that the isolated parameter transforms to the same expression, say E, wherever it occurs.
(4) Now disaggregate the inequalities into two groups: one for which the inequalities have the form $\dots \le E$, and the other one for which the inequalities read $E \le \dots$.
(5) Combine each inequality from the first group with each of the second group dropping the middle term E. Add to the set of inequalities those ones which have not been considered until now since they were E-free. □

In this way we could eliminate a particular parameter in inequalities which characterized some given set ending up with modified inequalities which characterize the same set.

We will apply these ideas in the proof of the following theorem on S_{sym} which modifies, reformulates and improves a result by Hladík [149] and which completes results in Alefeld, Kreinovich, and Mayer [28, 29, 30]. While Hladík's proof is based on results from linear programming we use different means and stay more elementary (cf. Mayer [217, 219]). In order to formulate the theorem we use the symbol $\prec_{\mathrm{lex}}$ which denotes strict lexicographic ordering of vectors $x, y \in \mathbb{R}^n$, i.e., $x \prec_{\mathrm{lex}} y$ if for some $k \le n$ we have $x_i = y_i$, $i < k$, and $x_k < y_k$. In addition, we remark that S_{sym} is empty if $[A] \in \mathbb{IR}^{n\times n}$ does not contain a symmetric matrix as an element. If $\underline{A} \ne \underline{A}^T$ or $\overline{A} \ne \overline{A}^T$, we can replace $[A]$ by the largest matrix $[B] \subseteq [A]$ with $[B] = [B]^T$ since $[A] \setminus [B]$ does not contain a symmetric matrix as an element and therefore does not influence S_{sym}. This is the reason why we will assume $[A] = [A]^T$, without loss of generality, from the start.

Theorem 5.2.6. *Let $[A] = [A]^T \in \mathbb{IR}^{n\times n}$, $[b] \in \mathbb{IR}^n$, $x \in \mathbb{R}^n$, $r = \check{b} - \check{A}x$. Then $x \in S_{\mathrm{sym}}$ if and only if*

$$x \in S \tag{5.2.14}$$

and if for all vectors $p, q \in \{0, 1\}^n$ with

$$0 \ne p \prec_{\mathrm{lex}} q \quad \text{and} \quad p^T q = 0 \tag{5.2.15}$$

one of the following four relations (5.2.16)–(5.2.19) *is satisfied.*

$$|x|^T\,|D_p \,\mathrm{rad}([A]) - \mathrm{rad}([A])D_q| \cdot |x| + |x|^T\,|D_p - D_q|\,\mathrm{rad}([b]) \ge |x^T(D_p - D_q)r|, \tag{5.2.16}$$

$$|x|^T(D_p \,\mathrm{rad}([A])D_{\overline{q}} + D_{\overline{p}}\,\mathrm{rad}([A])D_q)\,|x| + |x|^T(D_p + D_q)\,\mathrm{rad}([b]) \ge |x^T(D_p - D_q)r|, \tag{5.2.17}$$

$$(x^T D_p([b] - [A]D_{\overline{q}}x)) \cap (x^T D_q([b] - [A]D_{\overline{p}}x)) \ne \emptyset, \tag{5.2.18}$$

$$\left.\begin{aligned} &x^T D_p(b^- - A^+ D_{\overline{q}}x) \le x^T D_q(b^+ - A^- D_{\overline{p}}x),\\ \wedge\, &x^T D_q(b^- - A^+ D_{\overline{p}}x) \le x^T D_p(b^+ - A^- D_{\overline{q}}x), \end{aligned}\right\} \tag{5.2.19}$$

where $D_p = \mathrm{diag}(p)$, $D_q = \mathrm{diag}(q)$, $\overline{p} = e - p$, $\overline{q} = e - q$, and where $A^- = (a_{ij}^-)$, $A^+ = (a_{ij}^+)$, $b^- = (b_i^-)$, $b^+ = (b_i^+)$ are defined by the equalities

$$[a]_{ij} =: \begin{cases} [a_{ij}^-, a_{ij}^+] & \text{if } x_i \cdot x_j \ge 0,\\ [a_{ij}^+, a_{ij}^-] & \text{if } x_i \cdot x_j < 0, \end{cases} \qquad [b]_i =: \begin{cases} [b_i^-, b_i^+] & \text{if } x_i \ge 0,\\ [b_i^+, b_i^-] & \text{if } x_i < 0. \end{cases}$$

The set of of these relations with the restriction (5.2.15) *consists of* $(3^n - 2^{n+1} + 1)/2$ *inequalities, intersections, and double inequalities, respectively.*

Theorem 5.2.6 shows that the intersection $S_{\mathrm{sym}} \cap O$ is characterized by the intersection of hyperplanes and quadrics. In particular, its boundary can be curvilinear. The property '$x \in S$' in (5.2.14) can be replaced by any equivalent property of Theorem 5.2.2.

For $n = 2$ there is only one inequality (5.2.16), for $n = 3$ there are already six.

For the Example 5.2.4 Theorem 5.2.6 implies the equivalence

$$x \in S_{\mathrm{sym}} \Leftrightarrow \begin{cases} \frac{1}{2}|x_2| + 1 \geq |1 - x_1 - \frac{1}{2}x_2|, \\ \frac{1}{2}|x_1| + \frac{3}{2}|x_2| + 1 \geq |1 - \frac{1}{2}x_1 + \frac{5}{2}x_2|, \\ \frac{3}{2}x_2^2 + |x_1| + |x_2| \geq |x_1^2 - x_1 + \frac{5}{2}x_2^2 + x_2|, \end{cases} \tag{5.2.20}$$

where the first two inequalities result from the Oettli–Prager criterion in order to express '$x \in S$', and where the last inequality corresponds to (5.2.16) for $p = (0, 1)^T$, $q = (1, 0)^T$, the only vectors which satisfy (5.2.15).

Proof of Theorem 5.2.6. First we remark that the assumption $[A] = [A]^T$ implies $\check{A} = \check{A}^T$ and $\mathrm{rad}([A]) = (\mathrm{rad}([A]))^T$. We start with the equivalence

$$x \in S_{\mathrm{sym}} \Longleftrightarrow x \text{ satisfies (5.2.14)–(5.2.16)} \tag{5.2.21}$$

but restrict here ourselves to the direction '$\Rightarrow$' of the proof; the converse direction is lengthy and rather technical. It can be found in the Appendix F.

Let $x \in S_{\mathrm{sym}}$. Then $x \in S$, hence (5.2.14) holds. There is a symmetric matrix $\tilde{A} = \tilde{A}^T \in [A]$ and a vector $\tilde{b} \in [b]$ such that $\tilde{A}x = \tilde{b}$. This implies $x^T D_p \tilde{A} x = x^T D_p \tilde{b}$ and $x^T D_q \tilde{A}^T x = x^T D_q \tilde{b}$. Subtracting both equalities and introducing the representations

$$\tilde{A} = \check{A} + \Delta, \quad \tilde{b} = \check{b} + \delta,$$

yields

$$x^T D_p \Delta x - x^T D_q \Delta^T x - x^T (D_p - D_q)\delta = x^T (D_p - D_q) r.$$

Using $x^T D_q \Delta^T x = x^T \Delta D_q x$ and absolute values results in

$$|x^T (D_p - D_q) r| \leq |x|^T \cdot |D_p \Delta - \Delta D_q| \cdot |x| + |x|^T \cdot |D_p - D_q| \cdot |\delta|.$$

Since

$$\begin{aligned} |D_p \Delta - \Delta D_q|_{ij} &= |(D_p)_{ii} \Delta_{ij} - \Delta_{ij} (D_q)_{jj}| = |\Delta_{ij}| |(D_p)_{ii} - (D_q)_{jj}| \\ &\leq \mathrm{rad}([A])_{ij} |(D_p)_{ii} - (D_q)_{jj}| \\ &= |D_p \,\mathrm{rad}([A]) - \mathrm{rad}([A]) D_q|_{ij} \end{aligned}$$

and $|\delta| \leq \mathrm{rad}([b])$ we finally end up with (5.2.16) without any restrictions on $p, q \in \{0, 1\}^n$.

We show that p, q can be restricted to (5.2.15). With the complementary vectors $\overline{p} = e - p$, $\overline{q} = e - q$ we can transform the matrix $R(p, q) = |D_p \,\mathrm{rad}([A]) - \mathrm{rad}([A]) D_q| \in \mathbb{R}^{n \times n}$ in the following way.

$$\begin{aligned} R(p, q) &= |D_p \,\mathrm{rad}([A]) D_{\overline{q}} - D_{\overline{p}} \,\mathrm{rad}([A]) D_q| \\ &= D_p \,\mathrm{rad}([A]) D_{\overline{q}} + D_{\overline{p}} \,\mathrm{rad}([A]) D_q, \end{aligned} \tag{5.2.22}$$

where for the last equality we proceeded entrywise and exploited $p + \overline{p} = e$, $p \in \{0, 1\}^n$, so that either $p_i = 1$, $\overline{p}_i = 0$ or vice versa. Therefore, at most one summand of $(R(p, q))_{ij}$ in (5.2.22) differs from zero.

The first equality in (5.2.22) implies

$$\begin{aligned} R(\overline{p}, \overline{q}) &= |D_{\overline{p}} \operatorname{rad}([A]) D_q - D_p \operatorname{rad}([A]) D_{\overline{q}}| \\ &= |D_p \operatorname{rad}([A]) D_{\overline{q}} - D_{\overline{p}} \operatorname{rad}([A]) D_q| = R(p, q), \end{aligned}$$

and $q \prec_{\text{lex}} p$ is equivalent to $\overline{p} \prec_{\text{lex}} \overline{q}$. Therefore, with $D_{\overline{p}} - D_{\overline{q}} = -(D_p - D_q)$, the inequality (5.2.16) also holds for $q \prec_{\text{lex}} p$ if it is true for $p \prec_{\text{lex}} q$.

If $p = 0$, then (5.2.16) reduces to

$$|x|^T \operatorname{rad}([A]) D_q |x| + |x|^T D_q \operatorname{rad}([b]) \geq |x^T D_q r|$$

which follows also from (5.2.14), the Oettli–Prager criterion and $|x^T D_q r| \leq |x|^T D_q |r|$. The same is true for the case $q = 0$. Since $e = \overline{p}$ for $p = 0$ we can restrict to $p, q \in \{0, 1\}^n \setminus \{0, e\}$.

We show that all inequalities (5.2.16) with $p_i = q_i = 1$ for some indices i can be omitted. To this end choose $\hat{p}$, $\hat{q}$ such that

$$\hat{p}_i = \begin{cases} p_i, & \text{if } p_i q_i \neq 1 \\ 0 & \text{otherwise} \end{cases} \qquad \hat{q}_i = \begin{cases} q_i, & \text{if } p_i q_i \neq 1 \\ 0 & \text{otherwise} \end{cases}$$

and let $\hat{p}_r = p - \hat{p}$. Then

$$(\hat{p}_r)_i = \begin{cases} 1, & \text{if } p_i = q_i = 1 \\ 0 & \text{otherwise} \end{cases},$$

whence $\hat{p}_r = q - \hat{q}$, $\overline{p} = \overline{\hat{p}} - \hat{p}_r$, $\overline{q} = \overline{\hat{q}} - \hat{p}_r$. Hence

$$\begin{aligned} R(p, q) &= D_{\hat{p}} \operatorname{rad}([A]) D_{\overline{q}} + D_{\hat{p}_r} \operatorname{rad}([A]) D_{\overline{q}} + D_{\overline{p}} \operatorname{rad}([A]) D_{\hat{q}} + D_{\overline{p}} \operatorname{rad}([A]) D_{\hat{p}_r} \\ &= D_{\hat{p}} \operatorname{rad}([A]) D_{\overline{\hat{q}}} - D_{\hat{p}} \operatorname{rad}([A]) D_{\hat{p}_r} + D_{\hat{p}_r} \operatorname{rad}([A]) D_{\overline{q}} \\ &\quad + D_{\overline{\hat{p}}} \operatorname{rad}([A]) D_{\hat{q}} - D_{\hat{p}_r} \operatorname{rad}([A]) D_{\hat{q}} + D_{\overline{p}} \operatorname{rad}([A]) D_{\hat{p}_r}. \end{aligned}$$

From this and transformations like

$$\begin{aligned} |x|^T D_{\hat{p}} \operatorname{rad}([A]) D_{\hat{p}_r} |x| &= \left(|x|^T D_{\hat{p}} \operatorname{rad}([A]) D_{\hat{p}_r} |x|\right)^T \\ &= |x|^T D_{\hat{p}_r} (\operatorname{rad}([A]))^T D_{\hat{p}} |x| \\ &= |x|^T D_{\hat{p}_r} \operatorname{rad}([A]) D_{\hat{p}} |x| \in \mathbb{R} \end{aligned}$$

we get

$$\begin{aligned} |x|^T R(p, q) |x| &= |x|^T R(\hat{p}, \hat{q}) |x| + |x|^T D_{\hat{p}_r} \operatorname{rad}([A]) (I - D_{\hat{p}+q} + I - D_{p+\hat{q}}) |x| \\ &\geq |x|^T R(\hat{p}, \hat{q}) |x|, \end{aligned}$$

since $I - D_{\hat{p}+q} \geq O$ and $I - D_{p+\hat{q}} \geq O$. (Notice that $\hat{p} + q \leq e$ and $p + \hat{q} \leq e$ holds.) Furthermore, $D_p - D_q = D_{\hat{p}+\hat{p}_r} - D_{\hat{q}+\hat{p}_r} = D_{\hat{p}} + D_{\hat{p}_r} - (D_{\hat{q}} + D_{\hat{p}_r}) = D_{\hat{p}} - D_{\hat{q}}$. Therefore, the inequality (5.2.16) for p, q follows from that for $\hat{p}, \hat{q}$ and can be omitted. This shows that in (5.2.16) only vectors $p, q \in \{0, 1\}^n$ with $0 \neq p \prec_{\text{lex}} q$ and $p_i q_i = 0$, $i = 1, \ldots, n$, need to be considered.

There are 3^n possibilities for inequalities with $(p_i, q_i) \in \{(0, 0), (0, 1), (1, 0)\}$ for all $i = 1, \ldots, n$. Since 2^n is the number of possible vectors taken from the set $\{0, 1\}^n$, there are 2^n possibilities of inequalities with $p = 0$. The same amount of inequalities occurs for $q = 0$. Both numbers have to be subtracted from 3^n. Since hereby the case $p = q = 0$ is subtracted twice, we must add a 1 again. Notice that we did not allow the case $p_i = q_i = 1$ and excluded the cases $p = 0$ and $q = 0$. By virtue of $p^T q = 0$, this implies that we simultaneously excluded the cases $p = e$ and $q = e$. Taking into account $p \prec_{\text{lex}} q$, we finally end up with the number $(3^n - 2 \cdot 2^n + 1)/2$ as required.

Once we have proved the equivalence (5.2.21), the remaining part of the theorem is nearly straightforward:

Since (5.2.15) implies $|D_p - D_q| = D_p + D_q$, we get the equivalence of (5.2.16) and (5.2.17) by virtue of (5.2.22).

For the equivalence of (5.2.17) and (5.2.18) we apply the first equivalence of Theorem 2.4.8 (c), whence the relation (5.2.18) is equivalent to

$$\begin{aligned} &\left|\text{mid}(x^T D_p([b] - [A]D_{\bar{q}}x)) - \text{mid}\,(x^T D_q([b] - [A]D_{\bar{p}}x))\right| \\ &\qquad \leq \text{rad}(x^T D_p([b] - [A]D_{\bar{q}}x)) + \text{rad}(x^T D_q([b] - [A]D_{\bar{p}}x)). \end{aligned} \tag{5.2.23}$$

Now the left-hand side of (5.2.23) equals

$$\begin{aligned} &|x^T D_p(\check{b} - \check{A}D_{\bar{q}}x) - x^T D_q(\check{b} - \check{A}D_{\bar{p}}x)| \\ &\qquad = |x^T\{(D_p - D_q)\check{b} - D_p\check{A}D_{\bar{q}}x + D_q\check{A}D_{\bar{p}}x - D_p\check{A}D_q x + D_p\check{A}D_q x\}| \\ &\qquad = |x^T(D_p - D_q)(\check{b} - \check{A}x)| \end{aligned}$$

while the right-hand side can be written as

$$\begin{aligned} &|x|^T D_p \,\text{rad}([b] - [A]D_{\bar{q}}x) + |x|^T D_q \,\text{rad}([b] - [A]D_{\bar{p}}x) \\ &\qquad = |x|^T(D_p + D_q)\,\text{rad}([b]) + |x|^T D_p \,\text{rad}([A])D_{\bar{q}}|x| + |x|^T D_q \,\text{rad}([A])D_{\bar{p}}|x|. \end{aligned}$$

Thus we end up with the corresponding sides of (5.2.17).

The equivalence of (5.2.18) and (5.2.19) is an immediate consequence of the last equivalence in Theorem 2.4.8 (c). □

The omitted proof of the converse direction (see Appendix F) uses ideas which can be rediscovered in the proof of the following theorem on parameter elimination in more general systems of inequalities.

Theorem 5.2.7. *Let $f_{\lambda\mu}, g_\lambda, \lambda = 1, \ldots, k\ (\geq 2), \mu = 1, \ldots, m$, be real-valued functions of $x = (x_1, \ldots, x_n)^T$ on some subset $D \subseteq \mathbb{R}^n$. Assume that there is a positive integer $k_1 < k$ such that*

$$f_{\lambda 1}(x) \not\equiv 0 \quad \text{for all } \lambda \in \{1, \ldots, k\}, \tag{5.2.24}$$

$$f_{\lambda 1}(x) \geq 0 \quad \text{for all } x \in D \text{ and all } \lambda \in \{1, \ldots, k\}, \tag{5.2.25}$$

$$\begin{aligned} &\text{for each } x \in D \text{ there is an index } \beta^* = \beta^*(x) \in \{1, \ldots, k_1\} \text{ with } f_{\beta^* 1}(x) > 0 \\ &\text{and an index } \gamma^* = \gamma^*(x) \in \{k_1 + 1, \ldots, k\} \text{ with } f_{\gamma^* 1}(x) > 0. \end{aligned} \tag{5.2.26}$$

For m parameters $u_1, \ldots, u_m$ varying in $\mathbb{R}$ and for x varying in D define the sets S_1, S_2 by

$$\begin{aligned} S_1 &:= \{ x \in D \mid \exists u_\mu \in \mathbb{R},\ \mu = 1, \ldots, m\colon \text{(5.2.27), (5.2.28) holds} \}, \\ S_2 &:= \{ x \in D \mid \exists u_\mu \in \mathbb{R},\ \mu = 2, \ldots, m\colon \text{(5.2.29) holds} \}, \end{aligned}$$

where the inequalities (5.2.27)–(5.2.29) are given by

$$g_\beta(x) + \sum_{\mu=2}^{m} f_{\beta\mu}(x) u_\mu \leq f_{\beta 1}(x) u_1, \quad \beta = 1, \ldots, k_1, \tag{5.2.27}$$

$$f_{\gamma 1}(x) u_1 \leq g_\gamma(x) + \sum_{\mu=2}^{m} f_{\gamma\mu}(x) u_\mu, \quad \gamma = k_1 + 1, \ldots, k; \tag{5.2.28}$$

and

$$\begin{aligned} g_\beta(x) f_{\gamma 1}(x) + \sum_{\mu=2}^{m} f_{\beta\mu}(x) f_{\gamma 1}(x) u_\mu &\leq g_\gamma(x) f_{\beta 1}(x) + \sum_{\mu=2}^{m} f_{\gamma\mu}(x) f_{\beta 1}(x) u_\mu, \\ &\beta = 1, \ldots, k_1, \quad \gamma = k_1 + 1, \ldots, k. \end{aligned} \tag{5.2.29}$$

(Trivial inequalities such as $0 \leq 0$ can be omitted.)

Then

$$S_1 = S_2.$$

Proof. First we show $S_1 \subseteq S_2$. To this end let $S_1 \neq \emptyset$, fix $x \in S_1$ and let $u_1, \ldots u_m \in \mathbb{R}$ be such that the inequalities (5.2.27), (5.2.28) hold for x. Multiply (5.2.27) by $f_{\gamma 1}(x)$ and (5.2.28) by $f_{\beta 1}(x)$. This implies

$$\begin{aligned} &g_\beta(x) f_{\gamma 1}(x) + \sum_{\mu=2}^{m} f_{\beta\mu}(x) f_{\gamma 1}(x) u_\mu \\ &\qquad \leq f_{\beta 1}(x) f_{\gamma 1}(x) u_1 \leq g_\gamma(x) f_{\beta 1}(x) + \sum_{\mu=2}^{m} f_{\gamma\mu}(x) f_{\beta 1}(x) u_\mu \end{aligned}$$

for $\beta = 1, \ldots, k_1$ and $\gamma = k_1 + 1, \ldots, k$. Dropping the middle term results in (5.2.29), whence $S_1 \subseteq S_2$.

In order to prove the converse $S_1 \supseteq S_2$ let $S_2 \neq \emptyset$, fix $x \in S_2$ and let $u_2, \ldots, u_m \in \mathbb{R}$ be such that the inequalities (5.2.29) hold for x. Divide (5.2.29) by $f_{\beta 1}(x)$ if $f_{\beta 1}(x) > 0$, and by $f_{\gamma 1}(x)$ if $f_{\gamma 1}(x) > 0$. Unless $f_{\beta 1}(x) = 0$ and $f_{\gamma 1}(x) = 0$ (in this case (5.2.29) reads $0 \le 0$ and can be omitted) one gets equivalently

$$g_\beta(x) + \sum_{\mu=2}^{m} f_{\beta\mu}(x)u_\mu \le 0, \quad \text{if } f_{\beta 1}(x) = 0 \text{ and } f_{\gamma 1}(x) > 0, \tag{5.2.30}$$

$$0 \le g_\gamma(x) + \sum_{\mu=2}^{m} f_{\gamma\mu}u_\mu, \quad \text{if } f_{\beta 1}(x) > 0 \text{ and } f_{\gamma 1}(x) = 0, \tag{5.2.31}$$

$$\left(g_\beta(x) + \sum_{\mu=2}^{m} f_{\beta\mu}(x)u_\mu \right) / f_{\beta 1}(x) \le \left(g_\gamma(x) + \sum_{\mu=2}^{m} f_{\gamma\mu}(x)u_\mu \right) / f_{\gamma 1}(x), \quad \text{if } f_{\beta 1}(x) f_{\gamma 1}(x) > 0. \tag{5.2.32}$$

By virtue of (5.2.26) there exists at least one pair $(\beta^*, \gamma^*) \in \{1, \ldots, k_1\} \times \{k_1 + 1, \ldots, k\}$ such that $f_{\beta^* 1}(x) f_{\gamma^* 1}(x) > 0$. Let M_1 be the maximum of the left-hand sides of all inequalities (5.2.32) and let M_2 be the minimum of the right-hand sides of all inequalities (5.2.32). Then M_1, M_2 are attained for some indices $\beta = \beta_0$ and $\gamma = \gamma_0$, respectively. Since β and γ vary independently there is an inequality (5.2.32) with $\beta = \beta_0$ and $\gamma = \gamma_0$ simultaneously. This proves $M_1 \le M_2$. Choose $u_1 \in [M_1, M_2]$ and apply (5.2.30) and (5.2.32), respectively, with $\gamma = \gamma_0$ (which implies $f_{\gamma_0 1}(x) > 0$) and $\beta = 1, \ldots, k_1$. If $f_{\beta 1}(x) = 0$, then (5.2.30) yields the corresponding inequality in (5.2.27). If $f_{\beta 1}(x) > 0$, then

$$\left(g_\beta(x) + \sum_{\mu=2}^{m} f_{\beta\mu}(x)u_\mu \right) / f_{\beta 1}(x) \le M_1 \le u_1$$

implies the corresponding inequality in (5.2.27). By applying (5.2.31) and (5.2.32), respectively with $\beta = \beta_0$, the inequalities (5.2.28) can be seen analogously, whence $S_2 \subseteq S_1$. □

Notice that the parameter u_1 which occurs in the definition of S_1 is no longer needed in order to describe S_2. Therefore, we call the transition from the inequalities (5.2.27), (5.2.28) to the inequalities in (5.2.29) the *elimination of* u_1.

It is obvious that the assertion of Theorem 5.2.7 remains true if the inequalities in (5.2.27), (5.2.28) and the inequalities in (5.2.29) are supplemented by inequalities which do not contain the parameter u_1, as long as these inequalities are the same in both cases.

We next state a corollary which may facilitate the terms occurring in the Fourier–Motzkin elimination process.

Corollary 5.2.8. *With the notation and the assumptions of Theorem* 5.2.7 *let*

$$f_{\beta 1}(x) = h_{\beta\gamma}(x)\tilde{f}_{\beta 1}(x), \quad f_{\gamma 1}(x) = h_{\beta\gamma}(x)\tilde{f}_{\gamma 1}(x)$$

with nonnegative functions $\tilde{f}_{\beta 1}, \tilde{f}_{\gamma 1}, h_{\beta\gamma}$ defined on D. Then the assertion of Theorem 5.2.7 *remains true if $f_{\beta 1}(x), f_{\gamma 1}(x)$ are replaced in* (5.2.29) *by $\tilde{f}_{\beta 1}(x)$ and $\tilde{f}_{\gamma 1}(x)$, respectively.*

Proof. Define

$$S_3 = \{ x \in D \mid \exists u_\mu \in \mathbb{R},\ \mu = 2, \ldots, m\colon\ (5.2.34) \text{ holds} \}, \tag{5.2.33}$$

where the inequality (5.2.34) is given by

$$g_\beta(x)\tilde{f}_{\gamma 1}(x) + \sum_{\mu=2}^{m} f_{\beta\mu}(x)\tilde{f}_{\gamma 1}(x)u_\mu \le g_\gamma(x)\tilde{f}_{\beta 1}(x) + \sum_{\mu=2}^{m} f_{\gamma\mu}(x)\tilde{f}_{\beta 1}(x)u_\mu$$

$$\beta = 1, \ldots, k_1, \quad \gamma = k_1 + 1, \ldots, k. \tag{5.2.34}$$

Multiply (5.2.27) and (5.2.28) by $\tilde{f}_{\gamma 1}(x)$ and $\tilde{f}_{\beta 1}(x)$, respectively. This results in (5.2.34) and shows $S_1 \subseteq S_3$. In order to prove $S_3 \subseteq S_1$ fix $x \in S_3$ and choose $u_2, \ldots, u_m \in \mathbb{R}$ such that (5.2.34) holds for x. Multiply the corresponding inequality (5.2.34) by $h_{\beta\gamma}(x)$. This yields (5.2.29), hence $x \in S_2$. Now Theorem 5.2.7 implies $x \in S_1$. □

Corollary 5.2.8 is particularly useful if $f_{\beta 1} = f_{\gamma 1} \ge 0$, where $f \ge 0$ means $f(x) \ge 0$ for all $x \in D$. Then $h_{\beta\gamma} = f_{\beta 1} = f_{\gamma 1} \ge 0$, $\tilde{f}_{\beta 1} = \tilde{f}_{\gamma 1} = 1 > 0$ and the corresponding inequality in (5.2.29) reads

$$g_\beta(x) + \sum_{\mu=2}^{m} f_{\beta\mu}(x)u_\mu \le g_\gamma(x) + \sum_{\mu=2}^{m} f_{\gamma\mu}(x)u_\mu .$$

Another typical application of Corollary 5.2.8 occurs if the functions $f_{\lambda\mu}$, g_λ all are polynomials and if $f_{\beta 1}$ and $f_{\gamma 1}$ have a nonconstant polynomial as a common factor.

No topological assumption such as continuity of $f_{\lambda\mu}$, g_λ or connectivity of D is required in Theorem 5.2.2. The assumption (5.2.24) prevents $f_{\lambda 1}$ from being completely omitted in (5.2.27), (5.2.28) and (5.2.29). If $f_{\lambda 1}(x) \le 0$ on D, one can simply fulfill (5.2.25) by multiplying the corresponding inequality by -1. If neither $f_{\lambda 1}(x) \ge 0$ nor $f_{\lambda 1}(x) \le 0$ holds uniformly on D, one can split D in several appropriate subdomains D_i, with $\bigcup_i D_i = D$, for each of which the assumptions of Theorem 5.2.7 hold. As was shown in Alefeld, Kreinovich, Mayer [33] the restriction (5.2.26) cannot be dropped.

In our applications the parameters u_μ will be the matrix entries a_{ij} and the components b_i of the right-hand side b of a linear system $Ax = b$. They will be restricted to compact intervals by $A \in [A]$ and $b \in [b]$. This generates inequalities of the form $\underline{a} \le u_\mu \le \overline{a}$ with a corresponding function $f_{\lambda\mu} = 1$. Since such an inequality depends on a single u_μ, it is only used when this parameter is eliminated. Therefore, in the sequel the assumption (5.2.24) will be fulfilled for any domain D.

Theorem 5.2.7 and Corollary 5.2.8 can be applied repeatedly in order to eliminate some or all expressions in which the parameters u_μ occur linearly. But even in simple cases like S_{sym} the number of inequalities can increase exponentially.

In order to keep track we shortly summarize the steps to be executed when eliminating the parameters in the inequalities describing some set $S_1 \subseteq D$. These steps generalize those on page 165.

Elimination process. Given a domain $D \subseteq \mathbb{R}^n$ and a set of inequalities in $x \in D$ with parameters $u_1, \dots u_m$ which occur linearly, denote D together with this set of inequalities as a record and store it on a stack named Stack 1.

Step 1: Fetch the first record (i. e., the domain and the corresponding set of inequalities) from Stack 1, fix a parameter, say u_1, bring those inequalities into the form (5.2.27), (5.2.28) which contain u_1. (Renumber and rename if necessary, in order to have a domain named D, a parameter named u_1, and subsequent inequalities according to (5.2.27), (5.2.28).)

Step 2: Check the assumptions of Theorem 5.2.7 for the inequalities which contain u_1. If (5.2.25) is not satisfied, then multiply the corresponding inequality by -1. If this does not help, split D into appropriate subdomains D_i and replace the record with D by corresponding ones with D_i. If (5.2.26) is not fulfilled for each D_i, then stop. Otherwise put the records to a stack named Stack 2.

Step 3: As long as Stack 2 is not empty fetch from it the last record and eliminate u_1 according to Theorem 5.2.7 or Corollary 5.2.8. If the new record no longer contains any parameter u_μ, then store it into a file. Otherwise put it to Stack 1 as the last element. If Stack 1 is not empty, then go to Step 1. □

We want to apply Theorem 5.2.7 and, whenever possible, Corollary 5.2.8 in order to characterize S_{sym} once more. This is the second way which we mentioned on page 164. Again we assume $[A] = [A]^T$.

Let O be a fixed orthant. We start with $D = O$ and (5.2.4)–(5.2.6), this time reducing the amount of free parameters nearly to one half by using $a_{ij} = a_{ji}$. The elimination process for the b_i and the diagonal entries a_{ii} is the same as for S and is left to the reader. The elimination of the off-diagonal entries a_{ij}, $i < j$, $i, j = 1, \dots, n$, differs due to the dependency $a_{ij} = a_{ji}$. For instance, when handling a_{12} first, one gets the (nontrivial) new inequalities

$$\underline{b}_1 - \hat{a}_{11}^+ x_1 - \sum_{j=3}^n a_{1j} x_j \le \hat{a}_{12}^+ x_2, \quad \hat{a}_{12}^- x_2 \le \overline{b}_1 - \hat{a}_{11}^- x_1 - \sum_{j=3}^n a_{1j} x_j, \tag{5.2.35}$$

$$\underline{b}_2 - \hat{a}_{22}^+ x_2 - \sum_{j=3}^n a_{2j} x_j \le \hat{a}_{12}^+ x_1, \quad \hat{a}_{12}^- x_1 \le \overline{b}_2 - \hat{a}_{22}^- x_2 - \sum_{j=3}^n a_{2j} x_j, \tag{5.2.36}$$

$$b_1^- x_1 - a_{11}^+ x_1^2 - \sum_{j=3}^n a_{1j} x_1 x_j \le b_2^+ x_2 - a_{22}^- x_2^2 - \sum_{j=3}^n a_{2j} x_2 x_j, \tag{5.2.37}$$

$$b_2^- x_2 - a_{22}^+ x_2^2 - \sum_{j=3}^n a_{2j} x_2 x_j \le b_1^+ x_1 - a_{11}^- x_1^2 - \sum_{j=3}^n a_{1j} x_1 x_j, \tag{5.2.38}$$

where

$$\hat{a}_{ij}^- = \begin{cases} \underline{a}_{ij}, & \text{if } x_j \geq 0 \\ \overline{a}_{ij}, & \text{if } x_j < 0 \end{cases}, \qquad \hat{a}_{ij}^+ = \begin{cases} \overline{a}_{ij}, & \text{if } x_j \geq 0 \\ \underline{a}_{ij}, & \text{if } x_j < 0 \end{cases},$$

$$a_{ij}^- = \begin{cases} \underline{a}_{ij}, & \text{if } x_i x_j \geq 0 \\ \overline{a}_{ij}, & \text{if } x_i x_j < 0 \end{cases}, \qquad a_{ij}^+ = \begin{cases} \overline{a}_{ij}, & \text{if } x_i x_j \geq 0 \\ \underline{a}_{ij}, & \text{if } x_i x_j < 0 \end{cases},$$

$$b_i^- = \begin{cases} \underline{b}_i, & \text{if } x_i \geq 0 \\ \overline{b}_i, & \text{if } x_i < 0 \end{cases}, \qquad b_i^+ = \begin{cases} \overline{b}_i, & \text{if } x_i \geq 0 \\ \underline{b}_i, & \text{if } x_i < 0 \end{cases}.$$

The inequalities (5.2.35)–(5.2.36) coincide with those for S. The inequalities (5.2.37), (5.2.38) are new. They contain quadratic polynomials. When eliminating a_{1j} for $j = 3, \dots, n$ according to Corollary 5.2.8, the i-th inequality in (5.2.1) has to be multiplied by x_i for $i = 3, \dots, n$. Afterwards, no additional multiplication is needed in inequalities which have a form analogous to (5.2.37), (5.2.38). This is true because the function $f_{\lambda\mu}$ in front of a_{ij} reads $f_{\lambda\mu}(x) = x_i x_j$ in these inequalities, and in the remaining (nonquadratic) inequalities they are given by $f_{\lambda\mu}(x) = x_i$, $f_{\lambda\mu}(x) = x_j$ and $f_{\lambda\mu}(x) = 1$, respectively. Notice that the sign of the function $x_i x_j$ remains constant over a fixed orthant O. This is the reason why no splitting is needed for $D = O$ during the elimination process. Pursuing this process shows that the final inequalities for $S_{\text{sym}} \cap O$ consist of the inequalities in Theorem 5.2.2 (f) which characterize S, and quadratic inequalities. We thus get the following result which is already contained in Theorem 5.2.6.

Theorem 5.2.9. *Let $[A] = [A]^T \in \mathbb{IR}^{n\times n}$ (not necessarily regular) and let $[b] \in \mathbb{IR}^n$. Then in each orthant the symmetric solution set S_{sym} can be represented as the intersection of the solution set S and sets with quadrics as boundaries.*

For 2×2 matrices there are only two quadratic inequalities which supplement the linear ones of Theorem 5.2.2 (f). They are given by

$$\left.\begin{aligned} b_1^- x_1 - b_2^+ x_2 - a_{11}^+ x_1^2 + a_{22}^- x_2^2 &\leq 0 \\ -b_1^+ x_1 + b_2^- x_2 + a_{11}^- x_1^2 - a_{22}^+ x_2^2 &\leq 0 \end{aligned}\right\} \tag{5.2.39}$$

with $b_i^\pm$, $a_{ii}^\pm$ as above. They correspond to the only inequality (5.2.16). (Notice the absolute values there!)

Thus for S_{sym} of Example 5.2.4 we need the inequalities for S together with

$$4x_1^2 + (4x_2 + 1)^2 \geq 1, \qquad (x_1 - 1)^2 + x_2^2 \leq 1 \tag{5.2.40}$$

in the first quadrant O_1 and

$$x_1^2 + 4x_2^2 \geq 0, \quad (x_1 - 1)^2 + (x_2 + 1)^2 \leq 2 \tag{5.2.41}$$

in the forth quadrant O_4. The first inequality in (5.2.40) represents an ellipse and its exterior. Since the ellipse lies completely in the closed lower half-plane this inequality is not a restriction for $S_{\mathrm{sym}} \cap O_1$. The first inequality in (5.2.41) is always true. Therefore, only the respective second inequality which represents a closed disc is really relevant for S_{sym} as a subset of S. Although both look different from the last inequality in (5.2.20), they are equivalent. The proof is not quite trivial; we leave it to the reader as Exercise 5.2.2 or refer to Mayer [217].

Both solution sets are illustrated in Figure 5.2.1 on page 163.

Persymmetric matrices $A \in \mathbb{R}^{n\times n}$ are characterized by $EA = (EA)^T$ with

$$E = \begin{pmatrix} O & & 1 \\ & \cdot^{\cdot^{\cdot}} & \\ 1 & & O \end{pmatrix}, \tag{5.2.42}$$

or, equivalently, by $a_{ij} = a_{n+1-j,n+1-i}$ for all indices i, j. They are symmetric with respect to their counterdiagonal. The corresponding persymmetric solution set reads

$$S_{\mathrm{per}} = \{\, x \in \mathbb{R}^n \mid Ax = b,\ EA = (EA)^T \in E[A] = (E[A])^T,\ b \in [b]\,\}.$$

A description of S_{per} can be deduced directly from Theorem 5.2.6; cf. Exercise 5.2.6 or Mayer [217, 219].

Skew-symmetric matrices $A \in \mathbb{R}^{n\times n}$ are defined by $A = -A^T$, or, equivalently, by $a_{ij} = -a_{ji}$ for all indices i, j, whence $a_{ii} = 0$, $i = 1, \ldots, n$. The skew-symmetric solution set is given by

$$S_{\mathrm{skew}} = \{\, x \in \mathbb{R}^n \mid Ax = b,\ A = -A^T \in [A] = -[A]^T,\ b \in [b]\,\},$$

where we implicitly assume here $[a]_{ii} = 0$, $i = 1, \ldots, n$.

Similar to Theorem 5.2.6 and with the notation there, the following equivalence on S_{skew} can be proved (cf. Exercise 5.2.7, Hladík [149], Mayer [217, 219]).

Theorem 5.2.10. *Let $[A] = -[A]^T \in \mathbb{IR}^{n\times n}$, $[a]_{ii} = 0$ for $i = 1, \ldots, n$, $[b] \in \mathbb{IR}^n$, $x \in \mathbb{R}^n$, $r = \check{b} - \check{A}x$. Then $x \in S_{\mathrm{skew}}$ if and only if $x \in S$ and if for all $2^n - n - 1$ vectors*

$$p \in \{0,1\}^n \setminus \{0\}, \quad p \neq e^{(i)}, \quad i = 1, \ldots, n, \tag{5.2.43}$$

one of the following four relations (5.2.44)–(5.2.47) *is satisfied:*

$$\tfrac{1}{2}|x|^T \cdot |D_p \operatorname{rad}([A]) - \operatorname{rad}([A])D_p| \cdot |x| + |x|^T D_p \operatorname{rad}([b]) \geq |x^T D_p r|, \tag{5.2.44}$$

$$|x|^T D_p \operatorname{rad}([A]) D_{\overline{p}} |x| + |x|^T D_p \operatorname{rad}([b]) \geq |x^T D_p r|, \tag{5.2.45}$$

$$x^T D_p [b] \cap x^T D_p [A] D_{\overline{p}} x \neq \emptyset, \tag{5.2.46}$$

$$\left.\begin{aligned} x^T D_p b^- &\leq x^T D_p A^+ D_{\overline{p}} x, \\ \wedge\, x^T D_p A^- D_{\overline{p}} x &\leq x^T D_p b^+, \end{aligned}\right\}. \tag{5.2.47}$$

For the equivalence of (5.2.44) and (5.2.45) one can use (5.2.22) with $q = p$ and exploit the symmetry of $\mathrm{rad}([A])$ which follows from $\underline{A} = -\overline{A}^T$ and $\check{A} = -\check{A}^T$ for $[A]$ from Theorem 5.2.10.

For the equivalence of (5.2.45), (5.2.46), and (5.2.47) one applies Theorem 2.4.8 (c) and the equality

$$x^T D_p \check{A} D_p x = (x^T D_p \check{A} D_p x)^T = x^T D_p \check{A}^T D_p x = -x^T D_p \check{A} D_p x,$$

which proves $x^T D_p \check{A} D_p x = 0$. This implies

$$\begin{aligned} x^T D_p r &= x^T D_p(\check{b} - \check{A}(D_p + D_{\overline{p}})x) = x^T D_p \check{b} - x^T D_p \check{A} D_{\overline{p}} x \\ &= \mathrm{mid}(x^T D_p [b]) - \mathrm{mid}(x^T D_p [A] D_{\overline{p}} x). \end{aligned}$$

If $n = 2$ then there is only one inequality (5.2.45), if $n = 3$ there are four.

Solution sets with more general restrictions and further references can be found in Alefeld, Kreinovich, Mayer [33]. Methods by Jansson [153, 154] and Rohn [306] to enclose S_{sym} as well as some historical notes were given in Mayer [216].

Exercises

Ex. 5.2.1 (Rohn [308]). Show that $[A]x - [b] = \{Ax - b \mid A \in [A],\ b \in [b]\}$ holds and that $x \in S$ can also be described equivalently by the inequality $|\,\mathrm{mid}([A]x - [b])| \le \mathrm{rad}([A]x - [b])$.

The nice feature of this description is the fact that both sides of the inequality contain the same interval vector $[A]x - [b]$.

Ex. 5.2.2. Show that the two inequalities $(x_1 - 1)^2 + x_2^2 \le 1$ for $(x_1, x_2) \in O_1$ and $(x_1 - 1)^2 + (x_2 + 1)^2 \le 2$ for $(x_1, x_2) \in O_4$ are equivalent to the inequalities (5.2.20).

Hint: Use $S_{\mathrm{sym}} \subseteq S \subseteq O_1 \cup O_4$ and consider the cases $x \in O_1$, $x \in O_4$, $T = x_1 - x_1 + \frac{5}{2}x_2^2 + x_2 \ge 0$, and $T < 0$.

Ex. 5.2.3 (Mayer [219]). List all six pairs (p, q) of possible vectors according to (5.2.15) in Theorem 5.2.6 for the case $n = 3$. What do the intersections (5.2.18) look like in the cases $n = 2$ and $n = 3$?

Ex. 5.2.4 (Mayer [217]). Let

$$[A]_\varepsilon = \begin{pmatrix} 1 & [-1+\varepsilon, 1-\varepsilon] \\ [-1+\varepsilon, 1-\varepsilon] & -1 \end{pmatrix}, \quad 0 \le \varepsilon \le 1, \quad [b] = \begin{pmatrix} 2 \\ 2 \end{pmatrix}$$

and denote by S^ε, respectively $S^\varepsilon_{\mathrm{sym}}$, the solution set, respectively the symmetric solution set, of the interval linear system $[A]_\varepsilon x = [b]$.

(a) Derive the inequalities which characterize the sets S^ε and $S^\varepsilon_{\mathrm{sym}}$ and plot these sets for $\varepsilon = 1/2$. How do these sets behave when $\varepsilon > 0$ gets smaller? How do they look in the limit cases $\varepsilon = 0$ and $\varepsilon = 1$? (Notice that $[A]_0$ is singular while $[A]_\varepsilon$ is regular for $0 < \varepsilon \le 1$.)

(b) Find an interval vector $[z] \in \mathbb{IR}^2$ which encloses $S_{\mathrm{sym}}^{\varepsilon}$ for all $\varepsilon \in [0,1]$. What can be said about the distance $q([z], \square S^{\varepsilon})$ when ε tends to zero? Which conclusion could one draw from this observation?

Ex. 5.2.5. Plot S and S_{sym} for the data in Example 5.2.4. Proceed as follows: Start with a rectangular grid of sufficiently small mesh size which covers S. Test the Oettli–Prager criterion in the grid points. If the test is positive, plot the grid point. Proceed similarly for S_{sym} taking into account (5.2.16).

Ex. 5.2.6 (Mayer [219]). Characterize the persymmetric solution set similarly to Theorem 5.2.6.

Ex. 5.2.7. Prove Theorem 5.2.10 for the skew-symmetric solution set.
Hint: Replace $(\dot{\underline{L}})_p^q = (\dot{\underline{L}})_q^p$ by $(\dot{\underline{L}})_p^q = -(\dot{\overline{L}})_q^p$ in Lemma F.3 of Appendix F.

Ex. 5.2.8. Perskew-symmetric matrices $A \in \mathbb{R}^{n\times n}$ are defined by $EA = -(EA)^T$ with E from (5.2.42), or, equivalently, by $a_{ij} = -a_{n+1-j,n+1-i}$. Show that $a_{i,n+1-i} = 0$, $i = 1,\dots,n$, holds. How could one introduce a perskew-symmetric solution set? Characterize this set similarly to Theorem 5.2.10. Construct a perskew-symmetric 4×4 matrix.

Ex. 5.2.9. Centro-symmetric matrices $A \in \mathbb{R}^{n\times n}$ are defined by $EA = AE$ with E from (5.2.42), or, equivalently, by $a_{ij} = a_{n+1-i,n+1-j}$. Construct a centro-symmetric 3×3 and 4×4 matrix. How could one introduce a centro-symmetric solution set? Describe this set by inequalities. (Open problem!)

Ex. 5.2.10. Centroskew-symmetric matrices $A \in \mathbb{R}^{n\times n}$ are defined by $EA = -AE$ with E from (5.2.42), or, equivalently, by $a_{ij} = -a_{n+1-i,n+1-j}$. Construct a centroskew-symmetric 3×3 and 4×4 matrix. How could one introduce a centroskew-symmetric solution set? Describe this set by inequalities. (Open problem!)

5.3 Interval hull

The interval hull of a bounded subset of $\mathbb{R}^n$ was defined to be (with respect to inclusion) the smallest interval vector which encloses this subset. Theoretically it is possible to compute the interval hull $\square S$ of the solution set S of a regular interval linear system $[A]x = [b]$, but this computation can be costly unless the input data have some additional properties. Nevertheless we will present an algorithm for computing $\square S$. In addition, we will comment on particular cases. Supplementary information on the interval hull can be found in the last chapter of Neumaier [257]. We start this section with two auxiliary results. The first is due to Farkas [93] and is well known in linear programming. It provides a criterion which guarantees the existence of a nonnegative solution of a linear $m \times n$ system.

Lemma 5.3.1 (Farkas). *Let $A \in \mathbb{R}^{m\times n}$, $b \in \mathbb{R}^m$. Then $Ax = b$ has a nonnegative solution if and only if for each $y \in \mathbb{R}^n$ with $y^TA \geq 0$ one obtains $y^Tb \geq 0$.*

Proof. If $Ax = b$ has a nonnegative solution x^*, then $y^TA \geq 0$ implies $0 \leq y^TAx^* = y^Tb$.

Conversely, let $y^TA \geq 0$ imply $y^Tb \geq 0$ and assume that b is not contained in the convex set $M = \{Ax \mid x \geq 0\}$. Since M is closed, the distance $\delta = \inf_{x\in M}\|Ax - b\|_2 > 0$ is attained for some element $x^0 \in M$. Clearly, the intersection of M with the open ball $B(b, \delta)$ is empty. Let T be the hyperplane through x^0 which is tangential to the sphere $\partial B(b, \delta)$ and let H be the open half space which is bounded by T and which contains b. If $x^1 \in H \cap M$, the line between x^0 and x^1 would intersect $B(b, \delta)$, hence the convexity of M would imply the contradiction $M \cap B(b, \delta) \neq \emptyset$. Therefore, $H \cap M = \emptyset$. By moving T in parallel towards b by $\delta/2$ one ends up with some hyperplane

$$\hat{n}^Tz = \gamma, \tag{5.3.1}$$

which separates b from M. W.l.o.g. let

$$\hat{n}^Tb < \gamma \tag{5.3.2}$$

and

$$\hat{n}^Tz > \gamma \quad \text{for } z \in M. \tag{5.3.3}$$

(Otherwise multiply (5.3.1) by -1 and rename.) For $x = 0$ we get $z = Ax = 0 \in M$, whence $\gamma < 0$ by (5.3.3). Now we show

$$\hat{n}^TA_{*,j} \geq 0 \tag{5.3.4}$$

for each column of A. If $\hat{n}^TA_{*,j_0} < 0$ holds for some column A_{*,j_0}, then choose $x_{j_0} = \frac{\gamma}{\hat{n}^TA_{*,j_0}} > 0$ and $x_j = 0$ for $j \neq j_0$. Hence $z = \sum_{j=1}^n A_{*,j}x_j \in M$ and $\hat{n}^Tz = (\hat{n}^TA_{*,j_0})x_{j_0} = \gamma$, contradicting (5.3.3). This proves (5.3.4) from which we get $\hat{n}^TA \geq 0$. The assumption of the theorem (with $y = \hat{n}$) implies $\hat{n}^Tb \geq 0 > \gamma$ which contradicts (5.3.2). Therefore, $b \in M$, and the theorem is proved. □

Our second auxiliary result looks rather technical. It considers two regular matrices A, A' which map two vectors $x \neq x'$ to the same image. It says that these matrices differ in at least two columns with the same index j, while the components x_j, x'_j, with which these columns are multiplied, have the same sign.

Lemma 5.3.2. *Let $[A] \in \mathbb{IR}^{n\times n}$ be regular and let $Ax = A'x'$ for some $A, A' \in [A]$, $x \neq x'$. Then there exists an index $j \in \{1, \ldots, n\}$ such that $A_{*,j} \neq A'_{*,j}$ and $x_jx'_j > 0$.*

Proof. Assume that there exist $A, A' \in [A]$ and $x \neq x'$ such that $Ax = A'x'$ and for each j with $A_{*,j} \neq A'_{*,j}$ one has $x_jx'_j \leq 0$. Let $J = \{j \mid A_{*,j} \neq A'_{*,j},\ x_jx'_j < 0\}$ and define $\tilde{A} \in \mathbb{R}^{n\times n}$ as follows. If $j \in J$, then

$$\tilde{A}_{*,j} = \frac{x_j}{x_j - x'_j}A_{*,j} - \frac{x'_j}{x_j - x'_j}A'_{*,j}.$$

If $j \notin J$, choose $\tilde{A}_{*,j} \in \{A_{*,j}, A'_{*,j}\}$ such that $x_j A_{*,j} - x'_j A'_{*,j} = (x_j - x'_j)\tilde{A}_{*,j}$. Notice that in the case $j \notin J$ either $A_{*,j} = A'_{*,j}$, or $A_{*,j} \neq A'_{*,j}$ and $x_j x'_j = 0$. In the case $j \in J$ we have $x_j/(x_j - x'_j) = x_j^2/(x_j^2 - x_j x'_j) > 0$ and similarly $-x'_j/(x_j - x'_j) = -x_j x'_j/(x_j^2 - x_j x'_j) > 0$. Hence for $j \in J$ the column $\tilde{A}_{*,j}$ is a convex combination of $A_{*,j}$ and $A'_{*,j}$ and therefore in $[A]_{*,j}$. Hence $\tilde{A} \in [A]$ and $\tilde{A}(x - x') = \sum_{j=1}^n \tilde{A}_{*,j}(x_j - x'_j) = \sum_{j=1}^n (A_{*,j}x_j - A'_{*,j}x'_j) = Ax - A'x' = 0$. Since we assumed $x \neq x'$ this proves $\tilde{A}$ to be singular, which contradicts the assumption of the theorem. □

We consider Rohn's sign accord algorithm for solving the problem

$$(\check{A} - D\,\mathrm{rad}([A])\tilde{D})x = \check{b} + D\,\mathrm{rad}([b]), \quad \tilde{D}x \geq 0, \; |\tilde{D}| = I, \tag{5.3.5}$$

for a given signature matrix $D \in \mathbb{R}^{n\times n}$ (which was defined in Section 1.2 by $|D| = I$). It will turn out that (5.3.5) has a unique solution x^D and that $\square S$ can be represented by means of all such solutions if D varies in the set of the 2^n different signature matrices.

In order to get some insight in equation (5.3.5) we state several equivalent formulations of it before we present the algorithm. To this end we define the nonnegative vectors

$$x^+ = \max\{x, 0\}, \quad x^- = \max\{-x, 0\} \tag{5.3.6}$$

for $x \in \mathbb{R}^n$, where here and in the sequel max and min are applied to vectors componentwise. From (5.3.6) one immediately obtains $x = x^+ - x^-$ and $|x| = x^+ + x^-$.

First we bring the vector equation in (5.3.5) into the equivalent form

$$\check{A}x - \check{b} = D\,(\mathrm{rad}([A])|x| + \mathrm{rad}([b]))\,. \tag{5.3.7}$$

Introducing x^+, x^- results in

$$x^+ = (\check{A} - D\mathrm{rad}([A]))^{-1}(\check{A} + D\mathrm{rad}([A]))x^- + (\check{A} - D\mathrm{rad}([A]))^{-1}(\check{b} + D\mathrm{rad}([b])), \tag{5.3.8}$$

which is a linear complementarity problem (cf. Berman, Plemmons [68], Chapter 10) since $x^+ \geq 0$, $x^- \geq 0$, and $(x^+)^T x^- = 0$. Taking absolute values on both sides of (5.3.7) results in

$$|\check{A}x - \check{b}| = \mathrm{rad}([A])|x| + \mathrm{rad}([b]), \tag{5.3.9}$$

i.e., the Oettli–Prager criterion holds with equality (cf. Theorem 5.2.2 (d)). This implies that each solution x^D of (5.3.5) belongs to the solution set S. Rearranging (5.3.7) finally yields the equivalent fixed point form

$$x = \check{A}^{-1}D\,\mathrm{rad}([A])|x| + \check{A}^{-1}(\check{b} + D\,\mathrm{rad}([b])). \tag{5.3.10}$$

After these equivalences we formulate Rohn's algorithm in Rohn [301] for solving (5.3.5).

Algorithm 5.3.3 (Sign accord algorithm).
Step 0: Select $D' \in \mathbb{R}^{n\times n}$ with $|D'| = I$.
(Recommended: D' such that $D'(\check{A}^{-1}(\check{b} + D\,\mathrm{rad}([b]))) \geq 0$.)
Step 1: Solve $(\check{A} - D\,\mathrm{rad}([A])D')x = \check{b} + D\,\mathrm{rad}([b])$.
Step 2: If $D'x \geq 0$, set $\tilde{D} := D'$, $x^D := x$ and terminate.
Step 3: Find $k = \min\{j \mid d'_{jj}x_j < 0\}$.
Step 4: Set $d'_{kk} := -d'_{kk}$ and go to Step 1.

Notice that the signature matrix D is given and remains fixed during the whole algorithm. The second signature matrix D' which occurs can theoretically be chosen arbitrarily in Step 0, although it is recommended to define its diagonal entries by the signs of the components of some particular vector. If this vector has a zero component, put the corresponding diagonal entry d'_{ii} to one. The matrix D' is changed by one diagonal entry in each run of the loop. After the algorithm has terminated, D' is renamed as $\tilde{D} = D'$. It fulfills $\tilde{D}x^D = |x^D|$ for the solution x^D of the final linear system.

Our first theoretical result on the sign accord algorithm shows that it is finite, i.e., it stops after finitely many runs of its loop. Moreover, existence and uniqueness for the problem (5.3.5) is proved.

Theorem 5.3.4. *Let $[A] \in \mathbb{IR}^{n\times n}$ be regular and $[b] \in \mathbb{IR}^n$. Then the sign accord algorithm is finite for each signature matrix D and each starting signature matrix D'. It terminates with a signature matrix $\tilde{D}$ and a vector x^D which satisfy* (5.3.5). *The vector x^D does not depend on D' and is unique. The diagonal entries $\tilde{d}_{jj}$ of $\tilde{D}$ are unique, except for those which are multiplied by components $x^D_j = 0$. For each vector x^D there are at most 2^n linear systems to be solved.*

Proof. We show by induction that each k in Step 3 can occur there at most 2^{n-k} times.

Case $k = n$: Assume that n appears at least twice in Step 3 and let D', x, $\hat{D}'$, $\hat{x}$ correspond to its two nearest occurrences. Then $d'_{jj}x_j \geq 0$, $\hat{d}'_{jj}\hat{x}_j \geq 0$ for $j = 1, \ldots, n-1$, and $d'_{nn}\hat{d}'_{nn} = -1$, $d'_{nn}x_n < 0$, $\hat{d}'_{nn}\hat{x}_n < 0$; hence $d'_{jj}x_j\hat{d}'_{jj}\hat{x}_j \geq 0$ for $j = 1, \ldots, n$. For $j = n$ this implies $x_n \neq 0$, $\hat{x}_n \neq 0$ and $x_n\hat{x}_n < 0$, whence $x \neq \hat{x}$. Apply Lemma 5.3.2 to $(\check{A} - D\,\mathrm{rad}([A])D')x = \check{b} + D\,\mathrm{rad}([b]) = (\check{A} - D\,\mathrm{rad}([A])\hat{D}')\hat{x}$. There exists an index j such that $d'_{jj}\hat{d}'_{jj} = -1$ and $x_j\hat{x}_j > 0$. This implies the contradiction $d'_{jj}x_j\hat{d}'_{jj}\hat{x}_j < 0$.

Case $k < n$: Again let D', x, $\hat{D}'$, $\hat{x}$ correspond to two nearest occurrences of k so that $d'_{jj}x_j\hat{d}'_{jj}\hat{x}_j \geq 0$ for $j = 1, \ldots, k$. Then Lemma 5.3.2 implies the existence of an index j with $d'_{jj}\hat{d}'_{jj} = -1$ and $x_j\hat{x}_j > 0$. Hence $d'_{jj}x_j\hat{d}'_{jj}\hat{x}_j < 0$, so that $j > k$. By the induction hypothesis j can occur at most 2^{n-j} times for $j = k+1, k+2, \ldots, n$. This means that k cannot occur more than $1 + (2^{n-(k+1)} + 2^{n-(k+2)} + \cdots + 2 + 1) = 2^{n-k}$ times.

Therefore, the algorithm is finite and terminates with $\tilde{D}$, x^D which satisfy (5.3.5). Moreover, at most $(1 + 2 + 2^2 + \cdots + 2^{n-1}) + 1 = 2^n$ linear systems have to be solved.

It remains to prove uniqueness and independence of D'. To this end assume that (5.3.5) has two solutions $\tilde{D}$, x^D and $\tilde{\tilde{D}}$, $\hat{x}^D$ with $x^D \neq \hat{x}^D$. Since the right-hand side of

(5.3.5) is the same for both pairs, Lemma 5.3.2 assures the existence of an index j with $\tilde{d}_{jj}\tilde{\tilde{d}}_{jj} = -1$ and $x_j^D \hat{x}_j^D > 0$. This implies $\tilde{d}_{jj} x_j^D \tilde{\tilde{d}}_{jj} \hat{x}_j^D < 0$ contrary to $\tilde{d}_{jj} x_j^D > 0$ and $\tilde{\tilde{d}}_{jj} \hat{x}_j^D > 0$. Therefore, x^D is unique and so is $\tilde{D}$, except for the cases mentioned in the theorem. □

Next we show how to represent $\square S$ by means of x^D for varying signature matrices D. To this end we use the notation $\operatorname{conv} X$ for the convex hull of a set X, i.e., the smallest convex superset of X.

Theorem 5.3.5. *Let $[A] \in \mathbb{IR}^{n\times n}$ be regular, $[b] \in \mathbb{IR}^n$ and $X = \{x^D \mid |D| = I,\ x^D \text{solves (5.3.5)}\}$. Then*

$$\operatorname{conv} S = \operatorname{conv} X \tag{5.3.11}$$

and

$$\underline{x}^H = \min X, \quad \overline{x}^H = \max X, \tag{5.3.12}$$

where $[x]^H = \square S$. At most 2^{2n} linear systems have to be solved for computing $\square S$.

Proof. Since $x^D \in S$ we trivially get $\operatorname{conv} X \subseteq \operatorname{conv} S$. In order to prove the converse inclusion choose an arbitrary element $x \in S$. Then $A_0 x = b_0$ for some $A_0 \in [A]$, $b_0 \in [b]$. If we can show that the linear system

$$\left.\begin{aligned} \textstyle\sum_{|D|=I} \lambda_D (A_0 x^D) &= b_0 \\ \textstyle\sum_{|D|=I} \lambda_D &= 1 \end{aligned}\right\} \tag{5.3.13}$$

has a nonnegative solution $\lambda = (\lambda_D)_{|D|=I} \in \mathbb{R}^{2^n}$, then (5.3.13) implies

$$A_0 \Big(\textstyle\sum_{|D|=I} \lambda_D x^D\Big) = b_0,$$

hence by the regularity of A_0 we get $x = \sum_{|D|=I} \lambda_D x^D \in \operatorname{conv} X$. Since $x \in S$ was arbitrary, this proves $S \subseteq \operatorname{conv} X$ and $\operatorname{conv} S \subseteq \operatorname{conv} X$. Hence (5.3.11) follows.

We will apply the Farkas Lemma 5.3.1 to (5.3.13) in order to show the existence of $\lambda \ge 0$. Notice that the columns of the system (5.3.13) are

$$\begin{pmatrix} A_0 x^D \\ 1 \end{pmatrix}$$

and the right-hand side is

$$\begin{pmatrix} b_0 \\ 1 \end{pmatrix}.$$

In view of the lemma let $y \in \mathbb{R}^n$, $y_0 \in \mathbb{R}$ satisfy $y^T A_0 x^D + y_0 \ge 0$ for each signature matrix D. Choose D such that $|y| = -Dy$ holds. Then $|y^T(A_0 - \check{A})| \le |y^T| \operatorname{rad}([A]) = -y^T D \operatorname{rad}([A])$, whence $y^T(\check{A} + D \operatorname{rad}([A])) \le y^T A_0 \le y^T(\check{A} - D \operatorname{rad}([A]))$. Similarly,

$y^T(\check{b} + D\,\mathrm{rad}([b])) \le y^T b_0$. Then with $\tilde{D}x^D = |x^D|$ and (5.3.5) we get

$$\begin{aligned}-y_0 &\le y^T A_0 x^D = y^T A_0 (x^D)^+ - y^T A_0 (x^D)^- \\ &\le y^T(\check{A} - D\,\mathrm{rad}([A]))(x^D)^+ - y^T(\check{A} + D\,\mathrm{rad}([A]))(x^D)^- \\ &= y^T(\check{A} - D\,\mathrm{rad}([A])\tilde{D})x^D = y^T(\check{b} + D\,\mathrm{rad}([b])) \le y^T b_0.\end{aligned}$$

Hence $y^T b_0 + y_0 \ge 0$ so that the assumptions of the Farkas lemma are fulfilled and a nonnegative solution λ of (5.3.13) is guaranteed.

The assertion (5.3.12) follows from (5.3.11) since $\mathrm{conv}\, S$ is a convex polyhedron. From Theorem 5.3.4 at most 2^n linear systems have to be solved for a fixed signature matrix D. Since there are 2^n different matrices D, at most 2^{2n} linear systems have to be solved for computing $\square S$. □

Corollary 5.3.6. *Let* $[A] \in \mathbb{IR}^{n\times n}$ *be regular,* $[b] \in \mathbb{IR}^n$*, and* $E = \{\, x \mid Ax = b,\ A \in \partial[A],\ b \in \partial[b]\,\}$*. Then*

$$\mathrm{conv}\, S = \mathrm{conv}\, E \tag{5.3.14}$$

and

$$\underline{x}^H = \min E, \quad \overline{x}^H = \max E, \tag{5.3.15}$$

where $[x]^H = \square S$.

Proof. With X as in Theorem 5.3.5 the assertions follow immediately from $X \subseteq E \subseteq S$ and the same theorem. □

We present a second proof of Corollary 5.3.6 which is very elementary and independent of the sign accord algorithm. It follows the lines in Hartfiel [136]. To this end we call any entry $a_{ij} \in [a]_{ij} \setminus \{\underline{a}_{ij}, \overline{a}_{ij}\}$ a free entry in $A \in [A]$, and analogously we define $b_i \in [b]_i \setminus \{\underline{b}_i, \overline{b}_i\}$ to be a free entry in $b \in [b]$. By $\upsilon(A)$, $\upsilon(b)$ we denote the number of free entries in A, and b, respectively, and we use $\upsilon(A, b) = \upsilon(A) + \upsilon(b)$.

Second proof of Corollary 5.3.6. Since $E \subseteq S$, we get $\mathrm{conv}\, E \subseteq \mathrm{conv}\, S$. For the converse inclusion we proceed by induction on $\upsilon(A, b)$. If $\upsilon(A, b) = 0$, then $A \in \partial[A]$, $b \in \partial[b]$, hence $x = A^{-1}b \in E \subseteq \mathrm{conv}\, E$. Now suppose that $\upsilon(A, b) < m$ implies $x = A^{-1}b \in \mathrm{conv}\, E$, choose $A \in [A]$, $b \in [b]$ such that $\upsilon(A, b) = m$ and let $x = A^{-1}b$.

Case 1, $\upsilon(b) \ne 0$: Choose $b', b'' \in [b]$ with $b' \ne b''$ such that $b = \alpha b' + (1 - \alpha)b''$, where $\alpha \in (0, 1)$ and $\upsilon(b') = \upsilon(b'') < \upsilon(b)$. Then by the induction hypothesis $x' = A^{-1}b' \in \mathrm{conv}\, E$ and $x'' = A^{-1}b'' \in \mathrm{conv}\, E$, hence $x = \alpha x' + (1 - \alpha)x'' \in \mathrm{conv}\, E$.

Case 2, $\upsilon(A) \ne 0$: Let a_{ij} be a free entry in A and define $A(\theta) = A + \theta e^{(i)}(e^{(j)})^T$. If $x(\theta) = A(\theta)^{-1}b$ exists we can express $x(\theta)_j$ using Cramer's rule. Taking into account that θ occurs only once in $A(\theta)$ and calculating the resulting two determinants yields

$$x(\theta)_j = \frac{\alpha_j\theta + \beta_j}{\alpha\theta + \beta}$$

with constants $\alpha, \alpha_j, \beta, \beta_j$, which are independent of θ. Hence, $x(\theta)_j = c_j + \delta(\theta)d_j$ for some c_j, d_j and $\delta(\theta) = \theta$ if $\alpha = 0$ and $\delta(\theta) = \frac{1}{\alpha\theta+\beta}$ if $\alpha \ne 0$. Choose θ_1 and θ_2 so that $a_{ij}(\theta_1) = \underline{a}_{ij}$ and $a_{ij}(\theta_2) = \overline{a}_{ij}$. Then $\theta_1 < 0 < \theta_2$. Since $\delta(\theta)$ is independent of j the solutions $x(\theta)$ lie on a straight line and $x = x(0) = \gamma x(\theta_1) + (1-\gamma)x(\theta_2)$ for some $\gamma \in (0, 1)$. (Note that $\delta(\theta)$ is strictly monotone with respect to θ.) Since $\upsilon(A(\theta_1), b) = \upsilon(A(\theta_2), b) < \upsilon(A, b)$ the induction hypothesis implies $x(\theta_1) \in \operatorname{conv} E$, $x(\theta_2) \in \operatorname{conv} E$, whence $x \in \operatorname{conv} E$. Summarizing, we have $S \subseteq \operatorname{conv} E$, whence $\operatorname{conv} S \subseteq \operatorname{conv} E$. □

Corollary 5.3.6 suggests an upper bound of 2^{n^2+n} linear systems to be solved for computing $\square S$. This is tremendously more than the bound provided by Theorem 5.3.5. But even this bound can be reduced as the following theorem shows.

Theorem 5.3.7. *Let $[A] \in \mathbb{IR}^{n\times n}$ be regular, $[b] \in \mathbb{IR}^n$, $D \in \mathbb{R}^{n\times n}$ a fixed signature matrix and x the solution of the vector equation in* (5.3.5). *If*

$$|\check{A}^{-1} D \operatorname{rad}([A])|x|\,| < |x| \tag{5.3.16}$$

holds, then the sign accord algorithm solves only one linear system in Step (1) *when started as recommended, i.e., $x = x^D$ holds, where x^D is the solution of* (5.3.5).

Proof. Since x solves the vector equation in (5.3.5) it also solves the equivalent fixed point form (5.3.10) from which we obtain

$$|x - \check{A}^{-1}(\check{b} + D \operatorname{rad}([b]))| = |\check{A}^{-1} D \operatorname{rad}([A])|x|\,| < |x|. \tag{5.3.17}$$

Therefore, for each fixed j the j-th component of the two vectors on the left-hand side differ from zero and has the same sign. If one starts with D' as recommended and if x solves the linear system in Step 1 for the first time, one has $D'(\check{A}^{-1}(\check{b} + D \operatorname{rad}([b]))) \ge 0$ by the choice of D' and $D'x \ge 0$ by virtue of (5.3.17). This terminates the algorithm in Step 2 with $x = x^D$. □

At first glance Theorem 5.3.7 looks very theoretical. But if $\operatorname{rad}([A])$ and $\operatorname{rad}([b])$ tend to zero, then the left-hand side of (5.3.16) does the same while the right-hand side tends to $|\hat{x}|$, where $\hat{x} = \check{A}^{-1}\check{b}$. Hence if $|\hat{x}| > 0$ and $\operatorname{rad}([A])$, $\operatorname{rad}([b])$ are small then (5.3.16) holds for each signature matrix D. In addition, practical experience shows that in many cases the sign accord algorithm terminates after solving only one linear system if one follows the recommendation in Step 0 even if $\operatorname{rad}([A])$ and $\operatorname{rad}([b])$ are larger. In this ideal case only 2^n linear systems are necessary in order to determine $\square S$. Taking into account that the initialization in Step 0 requires the solution of an additional linear system $\check{A}z = \check{b} + D \operatorname{rad}([b])$, then there is a *total* of 2^{n+1} linear systems to be solved for $\square S$. This is nearly the square root of the upper bound given in Theorem 5.3.5.

The worst upper bound 2^n of Theorem 5.3.4 for the computation of a *single* x^D can be attained if D' is initialized improperly in Step 0 of the algorithm. This is demonstrated by the following example in Rohn [301].

Example 5.3.8. Define $[A] \in \mathbb{IR}^{n\times n}$ and $[b] \in \mathbb{IR}^n$, $n \ge 2$, by

$$\check{A} = I, \quad \operatorname{rad}([a]_{ij}) = \begin{cases} 2, & \text{if } j = i+1,\ 1 \le i \le n-1 \\ 0 & \text{otherwise} \end{cases}, \quad i, j = 1, \ldots, n,$$

$$\check{b} = 0, \quad \operatorname{rad}([b]_i) = 1, \quad i = 1, \ldots, n. \tag{5.3.18}$$

Denote by $p_D(D')$ the number of linear systems to be solved in Algorithm 5.3.3 in order to find x^D, $|D| = I$, for a starting signature matrix D' in Step 0. Then

$$p_D(D') = 1 + \sum_{j=1}^{n} \left(1 - \prod_{i=j}^{n} d_{ii}d'_{ii}\right) 2^{j-2}. \tag{5.3.19}$$

In order to prove this formula we first notice that the solution x of the linear system $(\check{A} - D\operatorname{rad}([A])D')x = \check{b} + D\operatorname{rad}([b])$ in Step 1 satisfies

$$x_j = d_{jj} \sum_{m=0}^{n-j} 2^m \prod_{i=j+1}^{j+m} d_{ii}d'_{ii}, \quad j = 1, \ldots, n. \tag{5.3.20}$$

This can be seen by backward substitution and by induction on $j = n, n-1, \ldots, 1$ taking into account the upper bidiagonal form $I - (2d_{ii}\delta_{i,j-1}d'_{jj})$ of the matrix above, whence $x_n = d_{nn}$, $x_j = d_{jj} + 2d_{jj}d'_{j+1,j+1}x_{j+1}$, $j = n-1, n-2, \ldots, 1$. From (5.3.20) we get

$$d'_{jj}x_j = \sum_{m=0}^{n-j} 2^m \prod_{i=j}^{j+m} d_{ii}d'_{ii}. \tag{5.3.21}$$

Since $\sum_{m=0}^{n-j-1} 2^m = 2^{n-j} - 1 < 2^{n-j}$ the last summand in (5.3.21) determines the sign of the left-hand side, hence

$$\operatorname{sign}(d'_{jj}x_j) = \operatorname{sign}\left(\prod_{i=j}^{n} d_{ii}d'_{ii}\right) = \prod_{i=j}^{n} d_{ii}d'_{ii}, \quad j = 1, \ldots, n.$$

Now we proceed by induction on the values which $p_D(D')$ can assume. If $p_D(D') = 1$ then Algorithm 5.3.3 terminates with $D'x \ge 0$ when reaching Step 2 for the first time. Hence for each j we have

$$\prod_{i=j}^{n} d_{ii}d'_{ii} = \operatorname{sign}(d'_{jj}x_j) = 1,$$

so that the formula holds. Now assume that (5.3.19) holds whenever $p_D(D') \le s$, $s \ge 1$, and choose two signature matrices D, D' such that $p_D(D') = s+1$. Let D'' be the updated value of D' when passing Step 4 for the first time. Then $d''_{kk} = -d'_{kk}$ while $d''_{jj} = d'_{jj}$ for $j \ne k$. From this we get

$$\prod_{i=j}^{n} d_{ii}d''_{ii} = \begin{cases} -\prod_{i=j}^{n} d_{ii}d'_{ii} = -\operatorname{sign}(d'_{jj}x_j) = -1, & \text{if } j < k, \\ -\operatorname{sign}(d'_{kk}x_k) = 1, & \text{if } j = k, \\ \prod_{i=j}^{n} d_{ii}d'_{ii}, & \text{if } j > k. \end{cases}$$

Using the induction hypothesis we finally obtain

$$p_D(D') = 1 + p_D(D'') = 2 + \sum_{j=1}^{n}\left(1 - \prod_{i=j}^{n} d_{ii}d''_{ii}\right)2^{j-2}$$

$$= 2 + \sum_{j=1}^{k-1} 2 \cdot 2^{j-2} + \sum_{j=k+1}^{n}\left(1 - \prod_{i=j}^{n} d_{ii}d'_{ii}\right)2^{j-2}$$

$$= 1 + 2^{k-1} + \sum_{j=k+1}^{n}\left(1 - \prod_{i=j}^{n} d_{ii}d'_{ii}\right)2^{j-2}$$

$$= 1 + \sum_{j=1}^{n}\left(1 - \prod_{i=j}^{n} d_{ii}d'_{ii}\right)2^{j-2},$$

which completes the proof of formula (5.3.19).

Now we consider $p_D(D')$ for several choices of D and D'.

(i) If $d_{ii} = d'_{ii}$ for all i with the exception of $i = n$, then $\prod_{i=j}^{n} d_{ii}d'_{ii} = d_{nn}d'_{nn} = -1$ for each j, whence (5.3.19) yields $p_D(D') = 1 + \sum_{j=1}^{n} 2^{j-1} = 2^n$.

(ii) If $d'_{ii} = \operatorname{sign}((\check{b} + D\operatorname{rad}([b]))_i)$, $i = 1, \dots, n$, as recommended, then $d'_{ii} = \operatorname{sign}(d_{ii})$ and $\prod_{i=j}^{n} d_{ii}d'_{ii} = 1$, whence $p_D(D') = 1$.

(iii) Let D and $s \in \{1, \dots, 2^n\}$ be given. Then there is a signature matrix D' such that $p_D(D') = s$. In order to see this, represent $s - 1$ as a binary number such that $s = 1 + \sum_{j=1}^{n} \beta_j 2^{j-1}$, $\beta_j \in \{0, 1\}$ and define D' inductively from $\prod_{i=j}^{n} d_{ii}d'_{ii} = 1 - 2\beta_j$, $j = n, n-1, \dots, 1$. Then $p_D(D') = s$.

(iv) Let D' and $s \in \{1, \dots, 2^n\}$ be given. Then there is a signature matrix D such that $p_D(D') = s$. This can be seen similarly as in (iii). □

In order to reduce work in Algorithm 5.3.3 one might hope that some of the solutions x^D coincide so that there is no need to consider all possibilities of a signature matrix D. The following result, however, is negative in this respect.

Theorem 5.3.9. *Let* $[A] \in \mathbb{IR}^{n\times n}$ *be regular and* $[b] \in \mathbb{IR}^n$. *If either of the assumptions*
(i) $\operatorname{rad}([b]) > 0$,
(ii) $\operatorname{rad}([A]) > O$ *and* $0 \notin [b]$
holds, then $D \neq \hat{D}$ *implies* $x^D \neq x^{\hat{D}}$ *for any two signature matrices* $D, \hat{D} \in \mathbb{R}^{n\times n}$.

Proof. First we note that in the case (ii) the solution set S does not contain $x = 0$ since $0 \notin [b]$. Therefore, each of the two assumptions implies $\operatorname{rad}([A])|x| + \operatorname{rad}([b]) > 0$ for $x \in S$. Let $x^D = x^{\hat{D}}$. Then (5.3.7) implies

$$D(\operatorname{rad}([A])|x^D| + \operatorname{rad}([b])) = \check{A}x^D - \check{b} = \check{A}x^{\hat{D}} - \check{b}$$
$$= \hat{D}(\operatorname{rad}([A])|x^{\hat{D}}| + \operatorname{rad}([b])) = \hat{D}(\operatorname{rad}([A])|x^D| + \operatorname{rad}([b])).$$

Since $\operatorname{rad}([A])|x| + \operatorname{rad}([b]) > 0$, we get $D = \hat{D}$. □

Since Theorem 5.3.9 is discouraging, we will now concentrate on particular classes of interval matrices. The first class consists of matrices whose radius has rank one. Then there exist nonnegative vectors $u, v \in \mathbb{R}^n$ such that $\mathrm{rad}([A]) = uv^T$. We will see that in this case the solution in Step 1 of the algorithm can be given explicitly.

Theorem 5.3.10. *Let $[A] \in \mathbb{IR}^{n\times n}$ be regular with $\mathrm{rad}([A]) = uv^T$ for some nonnegative vectors u, v, and let $[b] \in \mathbb{IR}^n$. Then for each of the signature matrices D, D' the solution of the system*

$$(\check{A} - D\,\mathrm{rad}([A])D')x = \check{b} + D\,\mathrm{rad}([b]) \tag{5.3.22}$$

in Step (1) of the sign accord algorithm is given by

$$x = \alpha_{DD'} z^D + b^D, \tag{5.3.23}$$

where $z^D = \check{A}^{-1}Du$, $b^D = \check{A}^{-1}(\check{b} + D\,\mathrm{rad}([b]))$ and $\alpha_{DD'} = \dfrac{v^TD'b^D}{1 - v^TD'z^D}$.

Proof. Let x solve (5.3.22) and define $\alpha = v^TD'x \in \mathbb{R}$. Use $\mathrm{rad}([A]) = uv^T$ and (5.3.22) in order to obtain $\check{A}x - Du\alpha = \check{b} + D\,\mathrm{rad}([b])$. Hence $x = \alpha z^D + b^D$, which is (5.3.23) with α instead of $\alpha_{DD'}$. Premultiplying this equation by v^TD' results in $\alpha = \alpha v^TD'z^D + v^TD'b^D$. Solving for α shows $\alpha = \alpha_{DD'}$. □

Notice that the denominator of $\alpha_{DD'}$ is positive so that $\alpha_{DD'}$ is defined. Otherwise we have $\beta = v^TD'z^D = v^TD'\check{A}^{-1}Du \geq 1$. Then $v \neq 0$ and $\tilde{A} = \check{A} - \frac{1}{\beta}Duv^TD' \in [A]$. In particular, $\tilde{A}$ is regular since we assumed $[A]$ to be so. Now we get the contradiction

$$0 \neq v^TD'\check{A}^{-1}\tilde{A} = v^TD' - \frac{1}{\beta}(v^TD'\check{A}^{-1}Du)v^TD' = 0.$$

We leave it to the reader to reformulate Algorithm 5.3.3 in an economical way for interval matrices with rank-one radius. For details see Rohn [301], Algorithm 3.2.

Now we consider the case where S is completely contained in a fixed orthant. It turns out that the loop in the algorithm has to be passed only once. Hence for $\square S$ at most 2^n linear systems have to be solved.

Theorem 5.3.11. *Let $[A] \in \mathbb{IR}^{n\times n}$ be regular and $[b] \in \mathbb{IR}^n$. Moreover, let the solution set S be contained in a fixed orthant O which contains the sign vector z defined by $|z| = e$. Then the signature matrix $D' = \mathrm{diag}(z_1, \ldots, z_n)$ satisfies the recommendation in Step (0) of the sign accord algorithm, and for each fixed signature matrix D this algorithm terminates after having passed Step (1) only once.*

Proof. Follows immediately from $S \subseteq O$ which effects $D'x \geq 0$ in Step 2. □

In Definition 3.3.6 we introduced inverse stable interval matrices. For such interval matrices at most $2n$ instead of 2^n vectors x^D have to be computed for $\square S$ as we will see in our next theorem.

Theorem 5.3.12. *Let $[A] \in \mathbb{IR}^{n\times n}$ be inverse stable and $[b] \in \mathbb{IR}^n$. For $i = 1, \ldots, n$ define the diagonal entries of the signature matrices $D(i)$ by*

$$(D(i))_{jj} = \begin{cases} 1, & \text{if } (A^{-1})_{ij} \geq 0 \text{ for each } A \in [A], \\ -1 & \text{otherwise.} \end{cases} \tag{5.3.24}$$

Then

$$\underline{x}_i^H = x_i^{-D(i)}, \quad \overline{x}_i^H = x_i^{D(i)}$$

for $[x]^H = \square S$ *and* $i = 1, \ldots, n$.

Proof. For shortness set $D = D(i)$. Choose any $x \in S$ and let $A \in [A]$, $b \in [b]$ such that $Ax = b$. Define $h = DA(x^D - x)$. From $-D(A - \check{A})x^D \le |D(A - \check{A})x^D| \le \mathrm{rad}([A])|x^D|$ and $-D(\check{b} - b) \le |D(\check{b} - b)| \le \mathrm{rad}([b])$ we get $h = D(Ax^D - b - (\check{A}x^D - \check{b}) + D(\mathrm{rad}([A]))|x^D| + \mathrm{rad}([b])) = D(A - \check{A})x^D + \mathrm{rad}([A])|x^D| + D(\check{b} - b) + \mathrm{rad}([b]) \ge 0$. Taking into account the definition of $D(i)$ we obtain $(x^D - x)_i = (A^{-1}Dh)_i \ge 0$. Hence $(x^D)_i \ge x_i$. Since $x \in S$ was arbitrary and since $x^D \in S$ we finally get $\overline{x}_i^H = x_i^{D(i)}$. Analogously one proves $\underline{x}_i^H = x_i^{-D(i)}$. □

We specialize our result further by considering only such inverse stable matrices $[A]$ whose element matrices satisfy

$$\hat{D}A^{-1}D \ge O \quad \text{for each } A \in [A], \tag{5.3.25}$$

where D, $\hat{D}$ are some fixed signature matrices. According to (5.3.25) the matrix $D[A]\hat{D}$ is inverse positive – see Definition 3.3.1 in combination with Definition 1.10.19. For the corresponding interval hull only the two vectors x^{-D} and x^D are needed.

Theorem 5.3.13. *Let* $[A] \in \mathbb{IR}^{n\times n}$, $[b] \in \mathbb{IR}^n$, $[x]^H = \square S$. *In addition, let* $D[A]\hat{D}$ *be inverse positive, where* D, $\hat{D}$ *are two signature matrices. Then*

$$\underline{x}^H = \min\{x^{-D}, x^D\}, \quad \overline{x}^H = \max\{x^{-D}, x^D\}.$$

In particular, if $\hat{D} = I$, *then* $\underline{x}^H = x^{-D}$, $\overline{x}^H = x^D$.

Proof. First we remark that $[A]$ is inverse stable. For the matrix $D(i)$ of Theorem 5.3.12 we have $D(i) = \hat{d}_{ii}D$ for each $i \in \{1, \ldots, n\}$. Therefore, $D(i) \in \{-D, D\}$, whence $x^{D(i)} \in \{x^{-D}, x^D\}$ by virtue of the uniqueness guaranteed in Theorem 5.3.4. Now Theorem 5.3.12 concludes the proof. □

Theorem 5.3.13 says nothing about the number of linear systems to be solved for x^{-D} and x^D. Modifying the sign accord algorithm slightly results in a method for inverse positive matrices $[A]$ which is due to Beeck [64] and Barth, Nuding [59] and needs at most $n + 1$ linear systems for computing $[x]^H$. Notice that for inverse positive matrices the signature matrices in Theorem 5.3.13 are now $D = I = \hat{D}$ so that $\underline{x}^H = x^{-I}$, $\overline{x}^H = x^I$ holds. For the modified sign accord algorithm, one pass of the loop changes the sign *for all* indices k which satisfy $d'_{kk}x_k < 0$. Moreover, the recommendation is now replaced by solving the midpoint equation $\check{A}x - \check{b}$ and by initializing D' by means of the signs of this solution.

We formulate the modification for $D = -I$ in a way which is more suited for the proof in our subsequent theorem. For $D = I$ one has to replace $\underline{A}$ by $\overline{A}$ in Step 0. Moreover, there is no need to solve the midpoint equation again if already solved for $D = -I$.

In Step 1 the inequality sign '>' is replaced by '≤', in Step 2 the vector $\underline{x}^H$ must read $\overline{x}^H$ and in Step 3 the vector $\underline{b}$ is then $\overline{b}$.

Algorithm 5.3.14 (Modified sign accord algorithm for $D = -I$).
Step 0: Put $k := 0$, $A^{(0)} := \underline{A}$, solve $\check{A}x = \check{b}$ and denote this solution by $x^{(0)}$.
Step 1: Define $A^{(k+1)}$ by

$$a_{ij}^{(k+1)} := \begin{cases} \overline{a}_{ij} & \text{if } x_j^{(k)} > 0 \\ \underline{a}_{ij} & \text{otherwise} \end{cases}, \quad i,j = 1,\ldots,n.$$

Step 2: If $k > 0$ and $A^{(k+1)} = A^{(k)}$ set $\underline{x}^H := x$ and terminate.
Step 3: Solve $A^{(k+1)}x^{(k+1)} = \underline{b}$.
Step 4: Set $k := k + 1$ and go to Step 1.

Theorem 5.3.15. *Let $[A] \in \mathbb{IR}^{n\times n}$ be inverse positive and $[b] \in \mathbb{IR}^n$, $[x]^H = \square S$. Then Algorithm 5.3.14 terminates with $x^{(k)} = x^{-I} = \underline{x}^H$ for some $k \le n$. If combined appropriately with the corresponding modified algorithm for computing $x^I = \overline{x}^H$, at most $n + 1$ linear systems have to be solved for $[x]^H$ (plus the initial system $\check{A}x = \check{b}$). In addition,*

$$[x]^H = \begin{cases} [\underline{A}^{-1}\underline{b}, \overline{A}^{-1}\overline{b}], & \text{if } [b] \le 0, \\ [\underline{A}^{-1}\underline{b}, \underline{A}^{-1}\overline{b}], & \text{if } [b] \ni 0, \\ [\overline{A}^{-1}\underline{b}, \underline{A}^{-1}\overline{b}], & \text{if } [b] \ge 0. \end{cases}$$

Proof. By construction of $A^{(k+1)}$ we have $A^{(k+1)}x^{(k)} \ge Ax^{(k)}$ for all $A \in [A]$, in particular $(A^{(k+1)} - A^{(k)})x^{(k)} \ge 0$ and $(A^{(1)} - \check{A})x^{(0)} \ge 0$. Therefore, $z := A^{(k+1)}(x^{(k)} - x^{(k+1)}) = A^{(k+1)}x^{(k)} - \underline{b} \ge (A^{(k+1)} - A^{(k)})x^{(k)} \ge 0$, where in the case $k = 0$ the matrix $A^{(k)} = A^{(0)} = \underline{A}$ has to be replaced by $\check{A}$ taking into account $\underline{b} \le \check{b} = \check{A}x^{(0)}$. Since we assumed $[A]$ to be inverse positive, we obtain $x^{(k)} - x^{(k+1)} = (A^{(k+1)})^{-1}z \ge 0$. This implies $x_j^{(k+1)} \le x_j^{(k)} \le x_j^{(0)}$ so that there is at most one k for which the j-th component of $x^{(k)}$ can change its sign. Moreover, if $x_j^{(0)}$ is negative or zero, the j-th column of $A^{(k)}$ remains fixed during the whole algorithm. In an analogous way, corresponding columns of $A^{(k)}$ are fixed when computing $\overline{x}^H$. This time $x_j^{(0)} \ge 0$ is decisive so that at least n columns (note the special case $x_j^{(0)} = 0$!) remain fixed when computing *both* $\underline{x}^H$ and $\overline{x}^H$. Due to the restricted change of signs, the algorithm terminates with $A^{(k+1)} = A^{(k)}$ for some k with $1 \le k \le n$. For this k we have $\underline{b} = \check{b} + (-I)\,\mathrm{rad}([b]) = A^{(k)}x^{(k)} = (\check{A} - (-I)\,\mathrm{rad}([A])\tilde{D})x^{(k)}$ with $\tilde{D}x^{(k)} \ge 0$, $|\tilde{D}| = I$. Hence $x^{(k)} = x^{-I}$, and by Theorem 5.3.13 we get $x^{-I} = \underline{x}^H$. The remaining part of the theorem follows immediately from $O \le \overline{A}^{-1} \le \underline{A}^{-1}$. □

The number $n + 1$ in Theorem 5.3.15 cannot be decreased in general as the case $n = 1$ shows: Apart from the initial equation $\check{A}x^{(0)} = \check{b}$ at least the two equations $A^{(1)}x^{(1)} = \underline{b}$ and $A^{(1)}x^{(1)} = \overline{b}$ have to be solved.

Algorithm 5.3.14 can also be used for sign stable matrices $[A]$ which fulfill (5.3.25). Since $Ax = b$ is equivalent to $(DA\hat{D})(\hat{D}x) = Db$, apply the algorithm to the inverse positive matrix $D[A]\hat{D}$ and the vector $D[b]$ and use $\square S = \hat{D}[x]^{S'}$, where $[x]^{S'}$ denotes the interval hull of the system $D[A]\hat{D}x = D[b]$.

Of course, Algorithm 5.3.14 can be applied to M-matrices $[A]$ as particular inverse positive matrices.

For our next result we assume that $[A]$ has the midpoint I and is an H-matrix. Then it will turn out that only one matrix must be inverted in order to compute $\square S$. At first glance the assumptions seem to be very special. But when preconditioning $[A]$ by its midpoint inverse $\check{A}^{-1}$ one immediately obtains the particular form $[A] = I + [-R, R]$ with $R \geq O$. The H-matrix property implies that R is not too large, since according to Theorem 1.10.16 it is equivalent to $\rho(R) < 1$.

Theorem 5.3.16. *Let $[A] = I + [-R, R] \in \mathbb{IR}^{n\times n}$, $[b] \in \mathbb{IR}^n$, $\rho(R) < 1$, $M = (I - R)^{-1}$, $[x]^H = \square S$. Then for each $i \in \{1, \dots, n\}$ we have*

$$\underline{x}_i^H = \min\{\underset{\sim}{x}_i^H, \nu_i \underset{\sim}{x}_i^H\}, \qquad \overline{x}_i^H = \max\{\tilde{x}_i^H, \nu_i \tilde{x}_i^H\}, \tag{5.3.26}$$

where

$$\left.\begin{aligned} \underset{\sim}{x}_i^H &= -x_i^* + m_{ii}(\check{b} + |\check{b}|)_i, \quad \tilde{x}_i^H = x_i^* + m_{ii}(\check{b} - |\check{b}|)_i, \\ x_i^* &= \left(M(|\check{b}| + \mathrm{rad}([b]))\right)_i = (M|[b]|)_i \end{aligned}\right\} \tag{5.3.27}$$

and

$$\nu_i = \frac{1}{2m_{ii} - 1} \in (0, 1].$$

Proof. Let $i \in \{1, \dots, n\}$ be fixed. We will carry out the proof in three major steps:

(i) We prove $x_i \leq \max\{\tilde{x}_i^H, \nu_i \tilde{x}_i^H\}$ for each $x \in S$.

(ii) We show $\tilde{x}_i^H = x_i'$, $\nu_i \tilde{x}_i^H = x_i''$ for some elements x', x'' of S which proves $\overline{x}_i^H = \max\{\tilde{x}_i^H, \nu_i \tilde{x}_i^H\}$.

(iii) We prove the formula for $\underline{x}_i$ using the result for $\overline{x}_i$.

Ad (i): Since we assumed $\rho(R) < 1$, we can use the Neumann series in order to see $M \geq O$ and $O \leq MR = RM = M - I$. This latter inequality implies $m_{ii} \geq 1$, whence $2m_{ii} - 1 \geq 1$ and $\nu_i \in (0, 1]$. Define the diagonal matrix $D \in \mathbb{R}^{n\times n}$ by

$$d_{jj} = \begin{cases} 1 & \text{if } j \neq i \text{ and } \check{b}_j \geq 0, \\ -1 & \text{if } j \neq i \text{ and } \check{b}_j < 0, \\ 1 & \text{if } j = i \end{cases}$$

and let

$$\tilde{b} = D\check{b} + \mathrm{rad}([b]).$$

Then $\tilde{b} = |\check{b}| + \mathrm{rad}([b]) + (\check{b}_i - |\check{b}_i|)e^{(i)} = |[b]| + (\check{b}_i - |\check{b}_i|)e^{(i)}$ and

$$(M\tilde{b})_i = x_i^* + m_{ii}(\check{b}_i - |\check{b}_i|) = \tilde{x}_i^H.$$

Choose any element $x \in S$ and let $A \in [A]$, $b \in [b]$ such that $Ax = b$. For

$$\hat{x} = |x| + (x_i - |x_i|)e^{(i)}$$

we obtain

$$\begin{aligned}\hat{x}_i = x_i = b_i + ((I - A)x)_i &\le \check{b}_i + \operatorname{rad}([b]_i) + (|I - A||x|)_i \\ = \tilde{b}_i + (|I - A||x|)_i &\le \tilde{b}_i + (R|x|)_i\end{aligned}$$

and, analogously,

$$\begin{aligned}\hat{x}_j = |x_j| = |b_j + ((I - A)x)_j| &\le |\check{b}_j| + \operatorname{rad}([b]_j) + (|I - A||x|)_j \\ &\le \tilde{b}_j + (R|x|)_j, \quad j \ne i.\end{aligned}$$

Hence $\hat{x} \le \tilde{b} + R|x|$, and $M\hat{x} \le M\tilde{b} + MR|x| = M\tilde{b} + (M - I)|x|$. This finally yields

$$M(\hat{x} - |x|) + |x| \le M\tilde{b}. \tag{5.3.28}$$

If $x_i \ge 0$, then $\hat{x} = |x|$, and from (5.3.28) we get $x_i = |x_i| \le (M\tilde{b})_i = \tilde{x}_i^H$. If $x_i < 0$, then (5.3.28) implies

$$(2m_{ii} - 1)x_i = m_{ii}(x_i - |x_i|) + |x_i| = (M(\hat{x} - |x|))_i + |x_i| \le (M\tilde{b})_i = \tilde{x}_i^H,$$

whence $x_i \le \nu_i \tilde{x}_i^H$. This completes (i).

Ad (ii): Choose

$$x' = DM\tilde{b}, \quad x'' = DM(\tilde{b} - 2\nu_i \tilde{x}_i^H Re^{(i)}).$$

From

$$(I - DRD)x' = DM\tilde{b} - D(M - I)\tilde{b} = D\tilde{b} = \check{b} + D\operatorname{rad}([b]) \in [b]$$

we see that $x' \in S$. Define the diagonal matrix D' by $d'_{ii} = -1$ and $d'_{jj} = d_{jj}$ for $j \ne i$. Then

$$\begin{aligned}(I - DRD')DM &= DM - DR(I - 2e^{(i)}(e^{(i)})^T)M \\ &= DM - D(M - I) + 2DRe^{(i)}(e^{(i)})^T M \\ &= D + 2DRe^{(i)}(e^{(i)})^T M,\end{aligned}$$

hence

$$\begin{aligned}(I - DRD')x'' &= (D + 2DRe^{(i)}(e^{(i)})^T M)(\tilde{b} - 2\nu_i \tilde{x}_i^H Re^{(i)}) \\ &= D\tilde{b} + 2\tilde{x}_i^H DRe^{(i)}(-\nu_i + 1 - 2\nu_i(m_{ii} - 1)) \\ &= D\tilde{b} = \check{b} + D\operatorname{rad}([b]) \in [b].\end{aligned}$$

This proves $x'' \in S$. From $(e^{(i)})^T D = (e^{(i)})^T$ we get

$$x'_i = (e^{(i)})^T x' = (e^{(i)})^T DM\tilde{b} = (e^{(i)})^T M\tilde{b} = (M\tilde{b})_i = \tilde{x}_i^H$$

and analogously

$$\begin{aligned} x''_i &= \tilde{x}_i^H - 2v_i\tilde{x}_i^H (e^{(i)})^T MRe^{(i)} = \tilde{x}_i^H - 2v_i\tilde{x}_i^H (e^{(i)})^T (M - I)e^{(i)} \\ &= \tilde{x}_i^H - 2v_i\tilde{x}_i^H (m_{ii} - 1) = v_i\tilde{x}_i^H . \end{aligned}$$

Together with (i) this proves (ii).

Ad (iii): Apply the formula for $\overline{x}_i^H$ to the interval linear system $[A]x = -[b]$ with the solution set $S^- = -S$. Then we obtain $\max\{x_i \mid x \in S^-\} = \max\{\tilde{\tilde{x}}_i, v_i\tilde{\tilde{x}}_i\}$, where $\tilde{\tilde{x}}_i = x_i^* + m_{ii}(-\check{b} - |\check{b}|)_i = -\underset{\sim}{x}_i^H$. The formula for $\underline{x}_i^H$ follows now from

$$\begin{aligned} \underline{x}_i^H &= \min\{x_i \mid x \in S\} = -\max\{-x_i \mid x \in S\} = -\max\{x_i \mid x \in S^-\} \\ &= -\max\{-\underset{\sim}{x}_i^H, -v_i\underset{\sim}{x}_i^H\} = \min\{\underset{\sim}{x}_i^H, v_i\underset{\sim}{x}_i^H\}. \end{aligned}$$ □

Exercises

Ex. 5.3.1. Compute $\square S$ for the interval linear system $[A]x = [b]$ with

(a) $[A] = \begin{pmatrix} [3,4] & [-2,-1] \\ [-3,-2] & [7,8] \end{pmatrix}$, $[b] = \begin{pmatrix} [15,30] \\ [30,60] \end{pmatrix}$.

(b) $[A] = \frac{1}{4}\begin{pmatrix} [2,6] & [-1,1] \\ [-1,1] & [2,6] \end{pmatrix}$, $[b] = \begin{pmatrix} [-3,-9] \\ [0,6] \end{pmatrix}$.

Ex. 5.3.2. Let $[A] \in \mathbb{IR}^{n\times n}$, $[b] \in \mathbb{IR}^n$, $\rho(|\check{A}^{-1}|\,\mathrm{rad}([A])) < 1$. Show that the solution set S of $[A]x = [b]$ satisfies $S \subseteq \square S \subseteq [-z, z]$ with $z = (I - |\check{A}^{-1}|\,\mathrm{rad}([A]))^{-1}(|\check{A}^{-1}\check{b}| + |\check{A}^{-1}|\,\mathrm{rad}([b])) \geq 0$.

Hint: Prove $S \subseteq [-z, z]$. To this end start with $Ax = b$, $A = \check{A} + \Delta \in [A]$, $b = \check{b} + \delta \in [b]$.

5.4 Direct methods

The interval Gaussian algorithm – basics

In numerical analysis the Gaussian algorithm is a popular method for solving linear systems of equations $Ax = b$ with regular, nonstructured and moderately-sized matrices A. Although we assume that the reader is familiar with the Gaussian algorithm, we repeat the essential ideas in order to apply them to interval linear systems later on. We first describe the algorithm in its standard form, i.e., without interchanging rows or columns. We start with a linear system in matrix-vector form

$$\begin{pmatrix} a_{11} & a_{12} & \dots & a_{1n} \\ a_{21} & a_{22} & \dots & a_{2n} \\ \vdots & \vdots & & \vdots \\ a_{n1} & a_{n2} & \dots & a_{nn} \end{pmatrix} \cdot \begin{pmatrix} x_1 \\ x_2 \\ \vdots \\ x_n \end{pmatrix} = \begin{pmatrix} b_1 \\ b_2 \\ \vdots \\ b_n \end{pmatrix}$$

in which we will eliminate the variable x_1 in all equations except the first one. To this end we multiply the first equation by $\frac{a_{i1}}{a_{11}}$ and subtract it from the i-th equation. This requires $a_{11} \neq 0$, of course, and results in a linear system

$$\begin{pmatrix} a_{11} & a_{12} & \dots & a_{1n} \\ 0 & a_{22}^{(2)} & \dots & a_{2n}^{(2)} \\ \vdots & \vdots & & \vdots \\ 0 & a_{n2}^{(2)} & \dots & a_{nn}^{(2)} \end{pmatrix} \cdot \begin{pmatrix} x_1 \\ x_2 \\ \vdots \\ x_n \end{pmatrix} = \begin{pmatrix} b_1 \\ b_2^{(2)} \\ \vdots \\ b_n^{(2)} \end{pmatrix}$$

with

$$\begin{aligned} a_{ij}^{(2)} &= a_{ij} - \frac{a_{i1}}{a_{11}} a_{1j}, \quad i, j = 2, \dots, n, \\ b_i^{(2)} &= b_i - \frac{a_{i1}}{a_{11}} b_1, \qquad i = 1, \dots, n. \end{aligned}$$

For a more systematic description we set in addition $A^{(1)} = (a_{ij}^{(1)}) = A$, $b^{(1)} = (b_i^{(1)}) = b$ and $a_{1j}^{(2)} = a_{1j}^{(1)}$ for $j = 1, \dots, n$, $a_{i1}^{(2)} = 0$ for $i = 2, \dots, n$, $b_1^{(2)} = b_1^{(1)}$. If no division by zero occurs, this elimination process can be repeated. It produces a sequence of equivalent linear systems $A^{(k)}x = b^{(k)}$, $k = 1, \dots, n$, i.e., systems with the same solution as the initial one, $Ax = b$. The matrix $A^{(n)}$ is upper triangular so that the last system can be easily solved starting with the last equation. It is obvious that this process is only feasible if

$$a_{kk}^{(k)} \neq 0, \quad k = 1, \dots, n. \tag{5.4.1}$$

Assume (5.4.1) for the moment and let

$$L := \begin{pmatrix} 1 & & & O \\ l_{21} & 1 & & \\ \vdots & \ddots & \ddots & \\ l_{n1} & \dots & l_{n,n-1} & 1 \end{pmatrix} \qquad U := A^{(n)}$$

with

$$l_{ik} = \frac{a_{ik}^{(k)}}{a_{kk}^{(k)}}, \quad i = k+1, \dots, n.$$

It is well known that $A = LU$ holds and that the matrices L, U are uniquely determined by this representation of A if one requires the upper, respective lower triangular form for L and U with the ones in the diagonal for L. The Gaussian algorithm can be briefly described in three steps.

Step 1: Compute L and U. (LU decomposition)
Step 2: Solve $Ly = b$. (Forward substitution)
Step 3: Solve $Ux = y$. (Backward substitution)

A closer look reveals $y = b^{(n)}$. If $a_{kk}^{(k)} = 0$ occurs, the elimination process must be modified. For regular matrices A, one can show that there is always an entry $a_{ik}^{(k)}$, $i \geq k$, which differs from zero. Interchanging the i-th and k-th equation allows the elimination to continue. The (eventually modified) entry $a_{kk}^{(k)}$ is called the pivot element. Choosing $a_{kk}^{(k)}$ according to some rule is called pivoting. One frequently selects as the pivot element the one that has the largest absolute value of the entries $a_{ik}^{(k)}$, $i = k, \ldots, n$. For numerical stability this is even done if originally $a_{kk}^{(k)} \neq 0$. The change of order of the equations can be described by means of a premultiplication with a permutation matrix P so that $PA = LU$ holds. Again L, U are unique if P is fixed. In passing we note that for *singular* systems matrices P, L, U always exist and can, but need not be unique for fixed P. The latter can be seen by the simple systems

$$\begin{cases} x_1 + x_2 = 0, \\ x_1 + x_2 = 0, \end{cases} \quad \text{and} \quad \begin{cases} 0x_1 + 2x_2 = 0, \\ 0x_1 + 4x_2 = 0. \end{cases}$$

Now we apply the Gaussian algorithm to interval linear systems $[A]x = [b]$ with a given regular interval matrix $[A] \in \mathbb{IR}^{n\times n}$ and a given interval vector $[b] \in \mathbb{IR}^n$. We define $[A]^{(k)} \in \mathbb{IR}^{n\times n}$, $[b]^{(k)} = ([b]_i^{(k)}) \in \mathbb{IR}^n$, $k = 1, \ldots, n$, and $[x]^G = ([x]_i^G) \in \mathbb{IR}^n$ analogously to the point case by

$$[A]^{(1)} = [A], \quad [b]^{(1)} = [b],$$

$$[a]_{ij}^{(k+1)} = \begin{cases} [a]_{ij}^{(k)}, & i = 1, \ldots, k, j = 1, \ldots, n, \\ [a]_{ij}^{(k)} - \dfrac{[a]_{ik}^{(k)} \cdot [a]_{kj}^{(k)}}{[a]_{kk}^{(k)}}, & i = k+1, \ldots, n, j = k+1, \ldots, n, \\ 0 & \text{otherwise} \end{cases}$$

$$[b]_i^{(k+1)} = \begin{cases} [b]_i^{(k)}, & i = 1, \ldots, k, \\ [b]_i^{(k)} - \dfrac{[a]_{ik}^{(k)}}{[a]_{kk}^{(k)}} \cdot [b]_k^{(k)}, & i = k+1, \ldots, n, \end{cases} \qquad k = 1, \ldots, n-1,$$

$$[x]_i^G = \Big([b]_i^{(n)} - \sum_{j=i+1}^{n} [a]_{ij}^{(n)} [x]_j^G \Big) / [a]_{ii}^{(n)}, \quad i = n, n-1, \ldots, 1. \tag{5.4.2}$$

The construction of $[x]^G$ is called the interval Gaussian algorithm (without pivoting). Obviously, $[x]^G$ exists if and only if $0 \notin [a]_{kk}^{(k)}$ for $k = 1, \ldots, n$. In this case we say that the algorithm is feasible. If emphasis is laid on the dependence on $[A]$ and $[b]$, we use the notation $[x]^G = \mathrm{IGA}([A], [b])$.

Now we represent $[x]^G$ as a product of some matrices and $[b]$. To this end we define the diagonal matrices $[D]^{(k)}$, $k = 1, \ldots, n$, the lower triangular matrices $[L]^{(k)}$

and the upper triangular matrices $[U]^{(k)}$, $k = 1, \dots, n-1$, by

$$[d]_{ij}^{(k)} := \begin{cases} 1 & \text{if } i = j \neq k, \\ \frac{1}{[a]_{kk}^{(k)}} & \text{if } i = j = k, \\ 0 & \text{otherwise,} \end{cases}$$

$$[l]_{ij}^{(k)} := \begin{cases} 1 & \text{if } i = j, \\ -\frac{[a]_{ik}^{(k)}}{[a]_{kk}^{(k)}} & \text{if } j = k < i, \\ 0 & \text{otherwise,} \end{cases} \qquad [u]_{ij}^{(k)} := \begin{cases} 1 & \text{if } i = j, \\ -[a]_{kj}^{(k)} & \text{if } i = k < j, \\ 0 & \text{otherwise.} \end{cases} \tag{5.4.3}$$

Following the elimination process one sees that

$$\begin{aligned} [x]^G = {} & [D]^{(1)}([U]^{(1)}([D]^{(2)}([U]^{(2)}(\dots([U]^{(n-1)}([D]^{(n)}([L]^{(n-1)}(\dots \\ & \dots([L]^{(2)}([L]^{(1)}[b]))\dots))))\dots)))) \end{aligned} \tag{5.4.4}$$

and

$$[b]^{(k)} = [L]^{(k)}([L]^{(k-1)}(\dots([L]^{(1)}[b]))\dots)$$

holds. The multiplication with the matrices $[L]^{(k)}$ describes the forward substitution, while the multiplication with the matrices $[D]^{(k)}$, $[U]^{(k)}$ determines the backward substitution.

For later reasons we introduce the matrices

$$\begin{aligned} \mathrm{IGA}([A]) = {} & [D]^{(1)}([U]^{(1)}([D]^{(2)}([U]^{(2)}(\dots([U]^{(n-1)}([D]^{(n)}([L]^{(n-1)}(\dots \\ & \dots([L]^{(2)}[L]^{(1)})\dots))))\dots)))) \end{aligned} \tag{5.4.5}$$

and

$$\begin{aligned} |[A]^G| = {} & |[D]^{(1)}| \cdot |[U]^{(1)}| \cdot |[D]^{(2)}| \cdot |[U]^{(2)}| \cdots |[U]^{(n-1)}| \cdot |[D]^{(n)}| \\ & \times |[L]^{(n-1)}| \cdots |[L]^{(2)}| \cdot |[L]^{(1)}| \end{aligned} \tag{5.4.6}$$

where $|[A]^G|$ coincides with the corresponding symbol in Neumaier [257]. It can be interpreted as a measure for the relative width of $\mathrm{IGA}([A], [b])$ when compared with $\alpha = \| \, |[b]| \, \|_\infty$. Indeed, by (5.4.4) we get

$$\mathrm{IGA}([A], [b]) \subseteq \mathrm{IGA}([A], \alpha[-1, 1]e) = \alpha \, \mathrm{IGA}([A], [-1, 1]e) = \alpha |[A]^G| \, [-1, 1]e,$$

whence $\mathrm{rad}(\mathrm{IGA}([A], [b]))/\alpha \leq |[A]|^G e$ if $[b]$ is not degenerate. In addition,

$$|\mathrm{IGA}([A])| \leq |[A]^G| \tag{5.4.7}$$

and

$$\mathrm{IGA}([A]) = \mathrm{IGA}(A) = A^{-1} \quad \text{if } [A] \equiv A \text{ is degenerate,} \tag{5.4.8}$$

where (5.4.8) follows from

$$(\mathrm{IGA}([A]))_{*,j} = \mathrm{IGA}([A], e^{(j)}) = \mathrm{IGA}(A, e^{(j)}) = (A^{-1})_{*,j}.$$

There is another way to access to $[x]^G$ due to Neumaier [257] using the Schur complement. To this end consider the block form

$$[A] = \begin{pmatrix} [a]_{11} & [c]^T \\ [d] & [A]' \end{pmatrix} \quad \text{with } [c], [d] \in \mathbb{IR}^{n-1}.$$

If $0 \notin [a]_{11}$, then $\Sigma([A]) = [A]' - \frac{1}{[a]_{11}}[d][c]^T \in \mathbb{IR}^{(n-1)\times(n-1)}$ is called the Schur complement of $[A]$. It is not defined if $0 \in [a]_{11}$.

We call $([L], [U])$ the triangular decomposition of $[A] \in \mathbb{IR}^{n\times n}$ if either $n = 1$, $[L] = 1$, $[U] = [A] \not\ni 0$ or

$$[L] = \begin{pmatrix} 1 & 0 \\ [d]/[a]_{11} & [L]' \end{pmatrix}, \qquad [U] = \begin{pmatrix} [a]_{11} & [c]^T \\ 0 & [U]' \end{pmatrix},$$

where $0 \notin [a]_{11}$ and $([L]', [U]')$ is the triangular decomposition of $\Sigma([A])$.

If this triangular decomposition exists, then

$$[x]^G = \mathrm{IGA}([U], \mathrm{IGA}([L], [b])),$$

which means solving two triangular systems as forward and backward substitution. With the entries from (5.4.2) we have

$$[L] = \begin{pmatrix} 1 & & & O \\ [l]_{21} & 1 & & \\ \vdots & \ddots & \ddots & \\ [l]_{n1} & \dots & [l]_{n,n-1} & 1 \end{pmatrix} \tag{5.4.9}$$

with $[l]_{ik} = \frac{[a]_{ik}^{(k)}}{[a]_{kk}^{(k)}}$ for $i > k$. Similarly,

$$[U] = \begin{pmatrix} [u]_{11} & \dots & [u]_{1n} \\ & \ddots & \vdots \\ O & & [u]_{nn} \end{pmatrix} \tag{5.4.10}$$

with $[u]_{kj} = [a]_{kj}^{(k)} = [a]_{kj}^{(n)}$ for $j \geq k$. Obviously,

$$[L] = [\hat{L}]^{(1)}([\hat{L}]^{(2)}(\dots([\hat{L}]^{(n-3)}([\hat{L}]^{(n-2)}[\hat{L}]^{(n-1)}))\dots)) \tag{5.4.11}$$

if one defines $[\hat{L}]^{(k)}$ as $[L]^{(k)}$ in (5.4.3) with the minus signs reversed to plus. For point matrices A, the corresponding matrix $\hat{L}^{(k)}$ is just the inverse of $L^{(k)}$ and L is the corresponding lower triangular matrix of the LU decomposition of A. Similarly,

$$[U] = [\hat{D}]^{(n)}([\hat{U}]^{(n-1)}([\hat{D}]^{(n-1)}(\dots([\hat{U}]^{(2)}([\hat{D}]^{(2)}([\hat{U}]^{(1)}[\hat{D}]^{(1)})))\dots))) \tag{5.4.12}$$

if one defines $[\hat{U}]^{(k)}$ as $[U]^{(k)}$ in (5.4.3), again with the minus signs reversed to plus, and if one changes $1/[a]_{kk}^{(k)}$ into $[a]_{kk}^{(k)}$ in $[D]^{(k)}$ in order to get $[\hat{D}]^{(k)}$. Since $\langle [L] \rangle = \prod_{k=1}^{n-1} \langle [\hat{L}]^{(k)} \rangle$ and $\langle [\hat{L}]^{(k)} \rangle^{-1} = |[L]^{(k)}|$ we have

$$\langle [L] \rangle^{-1} = \prod_{k=1}^{n-1} |[L]^{(n-k)}|.$$

Similarly, $\langle [U] \rangle = \langle [\hat{D}]^{(n)} \rangle \cdot \prod_{k=1}^{n-1} \left(\langle [\hat{U}]^{(n-k)} \rangle \cdot \langle [\hat{D}]^{(n-k)} \rangle \right)$, whence

$$\langle [U] \rangle^{-1} = \left(\prod_{k=1}^{n-1} \left(|[D]^{(k)}| \cdot |[U]^{(k)}| \right) \right) \cdot |[D]^{(n)}|.$$

This proves

$$|[A]^G| = \langle [U] \rangle^{-1} \langle [L] \rangle^{-1}, \tag{5.4.13}$$

which coincides with Theorem 4.5.1 (iii) in Neumaier [257].

For degenerate matrices A, the matrix U is the corresponding upper triangular matrix of the LU decomposition of A. But notice that for interval matrices $[A]$, only $[A] \subseteq [L][U]$ holds.

Now we list some elementary properties of $\mathrm{IGA}([A], [b])$ and $\mathrm{IGA}([A])$.

Theorem 5.4.1. *Let $[A] \in \mathbb{IR}^{n\times n}$, $[b], [c] \in \mathbb{IR}^n$, and let $D \in \mathbb{R}^{n\times n}$ be a regular diagonal matrix.*

(a) *If $\mathrm{IGA}([A], [b])$ exists and $[\hat{A}] \subseteq [A]$, $[\hat{b}] \subseteq [b]$, then $\mathrm{IGA}([\hat{A}], [\hat{b}])$ exists and is contained in $\mathrm{IGA}([A], [b])$.* *(Subset property)*

(b) *If $\mathrm{IGA}([A], [b])$ exists, then the Gaussian algorithm is feasible for each pair $(A, b) \in ([A], [b])$. In particular, the solution set S of $[A]x = [b]$ is enclosed by $\mathrm{IGA}([A], [b])$.* *(Enclosure property)*

(c) *The existence of $\mathrm{IGA}([A], [b])$ does not depend on $[b]$. It is guaranteed if and only if $\mathrm{IGA}([A])$ exists.*

(d) $\mathrm{IGA}([A], \alpha[b]) = \alpha\, \mathrm{IGA}([A], [b])$, $\alpha \in \mathbb{R}$. *(Homogeneity with respect to $[b]$)*

(e) $\mathrm{IGA}([A], [b] + [c]) \subseteq \mathrm{IGA}([A], [b]) + \mathrm{IGA}([A], [c])$. *(Subadditivity)*

(f) *If $\mathrm{IGA}([A], [b])$ exists, then $\mathrm{IGA}([A]D, [b])$ and $\mathrm{IGA}(D[A], [b])$ exist and satisfy*

$$\mathrm{IGA}([A]D, [b]) = D^{-1}\, \mathrm{IGA}([A], [b]), \quad \mathrm{IGA}(D[A], D[b]) = \mathrm{IGA}([A], [b]).$$

(Scaling)

(g) *If $[A] \equiv A \in \mathbb{R}^{n\times n}$, then $\mathrm{IGA}([A], [b]) = \mathrm{IGA}(A, [b]) \supseteq A^{-1}[b]$. If, in addition, $\mathrm{rad}([b]) \le \alpha\, \mathrm{rad}([c])$ holds for some $\alpha \in \mathbb{R}^+$, then $\mathrm{rad}(\mathrm{IGA}(A, [b])) \le \alpha\, \mathrm{rad}(\mathrm{IGA}(A, [c]))$ follows.*

(h) *If $[b] \equiv b \in \mathbb{R}^n$, then $\mathrm{IGA}([A], [b]) = \mathrm{IGA}([A], b) \subseteq \mathrm{IGA}([A]) \cdot b$.*

Proof. (a), (b) follow from the subset property of the interval arithmetic, (c) is immediate and (d)–(h) can be seen from (5.4.2)–(5.4.5), and properties of interval arithmetic; cf. Chapters 2 and 3. □

Theorem 5.4.2. *Let* $\mathrm{IGA}([A])$ *exist for some* $[A] \in \mathbb{IR}^{n\times n}$.

(a) *If* $[\hat{A}] \subseteq [A]$, *then* $\mathrm{IGA}([\hat{A}])$ *exists and satisfies* $\mathrm{IGA}([\hat{A}]) \subseteq \mathrm{IGA}([A])$. *In addition,* $|[\hat{A}]^G| \le |[A]^G|$ *holds.*

(b) *For* $\tilde{A} \in [A]$ *we get*

$$\tilde{A}^{-1} \in \mathrm{IGA}([A]) \quad \textit{and} \quad |\tilde{A}^{-1}| \le |\,\mathrm{IGA}([A])| \le |[A]^G|. \tag{5.4.14}$$

In particular, $[A]$ *is regular.*

(c) *If* $\rho(|I - \mathrm{IGA}([A]) \cdot [A]|) < 1$ *or* $\rho(d(\mathrm{IGA}([A])) \cdot |[A]|) < 1$, *then* $\mathrm{IGA}([A])$ *is regular.*

(d) $\mathrm{IGA}([\hat{A}])$ *is a Lipschitz continuous interval function of* $[\hat{A}] \subseteq [A]$. *In particular, there is a positive constant* γ *depending on* $[A]$ *only such that*

$$\|d(\mathrm{IGA}([\hat{A}]))\|_\infty \le \gamma \|d([\hat{A}])\|_\infty \tag{5.4.15}$$

holds for all matrices $[\hat{A}] \subseteq [A]$.

Proof. (a) follows from the inclusion monotony of the interval arithmetic.

(b) follows from (a), (5.4.7), and (5.4.8).

(c) Let $C \in \mathrm{IGA}([A])$, $\tilde{A} \in [A]$, $[M] = I - \mathrm{IGA}([A]) \cdot [A]$.

If $\rho(|[M]|) < 1$, then $\rho(I - C\tilde{A}) \le \rho(|[M]|) < 1$, hence the Neumann series of $I - C\tilde{A}$ is convergent and represents $(I - (I - C\tilde{A}))^{-1} = (C\tilde{A})^{-1}$. Therefore, neither of the factors C and $\tilde{A}$ can be singular. In particular, $\mathrm{IGA}([A])$ is regular.

Now let $\rho(d(\mathrm{IGA}([A])) \cdot |[A]|) < 1$. For fixed $i, j \in \{1, \dots, n\}$ there is an element $\tilde{m}_{ij} \in [m]_{ij}$ which satisfies $|\tilde{m}_{ij}| = |[m]_{ij}|$. By Theorem 4.1.5 one can find matrices $S \in \mathrm{IGA}([A])$, $T \in [A]$ which depend on i, j such that $\tilde{m}_{ij} = (I - ST)_{ij}$ holds. Hence

$$|[m]_{ij}| = |I - ST|_{ij} = |(T^{-1} - S)T|_{ij} \le (|T^{-1} - S| \cdot |T|)_{ij} \le (d(\mathrm{IGA}([A])) \cdot |[A]|)_{ij}$$

follows and implies $\rho(|[M]|) \le \rho(d(\mathrm{IGA}([A])) \cdot |[A]|) < 1$. The regularity of $\mathrm{IGA}([A])$ is now an immediate consequence of what we have already proven.

(d) As (5.4.2), (5.4.3), and (5.4.5) show, the entries of $\mathrm{IGA}([A])$ are rational functions of the entries $[a]_{ij}$ of $[A]$. Transforming $\mathrm{IGA}([\hat{A}])$ and $[\hat{A}]$ column by column equivalently into two vectors of dimension $m = n^2$ one can apply Theorem 4.1.16 in order to see that $\mathrm{IGA}([\hat{A}])$ is a Lipschitz continuous interval function of $[\hat{A}] \subseteq [A]$. In particular, by virtue of Theorem 4.1.18 (a) we get $\max_{i,j} d((\mathrm{IGA}([\hat{A}]))_{ij}) \le \gamma \max_{i,j} d([\hat{a}]_{ij})$, whence

$$\|d(\mathrm{IGA}([\hat{A}]))\|_\infty \le n \max_{i,j} d((\mathrm{IGA}([\hat{A}]))_{ij}) \le n\gamma \max_{i,j} d([\hat{a}]_{ij}) \le n\gamma \|d([\hat{A}])\|_\infty.$$

This proves (5.4.15) when replacing $n\gamma$ by γ. □

Notice that by the norm equivalence the row sum norm $\|\cdot\|_\infty$ in (5.4.15) can be replaced by any matrix norm.

Without the assumption in Theorem 5.4.2 (c) the matrix $\mathrm{IGA}([A])$ can be singular, as the example

$$[A] = \begin{pmatrix} [1, 2] & [-2, -1] \\ 1 & 1 \end{pmatrix}$$

shows; cf. Exercise 5.4.2.

The interval Gaussian algorithm – feasibility

We address the feasibility of the algorithm, i.e., we derive criteria which guarantee the existence of $[x]^G$. We start with a necessary and sufficient criterion which assumes $[A]$ to be degenerate so that according to Theorem 5.4.1 (b) we obtain a classical feasibility criterion.

Theorem 5.4.3. *Let $[A] \equiv A \in \mathbb{R}^{n\times n}$, i.e., let $[A]$ be degenerate, and let $[b] \in \mathbb{IR}^n$. Then $[x]^G$ exists if and only if each leading principal submatrix $A_k = (a_{ij})_{i,j=1,\dots,k}$, $k \le n$, is regular.*

Proof. The LU decomposition of the Gaussian algorithm applied to A_k produces the same entries $a_{ij}^{(k)}$, $i, j \ge k$, as the LU decomposition for A.

'$\Leftarrow$': Since $[x]^G$ exists we obtain $0 \ne a_{11}^{(1)} a_{22}^{(2)} \cdots a_{kk}^{(k)} = \det A_k$. Therefore, A_k is regular.

'$\Rightarrow$': Since $A_1 = (a_{11})$ is regular we have $0 \ne a_{11}$. Hence the matrix $A^{(1)}$ exists and satisfies $a_{11}^{(1)} \ne 0$. Assume now that $A^{(k-1)}$ exists and satisfies $a_{ii}^{(i)} \ne 0$ for $i = 1, \dots,$ $k-1$. Then $A^{(k)}$ exists. If $a_{kk}^{(k)} = 0$, then $\det A_k = a_{11}^{(1)} a_{22}^{(2)} \cdots a_{kk}^{(k)} = 0$ which contradicts the regularity of A_k. The proof concludes by induction. □

Corollary 5.4.4. *Let $[A] \equiv A \in \mathbb{R}^{n\times n}$ be an H-matrix and $[b] \in \mathbb{IR}^n$. Then $[x]^G$ exists.*

Proof. Since A is an H-matrix there is a positive vector $u \in \mathbb{R}^n$ such that $\langle A\rangle u > 0$. Shortening u correspondingly shows that each leading submatrix A_k fulfills an analogous inequality and is therefore an H-matrix. From Corollary 1.10.17 we know that A_k is regular, hence Theorem 5.4.3 concludes the proof. □

Since $\check{A}^{-1}[A] = I + [-R, R] =: [\hat{A}]$ with $R = |\check{A}^{-1}| \operatorname{rad}([A])$ we can precondition any regular interval linear system $[A]x = [b]$ by the midpoint inverse in order to end up with a new system with a matrix of the form $[\hat{A}]$. This is the reason why we are going to consider interval systems with such matrices. But notice that preconditioning by the midpoint inverse can increase the solution set since $[\hat{A}]$ usually does not coincide with the set $\{\check{A}^{-1}A \mid A \in [A]\}$. As an example choose

$$[A] = \begin{pmatrix} [1,3] & [-3,-1] \\ [1,3] & [1,3] \end{pmatrix}, \quad [b] = \begin{pmatrix} [0,2] \\ [0,2] \end{pmatrix}.$$

Then

$$\check{A}^{-1} = \frac{1}{4}\begin{pmatrix} 1 & 1 \\ -1 & 1 \end{pmatrix}, \quad \check{A}^{-1}[A] = \frac{1}{2}\begin{pmatrix} [1,3] & [-1,1] \\ [-1,1] & [1,3] \end{pmatrix},$$
$$\check{A}^{-1}[b] = \frac{1}{2}\begin{pmatrix} [0,2] \\ [-1,1] \end{pmatrix},$$

and $x = (-1,-1)^T$ satisfies $(\check{A}^{-1}[A])x \cap \check{A}^{-1}[b] \ne \emptyset$ but $[A]x \cap [b] = \emptyset$. Hence by Theorem 5.2.2 the vector x belongs to the solution set of the preconditioned system but not to that of the original one.

Theorem 5.4.5. *Let $O \le R \in \mathbb{R}^{n\times n}$, $[A] = I + [-R, R] \in \mathbb{IR}^{n\times n}$, $[b] \in \mathbb{IR}^n$. Then the following conditions are equivalent.*
(a) $[x]^G$ *exists,*
(b) $I - R$ *is an* M*-matrix,*
(c) $\rho(R) < 1$,
(d) $[A]$ *is regular,*
(e) $[A]$ *is strongly regular (as defined after Example* 3.1.3*),*
(f) $[A]$ *is an* H*-matrix.*

Proof. The equivalence of (b) and (c) follows from Theorem 1.10.21 with $A = I - R$, $(M, N) = (I, R)$. The equivalence of (d) and (e) is trivial since $\check{A} = I$.

'(b) ⇒ (f) ⇒ (d) ⇒ (c)': Let (b) hold. It implies $1 - r_{ii} > 0$, whence $\langle [A]\rangle = I - R$ is an M-matrix, i.e., $[A]$ is an H-matrix as in (f). Now let (f) hold and choose $\tilde{A} \in [A]$. By the Definition 3.3.1 $\tilde{A}$ is an H-matrix which is regular by virtue of Corollary 1.10.17. Hence $[A]$ is regular as stated in (d). In order to deduce (c) from (d) assume $\rho := \rho(R) \ge 1$. Choose $u \ge 0$ such that $Ru = \rho u$ holds. Then $(I - (1/\rho)R)u = 0$, hence $I - (1/\rho)R \in I + [-R, R] = [A]$ is singular and so is $[A]$, contradicting the assumption (d).

'(a) ⇒ (d)': To this end choose any matrix $\tilde{A} \in [A]$. Since $[x]^G$ exists for $[A]$ it also exists for $\tilde{A}$, hence $\tilde{A}$ is regular, and so is $[A]$. For the implication '(b) ⇒ (a)' first note that according to Corollary 1.10.5 the matrices $I - \alpha R$ are M-matrices for $0 \le \alpha \le 1$, hence by Corollary 5.4.4 the interval Gaussian algorithm is feasible for them. Denote by $(I - \alpha R)^{(k)}$ the matrices of this algorithm corresponding to $[A]^{(k)}$. Then the continuous function $f(\alpha) = ((I - \alpha R)^{(k)})_{kk}$ is positive. Otherwise $f(0) = 1 > 0$ implies that $f(\alpha) = 0$ for some $\alpha \in (0, 1]$, which contradicts the feasibility for the matrix $I - \alpha R$. Using the formulae in Theorem 2.3.3 and Theorem 2.4.2, the particular form of $[A]$ implies

$$\langle [A]^{(k)}\rangle = \langle [A]\rangle^{(k)} = (I - R)^{(k)}. \tag{5.4.16}$$

for $k = 1, \dots, n$ by induction. Hence $[x]^G$ exists. □

For general H-matrices Alefeld proved in [7] the following sufficient criterion.

Theorem 5.4.6. *If $[A] \in \mathbb{IR}^{n\times n}$ is an H-matrix, then $[x]^G$ exists for any $[b] \in \mathbb{IR}^n$, and $|[A]^G| \le \langle [A]\rangle^{-1}$ holds.*

Proof. Since $0 \notin [a]_{ii}$ the diagonal matrix $D = \operatorname{diag}(\check{a}_{11}, \dots, \check{a}_{nn}) \in \mathbb{R}^{n\times n}$ is regular. Define

$$[B] = D + [-\tilde{R}, \tilde{R}] \quad \text{with } \tilde{r}_{ij} = \begin{cases} \operatorname{rad}([a]_{ii}), & \text{if } i = j, \\ |[a]_{ij}|, & \text{if } i \ne j. \end{cases}$$

Since $[A] \subseteq [B]$, $\langle [A]\rangle = \langle [B]\rangle$ the matrix $[B]$ is an H-matrix, too. Hence $[B]$ is regular and the same holds for the matrix $D^{-1}[B]$. Since $D^{-1}[B] = I + [-R, R]$ with $R = |D^{-1}|\tilde{R}$, Theorem 5.4.5 applies and guarantees the existence of $\mathrm{IGA}(D^{-1}[B], D^{-1}[b])$. From Theorem 5.4.1 we obtain the existence of $\mathrm{IGA}([B], [b])$ and that of $[x]^G = \mathrm{IGA}([A], [b])$.

In order to prove the inequality of the theorem we remark first that the inclusion monotony of interval arithmetic guarantees $|[A]^G| \le |[B]^G|$. With $[\hat{B}] = [B]D^{-1} = I + [-\hat{R}, \hat{R}]$, $\hat{R} = \tilde{R}|D^{-1}|$ one gets $\langle [\hat{B}]^{(k)} \rangle = (I - \hat{R})^{(k)} = \langle [\hat{B}] \rangle^{(k)}$ by virtue of (5.4.16) with $[\hat{B}]$, $\hat{R}$ instead of $[A]$, R. Here the superscripts indicate the intermediate matrices of the interval Gaussian algorithm as defined in (5.4.2).

Now we compute $([L], [U])$ in (5.4.9), (5.4.10) for $[B]$, similarly $([\hat{L}], [\hat{U}])$ for $[\hat{B}]$, and $(\tilde{L}, \tilde{U})$ for $\langle [\hat{B}] \rangle = I - \hat{R}$. Then we obtain $[\hat{U}] = [\hat{B}]^{(n)}$, whence $\langle [\hat{U}] \rangle = \langle [\hat{B}]^{(n)} \rangle = \langle [\hat{B}] \rangle^{(n)} = (I - \hat{R})^{(n)} = \tilde{U}$. Similarly, $[\hat{l}]_{ik} = [\hat{b}]_{ik}^{(k)} / [\hat{b}]_{kk}^{(k)}$ for $i > k$, whence $-|[\hat{l}]_{ik}| = -|[\hat{b}]_{ik}^{(k)}| / \langle [\hat{b}]_{kk}^{(k)} \rangle = \langle [\hat{B}] \rangle_{ik}^{(k)}$, $i > k$, and $\langle [\hat{L}] \rangle = \tilde{L}$. Now (5.4.13) implies

$$|[\hat{B}]^G| = \langle [\hat{U}] \rangle^{-1} \langle [\hat{L}] \rangle^{-1} = \tilde{U}^{-1} \tilde{L}^{-1} = (\tilde{L}\tilde{U})^{-1} = (I - \hat{R})^{-1} = \langle [\hat{B}] \rangle^{-1}$$

since $\tilde{L}\tilde{U}$ is the LU decomposition of $I - \hat{R} = \langle [\hat{B}] \rangle$. From the formulae (5.4.2) of the interval Gaussian algorithm one sees as in the proof of Theorem 5.4.1 (f) that $[B] = [\hat{B}]D$ implies $[B]^{(k)} = [\hat{B}]^{(k)}D$, whence $[L] = [\hat{L}]$, $[U] = [\hat{U}]D$. Similar to that above, one ends up with

$$\begin{aligned} |[B]^G| &= \langle [U] \rangle^{-1} \langle [L] \rangle^{-1} = \langle [\hat{U}D] \rangle^{-1} \langle [\hat{L}] \rangle^{-1} = (\langle [\hat{U}] \rangle |D|)^{-1} \langle [\hat{L}] \rangle^{-1} \\ &= |D|^{-1} (\tilde{L}\tilde{U})^{-1} = |D|^{-1} \langle [\hat{B}] \rangle^{-1} = \langle [\hat{B}]D \rangle^{-1} = \langle [B] \rangle^{-1}, \end{aligned}$$

which terminates the proof. □

Theorem 5.4.6 at once implies the following corollary, since there all matrices are H-matrices by Theorem 3.3.5.

Corollary 5.4.7. *If $[A] \in \mathbb{IR}^{n \times n}$ and $[b] \in \mathbb{IR}^n$, then $[x]^G$ exists in each of the following cases.*

(a) *$[A]$ is an M-matrix.*
(b) *$\langle [A] \rangle$ is strictly diagonally dominant.*
(c) *$\langle [A] \rangle$ is irreducibly diagonally dominant.*
(d) *$\langle [A] \rangle$ is regular and diagonally dominant.*
(e) *$[A]$ is a regular triangular matrix.*

Recall that all matrices $[A]$ with $\langle [A] \rangle u > 0$ for some positive vector u are H-matrices, thus $[x]^G$ exists for them. If $\langle [A] \rangle$ is irreducible, by virtue of Theorem 1.10.6 this criterion can even be weakened to $\langle [A] \rangle u \ge 0$, if at least one component of this product is positive. If one weakens this condition further, additional assumptions are necessary for the existence of $[x]^G$. This will be shown now. We will consider matrices $[A]$ with an irreducible comparison matrix and $\langle [A] \rangle u = 0$ for some $u > 0$. According to Theorem 1.10.13 such matrices are no longer H-matrices, but are close to them because they belong to their boundary. We start with some auxiliary lemmas.

Lemma 5.4.8. *Let* $[a], [b], [c], [d] \in \mathbb{IR}$ *with* $0 \notin [d]$. *Define* $[a]' \in \mathbb{IR}$ *by*

$$[a]' = [a] - \frac{[b] \cdot [c]}{[d]}$$

and s *by* $s = \operatorname{sign}(\check{a}) \operatorname{sign}(\check{b}) \operatorname{sign}(\check{c}) \operatorname{sign}(\check{d})$. *Then the following assertions hold.*

(a) *The inequality*

$$|[a]'| \leq |[a]| + \frac{|[b]| \cdot |[c]|}{\langle [d] \rangle} \tag{5.4.17}$$

holds. Moreover, equality is valid in (5.4.17) if and only if

$$s \neq 1, \text{ i.e., } s \leq 0. \tag{5.4.18}$$

In this case

$$\operatorname{sign}(\check{a}') = \begin{cases} \operatorname{sign}(\check{a}), & \text{if } \check{a} \neq 0 \\ -\operatorname{sign}(\check{b}) \operatorname{sign}(\check{c}) \operatorname{sign}(\check{d}), & \text{if } \check{a} = 0. \end{cases} \tag{5.4.19}$$

(b) *The inequality*

$$\langle [a]' \rangle \geq \langle [a] \rangle - \frac{|[b]||[c]|}{\langle [d] \rangle} \tag{5.4.20}$$

holds. Moreover, if

$$\langle [a] \rangle \geq \frac{|[b]||[c]|}{\langle [d] \rangle}, \tag{5.4.21}$$

then equality is valid in (5.4.20) if and only if

$$s \neq -1, \quad \text{i.e., } s \geq 0. \tag{5.4.22}$$

In this case

$$\operatorname{sign}(\check{a}') = \operatorname{sign}(\check{a}), \quad \text{provided } [a]' \neq 0. \tag{5.4.23}$$

Proof. Let

$$[t] = \frac{[b][c]}{[d]}.$$

Then

$$|[t]| = \frac{|[b]|\,|[c]|}{\langle [d] \rangle}$$

by Theorem 2.4.2.

(a) The inequality follows directly from Theorem 2.4.2. By Theorem 2.4.4 we get

$$\begin{aligned} |[a]'| &= |\check{a}'| + \operatorname{rad}([a]') \\ &= |\check{a} - \check{t}| + \operatorname{rad}([a]) + \operatorname{rad}([t]) \\ &= |\check{a} - \check{t}| - (|\check{a}| + |\check{t}|) + |[a]| + |[t]|. \end{aligned}$$

Therefore, equality holds in (5.4.17) if and only if

$$|\check{a} - \check{t}| = |\check{a}| + |\check{t}|,$$

i.e.,

$$\operatorname{sign}(\check{a}) = 0, \text{ or } \operatorname{sign}(\check{t}) = 0, \text{ or } \operatorname{sign}(\check{a}) = -\operatorname{sign}(\check{t}) \neq 0. \tag{5.4.24}$$

By Theorem 2.4.4 we obtain $s = \operatorname{sign}(\check{a})\operatorname{sign}(\check{t})$, hence (5.4.24) implies $s \neq 1$, and vice versa.

Again by Theorem 2.4.4 we get

$$\operatorname{sign}(\check{a}') = \operatorname{sign}(\check{a} - \check{t}),$$

thus (5.4.24) proves (5.4.19).

(b) The inequality follows from Theorem 2.4.2. If $[t] = 0$, then $[a]' = [a]$, hence (5.4.21)–(5.4.23) hold. Therefore, let $[t] \neq 0$. Together with (5.4.21) this implies

$$\langle [a] \rangle \geq |[t]| > 0, \tag{5.4.25}$$

hence $0 \notin [a]$ and ($\underline{a} \geq |[t]|$ or $-\overline{a} \geq |[t]|$); thus $0 \notin \operatorname{int}([a] - [t]) = \operatorname{int}([a]')$. By Theorem 2.4.4 we get

$$\begin{aligned} \langle [a]' \rangle &= |\check{a}'| - \operatorname{rad}([a]') = |\check{a} - \check{t}| - (\operatorname{rad}([a]) + \operatorname{rad}([t])) \\ &= |\check{a} - \check{t}| - (|\check{a}| - |\check{t}|) + \langle [a] \rangle - |[t]|. \end{aligned}$$

Therefore, equality holds in (5.4.20) if and only if

$$|\check{a} - \check{t}| = |\check{a}| - |\check{t}|, \tag{5.4.26}$$

i.e., $\check{t} = 0$ or $\operatorname{sign}(\check{a}) = \operatorname{sign}(\check{t})$. This implies (5.4.22), and vice versa. In order to prove (5.4.23) for $[t] \neq 0$ (and equality in (5.4.20)) we first consider the case of two degenerate intervals $[a]$ and $[t]$. Then $|\check{a}| = \langle [a] \rangle \neq |\check{t}| = |[t]|$ because of $[a]' \neq 0$ and (5.4.22), hence $|\check{a}| > |\check{t}|$ follows from (5.4.25) in this case. If $[a]$ or $[t]$ are not degenerate, then $|\check{a}| > |\check{t}|$ follows directly from (5.4.25). Thus in both cases Theorem 2.4.4 (c) and (5.4.26) imply $\operatorname{sign}(\check{a}') = \operatorname{sign}(\check{a} - \check{t}) = \operatorname{sign}(\check{a})$. □

It is obvious that $\operatorname{sign}(\check{a})$ determines whether $|[a]| = |\underline{a}|$ or $|[a]| = |\overline{a}|$ holds.

Lemma 5.4.9. *Let $[A] \in \mathbb{IR}^{n \times n}$ have a comparison matrix which is irreducible and satisfies $\langle [A] \rangle u \geq 0$ for some positive vector u. Then*

(a) $\langle [a]_{ii} \rangle > 0$, $i = 1, \ldots, n$.

(b) *The Schur complement $\Sigma(\langle [A] \rangle)$ of $\langle [A] \rangle$ exists and is irreducible, provided that $n \geq 3$.*

Proof. (a) If $n = 1$, the assertion follows from the definition of the irreducibility. For $n > 1$ and any index i, there is an index $j = j(i) \neq i$ such that $[a]_{ij} \neq 0$, hence

$$0 \leq (\langle A \rangle u)_i \leq \langle [a]_{ii} \rangle u_i - |[a]_{ij}| u_j < \langle [a]_{ii} \rangle u_i.$$

(b) The existence of $\Sigma(\langle [A] \rangle)$ follows from (a). Choose any two indices $i, j \in \{2, \ldots, n\}$, assuming $i \neq j$. Then

$$(\langle [A] \rangle^{(2)})_{ij} = -|[a]_{ij}| - \frac{|[a]_{i1}|\,|[a]_{1j}|}{\langle [a]_{11} \rangle},$$

thus

$$(\langle [A]\rangle^{(2)})_{ij} < 0 \iff [a]_{ij} \neq 0 \text{ or } ([a]_{i1} \neq 0 \text{ and } [a]_{1j} \neq 0). \tag{5.4.27}$$

Choose a shortest path in the directed graph $G(\langle [A]\rangle)$ which connects the node i to the node j, say,

$$i =: i_1 \to i_2 \to \cdots \to i_{s-1} \to i_s := j. \tag{5.4.28}$$

Case 1, $i_l \neq 1$ for all $l \in \{1, \dots, s\}$: By (5.4.27) $(\langle [A]\rangle^{(2)})_{i_l i_{l+1}} \neq 0$, $l = 1, \dots, s-1$, hence (5.4.28) is a path in $G(\langle [A]\rangle^{(2)})$ which connects the node i to the node j.

Case 2, $i_l = 1$ for some index $l \in \{1, \dots, s\}$: Since the path of (5.4.28) is of minimal length, l is the only index for which $i_l = 1$. Then $l \notin \{1, s\}$ and $[a]_{i_{l-1}1} \neq 0$, $[a]_{1i_{l+1}} \neq 0$, hence $(\langle [A]\rangle^{(2)})_{i_{l-1}, i_{l+1}} < 0$ by (5.4.27). Therefore, the path

$$i = i_1 \to i_2 \to \cdots \to i_{l-1} \to i_{l+1} \to \cdots \to i_s = j$$

connects the node i to the node j in $G(\langle [A]\rangle^{(2)})$.

Thus we have shown that for any nodes $i, j \in \{2, \dots, n\}$, $i \neq j$, there is a path in $G(\langle [A]\rangle^{(2)})$ connecting i to j without containing the node 1. By concatenating the paths from i to any node $l \in \{2, \dots, n\}$ and from l to i, this assertion holds also for the case $i = j$. Hence, any two nodes are connected in $G(\Sigma_{\langle [A]\rangle})$ and thus $\Sigma_{\langle [A]\rangle}$ is irreducible. □

We are now ready to study matrices $[A]$ with a singular comparison matrix as described above.

Theorem 5.4.10. *Let $[A] \in \mathbb{IR}^{n\times n}$, $n \geq 2$, and let $[b] \in \mathbb{IR}^n$. Assume that $\langle [A]\rangle$ is irreducible and satisfies $\langle [A]\rangle u = 0$ for some positive vector u. If*

$$\operatorname{sign}(\check{a}_{i_0 j_0}) \cdot \operatorname{sign}(\check{a}_{i_0 k_0}) \cdot \operatorname{sign}(\check{a}_{k_0 k_0}) \cdot \operatorname{sign}(\check{a}_{k_0 j_0}) = \begin{cases} 1 & \text{if } i_0 \neq j_0 \\ -1 & \text{if } i_0 = j_0 \end{cases} \tag{5.4.29}$$

for at least one triple $i_0, j_0, k_0 \in \{1, \dots, n\}$, $k_0 < i_0, j_0$, then $[x]^G$ exists.

Proof. W.l.o.g. we may assume that all nondiagonal entries $[a]_{ij}$ of $[A]$ are zero-symmetric intervals as long as $(i, j) \notin \{(i_0, j_0), (i_0, k_0), (k_0, j_0)\}$. If they are not zero-symmetric from the start, replace them by $-[|[a]_{ij}|, |[a]_{ij}|]$. This does not change the assumptions and may enlarge $[A]$, but simplifies the proof. By virtue of Theorem 1.10.12 the matrix $\langle [A]\rangle$ is a singular M-matrix. According to Theorem 1.10.13, each of its $(n-1) \times (n-1)$ principal submatrices is a regular M-matrix, hence the corresponding submatrices of $[A]$ are H-matrices. Therefore, the matrices $\langle A\rangle^{(k)}$ and $[A]^{(k)}$ of the interval Gaussian algorithm exist for $k = 1, \dots, n$ and satisfy

$$0 = \langle [A]\rangle^{(n)}_{nn} \quad \text{while } 0 < \langle [A]\rangle^{(k)}_{jj} \text{ for } k < n,\ j = 1, \dots, n. \tag{5.4.30}$$

Moreover, by our additional assumptions and by induction one sees that for $k = 1, \dots, k_0$ the following properties hold:

(i) The signs of the entries of $\check{A}^{(k)}$ are the same as the corresponding ones of $\check{A}$, since $[a]_{ik}^{(k)}[a]_{kj}^{(k)}/[a]_{kk}^{(k)}$ is zero-symmetric for $k < k_0$.
(ii) $\langle [A]^{(k)}\rangle = \langle [A]\rangle^{(k)}$ by the same reason and Lemma 5.4.8. In addition, $\langle [A]^{(k)}\rangle u = 0$, since $\langle [A]\rangle^{(k)} u = L^{(k-1)}_{\langle [A]\rangle}\langle [A]^{(k-1)}\rangle u$. Here $L^{(l)}_{\langle [A]\rangle}$ is defined for $\langle [A]\rangle$ analogously to (5.4.3).
(iii) By Lemma 5.4.9 the Schur complements of the matrices $\langle [A]\rangle^{(k)}$ remain irreducible (with the restriction $k < k_0$ in the case $k_0 = n-1$).

Therefore, the assumptions of the theorem hold for the entries of $[A]^{(k_0)}$ if the irreducibility is required for the submatrix $\Sigma([A]^{(k_0-1)}) = ([a]_{ij}^{(k_0)})_{i,j=k_0,\ldots,n}$ only. If $k_0 < n-1$, then (5.4.29) (for $[A]^{(k_0)}$ instead of $[A]$) together with Lemma 5.4.8 implies $\langle [A]^{(k_0+1)}\rangle u \ge 0$, $(\langle [A]^{(k_0+1)}\rangle u)_{i_0} > 0$, hence $\Sigma([A]^{(k_0)})$ is an H-matrix by virtue of Theorem 1.10.6, and $[x]^G$ exists. Notice that (5.4.21) is fulfilled by (5.4.30) and property (ii) above. If $k_0 = n-1$, then $i_0 = j_0 = n$. From Lemma 5.4.8 together with (5.4.30) we get $\langle [a]_{nn}^{(n)}\rangle > 0$, whence $[x]^G$ exists again. □

Example 5.4.11. Define $[A]_\alpha$ by

$$[A]_\alpha = \begin{pmatrix} 2 & 1 & -1 \\ [\alpha, 1] & 2 & -1 \\ [-1, 1] & -1 & 2 \end{pmatrix}, \quad \text{where } \alpha \in [-1, 1].$$

$[A]_\alpha$ is not an H-matrix since $\langle [A]_\alpha\rangle$ is singular with $\langle [A]_\alpha\rangle e = 0$. For $\alpha \ne -1$ the sign matrix

$$S_{[A]} := (\operatorname{sign}(\check{a}_{ij})) \in \mathbb{R}^{3\times 3}$$

is given by

$$S_{[A]} = \begin{pmatrix} 1 & 1 & -1 \\ 1 & 1 & -1 \\ 0 & -1 & 1 \end{pmatrix}.$$

Then (5.4.29) is fulfilled for $(i_0, j_0, k_0) = (2, 3, 1)$, and by Theorem 5.4.10, $\mathrm{IGA}([A]_\alpha, [b])$ exists for $\alpha \in (-1, 1]$ and any $[b] \in \mathbb{IR}^n$.

For $\alpha = -1$, condition (5.4.29) is not fulfilled. Each element matrix $\tilde{A} \in [A]$ is regular since

$$\det\begin{pmatrix} 2 & 1 & -1 \\ \alpha & 2 & -1 \\ \beta & -1 & 2 \end{pmatrix} = 6 - \alpha + \beta \ge 4 \ne 0, \quad \alpha, \beta \in [-1, 1].$$

Moreover, the Gaussian algorithm is feasible for any matrix $\tilde{A} \in [A]_{-1}$ by virtue of Theorem 5.4.3. But $[x]^G = \mathrm{IGA}([A]_{-1}, [b])$ does not exist since $[A]_{-1}$ and

$$[B] := \begin{pmatrix} 2 & [-1, 1] & -1 \\ [-1, 1] & 2 & -1 \\ [-1, 1] & -1 & 2 \end{pmatrix}$$

produce the same Schur complement $\Sigma([A]_{-1}) = \Sigma([B])$. This is due to the equality

$$([A]_{-1})_{21}([A]_{-1})_{12} = [-1, 1] \cdot 1 = [-1, 1] \cdot [-1, 1] = [b]_{21}[b]_{12}.$$

The products occur when computing $([A]_{-1})^{(2)}$ and $[B]^{(2)}$, respectively. Since $\langle [B] \rangle \in [B]$ and $\langle [B] \rangle e = 0$, the matrix $[B]$ contains a singular matrix, hence IGA($[B]$, $[b]$), and therefore IGA($[A]_{-1}$, $[b]$) cannot exist. □

The second part of the preceding example suggests that the criterion (5.4.29) is necessary and sufficient for the existence of $[x]^G$.

On the other side the matrix

$$[A] = \begin{pmatrix} 2 & 2 & 0 \\ 0 & 2 & 2 \\ 2 & 0 & [2, 4] \end{pmatrix} \tag{5.4.31}$$

shows that this expectation is wrong. Difficulties arise if $\check{a}_{ij} = 0$. These zeros can disappear in a later stage of the algorithm. Therefore we will associate with $[A]$ an extended sign matrix S' which takes this change into account. In fact we update the signs s_{ij} whenever (i, j) becomes an edge for the first time in the directed graphs $G(\,\mathrm{mid}(\Sigma([A]^{(k)}))\,)$, which can be interpreted as some sort of elimination graph (cf., however, Definition 5.4.21 below).

Definition 5.4.12. Let $[A] \in \mathbb{IR}^{n\times n}$ and define the sign matrix S by $s_{ij} := \mathrm{sign}(\check{a}_{ij})$. Then the extended sign matrix S' is defined by the following process

$S' := S$
for $k := 1$ to $n-1$ do
  for $i := k+1$ to n do
    for $j := k+1$ to n do
      if $s'_{ij} = 0$ then $s'_{ij} := -s'_{ik}s'_{kk}s'_{kj}$.

Introducing this extended sign matrix S' in Theorem 5.4.10 yields the following equivalence.

Theorem 5.4.13. *Let $[A] \in \mathbb{IR}^{n\times n}$, $n \geq 2$, and let $[b] \in \mathbb{IR}^n$. Assume that $\langle [A] \rangle$ is irreducible and satisfies $\langle [A] \rangle u = 0$ for some positive vector u. Then $[x]^G$ exists if and only if*

$$s'_{ij}s'_{ik}s'_{kk}s'_{kj} = \begin{cases} 1, & \text{if } i \neq j \\ -1, & \text{if } i = j \end{cases} \tag{5.4.32}$$

holds for at least one triple $i_0, j_0, k_0 \in \{1, \ldots, n\}$, $k_0 < i_0, j_0$, with S' as in Definition 5.4.12.

The proof of this theorem can be found in Mayer, Pieper [220] together with Mayer [214]. It is essentially based on a relation between s'_{ij} and $\mathrm{sign}(\check{a}^{(k)}_{ij})$ and uses ideas involved in the proof of Theorem 5.4.10.

For the matrix in (5.4.31) we get

$$S' = \begin{pmatrix} 1 & 1 & 0 \\ 0 & 1 & 1 \\ 1 & -1 & 1 \end{pmatrix}.$$

Therefore, (5.4.32) holds for $(i_0, j_0, k_0) = (3, 3, 2)$, and $[x]^G$ exists.

Replacing $[a]_{33} = [2, 4]$ by $[a]_{33} = [-4, -2]$ results in

$$S' = \begin{pmatrix} 1 & 1 & 0 \\ 0 & 1 & 1 \\ 1 & -1 & -1 \end{pmatrix}$$

and (5.4.32) does not hold as one can easily check. Therefore, $[x]^G$ does not exist.

The Theorems 5.4.10 and 5.4.13 were formulated for $\langle [A] \rangle$ being irreducible. This is not a severe restriction as the following discussion shows.

Definition 5.4.14. Let $[A] \in \mathbb{IR}^{n\times n}$ and let π be a permutation with corresponding permutation matrix $P \in \mathbb{R}^{n\times n}$ such that $R_{[A]} = ([R]_{st})_{s,t=1,\dots,r} = P[A]P^T$ is the reducible normal form of $[A]$ as defined in Section 3.2. Let P be chosen such that the *order* of the rows in the individual diagonal blocks $[R]_{ss}$, $s = 1, \dots, r$, is the same as in $[A]$, i.e., if $i_1 < i_2$ holds for two row indices i_1, i_2 of $[A] = ([a]_{ij})$ with $\pi(i_1), \pi(i_2)$ belonging to the row indices of $[R]_{ss}$, then $\pi(i_1) < \pi(i_2)$. Then we say that π and P are 'order preserving within blocks'.

We illustrate this definition by the matrices

$$A = \begin{pmatrix} 7 & 6 & 5 \\ 0 & 4 & 3 \\ 0 & 2 & 1 \end{pmatrix}, \quad P_1 = \begin{pmatrix} 0 & 1 & 0 \\ 0 & 0 & 1 \\ 1 & 0 & 0 \end{pmatrix}, \quad P_2 = \begin{pmatrix} 0 & 0 & 1 \\ 0 & 1 & 0 \\ 1 & 0 & 0 \end{pmatrix}.$$

The permutation matrix P_1 is order preserving within blocks with

$$R_A = \begin{pmatrix} 4 & 3 & 0 \\ 2 & 1 & 0 \\ 6 & 5 & 7 \end{pmatrix},$$

whereas P_2 is not. Here

$$R_A = \begin{pmatrix} 1 & 2 & 0 \\ 3 & 4 & 0 \\ 5 & 6 & 7 \end{pmatrix}.$$

Notice that the order of the blocks $[R]_{st}$ themselves is not influenced by the property defined above.

With Definition 5.4.14 we are able to formulate a result which shows how the feasibility of the interval Gaussian algorithm for arbitrary interval matrices can be reduced to that for irreducible ones.

Theorem 5.4.15. *Let $[A] \in \mathbb{IR}^{n\times n}$, $[b] \in \mathbb{IR}^n$ and let $P \in \mathbb{R}^{n\times n}$ be a permutation matrix which is order preserving within blocks such that $R_{[A]} = ([R]_{st})_{s,t=1,\ldots,r} = P[A]P^T$ is the reducible normal form of $[A]$. Then the following properties are equivalent.*

(i) $\mathrm{IGA}([A], [b])$ *exists.*
(ii) $\mathrm{IGA}(P[A]P^T, P[b]) = \mathrm{IGA}(R_{[A]}, P[b])$ *exists.*
(iii) $\mathrm{IGA}([R]_{ss}, (P[b])_s)$ *exists for* $s = 1, \ldots, r$, *where* $(P[b])_s$ *denotes the part of* $P[b]$ *which corresponds to* $[R]_{ss}$.

Proof. The equivalence of (ii) and (iii) is trivial by virtue of the particular form of $[R] = R_{[A]}$ and the fact that the feasibility depends only on the diagonal entries $[r]_{kk}^{(k)}$.

For the implication '(i) ⇒ (iii)' let $I = \{i_1, i_2, \ldots, i_l\}$ be the row indices of $[A]$ which are transformed to those of any fixed block $[R]_{ss}$ and let $I' = \{i'_1, i'_2, \ldots, i'_{l'}\}$ be the row indices of $[A]$ which are transformed to those of any other arbitrary fixed block $[R]_{s's'}$. Without loss of generality assume $s < s'$. Denote by π the permutation corresponding to P. Then $[a]_{i_m i'_{m'}} = 0$ for all indices $i_m \in I$ and all indices $i'_{m'} \in I'$ since $[a]_{\pi(i_m)\pi(i'_{m'})} = 0$. Therefore, if the interval Gaussian algorithm is applied to $[A]$, the diagonal entry in the i_m-th column, $i_m \in I$, does not effect a change of the entries $[a]_{i'j'}$ if $(i', j') \in I' \times I'$, and the diagonal entry in the $i'_{m'}$-th column does not effect a change of the entries $[a]_{ij}$ if $(i, j) \in I \times I$ even if $i'_{m'} < i_m$. Taking into account that s, s' were arbitrary and that the order of the indices in I, I' is preserved, the feasibility of the interval Gaussian algorithm for $[A]$ implies that of $[R]_{ss}$, $[R]_{s's'}$. This proves the implication '(i) ⇒ (iii)'.

The converse implication can be seen analogously using essentially the fact that diagonal entries $[a]_{ii}$, $[a]_{i'i'}$ such that $\pi(i)$, $\pi(i')$ belong to different blocks in $R_{[A]}$ do not effect changes of each other when the interval Gaussian algorithm is applied. □

In Theorem 5.4.15 we required P to be order preserving within blocks. Without this assumption Theorem 5.4.15 can fail, of course, even if $[A]$ is irreducible as the example

$$[A] = \begin{pmatrix} 6 & -1 & [-2, 2] \\ -3 & 2 & [-2, 2] \\ -3 & 1 & 4 \end{pmatrix}$$

shows. Here, the interval Gaussian algorithm is feasible for $[A]$, while it is not for $P[A]P^T$ with

$$P = \begin{pmatrix} 0 & 0 & 1 \\ 0 & 1 & 0 \\ 1 & 0 & 0 \end{pmatrix}.$$

In both cases, however, the classical Gaussian algorithm is feasible for any matrix $\tilde{A} \in [A]$ and $\tilde{A} \in P[A]P^T$, respectively.

Notice also that we did not exploit the reducibility of $R_{[A]}$ when applying the interval Gaussian algorithm. So one might think of solving the smaller system $[R]_{11}x = [b]_1$ with the block $[b]_1$ first, then multiplying $[R]_{i1}$, $i = 2, \ldots, r$, by the solution $[x]_1^G$ and subtracting the result from the blocks $[b]_i$, $i = 2, \dots, r$, before repeating this procedure. This is the usual way when dealing with point systems. For interval matrices this can blow up $[x]^G$ as the following example shows, where $[x]^G_{\text{mod}}$ denotes the result of this modified Gaussian algorithm.

$$A = \begin{pmatrix} 1 & -1 & 0 \\ 1 & 1 & 0 \\ 1 & -1 & 1 \end{pmatrix}, \qquad [b] = \begin{pmatrix} 0 \\ [-2, 2] \\ 0 \end{pmatrix},$$

$$[x]^G = ([-1, 1], [-1, 1], 0)^T \subset ([-1, 1], [-1, 1], [-2, 2])^T = [x]^G_{\text{mod}}.$$

Now we consider upper interval Hessenberg matrices with particular sign patterns. To this end we generalize the definition of the sign of a real number onto intervals.

Definition 5.4.16. Let $[a] \in \mathbb{IR}$. Then the sign $\operatorname{sign}([a])$ of $[a]$ is defined by

$$\operatorname{sign}([a]) = \begin{cases} +1 & \text{if } \underline{a} > 0, \\ -1 & \text{if } \overline{a} < 0, \\ 0 & \text{otherwise,} \end{cases}$$

and the extended sign $\sigma([a])$ of $[a] \neq 0$ by

$$\sigma([a]) = \begin{cases} +1 & \text{if } \underline{a} \geq 0, \\ -1 & \text{if } \overline{a} \leq 0, \\ 0 & \text{otherwise.} \end{cases}$$

The following lemma on these signs can be seen immediately.

Lemma 5.4.17. *Let $[a], [b] \in \mathbb{IR}$. Then the following statements hold.*
(a) *If* $\operatorname{sign}([a]) = -\sigma([b]) \neq 0$, *then* $\operatorname{sign}([a] - [b]) = \operatorname{sign}([a])$.
(b) *If* $\sigma([a]) = -\sigma([b])$, *then* $\sigma([a] - [b]) = \sigma([a])$.
(c) $\sigma([a] \cdot [b]) = \sigma([a]) \cdot \sigma([b])$.
(d) *If* $0 \notin [b]$, *then* $\sigma([a]/[b]) = \sigma([a]) / \operatorname{sign}([b])$.

Definition 5.4.18. A matrix $[A] \in \mathbb{IR}^{n \times n}$ is an upper Hessenberg matrix if $[a]_{ij} = 0$ for $i > j + 1$, $j = 1, \dots, n - 2$.

Theorem 5.4.19. *Let $[A] \in \mathbb{IR}^{n \times n}$, $n \geq 2$, be an upper Hessenberg matrix with $0 \notin [a]_{ii}$, $i = 1, \dots, n$ and $[a]_{i+1,i} \neq 0$, $i = 1, \dots, n - 1$. Then $[x]^G$ exists for any right-hand side $[b]$ if for each $i = 1, \dots, n - 1$ and each $j = i + 1, \dots, n$ one of the following two (mutually exclusive) conditions holds.*
(i) $[a]_{ij} = 0 \Rightarrow [a]_{pj} = 0$, $p = 1, \dots, i - 1$;
(ii) $[a]_{ij} \neq 0 \Rightarrow \sigma([a]_{ii})\sigma([a]_{i+1,j}) = -\sigma([a]_{i+1,i})\sigma([a]_{ij})$.

Proof. With $[A]^{(0)} := [A] = [A]^{(1)}$ we prove the following statements by induction on k for $k = 1, \dots, n - 1$, $i = 1, \dots, n$:

(α) $\operatorname{sign}([a]_{ii}^{(k-1)}) = \operatorname{sign}([a]_{ii}^{(k)}) \neq 0$.

(β) If $[a]_{ij}^{(k-1)} = 0$ for some $j \in \{i+1, \ldots, n\}$, then $[a]_{ij}^{(k)} = 0$ for this j.

(γ) If $[a]_{ij}^{(k-1)} \neq 0$ for some $j \in \{i+1, \ldots, n\}$, then $\sigma([a]_{ij}^{(k-1)}) = \sigma([a]_{ij}^{(k)})$ for this j. In particular, $[a]_{ij}^{(k)} \neq 0$.

Applying (α) iteratively finally yields $\operatorname{sign}([a]_{ii}^{(n)}) = \operatorname{sign}([a]_{ii}) \neq 0$, $i = 1, \ldots, n$, whence the existence of $[x]^G$ follows.

For $k = 1$ the statements (α)–(γ) hold by virtue of the particular definition of $[A]^{(0)}$. Assume now that they hold up to some $k < n$. By the Hessenberg form and by the formulae of the Gaussian algorithm we have $[a]_{k+1,k}^{(k+1)} = 0$ and $[a]_{ij}^{(k+1)} = [a]_{ij}^{(k)}$ for $i \neq k+1$ and for $i = k+1$, $j < k$. Therefore, we can restrict ourselves to $i = k+1$, $j \geq k+1$. Again by the upper Hessenberg form we get

$$[a]_{k+1,j}^{(k+1)} = [a]_{k+1,j}^{(k)} - \frac{[a]_{k+1,k}^{(k)}[a]_{kj}^{(k)}}{[a]_{kk}^{(k)}} = [a]_{k+1,j} - \frac{[a]_{k+1,k}[a]_{kj}^{(k)}}{[a]_{kk}^{(k)}}. \tag{5.4.33}$$

If $[a]_{kj}^{(k)} = 0$ holds for some $j \geq k+1$, then $[a]_{k+1,j}^{(k+1)} = [a]_{k+1,j}^{(k)}$, whence (α)–(γ) are true for this j and for k being replaced by $k+1$. Assume now $[a]_{kj}^{(k)} \neq 0$ for some $j \geq k+1$. Then

$$\sigma([a]_{kj}^{(k)}) = \sigma([a]_{kj}) \tag{5.4.34}$$

by iterative application of the induction hypothesis. In particular,

$$[a]_{kj} \neq 0. \tag{5.4.35}$$

By the assumption of the theorem we have $[a]_{k+1,k} \neq 0$, and the induction hypothesis yields $\operatorname{sign}([a]_{kk}^{(k)}) = \operatorname{sign}([a]_{kk}) \neq 0$.

First we consider the case $j = k+1$. From (5.4.35) we know $[a]_{k,k+1} \neq 0$. Therefore, from (ii) we get

$$0 \neq \operatorname{sign}([a]_{kk}) \cdot \operatorname{sign}([a]_{k+1,k+1}) = \sigma([a]_{kk}) \cdot \sigma([a]_{k+1,k+1}) = -\sigma([a]_{k+1,k})\sigma([a]_{k,k+1}),$$

which implies

$$0 \neq \operatorname{sign}([a]_{k+1,k+1}) = -\sigma\left(\frac{[a]_{k+1,k}[a]_{k,k+1}^{(k)}}{[a]_{kk}^{(k)}}\right)$$

by virtue of (α), (γ), (5.4.34) and Lemma 5.4.17. By the same lemma and (5.4.33) this implies (α) for $k+1$.

Now we consider the case $j > k+1$. If $[a]_{k+1,j} = 0$, then (i) implies $[a]_{kj} = 0$, which contradicts (5.4.35). Therefore, both these entries differ from zero, and (ii) yields

$$\operatorname{sign}([a]_{kk}) \cdot \sigma([a]_{k+1,j}) = -\sigma([a]_{k+1,k})\sigma([a]_{kj}),$$

whence

$$\sigma([a]_{k+1,j}) = -\sigma\left(\frac{[a]_{k+1,k}[a]_{kj}^{(k)}}{[a]_{kk}^{(k)}}\right).$$

Hence (5.4.33) and Lemma 5.4.17 guarantee (γ) for $k+1$. (Use $\sigma([a]_{k+1,j}) = \operatorname{sign}([a]_{k+1,j}) \neq 0$ if $[a]_{k+1,j}$ is degenerate.) □

For upper Hessenberg matrices only one element must be eliminated in each elimination step k. Hence $[A]^{(k)}$ and $[A]^{(k+1)}$ differ only by the row $k+1$. Condition (i) says that with a zero entry all other entries further up in the same column are zero. Condition (ii) allows only particular sign patterns for the matrices $A \in [A]$. By virtue of (α)–(γ) of the previous proof this sign pattern remains fixed in the upper triangle during the whole elimination process.

If $[a]_{ij} \neq 0$ for some i, j with $i < j$, then (i) implies

$$[a]_{kj} \neq 0 \quad \text{for all } k \text{ with } i < k < j. \tag{5.4.36}$$

If, in addition, $\sigma([a]_{ij}) = 0$, then $\sigma([a]_{pj}) = 0$, $p = i+1, \ldots, j-1$, by virtue of (ii) and induction. In particular, $\sigma([a]_{j-1,j}) = 0$. This contradicts (ii) for $i = j-1$ since $\sigma([a]_{j-1,j-1})\sigma([a]_{jj}) \neq 0$. Hence if the assumptions of Theorem 5.4.19 are fulfilled, no entry in the upper triangle of $[A]$ contains zero in its interior, i.e., either $[a]_{ij} = 0$ or $\sigma([a]_{ij}) \neq 0$ holds for $i \leq j$. Using (ii) if $j = i+1$, and (ii) and (5.4.36) with $k = i+1$ if $j > i+1$, we also get $\sigma([a]_{i+1,i}) \neq 0$ provided that $[a]_{ij} \neq 0$. Thus $\sigma([a]_{i+1,i}) = 0$ can only occur if $[a]_{ij} = 0$, $j = i+1, \ldots, n$, whence by (i) the whole block $([a]_{kj})_{k=1,\ldots,i,\, j=i+1,\ldots,n}$ is zero.

Assume now $\sigma([a]_{ii}) = 1$, $i = 1, \ldots, n$ (in addition to $0 \notin [a]_{ii}$) which can always be obtained by an appropriate scaling. Then the nonzero sign pattern of $[A]$ is completely controlled by the signs of the entries in the first lower subdiagonal. This can be seen for instance when starting with the last row and filling up the sign pattern row by row. In order to illustrate the possible sign structure we assume for simplicity that no entry in the upper triangle is zero. Then $\sigma([a]_{i+1,i}) \neq 0$, $i = 1, \ldots, n-1$. Precondition now $[A]$ by a signature matrix $D = \operatorname{diag}(d_1, \ldots, d_n)$, $d_i = \pm 1$, from the left such that $\sigma([a]_{ii}) = 1$, $i = 1, \ldots, n$, as above. Then precondition from the left and the right by a signature matrix $\hat{D} = \operatorname{diag}(\hat{d}_1, \ldots, \hat{d}_n)$ in the following recursive way: $\hat{d}_1 = \pm 1$ is arbitrary. If $\hat{d}_1, \ldots, \hat{d}_i$ are known, choose $\hat{d}_{i+1} = \pm 1$ such that $\sigma(\hat{d}_{i+1} d_{i+1} [a]_{i+1,i} \hat{d}_i) = -1$. Consider now $[\hat{A}] = \hat{D}D[A]\hat{D}$. Since $\sigma([\hat{a}]_{ii}) = 1 = -\sigma([\hat{a}]_{i+1,i})$ condition (ii) reads $\sigma([a]_{i+1,j}) = \sigma([a]_{ij})$, $j = i+1, \ldots, n$, from which we necessarily get the sign pattern

$$\begin{pmatrix} + & + & + & \cdots & + \\ - & + & + & \cdots & + \\ 0 & - & + & \cdots & + \\ \vdots & \ddots & \ddots & \ddots & \vdots \\ 0 & \cdots & 0 & - & + \end{pmatrix}, \tag{5.4.37}$$

i.e., the signs of the entries of each $\tilde{A} \in [\hat{A}]$ are nonpositive in the first lower subdiagonal, positive in the diagonal, and nonnegative in the whole strict upper triangle. Therefore, if $[A]$ fulfills the assumptions of Theorem 5.4.19 and has no zero in the upper triangle, then the sign pattern of its element matrices grows out from (5.4.37) changing there the signs according to a multiplication of $[A]$ with a signature matrix D_1 from the left and a signature matrix D_2 from the right. Notice that by virtue of

Theorem 5.4.1, IGA($[A]$, $[b]$) exists if and only if IGA($D_1[A]D_2, D_1[b]$) exists, where D_1, D_2 are arbitrary nonsingular diagonal matrices from $\mathbb{R}^{n\times n}$.

For our next class of matrices we need the concept of an undirected graph which modifies Definition 1.7.11 slightly.

Definition 5.4.20. The undirected graph $\overline{G}([A]) = (N, E)$ of a matrix $[A] \in \mathbb{IR}^{n\times n}$ consists of the set $N = \{1, \dots, n\}$ of nodes and of the set $E = \{\{i, j\} \mid [a]_{ij} \neq 0,\ i \neq j\}$ of undirected edges.

The degree $\deg(i) = \deg_{\overline{G}([A])(i)}$ of a node i is the number of edges which contain i.

The sequence $(i_0, i_1, \dots, i_k)$ of nodes with $\{i_l, i_{l+1}\} \in E$ for $l = 0, \dots, k-1$ is called an undirected path of length k from i_0 to i_k. If $k \geq 3$ and $i_0 = i_k$ while $i_l \neq i_m$ for $l \neq m$, $l, m = 0, \dots, k-1$, we call this path a cycle. A graph is connected if for each pair of different nodes i, j there is a path from i to j.

A connected graph without cycles is a tree.

Notice that by our definition $E = \emptyset$ if $n = 1$ and that a cycle in an undirected graph has at least length 3. We next define a sequence of particular graphs.

Definition 5.4.21. Let $[A] \in \mathbb{IR}^{n\times n}$, $n \geq 2$. The elimination graphs $G^k = (N^k, E^k)$, $k = 1, \dots, n$, associated with $[A]$ are defined by $G^1 = \overline{G}([A])$ and

$$\begin{aligned} N^{k+1} &= \{k+1, \dots, n\}, \\ E^{k+1} &= \left\{\{i, j\} \mid i, j \in N^{k+1} \text{ and } \{i, j\} \in E^k\right\} \\ &\quad \cup \left\{\{i, j\} \mid i, j \in N^{k+1},\ i \neq j, \text{ and } \{i, k\}, \{j, k\} \in E^k\right\}. \end{aligned} \tag{5.4.38}$$

Again $E^n = \emptyset$. Denote by $[A]_k$ the trailing lower right $(n-k+1) \times (n-k+1)$ submatrices $([a]_{ij}^{(k)})_{i,j=k,\dots,n}$ of the matrices $[A]^{(k)}$ occurring in the interval Gaussian algorithm. Then the elimination graphs are the undirected graphs of $[A]_k$ if one assumes that $[a]_{ij}^{(k+1)} \neq 0$ holds for $i, j > k$ whenever $[a]_{ij}^{(k)} \neq 0$. This is a very weak assumption since only if $[a]_{ij}^{(k)} \equiv a_{ij}^{(k)} \neq 0$ is a point interval and if, in addition, $a_{ij}^{(k)} = ([a]_{ik}^{(k)}[a]_{kj}^{(k)})/[a]_{kk}^{(k)}$ holds, then this assumption will be violated. In any case $\overline{G}([A]_k)$ is a subgraph of G^k.

Definition 5.4.22. We say that the matrix $[A] \in \mathbb{IR}^{n\times n}$, is ordered by minimum degree if

$$\deg_{G^k}(k) = \min\{\deg_{G^k}(i) \mid i = k, \dots, n\}$$

holds for all $k = 1, \dots, n$.

Notice that the subscript G^k indicates that we mean the degree in the elimination graph G^k. Any matrix can be ordered by minimum degree using an appropriate simultaneous permutation of its rows and columns. See for instance George, Liu [118].

Lemma 5.4.23. *Let $[A] \in \mathbb{R}^{n\times n}$, $n \geq 2$. If $G([A])$ is a tree and if $[A]$ is ordered by minimum degree, then for $k = 1, \dots, n-1$ we have*

(a) *G^k is a tree.*

(b) $\deg_{G^k}(k) = 1$.

Proof. We first prove (b) provided that one already knows that G^k is a tree. Choose any node i_0 from G^k and consider the longest path in G^k which starts from i_0 and does not contain any node twice. Since G^k is assumed to be a tree, it is connected, hence such a path exists and has at least length one. The node at the end of this path has degree one since otherwise there is either a longer path with the same property or there is a cycle which is impossible in a tree. Moreover, a node with degree zero cannot occur in G^k since a tree is connected. Now (b) follows from Definition 5.4.22.

In order to prove (a) we proceed by induction. Assume that G^k, $k < n-1$, is a tree which is correct for $k = 1$. We first show that G^{k+1} is connected. To this end let $v, w \ge k+1$ be two different nodes in G^{k+1}. Since G^k is connected there is a path $(v = v_0, v_1, \dots, v_l = w)$ in G^k. Since by (b) we have $\deg_{G^k}(k) = 1$, there is exactly one node $j > k$ for which $\{k, j\} \in E^k$. Assume that our path contains the node k for some $i \in \{1, \dots, l-1\}$. Then $v_{i-1} = v_{i+1} = j \in N^{k+1}$ and $(v_0, \dots, v_{i-2}, j, v_{i+2}, \dots, v_l)$ is also a path which connects v to w in G^k. By repeating this process we can completely remove the node k from the path. The remaining one connects v and w in G^{k+1}.

Next we show that G^{k+1} does not have any cycles. Assume to the contrary that $(v = v_0, v_1, \dots, v_l = v)$ is a cycle in G^{k+1}. Since $\deg_{G^k}(k) = 1$, the second set of the union in (5.4.38) is empty which shows $E^{k+1} \subseteq E^k$. Therefore, $(v = v_0, v_1, \dots, v_l = v)$ is also a cycle in G^k, which is impossible since G^k is a tree. □

After these preparations we can state another necessary and sufficient criterion.

Theorem 5.4.24. *Let $[A] \in \mathbb{IR}^{n\times n}$, $[b] \in \mathbb{IR}^n$. Let $G([A])$ be a tree and $[A]$ be ordered by minimum degree. Then $[x]^G$ exists if and only if the Gaussian algorithm is feasible for all matrices $\tilde{A} \in [A]$.*

Proof. The necessity follows from Theorem 5.4.1 (b). In order to prove sufficiency we will show by induction that

$$[A]^{(k)} = \{ A^{(k)} \mid A \in [A] \}, \quad k = 1, \dots, n, \tag{5.4.39}$$

holds, whence

$$[a]_{kk}^{(k)} = \{ a_{kk}^{(k)} \mid A \in [A] \}, \quad k = 1, \dots, n.$$

Since by assumption this latter set cannot contain zero, $[x]^G$ exists.

Assume now that (5.4.39) is correct up to some $k < n$ which is certainly true for $k = 1$. According to Lemma 5.4.23 we have $\deg_{G^k}(k) = 1$ for $k < n$. We have already remarked above that the undirected graph of the trailing $(n-k+1) \times (n-k+1)$ submatrix $[A]_k$ of $[A]^{(k)}$ is a subgraph of G^k. Both facts together show that there is at most one index $l > k$ such that $[a]_{lk}^{(k)} \ne 0$ or $[a]_{kl}^{(k)} \ne 0$, whereas $[a]_{ik}^{(k)}[a]_{ki}^{(k)} = 0$ for all $i \in \{k+1, \dots, n\} \setminus \{l\}$. Therefore, the formula (5.4.2) for $[a]^{(k+1)}$ implies

$$[a]_{ij}^{(k+1)} = [a]_{ij}^{(k)} \quad \text{for all pairs } (i, j) \ne (l, l), \tag{5.4.40}$$

while for $(i, j) = (l, l)$ we have

$$[a]_{ll}^{(k+1)} = [a]_{ll}^{(k)} - [a]_{lk}^{(k)}[a]_{kl}^{(k)}/[a]_{kk}^{(k)}.$$

The right-hand side is a term in which each interval occurs only once. By Theorem 4.1.5 it equals the range

$$\{ a_{ll}^{(k)} - a_{lk}^{(k)} a_{kl}^{(k)} / a_{kk}^{(k)} \mid a_{ll}^{(k)} \in [a]_{ll}^{(k)},\ a_{lk}^{(k)} \in [a]_{lk}^{(k)},\ a_{kl}^{(k)} \in [a]_{kl}^{(k)},\ a_{kk}^{(k)} \in [a]_{kk}^{(k)} \}.$$

Together with (5.4.40) and the induction hypothesis this proves (5.4.39) for $k+1$. □

The next corollary deals with tridiagonal matrices and arrowhead matrices. These are matrices of the structure

$$\begin{pmatrix} \times & \times & 0 & 0 & 0 \\ \times & \times & \times & 0 & 0 \\ 0 & \times & \times & \times & 0 \\ 0 & 0 & \times & \times & \times \\ 0 & 0 & 0 & \times & \times \end{pmatrix} \quad \text{and} \quad \begin{pmatrix} \times & 0 & 0 & 0 & \times \\ 0 & \times & 0 & 0 & \times \\ 0 & 0 & \times & 0 & \times \\ 0 & 0 & 0 & \times & \times \\ \times & \times & \times & \times & \times \end{pmatrix},$$

respectively, where the crosses '×' denote zeros or nonzeros.

Corollary 5.4.25. *For the following classes of interval matrices $[A] \in \mathbb{IR}^{n\times n}$ the assumptions of Theorem 5.4.24 are fulfilled.*
(a) *$[A]$ is a tridiagonal matrix.*
(b) *$[A]$ is a northwest–southeast arrowhead matrix.*
(c) *$[A]$ is a 2×2 matrix.*
In particular, $[x]^G$ exists if and only if it exists for each element matrix $\tilde{A} \in [A]$.

Proof. The undirected graph of a triangular matrix or an arrowhead matrix in (b) is a tree or a union of trees. Moreover, $[A]$ is ordered by minimum degree. A 2×2 matrix is tridiagonal as well as an arrowhead matrix as in (b). □

In order to formulate an application of the preceding corollary we need the definition of additional classes of real matrices.

Definition 5.4.26. A matrix $A \in \mathbb{R}^{n\times n}$ is called totally positive (totally nonnegative) if each of its minors is positive (nonnegative). It is called oscillatory if it is totally nonnegative and if at least one of its powers A^k is totally positive.

By inspecting the minors of order one it is obvious that totally positive (totally nonnegative) matrices are, in particular, positive (nonnegative) matrices in the sense of Definition 1.9.1 (b).

Corollary 5.4.27. *Let $[A] \in \mathbb{IR}^{n\times n}$ be tridiagonal and $[b] \in \mathbb{IR}^n$. In each of the following cases $[x]^G$ exists.*
(a) *Each matrix $\tilde{A} \in [A]$ is totally positive.*
(b) *Each matrix $\tilde{A} \in [A]$ is regular and totally nonnegative.*
(c) *Each matrix $\tilde{A} \in [A]$ is oscillatory.*

Proof. Let $\tilde{A}_k$ be the $k \times k$ leading submatrix of any matrix $\tilde{A} \in [A]$. We will show that $\det \tilde{A}_k \neq 0$ holds. Then Theorem 5.4.3 together with Theorem 5.4.24 can be applied and Corollary 5.4.25 implies the assertion.

(a) $\det \tilde{A}_k > 0$ follows by Definition 5.4.26.

(b) Since $\tilde{A}$ is regular and totally nonnegative we must have $\det \tilde{A} > 0$; $\det A_k \neq 0$ follows by virtue of inequality (116) in Gantmacher [111], p. 443.

(c) Since $\det \tilde{A}^l > 0$ for some integer l, the matrix $\tilde{A}$ is regular, and (b) applies. □

We continue our studies of the interval Gaussian algorithm by a perturbation theorem due to Neumaier [255]. It provides a sufficient criterion for the feasibility of the Gaussian algorithm for all matrices $[B]$ which are supersets of a given matrix $[A]$ with existent vector $[x]^G$. It uses the matrix $|[A]^G|$ defined in (5.4.6) and needs a preparatory lemma.

Lemma 5.4.28. *Let* $[A] \subseteq [B] \in \mathbb{IR}^{n\times n}$, $[b] \in \mathbb{IR}^n$ *and assume that* $\mathrm{IGA}([A],[b])$ *exists with* $([L],[U])$ *denoting the triangular decomposition of* $[A]$. *If*

$$q([A],[B])u < \langle [L]\rangle \langle [U]\rangle u \tag{5.4.41}$$

holds for some positive vector u, *then* $\mathrm{IGA}([B],[b])$ *exists.*

Proof. If $n = 1$ we get

$$q([A],[B])u < \langle [a]_{11}\rangle u,$$

whence

$$\langle [a]_{11}\rangle - q([A],[B]) > 0.$$

Since $[A] \subseteq [B]$, this shows $0 < \langle [b]_{11}\rangle$ which proves the existence of $\mathrm{IGA}([B],[b])$.

Assume now that the statement is true for some dimension $n \geq 1$, and let (5.4.41) hold for

$$[A] = \begin{pmatrix} [a]_{11} & [c]^T \\ [d] & [A]' \end{pmatrix} \subseteq [B] = \begin{pmatrix} [b]_{11} & [e]^T \\ [f] & [B]' \end{pmatrix} \in \mathbb{IR}^{(n+1)\times(n+1)}. \tag{5.4.42}$$

First we show $\langle [b]_{11}\rangle > 0$. With

$$q([A],[B]) = (q_{ij}) = \begin{pmatrix} q_{11} & r^T \\ s & Q' \end{pmatrix} \geq O \tag{5.4.43}$$

we get from (5.4.41)

$$\sum_{j=1}^{n+1} q_{1j}u_j < \langle [a]_{11}\rangle\, u_1 - \sum_{j=2}^{n+1} |[a]_{1j}|\, u_j,$$

hence

$$\langle [a]_{11}\rangle - q_{11} > \left\{ \sum_{j=2}^{n+1} (q_{1j} + |[a]_{1j}|)\, u_j \right\} \Big/ u_1 \geq 0.$$

Together with (5.4.42) this implies $0 < \langle [b]_{11}\rangle$, whence the Schur complement $\Sigma([B]) \supseteq \Sigma([A])$ exists.

By our assumptions, the Schur complement $\Sigma([A])$ has a triangular decomposition $([L]', [U]')$. If we can show that

$$q(\Sigma([A]), \Sigma([B]))u' < \langle [L]' \rangle \langle [U]' \rangle u' \tag{5.4.44}$$

holds for some vector $u' > 0$, then $\Sigma([B])$ has a triangular decomposition, say $([\hat{L}]', [\hat{U}]')$, by the hypothesis of our induction, and with

$$[\hat{L}] := \begin{pmatrix} 1 & O \\ \frac{[f]}{[b]_{11}} & [\hat{L}]' \end{pmatrix}, \quad [\hat{U}] := \begin{pmatrix} [b]_{11} & [e]^T \\ O & [\hat{U}]' \end{pmatrix}$$

we obtain the triangular decomposition $([\hat{L}], [\hat{U}])$ of $[B]$.

In order to prove (5.4.44) we will apply Neumaier's β-function from Definition 2.5.8 componentwise, and we will use the notation from (5.4.42), (5.4.43). We then get

$$\begin{aligned} q(\Sigma([A]), \Sigma([B])) &= q([A]' - [d][c]^T[a]_{11}^{-1}, [B]' - [f][e]^T[b]_{11}^{-1}) \\ &\le Q' + q([d][c]^T[a]_{11}^{-1}, [f][e]^T[b]_{11}^{-1}) \\ &= Q' - |[d][c]^T[a]_{11}^{-1}| + \beta([d][c]^T[a]_{11}^{-1}, [f][e]^T[b]_{11}^{-1}) \\ &= Q' - |[d]|\,|[c]^T|\,\langle [a]_{11}\rangle^{-1} + \beta([d][c]^T[a]_{11}^{-1}, [f][e]^T[b]_{11}^{-1}) \\ &\le Q' - |[d]|\,|[c]^T|\,\langle [a]_{11}\rangle^{-1} + \beta([d], [f]) \cdot \beta([c]^T, [e]^T) \cdot \beta([a]_{11}^{-1}, [b]_{11}^{-1}) \\ &= Q' - |[d]|\,|[c]^T|\,\langle [a]_{11}\rangle^{-1} + (|[d]| + s)(|[c]| + r)^T \beta([a]_{11}^{-1}, [b]_{11}^{-1}). \end{aligned} \tag{5.4.45}$$

Now we want to apply Theorem 2.5.9 (b) to the last factor in (5.4.45). To this end we have to show

$$\langle [a]_{11} \rangle > q([a]_{11}, [b]_{11}) = q_{11}. \tag{5.4.46}$$

Therefore, we set

$$u = \begin{pmatrix} u_1 \\ u' \end{pmatrix}$$

in (5.4.41). With the notation (5.4.43) we obtain

$$\begin{pmatrix} q_{11} & r^T \\ s & Q' \end{pmatrix} \begin{pmatrix} u_1 \\ u' \end{pmatrix} < \begin{pmatrix} 1 & O \\ -\frac{|[d]|}{\langle [a]_{11} \rangle} & \langle [L]' \rangle \end{pmatrix} \begin{pmatrix} \langle [a]_{11} \rangle & -|[c]^T| \\ O & \langle [U]' \rangle \end{pmatrix} \begin{pmatrix} u_1 \\ u' \end{pmatrix},$$

whence

$$q_{11}u_1 + r^T u' < \langle [a]_{11} \rangle u_1 - |[c]^T| u' \tag{5.4.47}$$

and

$$s u_1 + Q' u' < -|[d]| u_1 + |[d]|\,|[c]^T|\,\langle [a]_{11} \rangle^{-1} u' + \langle [L]' \rangle \langle [U]' \rangle u'. \tag{5.4.48}$$

Since $u_1 > 0$, the inequality (5.4.47) implies (5.4.46), and Theorem 2.5.9 (b) and (5.4.45) yield

$$\begin{aligned} q(\Sigma([A]), \Sigma([B])) \le Q' &- |[d]|\,|[c]^T|\,\langle [a]_{11} \rangle^{-1} \\ &+ (|[d]| + s)(|[c]| + r)^T (\langle [a]_{11} \rangle - q_{11})^{-1}. \end{aligned}$$

Together with (5.4.46), (5.4.47), (5.4.48) this implies

$$\begin{aligned} q(\Sigma([A]), \Sigma([B]))u' &\le Q'u' - |[d]|\,|[c]^T|\langle [a]_{11}\rangle^{-1}u' \\ &\quad + (|[d]| + s)(\langle [a]_{11}\rangle - q_{11})^{-1}(|[c]| + r)^T u' \\ &< -su_1 - |[d]|u_1 + \langle [L]'\rangle\,\langle [U]'\rangle u' \\ &\quad + (|[d]| + s)(\langle [a]_{11}\rangle - q_{11})^{-1}(\langle [a]_{11}\rangle u_1 - q_{11}u_1) \\ &= \langle [L]'\rangle\,\langle [U]'\rangle u'. \end{aligned}$$

This proves (5.4.44) and terminates the induction. □

Theorem 5.4.29. *Let* $[A], [B] \in \mathbb{IR}^{n\times n}$, $[b] \in \mathbb{IR}^n$ *and assume that* IGA$([A], [b])$ *exists. If*

$$\rho(|[A]^G|\, q([A], [B]) < 1, \tag{5.4.49}$$

then IGA$([B], [b])$ *exists.*

Proof. Let $Q = q([A], [B])$, $[C] = [A] + [-Q, Q]$. Then $[B] \subseteq [C]$, and IGA$([B], [b])$ exists if IGA$([C], [b])$ does. By (5.4.49) the inverse of $I - |[A]^G|Q$ exists and can be represented as Neumann series

$$(I - |\,[A]^G|Q)^{-1} = \sum_{k=0}^{\infty}(|[A]^G|Q)^k \ge O.$$

With any $v \in \mathbb{R}^n$ satisfying $v > 0$ define

$$u = (I - |[A]^G|Q)^{-1}\,|[A]^G|v. \tag{5.4.50}$$

Since $|[A]^G| \ge O$ and $(I - |[A]^G|Q)^{-1} \ge O$ are regular (use (5.4.13)) each of their rows contains at least one positive entry. Therefore $|[A]^G|v > 0$ and $u > 0$. Now (5.4.50) yields

$$|[A]^G|Qu = u - |[A]^G|v,$$

whence

$$Qu = \langle [L]\rangle\langle [U]\rangle u - v < \langle [L]\rangle\langle [U]\rangle u,$$

with $([L], [U])$ being the triangular decomposition of $[A]$. Hence, Lemma 5.4.28 guarantees the feasibility of the interval Gaussian algorithm for $[C]$ and therefore also for $[B]$. □

We illustrate Theorem 5.4.29 with an example.

Example 5.4.30. Let

$$[B] = \begin{pmatrix} 4 & 2 & 2 \\ 2 & 4 & [0, 2] \\ 2 & [0, 2] & 4 \end{pmatrix}.$$

Then $[B]$ is not an H-matrix since $\langle [B]\rangle e = 0$, hence $\langle [B]\rangle$ is singular. Consider

$$[A] := \check{B} = \begin{pmatrix} 4 & 2 & 2 \\ 2 & 4 & 1 \\ 2 & 1 & 4 \end{pmatrix} \subseteq [B].$$

Since $\langle [A] \rangle$ is irreducibly diagonally dominant, the interval Gaussian algorithm is feasible for $[A]$. A simple computation yields

$$[L] = \begin{pmatrix} 1 & 0 & 0 \\ 1/2 & 1 & 0 \\ 1/2 & 0 & 1 \end{pmatrix}, \qquad [U] = \begin{pmatrix} 4 & 2 & 2 \\ 0 & 3 & 0 \\ 0 & 0 & 3 \end{pmatrix},$$

$$\langle [L] \rangle^{-1} = \begin{pmatrix} 1 & 0 & 0 \\ 1/2 & 1 & 0 \\ 1/2 & 0 & 1 \end{pmatrix}, \quad \langle [U] \rangle^{-1} = \begin{pmatrix} 1/4 & 1/6 & 1/6 \\ 0 & 1/3 & 0 \\ 0 & 0 & 1/3 \end{pmatrix},$$

and

$$|[A]^G| = \langle [U] \rangle^{-1} \langle [L] \rangle^{-1} = \frac{1}{12} \begin{pmatrix} 5 & 2 & 2 \\ 2 & 4 & 0 \\ 2 & 0 & 4 \end{pmatrix}.$$

From

$$q([A], [B]) = \begin{pmatrix} 0 & 0 & 0 \\ 0 & 0 & 1 \\ 0 & 1 & 0 \end{pmatrix}$$

we get the matrix

$$|[A]^G| q([A], [B]) = \begin{pmatrix} 0 & 1/6 & 1/6 \\ 0 & 0 & 1/3 \\ 0 & 1/3 & 0 \end{pmatrix}$$

which has the eigenvalues $-1/3, 0, 1/3$. Therefore, Theorem 5.4.29 applies. The matrices $[\hat{L}]$, $[\hat{U}]$ of the triangular decomposition of $[B]$ are

$$[\hat{L}] = \begin{pmatrix} 1 & 0 & 0 \\ 1/2 & 1 & 0 \\ 1/2 & [-1/3, 1/3] & 1 \end{pmatrix} \quad \text{and} \quad [\hat{U}] = \begin{pmatrix} 4 & 2 & 2 \\ 0 & 3 & [-1, 1] \\ 0 & 0 & [8/3, 10/3] \end{pmatrix}.$$

Example 5.4.30 also illustrates the following corollary which is an immediate consequence of Theorem 5.4.29 and of $q(\check{A}, [A]) = \text{rad}([A])$.

Corollary 5.4.31. *Let the Gaussian algorithm be feasible for the midpoint matrix $\check{A}$ of $[A] \in \mathbb{IR}^{n \times n}$ and assume that*

$$\rho(|\check{A}^G| \, \text{rad}([A])) < 1,$$

where $|\check{A}^G|$ is defined for $\check{A}$ analogously to (5.4.6). Then $[x]^G$ exists for $[A]$.

We continue this section with a nonexistence theorem that can indicate that $[x]^G$ does not exist for some interval matrix $[A]$ although it exists for each element matrix $\tilde{A} \in [A]$.

The idea is essentially based on the assumption that

$$I \in [A]. \tag{5.4.51}$$

This is not a severe restriction as long as $0 \notin [a]_{ii}$, $i = 1, \dots, n$, while $0 \in [a]_{ij}$, $i \neq j$. In this case define the diagonal matrix

$$D = \mathrm{diag}(\mathrm{sign}(\check{a}_{11})\langle [a]_{11}\rangle, \dots, \mathrm{sign}(\check{a}_{nn})\langle [a]_{nn}\rangle)$$

and consider $[B] = D^{-1}[A]$ which contains I. By Theorem 5.4.1 (f), $[x]^G$ does not exist for $[A]$ if and only if it does not exist for $[B]$. Trivially, (5.4.51) implies

$$I \in [A]^{(k)}, \tag{5.4.52}$$

provided that this matrix exists. In particular, we get $1 \in [a]_{kk}^{(k)}$. Therefore, the interval Gaussian algorithm breaks down, whenever $\underline{a}_{kk}^{(k)} \leq 0$. This is the basis of the subsequent theorem.

Theorem 5.4.32. *Let $I \in [A] \in \mathbb{IR}^{n \times n}$ with $\underline{a}_{ii} = 1$, $i = 1, \dots, n$, and let $[b] \in \mathbb{IR}^n$. Choose $c_{ik} \in \{\underline{a}_{ik}, \overline{a}_{ik}\}$, $c_{ki} \in \{\underline{a}_{ki}, \overline{a}_{ki}\}$ such that*

$$c_{ik}c_{ki} = \max\{\,\underline{a}_{ik}\underline{a}_{ki}, \overline{a}_{ik}\overline{a}_{ki}\,\} \quad \textit{for } i > k,\ k = 1, \dots, n-1.$$

Define $c_{kk}^{(k)}$, $k \in \{1, \dots, n\}$, by $c_{11}^{(1)} = 1$ and for $k \geq 2$ recursively by

$$c_{kk}^{(k)} = 1 - \sum_{j=1}^{k-1} \frac{c_{kj}c_{jk}}{c_{jj}^{(j)}} \tag{5.4.53}$$

as long as $c_{k-1,k-1}^{(k-1)} > 0$.

If the definition (5.4.53) stops with $c_{k_0k_0}^{(k_0)} \leq 0$ for some $k_0 \leq n$, then the interval Gaussian algorithm is not feasible.

Proof. Assume that the interval Gaussian algorithm is feasible when the definition (5.4.53) stops with $c_{k_0k_0}^{(k_0)} \leq 0$ for some $k_0 \leq n$. Then the matrices $[A]^{(k)}$, $k = 1, \dots, n$, exist, and (5.4.52) implies $\overline{a}_{kk}^{(k)} \geq 1$, hence $\underline{a}_{kk}^{(k)} > 0$, $k = 1, \dots, n$.

The assumption $I \in [A]$ implies $0 \in [a]_{st}$, $s \neq t$, whence $[a]_{ij} \subseteq [a]_{ij}^{(2)} = [a]_{ij} - \frac{[a]_{i1}[a]_{1j}}{[a]_{11}}$. With (5.4.52) and by induction we get

$$[a]_{ij} \subseteq [a]_{ij}^{(k)}, \quad i, j \geq k,\ k = 1, \dots, n. \tag{5.4.54}$$

Let

$$c_{ii}^{(1)} = 1, \quad c_{ii}^{(k)} = 1 - \sum_{j=1}^{k-1} \frac{c_{ij}c_{ji}}{c_{jj}^{(j)}}, \quad k = 2, \dots, k_0.$$

We show by induction on $k \leq k_0$ that

$$\underline{a}_{ii}^{(k)} \leq c_{ii}^{(k)}, \quad i \geq k. \tag{5.4.55}$$

If $k = 1$ this is trivial since $\underline{a}_{ii}^{(1)} = \underline{a}_{ii} = 1 = c_{ii}^{(1)}$ by assumption and by definition, respectively. Let (5.4.55) hold for some $k < k_0$. From the formulae of the interval Gaussian algorithm, (5.4.52), (5.4.54) and the induction hypothesis (5.4.55) we get

$$\begin{aligned}\underline{a}_{ii}^{(k+1)} &= \underline{a}_{ii}^{(k)} - \frac{\max([a]_{ik}^{(k)}[a]_{ki}^{(k)})}{\underline{a}_{kk}^{(k)}} \leq c_{ii}^{(k)} - \frac{\max([a]_{ik}[a]_{ki})}{\underline{a}_{kk}^{(k)}} \\ &\leq c_{ii}^{(k)} - \frac{c_{ik}c_{ki}}{c_{kk}^{(k)}} = 1 - \sum_{j=1}^{k} \frac{c_{kj}c_{jk}}{c_{jj}^{(j)}} = c_{ii}^{(k+1)}, \quad i = k+1, \ldots, n,\end{aligned}$$

which proves (5.4.55). Hence

$$\underline{a}_{kk}^{(k)} \leq c_{kk}^{(k)}, \quad k = 1, \ldots, k_0.$$

In particular,

$$\underline{a}_{k_0k_0}^{(k_0)} \leq c_{k_0k_0}^{(k_0)} \leq 0.$$

Therefore, the interval Gaussian algorithm breaks down with $0 \in [a]_{k_0k_0}^{(k_0)}$ in contrast to our assumption. □

We will apply this theorem to an example due to Reichmann [291] with which he demonstrated for the first time that the interval Gaussian algorithm breaks down although the given input matrix $[A]$ contains only matrices $\tilde{A}$ for which the usual Gaussian algorithm is feasible without pivoting.

Example 5.4.33. Let

$$[A] = \begin{pmatrix} 1 & [c] & [c] \\ [c] & 1 & [c] \\ [c] & [c] & 1 \end{pmatrix}, \qquad [c] = [0, x],\ x \in [0, 1).$$

First we show that the Gaussian algorithm is feasible for each element matrix

$$\tilde{A} = \begin{pmatrix} 1 & a_1 & a_2 \\ a_3 & 1 & a_4 \\ a_5 & a_6 & 1 \end{pmatrix}, \in [A],$$

if $x \leq \frac{2}{3}$, i.e., $a_i \in \left[0, \frac{2}{3}\right]$. The first two leading submatrices of $\tilde{A}$ have a determinant which is one, and $1 - a_1a_3 \geq 1 - x^2 > 0$. So it remains to study $\det \tilde{A}$ itself. To this end define $a = (a_1, a_2, \ldots, a_6)^T \geq 0$ and

$$\begin{aligned} f(a) = \det A &= 1 - a_4a_6 - a_1(a_3 - a_4a_5) + a_2(a_3a_6 - a_5) \\ &= 1 + a_1a_4a_5 + a_2a_3a_6 - a_4a_6 - a_1a_3 - a_2a_5. \end{aligned}$$

We look for $\tilde{a}$ such that

$$f(\tilde{a}) = \min\left\{ f(a) \mid a_i \in \left[0, \frac{2}{3}\right]\right\}. \tag{5.4.56}$$

Case 1, $\tilde{a}_i = 0$ for at least one index $i \in \{1, \dots, 6\}$: Then

$$f(\tilde{a}) \geq 1 - \frac{2}{3} \cdot \frac{2}{3} - \frac{2}{3} \cdot \frac{2}{3} > 0.$$

Case 2, $\tilde{a}_i = \frac{2}{3}$ for all $i \in \{1, \dots, 6\}$: Then

$$f(\tilde{a}) = 1 + 2 \cdot \frac{8}{27} - 3 \cdot \frac{4}{9} = \frac{7}{27} > 0.$$

Case 3, $\tilde{a}_i \in \left(0, \frac{2}{3}\right)$ for at least one $i \in \{1, \dots, 6\}$: W.l.o.g. let $i = 6$. The function $g(s) := f(\tilde{a}_1, \dots, \tilde{a}_5, s)$ satisfies $f(\tilde{a}) = g(\tilde{a}_6)$ and has a relative minimum if $g'(s) = 0$, i.e, if $\tilde{a}_2\tilde{a}_3 - \tilde{a}_4 = 0$. In this case

$$\begin{aligned} f(\tilde{a}) &= 1 + \tilde{a}_1\tilde{a}_4\tilde{a}_5 + (\tilde{a}_2\tilde{a}_3 - \tilde{a}_4)\tilde{a}_6 - \tilde{a}_1\tilde{a}_3 - \tilde{a}_2\tilde{a}_5 \\ &\geq 1 + 0 - \frac{4}{9} - \frac{4}{9} = \frac{1}{9} > 0. \end{aligned}$$

Hence $\det \tilde{A} \neq 0$ for all $\tilde{A} \in [A]$, and $\mathrm{IGA}(\tilde{A}, b)$ exists as long as $\tilde{a}_{ij} \in [0, 2/3]$ for $i \neq j$.

For the corresponding interval matrix $[A]$ we obtain

$$[a]_{11}^{(1)} = [1, 1],\ [a]_{22}^{(2)} = [1 - x^2, 1],\ [a]_{33}^{(3)} = \left[1 - x^2 - \frac{x^2}{1 - x^2}, 1 + \frac{x^3}{1 - x^2}\right]. \tag{5.4.57}$$

Since $0 \notin [a]_{22}^{(2)}$ the matrix $[A]^{(3)}$ always exists. If $x = \frac{1}{2}(-1 + \sqrt{5}) < \frac{2}{3}$ as in Reichmann [291], then $\underline{a}_{33}^{(3)} = 0$ and if $x = \frac{2}{3}$ as in Neumaier [257], then $\underline{a}_{33}^{(3)} < 0$. Since in both cases $\overline{a}_{33}^{(3)} > 0$ the interval Gaussian algorithm breaks down.

This is also predicted by our preceding nonexistence theorem, where $c_{11}^{(1)}$, $c_{22}^{(2)}$ and $c_{33}^{(3)}$ are just the lower bounds of the corresponding intervals in (5.4.57).

Example 5.4.33 with $[c] = [-x, x]$, $x \in [0, 1)$, shows that the criterion in Theorem 5.4.32 is not necessary. Here, the value of $c_{33}^{(3)}$ is the same as above while

$$\underline{a}_{33}^{(3)} = \frac{2x^2 + x - 1}{x - 1} \leq 0 \quad \text{for} \quad \frac{1}{2} \leq x < 1.$$

Therefore, the criterion in Theorem 5.4.32 does not indicate the breakdown if $x \in \left[\frac{1}{2}, \frac{-1+\sqrt{5}}{2}\right)$. Since $[A]$ contains the singular matrix

$$\tilde{A} = \begin{pmatrix} 1 & -1/2 & -1/2 \\ -1/2 & 1 & -1/2 \\ -1/2 & -1/2 & 1 \end{pmatrix} \in [A]$$

if $[c] = [-x, x]$, $x \in [1/2, 1)$, the breakdown is a must.

The interval Gaussian algorithm – explicit formulae

In the subsequent part of this section we consider the quality of enclosure obtained by $[x]^G \supseteq S$. We start with the class of M-matrices for which we already know by Corollary 5.4.7 that $[x]^G$ exists. By the particular structure of $[A]$ we are able to extend some of the preceding statements.

Theorem 5.4.34. *Let $[A] \in \mathbb{IR}^{n\times n}$ be an M-matrix and $[b] \in \mathbb{IR}^n$. Moreover, denote by $\underline{D}^{(k)}$, $\underline{L}^{(k)}$, $\underline{U}^{(k)}$ the matrices in (5.4.3) for $\underline{A}$ and let $\underline{A} = \underline{L}\,\underline{U}$ be the triangular decomposition of $\underline{A}$. The corresponding quantities for $\overline{A}$ are denoted by means of an upper bar. Then we have*

(a) *The matrices $\underline{D}^{(k)}$, $\underline{L}^{(k)}$, $\underline{U}^{(k)}$ and the corresponding overlined quantities are non-negative and satisfy*

$$[D]^{(k)} = [\overline{D}^{(k)}, \underline{D}^{(k)}], \quad [L]^{(k)} = [\overline{L}^{(k)}, \underline{L}^{(k)}], \quad [U]^{(k)} = [\overline{U}^{(k)}, \underline{U}^{(k)}].$$

In addition, the matrices of the triangular decomposition of $[A]$ can be represented as $[L] = [\underline{L}, \overline{L}]$, $[U] = [\underline{L}, \overline{L}]$.

(b) $\mathrm{IGA}([A]) = [\overline{A}^{-1}, \underline{A}^{-1}]$, $\quad |[A]^G| = \underline{A}^{-1} = \langle [A]\rangle^{-1} = |\,\mathrm{IGA}([A])|$.

(c) $\mathrm{IGA}([A], [b]) \subseteq \mathrm{IGA}([A]) \cdot [b]$.

(d) *If $[A] \equiv A \in \mathbb{R}^{n\times n}$, then* $\mathrm{IGA}([A], [b]) = \mathrm{IGA}(A, [b]) = A^{-1}[b]$.

(e) $\mathrm{IGA}([A], [b]) = \begin{cases} [\overline{A}^{-1}\underline{b}, \underline{A}^{-1}\overline{b}], & \text{if } \underline{b} \geq 0, \\ [\underline{A}^{-1}\underline{b}, \underline{A}^{-1}\overline{b}], & \text{if } 0 \in [b], \\ [\underline{A}^{-1}\underline{b}, \overline{A}^{-1}\overline{b}], & \text{if } \overline{b} \leq 0, \end{cases}$

i.e., $\mathrm{IGA}([A], [b])$ *equals the interval hull $\square S$ if $[b]$ satisfies the indicated sign conditions.*

Proof. (a) We prove the assertion for $\underline{D}^{(k)}$, $\underline{L}^{(k)}$, $\underline{U}^{(k)}$ by induction on the size n of $[A]$. Assume that it is correct for some n which is trivial for $n = 1$. Let

$$[A] = \begin{pmatrix} [a]_{11} & [c]^T \\ [d] & [A]' \end{pmatrix} \in \mathbb{IR}^{(n+1)\times(n+1)}$$

be an M-matrix with $[c]$, $[d] \in \mathbb{IR}^n$. Then the assertion can be seen for $k = 1$ from the sign structure of $[A]$. Moreover, $\underline{a}_{11} > 0$, $\overline{c} \leq 0$, $\overline{d} \leq 0$, hence

$$\Sigma([A]) = [A]' - \frac{[d][c]^T}{[a]_{11}} = [A]' - \left[\frac{\overline{d}\overline{c}^T}{\overline{a}_{11}}, \frac{\underline{d}\underline{c}^T}{\underline{a}_{11}}\right] = [\Sigma(\underline{A}), \Sigma(\overline{A})].$$

Since $\underline{A}$ is an M-matrix there is some positive vector $u = (u_1, u') > 0$, $u' \in \mathbb{R}^n$, such that $\underline{A}u = \underline{A}^{(1)}u > 0$. From $\underline{L}^{(1)} \geq I$ we get $\underline{L}^{(1)}\underline{A}u \geq I\underline{A}u > 0$, whence $\Sigma(\underline{A})u' > 0$. By inspecting the off-diagonal entries one sees immediately that $\Sigma(\underline{A}), \Sigma(\overline{A}) \in Z^{n\times n}$, hence $\Sigma(\underline{A})$ is an M-matrix and so is $\Sigma([A])$. By the induction hypothesis, the matrices $[D]_\Sigma^{(k)}$, $[L]_\Sigma^{(k)}$, $[U]_\Sigma^{(k)}$ for $\Sigma([A]) = [\Sigma(\underline{A}), \Sigma(\overline{A})]$ corresponding to those in (5.4.3) can be represented as $[\overline{D}_\Sigma^{(k)}, \underline{D}_\Sigma^{(k)}]$ etc., where the upper and lower interval bounds are the corresponding matrices for $\Sigma(\underline{A})$ and $\Sigma(\overline{A})$, respectively. In order to facilitate notation we assume that these matrices are numbered starting with $k = 2$ instead of $k = 1$. Then

$$\underline{D}^{(k)} = \begin{pmatrix} 1 & 0 \\ 0 & \underline{D}_\Sigma^{(k)} \end{pmatrix}, \quad \underline{L}^{(k)} = \begin{pmatrix} 1 & 0 \\ 0 & \underline{L}_\Sigma^{(k)} \end{pmatrix}, \quad \underline{U}^{(k)} = \begin{pmatrix} 1 & 0 \\ 0 & \underline{U}_\Sigma^{(k)} \end{pmatrix}$$

with a similar representation for $\overline{D}^{(k)}, \overline{L}^{(k)}, \overline{U}^{(k)}$. This proves the assertion.

The statement for $[L]$, $[U]$ follows from their representation by means of the matrices $[\hat{D}]^{(k)}$, $[\hat{L}]^{(k)}$, $[\hat{U}]^{(k)}$ in (5.4.11), (5.4.12) and the connection of these matrices with $[D]^{(k)}$, $[L]^{(k)}$, $[U]^{(k)}$.

(b)–(e) follow from (5.4.4) and the nonnegativity of the matrices therein. □

Next we derive a representation for $[x]^G$ if $[A] = I + [-R, R]$, $R \geq O$, $\rho(R) < 1$ as in Theorem 5.4.5 which guarantees the existence of $[x]^G$. For this representation we prove the following lemma which provides explicit formulae for the quantities arising in the interval Gaussian algorithm. We will define a signature matrix $D = \operatorname{diag}(d_1, \ldots, d_n)$ such that $|\check{b}| = D\check{b}$, whence

$$\max(d_i[b]_i + [a]) = |d_i[b]_i + [a]| = |[b]_i| + |[a]|$$

for any symmetric interval $[a]$.

Lemma 5.4.35. *Let $[A] = I + [-R, R] \in \mathbb{IR}^{n\times n}$, $R \geq O$, $\rho(R) < 1$ and let $[b] \in \mathbb{IR}^n$. Apply the Gaussian algorithm to the real matrix $C = I - R$ and the right-hand side $w = |\check{b}| + \operatorname{rad}([b])$ and define the diagonal matrix $D = \operatorname{diag}(d_1, \ldots, d_n) \in \mathbb{R}^{n\times n}$ by*

$$d_i = \begin{cases} 1 & \text{if } \check{b}_i \geq 0, \\ -1 & \text{if } \check{b}_i < 0. \end{cases}$$

Then the quantities of the interval Gaussian algorithm for $[A]$, $[b]$ can be represented as follows:

$$[A]^{(k)} = [C^{(k)}, 2I - C^{(k)}] = I + [-(I - C^{(k)}), I - C^{(k)}] \tag{5.4.58}$$

$$\langle [A]^{(k)} \rangle = C^{(k)} \tag{5.4.59}$$

$$\begin{aligned} [b]^{(k)} &= D[2|\check{b}| - w^{(k)}, w^{(k)}] = D(|\check{b}| + [-(w^{(k)} - |\check{b}|), w^{(k)} - |\check{b}|]) \\ &= \check{b} + [-(w^{(k)} - |\check{b}|), w^{(k)} - |\check{b}|] \end{aligned} \tag{5.4.60}$$

$$\max(D[b]^{(k)}) = |[b]^{(k)}| = w^{(k)} \geq 0 \tag{5.4.61}$$

$$\max(D[x]^G) = |[x]^G| = C^{-1}w =: x^* \geq 0 \tag{5.4.62}$$

Proof. As in the proof of Theorem 5.4.34 one can show that $C^{(k)}$, $k = 1, \ldots, n$, are M-matrices. By inspecting the formulae for the interval Gaussian algorithm one easily sees that all the vectors $w^{(k)}$ are nonnegative.

First we prove (5.4.58) and (5.4.59). For $k = 1$ these formulae are trivial. Let them hold for some $k < n$ and choose $i, j > k$. Then $\underline{A}^{(k)} = C^{(k)}$ and $\langle [A]^{(k)} \rangle = C^{(k)}$ by assumption, and

$$\underline{a}_{ij}^{(k+1)} = \underline{a}_{ij}^{(k)} - \max\left(\frac{[a]_{ik}^{(k)} [a]_{kj}^{(k)}}{[a]_{kk}^{(k)}} \right) = c_{ij}^{(k)} - \frac{|[a]_{ik}^{(k)}|\,|[a]_{kj}^{(k)}|}{\langle [a]_{kk}^{(k)} \rangle}$$

$$= c_{ij}^{(k)} - \frac{c_{ik}^{(k)} c_{kj}^{(k)}}{c_{kk}^{(k)}} = c_{ij}^{(k+1)}.$$

Hence $\underline{A}^{(k+1)} = C^{(k+1)}$. Since $\check{A}^{(k)} = I$ the same holds for $[A]^{(k+1)}$ as can be seen from Theorem 2.4.4. Therefore, $\mathrm{rad}([A]^{(k+1)}) = I - C^{(k+1)}$ which yields (5.4.58) for $k+1$. From $[a]_{ij}^{(k+1)} = [c_{ij}^{(k+1)}, -c_{ij}^{(k+1)}]$ for $i \neq j$, and from $\langle [a]_{ii}^{(k+1)} \rangle = c_{ii}^{(k+1)}$ we get (5.4.59).

Now we prove (5.4.60) and (5.4.61) which certainly are trivial if $k = 1$. Let them hold for some $k < n$. Then $\max(D[b]^{(k)}) = w^{(k)} = |D[b]^{(k)}| = |[b]^{(k)}|$ by assumption and by $|D| = I$. This implies

$$\max(d_i[b]_i^{(k+1)}) = \max(d_i[b]_i^{(k)}) - \min\left(d_i \frac{[a]_{ik}^{(k)}}{[a]_{kk}^{(k)}} [b]_k^{(k)} \right)$$

$$= w_i^{(k)} - \min\left(\frac{[a]_{ik}^{(k)}}{\langle [a]_{kk}^{(k)} \rangle} |[b]_k^{(k)}| \right)$$

$$= w_i^{(k)} - \min\left(\frac{[a]_{ik}^{(k)}}{\langle [a]_{kk}^{(k)} \rangle} w_k^{(k)} \right) = w_i^{(k)} + \frac{|[a]_{ik}^{(k)}|}{\langle [a]_{kk}^{(k)} \rangle} w_k^{(k)}$$

$$= w_i^{(k)} - \frac{c_{ik}^{(k)}}{c_{kk}^{(k)}} w_k^{(k)} = w_i^{(k+1)},$$

where we exploited the symmetry of $[a]_{ik}^{(k)}$ and (5.4.58). With Theorem 2.4.4 one sees that $\mathrm{mid}(d_i[b]_i^{(k+1)}) = \mathrm{mid}(d_i[b]_i^{(k)}) = |\check{b}_i|$ whence $\mathrm{rad}([b]_i^{(k+1)}) = \mathrm{rad}(d_i[b]_i^{(k+1)}) = w_i^{(k+1)} - |\check{b}_i|$. This yields (5.4.60) and (5.4.61) for $k+1$.

Now we address (5.4.62). For $k = n$ we get

$$\max(d_n[x]_n^G) = |d_n[b]_n^{(n)}| \,/\, \langle [a]_{nn}^{(n)} \rangle = |[b]_n^{(n)}| \,/\, \langle [a]_{nn}^{(n)} \rangle,$$

which equals $|[x]_n^G|$ as well as $w_n^{(n)}/c_{nn}^{(n)} = x_n^*$. Here we used (5.4.59) and (5.4.61). Assume now that $\max(d_j[x]_j^G) = |[x]_j^G| = x_j^*$ holds for $j = n, n-1, \ldots, i+1$. Then

$$\max(d_i[x]_i^G) = \max\left(\left(d_i[b]_i^{(n)} - \sum_{j=i+1}^{n} d_i[a]_{ij}^{(n)} [x]_j^G \right) / [a]_{ii}^{(n)} \right)$$

$$= \max\left(\left(d_i[b]_i^{(n)} - \sum_{j=i+1}^{n} [a]_{ij}^{(n)} |[x]_j^G| \right) / [a]_{ii}^{(n)} \right)$$

$$= \left(|d_i [b]_i^{(n)}| + \sum_{j=i+1}^{n} |[a]_{ij}^{(n)}|\, |[x]_j^G| \right) / \langle [a]_{ii}^{(n)} \rangle$$

$$= \left(w_i^{(n)} - \sum_{j=i+1}^{n} c_{ij}^{(n)} x_j^* \right) / c_{ii}^{(n)} = x_i^*,$$

where we exploited the symmetry of $[a]_{ij}^{(n)}$. □

Based on Lemma 5.4.35 we will now show that it is sufficient to solve the single real system $Cx = w$ in order to compute $\mathrm{IGA}([A], [b])$ for matrices of the form $[A] = I + [-R, R]$.

Theorem 5.4.36. *Let* $[A] = I + [-R, R] \in \mathbb{IR}^{n\times n}$, $R \geq O$, $\rho(R) < 1$ *and let* $[b] \in \mathbb{IR}^n$, $C := I - R$, $w := |\check{b}| + \mathrm{rad}([b])$, $x^* := C^{-1}w$. *With* $C^{(n)}$ *from the Gaussian algorithm define* $f_i := 1/c_{ii}^{(n)}$, $i = 1, \ldots, n$. *Then for each* $i \in \{1, \ldots, n\}$ *we have*

$$\underline{x}_i^G = \min\{\underset{\sim}{x}_i, \mu_i \underset{\sim}{x}_i\}, \quad \overline{x}_i^G = \max\{\tilde{x}_i, \mu_i \tilde{x}_i\}, \tag{5.4.63}$$

where

$$\underset{\sim}{x}_i := -x_i^* + f_i(\check{b} + |\check{b}|)_i, \quad \tilde{x}_i := x_i^* + f_i(\check{b} - |\check{b}|)_i, \tag{5.4.64}$$

and

$$\mu_i := \frac{1}{2f_i - 1} \in (0, 1].$$

Proof. From (5.4.58) we get $c_{ii}^{(n)} \leq 2 - c_{ii}^{(n)}$. Hence $c_{ii}^{(n)} \leq 1$, $2f_i - 1 \geq 1 > 0$ and $0 < \mu_i \leq 1$.

Let

$$[z]_i := [b]_i^{(n)} - \sum_{j=i+1}^{n} [a]_{ij}^{(n)} [x]_j^G.$$

With Lemma 5.4.35 we obtain

$$[z]_i = [b]_i^{(n)} - \sum_{j=i+1}^{n} [a]_{ij}^{(n)}\, |[x]_j^G| = [b]_i^{(n)} - \sum_{j=i+1}^{n} [a]_{ij}^{(n)} x_j^*$$

and

$$\overline{z}_i = \overline{b}_i^{(n)} - \sum_{j=i+1}^{n} c_{ij}^{(n)} x_j^* = \check{b}_i - |\check{b}_i| + w_i^{(n)} - \sum_{j=i+1}^{n} c_{ij}^{(n)} x_j^* = \check{b}_i - |\check{b}_i| + c_{ii}^{(n)} x_i^* = c_{ii}^{(n)} \tilde{x}_i.$$

In particular, $\mathrm{sign}(\overline{z}_i) = \mathrm{sign}(\tilde{x}_i)$.

If $\overline{z}_i \geq 0$, then $\tilde{x}_i \geq 0$ and

$$\overline{x}_i^G = \frac{\overline{z}_i}{\underline{a}_{ii}^{(n)}} = \frac{\overline{z}_i}{c_{ii}^{(n)}} = f_i(\check{b}_i - |\check{b}_i|) + x_i^* = \tilde{x}_i \geq \mu_i \tilde{x}_i.$$

If $\overline{z}_i < 0$, then $\tilde{x}_i < 0$ and by (5.4.58) we get

$$\overline{x}_i^G = \frac{\overline{z}_i}{\overline{a}_{ii}^{(n)}} = \frac{\overline{z}_i}{2 - c_{ii}^{(n)}} = \mu_i \frac{\overline{z}_i}{c_{ii}^{(n)}} = \mu_i \tilde{x}_i \geq \tilde{x}_i .$$

This proves the second equality in (5.4.63).

In order to prove the first one we replace $[b]$ by $-[b]$. Then $[y]^G = \text{IGA}([A], -[b]) = -\text{IGA}([A], [b]) = -[x]^G$. Applying the second equality of (5.4.63) to $\overline{y}_i^G$ yields $\underline{x}_i^G = -\overline{y}_i^G = -\max\{\tilde{y}_i, \mu_i \tilde{y}_i\} = \min\{-\tilde{y}_i, -\mu_i \tilde{y}_i\}$ with $\tilde{y}_i := x_i^* + f_i(-\check{b}_i - |\check{b}_i|) = -\underset{\sim}{x}_i$. □

Theorem 5.4.36 shows a remarkable analogy to Theorem 5.3.16 which provides formulae for the interval hull $[x]^H = \square S$ of the solution set S. Notice that x^* is the same vector in both theorems since $M = C^{-1}$. By means of these theorems it is easily seen that $[x]_i^G = [x]_i^H$ if $f_i = m_{ii}$. The following result is essentially based on this fact. It shows that at least one bound of each component $[x]_i^G$ coincides with the corresponding bound of $[x]_i^H$.

Theorem 5.4.37. *The assumptions of Theorem 5.4.36 imply*

$$[x]_n^G = [x]_n^H . \tag{5.4.65}$$

In addition, for any $i \in \{1, \dots, n\}$ we get

$$\overline{x}_i^G = \overline{x}_i^H = x_i^* \textit{ if } \check{b}_i \geq 0, \quad \textit{and} \quad \underline{x}_i^G = \underline{x}_i^H = -x_i^* \textit{ if } \check{b}_i \leq 0.$$

In particular, with x^* from Theorem 5.4.36 the equality $[x]_i^G = [x]_i^H = [-x_i^*, x_i^*]$ holds if $\check{b}_i = 0$.

Proof. With the notation of the two Theorems 5.3.16 and 5.4.36 the last column of M can be written as $y = C^{-1}e^{(n)}$. By Cramer's rule $y_n = \det(C')/\det(C)$ with C' being the $(n-1) \times (n-1)$ leading principal submatrix of C. Since $\det(C) = c_{11}^{(n)} \cdot c_{22}^{(n)} \cdots c_{nn}^{(n)} = \det(C') \cdot c_{nn}^{(n)}$ one gets $f_{nn} = m_{nn}$. Therefore, (5.4.65) holds. The remaining part of the theorem follows directly from the Theorems 5.3.16 and 5.4.36 with $\tilde{x}_i^H = x_i^* = \tilde{x}_i \geq 0$ if $\check{b}_i \geq 0$ and $\underset{\sim}{x}_i^H = -x_i^* = \underset{\sim}{x}_i \leq 0$ if $\check{b}_i \leq 0$ which implies $\overline{x}_i^G = \overline{x}_i^H = x_i^*$ if $\check{b}_i \geq 0$ and $\underline{x}_i^G = \underline{x}_i^H = -x_i^*$ if $\check{b}_i \leq 0$. □

Unfortunately, $m_{ii} = f_i$, $i = 1, \dots, n-1$, does not hold, in general. Hence $[x]^G \neq [x]^H$, i.e., the enclosure of S by $[x]^G$ is not optimal. This is even true in the 2×2 case and, therefore, for tridiagonal matrices, as the following example shows.

Example 5.4.38. Let

$$[A] = \begin{pmatrix} 1 & [-1, 1] \\ [-\frac{1}{2}, \frac{1}{2}] & 1 \end{pmatrix}, \quad [b] = \begin{pmatrix} -1 \\ 1 \end{pmatrix} = \check{b}.$$

Then

$$C = \begin{pmatrix} 1 & -1 \\ -\frac{1}{2} & 1 \end{pmatrix}, \quad C^{(n)} = \begin{pmatrix} 1 & -1 \\ 0 & \frac{1}{2} \end{pmatrix},$$

$$M = C^{-1} = \begin{pmatrix} 2 & 2 \\ 1 & 2 \end{pmatrix}, \quad x^* = M|\check{b}| = \begin{pmatrix} 4 \\ 3 \end{pmatrix},$$

$$[x]^G = \begin{pmatrix} [-4, 2] \\ [\frac{1}{3}, 3] \end{pmatrix} \neq [x]^H = \begin{pmatrix} [-4, 0] \\ [\frac{1}{3}, 3] \end{pmatrix}.$$

In order to complete Theorem 5.4.37 we need a result on the matrix $C = I - R$.

Lemma 5.4.39. *Let $C = I - R \in \mathbb{R}^{n\times n}$, $R \geq O$, $\rho(R) < 1$, and let $M = C^{-1}$. Denote by $C^{(k)}$ the matrices of the Gaussian algorithm. Then the following assertions hold.*

(a) *For arbitrary $i, j \in \{1, \ldots, n\}$ we have $m_{ij} \neq 0$ if and only if i is connected to j in the graph $G(C)$.*

(b) *Let (L, U) be the triangular decomposition of C. Then $c_{ij}^{(k)} \neq 0$ for $i \neq j$ and $k \leq m := \min\{i, j\}$ if and only if i is connected to j in the graph $G(C)$ such that all intermediate nodes i_s in the corresponding path satisfy $i_s < \min\{i, j, k\}$.*
In particular, $l_{ij} \neq 0$ for $i > j$ if and only if i is connected to j in $G(C)$ such that all intermediate nodes i_s in the corresponding path satisfy $i_s < j$.

Proof. (a) From $M = (I - R)^{-1} = \sum_{p=0}^{\infty} R^p$ and from the nonnegativity of R we get $m_{ii} \geq 1$. Since C is an M-matrix it also has positive diagonal entries. Therefore, the assertion is true for $i = j$. Assume now $i \neq j$. Then $m_{ij} \neq 0$ if and only if $(R^p)_{ij} > 0$ for at least one $p \in \mathbb{N}$. For $p > 1$ this holds if and only if in the representation

$$(R^p)_{ij} = \sum_{i_1=1}^{n} \sum_{i_2=1}^{n} \cdots \sum_{i_{p-1}=1}^{n} r_{i,i_1} \cdot r_{i_1,i_2} \cdots r_{i_{p-1},j}$$

at least one summand differs from zero. The corresponding indices determine the path required in the assertion, and vice versa. For $p = 1$ the path is $i \to j$.

(b) '⇒': Let $c_{ij}^{(k)} \neq 0$ hold with $k \leq m$. Then in $G(C^{(k)})$ there exists the path $i \to j$. If $m = 1$, this implies the assertion. If $m > 1$, we see from the formula

$$c_{ij}^{(k)} = c_{ij}^{(k-1)} - \frac{c_{i,k-1}^{(k-1)} c_{k-1,j}^{(k-1)}}{c_{k-1,k-1}^{(k-1)}} \tag{5.4.66}$$

and from the signs of the entries of the M-matrix $C^{(k-1)}$ that $c_{ij}^{(k)} \neq 0$ if and only if $c_{ij}^{(k-1)} \neq 0$ or both $c_{i,k-1}^{(k-1)}, c_{k-1,j}^{(k-1)} \neq 0$. Therefore, in $G(C^{(k-1)})$ we obtain the path $i \to j$ or $i \to (k-1) \to j$. Repeat the arguments while the upper index goes down to 1. This yields the path in $G(C^{(1)}) = G(C)$ as it was asserted.

'⇐': Let the path of the assertion exist. Without loss of generality assume that it contains none of the nodes twice. (Otherwise cut off the piece between the two equal nodes including one of them.) We proceed by induction on the length p of the path. If

$p = 1$, then the path reads $i \to j$, hence $c_{ij} = c_{ij}^{(1)} \neq 0$. Using the arguments following (5.4.66) we obtain $c_{ij}^{(k)} \neq 0$ for all $k \leq m$. Assume now that the assertion is true for all pairs of indices for which the corresponding path in $G(C)$ has a length which is less than or equal to some p. Let i, j be connected by a path of length $p+1$ and let i_s be the largest node among all intermediate nodes of this path. Then by the hypothesis of the induction $c_{i,i_s}^{(i_s)} \neq 0$ and $c_{i_s,j}^{(i_s)} \neq 0$ whence $c_{ij}^{(i_s+1)} \neq 0$ by (5.4.66). As above, we get $c_{ij}^{(k)} \neq 0$ for all k with $i_{s+1} \leq k \leq m$.

The assertion for l_{ij} follows immediately, since $l_{ij} = c_{ij}^{(j)}/c_{jj}^{(j)} \neq 0$ if and only if $c_{ij}^{(j)} \neq 0$. □

Theorem 5.4.40. *The assumptions of Theorem 5.4.36 imply*

$$[x]_i^G \neq [x]_i^H \tag{5.4.67}$$

if and only if the following two properties (i), (ii) *or, equivalently* (i), (iii) *hold.*

(i) $\check{b}_i \neq 0$.

(ii) *Using the matrices L from the LU decomposition of $C = I - R$ and $M = (I - R)^{-1}$ there is an index $k > i$ such that $m_{ik} \neq 0$ and $l_{ki} \neq 0$.*

(iii) *There is an index $k > i$ such that the node i is connected to k in the graph $G(C)$, and, vice versa, the node k is connected to i, where in this latter case the intermediate nodes i_j of the path have to satisfy $i_j < i$.*

Proof. We use again the notations from the Theorems 5.3.16 and 5.4.36. By these theorems and Theorem 5.4.37 the inequality (5.4.67) is equivalent to (i) and $m_{ii} \neq f_i$. In order to derive an equivalent condition for $m_{ii} - f_i = (M - (C^{(n)})^{-1})_{ii} \neq 0$ let $C = LU = LC^{(n)}$ be the LU decomposition of C. Then $(C^{(n)})^{-1} = C^{-1}L = ML$ implies

$$M - (C^{(n)})^{-1} = M(I - L) \geq O. \tag{5.4.68}$$

Hence $(M - (C^{(n)})^{-1})_{ii} \neq 0$ if and only if there is an index $k > i$ such that $m_{ik} \neq 0$ and $l_{ki} \neq 0$. Here, we exploited $M \geq O$, $I - L \geq O$; $k > i$ is required since $I - L$ is a strictly lower triangular matrix. This proves the equivalence of (5.4.67) with (i), (ii). Since by Lemma 5.4.39 (ii) is equivalent to (iii) the theorem is proved. □

Theorem 5.4.40 shows that $[x]^H = [x]^G$ is true for 2×2 matrices of the form $[A] = I + [-R, R]$ if and only if $\check{b}_1 = 0$ or $[a]_{12} = 0$ or $[a]_{21} = 0$. This is equivalent to '$\check{b}_1 = 0$ or C is reducible'. In particular, if $\check{b}_1 \neq 0$ and if C is irreducible, then $[x]_1^G \neq [x]_1^H$ holds. This can be generalized to $n \times n$ matrices.

Corollary 5.4.41. *Let the assumptions of Theorem 5.4.36 hold with $n > 1$, and let C be irreducible. Then for an arbitrary $i < n$*

$$[x]_i^G = [x]_i^H \text{ if and only if } \check{b}_i = 0. \tag{5.4.69}$$

Proof. Since C is irreducible and $i < n$ there are two paths in the graph $G(C)$ which connect i to $i+1$ and $i+1$ to i, respectively. Concatenate these paths to end up with

$i \to \cdots \to i+1 \to \cdots \to i$. Trace back this path starting with the right node i until you find the first node, say i_0, which is larger than i. (At latest, $i+1$ is such a node.) Then (iii) of Theorem 5.4.40 is fulfilled with $k = i_0$, i.e., (iii) always holds if C is irreducible. Thus (5.4.67) is equivalent to (i), which proves (5.4.69). □

Corollary 5.4.41 shows that for matrices of the form $[A] = I + [-R, R]$ the interval Gaussian algorithm often overestimates the interval hull of S, a phenomenon which is well known in the literature. (Cf. Neumaier [257], p. 160ff; Neumaier [258], Rohn [304], Rump [317]; or Wongwises [364], e.g.)

The interval Gaussian algorithm – pivoting

In classical numerical analysis pivoting is an essential feature when applying the Gaussian algorithm. We will do so similarly for the interval version of this algorithm. To this end we want to select the pivot so that the absolute value and the radius of the entries $[a]_{ij}^{(k+1)}$ do not increase too much when compared with the corresponding quantities for $[a]_{ij}^{(k)}$. The inequalities

$$|[a]_{ij}^{(k+1)}| \le |[a]_{ij}^{(k)}| + \frac{|[a]_{ik}^{(k)}[a]_{kj}^{(k)}|}{\langle [a]_{kk}^{(k)} \rangle}$$

and

$$\begin{aligned}\operatorname{rad}([a]_{ij}^{(k+1)}) &\le \operatorname{rad}([a]_{ij}^{(k)})\\ &\quad + \frac{1}{\langle [a]_{kk}^{(k)} \rangle} \operatorname{rad}([a]_{ik}^{(k)}[a]_{kj}^{(k)}) + \operatorname{rad}\left(\frac{1}{[a]_{kk}^{(k)}}\right) |\operatorname{mid}([a]_{ik}^{(k)}[a]_{kj}^{(k)})|\end{aligned}$$

suggest making $1/\langle [a]_{kk}^{(k)} \rangle$ and $\operatorname{rad}(1/[a]_{kk}^{(k)})$ small simultaneously. Since for intervals $[a]$ with $0 \notin [a]$ Theorem 2.4.5 implies

$$\begin{aligned}\operatorname{rad}\left(\frac{1}{[a]}\right) &= \frac{\operatorname{rad}([a])}{\langle [a] \rangle |[a]|} = \frac{\operatorname{rad}([a])}{\langle [a] \rangle (\langle [a] \rangle + 2\operatorname{rad}([a]))}\\ &= \frac{1}{2\langle [a] \rangle}(1 - \frac{\langle [a] \rangle}{\langle [a] \rangle + 2\operatorname{rad}([a])})\end{aligned}$$

the radius of $1/[a]$ decreases with $\operatorname{rad}([a])$ if $\langle [a] \rangle$ is fixed, and it does the same if $\langle [a] \rangle$ increases while $\operatorname{rad}([a])$ is fixed. This motivates the following pivot selection:

Choose $[a]_{i_0,k}^{(k)}$ as pivot such that the quotient

$$q([a]_{ik}^{(k)}) = \frac{1 + \operatorname{rad}([a]_{ik}^{(k)})}{\langle [a]_{ik}^{(k)} \rangle}, \quad i \ge k,\ 0 \notin [a]_{ik}^{(k)}, \tag{5.4.70}$$

is minimized.

Here, the one in the numerator is introduced in order to have an additional choice among degenerate entries.

Hebgen suggested in [138] merely to minimize

$$\frac{\operatorname{rad}([a]_{ik}^{(k)})}{\langle [a]_{ik}^{(k)} \rangle}, \quad i \ge k,\ 0 \notin [a]_{ik}^{(k)}, \tag{5.4.71}$$

supplemented by an additional selection for degenerate entries. This criterion is invariant against scaling, i.e., by multiplying rows by real numbers ($\neq 0$). According to (5.4.71) intervals with large radius and large mignitude are thus as good as intervals with small radius and small mignitude while in (5.4.70) the mignitude is slightly overemphasized. Note however, that in both situations it cannot be guaranteed that the selection yields the smallest width for $[a]_{ij}^{(k+1)}$. If, e.g.,

$$[A] = \begin{pmatrix} [1,3] & [-1,1] & [-1,1] \\ [2,12] & [-1,1] & [-1,1] \\ [-1,1] & [-1,1] & 10 \end{pmatrix},$$

then $q([1,3]) = 2 < q([2,12]) = 3$, whence

$$[a]_{33}^{(2)} = 10 - \begin{cases} [-1,1], & \text{if } [1,3] \text{ is the pivot,} \\ [-1/2, 1/2], & \text{if } [2,12] \text{ is the pivot,} \end{cases}$$

so that a selection according to (5.4.70) or (5.4.71) is certainly not the best.

The interval Cholesky method – basics

The Cholesky method of classical numerics was tailored to solve a linear system of equations with a symmetric and positive definite matrix A. It approximately halves the amount of work needed for the Gaussian algorithm and produces a unique lower triangular matrix L^C with positive diagonal entries such that $A = L^C(L^C)^T$ holds. By the following theorem and its proof we can see that A has a triangular decomposition (L, U) and how L, U are related to L^C.

Theorem 5.4.42. *Let $A = A^T \in \mathbb{R}^{n\times n}$ and denote by A_k the $k \times k$ leading principal submatrix of A. Then the following statements are equivalent.*

(a) *A is positive definite.*

(b) $\det A_k > 0$, $k = 1, \dots, n$.

(c) *There is a unique lower triangular matrix L^C with positive diagonal entries such that $L^C(L^C)^T$ holds.*

Proof. '(a) ⇒ (b)': By virtue of Theorem 1.8.11 all eigenvalues of A are positive, hence $\det A > 0$ holds by Theorem 1.8.1. For $x' = (x_1, \dots, x_k)^T \in \mathbb{R}^k \setminus \{0\}$ and

$x = (x'^T, 0, \dots, 0)^T \in \mathbb{R}^n$ we have $0 < x^T A x = x'^T A_k x$, hence A_k is also symmetric and positive definite, whence $\det A_k > 0$.

'(b) ⇒ (c)': According to Theorem 5.4.3 and (b) the matrix A has a triangular decomposition $A = LU$. Denote by L_k, U_k the $k \times k$ leading principal submatrices of L, U. Then $A_k = L_k U_k$ and $0 < \det A_k = \det L_k \det U_k = 1 \cdot \det U_k = \prod_{l=1}^k a_{ll}^{(l)}$. Therefore, $a_{11}^{(1)} = \det A_1 > 0$ and $a_{kk}^{(k)} = \det A_k / \det A_{k-1} > 0$, $k = 2, \dots, n$, so that

$$D = \operatorname{diag}(\sqrt{a_{11}^{(1)}}, \dots, \sqrt{a_{nn}^{(n)}})$$

exists. With $\hat{U} = D^{-2}U$ we obtain $A = LD^2\hat{U} = A^T = \hat{U}^T D^2 L^T$ whence by the uniqueness of the triangular decomposition of A we must have $L = \hat{U}^T$. With $L^C = LD$ we then get (c). Uniqueness can be derived either directly as in Theorem 1.8.13 or from the uniqueness of the triangular decomposition of A.

'(c) ⇒ (a)' follows as in Theorem 1.8.13. □

The equivalence of (a) and (c) in the preceding theorem was already stated as Theorem 1.8.13. It was proved there by different means. Notice that the properties (b) and (c) of Theorem 5.4.42 are *computable* criteria for positive definiteness if rounding errors are avoided.

The interval Cholesky method for $[A] = [A]^T \in \mathbb{IR}^{n\times n}$, $[b] \in \mathbb{IR}^n$ follows the definition of the classical Cholesky method which is organized in three steps like the Gaussian algorithm:

Step 1: Compute L^C. ($L^C(L^C)^T$ decomposition)
Step 2: Solve $L^C y = b$. (Forward substitution)
Step 3: Solve $(L^C)^T x = y$. (Backward substitution)

For intervals these three steps can be realized as follows:

$$[l]_{ij}^C = \begin{cases} \left([a]_{jj} - \sum_{k=1}^{j-1}([l]_{jk}^C)^2\right)^{1/2} & \text{if } i = j \\ \left([a]_{ij} - \sum_{k=1}^{j-1}[l]_{ik}^C[l]_{jk}^C\right) / [l]_{jj}^C, & \text{if } i = j+1, \dots, n, \end{cases} \qquad j = 1, \dots, n;$$

$$[y]_i = \left([b]_i - \sum_{j=1}^{i-1}[l]_{ij}^C[y]_j\right) / [l]_{ii}^C, \quad i = 1, \dots, n;$$

$$[x]_i^C = \left([y]_i - \sum_{j=i+1}^{n}[l]_{ji}^C[x]_j^C\right) / [l]_{ii}^C, \quad i = n, n-1, \dots, 1. \tag{5.4.72}$$

Here, according to Definition 2.6.1 we use

$$[a]^2 = \{a^2 \mid a \in [a]\} \subseteq [a] \cdot [a] \quad \text{with equality if and only if } 0 \notin \operatorname{int}([a]), \tag{5.4.73}$$

and

$$[a]^{1/2} = \sqrt{[a]} = \{\sqrt{a} \mid a \in [a]\}$$

for intervals $[a]$, where we assume $\underline{a} \ge 0$ in the latter case.

The construction of $[x]^C$ is called the interval Cholesky method. The vector $[x]^C$ exists if and only if the radicands of $[l]_{ii}^C$ have a positive lower bound for $i = 1, \dots, n$. In this case we say that the Cholesky method is feasible. We also use the notation $[x]^C = \mathrm{ICh}([A], [b])$.

Many properties of this algorithm are analogous to the interval Gaussian algorithm. Therefore, we can proceed faster than there. We start with a representation of $[x]^C$ as a product. For this purpose we use the quantities $[l]_{ij}^C$ from the Cholesky method and define the diagonal matrices $[D]_C^{(k)}$, $k = 1, \dots, n$, and the lower triangular matrices $[L]_C^{(k)}$, $k = 1, \dots, n-1$, by

$$
\begin{aligned}
([D]_C^{(k)})_{ij} &:= \begin{cases} 1 & \text{if } i = j \neq k, \\ 1/[l]_{kk}^C & \text{if } i = j = k, \\ 0 & \text{otherwise,} \end{cases} \\
([L]_C^{(k)})_{ij} &:= \begin{cases} 1 & \text{if } i = j, \\ -[l]_{ik}^C & \text{if } j = k < i, \\ 0 & \text{otherwise.} \end{cases}
\end{aligned}
\tag{5.4.74}
$$

Then the last two steps in (5.4.72) yield

$$
\begin{aligned}
[y] &= [D]_C^{(n)}([L]_C^{(n-1)}([D]_C^{(n-1)}(\dots([L]_C^{(1)}([D]_C^{(1)}[b]))\dots))), \\
[x]^C &= [D]_C^{(1)}(([L]_C^{(1)})^T(\dots(([L]^{(n-1)})^T([D]_C^{(n)}[y]))\dots)).
\end{aligned}
\tag{5.4.75}
$$

Let $[L]^C$ be the lower triangular matrix with the entries $[l]_{ij}^C$, $i \geq j$, from the interval Cholesky method. We leave it to the reader to show that

$$|[D]_C^{(n)}| \cdot |[L]_C^{(n-1)}| \cdot |[D]_C^{(n-1)}| \cdots |[L]_C^{(1)}| \cdot |[D]_C^{(1)}| = \langle [L]_C \rangle^{-1} \tag{5.4.76}$$

holds. Analogously to $|[A]^G|$ we introduce $|[A]^C|$ using (5.4.76) for shortness:

$$|[A]^C| := \langle ([L]^C)^T \rangle^{-1} \langle [L]^C \rangle^{-1}. \tag{5.4.77}$$

It has a similar interpretation as $|[A]^G|$.

By virtue of the formulae (5.4.72) or the representation (5.4.75) we get at once the following properties of which (b) with $S_{\mathrm{sym}} \subseteq [x]^C$ is most remarkable.

Theorem 5.4.43. *Let $[A] = [A]^T \in \mathbb{IR}^{n\times n}$, $[b], [c] \in \mathbb{IR}^n$ and let $D \in \mathbb{R}^{n\times n}$ be a diagonal matrix with positive diagonal entries.*

(a) *If $\mathrm{ICh}([A], [b])$ exists and $[\hat{A}] = [\hat{A}]^T \subseteq [A]$, $[\hat{b}] \subseteq [b]$, then $\mathrm{ICh}([\hat{A}], [\hat{b}])$ exists and is contained in $\mathrm{ICh}([A], [b])$.* *(Subset property)*

(b) *If $\mathrm{ICh}([A], [b])$ exists, then the Cholesky method is feasible for each pair $(A, b) \in ([A], [b])$ with $A = A^T$. In particular, the symmetric solution set S_{sym} of $[A]x = [b]$ is enclosed by $\mathrm{ICh}([A], [b])$.* *(Enclosure property)*

(c) *The existence of $\mathrm{ICh}([A], [b])$ does not depend on $[b]$.*

(d) $\mathrm{ICh}([A], \alpha[b]) = \alpha\,\mathrm{ICh}([A], [b]), \quad \alpha \in \mathbb{R}.$ *(Homogeneity with respect to $[b]$)*
(e) $\mathrm{ICh}([A], [b] + [c]) \subseteq \mathrm{ICh}([A], [b]) + \mathrm{ICh}([A], [c]).$ *(Subadditivity)*
(f) *If* $\mathrm{ICh}([A], [b])$ *exists, then* $\mathrm{ICh}(D[A]D, [b])$ *exists and satisfies*

$$\mathrm{ICh}(D[A]D, D[b]) = D^{-1}\,\mathrm{ICh}([A], [b]).$$

(Scaling)

(g) *If* $[A] \equiv A \in \mathbb{R}^{n\times n}$, *then* $\mathrm{ICh}([A], [b]) = \mathrm{ICh}(A, [b]) \supseteq A^{-1}[b]$.
(h) *If* $[b] \equiv b \in \mathbb{R}^n$, *then* $\mathrm{ICh}([A], [b]) = \mathrm{ICh}([A], b) \subseteq \mathrm{ICh}([A]) \cdot b$.

Before we give an alternative access to $[x]^C$ by a recursive definition using a Schur complement, we illustrate some of the phenomena of the interval Cholesky method by means of a simple example.

Example 5.4.44. Let

$$[A] = \begin{pmatrix} 4 & [-1, 1] \\ [-1, 1] & 4 \end{pmatrix}, \quad [b] = \begin{pmatrix} 6 \\ 6 \end{pmatrix}.$$

With $A = \begin{pmatrix} 4 & \alpha \\ \beta & 4 \end{pmatrix}$ for $A \in [A]$ we get

$$A^{-1}b = \frac{6}{16 - \alpha\beta}\begin{pmatrix} 4 - \alpha \\ 4 - \beta \end{pmatrix}, \quad \text{where } \alpha, \beta \in [-1, 1].$$

If $A = A^T \in [A]$, then $\beta = \alpha$ yields

$$A^{-1}b = \frac{6}{4 + \alpha}\begin{pmatrix} 1 \\ 1 \end{pmatrix}.$$

Thus

$$\Box S_{\mathrm{sym}} = \left(\left[\frac{18}{15}, 2\right], \left[\frac{18}{15}, 2\right]\right)^T, \qquad \Box S = \left(\left[\frac{18}{17}, 2\right], \left[\frac{18}{17}, 2\right]\right)^T$$
$$[x]^C = \left([1, 2], \left[\frac{18}{16}, 2\right]\right)^T, \qquad [x]^G = \left([1, 2], \left[\frac{18}{17}, 2\right]\right)^T$$

where $[x]^G$ denotes the vector resulting from the interval Gaussian algorithm. The sets

$$S_{\mathrm{sym}} = \left\{\frac{6}{4 + \alpha}\begin{pmatrix} 1 \\ 1 \end{pmatrix} \,\middle|\, -1 \le \alpha \le 1\right\} = \left\{\gamma \cdot \begin{pmatrix} 1 \\ 1 \end{pmatrix} \,\middle|\, \frac{6}{5} \le \gamma \le 2\right\}$$

and

$$S = \mathrm{conv}\left\{\left(\frac{6}{5}, \frac{6}{5}\right)^T, (2, 2)^T, \left(\frac{18}{17}, \frac{30}{17}\right)^T, \left(\frac{30}{17}, \frac{18}{17}\right)^T\right\}$$

can be seen in Figure 5.4.1.

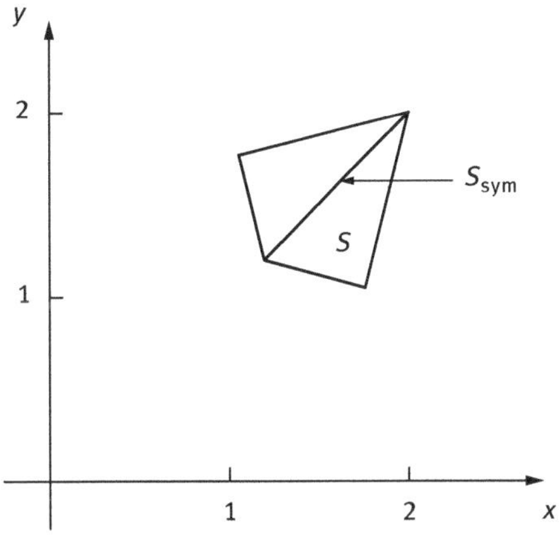

Fig. 5.4.1: The sets S and S_{sym}.

Example 5.4.44 shows that the following properties can occur.
(i) $\square S_{\mathrm{sym}} \neq \square S$ (cf. also Neumaier [257]),
(ii) $\square S_{\mathrm{sym}} \neq [x]^C$,
(iii) $\square S \neq [x]^G$,
(iv) $S \not\subseteq [x]^C$ (but $S_{\mathrm{sym}} \subseteq [x]^C$; cf. Theorem 5.4.43),
(v) $[x]^C \subseteq [x]^G$ with $[x]^C \neq [x]^G$.

Our next example illustrates another possible property.
(vi) $[x]^G \subseteq [x]^C$ with $[x]^G \neq [x]^C$.

So unfortunately neither $[x]^C \subseteq [x]^G$ nor the converse can be guaranteed in general.

Example 5.4.45. Let

$$[A] = \begin{pmatrix} [1,4] & [0,1] \\ [0,1] & 3 \end{pmatrix}, \quad [b] = \begin{pmatrix} 2 \\ [0,2] \end{pmatrix}.$$

Then $[x]^G = ([0.25, 3], [-1, 1])^T \subset [x]^C = ([0, 3], [-1, 1])^T$.

The reasons why the two examples above work, is best seen by expressing $[x]^C$ and $[x]^G$ in terms of the input data. One obtains

$$[x]_2^C = \frac{1}{[a]_{22} - \frac{[a]_{12}^2}{[a]_{11}}} \cdot \left\{ [b]_2 - \frac{[a]_{12}}{[a]_{11}} [b]_1 \right\}$$

$$[x]_1^C = \frac{1}{\sqrt{[a]_{11}}} \cdot \left\{ \frac{[b]_1}{\sqrt{[a]_{11}}} - \frac{[a]_{12}}{\sqrt{[a]_{11}}} [x]_2^C \right\}$$

$$[x]_2^G = \frac{1}{[a]_{22} - \frac{[a]_{12} \cdot [a]_{12}}{[a]_{11}}} \cdot \left\{ [b]_2 - \frac{[a]_{12}}{[a]_{11}} [b]_1 \right\}$$

$$[x]_1^G = \frac{1}{[a]_{11}} \cdot \left\{ [b]_1 - [a]_{12} [x]_2^G \right\}.$$

Hence, by (5.4.73), we always get $[x]_2^C \subseteq [x]_2^G$. If, however, $0 \notin \operatorname{int}([a]_{12})$, then $[x]_2^C = [x]_2^G$, and the subdistributivity causes $[x]_1^G \subseteq [x]_1^C$. Similar phenomena can appear in higher dimensions, too.

Now we present the alternative description of the Cholesky method which we have already announced.

Consider the matrix $[A] = [A]^T \in \mathbb{IR}^{n\times n}$ in the block form

$$[A] = \begin{pmatrix} [a]_{11} & [c]^T \\ [c] & [A]' \end{pmatrix} \quad \text{with } [c] \in \mathbb{IR}^{n-1}.$$

In connection with the Cholesky method we assume that the terms $[c]_i[c]_i$ in the Schur complement $\Sigma([A]) = [A]' - \frac{1}{[a]_{11}}[c][c]^T$ are evaluated as squares according to (5.4.73).

We call $([L]^C, ([L]^C)^T)$ the Cholesky decomposition of $[A] \in \mathbb{IR}^{n\times n}$ if $0 < \underline{a}_{11}$ and if either $n = 1$ and $[L]^C = (\sqrt{[a]_{11}})$ or

$$[L]^C = \begin{pmatrix} \sqrt{[a]_{11}} & 0 \\ [c]/\sqrt{[a]_{11}} & [L]' \end{pmatrix}, \tag{5.4.78}$$

where $([L]', ([L]')^T)$ is the Cholesky decomposition of $\Sigma([A])$.

If this triangular decomposition exists then

$$[x]^C = \mathrm{ICh}(([L]^C)^T, \mathrm{ICh}([L]^C, [b]))$$

which means solving two triangular systems as forward and backward substitution.

Theorem 5.4.46. *The matrix $[L]^C$ in (5.4.78) exists if and only if $[L]^C$ from (5.4.72) exists. In this case, both matrices are identical.*

Proof. We prove the assertion by induction with respect to the number n of rows/columns of $[A]$.

If $n = 1$ the assertion follows from $\sqrt{[a]_{11}}\sqrt{[a]_{11}} = [a]_{11}$ for $0 \le \underline{a}_{11}$.

Let the assertion be true for some n and choose $[A]$ from $\mathbb{IR}^{(n+1)\times(n+1)}$. By ease of argumentation we replace $[L]^C$, $[L]'$ in (5.4.78) by $[M]$, $[M]'$.

Assume first that $[L]^C$ exists, where $[L]^C$ is computed by the interval Cholesky method (5.4.72). We show that $[A]$ has the Cholesky decomposition $([M], [M]^T)$ satisfying $[M] = [L]^C$. Since $[L]^C$ exists, we obtain $\underline{a}_{11} > 0$. Hence $[l]_{i1}^C = [m]_{i1}$ for $i = 1, \ldots, n+1$.

For $j \geq 2$, the formulae in the interval Cholesky method can be reformulated by

$$\begin{cases} [l]^C_{jj} = \left(([a]_{jj} - ([l]^C_{j1})^2) - \sum_{k=2}^{j-1} ([l]^C_{jk})^2 \right)^{1/2} \\ \quad = \left(([a]_{jj} - \frac{[a]^2_{j1}}{\sqrt{[a]_{11}}\sqrt{[a]_{11}}}) - \sum_{k=2}^{j-1} ([l]^C_{jk})^2 \right)^{1/2} \\ \quad = \left(([a]_{jj} - \frac{[a]^2_{j1}}{[a]_{11}}) - \sum_{k=2}^{j-1} ([l]^C_{jk})^2 \right)^{1/2}, \\ [l]^C_{ij} = \left(([a]_{ij} - [l]^C_{i1}[l]^C_{j1}) - \sum_{k=2}^{j-1} [l]^C_{ik}[l]^C_{jk} \right) / [l]^C_{jj} \\ \quad = \left(([a]_{ij} - \frac{[a]_{i1}[a]_{j1}}{[a]_{11}}) - \sum_{k=2}^{j-1} [l]^C_{ik}[l]^C_{jk} \right) / [l]^C_{jj}. \end{cases} \tag{5.4.79}$$

These formulae can be interpreted as (5.4.72) applied to $\Sigma([A]) \in \mathbb{IR}^{n\times n}$ which results in a lower triangular matrix $[L]'$. By the hypothesis made for this induction the matrix $[M]'$ exists and equals $[L]'$. Thus $[M]$ exists and satisfies $[M] = [L]_C$.

Assume now conversely that $[M]$ exists. Then, again, $\underline{a}_{11} > 0$, $[l]^C_{i1} = [m]_{i1}$ for $i = 1, \dots, n+1$, and $[L]' = [M]'$ by the hypothesis and by (5.4.79). This concludes the proof. □

The interval Cholesky method – feasibility

As for the interval Gaussian algorithm, the interval Cholesky method can break down even if it is feasible for all symmetric element matrices contained in $[A] = [A]^T$. This can be seen from the Reichmann matrix $[A] = [A]^T$ in Example 5.4.33 with $x = 2/3$. As was already shown there, the leading principal submatrices of each matrix $\tilde{A} \in [A]$ have positive determinants, hence the symmetric ones are positive definite. According to Theorem 5.4.42 the Cholesky method is feasible for each $\tilde{A}$ but the interval version fails since $[l]^C_{11} = 1$, $[l]^C_{21} = [l]^C_{31} = [0, 2/3]$, $[l]^C_{22} = [\sqrt{5}/3, 1]$, $[l]^C_{32} = [-4/(3\sqrt{5}), 2/\sqrt{5}]$, and $[a]_{33} - ([l]^C_{31})^2 - ([l]^C_{32})^2 = [-11/45, 1]$ contains zero in its interior, i.e., $[l]^C_{33}$ does not exist.

As for the interval Gaussian algorithm, H-matrices are also appropriate for the interval Cholesky method. This is stated in our first criterion.

Theorem 5.4.47. *Let $[A] \in \mathbb{IR}^{n\times n}$ be an H-matrix satisfying $[A] = [A]^T$ and $0 < \underline{a}_{ii}$, $i = 1, \dots, n$. Then $[x]^C$ exists for any $[b] \in \mathbb{IR}^n$, and the matrix $[L]^C$ of the Cholesky decomposition is again an H-matrix.*

Proof. By the assumptions, $\hat{A} := \langle [A] \rangle$ is a Stieltjes matrix, in particular, it is symmetric and positive definite by Theorem 1.10.8. According to Theorem 5.4.42 $\hat{A}$ can be represented as $\hat{A} = \hat{L}\hat{L}^T$ by using the Cholesky method. From the formulae of this method it follows immediately that the triangular matrix $\hat{L}$ is contained in $Z^{n\times n}$ and has positive diagonal entries. Therefore, it is an M-matrix. We show by induction with respect to the column index j that the matrix $[L]^C$ exists and that

$$\hat{L} \le \langle [L]^C \rangle \tag{5.4.80}$$

holds. Then Corollary 1.10.5 guarantees that $[L]^C$ is an H-matrix.

For $j = 1$, $[l]_{11}^C = \sqrt{[a]_{11}}$ exists since we assumed $\underline{a}_{11} > 0$. We get $\langle [l]_{11}^C \rangle = \sqrt{\langle [a]_{11} \rangle} = \hat{l}_{11}$, $[l]_{i1}^C = [a]_{i1}/[l]_{11}^C$. Moreover,

$$|[l]_{i1}^C| = \left| \frac{[a]_{i1}}{[l]_{11}^C} \right| = \frac{|[a]_{i1}|}{\langle [l]_{11}^C \rangle} = -\hat{l}_{i1}, \quad i = 2, \ldots n.$$

Let all columns of $[L]^C$ exist which have an index less than $j > 1$. Assume that (5.4.80) holds for all these columns and define

$$[s] := \sum_{k=1}^{j-1} ([l]_{jk}^C)^2$$

and

$$[t] := [a]_{jj} - [s].$$

Then using $\underline{a}_{jj} > 0$ and the induction hypothesis we obtain

$$\begin{aligned} 0 < \hat{l}_{jj}^2 &= \langle [a]_{jj} \rangle - \sum_{k=1}^{j-1} \hat{l}_{jk}^2 \le \langle [a]_{jj} \rangle - \sum_{k=1}^{j-1} |[l]_{jk}^C|^2 \\ &= \underline{a}_{jj} - \overline{s} = \underline{t} = \langle [t] \rangle = \left\langle [a]_{jj} - \sum_{k=1}^{j-1} ([l]_{jk}^C)^2 \right\rangle. \end{aligned}$$

Hence $0 \notin [a]_{jj} - \sum_{k=1}^{j-1} ([l]_{jk}^C)^2$. Therefore, $[l]_{jj}^C$ exists and satisfies $\langle [l]_{jj}^C \rangle \ge \hat{l}_{jj}$.

For $i > j$ we get

$$\begin{aligned} |[l]_{ij}^C| &\le \left(|[a]_{ij}| + \sum_{k=1}^{j-1} |[l]_{ik}^C|\, |[l]_{jk}^C| \right) / \langle [l]_{jj}^C \rangle \\ &\le \left(|[a]_{ij}| + \sum_{k=1}^{j-1} \hat{l}_{ik} \hat{l}_{jk} \right) / \hat{l}_{jj} = -\hat{l}_{ij}. \end{aligned}$$

This implies $\hat{l}_{ij} \le -|[l]_{ij}^C|$. □

The following corollary is an immediate consequence of Theorem 5.4.47 because the matrices therein are H-matrices.

Corollary 5.4.48. *Let $[A] = [A]^T \in \mathbb{IR}^{n\times n}$ with $0 < \underline{a}_{ii}$, $i = 1, \dots, n$. Then in each of the following cases $[x]^C$ exists.*
(a) *$\langle [A] \rangle$ is strictly diagonally dominant.*
(b) *$\langle [A] \rangle$ is irreducibly diagonally dominant.*
(c) *$\langle [A] \rangle$ is regular and diagonally dominant.*
(d) *$\langle [A] \rangle$ is positive definite.*

Symmetric H-matrices are closely related to positive definite matrices as the following theorem shows.

Theorem 5.4.49. *Let $[A] \in \mathbb{IR}^{n\times n}$ be an H-matrix satisfying $[A] = [A]^T$ and $0 < \underline{a}_{ii}$, $i = 1, \dots, n$. Then each symmetric matrix $\tilde{A} \in [A]$ is positive definite.*

Proof. Since $\langle [A] \rangle$ is an M-matrix, $\langle \tilde{A} \rangle \geq \langle [A] \rangle$ is an M-matrix, too. Because of $\underline{a}_{ii} > 0$, the matrix $\tilde{A}$ has a nonnegative diagonal part D. Split $\tilde{A}$ into $\tilde{A} = D - B$. Then $\langle \tilde{A} \rangle = D - |B| = sI - (sI - D + |B|)$, $s \in \mathbb{R}$. By Theorem 1.10.4, s can be chosen such that

$$s > \rho(sI - D + |B|) \quad \text{and} \quad sI - D + |B| \geq O; \tag{5.4.81}$$

hence $sI \geq D$, and

$$|sI - \tilde{A}| = |sI - D + B| \leq |sI - D| + |B| = sI - D + |B|$$

implies

$$\rho(sI - \tilde{A}) \leq \rho(sI - D + |B|) < s.$$

Therefore, all eigenvalues λ of $\tilde{A}$ satisfy $|s - \lambda| < s$, whence $\lambda > 0$. This proves the assertion by Theorem 1.8.11. □

Notice that the converse of Theorem 5.4.49 is not true. This is shown by the Reichmann matrix in Example 5.4.33. Every symmetric matrix $\tilde{A} \in [A] = [A]^T$ is positive definite, and $[A]$ satisfies $\underline{a}_{ii} > 0$, $i = 1, \dots, n$. But $[A]$ is not an H-matrix since otherwise $[x]^C$ exists by Theorem 5.4.47.

Theorem 5.4.50. *Let $[A] = [A]^T \in \mathbb{IR}^{n\times n}$ be an M-matrix and let $[b] \in \mathbb{IR}^n$. Then the following statements hold.*
(a) *$[x]^C$ exists.*
(b) *If $(L^{(l)}, (L^{(l)})^T)$, $(L^{(u)}, (L^{(u)})^T)$ are the Cholesky decompositions of $\underline{A}$ and $\overline{A}$, respectively, then $L^{(l)}, L^{(u)}$ are M-matrices. The matrix $[L]^C$ of the Cholesky decomposition of $[A]$ can be represented as*

$$[L]^C = [L^{(l)}, L^{(u)}]; \tag{5.4.82}$$

in particular, $[L]^C$ is an M-matrix.
(c) *If $\underline{b} \geq 0$ or $0 \in [b]$ or $\overline{b} \leq 0$, then $[x]^C = \square S_{\text{sym}}$.*

Proof. (a) follows from Theorem 5.4.47.

(b) Since $\underline{A}$, $\overline{A}$ are Stieltjes matrices, the formulae in (5.4.72) show at once that $L^{(l)}, L^{(u)} \in Z^{n\times n}$. Theorem 5.4.47, applied to $\underline{A}$ and to $\overline{A}$, respectively, implies that they are M-matrices.

We prove (5.4.82) by induction with respect to the column index j.

For $j = 1$ we get at once

$$[l]^C_{11} = \sqrt{[a]_{11}} = [\sqrt{\underline{a}_{11}}, \sqrt{\overline{a}_{11}}] = [l^{(l)}_{11}, l^{(u)}_{11}]$$

and

$$[l]^C_{i1} = \frac{[a_{i1}]}{[l]_{11}} = \left[\frac{\underline{a}_{i1}}{l^{(l)}_{11}}, \frac{\overline{a}_{i1}}{l^{(u)}_{11}}\right] = [l^{(l)}_{i1}, l^{(u)}_{i1}], \quad i > 1, \text{ with } l^{(u)}_{i1} \leq 0,$$

where we took into account $\overline{a}_{i1} \leq 0$ for $i > 1$.

Assume now that (5.4.82) holds for all columns with an index less than $j > 1$. Then

$$[l]^C_{jj} = \left([a]_{jj} - \sum_{k=1}^{j-1} [(l^{(u)}_{jk})^2, (l^{(l)}_{jk})^2]\right)^{1/2} = [l^{(l)}_{jj}, l^{(u)}_{jj}]$$

and

$$[l]^C_{ij} = \left([a]_{ij} - \sum_{k=1}^{j-1} [l^{(u)}_{ik} l^{(u)}_{jk}, l^{(l)}_{ik} l^{(l)}_{jk}]\right) \cdot \left[\frac{1}{l^{(u)}_{jj}}, \frac{1}{l^{(l)}_{jj}}\right] = [l^{(l)}_{ij}, l^{(u)}_{ij}] \quad \text{for } i > j,$$

since $\overline{a}_{ij} \leq 0$ and $l^{(u)}_{ij} \leq 0$ for $i > j$. This proves the assertion.

(c) Denote by $D^{(l,s)}$, $L^{(l,s)}$ and $D^{(u,s)}$, $L^{(u,s)}$ the matrices in the representation (5.4.75) when the Cholesky method is applied to $\underline{A}$ and $\overline{A}$, respectively. By (b) and by (5.4.74), these matrices are nonnegative, and

$$[D]^{(s)} = [D^{(u,s)}, D^{(l,s)}], \quad [L]^{(s)} = [L^{(u,s)}, L^{(l,s)}].$$

Hence from (5.4.75) we get

$$[x]^C = \begin{cases} [\underline{A}^{-1}\underline{b}, \overline{A}^{-1}\overline{b}] & \text{if } \overline{b} \leq 0 \\ [\underline{A}^{-1}\underline{b}, \underline{A}^{-1}\overline{b}] & \text{if } 0 \in [b] \\ [\overline{A}^{-1}\underline{b}, \underline{A}^{-1}\overline{b}] & \text{if } \underline{b} \geq 0 \end{cases}. \qquad \square$$

Our final result on the interval Cholesky method is a perturbation theorem analogously to Theorem 5.4.29. Since the proof is similar to the one for that theorem we will skip it here. Its details can be found in Alefeld, Mayer [37].

Theorem 5.4.51. *Let* $[A], [B] \in \mathbb{IR}^{n\times n}$, $[A] = [A]^T$, $[B] = [B]^T$, $[b] \in \mathbb{IR}^n$. *Suppose that* $\mathrm{ICh}([A], [b])$ *exists. If*

$$\rho(|[A]^C|\, q([A], [B])) < 1, \tag{5.4.83}$$

then the Cholesky method is feasible for $[B]$.

Exercises

Ex. 5.4.1. Prove the details in Theorem 5.4.1.

Ex. 5.4.2. Compute IGA([A]) for

$$[A] = \begin{pmatrix} [1,2] & [-2,-1] \\ 1 & 1 \end{pmatrix}$$

and show that IGA([A]) contains the singular matrix

$$\tilde{A} = \frac{1}{3}\begin{pmatrix} -1 & 1 \\ -1 & 1 \end{pmatrix}.$$

Ex. 5.4.3. Show that the interval Gaussian algorithm is feasible for Hessenberg matrices with one of the following sign patterns, where a 'plus' means $\operatorname{sign}([a]_{ii}) = +1$ or $\sigma([a]_{ij}) = +1$, $i \neq j$, while a 'minus' means $\operatorname{sign}([a]_{ii}) = -1$ or $\sigma([a]_{ij}) = -1$, $i \neq j$. How can these matrices be transformed in order to obtain a sign pattern of the form (5.4.37)?

$$\begin{pmatrix} + & + & + & + & + \\ - & + & + & + & + \\ 0 & - & + & + & + \\ 0 & 0 & - & + & + \\ 0 & 0 & 0 & - & + \end{pmatrix}, \quad \begin{pmatrix} + & + & - & - & - \\ - & + & - & - & - \\ 0 & - & - & - & - \\ 0 & 0 & - & + & + \\ 0 & 0 & 0 & - & + \end{pmatrix},$$

$$\begin{pmatrix} + & - & - & 0 & + \\ + & + & + & 0 & - \\ 0 & + & - & 0 & + \\ 0 & 0 & + & + & + \\ 0 & 0 & 0 & - & + \end{pmatrix}, \quad \begin{pmatrix} + & + & - & + & - \\ + & - & + & - & + \\ 0 & + & + & - & + \\ 0 & 0 & - & - & + \\ 0 & 0 & 0 & + & + \end{pmatrix}.$$

Ex. 5.4.4. Show by the example

$$[A] = \begin{pmatrix} [1,3] & 4 \\ 4 & 4 \end{pmatrix}$$

in Neumaier [255] that $[A]^G$ exists, hence $[A]$ is regular. Show, in addition, that $\rho(|\check{A}^{-1}|\operatorname{rad}([A])) < 1$ holds, but $\rho(|\check{A}^G|\operatorname{rad}([A])) \geq 1$. Thus the example fulfills the assumption of Theorem 3.3.7 (a) but not that of Corollary 5.4.31.

5.5 Iterative methods

One of the earliest references to an iterative method for solving linear systems of equations is contained in a letter by Gauss to his student Christian Ludwig Gerling dated 26 December 1823 in the context of solving least squares problems via the normal equations. After briefly describing his method on a 4×4 example, Gauss wrote:

> I recommend this method to you for imitation. You will hardly ever again eliminate directly, at least not when you have more than two unknowns. The indirect procedure can be done while half asleep, or while thinking about other things.[1]

Although Gauss seems to be a little too optimistic, iterative methods are often applied in numerical analysis, especially for large linear systems. One way to end up with an iterative algorithm is to transform $Ax = b$ to a fixed point form and to use the standard fixed point iteration afterwards. This way will also be pursued in the following. There are essentially two classes of iterative interval methods for enclosing solutions of interval linear systems: One is based on a splitting $[A] = [M] - [N]$ leading – among others – to the interval Jacobi and the interval Gauss–Seidel method. The other one preconditions $Ax = b$ before splitting and ends up with the Krawczyk method. There are many variants of these methods, partly combined with intersections of an interval expression and the actual iterate, and partly formulated with some approximation $\tilde{x}$ of a solution and an enclosure of the error.

Iterative methods based on splittings

In order to derive traditional iterative methods for point systems

$$Ax = b \tag{5.5.1}$$

one splits $A \in \mathbb{R}^{n\times n}$ into $A = M - N$ and assumes M to be regular and 'simple'. Then (5.5.1) is equivalent to

$$Mx = Nx + b, \tag{5.5.2}$$

or

$$x = M^{-1}(Nx + b) = M^{-1}Nx + M^{-1}b \tag{5.5.3}$$

which induces the iterative process

$$Mx^{k+1} = Nx^k + b, \tag{5.5.4}$$

or

$$x^{k+1} = M^{-1}(Nx^k + b) = M^{-1}Nx^k + M^{-1}b. \tag{5.5.5}$$

If $\tilde{x}$ is some fixed vector, preferably a good approximation of a solution of (5.5.1), then this linear system can be rewritten as $A(x - \tilde{x}) = b - A\tilde{x}$ which leads to

$$M(x^{k+1} - \tilde{x}) = N(x^k - \tilde{x}) + b - A\tilde{x} \tag{5.5.6}$$

---

1 C. F. Gauss, *Über Stationsausgleichungen* (Correspondence Gauss to Gerling, Göttingen, 26 December 1823). In: Carl Friedrich Gauss, Werke, Vol. 9, pp. 278–281, edited by Königliche Gesellschaft der Wissenschaften zu Göttingen, Teubner, Leipzig, 1903. Translation by G. E. Forsythe, Notes, MTAC, Vol. 5, No. 36 (1951) pp. 255–258.

or

$$x^{k+1} = \tilde{x} + M^{-1}(N(x^k - \tilde{x}) + b - A\tilde{x}). \tag{5.5.7}$$

Since one rarely computes inverses in numerical analysis, the representations (5.5.5) and (5.5.7) are interesting for theoretical reasons only. In practice, the new iterates are computed solving the linear systems (5.5.4), and (5.5.6), respectively, which explains why M should be 'simple'.

For interval linear systems $[A]x = [b]$ one can proceed similarly, starting with a splitting $[A] = [M] - [N]$ and ending, for instance, with the iterations

$$[x]^{k+1} = \mathrm{IGA}([M], [N][x]^k + [b]) \tag{5.5.8}$$

and

$$[x]^{k+1} = \tilde{x} + \mathrm{IGA}([M], [N]([x]^k - \tilde{x}) + [b] - [A]\tilde{x}). \tag{5.5.9}$$

They are sometimes modified by iterating with intersections like

$$[x]^{k+1} = \{\mathrm{IGA}([M], [N][x]^k + [b])\} \cap [x]^k \tag{5.5.10}$$

and

$$[x]^{k+1} = \tilde{x} + \{\mathrm{IGA}([M], [N]([x]^k - \tilde{x}) + [b] - [A]\tilde{x}) \cap ([x]^k - \tilde{x})\} \tag{5.5.11}$$

with a breakdown if the intersections are empty. In (5.5.8)–(5.5.11) we assume that $\mathrm{IGA}([M])$ exists, which implies that $[M]$ is regular. If $[M] = I$ we get the (interval) Richardson iteration

$$[x]^{k+1} = (I - [A])[x]^k + [b]. \tag{5.5.12}$$

If $[M] = \mathrm{diag}([a]_{11}, \ldots, [a]_{nn})$, we obtain the (interval) Jacobi method

$$[x]_i^{k+1} = \Bigg([b]_i - \sum_{\substack{j=1 \\ j\neq i}}^{n} [a]_{ij}[x]_j^k\Bigg) \Big/ [a]_{ii}, \quad i = 1, \ldots, n. \tag{5.5.13}$$

If $[M]$ is the lower triangular part of $[A]$, the iteration (5.5.8) leads to the (interval) Gauss–Seidel method

$$[x]_i^{k+1} = \Bigg([b]_i - \sum_{j=1}^{i-1} [a]_{ij}[x]_j^{k+1} - \sum_{j=i+1}^{n} [a]_{ij}[x]_j^k\Bigg) \Big/ [a]_{ii}, \quad i = 1, \ldots, n. \tag{5.5.14}$$

This can be seen from the following lemma when substituting appropriately. We leave it to the reader to modify these particular methods along the lines (5.5.9)–(5.5.11). We also mention the Gauss–Seidel method with componentwise intersection

$$[x]_i^{k+1} = \Bigg\{\Bigg([b]_i - \sum_{j=1}^{i-1} [a]_{ij}[x]_j^{k+1} - \sum_{j=i+1}^{n} [a]_{ij}[x]_j^k\Bigg) \Big/ [a]_{ii}\Bigg\} \cap [x]_i^k, \quad i = 1, \ldots, n \tag{5.5.15}$$

which is not a special case of (5.5.10). As the subsequent lemma shows, an iteration with a componentwise intersection as in (5.5.15) can always be constructed if $[M]$ is a regular lower triangular matrix.

Lemma 5.5.1. *Let* $[A] \in \mathbb{IR}^{n\times n}$ *be a regular triangular matrix and* $[b] \in \mathbb{IR}^n$ *. Then*

$$[x] = \mathrm{IGA}([A], [b])$$

is equivalent to

$$[x]_i = \left([b]_i - \sum_{j=1}^{i-1} [a]_{ij}[x]_j \right) / [a]_{ii}, \quad i = 1, \ldots, n,$$

if $[A]$ *is lower triangular and to*

$$[x]_i = \left([b]_i - \sum_{j=i+1}^{n} [a]_{ij}[x]_j \right) / [a]_{ii}, \quad i = n, n-1, \ldots, 1,$$

if $[A]$ *is upper triangular.*

Proof. According to Corollary 5.4.7, $\mathrm{IGA}([A])$ exists. With the notation from Section 5.4 we show by induction on k that

$$[b]_i^{(k)} = [b]_i - \sum_{j=1}^{i-1} [a]_{ij}[x]_j \tag{5.5.16}$$

holds for $i \geq k$, if $[A]$ is lower triangular.

If $k = 1$, we have $[b]^{(1)} = [b]$ by definition. Assume that (5.5.16) holds for some $k < n$. Then for $i \geq k+1$ we obtain by one step of the interval Gaussian algorithm

$$[b]_i^{(k+1)} = [b]_i^{(k)} - [b]_k^{(k)} \cdot \frac{[a]_{ik}^{(k)}}{[a]_{kk}^{(k)}} = [b]_i^{(k)} - [b]_k^{(k)} \cdot \frac{[a]_{ik}}{[a]_{kk}}, \tag{5.5.17}$$

since by the lower triangular form of $[A]$ the k-th column of $[A]$ was not changed up to the k-th Gaussian step. Hence (5.5.17) implies

$$[b]_i^{(k+1)} = [b]_i^{(k)} - [x]_k[a]_{ik},$$

and by induction we get

$$[b]_i^{(k+1)} = [b]_i - \sum_{j=1}^{k-1} [a]_{ij}[x]_j - [a]_{ik}[x]_k$$

for $i \geq k+1$.

If $[A]$ is upper triangular, the assertion is trivial since the Gaussian algorithm consists of the backward substitution part only. □

Methods like (5.5.8) can be considered under several aspects, among them the following, where S denotes the solution set of $[A]x = [b]$ and $[x]^* = \lim_{k\to\infty}[x]^k$ in the case of convergence.

(a) Feasibility: Does $[x]^k$ exist for each starting vector $[x]^0$?
(b) Global convergence: Does each sequence $([x]^k)$ converge and is the limit $[x]^*$ independent of $[x]^0$?
(c) Inclusion monotony: Do the implications

$$\begin{aligned} [\hat{x}]^0 \subseteq [x]^1 \quad &\Rightarrow \quad [\hat{x}]^k \subseteq [x]^k, \\ [x]^1 \subseteq [x]^0 \quad &\Rightarrow \quad [x]^k \subseteq [x]^{k-1} \subseteq [x]^0, \\ [x]^1 \supseteq [x]^0 \quad &\Rightarrow \quad [x]^k \supseteq [x]^{k-1} \supseteq [x]^0 \end{aligned}$$

hold for $k = 0, 1, \ldots$?
(d) Inclusion property: Does $S \subseteq [x]^*$ hold in the case of global convergence?
(e) Permanence of enclosure: Does $S \subseteq [x]^0$ imply $S \subseteq [x]^k$?
(f) Quality of enclosure: What can be said about the distance $q(\square S, [x]^*)$ in the case of global convergence? When does $[x]^* = \square S$ hold?
(g) Speed of convergence: What can be said on the R-order of (5.5.8) as given in Definition 5.5.3 below?

We are going to answer these questions in detail. To this end we transfer some basic definitions of traditional numerics to interval analysis.

Definition 5.5.2. Let $[A] \in \mathbb{IR}^{n\times n}$. We call $([M], [N])$ a *splitting* of $[A]$ if $[A] = [M] - [N]$. We call such a splitting of $[A]$
(a) a triangular splitting, if $[M]$ is a regular lower or upper triangular matrix;
(b) a regular splitting, if $(\tilde{M}, \tilde{N})$ is a regular splitting for each pair $(\tilde{M}, \tilde{N}) \in [M] \times [N]$;
(c) an *M-splitting*, if $[M]$ is an M-matrix and $[N] \geq O$.

Definition 5.5.3 (Root convergence factors, R-order). Let $\|\cdot\|$ be any fixed monotone norm on $\mathbb{R}^n$ and let $\mathcal{J}$ be an iterative method which produces interval vectors $[x]^k \in \mathbb{IR}^n$, $k = 0, 1, \ldots$. Denote by $C(\mathcal{J}, [x]^*)$ the set of all output sequences of $\mathcal{J}$ which converge to some vector $[x]^* \in \mathbb{IR}^n$.
(a) We call $\mathcal{J}$ convergent to $[x]^*$, if $C(\mathcal{J}, [x]^*)$ contains all possible output sequences of $\mathcal{J}$.
(b) For $p \geq 1$ the root convergence factors, or R_p-factors, *of a sequence* $([x]^k) \in C(\mathcal{J}, [x]^*)$ are defined as

$$R_p([x]^k) = \begin{cases} \limsup_{k\to\infty} \|q([x]^k, [x]^*)\|^{1/k}, & \text{if } p = 1, \\ \limsup_{k\to\infty} \|q([x]^k, [x]^*)\|^{1/p^k}, & \text{if } p > 1. \end{cases}$$

(c) If $\mathcal{J}$ is convergent to $[x]^*$, then for $p \geq 1$ the R_p-factor of the method $\mathcal{J}$ is defined as

$$R_p(\mathcal{J}) = \sup\{\, R_p([x]^k) \mid ([x]^k) \in C(\mathcal{J}, [x]^*)\,\}.$$

(d) If $\mathcal{J}$ is convergent to $[x]^*$, then for $p \geq 1$ the R-order of the method $\mathcal{J}$ is defined as

$$O_R(\mathcal{J}) = \begin{cases} \infty, & \text{if } R_p(\mathcal{J}) = 0 \text{ for } p \geq 1, \\ \inf\{p \mid p \in [1, \infty),\ R_p(\mathcal{J}) = 1\} & \text{otherwise.} \end{cases}$$

From the equivalence of norms in $\mathbb{R}^n$ one can easily see that the definitions in (b)–(d) are independent of the particular monotone norm $\|\cdot\|$. For iterations of the form (5.5.8) or (5.5.9), the R_1-factor is sometimes called the asymptotic convergence factor. The R-order is a measure for the speed of convergence. It is useful when comparing two iterative methods. Cum grano salis: the larger the order, the faster the convergence. One can show (Exercise 5.5.2):

Theorem 5.5.4. *With the notation of Definition* 5.5.3 *and* $p \geq 1$ *the following statements hold if* $\mathcal{J}$ *is convergent to* $[x]^*$.

(a) *If* $\|q([x]^{k+1}, [x]^*)\| \leq \alpha \|q([x]^k, [x]^*)\|^p$, $k \geq k_0 = k_0((x^k))$, *for all sequences* $([x]^k) \in C(\mathcal{J}, [x]^*)$ *and fixed* $\alpha \in \mathbb{R}$, *then* $O_R(\mathcal{J}) \geq p$.

(b) *If* $\|q([x]^{k+1}, [x]^*)\| \geq \beta \|q([x]^k, [x]^*)\|^p > 0$, $k \geq k_0 = k_0((x^k))$, *for some sequence* $([x]^k) \in C(\mathcal{J}, [x]^*)$ *and some* $\beta \in \mathbb{R}$, *then* $O_R(\mathcal{J}) \leq p$.

(c) *If the assumptions of* (a) *and* (b) *hold simultaneously for one and the same* p, *then* $O_R(\mathcal{J}) = p$.

The splittings (a) and (c) in Definition 5.5.2 imply the existence of IGA($[M]$) and therefore the feasibility of (5.5.8). We also have the following result.

Theorem 5.5.5. *Let* $([M], [N])$ *be a splitting of* $[A] \in \mathbb{IR}^{n \times n}$ *and let* IGA($[M]$) *exist. Then the iteration* (5.5.8) *is feasible, inclusion monotone and satisfies permanence of enclosure. If, in addition,* (5.5.8) *is globally convergent with limit* $[x]^*$, *then* $S \subseteq [x]^*$ *for the solution set* S *of* $[A]x = [b]$.

Proof. The feasibility follows directly from the existence of IGA($[M]$). Inclusion monotony results from Theorem 5.4.1 (a). Assume now $S \subseteq [x]^0$. Choose $\tilde{x} \in S$. Then there are matrices $\tilde{A} \in [A]$, $\tilde{M} \in [M]$, $\tilde{N} \in [N]$ and a vector $\tilde{b} \in [b]$ such that $\tilde{A}\tilde{x} = \tilde{b}$, $\tilde{A} = \tilde{M} - \tilde{N}$. Hence $\tilde{M}\tilde{x} = \tilde{N}\tilde{x} + \tilde{b}$, which implies $\tilde{x} \in \text{IGA}([M], [N][x]^0 + [b]) = [x]^1$ and, finally, $S \subseteq [x]^1$. An inductive argument proves $S \subseteq [x]^k$, $k = 0, 1, \ldots$.

If (5.5.8) is globally convergent, we can start with $[x]^0 \equiv \tilde{x} \in S$. Similarly as above we can see $\tilde{x} \in [x]^k$, $k = 0, 1, \ldots$, whence $\tilde{x} \in [x]^*$. Since $\tilde{x} \in S$ was arbitrary we get $S \subseteq [x]^*$. □

Next we study global convergence of (5.5.8). We start with a general result which for the Richardson splitting goes back to Otto Mayer [226].

Theorem 5.5.6. *Let* $([M], [N])$ *be a splitting of* $[A] \in \mathbb{IR}^{n \times n}$ *and let* IGA($[M]$) *exist. Then the iteration* (5.5.8) *is globally convergent to some vector* $[x]^* \in \mathbb{IR}^n$ *if and only if* $\rho(|[M]^G| \cdot |[N]|) < 1$ *with* $|[M]^G|$ *as in* (5.4.6). *In this case the statements of Theo-*

rem 5.5.5 hold, the matrix $[A]$ *is regular, and* $\rho(|[M]^G| \cdot |[N]|)$ *is an upper bound for the* R_1*-factor of the iteration* (5.5.8).

Proof. Let $P = |[M]^G| \cdot |[N]|$, $[x], [y] \in \mathbb{IR}^n$, $[f]([x]) = \mathrm{IGA}([M], [N][x] + [b])$.

'$\Leftarrow$': From (5.4.4) we get

$$q([f]([x]), [f]([y])) \le Pq([x], [y]). \tag{5.5.18}$$

Since $\rho(P) < 1$, by assumption $[f]$ is a P-contraction, hence (5.5.8) is globally convergent by virtue of Theorem 4.2.6.

'$\Rightarrow$': From $d([f]([x])) \ge Pd([x])$ we get

$$d([x]^k) \ge P^k d([x]^0), \quad k = 0, 1, \dots \tag{5.5.19}$$

for the iteration (5.5.8). Assume that $\rho(P) \ge 1$. Choose $[x]^0$ such that $d([x]^0)$ is a non-negative eigenvector of P associated with the eigenvalue $\rho(P)$. W.l.o.g. assume that $d([x]^*_{i_0}) < d([x]^0_{i_0})$ holds for at least one index i_0; otherwise rescale $[x]^0$ appropriately. Then (5.5.19) implies $d([x]^k) \ge \rho(P)^k d([x]^0)$ from which we get the contradiction

$$d([x]^*_{i_0}) \begin{cases} = \infty & \text{if } \rho(P) > 1, \\ \ge d([x]^0_{i_0}) & \text{if } \rho(P) = 1. \end{cases}$$

Therefore, $\rho(P) < 1$.

Choose $\tilde{A} \in [A]$, $\tilde{M} \in [M]$, $\tilde{N} \in [N]$ such that $\tilde{A} = \tilde{M} - \tilde{N}$. Then $|\tilde{M}^{-1}| \le |[M]^G|$ by virtue of (5.4.14). Hence $\rho(\tilde{M}^{-1}\tilde{N}) \le \rho(P) < 1$ holds and $(I - \tilde{M}^{-1}\tilde{N})^{-1}\tilde{M}^{-1} = (\tilde{M} - \tilde{N})^{-1} = \tilde{A}^{-1}$ exists. Since $\tilde{A} \in [A]$ was arbitrary, this proves the regularity of $[A]$.

From (5.5.18) we get $q([x]^k, [x]^*) \le P^k q([x]^0, [x]^*)$, whence $\|q([x]^k, [x]^*)\| \le \|P\|^k \|q([x]^0, [x]^*)\|$ for any monotone norm of $\mathbb{R}^n$. Thus $\|P\|$ is an upper bound for the R_1-factor of the iteration (5.5.8), and Theorem 1.9.12 implies this property for $\rho(P)$, too. □

Notice that $\rho(|[M]^G| \cdot |[N]|)$ is only an upper bound for the R_1-factor. It can be sharp but does not need to be so as the following example shows.

Example 5.5.7.

(a) Let $[M] = I = 2[N] \in \mathbb{IR}^{1\times 1}$, $[A] = [M] - [N]$, $[b] = 0 \in \mathbb{IR}^1$. Then the iteration $\mathcal{J}$ in (5.5.8) yields

$$[x]^{k+1} = \frac{1}{2}[x]^k = \frac{1}{2^{k+1}}[x]^0$$

with limit $[x]^* = 0$. With $[x]^0 = 1$ one gets

$$\|q([x]^k, 0)\|^{1/k} = \frac{1}{2} \le R_1(\mathcal{J}, [x]^*) \le \rho(|[M]^G| \cdot |[N]|) = \frac{1}{2},$$

i.e., $R_1(\mathcal{J}, [x]^*) = \rho(|[M]^G| \cdot |[N]|) = 1/2$.

(b) Let

$$[M] = I, \quad [N] = \begin{pmatrix} 0 & [0, 1/2] \\ [0, 1/2] & 0 \end{pmatrix}, \quad [A] = [M] - [N], \quad [b] = \begin{pmatrix} 1 \\ -1 \end{pmatrix}.$$

Then

$$\rho(|[M]^G| \cdot |[N]|) = 1/2 < 1, \quad [x]^* = [1/2, 1] \begin{pmatrix} 1 \\ -1 \end{pmatrix}.$$

Starting the iteration $\mathfrak{I}$ in (5.5.8) with

$$[x]^0 = [0, 2] \begin{pmatrix} 1 \\ -1 \end{pmatrix} \quad \text{results in } [x]^1 = [0, 1] \begin{pmatrix} 1 \\ -1 \end{pmatrix} \text{ and } [x]^2 = [x]^*.$$

Starting with

$$[\hat{x}]^0 = \frac{3}{4} \begin{pmatrix} 1 \\ -1 \end{pmatrix} \quad \text{yields } [\hat{x}]^1 = [5/8, 1] \begin{pmatrix} 1 \\ -1 \end{pmatrix} \text{ and } [\hat{x}]^2 = [x]^*.$$

Now start $\mathfrak{I}$ with an arbitrary vector $[\tilde{x}]^0 \in \mathbb{IR}^2$. Since $\lim_{k \to \infty} [\tilde{x}]^k = [x]^*$ there is an integer k_0 such that $[\hat{x}]^0 \subseteq [\tilde{x}]^{k_0} \subseteq [x]^0$, whence

$$[x]^* = [\hat{x}]^2 \subseteq [\tilde{x}]^{k_0+2} \subseteq [x]^2 = [x]^*.$$

Therefore, $R_1(\mathfrak{I}) = 0 < \rho(|[M]^G| \cdot |[N]|) = 1/2$.

Normally the matrix $|[M]^G|$ in Theorem 5.5.6 is unknown. Therefore, we look for criteria which guarantee $\rho(|[M]^G| \cdot |[N]|) < 1$.

Theorem 5.5.8. *Let* $([M], [N])$ *be a splitting of* $[A] \in \mathbb{IR}^{n \times n}$. *Then*

$$\langle [M] \rangle - |[N]| \text{ is an } M\text{-matrix} \tag{5.5.20}$$

if and only if

$$[M] \text{ is an } H\text{-matrix with } \rho(\langle [M] \rangle^{-1} |[N]|) < 1. \tag{5.5.21}$$

If one of these two equivalent conditions is fulfilled, then

$$\rho(|[M]^G| \cdot |[N]|) \le \rho(\langle [M] \rangle^{-1} |[N]|) < 1$$

with $|[M]^G|$ *as in* (5.4.6), *and Theorem* 5.5.6 *can be applied. In particular, the iteration* (5.5.8) *is globally convergent.*

Proof. Let (5.5.20) hold. Then $\langle [M] \rangle$ is an M-matrix by virtue of Corollary 1.10.5. Hence $[M]$ is an H-matrix. From (5.5.20) we get $(\langle [M] \rangle - |[N]|)^{-1} \ge O$. Trivially, $(\langle [M] \rangle, |[N]|)$ is a regular splitting of $\langle [M] \rangle - |[N]|$ which implies the second property of (5.5.21) by virtue of Theorem 1.10.21.

Conversely, let (5.5.21) hold. Then the inverse $\langle [M]\rangle^{-1}$ exists and is nonnegative. From the second property of (5.5.21) we get

$$\begin{aligned}(\langle [M]\rangle - |[N]|)^{-1} &= (I - \langle [M]\rangle^{-1}|[N]|)^{-1}\langle [M]\rangle^{-1}\\ &= \sum_{k=0}^{\infty}(\langle [M]\rangle^{-1}|[N]|)^k\langle [M]\rangle^{-1} \ge O,\end{aligned}$$

whence $\langle [M]\rangle - |[N]| \in Z^{n\times n}$ is an M-matrix.

Let now (5.5.20) or, equivalently, (5.5.21) hold. Then $|[M]^G| \le \langle [M]\rangle^{-1}$ by virtue of Theorem 5.4.6, and (5.5.21) implies $\rho(|[M]^G| \cdot |[N]|) \le \rho(\langle [M]\rangle^{-1}|[N]|) < 1$. Therefore, Theorem 5.5.6 concludes the proof. □

If

$$\langle [A]\rangle = \langle [M]\rangle - |[N]|, \tag{5.5.22}$$

then $\langle [A]\rangle \in Z^{n\times n}$. The equality (5.5.22) holds for instance for so-called direct splittings $([M], [N])$ of $[A]$. These are splittings with $[a]_{ij} = [m]_{ij}$ or $[a]_{ij} = -[n]_{ij}$ for all $i, j \in \{1, \dots, n\}$. Therefore, if the assumptions of Theorem 5.5.8 hold, the matrix $[A]$ is an H-matrix.

Theorem 5.5.9. *Let $([M], [N])$ be a splitting of $[A] \in \mathbb{IR}^{n\times n}$ and let $\mathrm{IGA}([M])$ exist. If some starting vector $[x]^0$ of the iteration (5.5.8) satisfies*

$$d([x]^1) < d([x]^0) \tag{5.5.23}$$

or

$$[x]^1 \subseteq \operatorname{int}([x]^0), \tag{5.5.24}$$

then $\rho(|[M]^G| \cdot |[N]|) < 1$ with $|[M]^G|$ as in (5.4.6), i.e., Theorem 5.5.6 can be applied. In particular, the iteration (5.5.8) is globally convergent to some vector $[x]^* \in \mathbb{IR}^n$.

Proof. Define $P = |[M]^G| \cdot |[N]|$ and $[f]([x]) = \mathrm{IGA}([M], [N][x] + [b])$ as in the proof of Theorem 5.5.6, and let (5.5.23) hold. Together with (5.4.4) we get

$$d([x]^0) > d([x]^1) = d([f]([x]^0)) \ge P d([x]^0),$$

whence $\rho(P) < 1$ by virtue of Theorem 1.9.13 (a). Thus Theorem 5.5.6 concludes the proof in the case (5.5.23).

Since (5.5.24) implies (5.5.23), the theorem also holds in this case. □

Remark 5.5.1. In practical computation (5.5.24) has to be replaced by

$$[x]^1_{\mathrm{comp}} \subseteq \operatorname{int}([x]^0), \tag{5.5.25}$$

where $[x]^0$ is a machine interval (vector) and $[x]^k_{\mathrm{comp}}$ denotes the k-th iterate obtained on a computer using machine interval arithmetic. Here we assume that this arithmetic obeys the rules of Section 2.7, among them outward rounding and inclusion monotonicity. Since there are only finitely many machine numbers, these rules imply

$$[x]^{k+1}_{\text{comp}} = [x]^{k}_{\text{comp}} \tag{5.5.26}$$

for some k provided that (5.5.25) holds. Therefore, (5.5.26) can be used as a stopping criterion in practical computation, perhaps combined with an inequality $k \le k_{\max}$, where $k_{\max}$ is a maximum number of iterations in order to stop within an appropriate time.

Iterations based on triangular splittings

The Richardson iteration (5.5.12), the Jacobi method (5.5.13), and the Gauss–Seidel method (5.5.14) are based on triangular splittings. This is the reason why we are going to study such splittings in detail.

Theorem 5.5.10. *Let* $([M], [N])$ *be a triangular splitting of* $[A] \in \mathbb{IR}^{n\times n}$. *Then the iteration* (5.5.8) *is globally convergent to some vector* $[x]^* \in \mathbb{IR}^n$ *if and only if* $\rho(\langle [M]\rangle^{-1}|[N]|) < 1$. *In addition, the assertions of Theorem* 5.5.6 *hold.*

Proof. Since $[M]$ is a regular triangular matrix it is an H-matrix according to Theorem 3.3.5 (f), and global convergence follows from Theorem 5.5.8.

Conversely, let (5.5.8) be globally convergent to some limit $[x]^*$. Split the matrix $[M]$ into $[M] = [M]_D - [M]_T$, where $[M]_D$ denotes its diagonal part and $[M]_T$ is defined by the splitting. From Lemma 5.5.1 with $[M]$ and $[N][x] + [b]$ instead of $[A]$, $[b]$ we obtain

$$[x]^{k+1} = [M]_D^{-1}([N][x]^k + [b] + [M]_T[x]^{k+1}),$$

whence

$$d([x]^{k+1}) \ge |[M]_D^{-1}|\,(|[N]|d([x]^k) + |[M]_T|\, d([x]^{k+1})). \tag{5.5.27}$$

Since $|[M]_D^{-1}| = \langle [M]_D\rangle^{-1} \ge O$ has diagonal form, the inequality (5.5.27) and $\langle [M]_D\rangle \ge O$ imply

$$(\langle [M]_D\rangle - |[M]_T|)d([x]^{k+1}) = \langle [M]\rangle\, d([x]^{k+1}) \ge |[N]|\, d([x]^k)$$

and

$$d([x]^{k+1}) \ge Pd([x]^k) \quad \text{with } P = \langle [M]\rangle^{-1}|[N]| \ge O.$$

As in the proof of Theorem 5.5.6 choose $[x]^0$ such that $d([x]^0)$ is an eigenvector of P associated with the eigenvalue $\rho = \rho(P)$. W.l.o.g. assume $d([x]^0_{i_0}) > d([x]^*_{i_0})$ for at least one index i_0. Then

$$d([x]^k) \ge P^k d([x]^0) = \rho^k d([x]^0)$$

and

$$\limsup_{k\to\infty} \rho^k d([x]^0_{i_0}) \le d([x]^*_{i_0}) < d([x]^0_{i_0}),$$

whence $\rho < 1$ follows. The proof terminates by virtue of Theorem 5.5.6. □

Now we address particular classes of triangular splittings. To this end we need the following definition.

Definition 5.5.11. Let $[C] \in \mathbb{IR}^{n\times n}$ be a regular diagonal matrix. The set $S_{[C]}$ is defined as the set of all triangular splittings $([M], [N])$ of $[A] \in \mathbb{IR}^{n\times n}$ which satisfy

$$\left.\begin{aligned} [m]_{ii} &= [c]_{ii} \\ [m]_{ij}[n]_{ij} &= 0, \quad i \neq j \end{aligned}\right\} \quad i, j = 1, \dots, n.$$

Notice that $([M], [N]) \in S_{[C]}$ implies

$$[m]_{ij} = [a]_{ij},\ [n]_{ij} = 0 \quad \text{or} \quad [m]_{ij} = 0,\ [n]_{ij} = -[a]_{ij} \tag{5.5.28}$$

for each index pair (i, j) with $i \neq j$, i.e., $([M], [N])$ is 'nearly' a direct splitting of $[A]$.

In the sequel we will often use the splitting

$$[A] = [D] - [E] - [F] \quad \text{of } [A] \in \mathbb{IR}^{n\times n}, \tag{5.5.29}$$

where $[D] = \operatorname{diag}([a]_{11}, \dots, [a]_{nn})$ denotes the diagonal part of $[A]$, $-[E]$ its strictly lower triangular part, and $-[F]$ its strictly upper triangular part. We always assume that $[D]$ is regular.

Theorem 5.5.12. *Let $[A], [C] \in \mathbb{IR}^{n\times n}$, where $[C]$ is a regular diagonal matrix. If the iteration* (5.5.8) *is globally convergent for* some *triangular splitting* $([M], [N]) \in S_{[C]}$, *then for* each *splitting* $([\hat{M}], [\hat{N}]) \in S_{[C]}$. *In this case the limit $[x]^*$ is the same for each splitting from $S_{[C]}$ and*

$$|[N]| \leq |[\hat{N}]| \quad \textit{implies} \quad \rho(\langle [M]\rangle^{-1}|[N]|) \leq \rho(\langle [\hat{M}]\rangle^{-1}|[\hat{N}]|) < 1. \tag{5.5.30}$$

Proof. The global convergence together with Theorem 5.5.10 and the equivalence of (5.5.20) and (5.5.21) imply that the matrix $B = \langle [M]\rangle - |[N]|$ is an M-matrix. For $([\hat{M}], [\hat{N}]) \in S_{[C]}$ we have $[m]_{ii} = [\hat{m}]_{ii}$, hence $[n]_{ii} = [\hat{n}]_{ii}$ by virtue of the reduction rules in Theorem 2.3.2 (h). Using (5.5.28) we obtain $\langle [\hat{M}]\rangle - |[\hat{N}]| = B$, and Theorem 5.5.8 implies global convergence of (5.5.8) for $([\hat{M}], [\hat{N}])$, too.

In order to show equality of the limit $[x]^*$ we decompose $[M]$, $[\hat{M}]$ into $[M] = [M]_D - [M]_T$ and $[\hat{M}] = [\hat{M}]_D - [\hat{M}]_T$, where $[M]_D = [\hat{M}]_D = [C]$ is the diagonal part of the corresponding matrices, and $[M]_T$, $[\hat{M}]_T$ are defined by the splittings. With Lemma 5.5.1 we get

$$\begin{aligned} [x]^* &= [M]_D^{-1}([M]_T[x]^* + [N][x]^* + [b]) = [\hat{M}]_D^{-1}(([M]_T + [N])[x]^* + [b]) \\ &= [\hat{M}]_D^{-1}(([\hat{M}]_T + [\hat{N}])[x]^* + [b]) = [\hat{M}]_D^{-1}([\hat{M}]_T[x]^* + [\hat{N}][x]^* + [b]). \end{aligned}$$

Thus $[x]^*$ is also the limit of (5.5.8) for the splitting $([\hat{M}], [\hat{N}])$.

The inequality (5.5.30) follows from Theorem 1.10.22 with the regular splittings $(\langle [M]\rangle, |[N]|)$, $(\langle [\hat{M}]\rangle, |[\hat{N}]|)$ of B which is an M-matrix and therefore inverse positive.

□

Remark 5.5.2.

(a) In case of global convergence, the splitting $([C] - [E], [N]') \in S_{[C]}$ yields the smallest value of $\rho(\langle[M]\rangle^{-1}|[N]|)$ (= bound of the R_1-factor of (5.5.8)) for each splitting $([M], [N]) \in S_{[C]}$ with a *lower* triangular matrix $[M]$. In this case $[N]'$ is an upper triangular matrix. Correspondingly, $([C] - [F], [N]'') \in S_{[C]}$ yields the smallest value of $\rho(\langle[M]\rangle^{-1}|[N]|)$ for each splitting $([M], [N]) \in S_{[C]}$ with an *upper* triangular matrix $[M]$. In this case $[N]''$ is a lower triangular matrix.
The largest value of this spectral radius is obtained for $([C], [\tilde{N}]) \in S_{[C]}$.

(b) The Jacobi method and the Gauss–Seidel method are based on splittings from $S_{[D]}$. According to Theorem 5.5.12 both methods are globally convergent if and only if

$$\rho_J = \rho(\langle[D]\rangle^{-1}(|[E]| + |[F]|)) < 1.$$

The limit $[x]^*$ coincides. For each splitting $([M], [N]) \in S_{[D]}$ with a lower triangular matrix $[M]$ item (a) implies

$$\rho_{\mathrm{GS}} = \rho((\langle[D]\rangle - |[E]|)^{-1}|[F]|) \le \rho(\langle[M]\rangle^{-1}|[N]|) \le \rho_J,$$

i.e., the Gauss–Seidel method lets us expect the fastest convergence, the Jacobi method the slowest.
Notice that in the noninterval case, where only $x^0 \in \mathbb{R}$ is admitted, the convergence of one of these two methods does not necessarily imply the convergence of the other one.

Next we show that the limit of the Gauss–Seidel method is optimal in a certain sense. To this end we need part (a) of the following lemma.

Lemma 5.5.13. *Let* $[a], [b], [c], [d] \in \mathbb{IR}$ *with* $0 \notin [d]$, $0 \notin [d] - [c]$.

(a) $[a] \subseteq \dfrac{[b]}{[d] - [c]}$ *implies* $[a] \subseteq \dfrac{[b] + [a][c]}{[d]}$.

(b) *If, in addition,* $[c] \ge 0$, $[d] > 0$, $[d] - [c] > 0$, *then*

$$[a] = \frac{[b]}{[d] - [c]} \quad \textit{implies} \quad [a] = \frac{[b] + [a][c]}{[d]}.$$

Proof. (a) Let $\tilde{a} \in [a]$. By virtue of Theorem 4.1.5 there are real numbers $\tilde{b} \in [b]$, $\tilde{c} \in [c]$, $\tilde{d} \in [d]$ such that $\tilde{a} = \tilde{b}/(\tilde{d} - \tilde{c})$. This equation can be transformed to $\tilde{a} = (\tilde{b} + \tilde{a}\tilde{c})/\tilde{d} \in ([b] + [a][c])/[d]$ which proves the assertion.

(b) From $[a] = \dfrac{[b]}{[d] - [c]}$ and Corollary 2.3.9 we get

$$\begin{aligned}\frac{[b] + [a][c]}{[d]} &= \frac{[b]}{[d]}\left\{1 + \frac{[c]}{[d] - [c]}\right\} = \frac{[b]}{[d]}\left\{1 + \left[\frac{\underline{c}}{\overline{d} - \underline{c}}, \frac{\overline{c}}{\underline{d} - \overline{c}}\right]\right\} \\ &= \frac{[b]}{[d]}\left[\frac{\overline{d}}{\overline{d} - \underline{c}}, \frac{\underline{d}}{\underline{d} - \overline{c}}\right] = [b]\left[\frac{1}{\overline{d} - \underline{c}}, \frac{1}{\underline{d} - \overline{c}}\right] = \frac{[b]}{[d] - [c]} = [a].\end{aligned}$$

□

Theorem 5.5.14. *Assume that the Gauss–Seidel method is globally convergent to some limit* $[x]^*_{\mathrm{GS}}$, *and let the same hold with limit* $[x]^*$ *for* (5.5.8) *based on another lower triangular splitting* $([M], [N])$ *of* $[A] \in \mathbb{IR}^{n\times n}$. *Then*

$$[x]^*_{\mathrm{GS}} \subseteq [x]^* \tag{5.5.31}$$

and

$$\rho((\langle[D]\rangle - |[E]|)^{-1}|[F]|) \le \rho(\langle[M]\rangle^{-1}|[N]|) < 1. \tag{5.5.32}$$

Proof. Let $[M]_D$ be the diagonal part of $[M]$ and define $[M]_T$ by $[M] = [M]_D - [M]_T$. With the notation (5.5.29) and from $d([A]) = d([M]) + d([N]) \ge d([M])$ we get $d([D]) \ge d([M]_D)$, $d([E]) \ge d([M]_T)$. By virtue of Theorem 2.3.10 (a) there are a diagonal matrix $[\hat{D}]$ and a strictly lower triangular matrix $[\hat{E}]$ such that $[D] = [M]_D - [\hat{D}]$, $[E] = [M]_T + [\hat{E}]$. Then $[A] = [D] - [E] - [F] = [M] - [N]$ implies

$$[N] = [\hat{D}] + [\hat{E}] + [F]. \tag{5.5.33}$$

Using Lemma 5.5.13 (componentwise), Remark 5.5.2 (b) and the particular form of the matrices we get

$$\begin{aligned}
[x]^*_{\mathrm{GS}} &= [D]^{-1}([E][x]^*_{\mathrm{GS}} + [F][x]^*_{\mathrm{GS}} + [b]) \\
&\subseteq [M]_D^{-1}\{([M]_T + [\hat{E}])[x]^*_{\mathrm{GS}} + [\hat{D}][x]^*_{\mathrm{GS}} + [F][x]^*_{\mathrm{GS}} + [b]\} \\
&\subseteq [M]_D^{-1}\{([M]_T[x]^*_{\mathrm{GS}} + ([\hat{E}][x]^*_{\mathrm{GS}} + [\hat{D}][x]^*_{\mathrm{GS}} + [F][x]^*_{\mathrm{GS}}) + [b]\} \\
&= [M]_D^{-1}\{[M]_T[x]^*_{\mathrm{GS}} + [N][x]^*_{\mathrm{GS}} + [b]\} =: [y].
\end{aligned} \tag{5.5.34}$$

Start the iteration (5.5.8) for $([M], [N])$ with $[x]^0 = [x]^*_{\mathrm{GS}}$. Then

$$[x]^1_i = \left\{-\sum_{j=1}^{i-1}[m]_{ij}[x]^1_j + ([N][x]^*_{\mathrm{GS}} + [b])_i\right\} \Big/ [m]_{ii}, \quad i = 1, \dots n. \tag{5.5.35}$$

With (5.5.34) we obtain $([x]^*_{\mathrm{GS}})_1 \subseteq [y]_1 = [x]^1_1$. Assume now that $([x]^*_{\mathrm{GS}})_k \subseteq [x]^1_k$ holds for $k = 1, \dots, i-1 < n$. Together with (5.5.34) and (5.5.35) this implies $([x]^*_{\mathrm{GS}})_i \subseteq [y]_i \subseteq [x]^1_i$, whence $[x]^*_{\mathrm{GS}} = [x]^0 \subseteq [x]^1$ follows. By virtue of the inclusion monotony of the method (Theorem 5.5.5) we finally obtain

$$[x]^*_{\mathrm{GS}} = [x]^0 \subseteq [x]^1 \subseteq \cdots \subseteq [x]^k \quad \text{with} \lim_{k\to\infty}[x]^k = [x]^*.$$

This proves (5.5.31).

In order to show (5.5.32) we define ρ_ε as the spectral radius of the matrix $\langle[M]\rangle^{-1} \cdot (|[N]| + \varepsilon ee^T)$. This matrix is positive if $\varepsilon > 0$ and therefore has a Perron vector x_ε. Since $\rho_0 < 1$ by Theorem 5.5.10 we can choose $\varepsilon > 0$ so small that $\rho_\varepsilon < 1$ holds. In this case we have

$$\begin{aligned}
\langle[D] - [E]\rangle\rho_\varepsilon x_\varepsilon &- \langle[M]_D - [\hat{D}] - [M]_T - [\hat{E}]\rangle\rho_\varepsilon x_\varepsilon \\
&\ge \langle[M]\rangle\rho_\varepsilon x_\varepsilon - (|[\hat{D}]| + |[\hat{E}]|)\rho_\varepsilon x_\varepsilon \\
&= (|[N]| + \varepsilon ee^T)x_\varepsilon - (|[\hat{D}]| + |[\hat{E}]|)\rho_\varepsilon x_\varepsilon \\
&\ge (|[N]| - |[\hat{D}]| - |[\hat{E}]|)x_\varepsilon = |[F]|x_\varepsilon,
\end{aligned}$$

where the last equality follows from (5.5.33). Therefore, $\rho_\varepsilon x_\varepsilon \ge \langle [D] - [E] \rangle^{-1} |[F]| x_\varepsilon = (\langle [D] \rangle - |[E]|)^{-1} |[F]| x_\varepsilon$, and Theorem 1.9.13 (b) guarantees $\rho((\langle [D] \rangle - |[E]|)^{-1} |[F]|) \le \rho_\varepsilon$. The assertion follows with $\varepsilon \to 0$. □

According to Theorem 5.5.14 the Richardson iteration never yields a tighter enclosure of the solution set S than the Gauss–Seidel method. Sometimes however, this enclosure can be coarser as the following example shows: Let $[A] = ([1/2, 3/2] = [D] \in \mathbb{IR}^{1\times 1}$, $[E] = [F] = O$, $[b] = 1$. Then $[x]^*_{GS} = [2/3, 2]$ is the limit of the Gauss–Seidel method while $[x]^* = [0, 2]$ is the limit of the Richardson iteration.

Although the limit $[x]^*_{GS}$ of the Gauss–Seidel method is optimal in the sense of Theorem 5.5.14, it can differ from $\square S$. This can be seen by the following example which originates from J. Garloff.

Example 5.5.15. Let

$$[A] = \begin{pmatrix} 1 & -1/2 \\ 1/2 & 1 \end{pmatrix}, \quad [b] = \begin{pmatrix} [-1, 1] \\ [-1, 1] \end{pmatrix}.$$

Then

$$\square S \subseteq \mathrm{IGA}([A], [b]) = \frac{1}{5} \begin{pmatrix} [-8, 8] \\ [-6, 6] \end{pmatrix} \subset [x]^*_{GS} = \begin{pmatrix} [-2, 2] \\ [-2, 2] \end{pmatrix}.$$

Notice that $[A]$ is an H-matrix but not an M-matrix.

Iterations based on M-splittings

M-splittings are very restrictive but have many nice properties. In particular, they are regular splittings. This is the reason why we are going to study them.

Theorem 5.5.16. *Let* $([M], [N])$ *be an* M*-splitting of an* M*-matrix* $[A] \in \mathbb{IR}^{n\times n}$. *Then* $|[M]^G| \cdot |[N]| = \underline{M}^{-1}\overline{N} = \langle [M] \rangle^{-1} |[N]|$ *and* $\rho(|[M]^G| \cdot |[N]|) < 1$. *Hence Theorem 5.5.6 applies. In particular,* (5.5.8) *is globally convergent.*

Proof. Since $[M]$ is an M-matrix we have $|[M]^G| = \underline{M}^{-1} = \langle [M] \rangle^{-1}$ by Theorem 5.4.34 (b). The first statement follows now from $[N] \ge O$. From $\underline{A} = \underline{M} - \overline{N}$ one sees that $(\underline{M}, \overline{N})$ is a regular splitting of the inverse positive matrix $\underline{A}$. Hence $\rho(|[M]^G| \cdot |[N]|) = \rho(\underline{M}^{-1}\overline{N}) < 1$ by virtue of Theorem 1.10.21. □

Notice that Theorem 5.5.16 can also be proved with Theorem 5.5.8.

In order to prove our next result we need the following definition.

Definition 5.5.17. Let $([M], [N])$ be a splitting of $[A] \in \mathbb{IR}^{n\times n}$ and let the iteration (5.5.8) converge globally to some limit $[x]^*$. Then we define the matrices M^*, M^{**}, N^*, $N^{**} \in \mathbb{R}^{n\times n}$ by

$$m^*_{ij} = \begin{cases} \underline{m}_{ij}, & \text{if } \underline{x}^*_j \le 0 \\ \overline{m}_{ij}, & \text{if } \underline{x}^*_j > 0 \end{cases}, \qquad m^{**}_{ij} = \begin{cases} \overline{m}_{ij}, & \text{if } \overline{x}^*_j < 0 \\ \underline{m}_{ij}, & \text{if } \overline{x}^*_j \ge 0 \end{cases},$$

$$n_{ij}^{*} = \begin{cases} \overline{n}_{ij}, & \text{if } \underline{x}_j^{*} \le 0 \\ \underline{n}_{ij}, & \text{if } \underline{x}_j^{*} > 0 \end{cases}, \qquad n_{ij}^{**} = \begin{cases} \underline{n}_{ij}, & \text{if } \overline{x}_j^{*} < 0 \\ \overline{n}_{ij}, & \text{if } \overline{x}_j^{*} \ge 0 \end{cases}.$$

Theorem 5.5.18. *Let $([M],[N])$ be an M-splitting of an M-matrix $[A] \in \mathbb{IR}^{n\times n}$. If $[M]$ is a triangular matrix or a degenerate matrix $[M] \equiv M \in \mathbb{R}^{n\times n}$, then the limit $[x]^*$ of (5.5.8) is the interval hull of the solution set S of $[A]x = [b]$, i.e., $[x]^* = \square S$.*

Proof. W.l.o.g. assume that $[M]$ is a lower triangular M-matrix. For upper triangular M-matrices $[M]$ the proof proceeds similarly. According to Lemma 5.5.1 the iteration (5.5.8) is equivalent to

$$[x]_i^{k+1} = \frac{1}{[m]_{ii}}\left\{[b]_i + \sum_{j=1}^{i-1}(-[m]_{ij})[x]_j^{k+1} + \sum_{j=1}^{n}[n]_{ij}[x]_j^{k}\right\}, \quad i = 1,\dots,n.$$

Let $[x]^*$ be the limit of the iteration (5.5.8). This limit exists since (5.5.8) is globally convergent according to Theorem 5.5.16. It satisfies

$$[x]_i^{*} = \frac{1}{[m]_{ii}}\left\{[b]_i + \sum_{j=1}^{i-1}(-[m]_{ij})[x]_j^{*} + \sum_{j=1}^{n}[n]_{ij}[x]_j^{*}\right\}, \quad i = 1,\dots,n.$$

With M^*, N^* as in Definition 5.5.17 we get

$$\underline{x}_i^{*} = \frac{1}{m_{ii}^{*}}\left\{\underline{b}_i + \sum_{j=1}^{i-1}(-m_{ij}^{*})\underline{x}_j^{*} + \sum_{j=1}^{n} n_{ij}^{*}\underline{x}_j^{*}\right\}, \quad i = 1,\dots,n,$$

i.e., $(M^* - N^*)\underline{x}^* = \underline{b}$. Hence $\underline{x}^* \in S$. With the matrices M^{**}, N^{**} as in the same definition one similarly obtains $(M^{**} - N^{**})\overline{x}^* = \overline{b}$, which shows $\overline{x}^* \in S$. This proves the assertion in the first case.

Now let $[M] \equiv M \in \mathbb{R}^{n\times n}$. According to Theorem 5.4.34 (d) we have $[x]^* = \mathrm{IGA}(M, [N][x]^* + [b]) = M^{-1}([N][x]^* + [b])$, which implies $\underline{x}^* = M^{-1}(N^*\underline{x}^* + \underline{b})$ with N^* as in Definition 5.5.17. Therefore, $\underline{x}^* \in S$, and similarly we get $\overline{x}^* \in S$. □

If $[M]$ in Theorem 5.5.18 is a general M-matrix, then $[x]^* = \square S$ can no longer be guaranteed as the following example shows.

Example 5.5.19. Let

$$[A] = [M] = \begin{pmatrix} [2,4] & [-2,0] \\ [-1,0] & [2,4] \end{pmatrix}, \quad [N] = O, \quad [b] = \begin{pmatrix} [1,2] \\ [-2,2] \end{pmatrix}.$$

Then $\underline{A} \cdot (3/2, 1)^T = (1, 1/2)^T > 0$, hence $\underline{A}$ is an M-matrix by virtue of Theorem 1.10.4. Each matrix $\tilde{A} \in [A]$ satisfies $\underline{A} \le \tilde{A} \in Z^{n\times n}$, hence $\tilde{A}$ and therefore $[A] = [M]$ are M-matrices, too. Trivially, $([M],[N])$ is an M-splitting of the M-matrix $[A]$. The limit of the iteration (5.5.8) is $[x]^* = \mathrm{IGA}([A],[b]) = ([-3/2, 4], [-2, 3])^T$, and for $L = \{(x,y)^T \mid x = -5/4,\ y \in \mathbb{R}\}$ we get $S \cap L = \emptyset$ while $[x]^* \cap L \ne \emptyset$; cf. Exercise 5.5.5. Since $[A]$ is regular, the set S is compact and connected. Therefore it lies completely left or right of L and has a positive distance from it, whence $\square S \subset [x]^*$.

Our next theorem relates the limit of two methods of the form (5.5.8).

Theorem 5.5.20. *Let* $([M], [N])$, $([\hat{M}], [\hat{N}])$ *be two* M*-splittings of an* M*-matrix* $[A] \in \mathbb{IR}^{n\times n}$ *which satisfy*

$$[\hat{M}] \le [M], \quad d([\hat{M}]) \ge d([M]) \tag{5.5.36}$$

or, equivalently,

$$[\hat{N}] \le [N], \quad d([\hat{N}]) \le d([N]). \tag{5.5.37}$$

If $[x]^*$, $[\hat{x}]^*$ *are the corresponding limits of* (5.5.8), *then* $[x]^* \subseteq [\hat{x}]^*$.

Proof. We present here only selected parts of the lengthy proof. The details can be found in Mayer [203].

The equivalence of (5.5.36), (5.5.37) can be seen via

$$\begin{aligned}
&\underline{\hat{M}} \le \underline{M} \Leftrightarrow \overline{\hat{N}} = \underline{\hat{M}} - \underline{A} \le \underline{M} - \underline{A} = \overline{N},\\
&\overline{\hat{M}} \le \overline{M} \Leftrightarrow \underline{\hat{N}} \le \underline{N},\\
&d([A]) \ = \ d([M]) + d([N]) = d([\hat{M}]) + d([\hat{N}])\\
&\qquad\quad \Leftrightarrow d([\hat{M}]) - d([M]) = d([N]) - d([\hat{N}]).
\end{aligned}$$

Since $O \le d([N]) - d([\hat{N}]) = \overline{N} - \overline{\hat{N}} - (\underline{N} - \underline{\hat{N}})$ the matrix

$$[R] = [\underline{N} - \underline{\hat{N}}, \overline{N} - \overline{\hat{N}}]$$

exists and satisfies

$$[N] = [\hat{N}] + [R] \quad \text{with } [R] \ge O.$$

From

$$[A] = [M] - [N] = [M] - [\hat{N}] - [R] = [\hat{M}] - [\hat{N}]$$

we get

$$[\hat{M}] = [M] - [R].$$

Since $[\hat{N}] \ge O$, $[R] \ge O$ we can reformulate $([\hat{N}] + [R])[\hat{x}]^*$ as $[\hat{N}][\hat{x}]^* + [R][\hat{x}]^*$. In Mayer [203] we show the subset property in

$$\begin{aligned}
\mathrm{IGA}([M], [N][\hat{x}]^* + [b]) &= \mathrm{IGA}([M], [\hat{N}][\hat{x}]^* + [R][\hat{x}]^* + [b])\\
&\subseteq \mathrm{IGA}([M] - [R], [\hat{N}][\hat{x}]^* + [b])\\
&= \mathrm{IGA}([\hat{M}], [\hat{N}][\hat{x}]^* + [b]) = [\hat{x}]^*.
\end{aligned} \tag{5.5.38}$$

With $[f]([x]) = \mathrm{IGA}([M], [N][x] + [b])$ and $[x]^0 = [\hat{x}]^*$ we obtain $[x]^1 = [f]([x]^0) \subseteq [\hat{x}]^*$, and by induction $[x]^k \subseteq [\hat{x}]^*$, $k = 0, 1, \ldots$, which proves $[x]^* = \lim_{k\to\infty}[x]^k \subseteq [\hat{x}]^*$. □

Notice that (5.5.38) describes the transfer of $[R]$ from the right-hand side of the interval linear system to the coefficient matrix. For degenerate linear systems with solution x^* this can be seen as follows:

$$\begin{aligned} A &= M - N = M - \hat{N} - R = (M - R) - \hat{N} = \hat{M} - \hat{N} \\ &\Rightarrow \hat{M} = M - R, \\ x^* &= M^{-1}(Nx^* + b) = M^{-1}(\hat{N}x^* + Rx^* + b) \\ &\Rightarrow x^* = (M - R)^{-1}(\hat{N}x^* + b) = \hat{M}^{-1}(\hat{N}x^* + b). \end{aligned}$$

Since $[A] \le [\hat{M}]$ with $d([A]) \ge d([\hat{M}])$, Theorem 5.5.20 shows that among all M splittings $([\hat{M}], [\hat{N}])$ of an M-matrix $[A]$ the interval Gaussian algorithm (which results from the M-splitting $([A], O)$ and yields $[x]^*$ in one step, of course) delivers the coarsest enclosure $[x]^*$ of the solution set S of $[A]x = [b]$. We state this result as a corollary.

Corollary 5.5.21. *Let* $([M], [N])$ *be an* M*-splitting of an* M*-matrix* $[A]$*. Then the associated limit* $[x]^*$ *of the iteration* (5.5.8) *satisfies* $[x]^* \subseteq \mathrm{IGA}([A], [b])$*, where* $[b] \in \mathbb{IR}^n$.

Example 5.5.15 shows that this property can change if $[A]$ is not an M-matrix.

In Theorem 5.5.12 we were able to compare some bounds for the R_1-factor of (5.5.8) for two *triangular* splittings $([M], [N])$, $([\hat{M}], [\hat{N}])$ with additional restrictions. Now we want to prove a similar result for M-splittings for which $[M]$ and $[\hat{M}]$ are not necessarily triangular.

Theorem 5.5.22. *Let* $([M], [N])$, $([\hat{M}], [\hat{N}])$ *be two* M*-splittings of an* M*-matrix* $[A] \in \mathbb{IR}^{n\times n}$*. If* $\underline{\hat{M}} \le \underline{M}$ *or, equivalently,* $\overline{\hat{N}} \le \overline{N}$ *then*

$$\rho(\underline{\hat{M}}^{-1}\overline{\hat{N}}) = \rho(\langle\hat{M}\rangle^{-1}|[\hat{N}]|) \le \rho(\underline{M}^{-1}\overline{N}) = \rho(\langle M\rangle^{-1}|[N]|) < 1.$$

Proof. The equivalence of the assumptions follows from $\underline{A} = \underline{M} - \overline{N} = \underline{\hat{M}} - \overline{\hat{N}}$, the inequality of the spectral radii follows from Theorem 1.10.22 with (M, N), $(\hat{M}, \hat{N})$ being interchanged there. □

Example 5.5.23. Let

$$[A] = \begin{pmatrix} [2,4] & [-2,0] \\ [-1,0] & [2,4] \end{pmatrix}, \quad [b] = \begin{pmatrix} [1,2] \\ [-2,2] \end{pmatrix}, \quad [N] = \begin{pmatrix} [0,2] & [0,2] \\ [0,1] & [0,2] \end{pmatrix},$$
$$[M]_t = \overline{A} - t[N], \quad [N]_t = (1-t)[N], \qquad 0 \le t \le 1.$$

Then $([M]_t, [N]_t)$ is an M-splitting of the M-matrix $[A]$ for each $t \in [0, 1]$. Denote by $[x]_t^*$ the corresponding limit of (5.5.8) and by S the solution set of the interval linear system $[A]x = [b]$. Then $t \le \hat{t}$ implies $[M]_{\hat{t}} \le [M]_t$, $d([M]_{\hat{t}}) \ge d([M]_t)$, hence the Theorems 5.5.20 and 5.5.22 can be applied and yield

$$[x]_t^* \subseteq [x]_{\hat{t}}^* \quad \text{and} \quad \rho(\underline{M}_{\hat{t}}^{-1}\overline{N}_{\hat{t}}) \le \rho(\underline{M}_t^{-1}\overline{N}_t).$$

For $t = 0$ we have $[M]_0 = \text{diag}(4, 4) \in \mathbb{R}^{2\times 2}$, and Theorem 5.5.18 implies $[x]_0^* = \square S = ([-1, 4], [-1.5, 3])^T$. For $t = 1$ we obtain $[x]_1^* = \text{IGA}([A], [b]) = ([-1.5, 4], [-2, 3])^T$. An easy calculation (cf. Exercise 5.5.6) yields

$$\underline{M}_t^{-1}\overline{N}_t = \frac{1-t}{8-8t+t^2}\begin{pmatrix} 4-t & 4 \\ 2 & 4-t \end{pmatrix} \quad \text{and} \quad \rho(\underline{M}_t^{-1}\overline{N}_t) = 1 - \frac{3-\sqrt{8}}{4-\sqrt{8}-t}$$

where the spectral radius decreases monotonously. This is confirmed by Table 5.5.1 with the components of the iterate $[x]^{k_s}$ which fulfills the stopping criterion

$$|\underline{x}^{k+1} - \underline{x}^k| \le 10^{-10}|\underline{x}^k| \quad \text{and} \quad |\overline{x}^{k+1} - \overline{x}^k| \le 10^{-10}|\overline{x}^k| \tag{5.5.39}$$

for the first time starting with $[x]^0 = [b]$.

Tab. 5.5.1: Results of Example 5.5.23 with outward rounding to three digits after the decimal point.

t	$[x]_1^{k_s}$	$[x]_2^{k_s}$	k_s
0	[−1, 4]	[−1.5, 3]	128
0.7	[−1, 4]	[−1.5, 3]	47
0.8	[−1.001, 4]	[−1.5, 3]	35
0.85	[−1.076, 4]	[−1.576, 3]	29
0.9	[−1.176, 4]	[−1.676, 3]	22
0.95	[−1.312, 4]	[−1.812, 3]	15
1	[−1.5, 4]	[−2, 3]	1

The symmetric Gauss–Seidel method

The symmetric Gauss–Seidel method is a combination of two Gauss–Seidel steps (5.5.14), where in the second one the order of the newly computed components is just reversed. It reads

$$\begin{cases} [x]_i^{k+\frac{1}{2}} = \left\{[b]_i - \sum_{j=1}^{i-1}[a]_{ij}[x]_j^{k+\frac{1}{2}} - \sum_{j+1}^{n}[a]_{ij}[x]_j^k\right\} / [a]_{ii}, & i = 1, 2, \ldots, n, \\ [x]_i^{k+1} = \left\{[b]_i - \sum_{j=1}^{i-1}[a]_{ij}[x]_j^{k+\frac{1}{2}} - \sum_{j+1}^{n}[a]_{ij}[x]_j^{k+1}\right\} / [a]_{ii}, & i = n, n-1, \ldots, 1. \end{cases} \tag{5.5.40}$$

With the splitting $[A] = [D] - [E] - [F]$ of (5.5.29) three half-steps of (5.5.40) are represented by

$$\begin{aligned} [x]^{k+\frac{1}{2}} &= [D]^{-1}([E][x]^{k+\frac{1}{2}} + [F][x]^k + [b]), \\ [x]^{k+1} &= [D]^{-1}([E][x]^{k+\frac{1}{2}} + [F][x]^{k+1} + [b]), \\ [x]^{k+\frac{3}{2}} &= [D]^{-1}([E][x]^{k+\frac{3}{2}} + [F][x]^{k+1} + [b]). \end{aligned}$$

This reveals that the expressions $[E][x]^{k+\frac{1}{2}}$ and $[F][x]^{k+1} + [b]$ occur twice. Therefore, work can be saved when computing them only once and storing them for the second usage. This happens half-step by half-step. With the exception of the first and the last half-step of the iteration and of n additional divisions per step, the amount of work per iterate is thus the same as with the ordinary Gauss–Seidel method. While the limits of both methods will turn out to be the same, the bound for the R_1-factor can decrease, which lets us expect a faster convergence.

We start with an auxiliary result, where we use the following $n \times n$ matrices:

$$\begin{aligned} C_E &:= \langle [D] \rangle^{-1} |[E]| \quad & C_F &:= \langle [D] \rangle^{-1} |[F]|, \\ J &:= C_E + C_F, & G &:= (I - C_F)^{-1} C_E (I - C_E)^{-1} C_F. \end{aligned} \tag{5.5.41}$$

Lemma 5.5.24. *With the notation in* (5.5.29) *and* (5.5.41) *we have* $\rho(J) < 1$ *if and only if* $\rho(G) < 1$.

Proof. The matrices C_E and C_F are strict triangular matrices. Therefore, $\rho(C_E) = \rho(C_F) = 0$, whence $(I - C_E)^{-1} = \sum_{k=0}^{\infty} C_E^k \geq O$, $(I - C_F)^{-1} \geq O$.

Let $\rho(J) < 1$. Then similarly as above we can show $(I - J)^{-1} \geq O$. With $M = (I - C_E)(I - C_F)$, $N = C_E C_F$ we get

$$I - J = M - N. \tag{5.5.42}$$

Since (M, N) is a regular splitting of the inverse positive matrix $I - J$, Theorem 1.10.21 guarantees $\rho(M^{-1}N) < 1$. This implies $\rho(G) < 1$ because of

$$M^{-1}N = (I - C_F)^{-1}(I - C_E)^{-1} C_E C_F = (I - C_F)^{-1} C_E (I - C_E)^{-1} C_F = G.$$

Assume conversely that $\rho(G) < 1$ holds. Then

$$\begin{aligned} O \leq (I - G)^{-1} &= (I - (I - C_F)^{-1}(I - C_E)^{-1} C_E C_F)^{-1} \\ &= ((I - C_E)(I - C_F) - C_E C_F)^{-1}(I - C_E)(I - C_F) \\ &= (I - J)^{-1}(I - C_E)(I - C_F) \end{aligned}$$

holds, and multiplying with $(I - C_F)^{-1}(I - C_E)^{-1}$ implies $(I - J)^{-1} \geq O$. Since (I, J) is a regular splitting of the inverse positive matrix $I - J$ Theorem 1.10.21 proves $\rho(J) = \rho(I^{-1}J) < 1$. □

We need this lemma in our next result.

Theorem 5.5.25. *Let* $[A] \in \mathbb{IR}^{n \times n}$, $[b] \in \mathbb{IR}^n$. *Then the symmetric Gauss–Seidel method converges globally to some limit* $[x]^*$ *if and only if* $\rho(J) < 1$, *i.e., if and only if the Jacobi method and therefore the Gauss–Seidel method are globally convergent. In this case the limits coincide, the method is inclusion monotone and satisfies permanence of enclosure. In addition* $S \subseteq [x]^*$ *for the solution set* S *of* $[A]x = [b]$, *and* $\rho(G) < 1$ *is an upper bound of the* R_1*-factor.*

If $[A]$ *is an* M*-matrix and* $\rho(J) < 1$, *then* $[x]^* = \square S$ *holds for the common limit of the three methods mentioned above.*

Proof. Let $\rho(J) < 1$ hold and define $[f]\colon \mathbb{IR}^n \to \mathbb{IR}^n$ by

$$\begin{cases} [x]^{1/2} = [D]^{-1}([E][x]^{1/2} + [F][x] + [b]), \\ [f]([x]) = [D]^{-1}([E][x]^{1/2} + [F][f]([x]) + [b]), \end{cases}$$

where the number $1/2$ indicates the first half-step, not the power $1/2$. The evaluation proceeds componentwise according to (5.5.40) so that both left-hand sides are defined in a unique and straightforward way.

Using the standard properties of the Hausdorff distance we obtain for $[x]$, $[y] \in \mathbb{IR}^n$ and with the notation (5.5.41)

$$\begin{aligned} q([x]^{1/2}, [y]^{1/2}) &\le |[D]^{-1}| q([E][x]^{1/2} + [F][x] + [b], [E][y]^{1/2} + [F][y] + [b]) \\ &\le \langle [D] \rangle^{-1} \{ q([E][x]^{1/2}, [E][y]^{1/2}) + q([F][x], +[F][y]) \} \\ &\le C_E q([x]^{1/2}, [y]^{1/2}) + C_F q([x], [y]). \end{aligned}$$

Therefore, $q([x]^{1/2}, [y]^{1/2}) \le (I - C_E)^{-1} C_F q([x], [y])$ and

$$q([f]([x]), [f]([y])) \le (I - C_F)^{-1} C_E q([x]^{1/2}, [y]^{1/2}) \le G q([x], [y]).$$

Hence $[f]$ is a P-contraction with $P := G$, as we can see from the assumption and Lemma 5.5.24. Theorem 4.2.6 guarantees global convergence for (5.5.40) to a limit $[x]^*$, and as in the proof of Theorem 5.5.6 one can see that the R_1-factor is bounded by $\rho(G)$.

Let, conversely, (5.5.40) be globally convergent. Then

$$\begin{aligned} d([x]^{k+1/2}) &\ge |[D]^{-1}| \{ |[E]| d([x]^{k+1/2}) + |[F]| d([x]^k) \} \\ &= C_E d([x]^{k+1/2}) + C_F d([x]^k), \end{aligned}$$

i.e., $d([x]^{k+1/2}) \ge (I - C_E)^{-1} C_F d([x]^k)$. Similarly,

$$d([x]^{k+1}) \ge (I - C_F)^{-1} C_E d([x]^{k+1/2}) \ge G d([x]^k), \quad k = 0, 1, \ldots.$$

Now $\rho(G) < 1$ can be seen by contradiction as in the proof of Theorem 5.5.6, and Lemma 5.5.24 implies $\rho(J) < 1$. By virtue of Theorem 5.5.12 the limits of the Jacobi method, the Gauss–Seidel method, and the reverse Gauss–Seidel method $[x]^{k+1} = [D]^{-1}([E][x]^k + [F]^{k+1} + [b])$, $k = 0, 1, \ldots$, coincide. Starting the symmetric Gauss–Seidel method with this common limit $[\hat{x}]$ proves $[x]^* = [\hat{x}]^*$. The remaining statements of Theorem 5.5.25 are immediate or follow from Remark 5.5.2 (b) or Theorem 5.5.18. □

If $[A]$ is not an M-matrix, then $[x]^* = \square S$ is no longer true for the limit of the three methods mentioned in Theorem 5.5.25, as can be seen from Example 5.5.15.

Theorem 5.5.26. *Let the symmetric Gauss–Seidel method converge globally. Then*

$$\rho(G) \le \rho((I - C_E)^{-1} C_F) < 1, \tag{5.5.43}$$

i.e., the upper bound of the R_1-factor of the symmetric Gauss–Seidel method is not worse than that of the ordinary Gauss–Seidel method.

Proof. The global convergence of the symmetric Gauss–Seidel method guarantees that of the ordinary Gauss–Seidel method by virtue of Theorem 5.5.25. According to Theorem 5.5.10 or Remark 5.5.2 (b) this implies the right inequality of (5.5.43).

Define the matrix $M_\varepsilon = (I - C_E)^{-1}(C_F + \varepsilon ee^T)$, $\varepsilon \in \mathbb{R}$, and $\rho_\varepsilon = \rho(M_\varepsilon)$. Both are positive for $\varepsilon > 0$. Since $\rho_0 < 1$, we can choose $\varepsilon > 0$ sufficiently small such that $\rho_\varepsilon < 1$ holds. Let $x > 0$ be a Perron vector of M_ε. Then $M_\varepsilon x = \rho_\varepsilon x$ implies

$$(C_F + \varepsilon ee^T)x = (I - C_E)\rho_\varepsilon x,$$

hence

$$C_F x \le \frac{1}{\rho_\varepsilon}(C_F + \varepsilon ee^T)x = (I - C_E)x$$

follows. From this we get $C_E x \le (I - C_F)x$ which transforms into $(I - C_F)^{-1} C_E x \le x$. Therefore, $(I - C_F)^{-1} C_E M_\varepsilon x = \rho_\varepsilon (I - C_F)^{-1} C_E x \le \rho_\varepsilon x$. Theorem 1.9.13 proves $\rho((I - C_F)^{-1} C_E M_\varepsilon) \le \rho_\varepsilon$. Let ε tend to zero. This shows $\rho(G) \le \rho_0 = \rho((I - C_E)^{-1} C_F)$. □

Combining the Theorems 5.5.14 and 5.5.26 gives hope that the symmetric Gauss–Seidel method is faster than all methods (5.5.8) which are based on a triangular splitting.

We test various iterative methods by means of the discretized Love integral equation; cf. Love [195]. To this end we remind ourselves of the trapezoidal rule

$$T_n = \frac{h}{2}\left\{g(a) + 2\sum_{i=1}^{n-1} g(a + ih) + g(b)\right\}, \quad h = \frac{b-a}{n}, \tag{5.5.44}$$

which approximates the integral

$$\int_a^b g(x)\,dx, \quad g \in C^2[a, b]. \tag{5.5.45}$$

The sum (5.5.44) plus the remainder term $-\frac{h^2}{12} g''(\xi)$ represent the integral (5.5.45), where ξ is an appropriate element of $[a, b]$.

Example 5.5.27. The Love integral equation reads

$$u(x) + \frac{1}{\pi}\int_{-1}^{1} \frac{c}{c^2 + (x - t)^2} u(t)\,dt = f(x), \quad |x| \le 1, \tag{5.5.46}$$

with some given function f. This integral equation is a Fredholm integral equation of the second kind which occurs in the field of electrostatics. We consider (5.5.46) for $c = -1$ and $f(x) \equiv 1$. The trapezoidal rule applied to the equidistant decomposition

$$x_0 = -1 < x_1 = -1 + h < x_2 = -1 + 2h < \cdots < x_n = 1, \quad h = \frac{2}{n},$$

leads to the linear system

$$u_i - \frac{h}{\pi}\left\{\frac{u_0/2}{1 + (x_i + 1)^2} + \frac{u_n/2}{1 + (x_i - 1)^2} + \sum_{j=1}^{n-1} \frac{u_j}{1 + (x_i - x_j)^2}\right\} = 1, \quad i = 0, 1, \ldots, n. \tag{5.5.47}$$

Here u_i denotes the approximation for the value $u(x_i)$ of the unknown solution of (5.5.46). We write (5.5.47) in short form as $Az = e$ with $A \in \mathbb{R}^{(n+1)\times(n+1)}$,

$$a_{ij} = \delta_{ij} - \frac{1}{\pi n}\begin{cases} \dfrac{1}{1+(x_i-x_j)^2}, & \text{if } j \in \{0,n\}, \\ \dfrac{2}{1+(x_i-x_j)^2}, & \text{if } j \in \{1,\dots,n-1\}, \end{cases}$$

$z = (u_0, \dots, u_n)^T$. We show that A is strictly diagonally dominant, hence it is an M-matrix because $a_{ij} \le 0$ holds for $i \ne j$. To this end we define

$$T_n(x) = h\left\{\frac{1/2}{1+(x+1)^2} + \frac{1/2}{1+(x-1)^2} + \sum_{j=1}^{n-1}\frac{1}{1+(x-x_j)^2}\right\}, \tag{5.5.48}$$

which is the trapezoidal expression for $\int_a^b \frac{1}{1+(x-t)^2}\,dt$. We are going to show that

$$0 \le T_n(x) < \pi \tag{5.5.49}$$

holds for $|x| \le 1$. This implies $0 \le \frac{1}{\pi}T_n(x_i) < 1$, i.e., A is strictly diagonally dominant.

With the remainder term of the trapezoidal rule we get

$$\begin{aligned} 0 \le T_n(x) &= \int_{-1}^{1}\frac{1}{1+(x-t)^2}\,dt + \frac{h^2}{12}\cdot\frac{-2+6(x-\xi)^2}{(1+(x-\xi)^2)^3} \\ &\le \arctan(1-x) + \arctan(1+x) + \frac{1}{3n^2} \\ &\le 2\arctan 1 + \frac{1}{3n^2} = \frac{\pi}{2} + \frac{1}{3n^2} < \pi \end{aligned}$$

for $n \in \mathbb{N}$, $|x| \le 1$, $|x-\xi| \le 2$, since the fraction behind $h^2/12$ is less than one and the sum of the two arctangent terms increases for $-1 \le x \le 0$ and decreases for $0 \le x \le 1$. Next we show that $[0, 3e]$ contains the solution x^* of (5.5.47): If $Ax^* = e$, then $x^* = A^{-1}e > 0$. In addition,

$$(A(3e))_i = 3\left\{1 - \frac{1}{\pi}T_n(x_i)\right\} \ge 3 - \frac{3}{\pi}\left(\frac{\pi}{2} + \frac{1}{3n^2}\right) = \frac{3}{2} - \frac{1}{\pi n^2} > 1$$

holds for $n \in \mathbb{N}$ and $i = 0, 1, \dots, n$. This implies $A(3e) > e$, whence $3e > A^{-1}e = x^*$. Thus we can start our iterations with $[x]^0 = [0, 3e]$.

From the Theorems 5.5.16, 5.5.18, and 5.5.25 we know that the Richardson iteration, the Jacobi iteration, the Gauss–Seidel iteration, and the symmetric Gauss–Seidel iteration converge to x^*. Since the entries a_{ij} are certainly not machine numbers we must compute them using interval arithmetic. This results finally in an interval linear system

$$[A]x = e \tag{5.5.50}$$

to which our iterative methods are applied. With $[x]^0 = [0, 3e]$ we got $[x]^1 \subseteq \operatorname{int}([x]^0)$ on the computer for all the four methods just mentioned. Therefore, the Theorems 5.5.9

and 5.5.25 guarantee global convergence also when A is replaced by $[A]$. By this and the outward rounding of the machine interval arithmetic, the iterates $[x]^k$ do certainly not contract to x^* but contain x^* for all $k \in \mathbb{N}_0$. In our computations we chose $n = 64$ and stopped the iteration when the stopping criterion (5.5.26) was fulfilled for the first time. The results can be seen in Table 5.5.2, where we list k from the stopping criterion (5.5.26) and the largest diameter $d = \max_{0 \le i \le n} d([x]_i^k)$ of the corresponding components $[x]_i^k$.

In passing we note that the interval Gaussian algorithm applied to the computed matrix $[A]$ and the right-hand side e yields a superset of the final iterates above with $d = 4.662\,936\,703\,425\,658 \cdot 10^{-14}$. A method for verifying and enclosing the solution of the original, i.e., undiscretized Love integral equation (5.5.46) can be found for instance in Klein [166].

Tab. 5.5.2: Results of Example 5.5.27.

Method	k	d
Richardson	50	$1.154631945610163 \cdot 10^{-14}$
Jacobi	49	$1.154631945610163 \cdot 10^{-14}$
Gauss–Seidel	30	$4.884981308350689 \cdot 10^{-15}$
sym. Gauss–Seidel	19	$4.884981308350689 \cdot 10^{-15}$

Richardson iteration

The Richardson iteration (5.5.12) is a very simple splitting for which more results can be proved as for the general iteration (5.5.8). For convenience and w.l.o.g. we will consider the iteration

$$[x]^{k+1} = [A][x]^k + [b], \quad k = 0, 1, \ldots \tag{5.5.51}$$

in this subsection instead of (5.5.12). It can be interpreted as the Richardson iteration (5.5.12) applied to the interval linear system

$$(I - [A])x = [b] \tag{5.5.52}$$

with the Richardson splitting $(I, [A])$. Thus $\rho(|[M]^G| \cdot |[N]|) = \rho(|[A]|)$, which allows an easy reformulation of the previous theorems. With

$$[f]([x]) = [A][x] + [b] \tag{5.5.53}$$

the iteration (5.5.51) reads $[x]^{k+1} = [f]([x]^k)$ with global convergence if and only if $\rho(|[A]|) < 1$.

We start our studies with existence, uniqueness and representations of fixed points of f, even in the case $\rho(|[A]|) \ge 1$. Our first result is simply a reformulation of Theorem 5.5.6.

Theorem 5.5.28. *Let* $[A] \in \mathbb{IR}^{n\times n}$, $[b] \in \mathbb{IR}^n$. *If* $\rho(|[A]|) < 1$ *the function* $[f]$ *from* (5.5.53) *has a unique fixed point* $[x]^*$, *the iteration* (5.5.51) *converges globally to* $[x]^*$, *and a bound for its* R_1*-factor is* $\rho(|[A]|)$. *In addition,* $S \subseteq [x]^*$ *holds for the solution set* S *of the interval linear system* $(I - [A])x = [b]$.

Our second theorem lists some representations of fixed points for particular classes of matrices.

Theorem 5.5.29. *Let* $[A] \in \mathbb{IR}^{n\times n}$, $[b] \in \mathbb{IR}^n$ *with* $\rho(|[A]|) < 1$. *If one of the following three properties*
(i) $[A] = -[A]$,
(ii) $[A] \equiv A \in \mathbb{R}^{n\times n}$,
(iii) $[b] = -[b]$
is fulfilled, then the limit $[x]^*$ *of* (5.5.51) *can be represented as*

$$[x]^* = \check{x}^* + r_{x^*}[-1, 1] \quad \text{with } \check{x}^* = (I - \check{A})^{-1}\check{b},\ r_{x^*} = (I - |[A]|)^{-1}(r_A|\check{b}| + r_b). \tag{5.5.54}$$

Proof. (i) The assumption implies $\check{A} = O$, $|[A]| = r_A$. Taking midpoints in the limit equation

$$[x]^* = [A][x]^* + [b] \tag{5.5.55}$$

leads to $\check{x}^* = \check{b}$. Similarly, the radius operation yields

$$r_{x^*} = r_A|[x]^*| + r_b = r_A|\check{x}^*| + r_A r_{x^*} + r_b = r_A|\check{b}| + r_A r_{x^*} + r_b.$$

Solving for r_{x^*} results in (5.5.54).

(ii) Taking the radius on both sides of (5.5.55) results in $r_{x^*} = |A|r_{x^*} + r_b$ which – together with $r_A = O$ – leads to the expression for r_{x^*} in (5.5.54). The expression for $\check{x}^*$ follows from the midpoint operation applied to (5.5.55) with $A = \check{A}$.

(iii) Starting (5.5.51) with the zero-symmetric interval $[x]^0 = |[x]^0|[-1, 1]$ and taking into account $\check{b} = 0$ keeps symmetry in $[x]^1$ and, by induction, in all iterates $[x]^k$, $k = 0, 1, \ldots$. Therefore, $\check{x}^* = 0$ holds, and (5.5.54) follows from $[x]^* = |[A]|\,[x]^* + [b]$ taking the radius on both sides of (5.5.55). □

To our knowledge an explicit formula for $[x]^*$ in the case $[b] \equiv b \in \mathbb{R}^n$ is missing. Restricting $[A]$ slightly and generalizing $[b]$ in Theorem 5.5.29 (iii) yields the following result, where the Hadamard product $.*$ is used.

Theorem 5.5.30. *Let* $[A] \in \mathbb{IR}^{n\times n}$, $[b] \in \mathbb{IR}^n$, $0 \notin \mathrm{int}([a]_{ij})$ *for* $i, j = 1, \ldots, n$, $\rho(|[A]|) < 1$. *If* $0 \in [b]$, *then the limit* $[x]^*$ *of* (5.5.51) *contains* 0 *and can be represented as*

$$[x]^* = \check{x}^* + r_{x^*}[-1, 1] \quad \text{with } \check{x}^* = (I - S_{\check{A}} .* |[A]|)^{-1}\check{b},\quad r_{x^*} = (I - |[A]|)^{-1}r_b, \tag{5.5.56}$$

where $S_{\check{A}} = (\mathrm{sign}(\check{a}_{ij})) \in \mathbb{R}^{n\times n}$.

Proof. Since $0 = \tilde{A} \cdot 0 + 0$ we have $0 \in [x]^*$. Therefore, $|\check{x}^*| \le r_{x^*}$, and $r_{x^*}^- = |\check{x}^*|$, $r_{x^*}^+ = r_{x^*}$. The assertion follows now from $|\check{A}| + r_A = |[A]|$ and Theorem 3.1.5 (c) and (f). □

We apply this theorem to two examples.

Example 5.5.31.

(a) Consider (5.5.51) with

$$[A] = \frac{1}{4}\begin{pmatrix} [0,1] & 1 \\ -1 & [-1,0] \end{pmatrix} \quad \text{and} \quad [b] = \begin{pmatrix} [0,2] \\ [-2,8] \end{pmatrix}.$$

Here $\rho(|[A]|) = 1/2 < 1$, hence Theorem 5.5.30 applies. It yields $\check{x}^* = (2,2)^T$ and $r_{x^*} = (4,8)^T$, whence $[x]^* = ([-2,6],[-6,10])^T$.

(b) We apply Theorem 5.5.30 formally to

$$[A] = \begin{pmatrix} [2,4] & [0,2] \\ [-2,0] & [-3,-1] \end{pmatrix} \quad \text{and} \quad [b] = ([0,2],[0,4])^T.$$

This time we get $r_{x^*} = (I - |[A]|)^{-1} r_b = (1,-2)^T$. Since r_{x^*} cannot have negative components, the assumption $\rho(|[A]|) < 1$ must be violated. In fact we have $\rho(|[A]|) > 1$, and we will show below that the equation $[x] = [A][x] + [b]$ can only have a solution $[x]^*$ for the given matrix if $r_b = 0$. This is not the case in our example.

Now we take a closer look to matrices $[A] \ge O$. If $\rho(|[A]|) = \rho(\overline{A}) < 1$, the matrix $[B] = I - [A]$ is an M-matrix with $\overline{b}_{ii} \le 1$, $i = 1, \dots, n$, since $\underline{B} = I - \overline{A} \le I - \underline{A} = \overline{B} \in Z^{n\times n}$ and $\underline{B}^{-1} = (I - \overline{A})^{-1} = \sum_{k=0}^{\infty} \overline{A}^k \ge O$.

If, conversely, $[B] = I - (I - [B]) = I - [A] \in \mathbb{IR}^{n\times n}$ is an M-matrix with $\overline{b}_{ii} \le 1$, $i = 1, \dots, n$, then $\overline{B} = I - \underline{A} \in Z^{n\times n}$, whence $-\overline{b}_{ij} = \underline{a}_{ij} \ge 0$ for $i \ne j$. By our assumption on $\overline{b}_{ii}$ we have $\overline{b}_{ii} = 1 - \underline{a}_{ii} \le 1$, $i = 1, \dots, n$, which leads to $\underline{a}_{ii} \ge 0$, $i = 1, \dots, n$ and finally to $\underline{A} \ge O$. In addition, for the M-matrix $\underline{B}$ there is a positive vector u such that $0 < \underline{B}u = (I - \overline{A})u$, which implies $\overline{A}u < u$ for $\overline{A} \ge O$. Hence Theorem 1.9.13 guarantees $\rho(|[A]|) = \rho(\overline{A}) < 1$. Thus we have proved the following auxiliary result.

Lemma 5.5.32. *The matrix $[A] \in \mathbb{IR}^{n\times n}$ satisfies $[A] \ge O$ and $\rho(|[A]|) < 1$ if and only if $[B] = I - [A]$ is an M-matrix with $\overline{b}_{ii} \le 1$, $i = 1, \dots, n$.*

If $[B] \in \mathbb{IR}^{n\times n}$ is an M-matrix without any restrictions to its diagonal entries, then $D^{-1}[B]$ with $D = \mathrm{diag}(\overline{b}_{11}, \dots, \overline{b}_{nn}) \in \mathbb{R}^{n\times n}$ is an M-matrix to which Lemma 5.5.32 applies with $D^{-1}[B]$ instead of $[B]$. Thus iterations (5.5.51) with $[A] \ge O$ and interval linear systems $[B]x = [b]$ with an M-matrix $[B]$ are closely related. In particular, if $[A] \ge O$ and $\rho(|[A]|) < 1$ hold, then the limit $[x]^*$ of (5.5.51) is an optimal enclosure of the solution set S of the interval linear system $([I - [A])x = [b]$, i.e., $[x]^* = \square S$. This follows from Theorem 5.5.18 since $(I, [A])$ is an M-splitting of the M-matrix $[B] = I - [A]$, which is obviously inverse positive. We state this result separately in our next theorem in which we repeated the representation of $[x]^* = [x]^H$ in Theorem 5.3.15 replacing there $[A]$ by $I - [A]$.

Theorem 5.5.33. *Let $[b] \in \mathbb{IR}^n$, $[A] \in \mathbb{IR}^{n\times n}$, $[A] \geq O$, $\rho(|[A]|) < 1$. Then the limit $[x]^*$ of (5.5.51) is the interval hull $[x]^H$ of the solution set of $(I - [A])x = [b]$. It can be represented as*

$$[x]^* = \begin{cases} [(I-\overline{A})^{-1}\underline{b}, (I-\underline{A})^{-1}\overline{b}] \leq 0, & \textit{if } [b] \leq 0, \\ [(I-\overline{A})^{-1}\underline{b}, (I-\overline{A})^{-1}\overline{b}] \ni 0, & \textit{if } [b] \ni 0, \\ [(I-\underline{A})^{-1}\underline{b}, (I-\overline{A})^{-1}\overline{b}] \geq 0, & \textit{if } [b] \geq 0. \end{cases}$$

We do not know how to represent the limit $[x]^* = [x]^H$ by a *single* expression if $[b] \in \mathbb{IR}^n$ is *arbitrary* and $[A] \geq O$, $\rho(|[A]|) < 1$. But we can construct it by means of finitely many iterations using the modified sign accord algorithm 5.3.14; cf. Theorem 5.3.15, its proof and the text preceding Algorithm 5.3.14.

Limits $[x]^*$ of (5.5.51) are obviously algebraic solutions of the interval equation $[x] = [A][x] + [b]$ and, equivalently, fixed points of the interval function

$$[f]([x]) = [A][x] + [b]. \tag{5.5.57}$$

If $\rho(|[A]|) < 1$, we already know from Theorem 5.5.28 that $[f]$ has a unique fixed point. If $\rho(|[A]|) \geq 1$, the iteration (5.5.51) is no longer globally convergent, but fixed points of $[f]$ can exist as the following example shows.

Example 5.5.34.

(a) For each fixed $r \in [-1, 1]$ the interval function $[f]([x]) = [-1, r][x]$ has the infinitely many fixed points $[x]^* = [-t, t]$, $t \geq 0$, and no other ones. If $-1 \leq r < 1$, then the solution set $S_r = \{ x \in \mathbb{R} \mid (1-a)x = 0,\ a \in [-1, r] \} = \{0\}$ is contained in every fixed point $[x]^*$. If $r = 1$, then $S_1 = \mathbb{R}$, since any real number solves the particular equation $x = 1 \cdot x$. Therefore no fixed point of $[f]([x]) = [-1, 1][x]$ can contain the complete solution set S_1. Notice that $\rho(|[A]_r|) = 1$ for $[A]_r = ([-1, r]) \in \mathbb{IR}^{1\times 1}$, $r \in [-1, 1]$.

(b) For each fixed $r \in [-2, 2]$ the interval function $[f]([x]) = [-2, r][x]$ has the unique fixed point $[x]^* = 0$. If $-2 \leq r < 1$, then this fixed point contains the solution set $S_r := \{ x \in \mathbb{R} \mid (1-a)x = 0,\ a \in [-2, r] \} = \{0\}$ while $[x]^* \not\supseteq S_r = \mathbb{R}$ in the case $1 \leq r \leq 2$. Here, $\rho(|[A]_r|) = 2 > 1$ holds for $[A]_r = ([-2, r]) \in \mathbb{IR}^{1\times 1}$, $r \in [-2, 2]$.

This example shows also that the solution set S of the interval linear system $(I - [A])x = [b]$ does not need to be contained in a fixed point $[x]^*$ of $[f]$ from (5.5.57). Our next result states, however, that there is at least some connection between S and $[x]^*$.

Theorem 5.5.35. *Let $[A] \in \mathbb{IR}^{n\times n}$, $[b] \in \mathbb{IR}^n$. Denote by S the solution set of the interval linear system $(I - [A])x = [b]$, and let $[x]^*$ be a fixed point of $[f]$ defined in (5.5.57). Then for any linear system $(I - A)x = b$ with $A \in [A]$, $b \in [b]$, there is at least one solution which is contained in $[x]^*$, and $S \subseteq [x]^*$ holds if and only if $I - [A]$ is regular, i.e., if $I - [A]$ does not contain any singular matrix.*

Proof. Define the function h by $h(x) = Ax + b$ for any fixed $A \in [A]$, $b \in [b]$. Then $h(x) \in [A][x]^* + [b] = [x]^*$ holds for any $x \in [x]^*$. By Brouwer's fixed point theorem there is a vector $\tilde{x} \in [x]^*$ which satisfies $\tilde{x} = A\tilde{x} + b$. Since $\tilde{x} \in S$ this terminates the first part of the proof. If $S \subseteq [x]^*$ then S is bounded. Therefore, since we have already proved that any linear system $(I - A)x = b$ with $A \in [A]$ and $b \in [b]$ has at least one solution, $I - A$ must be regular for any $A \in [A]$. If, conversely, $I - [A]$ is regular then $(I - A)x = b$ is uniquely solvable for any $A \in [A]$, $b \in [b]$, and the first part of the theorem guarantees $S \subseteq [x]^*$. □

Now we want to generalize the results of Example 5.5.34. For simplicity we will restrict ourselves to matrices $[A]$ with irreducible absolute value. The reducible case is much more complicated and is studied in Arndt, Mayer [55].

Lemma 5.5.36. *Let $[b], [x]^* \in \mathbb{IR}^n$, $[A] \in \mathbb{IR}^{n\times n}$, $|[A]|$ irreducible, and $[x]^* = [A][x]^* + [b]$. Then either $d([x]^*) > 0$ or $d([x]^*) = 0$. If $d([x]^*) = 0$, then $d([b]) = 0$; if $d([b]) \neq 0$, then $d([x]^*) > 0$; if $\rho(|[A]|) > 1$, then $d([x]^*) = d([b]) = 0$.*

Proof. From the fixed point equation we get

$$d([x]^*) \geq |[A]|d([x]^*) + d([b]) \geq |[A]|d([x]^*) \geq |[A]|^k d([x]^*), \quad k = 1, 2, \ldots. \tag{5.5.58}$$

If $d([x]_j^*) > 0$ holds for some j and if i is arbitrary, then by Theorem 1.9.2 there is a power $|[A]|^k = (c_{ij})$ such that $c_{ij}d([x]_j^*) > 0$, which implies $d([x]_i^*) \geq c_{ij}d([x]_j^*) > 0$. Hence $d([x]^*) > 0$ follows. From (5.5.58) we get $d([b]) = 0$ if $d([x]^*) = 0$, and $d([x]^*) \neq 0$ if $d([b]) \neq 0$, whence $d([x]^*) > 0$ by what we have proved up to now. If $\rho(|[A]|) > 1$ and $d([x]^*) > 0$ hold, then define the diagonal matrix $D \in \mathbb{R}^{n\times n}$ by $d_{ii} = d([x]_i^*)$, $i = 1, \ldots, n$. From (5.5.58) we get the contradiction

$$1 \geq \max_{1\leq i\leq n} \frac{(|[A]|\, d([x^*]))_i}{d([x]_i^*)} = \|D^{-1}|[A]|\, D\|_\infty \geq \rho(|[A]|) > 1.$$ □

The example $[-1/2, 1/2] \cdot [0, 2] + 1 = [0, 2]$ shows that $d([b]) = 0$ does not necessarily imply $d([x]^*) = 0$.

From the preceding lemma we know that in the case $\rho(|[A]|) > 1$, $|[A]|$ irreducible, a fixed point of $[f]$ from (5.5.57) can only exist if $d([b]) = 0$. In addition, $[x]^*$ must be degenerate. Therefore, we will study degenerate fixed points of $[f]$ first, even if $\rho(|[A]|) \leq 1$.

Theorem 5.5.37. *Let $|[A]|$ be irreducible and let $\hat{A} \in \mathbb{R}^{n\times n}$ be the matrix which arises from $[A] \in \mathbb{IR}^{n\times n}$ by replacing the nondegenerate columns there by the corresponding columns of the identity matrix $I \in \mathbb{R}^{n\times n}$. In addition, let $[f]$ be defined as in (5.5.57).*

(a) *The interval function $[f]$ has a degenerate fixed point if and only if $[b]$ is degenerate, i.e., $[b] \equiv b \in \mathbb{R}^n$, and the linear system*

$$x = \hat{A}x + b \tag{5.5.59}$$

is solvable, i.e., b can be represented as a linear combination of the degenerate columns of $I - [A]$. In this case there is at least one solution x^* of (5.5.59) which *satisfies*

$$x_i^* = 0 \text{ for all } i \in M, \tag{5.5.60}$$

where M denotes the set of indices for which the columns of $[A]$ are nondegenerate. The degenerate fixed points $[x]^* \equiv x^*$ of $[f]$ are just the solutions of (5.5.59) *satisfying* (5.5.60).

(b) *If $[A]$ has no degenerate columns, then $[f]$ has a degenerate fixed point $[x]^* \equiv x^* \in \mathbb{R}^n$ if and only if $[b] \equiv 0 \in \mathbb{R}^n$. In this case $x^* = 0$; it is unique in $\mathbb{R}^n$, i.e., there are no additional degenerate fixed points of $[f]$.*

(c) *A degenerate solution $[x]^* \equiv x^* \in \mathbb{R}^n$ of $[f]$ is unique in $\mathbb{R}^n$ if and only if either $[A]$ has no degenerate columns – then $b = x^* = 0$ (cf.* (b)*) – or the degenerate columns of $I - [A]$ are linearly independent.*

(d) *A degenerate solution $[x]^* \equiv x^* \in \mathbb{R}$ of* (5.5.57) *is unique in $\mathbb{IR}^n$, i.e., there are no additional possibly nondegenerate fixed points of $[f]$, if and only if one of the following conditions hold:*
   (i) $\rho(|[A]|) < 1$;
   (ii) $\rho(|[A]|) > 1$ *and $[x]^* \equiv x^*$ is unique in $\mathbb{R}^n$ (which is studied in* (c)*).*
   In particular, $[x]^* \equiv x^*$ is not unique in $\mathbb{IR}^n$ if $\rho(|[A]|) = 1$.

Proof. (a) Let $[x]^* \equiv x^* \in \mathbb{R}^n$ be a fixed point of $[f]$. From $d([b]) \le d([b]) + d([A]x^*) = d(x^*) = 0$ we get $[b] \equiv b \in \mathbb{R}^n$ and $d([A]x^*) = 0$. This latter equality implies $x_i^* = 0$ for $i \in M$. Therefore, the fixed point equation $x^* = [A]x^* + b$ together with (5.5.60) proves the only-if-part of (a).

In order to verify the if-part of (a), choose any solution y^* of (5.5.59), replace the components y_i^* by 0 for every $i \in M$ and denote the resulting vector by x^*. Since (5.5.59) is equivalent to $(I - \hat{A})x = b$ and since by the particular form of $\hat{A}$ the i-th column of $I - \hat{A}$ is zero for $i \in M$, the vector x^* remains a solution of (5.5.59) and is a fixed point of $[f]$.

(b) If $[A]$ has no degenerate column, then (5.5.60) implies $x^* = 0$, and $b = 0$ follows from (5.5.59).

(c) If $[A]$ has no degenerate columns, the assertion is proved by (b). Otherwise it follows from (a) by using (5.5.59) and (5.5.60).

(d) In the case $\rho(|[A]|) < 1$ the assertion follows from Theorem 5.5.28, in the case $\rho(|[A]|) > 1$ it follows from Lemma 5.5.36. In the case $\rho(|[A]|) = 1$ no fixed point is unique in $\mathbb{IR}^n$ as we shall see later on. □

Our next result is a direct consequence of Lemma 5.5.36 and Theorem 5.5.37.

Theorem 5.5.38. *Let $|[A]|$ be irreducible with $\rho(|[A]|) > 1$, and let $[f]$ be defined as in* (5.5.57).

(a) *If $d([b]) \neq 0$, then $[f]$ has no fixed point.*

(b) *If $d([b]) = 0$, then every fixed point $[x]^*$ of $[f]$ is degenerate, and existence and uniqueness of $[x]^*$ are completely handled by Theorem* 5.5.37.

In order to prove the main result on (5.5.57) for $\rho(|[A]|) = 1$ we need some basic results which we gather in the subsequent lemma.

Lemma 5.5.39. *Let $|[A]|$ be irreducible with $\rho(|[A]|) = 1$ and let $[x]^* \in \mathbb{IR}^n$ be a fixed point of $[f]$ from (5.5.57) satisfying $d([x]^*) > 0$. Then the following properties hold.*

(a) *The vector $[b]$ is degenerate, i.e., $[b] \equiv b \in \mathbb{R}^n$, and $d([x]^*) = |[A]|d([x]^*)$, i.e., $d([x]^*)$ is a Perron vector of $|[A]|$. In particular*
$$[x]^* = \check{x}^* + [-v, v] \tag{5.5.61}$$
with the Perron vector $v := d([x]^*)/2 = \mathrm{rad}([x]^*)$. If $\dot{A} \in [A]$ is such that $|\dot{A}| = |[A]|$, then
$$\check{x}^* = \dot{A}\check{x}^* + b \tag{5.5.62}$$
and
$$[x]^* = \dot{A}[x]^* + b = [A][x]^* + b, \tag{5.5.63}$$
in particular,
$$\dot{A}[x]^* = [A][x]^*. \tag{5.5.64}$$
For each nondegenerate symmetric entry $[a]_{i_0j_0}$ of $[A]$ we have $\check{x}^*_{j_0} = 0$.

(b) *If $[y]^* \neq [x]^*$ is another fixed point of $[f]$, then*
$$q([x]^*, [y]^*) = |[A]|q([x]^*, [y]^*),$$
i.e., $q([x]^*, [y]^*)$ is a Perron vector of $|[A]|$.
The vector $[y]^*$ can be represented as $[y]^* = \check{y}^* + [-w, w]$ with $w = 0$ or w being a Perron vector of $|[A]|$, and $|\check{x}^* - \check{y}^*| = |[A]|\,|\check{x}^* - \check{y}^*|$ holds, i.e., either $\check{x}^* = \check{y}^*$ or $|\check{x}^* - \check{y}^*|$ is again a Perron vector of $|[A]|$.

Proof. (a) Let $d([b]) \neq 0$. Then $d([x]^*) = d([A][x]^*) + d([b]) \geq |[A]|d([x]^*)$ with inequality in at least one component. Hence
$$1 \geq \max_{1<i<n} \frac{(|[A]|d([x]^*))_i}{d([x]_i^*)}$$
and
$$1 > \min_{1\leq i\leq n} \frac{(|[A]|d([x]^*))_i}{d([x]_i^*)}.$$
By virtue of Theorem 1.9.8 we get $\rho(|A|) < 1$, which contradicts our assumption. Therefore, $d([b]) = 0$, $[b] \equiv b \in \mathbb{R}^n$, and the same arguments apply for showing $d([x]^*) = |[A]|\, d([x]^*)$.

The representation $[x]^* = \check{x}^* + [-v, v]$ with a Perron vector v of $|[A]|$ is a trivial consequence of what we have already proved.

Let $\dot{A} \in [A]$ be such that $|\dot{A}| = |[A]|$ holds. Then we get
$$\begin{aligned}\dot{A}\check{x}^* + [-v, v] + b &= \dot{A}\check{x}^* + |\dot{A}|\,[-v, v] + b = \dot{A}\check{x}^* + \dot{A}[-v, v] + b\\ &= \dot{A}(\check{x}^* + [-v, v]) + b = \dot{A}[x]^* + b\\ &\subseteq [A][x]^* + b = [x]^* = \check{x}^* + [-v, v].\end{aligned}$$

Since the vector on the left-hand side and the vector on the right-hand side have the same diameter $2v$ we can replace '$\subseteq$' by '$=$'. This implies immediately (5.5.62), (5.5.63) and (5.5.64).

If $[a]_{i_0j_0} = [-a_{i_0j_0}, a_{i_0j_0}]$ with some $a_{i_0j_0} > 0$, then $\dot{a}_{i_0j_0} = a_{i_0j_0}$ or $\dot{a}_{i_0j_0} = -a_{i_0j_0}$. Change the sign of this entry in $\dot{A}$ and denote the resulting matrix by $\ddot{A}$. Then $\ddot{A} \in [A]$ and $|\ddot{A}| = |[A]|$, hence $\check{x}^* = \ddot{A}\check{x}^* + b$ by (5.5.62). Subtracting this equation from (5.5.62) yields $(\dot{A} - \ddot{A})\check{x}^* = 0$. Since $\dot{a}_{ij} - \ddot{a}_{ij} = 0$ for $(i,j) \neq (i_0, j_0)$ while $\dot{a}_{i_0j_0} - \ddot{a}_{i_0j_0} \neq 0$ we must have $\check{x}^*_{j_0} = 0$.

The representation of $[y]^*$ is either trivial or follows from (5.5.61). Define $q := q([x]^*, [y]^*) + u$ where u is any Perron vector of $|[A]|$. Then $q > 0$ and $|[A]|u = u$ leads to

$$\begin{aligned} q &= q([A][x]^* + [b], [A][y]^* + [b]) + u = q([A][x]^*, [A][y]^*) + u \\ &\le |[A]|q([x]^*, [y]^*) + u = |[A]|q([x]^*, [y]^*) + |[A]|u = |[A]|q. \end{aligned} \tag{5.5.65}$$

If $q([x]^*, [y]^*) \neq |[A]|q([x]^*, [y]^*)$, then strict inequality holds in (5.5.65) at least for one component. Hence

$$1 < \max_{1\le i\le n} \frac{(|[A]|q)_i}{q_i}$$

and

$$1 \le \min_{1\le i\le n} \frac{(|[A]|q)_i}{q_i}.$$

This yields the contradiction $\rho(|[A]|) > 1$ by the same theorem as above.

Let $[y]^* = \check{y}^* + [-w, w]$ with $w = 0$ or w being a Perron vector of $|[A]|$. W.l.o.g. let $w \le v$. (Otherwise interchange the roles of $[x]^*$ and $[y]^*$.) Then $q([x]^*, [y]^*) = |\check{x}^* - \check{y}^*| + v - w$, and from $q([x]^*, [y]^*) = |[A]|q([x]^*, [y]^*)$ we obtain $|\check{x}^* - \check{y}^*| = |[A]|\,|\check{x}^* - \check{y}^*|$. □

We illustrate Lemma 5.5.39 by an example.

Example 5.5.40. Let

$$[A] = \begin{pmatrix} \frac{1}{2} & [0, \frac{1}{2}] \\ \frac{1}{2} & \frac{1}{2} \end{pmatrix}, \quad [b] \equiv b = \begin{pmatrix} 1 \\ -1 \end{pmatrix}.$$

Then

$$|[A]| = \begin{pmatrix} \frac{1}{2} & \frac{1}{2} \\ \frac{1}{2} & \frac{1}{2} \end{pmatrix}, \quad \text{and} \quad v = \begin{pmatrix} 1 \\ 1 \end{pmatrix}$$

is a Perron vector of $|[A]|$. The only matrix $\dot{A} \in [A]$ with $|\dot{A}| = |[A]|$ is

$$\dot{A} = \begin{pmatrix} \frac{1}{2} & \frac{1}{2} \\ \frac{1}{2} & \frac{1}{2} \end{pmatrix}.$$

Therefore,

$$\check{x}^* = \dot{A}\check{x}^* + b \Leftrightarrow \begin{pmatrix} \frac{1}{2} & -\frac{1}{2} \\ -\frac{1}{2} & \frac{1}{2} \end{pmatrix} \check{x}^* = \begin{pmatrix} 1 \\ -1 \end{pmatrix}$$

$$\Leftrightarrow \check{x}^* = \begin{pmatrix} 1 \\ -1 \end{pmatrix} + sv = \begin{pmatrix} 1+s \\ -1+s \end{pmatrix}, \qquad s \in \mathbb{R}.$$

Hence a fixed point $[x]^*$ of $[f]$ from (5.5.57) must have the form

$$[x]^* = \check{x}^* + t[-v, v] = \begin{pmatrix} [1+s-t, 1+s+t] \\ [-1+s-t, -1+s+t] \end{pmatrix}, \qquad t \geq 0. \tag{5.5.66}$$

Since the second row of $[A]$ is degenerate, one can easily see that an interval vector of the form (5.5.66) satisfies $[x]^* = [A][x]^* + [b]$ in the second component. In order to fulfill it in its first component we must have

$$\tfrac{1}{2}[1+s-t, 1+s+t] + \left[0, \tfrac{1}{2}\right][-1+s-t, -1+s+t] = [s-t, s+t]. \tag{5.5.67}$$

If $-1+s+t < 0$, then the upper bound of (5.5.67) reads $\frac{1}{2}(1+s+t) + 0 = s+t$ and leads to the contradiction $-1+s+t = 0$. If $-1+s-t > 0$, then the lower bound of (5.5.67) reads $\frac{1}{2}(1+s-t) + 0 = s-t$ and leads to the contradiction $-1+s-t = 0$. If

$$-1+s+t \geq 0 \quad \text{and} \quad -1+s-t \leq 0, \tag{5.5.68}$$

then (5.5.67) reads

$$\tfrac{1}{2}[1+s-t, 1+s+t] + \tfrac{1}{2}[-1+s-t, -1+s+t] = [s-t, s+t],$$

i.e., (5.5.67) is fulfilled. Since (5.5.68) is equivalent to $|1-s| \leq t$ we end up with the following result: The interval vector $[x]^*$ is a fixed point of $[f]$ from (5.5.57) if and only if

$$[x]^* = \begin{pmatrix} 1 \\ -1 \end{pmatrix} + sv + tv[-1, 1] \quad \text{with } t \geq |1-s|. \tag{5.5.69}$$

The vector $[x]^*$ is degenerate if and only if $t = 0$, hence $s = 1$ and

$$[x]^* \equiv x^* = \begin{pmatrix} 2 \\ 0 \end{pmatrix}.$$

This confirms Theorem 5.5.37 with

$$\hat{A} = \begin{pmatrix} \frac{1}{2} & 0 \\ \frac{1}{2} & 1 \end{pmatrix} \quad \text{and} \quad \hat{A}x^* + b = x^*.$$

It also confirms Lemma 5.5.39 (b). □

Theorem 5.5.41. *Let $|[A]|$ be irreducible with $\rho(|[A]|) = 1$ and let $[f]$ be defined as in* (5.5.57).

(a) *The interval function* $[f]$ *has a fixed point if and only if the following two properties hold:*
   (i) $[b] \equiv b \in \mathbb{R}^n$.
   (ii) *There is a vector* $\check{x} \in \mathbb{R}^n$ *such that*

$$\check{x} = \dot{A}\check{x} + b \quad \text{for all } \dot{A} \in [A] \text{ satisfying } |\dot{A}| = |[A]|. \tag{5.5.70}$$

(b) *If* $[f]$ *has a fixed point, then for all sufficiently large real numbers* $t > 0$ *the vector* $[x]^* = \check{x} + tv[-1, 1]$ *(*$\check{x}$ *from (5.5.70),* v *any fixed Perron vector of* $|[A]|$*) is also a fixed point of* $[f]$*. In particular, if* $[f]$ *has a fixed point, then there are infinitely many ones.*

Proof. (a) Let $[x]^* \in \mathbb{IR}^n$ be a fixed point of $[f]$. Then (i) follows from the Lemmas 5.5.36 (a) and 5.5.39 (a) together with Theorem 5.5.37 (a). In order to deduce (ii) we first assume $d([x]^*) = 0$, i.e., $[x]^* \equiv x^* \in \mathbb{R}^n$. By (5.5.59) we have $x^* = \hat{A}x^* + b$ with $\hat{A}$ from Theorem 5.5.37, and by (5.5.60) we obtain $x_i^* = 0$ for all indices i which number nondegenerate columns of $[A]$. Therefore, these columns can be replaced by the corresponding ones of any $\dot{A} \in [A]$ with $|\dot{A}| = |[A]|$ without changing the result $\hat{A}x^*$. Since the remaining columns of $\hat{A}$ are degenerate and thus necessarily coincide with those of $\dot{A}$, we get $\hat{A}x^* = \dot{A}x^*$ and finally $x^* = \dot{A}x^* + b$ for all $\dot{A} \in [A]$ with $|\dot{A}| = |[A]|$. This proves (ii) with $\check{x} = x^*$. In the case $d([x]^*) > 0$ the assertion (ii) follows from Lemma 5.5.39 (a) with $\check{x} = \check{x}^*$.

In order to prove the converse, let (i) and (ii) hold and define $\check{x}^* = \check{x}$ with $\check{x}$ from (ii). Let $v > 0$ be any Perron vector of $|[A]|$ and let $[x]^* = \check{x}^* + tv[-1, 1]$ with $t > 0$. If (5.5.64) holds for at least one matrix $\dot{A} \in [A]$ such that $|\dot{A}| = |[A]|$, then

$$\begin{aligned} [A][x]^* + b &= \dot{A}[x]^* + b = \dot{A}(\check{x}^* + tv[-1, 1]) + b \\ &= \dot{A}\check{x}^* + |\dot{A}|tv[-1, 1] + b = \check{x}^* + tv[-1, 1] = [x]^*, \end{aligned} \tag{5.5.71}$$

i.e., $[x]^*$ is a fixed point of $[f]$. Now we will prove that (5.5.64) holds for all $t > 0$ sufficiently large.

By virtue of $v > 0$ we can choose $t \ge 0$ such that $\underline{x}^* \le 0 \le \overline{x}^*$. If $\overline{a}_{ij} \le 0$, then $\dot{a}_{ij} := \underline{a}_{ij} = -|[a]_{ij}| \in [a]_{ij}$, and

$$[a]_{ij}[x]_j^* = \dot{a}_{ij}[x]_j^*. \tag{5.5.72}$$

If $\underline{a}_{ij} \ge 0$, then $\dot{a}_{ij} := \overline{a}_{ij} = |[a]_{ij}| \in [a]_{ij}$, and (5.5.72) holds again.

Now let $\underline{a}_{ij} < 0 < \overline{a}_{ij}$ hold.

Case 1, $\check{a}_{ij} = 0$, i.e., $[a]_{ij} = [-a_{ij}, a_{ij}]$ with some $a_{ij} > 0$: As at the end of the proof of Lemma 5.5.39 (a) we get by (ii) $\check{x}_j^* = \check{x}_j = 0$ whence $\underline{x}_j^* = -\overline{x}_j^*$ and $[a]_{ij}[x]_j^* = |[a]_{ij}|\,[x]_j^* = a_{ij}[x]_j^*$. Therefore, (5.5.72) holds with $\dot{a}_{ij} := a_{ij} = |[a]_{ij}| = \overline{a}_{ij} \in [a]_{ij}$.

Case 2, $\check{a}_{ij} > 0$, i.e., $\overline{a}_{ij} = |[a]_{ij}|$: If $\check{x}_j^* = 0$, then (5.5.72) holds with $\dot{a}_{ij} = \overline{a}_{ij}$.

If $\check{x}_j^* > 0$, then $[a]_{ij}[x]_j^* = [\min\{\underline{a}_{ij}\overline{x}_j^*, \overline{a}_{ij}\underline{x}_j^*\}, \overline{a}_{ij}\overline{x}_j^*]$. In order to fix the lower bound we remark that

$$\begin{aligned}
\underline{a}_{ij}\overline{x}_j^* \geq \overline{a}_{ij}\underline{x}_j^* &\Leftrightarrow (\check{a}_{ij} - \mathrm{rad}[a]_{ij})(\check{x}_j^* + tv_j) \geq (\check{a}_{ij} + \mathrm{rad}[a]_{ij})(\check{x}_j^* - tv_j) \\
&\Leftrightarrow -\mathrm{rad}[a]_{ij}\check{x}_j^* + t\check{a}_{ij}v_j \geq 0 \\
&\Leftrightarrow t \geq \frac{\mathrm{rad}([a]_{ij})}{|\check{a}_{ij}|} \cdot \frac{|\check{x}_j^*|}{v_j},
\end{aligned} \tag{5.5.73}$$

which is true for $t \geq 0$ sufficiently large. Hence (5.5.72) holds with $\dot{a}_{ij} := \overline{a}_{ij} = |[a]_{ij}| \in [a]_{ij}$.

If $\check{x}_j^* < 0$, then $[a]_{ij}[x]_j^* = -([a]_{ij}(-[x]_j^*)) = -(\dot{a}_{ij}(-[x]_j^*)) = \dot{a}_{ij}[x]_j^*$ as in the case $\check{x}_j^* > 0$ provided that (5.5.73) holds.

Case 3, $\check{a}_{ij} < 0$: Here, $[a]_{ij}[x]_j^* = -((-[a]_{ij})[x]_j^*) = -\tilde{a}_{ij}[x]_j^*$ with $\tilde{a}_{ij} := -\underline{a}_{ij} = |[a]_{ij}|$ provided that (5.5.73) is true. Setting $\dot{a}_{ij} := -\tilde{a}_{ij} = \underline{a}_{ij} = -|[a]_{ij}| \in [a]_{ij}$ results in (5.5.72).

(b) follows from the proof of (a). □

In Theorem 5.5.41 we showed that there are fixed points of the form

$$[x]^* = \check{x}^* + tv[-1, 1] \tag{5.5.74}$$

provided that $[f]$ has a fixed point at all. In our next theorem we prove that *all* fixed points of $[f]$ must have this form, and we derive a sharp lower bound for t in (5.5.74) such that, in fact, $[x] = [f]([x])$ holds. In addition, we study $\check{x}^*$ in view of uniqueness.

Theorem 5.5.42. *Let $|[A]|$ be irreducible with $\rho(|[A]|) = 1$ and choose any Perron vector $v > 0$ of $|[A]|$. Denote by M_{sym} the set of all indices for which the columns of $[A]$ contain at least one nondegenerate symmetric entry. Construct $[B] \in \mathbb{IR}^{n\times n}$ from $[A]$ by replacing the j-th column of $[A]$ by the j-th column of the identity matrix I for all $j \in M_{\mathrm{sym}}$ and let $\mathring{A} \in [B]$ be the unique matrix which satisfies $|\mathring{A}| = |[B]|$. In addition, let $[f]$ be defined as in (5.5.57).*

(a) *The interval function $[f]$ has a fixed point if and only if $[b]$ is degenerate, i.e., $[b] \equiv b \in \mathbb{R}^n$, and*
$$x = \mathring{A}x + b \tag{5.5.75}$$
is solvable. In this case, there is at least one solution $\check{x}$ of (5.5.75) which satisfies
$$\check{x}_i = 0 \quad \textit{for all } i \in M_{\mathrm{sym}}. \tag{5.5.76}$$

(b) *If $[b]$ is degenerate, i.e., $[b] \equiv b \in \mathbb{R}^n$, then for any solution $\check{x}$ of (5.5.75) satisfying (5.5.76) and for any real number t with*
$$t \geq m := \max\left\{0,\ \frac{\mathrm{rad}([a]_{ij})}{|\check{a}_{ij}|} \cdot \frac{|\check{x}_j|}{v_j},\ \frac{|\check{x}_j|}{v_j}\ \middle|\ 1 \leq i, j \leq n,\ \check{a}_{ij} \neq 0,\ \mathrm{rad}([a]_{ij}) \neq 0\right\} \tag{5.5.77}$$
the interval vector $[x]^* = \check{x} + tv[-1, 1]$ is a fixed point of $[f]$.

Conversely, if $[x]^*$ *is any fixed point of* $[f]$, *then* $[b]$ *is degenerate, i.e.,* $[b] \equiv b \in \mathbb{R}^n$, *and* $[x]^*$ *can be written in the form* $[x]^* = \check{x}^* + tv[-1, 1]$ *where* $\check{x}^*$ *solves* (5.5.75), (5.5.76) *and* t *satisfies* (5.5.77) *with* $\check{x} = \check{x}^*$.

(c) *If* $M_{\text{sym}} \neq \emptyset$, *i.e., if there are at least two different matrices* $\dot{A}, \ddot{A} \in [A]$ *with* $|\dot{A}| = |\ddot{A}| = |[A]|$, *then* (5.5.75) *has at most one solution which satisfies* (5.5.76).

(d) *If* $M_{\text{sym}} = \emptyset$, *i.e., if there is exactly one matrix* $\dot{A} \in [A]$ *with* $|\dot{A}| = |[A]|$, *then* $\dot{A} = \mathring{A}$, (5.5.76) *is trivially true and one of the following mutually exclusive cases occurs:*
   (i) $\rho(\dot{A}) < 1$, *whence* (5.5.75) *has a unique solution.*
   (ii) $\rho(\dot{A}) = 1$ *and* $\dot{A} \neq D^{-1}|[A]|D$ *for every signature matrix* $D = \operatorname{diag}(\sigma_1, \ldots, \sigma_n)$ *with* $\sigma_i \in \{-1, 1\}$, *whence* (5.5.75) *has a unique solution.*
   (iii) $\rho(\dot{A}) = 1$ *and* $\dot{A} = D^{-1}|[A]|D$ *for some signature matrix* $D = \operatorname{diag}(\sigma_1, \ldots, \sigma_n)$ *with* $\sigma_i \in \{-1, 1\}$. *Here,* (5.5.75) *has no solution if and only if* b *cannot be represented as linear combination of the column vectors of* $I - \dot{A}$. *Otherwise it has infinitely many solutions. They are given by*

$$\check{x}^* = \check{x} + sD^{-1}v, \tag{5.5.78}$$

where $\check{x}$ *is any fixed particular solution of* (5.5.75) *and* s *is any real number.*

Proof. (a) Let $[f]$ have a fixed point $[x]^*$. Then Theorem 5.5.41 implies $[b] \equiv b$ and

$$\check{x}^* = \dot{A}\check{x}^* + b \tag{5.5.79}$$

for all $\dot{A} \in [A]$ with $|\dot{A}| = |[A]|$. Define M as in Theorem 5.5.37 (a). Since

$$\check{x}^*_{j_0} = 0 \quad \text{for } j_0 \in M_{\text{sym}} \subseteq M \tag{5.5.80}$$

(cf. Theorem 5.5.37 (a) for $d([x]^*) = 0$ and Lemma 5.5.39 for $d([x]^*) > 0$) and since the i-th columns of $\dot{A}$ and $\mathring{A}$ coincide for $i \notin M_{\text{sym}}$, we get $\dot{A}\check{x}^* = \mathring{A}\check{x}^*$, and (5.5.75), (5.5.76) follow from (5.5.79), (5.5.80) with $x = \check{x} := \check{x}^*$.

Let, conversely, $[b] \equiv b$ be degenerate and let $\check{x}$ be a solution of (5.5.75) satisfying (5.5.76). Then $\dot{A}\check{x} = \mathring{A}\check{x}$ by the same arguments as above, and Theorem 5.5.41 (a) implies the existence of a fixed point of $[f]$.

(b) '$\Leftarrow$': Let $[x]^*$ be a fixed point of $[f]$. Then $[x]^*$ has the form

$$[x]^* = \check{x}^* + tv[-1, 1], \quad t \geq 0. \tag{5.5.81}$$

For $d([x]^*) > 0$ this follows from Lemma 5.5.39 since, by virtue of Theorem 1.9.4, the Perron vector $d([x]^*)/2$ of this lemma can be written as a positive multiple of our arbitrary Perron vector v. For $d([x]^*) = 0$ the representation (5.5.81) follows at once with $t = 0$.

Now we want to prove (5.5.77) for t in (5.5.81).

If $m = 0$, then (5.5.77) holds trivially.

If $m > 0$, then by the definition of m there is some pair (i, j) such that

$$\check{x}^*_j \neq 0 \quad \text{and} \quad \check{a}_{ij} \neq 0 \quad \text{and} \quad \operatorname{rad}([a]_{ij}) \neq 0 \tag{5.5.82}$$

hold together with

$$m = \frac{\operatorname{rad}([a]_{ij})}{|\check{a}_{ij}|} \cdot \frac{|\check{x}_j^*|}{v_j} \quad \text{or} \quad m = \frac{|\check{x}_j^*|}{v_j}.$$

If $d([x]^*) = 0$, then $\operatorname{rad}([a]_{ij}) \neq 0$ implies the contradiction $\check{x}_j^* = 0$ by virtue of (5.5.60). Therefore, $d([x]^*) > 0$, $t > 0$, and (5.5.64) implies

$$\dot{A}[x]^* = [A][x]^* \tag{5.5.83}$$

for any $\dot{A} \in [A]$ with $|\dot{A}| = |[A]|$. Since $\dot{a}_{ij}[x]_j^* \subset [a]_{ij}[x]_j^*$ would contradict (5.5.83) we get

$$\dot{a}_{ij}[x]_j^* = [a]_{ij}[x]_j^*. \tag{5.5.84}$$

From $d([x]_j^*) > 0$, (5.5.82), (5.5.84) and the multiplication table (Tab. 2.2.1) in Section 2.2 we must have $\underline{x}_j^* \le 0 \le \overline{x}_j^*$, i.e., $\check{x}_j^* - tv_j \le 0 \le \check{x}_j^* + tv_j$ whence

$$t \ge \frac{|\check{x}_j^*|}{v_j}. \tag{5.5.85}$$

If $\overline{a}_{ij} < 0$ or $\underline{a}_{ij} > 0$, then $\operatorname{rad}([a]_{ij}) < |\check{a}_{ij}|$, and together with (5.5.85) we obtain

$$\frac{\operatorname{rad}([a]_{ij})}{|\check{a}_{ij}|} \cdot \frac{|\check{x}_j^*|}{v_j} < \frac{|\check{x}_j^*|}{v_j} = m \le t.$$

If $0 \in [a]_{ij}$, then $|\check{a}_{ij}| \le \operatorname{rad}[a]_{ij}$ whence

$$m = \frac{\operatorname{rad}([a]_{ij})}{|\check{a}_{ij}|} \cdot \frac{|\check{x}_j^*|}{v_j} \ge \frac{|\check{x}_j^*|}{v_j}.$$

In addition, (5.5.82), (5.5.84) and the multiplication table mentioned above (Table 2.2.1) reveal the restrictions

$$\left.\begin{array}{ll} \overline{a}_{ij}\underline{x}_j^* \le \underline{a}_{ij}\overline{x}_j^* & \text{if } \underline{a}_{ij} \le 0 < \check{a}_{ij},\ \check{x}_j^* > 0, \\ \underline{a}_{ij}\underline{x}_j^* \le \overline{a}_{ij}\overline{x}_j^* & \text{if } \underline{a}_{ij} \le 0 < \check{a}_{ij},\ \check{x}_j^* < 0, \\ \overline{a}_{ij}\overline{x}_j^* \le \underline{a}_{ij}\underline{x}_j^* & \text{if } \check{a}_{ij} < 0 \le \overline{a}_{ij},\ \check{x}_j^* > 0, \\ \underline{a}_{ij}\overline{x}_j^* \le \overline{a}_{ij}\underline{x}_j^* & \text{if } \check{a}_{ij} < 0 \le \overline{a}_{ij},\ \check{x}_j^* < 0. \end{array}\right\} \tag{5.5.86}$$

Expressing the bounds of the intervals $[a]_{ij}$, $[x]_j^*$ by means of their midpoint and radius we can rewrite the first inequality in (5.5.86) by

$$(\check{a}_{ij} + \operatorname{rad}([a]_{ij}))(\check{x}_j^* - tv_j) \le (\check{a}_{ij} - \operatorname{rad}([a]_{ij}))(\check{x}_j^* + tv_j),$$

which is equivalent to

$$\operatorname{rad}([a]_{ij})\check{x}_j^* \le \check{a}_{ij} t v_j$$

and therefore to

$$m = \frac{\operatorname{rad}([a]_{ij})}{|\check{a}_{ij}|} \cdot \frac{|\check{x}_j^*|}{v_j} \le t. \tag{5.5.87}$$

The remaining three inequalities of (5.5.86) are also equivalent to (5.5.87). This proves (5.5.77).

'⇒': Now let $[b] \equiv b$ be degenerate and $[x]^* = \check{x} + tv[-1, 1]$ where $\check{x}$ satisfies (5.5.75), (5.5.76) and t satisfies (5.5.77). If $t = 0$, then $m = 0$, $[x]^* \equiv \check{x}$, and the definition of m together with (5.5.76) imply $\check{x}_j = 0$ for $\mathrm{rad}([a]_{ij}) \neq 0$. Hence (5.5.59) follows from (5.5.75) with $x = \check{x}$, and (5.5.60) holds, too. By Theorem 5.5.37 (a) the vector $[x]^* \equiv \check{x}$ is a fixed point of $[f]$. If $t > 0$, then $d([x]^*) > 0$. We first construct a matrix $\dot{A}$ such that

$$\dot{A}[x]^* = [A][x]^*, \quad \dot{A} \in [A], \quad |\dot{A}| = |[A]| \tag{5.5.88}$$

holds which, by virtue of $\dot{A}[x]^* \subseteq [A][x]^*$, $d(\dot{A}[x]^*) = d([A][x]^*)$, is equivalent to

$$\dot{a}_{ij}[x]_j^* = [a]_{ij}[x]_j^*, \quad \dot{a}_{ij} \in [a]_{ij}, \quad |\dot{a}_{ij}| = |[a]_{ij}| \quad \text{for } i, j = 1, \ldots, n. \tag{5.5.89}$$

If (5.5.82) does not hold, then $\dot{a}_{ij}$ from (5.5.89) can easily be found using (5.5.76) in the case $\check{a}_{ij} = 0$, $\mathrm{rad}([a]_{ij}) > 0$. Assume now that (5.5.82) is true (which, by the way, can only happen if $m > 0$). Then $t \geq m \geq \frac{|\check{x}_j|}{v_j}$, i.e., $|\check{x}_j| \leq tv_j$, whence $0 \in [x]_j^*$. In this case $\dot{a}_{ij}$ can be constructed as in the proof of Theorem 5.5.41. (Note that by virtue of (5.5.82) not all cases must be considered there.)

From (5.5.75), (5.5.76) we get $\dot{A}\check{x} = \mathring{A}\check{x}$. Using (5.5.75) and (5.5.88) the proof terminates now as in (5.5.71).

(c) Assume that there are two solutions $\hat{x}$ and $\tilde{x}$ of (5.5.75) which satisfy (5.5.76). Then $y = \hat{x} - \tilde{x}$ fulfills $y = \mathring{A}y$ and $y_i = 0$ for $i \in M_{\mathrm{sym}}$. Construct $\mathring{B}$ from $\mathring{A}$ by replacing the i-th column of $\mathring{A}$ by the zero vector for every $i \in M_{\mathrm{sym}}$. Then $|\mathring{B}| \leq |[A]|$ with inequality in at least one entry. By virtue of Theorem 1.9.7 we have $\rho(\mathring{B}) < \rho(|[A]|) = 1$. Since $y_i = 0$ for $i \in M_{\mathrm{sym}}$ we obtain $y = \mathring{A}y = \mathring{B}y$, whence $y = 0$.

(d) follows from Theorem 1.9.7 taking into account that the kernel of $I - \dot{A}$ is spanned by $D^{-1}v$ in the last case of (iii) since $\lambda = 1$ is a simple eigenvalue of $|[A]|$ and therefore of $\dot{A} = D^{-1}|[A]|D$. Notice that in Theorem 1.9.7 we must set $\phi = 0$ since $\lambda = e^{i\phi}\rho(C)$ has to be one in our situation, and the complex signature matrix D can be chosen to be real since $B := \dot{A}$ is real. (This follows from $B = D^{-1}CD = (b_{kl}) = (e^{i(\sigma_l-\sigma_k)}c_{kl}) \in \mathbb{R}^{n\times n} \Leftrightarrow e^{i(\sigma_l-\sigma_k)} \in \mathbb{R}$ for all $k, l \in \{1, \ldots, n\}$. Choosing $k = 1$ and $l = 2, \ldots, n$ yields $e^{i\sigma_l} = \tau_l e^{i\sigma_1}$ with $\tau_l \in \{-1, 1\}$. Hence $D = e^{i\sigma_1}\,\mathrm{diag}(1, \tau_2, \ldots, \tau_n)$, and σ_1 can be chosen to be zero since it has no influence on the representation of B.) □

Example 5.5.40 is an illustration of the last case in Theorem 5.5.42 (d) (iii) with $D = I$. It also confirms Theorem 5.5.42 (a).

Case (ii) of Theorem 5.5.42 (d) occurs, e.g., if $|[A]|$ is primitive and if, in addition, $[A]$ has the form $[A] = [-|[A]|, O]$. In this case $\dot{A} = -|[A]|$ has no eigenvalue equal to one, hence $\dot{A} \neq D^{-1}|[A]|D$ for every signature matrix D, and $I - \dot{A}$ is regular.

We close this subsection with a remark on the R_1-factor of (5.5.51). From Theorem 5.5.28 we know that $\rho(|[A]|) < 1$ is an upper bound for this factor. If $[A] \geq O$, we can even state a necessary and sufficient criterion for this bound to be sharp. To this end we need the following lemma.

Lemma 5.5.43. *Let* $[b] \in \mathbb{IR}^n$, $[A] \in \mathbb{IR}^{n\times n}$, $\rho(|[A]|) < 1$, *i.e., let the iteration* $\mathcal{J}$ *in* (5.5.51) *converge globally to a limit* $[x]^* \in \mathbb{IR}^n$.

(a) *If* $d([x]^*) > 0$ *and if the sequences* $([y]^k)$, $([z]^k)$ *are generated by* (5.5.51) *with*

$$[y]^0 \subseteq \operatorname{int}([x]^*) \subseteq [x]^* \subseteq \operatorname{int}([z]^0), \tag{5.5.90}$$

then $R_1(\mathcal{J}) = \max\{R_1([y]^k), R_1([z]^k)\}$.

(b) *If* $d([x]^*) = 0$ *and if the sequence* $([z]^k)$ *is generated by* (5.5.51) *with*

$$[x]^* \subseteq \operatorname{int}([z]^0), \tag{5.5.91}$$

then $R_1(\mathcal{J}) = R_1([z]^k)$.

Proof. (a) Let $([x]^k)$ be an arbitrary sequence generated by (5.5.51). By the assumptions, there is an index k_0 such that $[y]^0 \subseteq [x]^{k_0} \subseteq [z]^0$, whence $[y]^k \subseteq [x]^{k_0+k} \subseteq [z]^k$. Theorem 2.5.7 (p) implies

$$q([x]^{k_0+k}, [x]^*) \le \max\{\, q([y]^k, [x]^*),\ q([z]^k, [x]^*)\,\}.$$

If $\|\cdot\|$ denotes any monotone norm one obtains

$$\|q([x]^{k_0+k}, [x]^*)\| \le \max\{\, \|q([y]^k, [x]^*)\|,\ \|q([z]^k, [x]^*)\|\,\}$$

and

$$\begin{aligned}\|q([x]^{k_0+k}, [x]^*)\|^{1/k} &= \left(\|q([x]^{k_0+k}, [x]^*)\|^{1/(k+k_0)}\right)\\ &\quad\times \left(\left(\|q([x]^{k_0+k}, [x]^*)\|^{1/(k+k_0)}\right)^{1/k}\right)^{k_0}\\ &\le \max\{\, \|q([y]^k, [x]^*)\|^{1/k},\ \|q([z]^k, [x]^*)\|^{1/k}\,\}.\end{aligned}$$

From $0 \le R_1([x]^k) \le \rho(|[A]|) < 1$ one therefore gets

$$R_1([x]^k) \le \max\{\, R_1([y]^k), R_1([z]^k)\,\},$$

and the definition of the R_1-factor of $\mathcal{J}$ yields the assertion.

(b) is proved similarly. □

Theorem 5.5.44. *Let* $[b] \in \mathbb{IR}^n$, $[A] \in \mathbb{IR}^n$, $[A] \ge O$, $|[A]|$ *irreducible,* $\rho = \rho(|[A]|) < 1$. *Denote by* $\mathcal{J}$ *the iteration* (5.5.51). *Then* $R_1(\mathcal{J}) < \rho$ *if and only if there are indices* s, t *such that*

$$d([A]_{*s}) \ne 0,\ \ d([A]_{*t}) \ne 0,\ \ (((I - |[A]|)^{-1}\underline{b})_s > 0,\ \ ((I - |[A]|)^{-1}\overline{b})_t < 0. \tag{5.5.92}$$

In all other cases $R_1(\mathcal{J}) = \rho$ *holds.*

Proof. Let $[x]^*$ be the limit of (5.5.51) and $J = \{j \mid d([A]_{*j}) \neq 0\}$, $J' = \{1, \dots, n\} \setminus J$. Define A^*, A^{**} by

$$a^*_{ij} := \begin{cases} \underline{a}_{ij}, & \text{if } \underline{x}^*_j > 0 \\ \overline{a}_{ij}, & \text{if } \underline{x}^*_j \leq 0 \end{cases}, \quad a^{**}_{ij} := \begin{cases} \overline{a}_{ij}, & \text{if } \overline{x}^*_j \geq 0 \\ \underline{a}_{ij}, & \text{if } \overline{x}^*_j < 0 \end{cases}. \tag{5.5.93}$$

Then $A^*\underline{x}^* + \underline{b} = \underline{x}^*$, $A^{**}\overline{x}^* + \overline{b} = \overline{x}^*$.

'⇒': Let $R_1(\mathcal{I}) < \rho$, $x' = (I - |[A]|)^{-1}\underline{b}$, and assume $x'_j \leq 0$ for all $j \in J$. Since $|[A]| \in [A]$ we have $\underline{x} \leq x'$ and $\underline{x}_j \leq x'_j \leq 0$ for $j \in J$. This implies $A^*_{*j} = \overline{A}_{*j}$ for these indices j, and $A^*_{*j} = \underline{A}_{*j} = \overline{A}_{*j}$ for $j \in J'$. Thus we have $A^* = |[A]|$. Choose v as a Perron vector of A^* and define $[z]^k$ by (5.5.51) starting with $[z]^0 = [x]^* + v[-1, 1]$. Then

$$\begin{aligned} \underline{z}^1 &= \min\left([A]([x]^* + v[-1, 1]) + [b]\right) \\ &= \left(\sum\nolimits_{j\in J}[a]_{ij}([x]^*_j + v_j[-1, 1]) + \sum\nolimits_{j\in J'}[a]_{ij}([x]^*_j + v_j[-1, 1]) + \underline{b}_i\right) \\ &= \left(\sum\nolimits_{j\in J}|[a]_{ij}|(\underline{x}^*_j - v_j) + \sum\nolimits_{j\in J'}|[a]_{ij}|(\underline{x}^*_j - v_j) + \underline{b}_i\right) \\ &= A^*(\underline{x}^* - v) + \underline{b} = \underline{x}^* - \rho v \end{aligned}$$

holds. By induction we get $\underline{z}^k = \underline{x}^* - \rho^k v$, $k = 0, 1, \dots$. Therefore, $q([z]^k, [x]^*) \geq \rho^k v$, whence $R_1(\mathcal{I}) \geq R([z]^k) \geq \rho$ contradicting the hypothesis.

Similarly, one gets a contradiction if $R_1(\mathcal{I}) < \rho$ and $((I - |[A]|)^{-1}\overline{b})_j \geq 0$ for all $j \in J$. In this case $\overline{x}^*_j \geq 0$ for $j \in J$, $A^{**} = |[A]|$ and $\overline{z}^k = \overline{x}^* + \rho^k v$, $k = 0, 1, \dots$.

'⇐': Let s, t be as in (5.5.92). First we show that neither A^* nor A^{**} equals $|[A]|$. In contrast, assume $A^* = |[A]|$, for instance. Then $\underline{x}^* = (I - |[A]|)^{-1}\underline{b}$ and (5.5.92) implies $\underline{x}_s > 0$. Since $s \in J$ there is a nondegenerate entry $[a]_{i_0 s}$. By construction of A^* we obtain the contradiction $a^*_{i_0 s} = \underline{a}_{i_0 s} < \overline{a}_{i_0 s} = |[a_{i_0 s}]|$. Similarly, $A^{**} \neq |[A]|$.

From Theorem 1.9.7 we get $\rho_1 := \rho(A^*) < \rho$ and $\rho_2 := \rho(A^{**}) < \rho$. Choose $\varepsilon > 0$ so small that $\rho_1 \leq \rho_1^{(\varepsilon)} := \rho(A^* + \varepsilon ee^T) < \rho$ and let $u^{(\varepsilon)}$ be a Perron vector of the positive matrix $A^* + \varepsilon ee^T$. Then $A^* u^{(\varepsilon)} < (A^* + \varepsilon ee^T)u^{(\varepsilon)} = \rho_1^{(\varepsilon)} u^{(\varepsilon)}$. Assume that $u^{(\varepsilon)}$ is scaled such that $\operatorname{sign}(\underline{x}^*_i - u_i^{(\varepsilon)}) = \operatorname{sign} \underline{x}^*_i$ holds for all i with $\operatorname{sign} \underline{x}^*_i \neq 0$. Analogously, let $v^{(\varepsilon)}$ be a Perron vector of $A^{**} + \varepsilon ee^T$ with $\rho_2 \leq \rho_2^{(\varepsilon)} := \rho(A^{**} + \varepsilon ee^T) < \rho$, $A^{**}v^{(\varepsilon)} < \rho_2^{(\varepsilon)} v^{(\varepsilon)}$ and $\operatorname{sign}(\overline{x}^*_i + v_i^{(\varepsilon)}) = \operatorname{sign} \overline{x}^*_i$ for all i with $\operatorname{sign} \overline{x}^*_i \neq 0$. Start (5.5.51) with $[z]^0 = [x]^* + [-u^{(\varepsilon)}, v^{(\varepsilon)}]$. Then

$$\begin{aligned} [z]^1 &= [A^*(\underline{x}^* - u^{(\varepsilon)}) + \underline{b}, A^{**}(\overline{x}^* + v^{(\varepsilon)}) + \overline{b}] \\ &= [\underline{x}^* - A^* u^{(\varepsilon)}, \overline{x}^* + A^{**} v^{(\varepsilon)}] \\ &\subseteq [\underline{x}^* - \rho_1^{(\varepsilon)} u^{(\varepsilon)}, \overline{x}^* + \rho_2^{(\varepsilon)} v^{(\varepsilon)}] \subseteq [z]^0. \end{aligned}$$

The inclusion monotony of interval arithmetic and induction finally imply $[z]^{k+1} \subseteq [z]^k$, $k = 0, 1, \dots$, and

$$[z]^k \subseteq [\underline{x}^* - \rho_1^{(\varepsilon)} u^{(\varepsilon)}, \overline{x}^* + \rho_2^{(\varepsilon)} v^{(\varepsilon)}]. \tag{5.5.94}$$

From $[x]^* \subseteq [z]^0$ we get $[x]^* \subseteq [z]^k$, $k = 0, 1, \dots$, and (5.5.94) guarantees

$$q([z]^k, [x]^*) \le \sup\{(\rho_1^{(\varepsilon)})^k u^{(\varepsilon)}, (\rho_2^{(\varepsilon)})^k v^{(\varepsilon)}\}$$

and

$$R_1([z]^k) \le \max\{\rho_1^{(\varepsilon)}, \rho_2^{(\varepsilon)}\} < \rho. \tag{5.5.95}$$

In particular, the sequence $([z]^k)$ plays the same role as in Lemma 5.5.43. The assumption $d([x]^*) > 0$ or $d([x]^*) = 0$ of this lemma is fulfilled because of Lemma 5.5.36.

If $d([x]^*) = 0$, Lemma 5.5.43 (b) implies $R_1(\mathfrak{J}) \le \max\{\rho_1^{(\varepsilon)}, \rho_2^{(\varepsilon)}\} < \rho$. Since $\varepsilon > 0$ can be chosen arbitrarily small we get

$$R_1(\mathfrak{J}) \le \max\{\rho_1, \rho_2\} < \rho. \tag{5.5.96}$$

(If $\tilde{u}$, $\tilde{v}$ denote sufficiently small nonnegative eigenvectors of A^* and A^{**}, respectively, associated with the eigenvalues ρ_1 respective ρ_2, then (5.5.51) with $[x]^0 = [x]^* + [-\tilde{u}, \tilde{v}]$ shows that even equality holds in (5.5.96).)

Now let $d([x]^*) > 0$. In order to construct a sequence $([y]^k)$ as in Lemma 5.5.43 (a) we define the matrices B^*, B^{**} by means of

$$b_{ij}^* = \begin{cases} \underline{a}_{ij}, & \text{if } \underline{x}_j^* \ge 0 \\ \overline{a}_{ij}, & \text{if } \underline{x}_j^* < 0 \end{cases}, \quad b_{ij}^{**} = \begin{cases} \overline{a}_{ij}, & \text{if } \overline{x}_j^* > 0 \\ \underline{a}_{ij}, & \text{if } \overline{x}_j^* \le 0 \end{cases}.$$

Then

$$B^* \underline{x}^* + \underline{b} = \underline{x}^*, \quad B^{**}\overline{x}^* + \overline{b} = \overline{x}^*, \quad B^* \le A^*, \quad B^{**} \le A^{**}$$

holds and implies

$$\rho_1^{(\varepsilon)} u^{(\varepsilon)} = (A^* + \varepsilon e e^T) u^{(\varepsilon)} > B^* u^{(\varepsilon)}, \quad \rho_2^{(\varepsilon)} v^{(\varepsilon)} = (A^{**} + \varepsilon e e^T) v^{(\varepsilon)} > B^{**} v^{(\varepsilon)} \tag{5.5.97}$$

with the same quantities as above, eventually rescaled such that

$$\begin{cases} \operatorname{sign}(\underline{x}_i^* + u_i^{(\varepsilon)}) = \operatorname{sign} \underline{x}_i^* & \text{if } \operatorname{sign} \underline{x}_i^* \ne 0, \\ \operatorname{sign}(\overline{x}_i^* - v_i^{(\varepsilon)}) = \operatorname{sign} \overline{x}_i^* & \text{if } \operatorname{sign} \overline{x}_i^* \ne 0, \\ \underline{x}^* + u^{(\varepsilon)} < \overline{x}^* - v^{(\varepsilon)}. \end{cases}$$

The last inequality is possible since we assumed $d([x]^*) > 0$. With $[y]^0 = [\underline{x}^* + u^{(\varepsilon)}, \overline{x}^* - v^{(\varepsilon)}]$ we get similarly as above

$$\begin{aligned} [x]^* &= [A][x]^* + [b] \supseteq [A][y]^0 + [b] = [y]^1 \\ &= [B^*(\underline{x}^* + u^{(\varepsilon)}), B^{**}(\overline{x}^* - v^{(\varepsilon)})] + [b] \\ &\supset [\underline{x}^* + \rho_1^{(\varepsilon)} u^{(\varepsilon)}, \overline{x}^* - \rho_2^{(\varepsilon)} v^{(\varepsilon)}] \supseteq [y]^0, \end{aligned}$$

and by induction

$$[x]^* \supseteq [y]^k \supseteq [\underline{x}^* + (\rho_1^{(\varepsilon)})^k u^{(\varepsilon)}, \overline{x}^* - (\rho_2^{(\varepsilon)})^k v^{(\varepsilon)}], \quad k = 0, 1, \dots.$$

Hence $R_1([y]^k) \le \max\{\rho_1^{(\varepsilon)}, \rho_2^{(\varepsilon)}\}$, and (5.5.95), Lemma 5.5.43 and $\varepsilon \to 0$ imply $R_1(\mathfrak{J}) \le \max\{\rho_1, \rho_2\} < \rho$. (With $[x]^0$ as above we even obtain $R_1(\mathfrak{J}) = \max\{\rho_1, \rho_2\}$.) □

Krawczyk method for interval linear systems

As in the first subsection we start with the point system (5.5.1). This time we precondition it with some matrix C which, for the moment, we assume to be regular. Then (5.5.1) is equivalent to the fixed point form

$$x = Cb + (I - CA)x \tag{5.5.98}$$

or

$$x = \tilde{x} + C(b - A\tilde{x}) + (I - CA)(x - \tilde{x}) \tag{5.5.99}$$

with a fixed vector $\tilde{x}$. This induces the iterative methods

$$x^{k+1} = Cb + (I - CA)x^k, \quad k = 0, 1, \ldots, \tag{5.5.100}$$

and

$$x^{k+1} = \tilde{x} + C(b - A\tilde{x}) + (I - CA)(x^k - \tilde{x}), \quad k = 0, 1, \ldots, \tag{5.5.101}$$

respectively, where the latter one can be considered as an iterative refinement of $\tilde{x}$. If A, b are replaced by interval quantities $[A]$, $[b]$, one immediately gets interval iterations for the interval linear system $[A]x = [b]$. The point matrix C can be thought of as an approximate inverse of the midpoint $\check{A}$, and $\tilde{x}$ as an approximate solution of $\check{A}x = \check{b}$. We refer to each of the iterations

$$[x]^{k+1} = C[b] + (I - C[A])[x]^k, \quad k = 0, 1, \ldots, \tag{5.5.102}$$

and

$$[x]^{k+1} = \tilde{x} + C([b] - [A]\tilde{x}) + (I - C[A])([x]^k - \tilde{x}), \quad k = 0, 1, \ldots \tag{5.5.103}$$

as Krawczyk method, where the second iteration is sometimes more precisely called *residual Krawczyk iteration* as in Neumaier [257]. They will be modified and generalized to nonlinear systems in Chapter 6. They can be interpreted as the Richardson iteration applied to $(C[A])x = C[b]$ and $(C[A])(x - \tilde{x}) = C([b] - [A]\tilde{x})$, respectively, i.e., to a preconditioned system $[A]x = [b]$. For $\tilde{x} = 0$ the iteration (5.5.103) reduces to (5.5.102) while, conversely, (5.5.103) can be obtained from (5.5.102) by replacing $[b]$, $[x]^k$, $[x]^{k+1}$ by $[b] - [A]\tilde{x}$, $[x]^k - \tilde{x}$, $[x]^{k+1} - \tilde{x}$, so that results on (5.5.102) can be easily transferred to (5.5.103) and vice versa.

As in (5.5.10), (5.5.11) one can also iterate with intersections, for instance

$$[x]^{k+1} = \{\, C[b] + (I - C[A])[x]^k \,\} \cap [x]^k, \quad k = 0, 1, \ldots. \tag{5.5.104}$$

Since (5.5.100), (5.5.101) are contained in (5.5.102) and (5.5.103), respectively, we will consider only the latter ones, pointing out particularities of the previous ones if necessary.

We start with the simple iteration (5.5.102).

Theorem 5.5.45. *Let* $[x]^0 \in \mathbb{IR}^n$ *and denote by* $[x]^k$ *the iterates of* (5.5.102) *for some matrices* $[A] \in \mathbb{IR}^{n\times n}$, $C \in \mathbb{R}^{n\times n}$ *and some vector* $[b] \in \mathbb{IR}^n$. *If*

$$[x]_i^1 \subset [x]_i^0, \quad i = 1, \dots, n, \tag{5.5.105}$$

then the following properties hold.

(a) *The matrices* $[A]$, C *are regular.*

(b) *The iteration* (5.5.102) *is globally convergent to some limit* $[x]^*$ *which contains the solution set* S *of the interval linear system* $[A]x = [b]$.

(c) *The iterates* $[x]^k$ *satisfy*

$$[x]^* \subseteq [x]^k \subseteq [x]^{k-1} \subseteq \dots \subseteq [x]^0,$$

with $\lim_{k\to\infty}[x]^k = [x]^*$, $[x]^*$ *as in* (b).
If $[A] \equiv A \in \mathbb{R}^{n\times n}$, $[b] \equiv b \in \mathbb{R}^n$, *then* $[x]^* \equiv x^* \in \mathbb{R}^n$ *with* $Ax^* = b$.

Proof. (a), (b) follow from the Theorems 5.5.9, 5.5.6, and 5.5.5 applied to $C[A]$, $C[b]$ instead of $[A]$, $[b]$ and to $([M], [N]) = (I, I - C[A])$. Notice that Theorem 5.5.6 implies regularity of $C[A]$ and therefore also of C and $[A]$.

(c) The proof of Theorem 5.5.6 shows that

$$[f]([x]) := C[b] + (I - C[A])[x] = \mathrm{IGA}(I, (I - C[A])[x] + C[b]) \tag{5.5.106}$$

is a P-contraction with $P = |I - C[A]|$. The assertion follows now from Theorem 4.2.7 (c) and (d). □

The assumption (5.5.105) does not need to be fulfilled from the start. Then an iteration with ε-inflation as in Section 4.3 may help. It leads to $[x]^{k+1} = [f]([x]_\varepsilon^k) \subseteq [x]_\varepsilon^k$ after finitely many iterations as the Example 4.3.3 and Theorem 4.3.2 show if adapted appropriately.

Our next theorem starts with an a priori restriction of $I - C[A]$.

Theorem 5.5.46. *Let* $\tilde{x} \in \mathbb{R}^n$, $[x]^0 \in \mathbb{IR}^n$ *and denote by* $[x]^k$ *the iterates of* (5.5.102) *for some matrices* $[A] \in \mathbb{IR}^{n\times n}$, $C \in \mathbb{R}^{n\times n}$ *and some vector* $[b] \in \mathbb{IR}^n$. *If*

$$\|\,|I - C[A]|\,\|_\infty < 1, \tag{5.5.107}$$

then the following properties hold.

(a) *The matrices* $[A]$, C *are regular.*

(b) *The iteration* (5.5.102) *is globally convergent to some limit* $[x]^*$ *which satisfies*

$$[x]^* \subseteq [\tilde{x}] := \tilde{x} + [-\alpha, \alpha]e, \quad \text{where } \alpha := \frac{\|\,|C([b] - [A]\tilde{x})|\,\|_\infty}{1 - \|\,|I - C[A]|\,\|_\infty}.$$

If one starts (5.5.102) *with* $[x]^0 \supseteq [\tilde{x}]$, *then*

$$[x]^* \subseteq [x]^k, \quad k = 0, 1, \dots, \quad \text{and} \quad \lim_{k\to\infty}[x]^k = [x]^*.$$

Proof. The assumption implies $\rho(P) < 1$ for $P = |I - C[A]|$. Therefore, part (a) and the global convergence in (b) follow from Theorem 5.5.6 applied as in the previous proof.

Start the iteration (5.5.102) with $[x]^0 = \tilde{x}$. Then

$$q([x]^1, [x]^0) = q(C[b] + (I - C[A])\tilde{x}, \tilde{x}) = q(C[b] - (C[A])\tilde{x}, 0) = |C([b] - [A]\tilde{x})|,$$

and $[x]^* \subseteq [\tilde{x}]$ results from Theorem 4.2.7 (b). With $[f]$ as in (5.5.106) the remaining part of the theorem follows from $[x]^* = [f]([x]^*) \subseteq [f]([\tilde{x}]) \subseteq [f]([x]^0) = [x]^1$ and an obvious induction. □

The assumption (5.5.107) is certainly fulfilled if C is a sufficiently good approximation of the inverse of a matrix $\tilde{A} \in [A]$ and if the diameter of $[A]$ is sufficiently small. This follows from

$$\begin{aligned} I - C[A] &\subseteq I - C\left(\tilde{A} + d([A])[-1,1]\right) = I - C\tilde{A} + Cd([A])[-1,1] \\ &\approx I - \tilde{A}^{-1}\tilde{A} + |C|d([A])[-1,1] = |C|d([A])[-1,1]. \end{aligned}$$

The Theorems 5.5.45 and 5.5.46 can be easily reformulated and proved for the iteration (5.5.103), too. As was mentioned in Neumaier [257], p. 125f, the limit $[y]^*$ of the residual iteration (5.5.103) is not better than the limit $[x]^*$ of (5.5.102). We extend this result in Theorem 5.5.47.

Theorem 5.5.47. *Let* (5.5.102) *and* (5.5.103) *be globally convergent to some limits* $[x]^*$ *and* $[y]^*$*, respectively. Then*

$$[x]^* \subseteq [y]^* \tag{5.5.108}$$

holds with equality if $\tilde{x} = \check{x}^*$, $C = \check{A}^{-1}$. *For arbitrary choices* $\tilde{x} \in [x]^*$, $C^{-1} \in [A]$ *inequality is possible in* (5.5.108).

Proof. Start (5.5.102) with $[x]^0 = [y]^*$. Then

$$\begin{aligned} [x]^1 &= C[b] + (I - C[A])([y]^* - \tilde{x} + \tilde{x}) \\ &\subseteq C[b] + (I - C[A])([y]^* - \tilde{x}) + (I - C[A])\tilde{x} \\ &= \tilde{x} + C[b] - (C[A])\tilde{x} + (I - C[A])([y]^* - \tilde{x}) \\ &= \tilde{x} + C([b] - [A]\tilde{x}) + (I - C[A])([y]^* - \tilde{x}) = [y]^* \end{aligned}$$

holds, and an inductive argument implies $[x]^* \subseteq [y]^*$.

If $\tilde{x} = \check{x}^*$, $C = \check{A}^{-1}$ we get

$$\begin{aligned} [x]^* &= \check{A}^{-1}[b] + (I - \check{A}^{-1}[A])[x]^* \\ &= \check{A}^{-1}[b] + \check{A}^{-1}\,\mathrm{rad}([A])[x]^*[-1,1] \\ &= \check{A}^{-1}[b] + \check{A}^{-1}\,\mathrm{rad}([A])|[x]^*|[-1,1] \end{aligned}$$

and

$$\begin{aligned}[y]^* &= \check{x}^* + \check{A}^{-1}([b] - [A]\check{x}^*) + \check{A}^{-1}\operatorname{rad}([A])|[x]^* - \check{x}^*|[-1,1] \\ &= \check{x}^* + \check{A}^{-1}[b] - (I - \check{A}^{-1}\operatorname{rad}([A])[-1,1])\check{x}^* \\ &\quad + \check{A}^{-1}\operatorname{rad}([A])\operatorname{rad}([x]^*)[-1,1] \\ &= \check{A}^{-1}[b] + \check{A}^{-1}\operatorname{rad}([A])(|\check{x}^*| + \operatorname{rad}([x]^*))[-1,1] \\ &= \check{A}^{-1}[b] + \check{A}^{-1}\operatorname{rad}([A])|[x^*]|[-1,1] = [x]^*.\end{aligned}$$

The remaining part of the proof follows from the counterexample $[A] = [1,2]$, $C = 2/3$, $[b] = [-1,0]$, $\tilde{x} = -1$. Here, $[x]^* = [-1, 1/3]$, and

$$\begin{aligned}[y]^* &= \tilde{x} + C([b] - [A]\tilde{x}) + (I - C[A])([y]^* - \tilde{x}) \\ &\supseteq \tilde{x} + C([b] - [A]\tilde{x}) + (I - C[A])([x]^* - \tilde{x}) \\ &= [-13/9, 7/9] \supset [x]^*.\end{aligned}$$

□

In order to evaluate the quality of an enclosure $[x]$ of the solution set S of an interval linear system $[A]x = [b]$ it is advantageous to have a so-called inner enclosure $[y]$ of S which is defined as any interval vector contained in S. In this connection we sometimes more precisely call an enclosure $[x]$ an outer enclosure. If $[x]$ and $[y]$ are known, the Hausdorff distance $q([x],[y])$ is certainly an upper bound for the overestimation of $[x]$ with respect to S. If this bound is small, the enclosure $[x]$ of S is a good one.

Theorem 5.5.48. *Let $[x]^0 \in \mathbb{IR}^n$ and denote by $[x]^1$ the first iterate of (5.5.102) for some matrices $[A] \in \mathbb{IR}^{n\times n}$, $C \in \mathbb{R}^{n\times n}$ and some vector $[b] \in \mathbb{IR}^n$. Denote by S_i the projection of the solution set S of $[A]x = [b]$ onto the i-th coordinate axis. If*

$$[x]_i^1 \subset [x]_i^0, \quad i = 1, \ldots, n, \tag{5.5.109}$$

holds as in (5.5.105), then the following properties hold with

$$[z] = C[b], \quad [\Delta] = (I - C[A])[x]^0.$$

(a) *The component $[z]_i$ satisfies*

$$[z]_i = \{\, (C\tilde{b})_i \mid \tilde{b} \in [b] \,\}, \quad i = 1, \ldots, n.$$

(b) *The interval $[x]_i^1$ is an outer enclosure of S_i, i.e.,*

$$\underline{x}_i^1 = \underline{z}_i + \underline{\Delta}_i \le \min(S_i) \le \max(S_i) \le \overline{x}_i^1 = \overline{z}_i + \overline{\Delta}_i, \quad i = 1, \ldots, n.$$

Obviously, $d([x]^1) = d([z]) + d([\Delta])$ holds.

(c) *With $\underline{y} = \underline{z} + \overline{\Delta}$, $\overline{y} = \overline{z} + \underline{\Delta}$, one gets*

$$\min(S_i) \le \underline{y}_i, \quad \overline{y}_i \le \max(S_i), \quad i = 1, \ldots, n.$$

If $\underline{y} \le \overline{y}$, then $[y] = [\underline{y}, \overline{y}]$ is an inner enclosure of S which satisfies $d([y]) = d([z]) - d([\Delta]) = d([x]^1) - 2d([\Delta])$.

Proof. (a) follows from Theorem 4.1.5, the assertion in (b) results from Theorem 5.5.45 or is trivial.

(c) For each $x \in S$ there are $A \in [A]$, $b \in [b]$ such that $Ax = b$ holds, whence

$$x = Cb + (I - CA)x \in [z] + [\Delta]$$

follows. Therefore, by virtue of (a) and for any fixed index i there are vectors $x', x'' \in S$ which satisfy

$$x'_i = \underline{z}_i + \Delta'_i, \quad x''_i = \overline{z}_i + \Delta''_i$$

for appropriate vectors $\Delta', \Delta'' \in [\Delta]$. This implies $x'_i \le \underline{y}_i$ and $x''_i \ge \overline{y}_i$ and finishes the nontrivial part of the proof. □

Since the inequalities $\underline{z}_i + \underline{\Delta}_i \le \min(S_i) \le \underline{z}_i + \overline{\Delta}_i$ and $\overline{z}_i + \underline{\Delta}_i \le \max(S_i) \le \overline{z}_i + \overline{\Delta}_i$ hold, the diameter $d([\Delta])$ is a measure for the overestimation of S by $[x]^1$. If $[y]$ exists, then $d([\Delta]) = q([x]^1, [y])$ holds.

Notice that Theorem 5.5.48 holds also for the residual iteration (5.5.103). Here, $[z] = C([b] - [A]\tilde{x})$ and $[\Delta] = (I - C[A])([x]^0 - \tilde{x})$ and a summand $\tilde{x}_i$ has to be added on the right-hand sides in (b) and a summand $\tilde{x}$ has to be added in the definition of $\underline{y}, \overline{y}$ in (c). If $d([A])$, $d([b])$, and $d([x]^0)$ are small and if $\tilde{x} \in [x]^0$, $C = \check{A}^{-1}$, then one can expect that $d([z])$ is small and $d([\Delta])$ is much smaller than $d([z])$ because of

$$d([\Delta]) = d((I - C[A])([x]^0 - \tilde{x})) \le d(I - C[A])d([x]^0 - \tilde{x}) = |C|d([A])d([x]^0).$$

Here we used Theorem 3.1.5 (i). Thus $d([\Delta])$ can be thought to be 'small of second order' in this case.

The practical computation of an inner computation $[y]$ along the lines of Theorem 5.5.48 must be implemented carefully. For instance, for $\underline{y}_i$ we need an upper bound of $\underline{z}_i$ and of $\overline{\Delta}_i$. The latter can be obtained evaluating $(I - C[A])[x]^0$ in standard machine interval arithmetic and choosing the upper bound of the computed result. For the evaluation of $\underline{z}_i = \min(C[b])_i$ one needs a computation with upward rounding. The matrix C and – if needed – $\tilde{x}$ can be computed using floating point arithmetic. The assumption (5.5.109) can be fulfilled when iterating with ε-inflation; cf. again Theorem 4.3.2 and Example 4.3.3.

We mention that Theorem 5.5.48 remains true if assumption (5.5.109) is replaced by

$$[x]^1 \subseteq [x]^0 \quad \text{and} \quad S \subseteq [x]^0. \tag{5.5.110}$$

The simple proof of this modified Theorem 5.5.48 is based on Brouwer's fixed point theorem and is left to the reader as Exercise 5.5.12. It remains true for the residual iteration (5.5.103) if $[z]$ and $[\Delta]$ are altered as described above. If (5.5.110) holds, then part (b) of the theorem and an induction implies $S \subseteq [x]^k$, $k = 1, 2, \ldots$, whence $S \subseteq [x]^* = \lim_{k\to\infty}[x]^k$ follows. This remark can be applied as follows: If (5.5.109) is assumed and if the iteration (5.5.102) is started with $[x]^0 = [\tilde{x}]$ from Theorem 5.5.46, then

$S \subseteq [x]_{\text{comp}}$, where $[x]_{\text{comp}}$ denotes the *computed* fixed point of the iteration (5.5.102) similarly as $[x]^k_{\text{comp}}$ in (5.5.26) of Remark 5.5.1. Therefore, if $[f]$ is defined as in (5.5.106), then we get $[f]([x]_{\text{comp}}) = [x]_{\text{comp}}$ on the computer but only $[f]([x]_{\text{comp}}) \subseteq [x]_{\text{comp}}$ theoretically because of the outward rounding of the machine interval arithmetic. This allows the application of the modified Theorem 5.5.48 with $[x]^0 = [x]_{\text{comp}}$ which we will do in our example below.

Now we modify the residual iteration (5.5.103) slightly in order to enclose – hopefully better – the symmetric solution set S_{sym} of the interval linear system $[A]x = [b]$ with $[A] = [A]^T$. For $A = A^T \in [A]$, $b \in [b]$ we have

$$\begin{aligned}
(C(b - A\tilde{x}))_i &= \sum_{j=1}^{n} c_{ij}\Bigl(b_j - \sum_{\ell=1}^{n} a_{j\ell}\tilde{x}_\ell\Bigr) \\
&= \sum_{j=1}^{n} c_{ij}(b_j - a_{jj}\tilde{x}_j) - \sum_{j=1}^{n}\sum_{\ell=1}^{j-1}(c_{ij}\tilde{x}_\ell + c_{i\ell}\tilde{x}_j)a_{j\ell} \\
&\in \sum_{j=1}^{n} c_{ij}([b]_j - [a]_{jj}\tilde{x}_j) - \sum_{j=1}^{n}\sum_{\ell=1}^{j-1}(c_{ij}\tilde{x}_\ell + c_{i\ell}\tilde{x}_j)[a]_{j\ell} \\
&=: [z]_i^{\text{sym}} = \{\, (C(b - A\tilde{x}))_i \mid A = A^T \in [A],\ b \in [b] \,\},
\end{aligned} \tag{5.5.111}$$

where $i = 1, \ldots, n$ and where we exploited the symmetry of A and Theorem 4.1.5. Therefore we can iterate according to

$$[x]^{k+1} = \tilde{x} + [z]^{\text{sym}} + (I - C[A])([x]^k - \tilde{x}), \quad k = 0, 1, \ldots. \tag{5.5.112}$$

For this iteration there is a similar theorem to Theorem 5.5.48. Assume there $[A] = [A]^T$ and replace $[z]$ by $[z]^{\text{sym}}$, S by S_{sym} and (a) by (5.5.111). The Theorems 5.5.45–5.5.47 hold also for (5.5.112) when reformulated appropriately.

We illustrate the iterations (5.5.103) and (5.5.112) by an example which originates from Jansson [153].

Example 5.5.49. Let $\check{A} \in \mathbb{R}^{4\times 4}$ be defined by

$$\check{A} = \begin{pmatrix}
-758.0284 & 8.971284 & -507.7297 & -260.2576 \\
8.971284 & -507.7118 & 7.705539 & 508.9875 \\
-507.7297 & 7.705539 & -5.192805 & -510.2374 \\
-260.2576 & 508.9875 & -510.2374 & -259.0101
\end{pmatrix}$$

and $[A](t) = [A](t)^T \in \mathbb{IR}^{4\times 4}$ by $[a]_{ij}(t) = \check{a}_{ij} + \delta_{1,|i-j|} \cdot t \cdot [-1, 1]$ using the Kronecker symbol δ_{ij} and a very small positive parameter t. Let $\tilde{x} = (1, -1, 1, -1)^T$, $[b] \equiv b = \check{A} \cdot \tilde{x}$, and $C = \check{A}^{-1}$. Compute $\check{A}$, C, b in INTLAB using ordinary floating point arithmetic and the MATLAB function `inv()`. This changes the theoretical quantities slightly but does not change the effect of the example severely. Check the condition (5.5.107) of Theorem 5.5.46 and start the iterations (5.5.103) and (5.5.112) with $[\tilde{x}]$ from this theorem. Stop the iterations when two successive computed iterates $[x]^k$, $[x]^{k+1}$ coincide for the first time taking into account Remark 5.5.1. Denote by $[x]$ and $[x]^{\text{sym}}$ the *computed* final outer enclosures resulting from (5.5.103) and (5.5.112), respectively, and by $[y]$ and $[y]^{\text{sym}}$ the corresponding *computed* final inner enclosures according to Theorem 5.5.48 and its modification for S_{sym}. Some expressive quotients are listed for $t = 10^{-7}$ in Table 5.5.3, where the results are rounded to 5 digits by chopping. They were obtained after $k + 1 = 5$, respectively $k + 1 = 7$ iterations.

Tab. 5.5.3: Ratio of diameters for $t = 10^{-7}$.

i	$\frac{d([x]_i)}{d([x]_i^{\text{sym}})}$	$\frac{d([y]_i)}{d([x]_i)}$	$\frac{d([y]_i^{\text{sym}})}{d([x]_i^{\text{sym}})}$
1	$1.3123 \cdot 10^5$	0.99959	0.97321
2	$1.0166 \cdot 10^3$	0.99959	0.99979
3	$2.0278 \cdot 10^3$	0.99959	0.99958
4	$1.0115 \cdot 10^3$	0.99959	0.99979

The third and forth column indicate that $[x]_i$ and $[x]_i^{\text{sym}}$ are tight enclosures of S_i, and $(S_{\text{sym}})_i$, respectively. The second column shows then that $(S_{\text{sym}})_i$ is essentially smaller than S_i. Therefore, in this example it is advantageous to use the iteration (5.5.112) if an enclosure for S_{sym} is needed. Unfortunately, such a gain of enclosure does not always happen as the example in Exercise 5.5.14 shows. Trivially, for decreasing t the solution set S and the symmetric solution set S_{sym} must approach each other in Example 5.5.49. In fact, $S = S_{\text{sym}}$ holds for $t = 0$. The behavior of S with respect to S_{sym} is indicated by the second column of Table 5.5.4, where the ratios decrease nearly to one. The last two columns list the smallest indices k, and k_{sym} respectively, with which the stopping criterion (5.5.26) is fulfilled.

Tab. 5.5.4: Minimal ratio of diameters for decreasing t.

t	$\min_{i=1,\ldots,4} \frac{d([x]_i)}{d([x]_i^{\text{sym}})}$	$\min_{i=1,\ldots,4} \frac{d([y]_i)}{d([x]_i)}$	$\min_{i=1,\ldots,4} \frac{d([y]_i^{\text{sym}})}{d([x]_i^{\text{sym}})}$	k	k_{sym}
10^{-7}	$1.0115 \cdot 10^3$	0.99959	0.97321	4	6
10^{-8}	$1.0114 \cdot 10^3$	0.99995	0.99729	3	5
10^{-9}	$1.0111 \cdot 10^3$	0.99999	0.99973	3	4
10^{-10}	$1.0084 \cdot 10^3$	0.99999	0.99998	3	4
10^{-11}	$9.7787 \cdot 10^2$	0.99999	0.99999	2	3
10^{-12}	$7.5349 \cdot 10^2$	0.99999	0.99999	2	3
10^{-13}	$2.2988 \cdot 10^2$	0.99999	0.99999	2	2
10^{-14}	$2.7169 \cdot 10^1$	0.99999	0.99999	2	2
10^{-15}	$3.9427 \cdot 10^0$	0.99999	0.99999	2	2
10^{-16}	$1.6475 \cdot 10^0$	0.99999	0.99999	2	2

Exercises

Ex. 5.5.1. Reprove Lemma 5.5.1 starting first with the trivial case of a regular upper triangular matrix $[A]$. In the case of a regular lower triangular matrix $[A]$ apply the preceding result to the upper triangular matrix $P[A]P^T$ and the vector $P[b]$, where P is the permutation matrix in (5.2.42) (denoted by E there). Use $P = P^T$ and Theorem 5.4.1 (f) in order to see $[x] = \mathrm{IGA}([A], [b]) = P \cdot \mathrm{IGA}(P[A]P^T, P[b])$ and apply an index transformation $n + 1 - i \to i$ in the arising sum.

Ex. 5.5.2. Prove Theorem 5.5.4.

Ex. 5.5.3. Consider (5.5.8) only for real matrices A, M, N and real vectors b, x^0. Define D, E, F for $A \in \mathbb{R}^{n\times n}$ as in (5.5.29) and $\rho_J, \rho_{\mathrm{GS}}$ as in Remark 5.5.2 (b). Show that $\rho_{\mathrm{GS}}^2 = \rho_J$ holds if $n = 2$. (This implies $\rho_{\mathrm{GS}} < \rho_J$ if $\rho_J < 1$.) Find a matrix $A \in \mathbb{R}^{3\times 3}$ such that $\rho_J < \rho_{\mathrm{GS}} < 1$ holds.

Ex. 5.5.4. Prove the details of Example 5.5.15.

Ex. 5.5.5. Prove the details of Example 5.5.19.

Ex. 5.5.6. Prove the details of Example 5.5.23.

Ex. 5.5.7. Prove the details of Example 5.5.27.

Ex. 5.5.8. Let $[b] \in \mathbb{IR}^n$, $[A] \in \mathbb{IR}^{n\times n}$, $\rho(|[A]|) < 1$, and let $[x]^*$ be the limit of the Richardson iteration $[x]^{k+1} = [A][x]^k + [b]$. Show that $[x]^* = -[x]^*$ holds if and only if $[b] = -[b]$. (Cf. Theorem 5.5.29.)

Ex. 5.5.9. Let $[b] \in \mathbb{IR}^n$, $[A] \in \mathbb{IR}^{n\times n}$, $\rho(|[A]|) < 1$ and let $[x]^*$ be the limit of the Richardson iteration $[x]^{k+1} = [A][x]^k + [b]$.

(a) Show that there are matrices $\underset{\sim}{A}, \tilde{A} \in \partial[A]$ such that

$$[A][x]^* = [\underset{\sim}{A}\check{x}^*, \tilde{A}\check{x}^*] + [-|\underset{\sim}{A}|, |\tilde{A}|]r_{x^*}.$$

Hint: Prove this equation first for $n = 1$.

(b) Show by means of (a) that $[x]^* = [x]$ is the limit of the Richardson iteration (5.5.51) if $[x]$ satisfies the coupled linear systems

$$\check{x} = \tfrac{1}{2}(\tilde{A} + \underset{\sim}{A})\check{x} + \tfrac{1}{2}(|\tilde{A}| - |\underset{\sim}{A}|)r_x + \check{b}$$

$$r_x = \tfrac{1}{2}(\tilde{A} - \underset{\sim}{A})\check{x} + \tfrac{1}{2}(|\tilde{A}| + |\underset{\sim}{A}|)r_x + r_b$$

or, equivalently,

$$\underline{x} = \tfrac{1}{2}(\underset{\sim}{A} + |\underset{\sim}{A}|)\underline{x} + \tfrac{1}{2}(\underset{\sim}{A} - |\underset{\sim}{A}|)\overline{x} + \underline{b}$$

$$\overline{x} = \tfrac{1}{2}(\tilde{A} - |\tilde{A}|)\underline{x} + \tfrac{1}{2}(\tilde{A} + |\tilde{A}|)\overline{x} + \overline{b}$$

with $\underset{\sim}{A}, \tilde{A}$ as in (a).

Conversely, the limit $[x]^*$ of (5.5.51) satisfies the preceding linear systems.

Ex. 5.5.10. Apply the modified sign accord algorithm 5.3.14 to

$$[A] = \begin{pmatrix} [1/8, 1/4] & [0, 1/4] \\ [0, 1/2] & [1/8, 1/4] \end{pmatrix}, \quad [b] = \begin{pmatrix} [-10, -1] \\ [2, 10] \end{pmatrix}$$

in order to compute the solution $[x]^*$ of the interval equation $[x] = [A][x] + [b]$.

Ex. 5.5.11. Denote by C^+ the Moore–Penrose inverse of a matrix C. For C^+ it is well known that $C^+b \in \mathbb{R}^n$ is the least squares solution of $Cx = b$ of minimal Euclidean norm (cf. Stoer, Bulirsch [348], p. 220, for example). Show that if $[b] \equiv b \in \mathbb{R}^n$ and if (5.5.59) is solvable, then $x^* = (I - \hat{A})^+ b$ is a fixed point of $[f]$ from (5.5.57).

Ex. 5.5.12. Prove the assertions of Theorem 5.5.48 under the assumptions (5.5.110).

Ex. 5.5.13. Replace the matrix $\hat{A}$ in Example 5.5.49 by the Pascal matrix

$$P_n = \begin{pmatrix} i+j \\ i \end{pmatrix} = \left(\frac{(i+j)!}{i!\,j!} \right) \in \mathbb{R}^{n \times n},$$

modify P_n correspondingly in order to obtain an interval matrix $[A](t)$, and draw up a similar table to Table 5.5.4.

Ex. 5.5.14 (Mayer, [216]). Let

$$[A](t) = \begin{pmatrix} 1 & t[-1, 1] \\ t[-1, 1] & -1 \end{pmatrix}, \quad 0 \le t < 1, \ [b] = (2, 2)^T$$

and consider the interval linear system $[A]x = [b]$.

(a) Show that the solution set S lies completely in the fourth quadrant O_4 and has the form of a kite which lies symmetric with respect to the line $x_2 = -x_1$ and has a trailing end at $(2/(1-t), -2/(1-t))$.

(b) Show that the symmetric solution set S_{sym} is a connected part of the circle $(x_1 - 1)^2 + (x_2 + 1)^2 = 2$ which lies symmetric to the line $x_2 = -x_1$ and contains the point $(2, -2)$.

(c) Show that the limits of the iterations (5.5.103) and (5.5.112) coincide for every $t \in [0, 1)$. Show in addition that they cover the whole plane $\mathbb{R}^2$ if t tends to one. What do S and S_{sym} look like in the singular limit case $t = 1$?

Notes to Chapter 5

To 5.1: Details on Leontief's model can be found in Leontief [191]. Varying input data in this model are considered in Rohn [297].

To 5.2: An if and only if criterion for the nonconvexity of the solution set S is given in Rohn [298, 300]. The tolerance solution set and the control solution set are studied intensively in Shary [337, 338, 339, 340, 342] and Sharaya [336], where these sets are partly named tolerable solution set, and controllable solution set, respectively; cf. also Fiedler et al. [95], Theorem 2.28 and Theorem 2.29 from where we took our notation. A generalized version of the Oettli–Prager theorem was considered in Heindl [140].

Historical remarks on the symmetric solution set S_{sym} and additional references for the description and the enclosure of this set can be found in Mayer [216]. The problem of considering the hull of S_{sym} seems to be first mentioned by Neumaier in a Christmas letter addressed to Rohn dated December 23, 1985; cf. Rohn [305, 306]. Inspired by this letter Rohn [306] invented a first method to enclose S_{sym} in 1986. We considered Rohn's method in our survey [216]. The curvilinear boundary of S_{sym} was remarked experimentally by Jansson in his 1990 talk at the SCAN conference in Albena, Bulgaria. A first proof of this form succeeded in 1995 for 2×2 matrices by Alefeld, Mayer [37] and a year later for the general case by Alefeld, Kreinovich, Mayer [28, 30]. Another description using intersections of intervals was given in Mayer [213]. A first description in a closed form by a variety of inequalities was stated and proved by Hladík [149]. The form of Theorem 5.2.6 including the various equivalences and the presented proof goes back to Mayer [217, 219] based on Hladík's result in [149]. More general dependencies are considered in Alefeld, Kreinovich, Mayer [31, 33], and Popova [276]. The crucial Theorem 5.2.7 and Corollary 5.2.8 were first presented in Alefeld, Kreinovich, Mayer [31].

To 5.3: Farkas' lemma and variants of it including proofs can be found in Schrijver [332]. Its proof here is taken from Suchowitzki, Awdejewa [350], p. 381ff. Nearly all other results of Section 5.3 are results from Rohn [301]. Theorem 5.3.16 is based on a result of Hansen [131]. In its present form it goes back to Rohn [302]. It was generalized in Ning, Kearfott [263]. Additional remarks on the interval hull of the solution set were given in Neumaier [257].

To 5.4: For the 2×2 case the formulae of the interval Gaussian algorithm can be found already in the pioneering book of Moore [232]. In the general case they are contained in Alefeld [7]. The representation (5.4.4) of $[x]^G$ originates from Schwandt [335]. A survey on criteria for the feasibility of the interval Gaussian algorithm up to 1990 was given in Mayer [205]. The sufficient criterion in Theorem 5.4.10 is published in Frommer, Mayer [109]. Its generalization in Theorem 5.4.13 goes back to Mayer, Pieper [220]. The criterion for Hessenberg matrices in Theorem 5.4.19 originates from Mayer [215] and replaces the more extensive one in Reichmann [290]. Lemma 5.4.23, Theorem 5.4.24, and Corollary 5.4.25 are presented in Frommer [103]. Another criterion is derived in Frommer, Mayer [110]. The auxiliary result in Lemma 5.4.28 as well as the succeeding perturbation results in Theorem 5.4.29 and Corollary 5.4.31 are from Neumaier and can be found in Neumaier [255]. The negative result in Theorem 5.4.32 is due to Mayer [214], the Theorems 5.4.36–5.4.40 and Corollary 5.4.41 come from Mayer, Rohn [221]. In Garloff [115] and Jan Mayer [225] modifications of the interval Gaussian algorithm are suggested for improving its feasibility. Block methods are presented in Garloff [114]. Pivoting is studied in Hebgen [138], Wongwises [364].

The interval Cholesky method together with some basic properties was introduced in Alefeld, Mayer [35]. Additional conditions for its feasibility and comparisons with the vector resulting from the interval Gaussian algorithm can be found in Alefeld, Mayer [41]. All our results on this method are taken from these two papers and from Alefeld, Mayer [37]. Garloff [116, 117] suggests pivot tightening for the interval Cholesky method in order to improve its feasibility. A block version of the interval Cholesky method is considered in Schäfer [326, 327]. A sophisticated floating point Cholesky method is presented in Rump [316].

To 5.5: Most of the results presented in this section are meanwhile standard in interval analysis and can already be found in the books of Alefeld, Herzberger [26] and Neumaier [257]; see also Neumaier [254, 255] and Mayer [204]. It is an open question under which conditions $R_1(\mathcal{I}) = \rho(|[M]^G| \cdot |[N]|)$ holds in Theorem 5.5.6. Some basic ideas can be found in Mayer [201, 202]. Enclosures for at least one solution of a singular system are considered in Alefeld, Mayer [39, 40]. The algebraic solution of the fixed point equation $[x] = [A][x] + [b]$ is studied in Shary [341]. The discussion of its form in the case $\rho(|[A]|) \geq 1$ originates from Mayer, Warnke [222, 223, 224]. A generalization to reducible matrices $|[A]|$ can be found in Arndt, Mayer [55]. An extension to complex matrices is considered in Arndt [53].

The contents of Theorem 5.5.48 can be found in Rump [317], its symmetric variant for the iteration (5.5.112) in Jansson [153]. See also Rump [313, 318]. A way to check nonsingularity of a matrix A in a floating point system with directed rounding is presented in Oishi, Rump [265]. In this paper it is also shown how rigorous bounds for the error $\|x^* - \tilde{x}\|_\infty$ can be computed in a fast way, where x^* solves $Ax = b$ exactly and is unknown, while $\tilde{x}$ is only an approximation of x^* and is known.

6 Nonlinear systems of equations

A nonlinear system of n equations with n variables $x_1, \ldots, x_n$ is considered here either as a zero problem $f(x) = 0$ of a function $f: D \subseteq \mathbb{R}^n \rightarrow \mathbb{R}^n$ or as a fixed point problem $x = g(x)$ with $g: D \subseteq \mathbb{R}^n \rightarrow \mathbb{R}^n$. Newton's method and the general iteration method are two traditional methods to approximate a zero of f, and a fixed point of g, respectively. We will extend them to interval iterations verifying thus a zero or a fixed point within some iterate. We will get to know additional methods in this chapter. To this end we assume that f and g always have the required smoothness – at least continuity. Mostly it is sufficient that $f, g \in C^2(D)$ holds.

We will start with the interval Newton method with which we will be able to verify zeros of f and with which we can also decide whether some interval vector does not contain any zero of f.

6.1 Newton method – one-dimensional case

The traditional one-dimensional Newton method can be derived in several ways, where $x_0 \in D \subseteq \mathbb{R}$ is the starting value and $f'(x) \neq 0$ is assumed whenever it is necessary, in particular, $f'(x_0) \neq 0$.

(i) Linearization: Replace $f(x)$ by its Taylor polynomial $f(x_0) + f'(x_0)(x - x_0)$ of degree 1 and substitute the equation $f(x) = 0$ by $f(x_0) + f'(x_0)(x - x_0) = 0$. Solve the latter for x and rename x with x_1.

(ii) Geometric way: Intersect the tangent at $(x_0, f(x_0))$ of the graph of f with the x-axis and denote the resulting point by $(x_1, 0)$.

(iii) Fixed point iteration: Write the zero problem $f(x) = 0$ as the equivalent fixed point equation $x = x - f(x)/f'(x) =: g(x)$ and start the usual fixed point iteration for g with $x = x_0$.

In all the three derivations replace x_0, x_1 by x_k, x_{k+1} ending up with the Newton method

$$x_{k+1} = x_k - f(x_k)/f'(x_k), \quad k = 0, 1, \ldots. \tag{6.1.1}$$

It is well known that this method is locally quadratically convergent provided that $f \in C^2(D)$ has a zero x^* with $f'(x^*) \neq 0$.

In order to introduce the interval Newton method we consider f only on the interval $[x]^0$, assume $x^* \in [x]^0$, choose $\tilde{x}^0 \in [x]^0$ and use the Taylor theorem. Then

$$0 = f(x^*) = f(\tilde{x}^0) + f'(\xi)(x^* - \tilde{x}^0), \quad \xi \in [x]^0 \text{ appropriately,}$$

whence

$$x^* = \tilde{x}^0 - f(\tilde{x}^0)/f'(\xi) \in (\tilde{x}^0 - f(\tilde{x}^0)/f'([x]^0)) \cap [x]^0.$$

DOI 10.1515/9783110499469-007

This leads to the following interval Newton method:

$$\left.\begin{aligned} &\text{choose } \tilde{x}^k \in [x]^k, \\ &[n]^k := \tilde{x}^k - f(\tilde{x}^k)/f'([x]^k), \\ &[x]^{k+1} := [n]^k \cap [x]^k, \end{aligned}\right\} \quad k = 0, 1, \ldots . \tag{6.1.2}$$

Stop if the intersection is empty.

With $[m] = f'([x]^k)$ the method is illustrated in Figure 6.1.1.

The intersection in (6.1.2) leads to a sequence of iterates which is monotonously decreasing with respect to '$\subseteq$'.

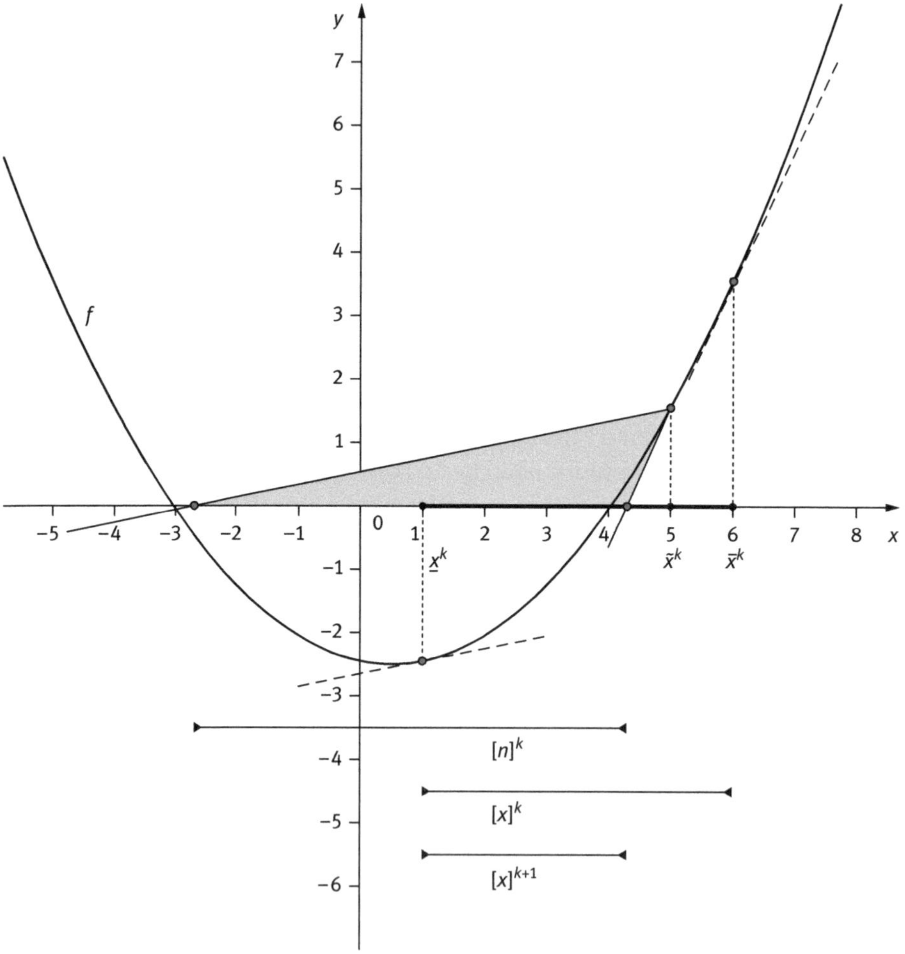

Fig. 6.1.1: One step of the Newton method.

Theorem 6.1.1. *Let D be an open subset of $\mathbb{R}$, $f \in C^1(D)$, $[x]^0 \subseteq D$, $0 \notin f'([x]^0)$, $f' \in L([x]^0)$ with Lipschitz constant $l_{f'} > 0$.*

(a) *If $[x]^0$ contains a zero x^* of f, then the following properties hold for f and the Newton method $\mathcal{I}$ in (6.1.2).*
   (i) *x^* is the only zero of f in $[x]^0$,*
   (ii) *x^* is contained in each iterate $[x]^k$ of (6.1.2),*
   (iii) *$[x]^{k+1} \subseteq [x]^k \subseteq \cdots \subseteq [x]^1 \subseteq [x]^0$,*
   (iv) *$d([x]^{k+1}) \le (1 - \langle f'([x]^k)\rangle / |f'([x]^k)|) d([x]^k) \le c(d([x]^k))^2$, $k = 0, 1, \ldots$, with $c > 0$ independent of k,*
   (v) *$d([x]^{k+1}) \le \frac{1}{2}(1 - \langle f'([x]^k)\rangle / |f'([x]^k)|) d([x]^k) \le \frac{c}{2}(d([x]^k))^2$, $k = 0, 1, \ldots$, if $\tilde{x}^k = \check{x}^k$, with $c > 0$ as in (iv),*
   (vi) *$\lim_{k\to\infty} [x]^k = x^*$,*
   (vii) *$O_R(\mathcal{I}) \ge 2$, i.e., the interval Newton method is at least quadratically convergent on $[x]^0$.*

(b) *The starting interval $[x]^0$ contains no zero of f if and only if (6.1.2) stops after finitely many steps with an empty intersection. In this case there are constants $a, b > 0$, which depend on $[x]^0$ (and on f) and satisfy the inequality*

$$\min\{\overline{n}^k - \underline{x}^k,\ \overline{x}^k - \underline{n}^k\} \le a(d([x]^k))^2 - b, \quad k = 0, 1, \ldots. \tag{6.1.3}$$

(c) *Unless $f(\tilde{x}^k) = 0$ in case (a) we have $\tilde{x}^k \notin [x]^{k+1}$ in both cases* (a) *and* (b). *In particular, if $\tilde{x}^k = \check{x}^k$, then in each Newton step $d([x]^k)$ is more than halved.*

Proof. (a)

(i) follows from $0 \notin f'([x]^0)$ whence f is strictly monotone on $[x]^0$.

(ii) Let $x^* \in [x]^k$ for some $k \in \mathbb{N}_0$. Then $0 = f(x^*) = f(\tilde{x}^k) + f'(\xi)(x^* - \tilde{x}^k)$ holds for some $\xi \in [x]^k$, and $x^* = \tilde{x}^k - f(\tilde{x}^k)/f'(\xi) \in [x]^{k+1}$ follows.

(iii) is obvious by virtue of the intersection in (6.1.2).

(iv) W.l.o.g. let $[\underline{f}'_k, \overline{f}'_k] = f'([x]^k) > 0$. Otherwise consider $-f$ and $-f'([x]^k)$. If $f(\tilde{x}^k) = 0$, then $[x]^{k+1} \equiv \tilde{x}^k$. This proves the assertion with $d([x]^{k+\ell}) = 0$, $\ell - 1, 2, \ldots$. If $f(\tilde{x}^k) > 0$, then

$$\overline{x}^{k+1} = \tilde{x}^k - f(\tilde{x}^k)/\overline{f}'_k < \tilde{x}^k, \tag{6.1.4}$$

and $\underline{x}^{k+1} = \max\{\underline{x}^k, \tilde{x}^k - f(\tilde{x}^k)/\underline{f}'_k\} \ge \tilde{x}^k - f(\tilde{x}^k)/\underline{f}'_k$. The last inequality implies $f(\tilde{x}^k) \ge \underline{f}'_k(\tilde{x}^k - \underline{x}^{k+1})$ and

$$\begin{aligned} d([x]^{k+1}) &= (\tilde{x}^k - f(\tilde{x}^k)/\overline{f}'_k) - \underline{x}^{k+1} \\ &\le \tilde{x}^k - (\underline{f}'_k/\overline{f}'_k)(\tilde{x}^k - \underline{x}^{k+1}) - \underline{x}^{k+1} \\ &= (1 - \underline{f}'_k/\overline{f}'_k)(\tilde{x}^k - \underline{x}^{k+1}) \le (1 - \underline{f}'_k/\overline{f}'_k) d([x]^k) \\ &= (1 - \langle f'([x]^k)\rangle/|f'([x]^k)|) d([x]^k) = \frac{1}{|f'([x]^k)|} d(f'([x]^k)) d([x]^k) \\ &\le \frac{l_{f'}}{|f'([x]^k)|}(d([x]^k))^2 \le \frac{l_{f'}}{\langle f'([x]^0)\rangle}(d([x]^k))^2. \end{aligned} \tag{6.1.5}$$

The assertion follows with $c := l_{f'}/\langle f'([x]^0)\rangle$. If $f(\tilde{x}^k) < 0$, the proof proceeds similarly and is left to the reader.

(v) If $\tilde{x}^k = \check{x}^k$, then $\tilde{x}^k - \underline{x}^{k+1} \le \tilde{x}^k - \underline{x}^k = \frac{1}{2}d([x]^k)$. Now the assertion follows from (iv) with $d([x]^k)$ being replaced by $\frac{1}{2}d([x]^k)$ in (6.1.5).

(vi) Let $\gamma = 1 - \langle f'([x]^0)\rangle/|f'([x]^0)|$. Then $\gamma < 1$ since $0 \notin f'([x]^0)$. From $[x]^k \subseteq [x]^0$ and the first inequality in (iv) one obtains

$$\begin{aligned} d([x]^{k+1}) &\le (1 - \langle f'([x]^0)\rangle/|f'([x]^0)|)d([x]^k) \\ &= \gamma d([x]^k) \le \gamma^{k+1} d([x]^0) \to 0 \text{ for } k \to \infty. \end{aligned} \tag{6.1.6}$$

Together with (ii) this implies the assertion.

(vii) follows from Theorem 5.5.4 and (iv) since $q([x]^{k+1}, x^*) \le d([x]^{k+1}) \le c d([x]^k)^2 \le 4c(q([x]^k, x^*))^2$. For the last inequality we used Theorem 2.5.7 (j) with $[b] = [x]^k$.

(b) If (6.1.2) stops with empty intersection, then (a) (ii) shows that $[x]^0$ cannot contain a zero x^* of f.

Assume conversely that $[x]^0$ does not contain a zero x^* of f and that (6.1.2) does not stop with empty intersection. Then (iii) and (iv) of (a) still hold, hence the sequence $([x]^k)$ converges to a point interval $[x, x] \subseteq [x]^0$. This can be seen as in the proof of (vi). In particular, $\lim_{k\to\infty} \tilde{x}^k = x$ holds. Taking the limit $k \to \infty$ in (6.1.2) and exploiting the continuity of the intersection operation yields

$$[x, x] = (x - f(x)/f'([x, x])) \cap [x, x],$$

hence $0 \in f(x)/f'([x, x])$, i.e., $f(x) = 0$. This contradicts the assumption.

In order to prove (6.1.3) we first assume $[\underline{f}'_k, \overline{f}'_k] = f'([x]^k) > 0$ and $f(x) > 0$ on $[x]^0$. Then $\overline{n}^k = \tilde{x}^k - f(\tilde{x}^k)/\overline{f}'_k$, and with $|f|_{\min} := \min\{|f(x)| \mid x \in [x]^0\}$ and $f(\tilde{x}^k) = f(\underline{x}^k) + f'(\xi)(\tilde{x}^k - \underline{x}^k)$ for some $\xi \in [x]^k$ we get

$$\begin{aligned} \overline{n}^k - \underline{x}^k &= \tilde{x}^k - f(\tilde{x}^k)/\overline{f}'_k - \underline{x}^k = (1 - f'(\xi)/\overline{f}'_k)(\tilde{x}^k - \underline{x}^k) - f(\underline{x}^k)/\overline{f}'_k \\ &= (\overline{f}'_k - f'(\xi))(\tilde{x}^k - \underline{x}^k)/\overline{f}'_k - f(\underline{x}^k)/\overline{f}'_k \\ &\le d(f'([x]^k))d([x]^k)/\overline{f}'_k - f(\underline{x}^k)/\overline{f}'_k \\ &\le (l_{f'}/\langle f'([x]^0)\rangle)(d([x]^k))^2 - |f|_{\min}/|f'([x]^0)|, \end{aligned} \tag{6.1.7}$$

where we used Theorem 4.1.18 (a) for the last inequality. This relation holds also for $f'([x]^0) < 0$ and $f(x) < 0$ for all $x \in [x]^0$ as can be seen by replacing f, f' by $-f, -f'$. If $f'([x]^0) < 0$ and $f(x) > 0$ for all $x \in [x]^0$, we get

$$\begin{aligned} \overline{x}^k - \underline{n}^k &= \overline{x}^k - \tilde{x}^k + f(\tilde{x}^k)/\underline{f}'_k = (1 - f'(\xi)/\underline{f}'_k)(\overline{x}^k - \tilde{x}^k) + f(\overline{x}^k)/\underline{f}'_k \\ &= (1 - |f'(\xi)|/|\underline{f}'_k|)(\overline{x}^k - \tilde{x}^k) + f(\overline{x}^k)/\underline{f}'_k \\ &\le d(f'([x]^k))d([x]^k)/|\underline{f}'_k| - f(\overline{x}^k)/|\underline{f}'_k| \\ &\le (l_{f'}/\langle f'([x]^0)\rangle)(d([x]^k))^2 - |f|_{\min}/|f'([x]^0)|, \end{aligned}$$

which holds also for $f'([x]^0) > 0$ and $f(x) < 0$ for all $x \in [x]^0$. Together with (6.1.7) this proves (6.1.3) with $a = l_{f'}/\langle f'([x]^0)\rangle$ and $b = |f|_{\min}/|f'([x]^0)|$.

(c) follows directly from $\overline{n}^k - \tilde{x}^k = -f(\tilde{x}^k)/\overline{f}'_k < 0$ if $f(\tilde{x}^k) \cdot f'([x]^k) > 0$ and from $\tilde{x}^k - \underline{n}^k = f(\tilde{x}^k)/\underline{f}'_k < 0$ if $f(\tilde{x}^k) \cdot f'([x]^k) < 0$. □

We comment on part (b) of Theorem 6.1.1 assuming that f has no zero in $[x]^0$. Since in this case (6.1.6) also holds, the diameter $d([x]^k)$ gets small if there are sufficiently many iterates. Now (6.1.3) implies $\overline{n}^k < \underline{x}^k$ or $\overline{x}^k < \underline{n}^k$ if $d([x]^k)$ is small enough, which is equivalent to $[n]^k \cap [x]^k = \emptyset$. As in Alefeld [15], property (6.1.3) can be denoted as *quadratic divergence* of the Newton method if f has no zero in $[x]^0$.

Remark 6.1.1. As for the traditional simplified Newton method, one can replace $f'([x]^k)$ in (6.1.2) by any fixed interval $[m]$ with $0 \notin [m]$. This leads to the Newton-like method

$$[x]^{k+1} = (\tilde{x}^k - f(\tilde{x}^k)/[m]) \cap [x]^k, \quad k = 0, 1, \dots . \tag{6.1.8}$$

One can even iterate by

$$[x]^{k+1} = \left(\tilde{x}^k - \frac{f(\tilde{x}^k)}{[m] \cap f'([x]^0)} \right) \cap [x]^k, \quad k = 0, 1, \dots, \tag{6.1.9}$$

or

$$[x]^{k+1} = \left(\tilde{x}^k - \frac{f(\tilde{x}^k)}{[m] \cap f'([x]^k)} \right) \cap [x]^k, \quad k = 0, 1, \dots, \tag{6.1.10}$$

where in all three iterations $\tilde{x}_k \in [x]^k$ is required. Notice that $0 \in f'([x]^0)$ is allowed in (6.1.9) and (6.1.10), but $0 \notin [m]$ has still to hold.

If $\tilde{x}^k$ coincides in (6.1.8) and (6.1.9) and if the iterates in (6.1.9) are denoted by $[y]^k$ in order to distinguish them from those in (6.1.8), then $[y]^k \subseteq [x]^k$ can be shown provided that $[x]^0 = [y]^0$; cf. Exercise 6.1.1.

The iteration (6.1.10) converges at least quadratically. The other results of Theorem 6.1.1 hold for (6.1.10) similarly. The iterations (6.1.8), (6.1.9) have at least the properties (a) (i)–(vi) and (b) without (6.1.3) of Theorem 6.1.1. □

Now we will see what happens if $0 \in f'([x]) = [\underline{f}', \overline{f}']$ is allowed. In addition to this we assume $x^*, \tilde{x} \in [x]$, $f(x^*) = 0 \neq f(\tilde{x})$. By virtue of the mean value theorem we get

$$f(x^*) = 0 = f(\tilde{x}) + f'(\xi)(x^* - \tilde{x}), \quad \xi \in [x],$$

hence $f'(\xi) \neq 0$ and

$$x^* = \tilde{x} - \frac{f(\tilde{x})}{f'(\xi)} \in (M_1 \cup M_2) \cap [x]$$

with

$$M_1 = \left\{ \tilde{x} - \frac{f(\tilde{x})}{s} \;\middle|\; s \in [\underline{f}', 0) \right\}, \quad M_2 = \left\{ \tilde{x} - \frac{f(\tilde{x})}{s} \;\middle|\; s \in (0, \overline{f}'] \right\},$$

cf. Figure 6.1.2.

This suggests the following modification of the Newton method using a stack.

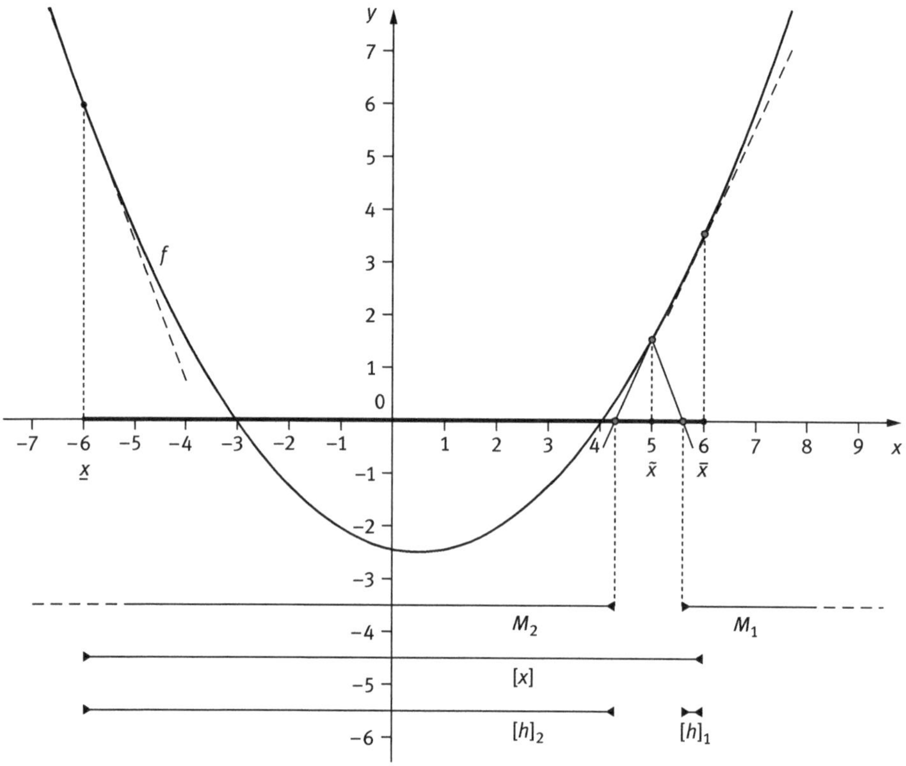

Fig. 6.1.2: Modified Newton method if $0 \in f'([x])$.

Modified Newton method ($0 \in f'([x]^0)$ allowed)

Let f, f', $[x]^0$ as in Theorem 6.1.1 with the exception of $0 \notin f'([x]^0)$. Assume that $[x]^0$ contains at most finitely many zeros of f and f'. Let $[v]$ be a vector of dynamic length n called a stack, and denote by $[h]_1$, $[h]_2$ intervals or the empty set.

(1) $n := 1$, $[v]_1 := [x]^0$

(2) if $n < 1$ then stop with empty stack else $[x] := [v]_n$, $n := n - 1$

(3) if $0 \notin f([x])$ then goto (2) % $[x]$ does not contain any zero of f

(4) choose $\tilde{x} \in [x]$ % for instance $\tilde{x} = \check{x}$

(5) if $f(\tilde{x}) = 0$ then stop with the zero $x^* := \tilde{x}$

(6) $j := 1$

  (a) if $0 \notin f'([x])$ then $[h]_j := (\tilde{x} - f(\tilde{x})/f'([x])) \cap [x]$

  (b) if $0 \in f'([x]) =: [\underline{f}', \overline{f}']$ then

    case $f(\tilde{x}) > 0$:

      if $\underline{f}' \neq 0$ then $[h]_1 := [\tilde{x} - f(\tilde{x})/\underline{f}', \infty) \cap [x]$

if $\overline{f}' \neq 0$ then
  if $\underline{f}' \neq 0$ then $j := j + 1$
  $[h]_j := (-\infty, \tilde{x} - f(\tilde{x})/\overline{f}'] \cap [x]$

case $f(\tilde{x}) < 0$:
  if $\overline{f}' \neq 0$ then $[h]_1 := [\tilde{x} - f(\tilde{x})/\overline{f}', \infty) \cap [x]$
  if $\underline{f}' \neq 0$ then
    if $\overline{f}' \neq 0$ then $j := j + 1$
    $[h]_j := (-\infty, \tilde{x} - f(\tilde{x})/\underline{f}'] \cap [x]$

(7) for $i = 1$ to j do
  if $[h]_i \neq \emptyset$ then $n := n + 1$, $[v]_n := [h]_i$

(8) goto (2).

Now we will comment on this algorithm.

Remark 6.1.2.

(a) Step (6) (a) is a Newton step.

(b) Step (6) (b) is based on the following equality: if $\tilde{b} \in \mathbb{R} \setminus \{0\}$, $0 \in [a] \in \mathbb{IR}$, $d([a]) > 0$, then

$$\left\{ \frac{\tilde{b}}{\tilde{a}} \;\middle|\; \tilde{a} \in [a] \setminus \{0\} \right\} = \begin{cases} \left(-\infty, \frac{\tilde{b}}{\underline{a}}\right] \cup \left[\frac{\tilde{b}}{\overline{a}}, \infty\right), & \text{if } \tilde{b} > 0, \\ \left(-\infty, \frac{\tilde{b}}{\overline{a}}\right] \cup \left[\frac{\tilde{b}}{\underline{a}}, \infty\right), & \text{if } \tilde{b} < 0. \end{cases}$$

If $\underline{a} = 0$ or $\overline{a} = 0$, then that interval disappears whose bound is undefined.

(c) In step (6) we cannot have $\underline{f}' = \overline{f}' = 0$ simultaneously. In step (6) (a) this is part of the assumption. In step (6) (b) $f'([x]) \equiv 0$ implies that f' has infinitely many zeros in $[x]$ because of $d([x]) > 0$. (If $d([x]) = 0$ then step (6) is not reached because of the steps (3) and (5)!) The final value of j in step (6) is only 1 or 2.

(d) In a practical realization, step (5) of the algorithm has to be modified and a stopping criterion has to be added, for instance in step (8). Thus one could stop if $[v]_n$ has a sufficiently small width. If one is interested in several zeros in $[x]^0$ and if the stopping criterion is true, then the momentary final element $[v]_n$ of $[v]$ could be stored in a second vector $[w]$ and n is replaced by $n - 1$ in order to study the remaining entries of $[v]$ for zeros. At the end, the vector $[w]$ contains candidates of small width for zeros of f. Whether these candidates really contain zeros must be studied separately, for instance in a more refined process. One can also think of changing the order with which the entries of $[v]$ are considered. So one could fetch the widest entry of $[v]$ first instead of the last one. In this case one should replace the terminology 'stack' by 'list' or another appropriate data structure because of the 'last in – first out' (LIFO) handling of a stack. □

In order to prove that the modified Newton method either verifies a zero of f in $[x]^0$ or shows that there cannot be any we need an auxiliary result which is closely related to the algorithm above and uses its notation without further reference.

Lemma 6.1.2. *Choose $\tilde{x} = \check{x}$ in the modified Newton method.*

(a) *If $[x]$ does not contain a zero of f when executing step (6) of this method, then either $[h]_j$ is empty or*

$$d([h]_j) \le \frac{1}{2} d([x]) - \Delta([x]) \quad \text{with } \Delta([x]) := \frac{\langle R_f([x]) \rangle}{|f'([x])|} \tag{6.1.11}$$

holds.

(b) *If the last component $[v]_n$ of the stack does not contain a zero of f, then after finitely many steps the stack does not contain any subset of $[v]_n$.*

Proof. (a) Let $[h]_j \ne \emptyset$. We first consider step (6) (a). If there $f(\check{x}) > 0$ and $f'([x]^0) > 0$ hold, then f increases on $[x]$. Hence

$$[h]_1 = \left[\check{x} - \frac{f(\check{x})}{\underline{f'}}, \check{x} - \frac{f(\check{x})}{\overline{f'}}\right] \cap [x] \subseteq \left[\underline{x}, \check{x} - \frac{f(\check{x})}{\overline{f'}}\right]$$

and

$$d([h]_1) \le \check{x} - \underline{x} - \frac{f(\check{x})}{\overline{f'}} \le \frac{1}{2} d([x]) - \Delta([x])$$

follows. Similarly, if $f(\check{x}) < 0$ and $f'([x]^0) > 0$, then $[h]_1 \subseteq [\check{x} - f(\check{x})/\overline{f'}, \overline{x}]$ and $d([h]_1) \le \frac{1}{2} d([x]) - \Delta([x])$. The remaining cases are left to the reader.

Next we address step (6) (b). Let $f(\check{x}) > 0$, $\underline{f'} \ne 0$. Then $\underline{f'} < 0$ since

$$0 \in f'([x]) \quad \text{and} \quad [h]_1 \subseteq \left[\check{x} - \frac{f(\check{x})}{\underline{f'}}, \overline{x}\right]$$

holds, which implies $d([h]_1) \le \frac{1}{2} d([x]) - \Delta([x])$ again. This result can be easily shown for the remaining cases, too. Notice that $f'([x]) \ne 0$ in step (6) as was stated in Remark 6.1.2 (c).

(b) Let $[y]$ be a subset of a former interval $[x] := [v]_n$. If $d([y]) = 0$, then $[y]$ disappears because of the steps (3) or (5). If $d([y]) \ne 0$, then $f'([y]) \ne 0$ holds according to Remark 6.1.2 (c). Since $[y] \subseteq [x]$, we get $\Delta([y]) \ge \Delta([x])$ with $\Delta([x])$ defined as in (6.1.11). If $[y]$ is replaced by some interval $[h]_i \ne 0$ in step (7), then i is at most 2 and the total length of such intervals $[h]_i$ is bounded by

$$2 \cdot \left(\tfrac{1}{2} d([y]) - \Delta([y])\right) \le d([y]) - \Delta([y]) \le d([y]) - \Delta([x]) = d([y]) - \Delta([v]_n)$$

according to (a). If $\Delta([v]_n) = 0$, then $0 \in R_f([v]_n)$ in contrast to our assumption. Therefore, $\Delta([v]_n) > 0$ holds and a replacement process as just described can occur at most $\lceil \frac{d([v]_n)}{\Delta([v]_n)} \rceil$ times. □

Now we are ready to prove our main result on the modified Newton method.

Theorem 6.1.3. *In the modified Newton method there occur exactly three cases:*

(1) *$[x]^0$ does not contain any zero of f. The method stops after finitely many iterations.*

(2) *$[x]^0$ contains a zero of f. This is verified after finitely many iterations in step (5).*

(3) $[x]^0$ *contains a zero of* f *and the method does not stop after finitely many iterations. If one denotes by* $[x]^k$, $k = 0, 1, \dots,$ *every last element of the stack which is fetched in step* (2), *then the sequence* $(\underline{x}^k)$ *converges monotonously increasing to the smallest zero* x^* *of* f *in* $[x]^0$.

Proof. We prove the theorem only for $\tilde{x} = \check{x}$.

(1) follows immediately from Lemma 6.1.2 (b). Assume now that neither (1) nor (2) is valid. As in the proof of Lemma 6.1.2 (a) one can see that

$$d([h]_j) \le \frac{1}{2} d([x]) \tag{6.1.12}$$

holds. According to the construction of $[x]^k$ we have $\underline{x}^k \le \underline{x}^{k+1} \le \overline{x}^0$, hence $(\underline{x}^k)$ converges to some $z \in [x]^0$. Lemma 6.1.2 (b) guarantees $x^* \in [x]^{k_\nu}$, $\nu = 1, 2, \dots$ for some subsequence $([x]^{k_\nu})$ of $([x]^k)$. From (6.1.12) we get $d([x]^{k_{\nu+1}}) \le d([x]^{k_\nu})/2$ even if $k_{\nu+1} \ne k_\nu + 1$; cf. Figure 6.1.3.

Therefore, $\lim_{\nu\to\infty} [x]^{k_\nu} = x^*$ holds and implies $\lim_{\nu\to\infty} \underline{x}^{k_\nu} = x^*$, hence

$$\lim_{k\to\infty} \underline{x}^k = z = x^*. \qquad \square$$

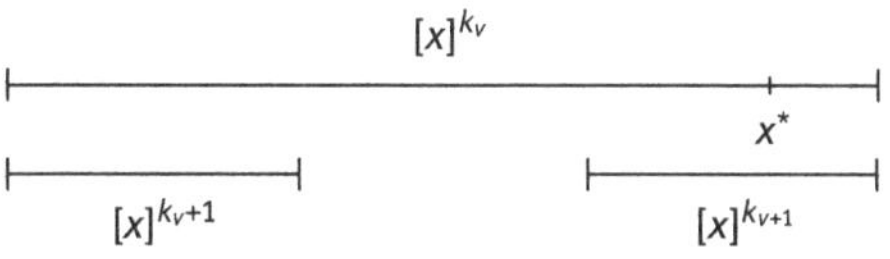

Fig. 6.1.3: Iterates $[x]^{k_\nu}$, $[x]^{k_\nu+1}$, and $[x]^{k_{\nu+1}}$.

Exercises

Ex. 6.1.1. Prove the statements of Remark 6.1.1.

6.2 Newton method – multidimensional case

In order to construct the multidimensional interval Newton method we proceed analogously to the one-dimensional case. To this end we use the multidimensional mean value theorem 1.6.1 and represent $f(x^*) = 0$ for $f \in C^1(D)$ as

$$0 = f(x^*) = f(\tilde{x}) + J(x^*, \tilde{x})(x^* - \tilde{x}) \tag{6.2.1}$$

with $x^*, \tilde{x} \in D$, $J(x, y) = \int_0^1 f'(x + t(y - x))dt \in \mathbb{R}^{n\times n}$. Assuming J to be regular we get

$$x^* = \tilde{x} - J(x^*, \tilde{x})^{-1} f(\tilde{x}) = \tilde{x} - \mathrm{IGA}(J(x^*, \tilde{x}), f(\tilde{x})) \in \tilde{x} - \mathrm{IGA}(f'([x]^0), f(\tilde{x})), \tag{6.2.2}$$

provided that $x^*, \tilde{x} \in [x]^0 \subseteq D$. With the Newton operator

$$N([x], \tilde{x}) = \tilde{x} - \mathrm{IGA}(f'([x]), f(\tilde{x})) \tag{6.2.3}$$

this induces the interval Newton method

$$[x]^{k+1} = N([x]^k, \tilde{x}^k) \cap [x]^k, \quad k = 0, 1, \dots, \tag{6.2.4}$$

with $\tilde{x}^k \in [x]^k$, $[x]^0 \subseteq D$. If the intersection in (6.2.4) is empty, the iteration is stopped.

In order to guarantee at least the first step of the method it is tacitly assumed that $\mathrm{IGA}(f'([x]^0))$ exists. This guarantees that also the succeeding iterates exist unless the intersection is empty. In addition, it implies the regularity of $J(x^*, \tilde{x}) \in f'([x]^0)$; cf. (5.4.14). This does, however, not guarantee that $\mathrm{IGA}(f'([x]^0))$ itself is regular; cf. $f(x, y) = (0.5x^2 - 0.5y^2, x + y)^T$, $[x]^0 = ([1, 2], [1, 2])^T$, and Exercise 5.4.2.

For (6.2.4) we obtain the following result.

Theorem 6.2.1. *Let D be an open convex subset of $\mathbb{R}^n$, $f : D \subseteq \mathbb{R}^n \to \mathbb{R}^n$, $f \in C^1(D)$. Let the entries f'_{ij} of f' be represented as expressions from $\mathbb{E}$ as in Definition 4.1.2 such that the interval arithmetic evaluation $f'([x])$ exists for $[x] \in \mathbb{I}(D)$. In addition, assume that $\mathrm{IGA}(f'([x]^0))$ exists for some starting vector $[x]^0 \subseteq D$.*

(a) *If f has a zero $x^* \in [x]^0$, then $[x]^k$, $k = 0, 1, \dots$, and $[x]^* := \lim_{k\to\infty}[x]^k$ exist and*

$$x^* \in [x]^* \subseteq [x]^{k+1} \subseteq [x]^k \subseteq [x]^0, \quad k = 0, 1, \dots, \tag{6.2.5}$$

holds.

(b) *There is no zero $x^* \in [x]^0$ if (6.2.4) stops because of empty intersection after finitely many iterations.*

Proof. (a) From (6.2.2) we get $x^* \in [x]^1$, and induction and the intersection in (6.2.4) imply the assertion.

(b) follows by contradiction immediately from (a). □

It is unknown whether the converse in (b) holds similarly to in (b) of Theorem 6.1.1. The convergence of $([x]^k)$ to a point interval as in the one-dimensional case is not mandatory as the Example 6.2.2 shows. This example is due to Schwandt [335], p. 85ff; its proof is left to the reader as Exercise 6.2.1.

Example 6.2.2. Let

$$f(x, y) = \begin{pmatrix} -x^2 + y^2 - 1 \\ x^2 - y \end{pmatrix}, \quad [x]^0 = \frac{1}{10}\begin{pmatrix} [11, 19] \\ [11, 19] \end{pmatrix},$$

$\tilde{x}^k = \check{x}^k$, $k = 0, 1, \dots$. Then f has the unique zero

$$x^* = \left(\sqrt{\frac{1 + \sqrt{5}}{2}}, \frac{1 + \sqrt{5}}{2} \right)^T$$

in $[x]^0$ and $[x]^0 = [x]^k$ holds for the iterates $[x]^k$, $k = 0, 1, \dots$, of the Newton method. In particular, $[x]^* = [x]^0$ is not degenerate.

Unfortunately, Theorem 6.2.1 does not verify a zero $x^* \in [x]^0$ of f nor does it guarantee uniqueness of such a zero. This changes if $[x]^0$ satisfies additional properties.

Theorem 6.2.3. *Let the assumptions of Theorem 6.2.1 be fulfilled. If*

$$N([x]^0, \tilde{x}^0) \subseteq [x]^0 \tag{6.2.6}$$

holds, then f has exactly one zero $x^* \in [x]^0$. We call (6.2.6) the Newton test.

Proof. Let $x \in [x]^0$. Then $g(x) = x - (J(x, \tilde{x}^0))^{-1} f(x)$ satisfies

$$\begin{aligned} g(x) &= x - (J(x, \tilde{x}^0))^{-1}(f(\tilde{x}^0) + J(x, \tilde{x}^0)(x - \tilde{x}^0)) \\ &= \tilde{x}^0 - (J(x, \tilde{x}^0))^{-1} f(\tilde{x}^0) \in N([x]^0, \tilde{x}^0) \subseteq [x]^0 \end{aligned}$$

by assumption. Therefore, g maps $[x]^0$ into itself. Since it is continuous, Brouwer's fixed point theorem 1.5.8 applied to g and $[x]^0$ guarantees the existence of a fixed point $x^* \in [x]^0$ of g. This implies $f(x^*) = 0$. In order to prove uniqueness of x^* let $y^* \in [x]^0$ be a second zero of f which differs from x^*. Replace $\tilde{x}$ in (6.2.2) by y^*. Then $x^* = y^* - J(x^*, y^*)^{-1} f(y^*) = y^*$ in contrast to $x^* \neq y^*$. □

Our next theorem provides conditions such that the Newton method either generates a sequence $([x]^k)$ which contracts to a point vector x^* or stops after finitely many iterations with an empty intersection. In the first case, x^* is the unique zero of f in $[x]^0$, in the second there is no zero of f in $[x]^0$.

Theorem 6.2.4. *Let the assumptions of Theorem 6.2.1 be fulfilled. In addition, let*

$$\rho(A) < 1 \quad \text{hold for } A = |I - \mathrm{IGA}(f'([x]^0)) \cdot f'([x]^0)|$$

or

$$\rho(B) < 1 \quad \text{for } B = d(\mathrm{IGA}(f'([x]^0))) \cdot |f'([x]^0)|.$$

(a) $\mathrm{IGA}(f'([x]^0))$ *is regular.*
(b) *Either* $f(\tilde{x}^k) = 0$ *or* $\tilde{x}^k \notin [x]^{k+1}$.
(c) *If f has a zero $x^* \in [x]^0$, then $\lim_{k\to\infty}[x]^k = x^*$ for the iterates $[x]^k$ of (6.2.4). In particular, x^* is the unique zero of f within $[x]^0$.*
(d) *The starting interval vector $[x]^0$ contains no zero of f if and only if (6.2.4) stops after finitely many steps with an empty intersection. In this case there are constants $a, b > 0$, which depend on $[x]^0$ (and f) only, and an index i_0 which may depend on the iterate $[x]^k$ such that the inequality*

$$\min\{\, \overline{N}([x]^k, \tilde{x}^k)_{i_0} - \underline{x}^k_{i_0},\ \overline{x}^k_{i_0} - \underline{N}([x]^k, \tilde{x}^k)_{i_0} \,\} \le a\|d([x]^k)\|_\infty^2 - b, \quad k = 0, 1, \ldots \tag{6.2.7}$$

holds, provided that $f' \in L([x]^0)$.

Proof. (a) follows immediately from Theorem 5.4.2 (c).

(b) If $\tilde{x}^k \in [x]^{k+1}$, then $\tilde{x}^k \in N([x]^k, \tilde{x}^k) = \tilde{x}^k - \mathrm{IGA}(f'([x]^k), f(\tilde{x}^k))$, whence $0 \in \mathrm{IGA}(f'([x]^k), f(\tilde{x}^k)) \subseteq \mathrm{IGA}(f'([x]^0)) \cdot f(\tilde{x}^k)$. Theorem 4.1.5 guarantees that there is some matrix $C \in \mathrm{IGA}(f'([x]^0))$ such that $0 = Cf(\tilde{x}^k)$. By virtue of (a) the matrix C is regular which implies $f(\tilde{x}^k) = 0$.

(c) Let $[x]^* = \lim_{k\to\infty}[x]^k$ as in Theorem 6.2.1 (a). Since the real vectors $\tilde{x}^k$ of the Newton method are contained in the bounded set $[x]^0$, there is a subsequence $(\tilde{x}^{k_\nu})$ with $\tilde{x}^{k_\nu} \in [x]^{k_\nu}$, $\nu = 0, 1, \ldots$, which converges to some vector $\tilde{x}^* \in [x]^* \subseteq [x]^0$. Replace k in (6.2.4) by k_ν and let ν tend to infinity. Then

$$x^*, \tilde{x}^* \in [x]^* = N([x]^*, \tilde{x}^*) \cap [x]^* \subseteq \tilde{x}^* - \mathrm{IGA}(f'([x]^*)) \cdot f(\tilde{x}^*) \tag{6.2.8}$$

by virtue of Theorem 5.4.1 (h). Hence $0 \in \mathrm{IGA}(f'([x]^0)) \cdot f(\tilde{x}^*)$ holds, which implies $f(\tilde{x}^*) = 0$ as in (b). Hence $N([x]^*, \tilde{x}^*) = \tilde{x}^*$ and $[x]^* = \lim_{k\to\infty}[x]^k = \lim_{\nu\to\infty}[x]^{k_\nu} = \tilde{x}^*$ follow. Together with (6.2.8) this yields $x^* = \tilde{x}^*$. Uniqueness of the zero follows as in the proof of Theorem 6.2.3.

(d) By virtue of Theorem 6.2.1 (b) there remains only one direction to be proved. Therefore, assume that $[x]^0$ does not contain any zero of f but (6.2.4) does not stop with empty intersection. Then as in (c) one can construct a vector $\tilde{x}^* \in [x]^0$ which satisfies $f(\tilde{x}^*) = 0$, contradicting the assumption.

In order to prove (6.2.7) we first mention that by virtue of the Theorems 4.1.18 and 5.4.2 (d) there are positive constants γ, κ depending only on $[x]^0$ such that

$$\|d(\mathrm{IGA}(f'([x])))\|_\infty \le \gamma \|d(f'([x]))\|_\infty \le \kappa \|d([x])\|_\infty \tag{6.2.9}$$

holds for all $[x] \subseteq [x]^0$, whence

$$d(\mathrm{IGA}(f'([x]))) \le \kappa e e^T \|d([x])\|_\infty.$$

Define g_C by

$$g_C(x) = \max_{1\le i\le n} |Cf(x)|_i, \quad x \in [x]^0,\ C \in \mathrm{IGA}(f'([x]^0)). \tag{6.2.10}$$

Then g_C is continuous on $[x]^0$, and so is

$$g(x) = \min\{\, g_C(x) \mid C \in \mathrm{IGA}(f'([x]^0)) \,\}.$$

Define the constant $g_{\min} = \min_{x\in[x]^0} g(x)$. Then $g_{\min}$ is positive. Otherwise there are $\hat{x} \in [x]^0$, $\hat{C} \in \mathrm{IGA}(f'([x]^0))$ such that $g_{\hat{C}}(\hat{x}) = 0$ which implies $\hat{C}f(\hat{x}) = 0$ and the contradiction $f(\hat{x}) = 0$.

Let $x, y \in [x] \subseteq [x]^0$, $C, \tilde{C} \in \mathrm{IGA}(f'([x]))$. With J from the multidimensional mean value theorem 1.6.1 we get

$$\begin{aligned} |Cf(x) - \tilde{C}f(y)| &\le |C - \tilde{C}| \cdot |f(x)| + |\tilde{C}| \cdot |f(x) - f(y)| \\ &\le d(\mathrm{IGA}(f'([x]))) \cdot |f([x]^0)| + |\mathrm{IGA}(f'([x]^0))| \cdot |J(x, y)| \cdot |x - y| \\ &\le \{\, \kappa e e^T |f([x]^0)| + |\mathrm{IGA}(f'([x]^0))| \cdot |f'([x]^0)| e \,\} \|d([x])\|_\infty \\ &\le \beta e \,\|d([x])\|_\infty, \end{aligned} \tag{6.2.11}$$

where $\beta \in \mathbb{R}$ is some positive constant which depends only on $[x]^0$. Let $[x] \subseteq [x]^0$ such that

$$\|d([x])\|_\infty \le g_{\min}/(2\beta) \tag{6.2.12}$$

and choose $\tilde{x}, \tilde{y} \in [x]$, $C, \tilde{C} \in \mathrm{IGA}(f'([x]))$. By virtue of (6.2.11) we obtain

$$|Cf(\tilde{x}) - \tilde{C}f(\tilde{y})| \le \frac{g_{\min}}{2}\, e. \tag{6.2.13}$$

If $i_0 \in \{1, \dots, n\}$ satisfies $g_C(\tilde{x}) = |Cf(\tilde{x})|_{i_0}$ (cf. (6.2.10)), then

$$|Cf(\tilde{x})|_{i_0} \ge g_{\min} > 0 \tag{6.2.14}$$

holds, and together with (6.2.13) we get

$$\begin{aligned}(\tilde{C}f(\tilde{y}))_{i_0} &= (Cf(\tilde{x}))_{i_0} + (\tilde{C}f(\tilde{y}) - Cf(\tilde{x}))_{i_0} \\ &\ge |Cf(\tilde{x})|_{i_0} - |Cf(\tilde{x}) - \tilde{C}f(\tilde{y})|_{i_0} \ge g_{\min} - g_{\min}/2 \\ &= g_{\min}/2 > 0 \quad \text{if } (Cf(\tilde{x}))_{i_0} > 0 \end{aligned} \tag{6.2.15}$$

and, similarly,

$$\begin{aligned}(\tilde{C}f(\tilde{y}))_{i_0} &\le -|Cf(\tilde{x})|_{i_0} + |Cf(\tilde{x}) - \tilde{C}f(\tilde{y})|_{i_0} \le -g_{\min} + g_{\min}/2 \\ &= -g_{\min}/2 < 0 \quad \text{if } (Cf(\tilde{x}))_{i_0} < 0. \end{aligned} \tag{6.2.16}$$

Hence

$$\operatorname{sign}((Cf(\tilde{x}))_{i_0}) = \operatorname{sign}((\tilde{C}f(\tilde{y}))_{i_0}) \ne 0 \tag{6.2.17}$$

follows. With $\tilde{x} \in [x] \subseteq [x]^0$, $N([x], \tilde{x}) = [\underline{n}, \overline{n}]$ and $N'([x], \tilde{x}) = [\underline{n}', \overline{n}'] := \tilde{x} - \mathrm{IGA}(f'([x])) \cdot f(\tilde{x})$ we have

$$\underline{n}' \le \underline{n} \le \overline{n} \le \overline{n}'.$$

By Theorem 4.1.5 and (a) there are regular matrices $C, \tilde{C} \in \mathrm{IGA}(f'([x]))$ such that

$$\underline{n}' = \tilde{x} - Cf(\tilde{x}), \quad \overline{n}' = \tilde{x} - \tilde{C}f(\tilde{x}).$$

With $J(x, y) \in \mathbb{R}^{n \times n}$ as above we get

$$\begin{aligned} x - \underline{n} \le x - \underline{n}' &= x - x + Cf(x) \\ &= \overline{x} - \tilde{x} + C\{f(\overline{x}) + J(\tilde{x}, \overline{x})(\tilde{x} - \overline{x})\} \\ &= (I - CJ(\tilde{x}, \overline{x}))(\overline{x} - \tilde{x}) + Cf(\overline{x}). \end{aligned}$$

Since $J(\tilde{x}, \overline{x}) \in f'([x]^0)$ this matrix is regular and

$$\begin{aligned} |(I - CJ(\tilde{x}, \overline{x}))(\overline{x} - \tilde{x})| &= |(J(\tilde{x}, \overline{x})^{-1} - C)J(\tilde{x}, \overline{x})(\overline{x} - \tilde{x})| \\ &\le d(\mathrm{IGA}(f'([x])) \cdot |J(\tilde{x}, \overline{x})| \cdot d([x]) \\ &\le \kappa e e^T |f'([x]^0)|\, e\, \|d([x])\|_\infty^2 \le a e\, \|d([x])\|_\infty^2 \end{aligned}$$

follows with some appropriate positive constant $a \in \mathbb{R}$ which depends only on $[x]^0$. This implies

$$\overline{x} - \underline{n} \le a e\, \|d([x])\|_\infty^2 + Cf(\overline{x}). \tag{6.2.18}$$

Similarly,

$$\overline{n} - \underline{x} \le ae \,\|d([x])\|_\infty^2 - \check{C}f(\underline{x}). \tag{6.2.19}$$

If $d([x])$ satisfies (6.2.12) and $\|d([x])\|_\infty \le \frac{1}{2}\sqrt{\frac{g_{\min}}{a}}$, then (6.2.13)–(6.2.19) with $\tilde{x} = \overline{x}$, $\tilde{y} = \underline{x}$ imply

$$\overline{n}_{i_0} - \underline{x}_{i_0} \le a\,\|d([x])\|_\infty^2 - g_{\min}/2 < 0$$

if $(Cf(\overline{x}))_{i_0} > 0$, and

$$\overline{x}_{i_0} - \underline{n}_{i_0} \le a\,\|d([x])\|_\infty^2 - g_{\min} \le a\,\|d([x])\|_\infty^2 - g_{\min}/2 < 0$$

if $(Cf(\overline{x}))_{i_0} < 0$. This proves (6.2.7) with $b = g_{\min}/2$. □

As in Section 6.1, the inequality (6.2.7) means that $N([x]^k, \tilde{x}^k) \cap [x]^k = \emptyset$ if f does not contain a zero in $[x]^0$ and if $\|d([x]^k)\|_\infty$ is small enough.

If $\tilde{x}^k = \check{x}^k$, then by virtue of Theorem 6.2.4 (b) at least one component of $[x]^k$ is more than halved in $[x]^{k+1}$. So one can hope that $\|d([x]^k)\|_\infty$ also becomes small in a few steps and that the iteration stops soon with empty intersection if $[x]^0$ does not contain a zero of f. As in Section 6.1 and Alefeld [17], one can again speak of the quadratic divergence of the Newton method if f has no zero in $[x]^0$.

Corollary 6.2.5. *Theorem 6.2.4 (c) holds also if $\rho(B) < 1$ is replaced by $\rho(B) < 2$, $\tilde{x}^k = \check{x}^k$, $k = 0, 1, \ldots$.*

Proof. Using the multidimensional mean value theorem and the notation there we obtain

$$\begin{aligned} d([x]^{k+1}) &\le d(N([x]^k, \check{x}^k)) \le d(\mathrm{IGA}(f'([x]^k))) \cdot |f(\check{x}^k)| \\ &\le d(\mathrm{IGA}(f'([x]^0))) \cdot |f(x^*) + J(\check{x}^k, x^*)(\check{x}^k - x^*)| \\ &\le d(\mathrm{IGA}(f'([x]^0))) \cdot |J(\check{x}^k, x^*)| \frac{1}{2} d([x]^k) \le \frac{1}{2} B d([x]^k) \\ &\le \cdots \le \left(\frac{1}{2}B\right)^{k+1} d([x]^0). \end{aligned}$$

This results in $\lim_{k\to\infty} d([x]^k) = 0$ and implies $\lim_{k\to\infty} [x]^k = x^*$. □

Unfortunately, $\mathrm{IGA}(f'([x]^0))$ does not always exist even if $f'([x]^0)$ contains only regular matrices. In addition, the assumptions of Theorem 6.2.4 cannot easily be checked since $\mathrm{IGA}(f'([x]^0))$ is rarely available. If, however, $[x]^0$ has small width, then $\mathrm{IGA}(f'([x]^0)) \cdot f'([x]^0)$ is approximately the identity matrix which implies $\rho(A) \ll 1$ and $\rho(B) \ll 1$, i.e., the assumptions of Theorem 6.2.4 can be expected to hold.

As a modification of the Newton method we can iterate according to

$$[x]^{k+1} = N([x]^0, \tilde{x}^k) \cap [x]^k, \quad k = 0, 1, \ldots \quad (\tilde{x}^k \in [x]^k) \tag{6.2.20}$$

which fixes the matrix $f'([x]^0)$ and is called the simplified Newton method. We leave it to the reader in Exercise 6.2.3 to show that Theorem 6.2.4 (with the exception of the inequality (6.2.7)) holds also for the simplified Newton method.

A possibility to avoid the application of the interval Gaussian algorithm and the computation of $\mathrm{IGA}(f'([x]^0))$ consists of iterating by means of a superset $[V]$ of the set $\{J^{-1}(x,y) \mid x,y \in [x]^0\}$ via

$$[x]^{k+1} = \{\tilde{x}^k - [V]\cdot f(\tilde{x}^k)\} \cap [x]^k, \quad k = 0,1,\ldots \quad (\tilde{x}^k \in [x]^k). \tag{6.2.21}$$

If $[V]$ is regular, the Theorems 6.2.3 and 6.2.4 (with the exception of the inequality (6.2.7)) hold similarly. For details on this Newton-like method we refer to § 19 in Alefeld, Herzberger [26] and to Exercise 6.2.3.

Exercises

Ex. 6.2.1. Prove the statements in Example 6.2.2.

Ex. 6.2.2. Show that for the matrices A, B in Theorem 6.2.4 the implication $\rho(A) < 1 \Rightarrow \rho(B) < 2$ holds.

Ex. 6.2.3. Prove as many results in this section on the Newton method as possible also for the simplified Newton method and the Newton-like method (6.2.21).

6.3 Krawczyk method

The Newton method in Section 6.2 depends heavily on the feasibility of the interval Gaussian algorithm for $f'([x]^0)$. As we learnt in Section 5.4, this feasibility is guaranteed for particular classes of matrices but may fail even if $f'([x]^0)$ is regular and pivoting is allowed. Therefore, we look for alternatives in order to verify a zero x^* of $f\colon D \subseteq \mathbb{R}^n \to \mathbb{R}^n$, $f \in C^1(D)$. One such alternative is given by the so-called Krawczyk method which we will derive now. To this end, choose $C \in \mathbb{R}^{n\times n}$, $[x] \subseteq D$, $\tilde{x} \in [x]$ and assume for the moment that $[x]$ contains a zero x^* of f. With the multidimensional mean value theorem 1.6.1 applied to $-Cf(x)$ and the notation there we obtain

$$0 = -Cf(x^*) = -Cf(\tilde{x}) - CJ(x^*,\tilde{x})(x^* - \tilde{x}),$$

whence

$$\begin{aligned} x^* &= \tilde{x} - Cf(\tilde{x}) + (I - CJ(x^*,\tilde{x}))(x^* - \tilde{x}) \\ &\in \tilde{x} - Cf(\tilde{x}) + (I - Cf'([x]))([x] - \tilde{x}) =: K([x],\tilde{x},C) \end{aligned} \tag{6.3.1}$$

follows. The function K is sometimes called the Krawczyk operator. This relation is the starting point of the Krawczyk method for nonlinear systems of equations

$$\begin{aligned}[x]^{k+1} &:= \{\tilde{x}^k - Cf(\tilde{x}^k) + (I - Cf'([x]^k))([x]^k - \tilde{x}^k)\} \cap [x]^k \\ &= K([x]^k, \tilde{x}^k, C) \cap [x]^k, \quad k = 0, 1, \ldots \end{aligned} \tag{6.3.2}$$

with $\tilde{x}^k \in [x]^k$. If the intersection in (6.3.2) is empty, the iteration is stopped. If it does not stop with empty intersection, we call it feasible for $[x]^0$. For C, often an approximation of $(\mathrm{mid}(f'([x]^0)))^{-1}$ is used. In this way the last expression of K can be considered to be quadratically small while the middle term is a correction of $\tilde{x}^k$ that decreases linearly.

Results can be proved for the method which are similar to those in Section 6.2 for the Newton method.

Theorem 6.3.1. *Let D be an open convex subset of $\mathbb{R}^n$, $f : D \subseteq \mathbb{R}^n \to \mathbb{R}^n$, $f \in C^1(D)$, $[x]^0 \subseteq D$. Let the entries f'_{ij} of f' be represented as expressions from $\mathbb{E}$ as in Definition 4.1.2 such that the interval arithmetic evaluation $f'([x])$ exists for $[x] \in \mathbb{I}(D)$. Then for the Krawczyk method (6.3.2) the following properties hold.*

(a) *If $x^* \in [x]^0$ is a zero of f and if C is regular, then the Krawczyk method is feasible for $[x]^0$ and satisfies*

$$x^* \in [x]^* := \lim_{k\to\infty} [x]^k \subseteq [x]^k \subseteq [x]^{k-1} \subseteq \cdots \subseteq [x]^0 \tag{6.3.3}$$

for $k = 0, 1, \ldots$. In particular, each iterate $[x]^k$ contains all zeros $x^* \in [x]^0$ of f and the same holds for the limit $[x]^*$.

(b) *If $K([x]^0, \tilde{x}^0, C) \cap [x]^0 = \emptyset$, then f has no zero in $[x]^0$.*

(c) *If the Krawczyk test*

$$K([x]^0, \tilde{x}^0, C) \subseteq [x]^0 \tag{6.3.4}$$

is valid and if C is regular, then there is at least one zero $x^* \in [x]^0$ of f and (a) *holds.*

(d) *If the strong Krawczyk test*

$$(K([x]^0, \tilde{x}^0, C))_i \subset [x]_i^0, \quad i = 1, \ldots, n, \tag{6.3.5}$$

is valid, then the matrices $f'([x]^0)$ and C are regular, f has exactly one zero x^* in $[x]^0$, and (a) *holds. If, in addition,*

$$\tilde{x}^k = \check{x}^k \tag{6.3.6}$$

is chosen in (6.3.2)*, then the iterates contract to $[x]^* \equiv x^*$; this holds also if one replaces* (6.3.6) *by*

$$\rho(|I - Cf'([x]^0)|) < \frac{1}{2}. \tag{6.3.7}$$

Proof. (a) Choose $x \in [x]^0$ and define g by $g(x) = x - Cf(x)$. Then we get

$$\begin{aligned} g(x) &= x - \tilde{x}^0 + \tilde{x}^0 - C\{f(\tilde{x}^0) + J(x, \tilde{x}^0)(x - \tilde{x}^0)\} \\ &= \tilde{x}^0 - Cf(\tilde{x}^0) + (I - CJ(x, \tilde{x}^0))(x - \tilde{x}^0) \\ &\in K([x]^0, \tilde{x}^0, C). \end{aligned} \tag{6.3.8}$$

If $x^* \in [x]^0$ is a zero of f, then $x^* = g(x^*) \in K([x]^0, \tilde{x}^0, C)$, hence $x^* \in [x]^1$, and (6.3.3) follows by induction and the intersection in (6.3.2).

(b) If f has a zero $x^* \in [x]^0$, then (6.3.1) implies $x^* \in K([x]^0, \tilde{x}^0, C)$, which yields the contradiction $x^* \in K([x]^0, \tilde{x}^0, C) \cap [x]^0 = \emptyset$.

(c) From (6.3.8) and (6.3.4) we get $g(x) \in K([x]^0, \tilde{x}^0, C) \subseteq [x]^0$ for arbitrary $x \in [x]^0$. Therefore, Brouwer's fixed point theorem 1.5.8 guarantees a fixed point $x^* \in [x]^0$ of g and thus a zero of f, which implies (a).

(d) The assumption implies

$$\{-Cf(\tilde{x}^0) + (I - Cf'([x]^0))([x]^0 - \tilde{x}^0)\}_i \subset ([x]^0 - \tilde{x}^0)_i, \tag{6.3.9}$$

hence C and $f'([x]^0)$ are regular by virtue of Theorem 3.1.6 (b), and (c) guarantees the existence of a zero $x^* \in [x]^0$ as well as (a). In order to prove uniqueness of this zero, assume that there is another zero $y^* \in [x]^0$ of f. Then the multidimensional mean value theorem yields $f(x^*) = 0 = f(y^*) + J(x^*, y^*)(x^* - y^*)$, and the regularity of $J(x^*, y^*) \in f'([x]^0)$ implies $x^* = y^*$.

If $\tilde{x}^k = \check{x}^k$ holds, then $\tilde{x}^* := \lim_{k\to\infty} \tilde{x}^k = \lim_{k\to\infty}(\underline{x}^k + \overline{x}^k)/2 = (\underline{x}^* + \overline{x}^*)/2 = \check{x}^*$ follows from (a), whence

$$\begin{aligned} d([x]^*) &\le d(\check{x}^* - Cf(\check{x}^*) + (I - Cf'([x]^*))([x]^* - \check{x}^*)) \\ &= d(|I - Cf'([x]^*)| \cdot \operatorname{rad}([x]^*) \cdot [-1, 1]) \\ &= |I - Cf'([x]^*)|\, d([x]^*) \\ &\le |I - Cf'([x]^0)|\, d([x]^*) \\ &\le \cdots \le |I - Cf'([x]^0)|^k d([x]^*). \end{aligned}$$

The limit $k \to \infty$ implies $d([x]^*) = 0$ since $\rho(\,|I - Cf'([x]^0)|\,) < 1$ by virtue of (6.3.9) and Theorem 3.1.6 (a). This proves $[x]^* = x^*$ in the case $\tilde{x}^k = \check{x}^k$.

In the case $\rho(|I - Cf'([x]^0)|) < 1/2$ we choose a convergent subsequence $(\tilde{x}^{k_j})$ with limit $\tilde{x}^*$. Then $\tilde{x}^* \in [x]^*$ and $[x]^* - \tilde{x}^* \subseteq d([x]^*)[-1, 1]$ holds, and similar to above we get

$$\begin{aligned} d([x]^*) &\le |I - Cf'([x]^*)|\, d([x]^*) \cdot 2 \\ &\le |I - Cf'([x]^0)| \cdot 2 \cdot d([x]^*) \\ &\le \cdots \le |I - Cf'([x]^0)|^k \cdot 2^k \cdot d([x]^*), \end{aligned}$$

again with the limit 0 if k tends to infinity. The assertion follows now as in the previous case. □

If the Krawczyk test (6.3.4) holds but C is not regular, then nothing can be guaranteed, as can be seen by the example $[x]^0 = -[x]^0 = [-r, r] \in \mathbb{IR}^n$, $r \geq 0$, $\tilde{x}^0 = \check{x}^0 = 0$, $C = O$, f arbitrary. If this test holds and if C is regular, then uniqueness of a zero cannot be deduced. This is demonstrated by the example $[x]^0 = -[x]^0 = [-r, r] \in \mathbb{IR}^n$, $r > 0$, $\tilde{x}^0 = \check{x}^0 = 0$, $C = I$, $f = 0$.

Notice that the Krawczyk tests (6.3.4) and (6.3.5) are frequently called Moore tests in the literature. We will use this latter terminology, too, but in a more general setting.

Our next result shows that the Krawczyk operator $K([x], \tilde{x}, C)$ behaves optimally in some sense if $\tilde{x} = \check{x}$ and $C = \hat{C} := (\mathrm{mid}(f'([x])))^{-1}$.

Theorem 6.3.2. *With the notation and the assumptions of Theorem* 6.3.1 *the following properties hold for arbitrary regular* $C \in \mathbb{R}^{n\times n}$.

(a) $\mathrm{rad}(K([x], \check{x}, C)) \leq \mathrm{rad}(K([x], \tilde{x}, C))$.

(b) *If* $K([x], \tilde{x}, C) \subseteq [x]$ *and* $\mathrm{mid}(f'([x]))$ *is regular, then* $K([x], \check{x}, \hat{C}) \subseteq K([x], \tilde{x}, C)$, *where* $\hat{C} = (\mathrm{mid}(f'([x])))^{-1}$.

Proof. (a) follows from

$$\begin{aligned}\mathrm{rad}(K([x], \check{x}, C)) &= |I - Cf'([x])|\,\mathrm{rad}([x] - \check{x})\\ &= |I - Cf'([x])|\,\mathrm{rad}([x] - \tilde{x})\\ &\leq \mathrm{rad}\big((I - Cf'([x]))([x] - \tilde{x})\big) = \mathrm{rad}\big(K([x], \tilde{x}, C)\big).\end{aligned}$$

(b) Let $[y] = K([x], \check{x}, C) \subseteq [x]$ hold. Then

$$\check{y} = \check{x} - Cf(\check{x}) \quad \text{and} \quad \mathrm{rad}([y]) = R\,\mathrm{rad}([x]) \tag{6.3.10}$$

with $R = |I - Cf'([x])|$. Taking into account Theorem 2.4.8 (a), the Krawczyk test and (6.3.10) imply

$$|Cf(\check{x})| = |\check{x} - \check{y}| \leq \mathrm{rad}([x]) - \mathrm{rad}([y]) = (I - R)\,\mathrm{rad}([x]). \tag{6.3.11}$$

Define $\hat{R}, \Delta$ by

$$\begin{aligned}\hat{R} &= |I - \hat{C}f'([x])| = |\hat{C}(\mathrm{mid}(f'([x])) - f'([x]))|\\ &= |\hat{C}|\,\mathrm{rad}(f'([x])) \leq |\hat{C}C^{-1}| \cdot |C|\,\mathrm{rad}(f'([x]))\end{aligned}$$

and

$$\Delta = |I - \hat{C}C^{-1}| = |\hat{C}C^{-1}(C\hat{C}^{-1} - I)| \leq |\hat{C}C^{-1}| \cdot |I - C\hat{C}^{-1}|.$$

By means of Theorem 2.4.4 (a) we get

$$\begin{aligned}\Delta + \hat{R} &\leq |\hat{C}C^{-1}|\,(|I - C\hat{C}^{-1}| + |C|\,\mathrm{rad}(f'([x])))\\ &= |\hat{C}C^{-1}|R = |I - (I - \hat{C}C^{-1})|R \leq (I + \Delta)R,\end{aligned}$$

hence $\Delta(I - R) \le R - \hat{R}$. With $[\hat{y}] = K([x], \check{x}, \hat{C})$ and (6.3.11) the relation

$$\begin{aligned} |\check{y} - \mathrm{mid}([\hat{y}])| &= |(\hat{C} - C)f(\check{x})| = |(\hat{C}C^{-1} - I)Cf(\check{x})| \\ &\le \Delta|Cf(\check{x})| \le \Delta(I - R)\,\mathrm{rad}([x]) \\ &\le (R - \hat{R})\,\mathrm{rad}([x]) = \mathrm{rad}([y]) - \mathrm{rad}([\hat{y}]) \end{aligned}$$

follows. This implies $[\hat{y}] \subseteq [y]$ and therefore (b). □

For this optimal choice of $\check{x}$ and C we present now conditions which guarantee that the Krawczyk test is positive. They were stated in Rall [279] and remind of the Newton–Kantorovich Theorem 1.5.10.

Theorem 6.3.3. *With the notation and the assumptions of Theorem 6.3.1 let $f'([x])$ be Lipschitz continuous on $[x]^0 \in \mathbb{IR}^n$ with Lipschitz constant $\gamma > 0$, i.e.,*

$$\|q(f'([x]), f'([y]))\|_\infty \le \gamma \|q([x], [y])\|_\infty \quad \textit{for all } [x], [y] \subseteq [x]^0.$$

Define $[x]_\rho = \check{x}^0 - \rho[-e, e]$, $C = (f'(\check{x}^0))^{-1}$ (which we assume to exist), and choose $\beta > 0$, $\eta \ge 0$ such that $\eta \ge \|(f'(\check{x}^0))^{-1} f(\check{x}^0)\|_\infty$, $\beta \ge \|C\|_\infty$. If

$$\alpha = \beta\gamma\eta \le 1/4, \tag{6.3.12}$$

then the Krawczyk test

$$K([x]_\rho, \check{x}^0, C) \subseteq [x]_\rho$$

is successful for each ρ such that

$$\rho \le \mathrm{rad}([x]^0) \quad \textit{and} \quad \rho^- := \frac{1 - \sqrt{1 - 4\alpha}}{2\beta\gamma} \le \rho \le \frac{1 + \sqrt{1 - 4\alpha}}{2\beta\gamma} =: \rho^+. \tag{6.3.13}$$

Proof. The assumptions (6.3.12) and (6.3.13) imply

$$\eta + \beta\gamma\rho^2 \le \rho \Leftrightarrow \rho - \beta\gamma\rho^2 - \eta \ge 0 \Leftrightarrow \rho^- \le \rho \le \rho^+. \tag{6.3.14}$$

From

$$\begin{aligned} |K([x]_\rho, \check{x}^0, C) - \check{x}^0| &\le |Cf(\check{x}^0)| + |I - Cf'([x]_\rho)|\rho e \\ &= |(f'(\check{x}^0))^{-1} f(\check{x}^0)| + |C| \cdot |f'(\check{x}^0) - f'([x]_\rho)|\rho e \end{aligned}$$

and (6.3.14) we get

$$\|\,|K([x]_\rho, \check{x}^0, C) - \check{x}^0|\,\|_\infty \le \eta + \beta\gamma\rho^2 \le \rho,$$

if $\rho \le \mathrm{rad}([x]^0)$. Hence $K([x]_\rho, \check{x}^0, C) \subseteq [x]_\rho$ follows. □

Unfortunately, the bound $1/4$ in Theorem 6.3.3 cannot be enlarged in general as the following example in Rall [279] shows.

Example 6.3.4. Let $f(x) = x^2 - \varepsilon$, $0 \le \varepsilon \le 1$, $\check{x}^0 = 1$, $C = (f'(\check{x}^0))^{-1} = 1/2$, $\beta = |C| = 1/2$, $\eta = (f'(\check{x}^0))^{-1} f(\check{x}^0) = (1-\varepsilon)/2 \ge 0$, $[x]^0 = \check{x}^0 + 2\eta[-1, 1] = [\varepsilon, 2-\varepsilon]$. Since $q(f'([x]), f'([y])) = 2q([x], [y])$ we can choose $\gamma = 2$. Then $\alpha = \beta\gamma\eta = (1-\varepsilon)/2 \le 1/2$, and the theorem of Newton–Kantorovich guarantees existence and uniqueness of the zero $x^* = \sqrt{\varepsilon} \in [x]^0$. Since

$$K([x]^0, \check{x}^0, C) = \left[-\frac{1}{2} + \frac{5}{2}\varepsilon - \varepsilon^2, \frac{3}{2} - \frac{3}{2}\varepsilon + \varepsilon^2\right] \subseteq [x]^0 \Leftrightarrow \frac{1}{2} \le \varepsilon \le 1,$$

i.e., $0 \le \alpha \le 1/4$, the bound $1/4$ cannot be increased.

For a comparison with an interval analysis based version of the Newton–Kantorovich theorem in Section 6.5, we notice here already that this latter theorem guarantees the existence of the solution $x^* = \sqrt{\varepsilon}$ in $[x]^0$ for the *full* range $[0, 1]$ of ε.

Our next result indicates some sort of quadratic convergence of the Krawczyk method.

Theorem 6.3.5. *Let the assumptions of Theorem 6.3.1 hold. If f' satisfies*

$$\|d(f'([x]))\|_\infty \le \hat{\gamma}\|d([x])\|_\infty \tag{6.3.15}$$

for all $[x] \subseteq D$ and an appropriate positive number $\hat{\gamma}$, then the inequality

$$\|d(K([x], \tilde{x}, C))\|_\infty \le \hat{\alpha}\|d([x])\|_\infty^2, \quad \text{with } \hat{\alpha} = \hat{\gamma}\|C\|_\infty \tag{6.3.16}$$

follows provided that $\tilde{x} \in [x] \subseteq D$, C regular, $C^{-1} \in f'([x])$.

Proof. By means of Theorem 3.1.5 (i) the assertion follows from

$$\begin{aligned} d(K([x], \tilde{x}, C)) &= d(\tilde{x} - Cf(\tilde{x}) + (I - Cf'([x]))([x] - \tilde{x})) \\ &= d((I - Cf'([x]))([x] - \tilde{x})) \\ &\le d(I - Cf'([x])) \cdot d([x] - \tilde{x}) \\ &= d(C(C^{-1} - f'([x])))\, d([x]) \\ &= |C| d(f'([x]))\, d([x]). \end{aligned}$$

□

By virtue of the Theorems 4.1.16 and 4.1.18 the assumption (6.3.15) is nearly always fulfilled in contrast to the assumption $C^{-1} \in f'([x]^k)$ for all $k = 0, 1, \ldots$ during the iteration (6.3.2).

For $f(x) = Ax - b$, $A \in \mathbb{R}^{n\times n}$, $b \in \mathbb{R}^n$ one gets $f'(x) = A$. Substituting $\tilde{x}^k$, $f'([x]^k)$ in (6.3.2) by $\tilde{x}$, A and dropping the intersection results in the Krawczyk residual iteration (5.5.103) for $[A] \equiv A$, $[b] \equiv b$. This justifies the terminology ‘Krawczyk iteration’ in Section 5.5. Notice that there $\tilde{x} \in [x]^k$ is no longer required.

The number of multiplications for computing K in (6.3.1) is apparently proportional to n^3 per iteration. Therefore, one can look for a reduction of this amount of work. This yields the modification of K due to Alefeld which starts with the representation

$$\begin{aligned}x^* &= \tilde{x} - C\{f(\tilde{x}) - (C^{-1} - J(x^*, \tilde{x}))(x^* - \tilde{x})\}\\ &= \tilde{x} - \tilde{C}^{-1}\{f(\tilde{x}) - (\tilde{C} - J(x^*, \tilde{x}))(x^* - \tilde{x})\}\\ &\in \tilde{x} - \tilde{C}^{-1}\{f(\tilde{x}) - (\tilde{C} - f'([x]))([x] - \tilde{x})\}\\ &\subseteq \tilde{x} - \mathrm{IGA}(\tilde{C}, f(\tilde{x}) - (\tilde{C} - f'([x]))([x] - \tilde{x}))\\ &=: A([x], \tilde{x}, \tilde{C})\end{aligned}$$

with $\tilde{C} = C^{-1}$ and the Alefeld operator A. Here, we tacitly assume that C is regular and $\tilde{x}, x^* \in [x]$, $f(x^*) = 0$. This suggests the Alefeld method

$$\begin{aligned}[x]^{k+1} &= \{\tilde{x}^k - \mathrm{IGA}(\tilde{C}, f(\tilde{x}^k) - (\tilde{C} - f'([x]^k))([x]^k - \tilde{x}^k))\} \cap [x]^k\\ &= A([x]^k, \tilde{x}^k, \tilde{C}) \cap [x]^k, \quad k = 0, 1, \ldots\end{aligned} \tag{6.3.17}$$

with $\tilde{x}^k \in [x]^k$. As with the Krawczyk method the iteration is stopped if the intersection is empty. Otherwise it is called feasible for $[x]^0$.

For (6.3.17), Theorem 6.3.1 holds similarly with the exception of (6.3.7) and the regularity of $\tilde{C}$, which must be assumed in advance.

Theorem 6.3.6. *Let the assumptions of Theorem 6.3.1 hold and let $\tilde{C} \in \mathbb{R}^{n\times n}$ be regular.*

(a) *If $x^* \in [x]^0$ is a zero of f then the Alefeld method (6.3.17) is feasible for $[x]^0$ and satisfies*

$$x^* \in [x]^* := \lim_{k\to\infty} [x]^k \subseteq [x]^k \subseteq [x]^{k-1} \subseteq \cdots \subseteq [x]^0 \tag{6.3.18}$$

for $k = 0, 1, \ldots$. In particular, each iterate $[x]^k$ contains all *zeros $x^* \in [x]^0$ of f and the same holds for the limit $[x]^*$.*

(b) *If $A([x]^0, \tilde{x}^0, \tilde{C}) \cap [x]^0 = \emptyset$, then there is no zero $x^* \in [x]^0$ of f.*

(c) *If the Alefeld test*

$$A([x]^0, \tilde{x}^0, \tilde{C}) \subseteq [x]^0 \tag{6.3.19}$$

is valid, then f has at least one zero $x^* \in [x]^0$ and (a) *holds.*

(d) *If the strong Alefeld test*

$$(A([x]^0, \tilde{x}^0, \tilde{C}))_i \subset [x]_i^0, \quad i = 1, \ldots, n \tag{6.3.20}$$

is valid, then $f'([x]^0)$ is regular, f has exactly one zero x^* in $[x]^0$, and (a) *holds. Furthermore, if*

$$\tilde{x}^k = \check{x}^k$$

is chosen then the iterates contract to $[x]^* \equiv x^*$.

Proof. (a), (b), and the existence and uniqueness of x^* in (c) follow similarly to Theorem 6.3.1 with $g(x) = x - \tilde{C}^{-1}f(x)$. The regularity of $f'([x]^0)$ in (d) results from Theorem 3.1.6 (b) and the subset property

$$\begin{aligned} &-\tilde{C}^{-1}\cdot f(\tilde{x}^0)+(I-\tilde{C}^{-1}f'([x]^0))([x]^0-\tilde{x}^0)\\ &\qquad = -\tilde{C}^{-1}f(\tilde{x}^0)+\{\tilde{C}^{-1}(\tilde{C}-f'([x]^0))\}([x]^0-\tilde{x}^0)\\ &\qquad \subseteq -\tilde{C}^{-1}f(\tilde{x}^0)+\tilde{C}^{-1}\{(\tilde{C}-f'([x]^0))([x]^0-\tilde{x}^0)\}\\ &\qquad = -\tilde{C}^{-1}\{f(\tilde{x}^0)-(\tilde{C}-f'([x]^0))([x]^0-\tilde{x}^0)\}+\tilde{x}^0-\tilde{x}^0\\ &\qquad \subseteq A([x]^0,\tilde{x}^0,\tilde{C})-\tilde{x}^0 \subset [x]^0-\tilde{x}^0. \end{aligned}$$

Here, '$\subset$' holds componentwise.

In order to prove the convergence to x^* in (d) and therefore uniqueness of x^*, we note that the assumption (6.3.20) guarantees the existence of some real number $\alpha \in [0,1)$ such that $d(A([x]^0,\check{x}^0,\tilde{C})) \le \alpha\, d([x]^0)$ is valid. This implies $d([x]^1) \le \alpha\, d([x]^0)$. Therefore, the 'right-hand side' of the interval Gaussian algorithm satisfies

$$\begin{aligned} d(f(\check{x}^1)-(\tilde{C}-f'([x]^1))([x]^1-\check{x}^1)) &\le d((\tilde{C}-f'([x]^0))[-1,1]\cdot \mathrm{rad}([x]^1))\\ &\le \alpha d(f(\check{x}^0)-(\tilde{C}-f'([x]^0))([x]^0-\check{x}^0)). \end{aligned}$$

With Theorem 5.4.1 (g) we finally get

$$d([x]^2) \le d(A([x]^1,\check{x}^1,\tilde{C})) \le \alpha d(A([x]^0,\check{x}^0,\tilde{C})) \le \alpha^2 d([x]^0),$$

and an induction yields $d([x]^k) \le \alpha^k d([x]^0)$. Thus the limit $[x]^*$ satisfies $d([x]^*) = 0$, and (6.3.18) implies $[x]^* \equiv x^*$. □

Another modification of the Krawczyk method originates from Hansen and Sengupta [132] and is considered by Moore and Qi in [239]. It is a Gauss–Seidel-like method which is represented as

$$[x]^{k+1} = M'([x]^k,\tilde{x}^k,C), \quad k=0,1,\ldots, \tag{6.3.21}$$

where $M'([x],\tilde{x},C)$ is defined for $\tilde{x}\in[x]$ and $i=1,2,\ldots,n$ by

$$\begin{aligned} M_i([x],\tilde{x},C) &= \tilde{x}_i-(Cf(\tilde{x}))_i+\sum_{j=1}^{i-1}(I-Cf'([x]))_{ij}(M'_j([x],\tilde{x},C)-\tilde{x}_j)\\ &\quad +\sum_{j=i}^{n}(I-Cf'([x]))_{ij}([x]_j-\tilde{x}_j),\\ M'_i([x],\tilde{x},C) &= M_i([x],\tilde{x},C)\cap[x]_i\,. \end{aligned} \tag{6.3.22}$$

If the intersection in (6.3.22) is empty for some i, then M' is not defined. If it is defined, we will call (6.3.21) a Gauss–Seidel-like Krawczyk method and M the Moore–Qi operator. Although the interval function M' seems to be more important in view of the iteration (6.3.21), the Moore–Qi operator M will play a crucial role in the subsequent Theorem 6.3.7.

Theorem 6.3.7. *Let the assumptions of Theorem 6.3.1 hold. In addition, let C be regular.*

(a) *If $x^* \in [x]^0$ is a zero of f, then the Gauss–Seidel-like Krawczyk method* (6.3.21) *is feasible for $[x]^0$ and satisfies*

$$x^* \in [x]^* := \lim_{k\to\infty} [x]^k \subseteq [x]^k \subseteq [x]^{k-1} \subseteq \cdots \subseteq [x]^0 \tag{6.3.23}$$

for $k = 0, 1, \ldots$. In particular, each iterate $[x]^k$ contains all *zeros $x^* \in [x]^0$ of f, and the same holds for the limit $[x]^*$.*

(b) *If $M'([x]^0, \tilde{x}^0, C)$ is not defined, then f has no zero in $[x]^0$.*

(c) *If $M'([x]^0, \tilde{x}^0, C)$ is defined and if the Moore–Qi test*

$$M([x]^0, \tilde{x}^0, C) \subseteq [x]^0 \tag{6.3.24}$$

is valid, then there is at least one zero $x^* \in [x]^0$ of f, and (a) *holds.*

(d) *Let $M'([x]^0, \tilde{x}^0, C)$ be defined. If the strong Moore–Qi test*

$$(M([x]^0, \tilde{x}^0, C))_i \subset [x]_i^0, \quad i = 1, \ldots, n, \tag{6.3.25}$$

is valid, then $Cf'([x]^0)$ is an H-matrix. In particular, $f'([x]^0)$ and C are regular, even if this is not assumed for C from the beginning. In addition, f has exactly one zero x^* in $[x]^0$, and (a) *holds.*

(e) *If the assumptions of* (d) *hold and if*

$$\rho(|I - Cf'([x]^0)|) < 1 \quad \text{and} \quad \tilde{x}^k = \check{x}^k, \quad k = 0, 1, \ldots \tag{6.3.26}$$

or

$$\rho(|I - Cf'([x]^0)|) < \frac{1}{2}, \tag{6.3.27}$$

then f has exactly one zero x^* in $[x]^0$, and the sequence in (a) *contracts to $[x]^* \equiv x^*$.*

Proof. (a) Let $x^* \in [x]^0$ be a zero of f and define g by $g(x) = x - Cf(x)$. Then by virtue of (6.3.8) we have $x^* = g(x^*) \in K([x]^0, \tilde{x}^0, C)$, hence $x_1^* \in K_1([x]^0, \tilde{x}^0, C) = M_1([x]^0, \tilde{x}^0, C)$ and $x_1^* \in M_1'([x]^0, \tilde{x}^0, C) = [x]_1^1$. Assume that $x_j^* \in M_j'([x]^0, \tilde{x}^0, C) = [x]_j^1$ for $j = 1, \ldots, i-1$ $(\le n-1)$. Then by this induction hypothesis and by (6.3.8) we get

$$\begin{aligned} x_i^* &= g_i(x^*) = \{\tilde{x}^0 - Cf(\tilde{x}^0) + (I - CJ(x^*, \tilde{x}^0))(x^* - \tilde{x}^0)\}_i \\ &\in \tilde{x}_i^0 - (Cf(\tilde{x}^0))_i + \sum_{j=1}^{n} (I - Cf'([x]^0))_{ij}(x_j^* - \tilde{x}_j^0) \\ &\subseteq \tilde{x}_i^0 - (Cf(\tilde{x}^0))_i + \sum_{j=1}^{i-1} (I - Cf'([x]^0))_{ij}(M_j'([x]^0, \tilde{x}^0, C) - \tilde{x}_j^0) \\ &\quad + \sum_{j=i}^{n} (I - Cf'([x]^0))_{ij}([x]_j^0 - \tilde{x}_j^0) = M_i([x]^0, \tilde{x}^0, C) \end{aligned}$$

which yields $x_i^* \in M_i'([x]^0, \tilde{x}^0, C) = [x]_i^1$. This proves $x^* \in [x]^1$, and (6.3.23) follows by induction and the intersection in (6.3.22).

(b) If f has a zero $x^* \in [x]^0$, then (a) implies $x^* \in M'([x]^0, \tilde{x}^0, C)$, contradicting the fact that $M'([x]^0, \tilde{x}^0, C)$ is not defined.

(c) The assumptions of (c) imply $M([x]^0, \tilde{x}^0, C) = M'([x]^0, \tilde{x}^0, C) \neq \emptyset$. Choose $x \in M([x]^0, \tilde{x}^0, C)$ arbitrarily. By virtue of (6.3.8) and (6.3.24) we obtain similarly to in (a) with g from there

$$\begin{aligned} g_i(x) &\in \tilde{x}_i^0 - (Cf(\tilde{x}^0))_i + \sum_{j=1}^{n} (I - Cf'([x]^0))_{ij}(M_j([x]^0, \tilde{x}^0, C) - \tilde{x}_j^0) \\ &\subseteq \tilde{x}_i^0 - (Cf(\tilde{x}^0))_i + \sum_{j=1}^{i-1} (I - Cf'([x]^0))_{ij}(M_j'([x]^0, \tilde{x}^0, C) - \tilde{x}_j^0) \\ &\quad + \sum_{j=i}^{n} (I - Cf'([x]^0))_{ij}([x]_j^0 - \tilde{x}_j^0) = M_i([x]^0, \tilde{x}^0, C). \end{aligned}$$

Therefore, the continuous function g maps $M([x]^0, \tilde{x}^0, C)$ into itself, hence Brouwer's fixed point theorem guarantees a fixed point $x^* \in M([x]^0, \tilde{x}^0, C) \subseteq [x]^0$ of g and thus a zero of f.

(d) Since (6.3.25) implies (6.3.24), the existence of a zero x^* as well as (a) are proved. We will show that $Cf'([x]^0)$ is an H-matrix. Then each matrix $C\tilde{f}'$ with $\tilde{f}' \in f'([x]^0)$ is an H-matrix and therefore regular. This implies the regularity of C and $\tilde{f}'$, and finally of $f'([x]^0)$. The uniqueness of x^* follows then as in the proof of Theorem 6.3.1.

In order to show that $Cf'([x]^0)$ is an H-matrix we choose $\tilde{P} \in Cf'([x]^0)$ such that $\langle \tilde{P} \rangle = \langle Cf'([x]^0) \rangle$ and consider the auxiliary sequence $([y]^k)$ which is defined by

$$\begin{aligned} &[y]^0 = [x]^0, \quad [y]^1 = [x]^1, \\ &[y]_i^{k+1} = \tilde{x}_i^0 - (Cf(\tilde{x}^0))_i + \sum_{j=1}^{i-1} (I - \tilde{P})_{ij}([y]_j^{k+1} - \tilde{x}_j^0) \\ &\qquad + \sum_{j=i}^{n} (I - \tilde{P})_{ij}([y]_j^k - \tilde{x}_j^0), \quad i = 1, \ldots, n; \ k = 1, 2, \ldots . \end{aligned} \tag{6.3.28}$$

We show by induction that $[y]^{k+1} \subseteq [y]^k$ holds for $k = 0, 1, \ldots$. For $k = 0$ this follows from (6.3.25), since then $[y]^1 = [x]^1 = M([x]^0, \tilde{x}^0, C) \subseteq [x]^0 = [y]^0$. Assume now that $[y]^k \subseteq [y]^{k-1}$ holds for some $k \geq 1$ and $[y]_i^{k+1} \subseteq [y]_i^k$ for $i = 1, \ldots, s-1$ and some $s \in \{1, \ldots, n\}$. Then (6.3.28) implies

$$\begin{aligned} [y]_s^{k+1} &= \tilde{x}_s^0 - (Cf(\tilde{x}^0))_s + \sum_{j=1}^{s-1} (I - \tilde{P})_{sj}([y]_j^{k+1} - \tilde{x}_j^0) + \sum_{j=s}^{n} (I - \tilde{P})_{sj}([y]_j^k - \tilde{x}_j^0) \\ &\subseteq \tilde{x}_s^0 - (Cf(\tilde{x}^0))_s + \sum_{j=1}^{s-1} (I - \tilde{P})_{sj}([y]_j^k - \tilde{x}_j^0) + \sum_{j=s}^{n} (I - \tilde{P})_{sj}([y]_j^{k-1} - \tilde{x}_j^0) \\ &= [y]_s^k. \end{aligned}$$

This concludes the induction with $s = n$ and guarantees the existence of the limit $[y]^* = \lim_{k\to\infty}[y]^k$. This limit satisfies $[y]^* \subseteq [y]^1 \subseteq [x]^0$, $d([y]^*) \leq d([y]^1) = d([x]^1) < d([x]^0)$, and

$$[y]^* = \tilde{x}^0 - Cf(\tilde{x}^0) + (I - \tilde{P})([y]^* - \tilde{x}^0)$$

from which we get

$$d([y]^*) = |I - \tilde{P}|d([y]^*). \tag{6.3.29}$$

Assume now that there is some vector u which satisfies $\langle \tilde{P}\rangle u \leq 0 \leq u$. Then

$$(1 - |1 - \tilde{p}_{ii}|)u_i \leq |\tilde{p}_{ii}|u_i \leq \sum_{j\neq i} |\tilde{p}_{ij}|u_j,$$

holds for $i = 1, \dots, n$, whence

$$u \leq |I - \tilde{P}|u. \tag{6.3.30}$$

Choose $\alpha > 0$ and a signature matrix D such that

$$[z] = [y]^* + [0, \alpha]Du \subseteq [x]^0 = [y]^0 \tag{6.3.31}$$

holds. Then $d([z]) = d([y]^*) + \alpha u$. We will prove by induction

$$d([z]) \leq d([y]^k), \quad k = 0, 1, \dots. \tag{6.3.32}$$

For $k = 0$ the inequality follows from (6.3.31). Assume that it holds for some $k \geq 0$ and that $d([z]_i) \leq d([y]_i^{k+1})$ is valid for $i = 1, \dots, s-1$ and some $s \in \{1, \dots, n\}$. Then for $i = s$ we get

$$\begin{aligned}
d([y]_s^{k+1}) &= \sum_{j=1}^{s-1} |I - \tilde{P}|_{sj} d([y]_j^{k+1}) + \sum_{j=s}^{n} |I - \tilde{P}|_{sj} d([y]_j^k) \\
&\geq \sum_{j=1}^{s-1} |I - \tilde{P}|_{sj} d([z]_j) + \sum_{j=s}^{n} |I - \tilde{P}|_{sj} d([z]_j) \\
&= \left(|I - \tilde{P}|d([z])\right)_s = \left(d([y]^*) + \alpha|I - \tilde{P}|u\right)_s \\
&\geq d([y]_s^*) + \alpha u_s = d([z]_s),
\end{aligned}$$

where we used (6.3.29)–(6.3.31). This proves (6.3.32).

With $k \to \infty$ in (6.3.32) we obtain $d([y]^*) + \alpha u = d([z]) \leq d([y]^*)$, which implies $u = 0$ since α was positive. Therefore, by virtue of Theorem 1.10.16 (d), the matrix $\tilde{P}$ is an H-matrix, and the same holds for $Cf'([x]^0)$, too.

(e) The proof proceeds similarly to that for Theorem 6.3.1 (d): from (6.3.22) with $[x] = [x]^k$, $\check{x} = \check{x}^k$, and $k \to \infty$ we get

$$\begin{aligned}
[x]^* &= M'([x]^*, \check{x}^*, C) \subseteq M([x]^*, \check{x}^*, C) \\
&= \check{x}^* - Cf(\check{x}^*) + (I - Cf'([x]^*))([x]^* - \check{x}^*) \\
&= \check{x}^* - Cf(\check{x}^*) + |I - Cf'([x]^*)|([x]^* - \check{x}^*) \\
&\subseteq \check{x}^* - Cf(\check{x}^*) + |I - Cf'([x]^0)|([x]^* - \check{x}^*),
\end{aligned}$$

whence

$$d([x]^*) \le |I - Cf'([x]^0)|d([x]^*) \le \cdots \le |I - Cf'([x]^0)|^k d([x]^*).$$

Therefore, $d([x]^*) = 0$ in the case (6.3.26).

In the case (6.3.27) we choose a convergent subsequence $\{\tilde{x}^{k_j}\}$ with limit $\tilde{x}^*$. Then we proceed similarly to above using $[x]^* - \tilde{x}^* \subseteq d([x]^*)[-1, 1]$. □

Notice that the intersection in (6.3.22) implies $M_i'([x], \tilde{x}, C) \subseteq [x]_i$ so that $M_i([x], \tilde{x}, C) \subseteq K_i([x], \tilde{x}, C)$ holds for all indices i. This means

$$M([x], \tilde{x}, C) \subseteq K([x], \tilde{x}, C) \tag{6.3.33}$$

and shows that the Moore–Qi test is more general than the Krawczyk test, i.e., it certainly holds if the Krawczyk test is satisfied. The following example by Moore and Qi [239] indicates that there are cases in which the Krawczyk test fails while the Moore–Qi test is successful.

Example 6.3.8.

$$f(x) = \begin{pmatrix} x_1^2 + 0.25x_2 - 0.1725 \\ x_2^2 - 3x_1 + 0.46 \end{pmatrix}, \quad [x] = \begin{pmatrix} [0, 0.5] \\ [0, 1] \end{pmatrix},$$

$$\tilde{x} = \begin{pmatrix} 0.25 \\ 0.5 \end{pmatrix}, \quad C = \begin{pmatrix} 0.8 & -0.2 \\ 2.4 & 0.4 \end{pmatrix}.$$

This yields

$$K([x], \tilde{x}, C) = \begin{pmatrix} [0.03, 0.43] \\ [-0.02, 0.98] \end{pmatrix} \not\subseteq [x],$$

but

$$M([x], \tilde{x}, C) = \begin{pmatrix} [0.03, 0.43] \\ [0.016, 0.944] \end{pmatrix} \subseteq \operatorname{int}([x]).$$

Because of outward rounding in practical interval computations one does not obtain $f'([x]^k)$ exactly but only a superset of it. In addition, the evaluation of f' per step may be very costly. Therefore, one is often content with an iteration of the form

$$\begin{aligned} [x]^{k+1} &= \{\tilde{x}^k - Cf(\tilde{x}^k) + (I - C[Y])([x]^k - \tilde{x}^k)\} \cap [x]^k \\ &= K_s([x]^k, \tilde{x}^k, [Y], C) \cap [x]^k, \quad k = 0, 1, \ldots, \end{aligned} \tag{6.3.34}$$

with the slope-based Krawczyk operator

$$K_s([x], \tilde{x}, [Y], C) := \tilde{x} - Cf(\tilde{x}) + (I - C[Y])([x] - \tilde{x}), \quad \tilde{x} \in [x]. \tag{6.3.35}$$

Here, $[Y]$ is an $n \times n$ interval matrix which satisfies

$$f(x) - f(y) \in [Y](x - y) \quad \text{for all } x, y \in [x]. \tag{6.3.36}$$

As in Neumaier [257] we call $[Y]$ an (interval) slope matrix generalizing the slope in Definition 4.1.7. But notice that there $s(x, y)$ was introduced as a real valued function of $x, y \in \mathbb{R}^n$ while here $[Y]$ is an interval matrix which may depend on $[x] \in \mathbb{IR}^n$. Although not necessary, we assume that in the iteration (6.3.34) the slope matrix $[Y]$ only depends on the starting vector $[x]^0$ and remains fixed for the succeeding iterates. We call the iteration (6.3.34) the slope-based Krawczyk iteration.

Obviously,

$$J(x, y)(x - y) \in [Y](x - y) \tag{6.3.37}$$

holds for all $x, y \in [x]$, where J is again the matrix of the multidimensional mean value theorem. Trivially, a candidate for $[Y]$ is $f'([x])$ but every superset of it will do, too. Such a superset is obtained automatically, for instance, if one computes $f'([x])$ using machine interval arithmetic.

Theorem 6.3.1 holds again with K being replaced by K_S and with $f'([x]^0)$ in (c) being substituted by $[Y]$. The Krawczyk test is replaced by the Moore test, respectively strong Moore test,

$$K_S([x]^0, \tilde{x}^0, [Y], C) \subseteq [x]^0, \quad \text{resp.} \quad K_S([x]^0, \tilde{x}^0, [Y], C)_i \subset [x]_i^0, \quad i = 1, \dots, n. \tag{6.3.38}$$

The proof follows the lines of that of Theorem 6.3.1 and is left to the reader. Only the uniqueness in (d) requires some care: If there are two zeros $x^*, y^* \in [x]^0$ of f, then (6.3.36) implies $0 = f(x^*) - f(y^*) \in [Y](x^* - y^*)$. By virtue of Theorem 4.1.5 there is a matrix $\tilde{Y} \in [Y]$ such that $0 = f(x^*) - f(y^*) = \tilde{Y}(x^* - y^*)$ holds. Since the second (= strict) subset property in (6.3.38) guarantees that $[Y]$ is regular, the zeros x^* and y^* must coincide.

Newton's method as well as Krawczyk's method primarily verify fixed points of some particular function g associated with a given function f whose zeros are finally to be looked for. If g is chosen independently of these two methods such that $x = g(x)$ is equivalent to $f(x) = 0$ and if g is a P-contraction, one can also iterate via the general fixed point iteration

$$[x]^{k+1} = g([x]^k), \quad k = 1, 2, \dots \tag{6.3.39}$$

as considered in Section 4.2, where properties on this method have also been stated; cf. particularly the Theorems 4.2.6 and 4.2.7.

In practical realizations of all these methods the crucial assumption $g([x]^0) \subseteq [x]^0$ or, even more restrictively, $g([x]^0)_i \subset [x]_i^0$, $i = 1, \dots, n$, will usually not be fulfilled in advance. Therefore an iterative process without intersection but with ε-inflation as described in Section 4.3 can help. If a fixed point x^* of g is assumed to exist, another possibility consists in iterating by a traditional (i.e. noninterval) method until a sufficiently good approximation $\tilde{x}$ of x^* is attained. Then $\tilde{x}$ is inflated to some interval vector $[x]$ with $\check{x} = \tilde{x}$ and the subset property is tested. A way to find $\tilde{x}$ was given by Alefeld, Gienger, and Potra in [23]. They start with the classical Newton method at some point $x^0 \in D$ and refer to the theorem of Newton–Kantorovich which was stated as Theorem 1.5.10 in Section 1.5. This theorem requires – among others – the assumptions

(i) $\|f'(x) - f'(y)\| \le \gamma\|x - y\|$ for all $x, y \in D$, $\gamma > 0$,
(ii) $\|f'(x^0)^{-1}\| \le \beta$, $\beta > 0$,
(iii) $\|f'(x^0)^{-1}f(x^0)\| \le \eta$,
(iv) $0 \le \alpha := \beta\gamma\eta \le 1/2$,

in order to construct the radius r of a ball $\overline{B}(x^0, r) \subseteq D$ which contains a unique zero x^* of f. The difficulty in applying this theorem consists in computing the constants β and γ, while η can be obtained from the first Newton step. Our aim is to get rid of the first two constants. To this end we make the following hypotheses on the Newton sequence (x^k) arising from the starting vector x^0.

(1) The sequence (x^k) converges to a zero x^* of f.
(2) The assumptions of the theorem of Newton–Kantorovich hold for some fixed x^k instead of x^0. (Thus the constants $\beta, \eta, \ldots$ refer to x^k.)
(3) The norm in this theorem is the maximum norm.

Since

$$\eta = \eta_k = \|f'(x^k)^{-1}f(x^k)\|_\infty = \|x^k - f'(x^k)^{-1}f(x^k) - x^k\|_\infty = \|x^{k+1} - x^k\|_\infty$$

we may assume $\eta > 0$ which implies $\alpha > 0$. Otherwise $f(x^k) = 0$ and we are done. From (1.5.11) (with x^k instead of x^0) we get

$$\begin{aligned} \|x^k - x^*\|_\infty &\le \frac{1}{\beta\gamma 2^0}(2\alpha)^{2^0} = \frac{1}{\beta\gamma\eta}2\alpha\eta = 2\|x^{k+1} - x^k\|_\infty = 2\eta_k, \\ \|x^{k+1} - x^*\|_\infty &\le \frac{1}{\beta\gamma \cdot 2}(2\alpha)^{2^1} = 2\alpha\eta \le \eta = \|x^{k+1} - x^k\|_\infty = \eta_k, \end{aligned}$$

see Figure 6.3.1. Define

$$\left.\begin{aligned} [x] &= x^{k+1} + \eta_k[-1, 1]e\,, \\ [x]_K &= K([x], x^{k+1}, f'(x^k)^{-1}) \end{aligned}\right\}. \tag{6.3.40}$$

Since $x^k \in [x]$, the assumption $C^{-1} \in f'([x])$ of Theorem 6.3.5 is fulfilled if $C = f'(x^k)^{-1}$. If the remaining assumptions of this theorem hold for $[x]$ from (6.3.40), then it implies

$$\|d([x]_K)\|_\infty \le \hat{\alpha}\,\|d([x])\|_\infty^2 = (4\hat{\alpha}\eta_k) \cdot \eta_k\,.$$

Now $x^* \in [x]$ guarantees

$$x^* \in [x]_K, \tag{6.3.41}$$

from which we get $[x]_K \subseteq x^* + \eta_k[-1, 1]e \cdot (4\hat{\alpha}\eta_k)$. If $x^* \in \mathrm{int}([x])$ holds and if η_k is sufficiently small, then one can expect that the crucial assumption

$$([x]_K)_i \subset [x]_i, \quad i = 1, \ldots, n, \tag{6.3.42}$$

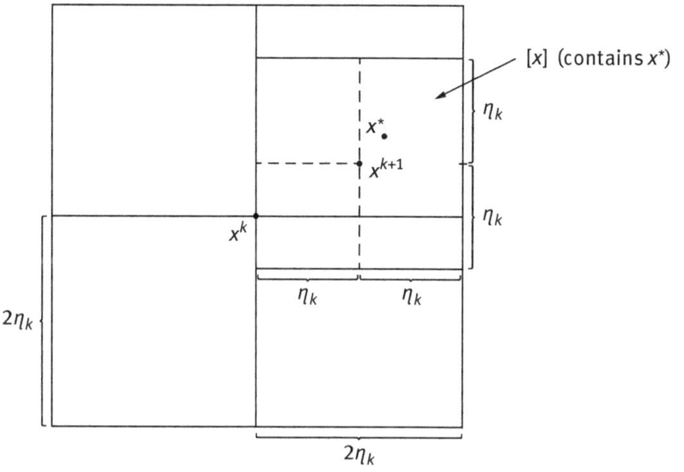

Fig. 6.3.1: Position of two successive iterates.

of Theorem 6.3.1 (d) is fulfilled. In this way one has verified 'a posteriori' and 'constructively' existence and uniqueness of a zero x^* of f in $[x]$ from (6.3.40). As a tool one has only to construct the Newton sequence (x^k) and to use an appropriate stopping criterion such that $\eta_k = \|x^{k+1} - x^k\|_\infty$ is 'sufficiently small'. In order to present such a stopping criterion we require

$$\frac{\|d([x]_K)\|_\infty}{\|x^{k+1}\|_\infty} \le \text{eps} = 2^{-t}, \tag{6.3.43}$$

where t is the length of the mantissa of the underlying floating point system with basis $b = 2$ and eps is the machine precision, i.e., eps $= \frac{1}{2}b^{-t+1}$; cf. Section 2.7, where $\ell' = t - 1$.

If the hypotheses above are valid then (6.3.41), (6.3.43) and $y \in [x]_K$ imply

$$\frac{\|x^* - y\|_\infty}{\|x^*\|_\infty} \le \frac{\|d([x]_K)\|_\infty}{\|x^*\|_\infty} \approx \frac{\|d([x]_K)\|_\infty}{\|x^{k+1}\|_\infty} \le \text{eps}, \tag{6.3.44}$$

i.e., the relative error with which $y \in [x]_K$ approximates the zero x^* is close to the machine precision (provided that $x^* \ne 0$). With

$$\tilde{\gamma} = \max\{\hat{\gamma}, \gamma\} > 0 \quad (\hat{\gamma} \text{ from Theorem 6.3.5}, \gamma \text{ from Theorem 1.5.10})$$

and $C = f'(x^k)^{-1}$ Theorem 6.3.5 yields

$$\|d([x])_K\|_\infty \le \|C\|_\infty \tilde{\gamma}\, \|d([x])\|_\infty^2 \le \|C\|_\infty \tilde{\gamma} \cdot 4\eta_k^2 .$$

Therefore, (6.3.44) holds (approximately) if

$$\frac{\|C\|_\infty \tilde{\gamma} \cdot 4\eta_k^2}{\|x^{k+1}\|_\infty} \le \text{eps} \tag{6.3.45}$$

is valid.

In order to eliminate C and $\tilde{y}$ from (6.3.45) we represent 0 as

$$\begin{aligned} 0 &= -f(x^{k-1}) - f'(x^{k-1})(-f'(x^{k-1})^{-1}f(x^{k-1})) \\ &= -f(x^{k-1}) - f'(x^{k-1})(x^k - x^{k-1}) \end{aligned}$$

from which we get

$$-f'(x^k)^{-1}\{f(x^k) - f(x^{k-1}) - f'(x^{k-1})(x^k - x^{k-1})\} = x^{k+1} - x^k. \tag{6.3.46}$$

By virtue of the multidimensional mean value theorem we obtain

$$\begin{aligned} &\|f(x^k) - f(x^{k-1}) - f'(x^{k-1})(x^k - x^{k-1})\|_\infty \\ &\quad = \left\| \int_0^1 \{f'(x^{k-1} + t(x^k - x^{k-1})) - f'(x^{k-1})\}\, dt\, (x^k - x^{k-1}) \right\|_\infty \\ &\quad \le \int_0^1 \|f'(x^{k-1} + t(x^k - x^{k-1})) - f'(x^{k-1})\|_\infty \, dt\, \|x^k - x^{k-1}\|_\infty \\ &\quad \le \gamma \int_0^1 t\, dt\, \eta_{k-1}^2 = \gamma \frac{1}{2} t^2 \Big|_0^1 \eta_{k-1}^2 \le \frac{1}{2} \tilde{\gamma}\, \eta_{k-1}^2 . \end{aligned}$$

Together with (6.3.46) this yields $\eta_k \le \|C\|_\infty \tilde{\gamma} \eta_{k-1}^2/2$, whence $\|C\|_\infty \tilde{\gamma} \ge 2\eta_k/\eta_{k-1}^2$. Since all our considerations are based on heuristics, we replace '$\ge$' by '$=$' in the last inequality. This allows us to substitute $\|C\|_\infty = 2\eta_k/(\tilde{\gamma}\eta_{k-1}^2)$ in (6.3.45) in order to get

$$\frac{8\eta_k^3}{\|x^{k+1}\|_\infty \, \eta_{k-1}^2} \le \text{eps}. \tag{6.3.47}$$

The inequality (6.3.47) is used as a stopping criterion for the Newton method. If it is fulfilled for some $k = k_0$, then one constructs $[x]$ from (6.3.40) with x^{k_0+1}, x^{k_0} and checks (6.3.42). Practical experiments show that sometimes η_{k_0} is already too small in order to guarantee (6.3.42). Then one constructs $[x]$ in (6.3.40) with $\sqrt{\eta_{k_0}\, \eta_{k_0-1}}$ instead of η_{k_0}, where the geometric mean is based on experience.

We illustrate the method with an example which is taken from Alefeld, Gienger, Potra [23].

Example 6.3.9 (Discretisation of a nonlinear bvp). Discretize

$$\left.\begin{aligned} 3y''y + (y')^2 &= 0 \\ y(0) = 0,\ y(1) &= 20 \end{aligned}\right\} \tag{6.3.48}$$

by means of

$$\left.\begin{aligned} t_i &= \frac{i}{n+1} \\ y_i &= y(t_i) \end{aligned}\right\} \quad i = 0, 1, \ldots, n+1$$

with the boundary values $y_0 = 0$, $y_{n+1} = 20$, and by means of

$$\left.\begin{aligned} y'(t_i) &= \frac{y_{i+1} - y_{i-1}}{2h} + O(h^2) \\ y''(t_i) &= \frac{y_{i+1} - 2y_i + y_{i-1}}{h^2} + O(h^2) \end{aligned}\right\}$$

with $h = \frac{1}{n+1} \to 0$.

Multiply (6.3.48) by h^2, delete $O(h^2)$, and replace y_i by x_i with $x_0 := y_0 = 0$, $x_{n+1} := y_{n+1} = 20$. Introduce $x = (x_1, \dots, x_n)^T$ in order to obtain

$$\begin{aligned} f_1(x) &= 3(x_2 - 2x_1)x_1 + \tfrac{1}{4}x_2^2 = 0, \\ f_i(x) &= 3(x_{i+1} - 2x_i + x_{i-1})x_i + \tfrac{1}{4}(x_{i+1} - x_{i-1})^2 = 0, \quad i = 2, \dots, n-1, \\ f_n(x) &= 3(20 - 2x_n + x_{n-1})x_n + \tfrac{1}{4}(20 - x_{n-1})^2 = 0. \end{aligned}$$

Choose the starting vector $x^0 = 10e$ for the Newton method applied to $f = (f_i)$. With the notation

$$\begin{aligned} x^k \quad & k\text{-th Newton iterate} \\ \eta^k &= \|x^{k+1} - x^k\|_\infty \\ g_k &= \sqrt{\eta_k \eta_{k-1}} \quad \text{(geometric mean)} \\ [x]_K \quad & \text{as in (6.3.40)} \\ r_k &= \frac{\|d([x]_K)\|_\infty}{\|x^{k+1}\|_\infty} \quad \text{(approximation of the relative error)} \end{aligned}$$

we got the results shown in Table 6.3.1. To this end we used INTLAB with eps $= 2^{-53} = (2^{-10})^{5.3} \approx (10^{-3})^{5.3} \approx 10^{-16}$ which is half of the MATLAB constant eps; cf. Section 2.7 for a discussion of MATLAB's eps.

Tab. 6.3.1: Results of Example 6.3.9 rounded appropriately.

n	k	η_k	η_{k-1}	g_k	stop with	$\|d([x]_K)\|_\infty$	r_k
10	7	$6.89 \cdot 10^{-13}$	$3.74 \cdot 10^{-6}$	$1.61 \cdot 10^{-9}$	η_k	$1.07 \cdot 10^{-14}$	$5.73 \cdot 10^{-16}$
20	7	$8.91 \cdot 10^{-11}$	$4.31 \cdot 10^{-5}$	$6.20 \cdot 10^{-8}$	η_k	$1.42 \cdot 10^{-14}$	$7.37 \cdot 10^{-16}$
50	8	$6.32 \cdot 10^{-13}$	$8.11 \cdot 10^{-7}$	$7.16 \cdot 10^{-10}$	η_k	$1.07 \cdot 10^{-14}$	$5.41 \cdot 10^{-16}$
100	9	$1.78 \cdot 10^{-15}$	$1.75 \cdot 10^{-8}$	$5.57 \cdot 10^{-12}$	g_k	$1.07 \cdot 10^{-14}$	$5.40 \cdot 10^{-16}$

The last two columns differ slightly from the results in Alefeld, Gienger, Potra [23] but show essentially the same magnitude. In the case $n = 100$ we could only prove $[x]_K \subseteq [x]$ instead of the stronger strict subset relation (6.3.42). The solution of the original nonlinear boundary value problem (6.3.48) is $y(t) = 20t^{3/4}$.

Exercises

Ex. 6.3.1. Work out the details of Example 6.3.4.

Ex. 6.3.2. Restate and prove Theorem 6.3.1 for the slope-based Krawczyk iteration (6.3.34).

Ex. 6.3.3. Recompute Example 6.3.8.

6.4 Hansen–Sengupta method

In [132] Hansen and Sengupta presented a method for verifying and enclosing a zero x^* of a function $f\colon D \subseteq \mathbb{R}^n \to \mathbb{R}^n$ which is a nonlinear version of the interval Gauss–Seidel method applied to the interval linear system

$$(Cf'([x]))(x - \tilde{x}) = -Cf(\tilde{x}). \tag{6.4.1}$$

Thus it is defined similarly to the Gauss–Seidel-like Krawczyk method (6.3.21) but differs from it slightly. It is given by

$$[x]^{k+1} = H'([x]^k, \tilde{x}^k, C), \quad k = 0, 1, \ldots, \tag{6.4.2}$$

where $H'([x], \tilde{x}, C)$ is defined for $\tilde{x} \in [x]$ and $i = 1, 2, \ldots, n$ by

$$\begin{aligned} H_i([x], \tilde{x}, C) &= \tilde{x}_i - \Big\{ (Cf(\tilde{x}))_i + \sum_{j=1}^{i-1} (Cf'([x]))_{ij} (H_j'([x], \tilde{x}, C) - \tilde{x}_j) \\ &\qquad + \sum_{j=i+1}^{n} (Cf'([x]))_{ij} ([x]_j - \tilde{x}_j) \Big\} \Big/ (Cf'([x]))_{ii}, \\ H_i'([x], \tilde{x}, C) &= H_i([x], \tilde{x}, C) \cap [x]_i. \end{aligned} \tag{6.4.3}$$

Here, we assume $0 \notin (Cf'([x]))_{ii}$ for $i = 1, \ldots, n$. If the intersection in (6.4.3) is empty for some i, then H' is not defined. If it is defined, we will call (6.4.2) the Hansen–Sengupta method and H the Hansen–Sengupta operator.

Theorem 6.4.1. *Let the assumptions of Theorem* 6.3.1 *hold. In addition, let C be regular and $0 \notin (CF'([x]^0))_{ii}$ for $i = 1, \ldots, n$.*

(a) *If $x^* \in [x]^0$ is a zero of f, then the Hansen–Sengupta method* (6.4.2) *is feasible for $[x]^0$ and satisfies*

$$x^* \in [x]^* = \lim_{k\to\infty} [x]^k \subseteq [x]^k \subseteq [x]^{k-1} \subseteq \cdots \subseteq [x]^0 \tag{6.4.4}$$

for $k = 0, 1, \ldots$. In particular, each iterate $[x]^k$ contains all *zeros $x^* \in [x]^0$ of f, and the same holds for the limit $[x]^*$.*

(b) *If $H'([x]^0, \tilde{x}^0, C)$ is not defined, then f has no zero in $[x]^0$.*

(c) *If* $H'([x]^0, \tilde{x}^0, C)$ *is defined and if the Hansen–Sengupta test*

$$H([x]^0, \tilde{x}^0, C) \subseteq [x]^0 \tag{6.4.5}$$

is valid, then there is at least one zero $x^* \in [x]^0$ *of* f, *and* (a) *holds.*

(d) *Let* $H'([x]^0, \tilde{x}^0, C)$ *be defined. If the strong Hansen–Sengupta test*

$$(H([x]^0, \tilde{x}^0, C))_i \subset [x]_i^0, \quad i = 1, \dots, n, \tag{6.4.6}$$

is valid, then $Cf'([x]^0)$ *is an* H*-matrix. In particular,* $f'([x]^0)$ *and* C *are regular even if this is not assumed for* C *from the start. In addition,* f *has exactly one zero* x^* *in* $[x]^0$, *and* (a) *holds.*

(e) *If the assumptions of* (d) *hold and if*

$$\tilde{x}^k = \check{x}^k, \quad k = 0, 1, \dots, \tag{6.4.7}$$

then f *has exactly one zero* x^* *in* $[x]^0$, *and the sequence in* (a) *contracts to* $[x]^* \equiv x^*$.

Proof. Since many steps of the proof are similar to those in the proof of Theorem 6.3.7 we will shorten them a bit.

(a) Let $x^* \in [x]^0$ be a zero of f and define g by $g(x) = -Cf(x)$. Then the multidimensional mean value theorem implies

$$0 = g(x^*) = g(\tilde{x}) - CJ(x^*, \tilde{x})(x^* - \tilde{x}).$$

Assume that $x_i^* \in [x]_i^1$ for $i = 1, \dots, s-1$ and some $s \in \{1, \dots, n\}$. Then

$$\begin{aligned}
x_s^* &= \tilde{x}_s^0 + \Big\{g_s(\tilde{x}^0) - \sum_{j=1}^{s-1} (CJ(x^*, \tilde{x}^0))_{sj}(x_j^* - \tilde{x}_j^0) \\
&\quad - \sum_{j=s+1}^{n} (CJ(x^*, \tilde{x}^0))_{sj}(x_j^* - \tilde{x}_j^0)\Big\} \,/\, (CJ(x^*, \tilde{x}^0))_{ss} \\
&\in \tilde{x}_s^0 - \Big\{(Cf(\tilde{x}^0))_s + \sum_{j=1}^{s-1} (Cf'([x]^0)_{sj}([x]_j^1 - \tilde{x}_j^0) \\
&\quad + \sum_{j=s+1}^{n} (Cf'([x]^0))_{ij}([x]_j^0 - \tilde{x}_j^0)\Big\} \,/\, (Cf'([x]^0))_{ss}.
\end{aligned}$$

This shows that $x_s^* \in H_s'([x]^0, \tilde{x}^0, C) = [x]_s^1$ holds, and finally $x^* \in [x]^1$. Now (6.4.4) follows by induction.

(b) follows from (a) as in the proof of Theorem 6.3.7.

(c) The assumptions of (c) imply $H([x]^0, \tilde{x}^0, C) = H'([x]^0, \tilde{x}^0, C) \neq \emptyset$. Define the function $h\colon [x]^0 \to \mathbb{R}^n$ componentwise by

$$\begin{aligned}
h_i(x) &= \tilde{x}_i^0 + \Big\{g_i(\tilde{x}^0) - \sum_{i=1}^{i-1} (CJ(x, \tilde{x}^0))_{ij}(h_j(x) - \tilde{x}_j^0) \\
&\quad - \sum_{j=i+1}^{n} (CJ(x, \tilde{x}^0))_{ij}(x_j - \tilde{x}_j^0)\Big\} \,/\, (CJ(x, \tilde{x}^0))_{ii},
\end{aligned} \tag{6.4.8}$$

where $J(x, \tilde{x}^0)$ is again the matrix from the multidimensional mean value theorem applied to f. Since $f \in C^1(D)$, by assumption this matrix depends continuously on x, and so does h. Obviously, $h(x) \in H([x]^0, \tilde{x}^0, C) \subseteq [x]^0$ holds for $x \in [x]^0$, hence Brouwer's fixed point theorem guarantees a fixed point $x^* \in [x]^0$ of h. A simple reformulation of (6.4.8) with $x = x^* = h(x^*)$ yields

$$0 = g(\tilde{x}^0) - CJ(x^*, \tilde{x}^0)(x^* - \tilde{x}^0) = g(x^*),$$

whence $f(x^*) = 0$.

(d) We only show that $Cf'([x]^0)$ is an H-matrix. The proof is then finished as in Theorem 6.3.7.

Choose $\tilde{P} \in Cf'([x]^0)$ such that $\langle \tilde{P} \rangle = \langle Cf'([x]^0) \rangle$ and consider the auxiliary sequence $([y]^k)$ which is defined by

$$\begin{aligned} &[y]^0 = [x]^0, \quad [y]^1 = [x]^1, \\ &[y]_i^{k+1} = \tilde{x}_i^0 - \left\{ (Cf(\tilde{x}^0))_i + \sum_{j=1}^{i-1} \tilde{p}_{ij}([y]_j^{k+1} - \tilde{x}_j^0) + \sum_{j=i+1}^{n} \tilde{p}_{ij}([y]_j^k - \tilde{x}_j^0) \right\} / \tilde{p}_{ii}, \\ &\qquad i = 1, \dots, n; \quad k = 1, 2, \dots . \end{aligned} \tag{6.4.9}$$

Similarly to the proof of Theorem 6.3.7 (d) we see that $[y]^{k+1} \subseteq [y]^k$ holds for $k = 0, 1, \dots$. This guarantees the existence of the limit $[y]^* = \lim_{k\to\infty}[y]^k$ which satisfies $[y]^* \subseteq [y]^1 \subseteq [x]^0$, $d([y]^*) \le d([y]^1) = d([x]^1) < d([x]^0)$, and

$$[y]_i^* = \tilde{x}_i^0 - \left\{ (Cf(\tilde{x}^0))_i + \sum_{j\ne i} \tilde{p}_{ij}([y]_j^* - \tilde{x}_j^0) \right\} / \tilde{p}_{ii}. \tag{6.4.10}$$

For the diameter in (6.4.10) we get after some rearrangements

$$\langle \tilde{P} \rangle d([y]^*) = 0. \tag{6.4.11}$$

Assume now that there is some vector u which satisfies $\langle \tilde{P} \rangle u \le 0 \le u$. Then

$$|\tilde{p}_{ii}| u_i \le \sum_{j\ne i} |\tilde{p}_{ij}| u_j, \quad i = 1, \dots, n, \tag{6.4.12}$$

holds. Choose $\alpha > 0$ and a signature matrix D such that

$$[z] = [y]^* + [0, \alpha] D u \subseteq [x]^0 = [y]^0 \tag{6.4.13}$$

holds. Then $d([z]) = d([y]^*) + \alpha u$. We will prove by induction

$$d([z]) \le d([y]^k), \quad k = 0, 1, \dots . \tag{6.4.14}$$

For $k = 0$ the inequality follows from (6.4.13). Assume that it holds for some $k \ge 0$ and that $d([z]_i) \le d([y]_i^{k+1})$ is valid for $i = 1, \dots, s-1$ and some $s \in \{1, \dots, n\}$. Then for

$i = s$ we get

$$\begin{aligned} d([y]_s^{k+1}) &= \left(\sum_{j=1}^{s-1} |\tilde p_{sj}| d([y]_j^{k+1}) + \sum_{j=s+1}^{n} |\tilde p_{sj}| d([y]_j^{k})\right) / |\tilde p_{ss}| \\ &\ge \left(\sum_{j=1}^{s-1} |\tilde p_{sj}| d([z]_j) + \sum_{j=s+1}^{n} |\tilde p_{sj}| d([z]_j)\right) / |\tilde p_{ss}| \\ &= \left(\sum_{j\neq s} |\tilde p_{sj}| (d([y]_j^*) + \alpha u_j)\right) / |\tilde p_{ss}| \\ &\ge d([y]_s^*) + \alpha u_s = d([z]_s), \end{aligned}$$

where we used (6.4.10)–(6.4.13). This proves (6.4.14).

With $k \to \infty$ in (6.4.14) we obtain $d([y]^*) + \alpha u = d([z]) \le d([y]^*)$, which implies $u = 0$. Therefore, $\tilde P$ is an H-matrix by virtue of Theorem 1.10.16 (d), and the same holds for $Cf'([x]^0)$.

(e) W.l.o.g. assume that $0 < (Cf'([x]^*))_{ii}$ holds. Similarly to in Theorem 6.3.1 (c) or Theorem 6.3.7 (e) we get

$$\begin{aligned} [x]_i^* &= H_i'([x]^*, \check x^*, C) = H_i([x]^*, \check x^*, C) \\ &= \check x_i^* - \left\{(Cf(\check x^*))_i + \textstyle\sum_{j\neq i} (Cf'([x]^*))_{ij}([x]_j^* - \check x_j^*)\right\} / (Cf'([x]^*))_{ii}. \end{aligned}$$

Since $\check x_i^*$ is the midpoint of the left-hand side $[x]_i^*$ and since $0 \notin (Cf'([x]^*))_{ii}$, the expression within braces must be a zero-symmetric interval. This implies $\{\ldots\}/(Cf'([x]^*))_{ii} = \{\ldots\}/\langle(Cf'([x]^*))_{ii}\rangle$ whence

$$[x]_i^* = \check x_i^* - \left\{(Cf(\check x^*))_i + \textstyle\sum_{j\neq i} |(Cf'([x]^*))_{ij}| ([x]_j^* - \check x_j^*)\right\} / \langle(Cf'([x]^*))_{ii}\rangle.$$

Equating the midpoints on the left- and right-hand sides shows that $-Cf(\check x^*) = 0$, whence $f(\check x^*) = 0$. Thus

$$[x]_i^* - \check x_i^* = -\left\{\textstyle\sum_{j\neq i} |(Cf'([x]^*))_{ij}| ([x]_j^* - \check x_j^*)\right\} / \langle(Cf'([x]^*))_{ii}\rangle \tag{6.4.15}$$

and

$$\langle Cf'([x]^*)\rangle d([x]^*) = 0.$$

Theorem 1.10.16 (d) implies $d([x]^*) = 0$ which proves the theorem. □

We want to compare the Hansen–Sengupta test with the Moore–Qi test. It will turn out that the former is more general than the latter which, by virtue of (6.3.33), is more general than the Krawczyk test. We start with an auxiliary result.

Lemma 6.4.2. *Let* $[a], [b], [c] \in \mathbb{IR}$, $0 \notin [b]$. *If* $[a] + (1 - [b])[c] \subseteq [c]$, *then* $[a]/[b] \subseteq [a] + (1 - [b])[c]$.

Proof. Let $\tilde{z} \in [a]/[b]$ be arbitrary. Then there exist $\tilde{a} \in [a]$, $\tilde{b} \in [b]$ such that $\tilde{z} = \tilde{a}/\tilde{b}$ holds. Since by the hypothesis we have $h(c) := \tilde{a} + (1 - \tilde{b})c \in [c]$ for $c \in [c]$, Brouwer's fixed point theorem guarantees a fixed point $c^* \in [c]$ of h. Hence $\tilde{a} + (1 - \tilde{b})c^* = c^*$ holds and $\tilde{z} = \tilde{a}/\tilde{b} = c^* \in [c]$ follows. This implies the assertion. □

As the example $[a] = [c] = [-1, 1]$, $[b] = 1/2$ shows, the subset hypothesis of Lemma 6.4.2 is necessary.

Theorem 6.4.3. *Let the assumptions of Theorem 6.4.1 hold. If the Moore–Qi test $M([x], \tilde{x}, C) \subseteq [x]$ is satisfied, then $H([x], \tilde{x}, C) \subseteq M([x], \tilde{x}, C)$ follows. In particular, the Hansen–Sengupta test is fulfilled.*

Proof. Assume that

$$H_i([x], \tilde{x}, C) = H_i'([x], \tilde{x}, C) \subseteq M_i'([x], \tilde{x}, C) = M_i([x], \tilde{x}, C) \subseteq [x]_i \tag{6.4.16}$$

holds for $i = 1, \dots, s-1$ and some $s \le n$. By virtue of Lemma 6.4.2 this is certainly true for $s = 1$. For $s > 1$ we have

$$\begin{aligned} H_s([x], \tilde{x}, C) = \tilde{x}_s - \Bigg\{ & (Cf(\tilde{x}))_s + \sum_{j=1}^{s-1} (Cf'([x]))_{sj}(H_j'([x], \tilde{x}, C) - \tilde{x}_j) \\ & + \sum_{j=s+1}^{n} (Cf'([x]))_{sj}([x]_j - \tilde{x}_j) \Bigg\} \Big/ (Cf'([x]))_{ss} \\ \subseteq \tilde{x}_s - \Bigg\{ & (Cf(\tilde{x}))_s + \sum_{j=1}^{s-1} (Cf'([x]))_{sj}(M_j'([x], \tilde{x}, C) - \tilde{x}_j) \\ & + \sum_{j=s+1}^{n} (Cf'([x]))_{sj}([x]_j - \tilde{x}_j) \Bigg\} \Big/ (Cf'([x]))_{ss}. \end{aligned}$$

If the right-hand side is abbreviated as $\tilde{x}_s + [a]/[b]$, the Moore–Qi test guarantees $\tilde{x}_s + [a] + (1 - [b])([x]_s - \tilde{x}_s) \subseteq [x]_s$. Now Lemma 6.4.2 with $[c] = [x]_s - \tilde{x}_s$ implies

$$\tilde{x}_s + [a]/[b] \subseteq \tilde{x}_s + [a] + (1 - [b])([x]_s - \tilde{x}_s) = M_s([x], \tilde{x}, C),$$

hence (6.4.16) follows for s. □

In Goldstejn [120] the following Jacobi-like modification H^{mod} of the Hansen–Sengupta operator H is considered.

$$H_i^{\text{mod}}([x], \tilde{x}, C) = \tilde{x}_i - \Big\{ (Cf(\tilde{x}))_i + \textstyle\sum_{j \ne i} (Cf'([x]))_{ij}([x]_j - \tilde{x}_j) \Big\} \Big/ (Cf'([x]))_{ii}, \qquad i = 1, \dots, n,$$

$$H^{\text{mod}'}([x], \tilde{x}, C) = H^{\text{mod}}([x], \tilde{x}, C) \cap [x], \tag{6.4.17}$$

where $C \in \mathbb{R}^{n \times n}$ is regular, $\tilde{x} \in [x]$, and $0 \notin (Cf'([x]))_{ii}$, $i = 1, \dots, n$. As in (6.4.3), $H^{\text{mod}'}$ is not defined if the intersection in (6.4.17) is empty. If it is defined, we will call the iteration

$$[x]^{k+1} = H^{\text{mod}'}([x]^k, \tilde{x}^k, C), \quad k = 0, 1, \dots, \tag{6.4.18}$$

the modified Hansen–Sengupta method and H^{mod} the modified Hansen–Sengupta operator. It is left to the reader to show that (6.4.18) has properties similar to those of (6.4.2) as stated in Theorem 6.4.1.

It is easy to see that $H([x], \tilde{x}, C) \subseteq H^{\text{mod}}([x], \tilde{x}, C)$ and $H'([x], \tilde{x}, C) \subseteq H^{\text{mod}'}([x], \tilde{x}, C)$ hold. This guarantees that the modified Hansen–Sengupta test

$$H^{\text{mod}}([x]^0, \tilde{x}^0, C) \subseteq [x]^0 \tag{6.4.19}$$

always implies the Hansen–Sengupta test (6.4.5).

Exercises

Ex. 6.4.1. State and prove a theorem similar to Theorem 6.4.1 also for the modified Hansen–Sengupta method (6.4.18).

6.5 Further existence tests

In Sections 6.3 and 6.4 we considered various methods – called tests – for verifying existence and uniqueness of zeros of a given continuously differentiable function $f : D \subseteq \mathbb{R}^n \to \mathbb{R}^n$. These tests contained as common condition the subset property

$$T([x], \tilde{x}, C) \subseteq [x], \tag{6.5.1}$$

respectively

$$T_i([x], \tilde{x}, C) \subset [x]_i, \quad i = 1, \dots, n, \tag{6.5.2}$$

where $\tilde{x} \in [x]$ and $T \in \{A, H, H^{\text{mod}}, K, K_S, M\}$. Here A stands for Alefeld, H for Hansen–Sengupta, H^{mod} for modified Hansen–Sengupta, K for Krawczyk, K_S for Moore, and M for Moore–Qi. In the case (6.5.1) there exists at least one zero in $[x]$, in the case (6.5.2) exactly one. In the present section we will extend these tests, and we will compare them. We start with the Moore–Kioustelidis test and the Miranda test which is a modification of the preceding one. Both tests are based on Miranda's Theorem 1.5.9. We will use the notation $[x]^{i,\pm}$ for the i-th faces of an interval vector $[x] \in \mathbb{IR}^n$, i.e.,

$$[x]^{i,-} = \{\tilde{x} \in [x] \mid \tilde{x}_i = \underline{x}_i\},$$
$$[x]^{i,+} = \{\tilde{x} \in [x] \mid \tilde{x}_i = \overline{x}_i\}.$$

It is obvious that the union $\bigcup_{i=1}^n [x]^{i,\pm}$ forms the surface of the n-dimensional box $[x]$.

Since we cannot expect that the crucial Miranda assumption $f_i(x) \cdot f_i(y) \le 0$, $i = 1, \dots n$, holds for all $x \in [x]^{i,+}$, $y \in [x]^{i,-}$ – see Exercise 6.5.1 – we will consider the function $g(x) = Cf(x)$ with $C \approx (f'(x^*))^{-1}$, where x^* denotes a zero of f and where we assume for the moment that f is sufficiently smooth. Expanding f at x^* yields

$$f(x) = f(x^*) + f'(x^*)(x - x^*) + o(\|x - x^*\|_\infty),$$

whence $g(x) \approx x - x^*$ near x^* so that Miranda's assumptions can be expected to be fulfilled for an interval vector $[x]$ with $x^* \in \mathrm{int}([x])$. This justifies considering g instead of f.

Of course, x^* is unknown – one does not know if it exists at all! Therefore, the usual procedure is to iterate with a traditional method in order to approximate a zero of f, to stop with some vector $\tilde{x}$, to compute an approximate inverse C of $f'(\tilde{x})$ (which we assume to be regular) and to enclose $\tilde{x}$ by an interval vector $[x]$ with $d([x]) > 0$ and $\check{x} = \tilde{x}$.

The following result in Moore, Kioustelidis [238] guarantees a zero of f.

Theorem 6.5.1. *Let $f\colon D \subseteq \mathbb{R}^n \to \mathbb{R}^n$, $f \in C^1(D)$, $0 \le r \in \mathbb{R}^n$, $\check{x} \in D$, $[x] = [\check{x} - r, \check{x} + r] \subseteq D$. Let $f([x])$ and $f'([x])$ exist and $C \in \mathbb{R}^{n\times n}$ be regular. If for $g(x) = Cf(x)$ the conditions*

$$g_i(\check{x} + r_i e^{(i)}) \cdot g_i(\check{x} - r_i e^{(i)}) \le 0 \tag{6.5.3}$$

$$\sum_{j\ne i} |(Cf'([x]^{i,+}))_{ij}| r_j \le |g_i(\check{x} + r_i e^{(i)})| \tag{6.5.4}$$

$$\sum_{j\ne i} |(Cf'([x]^{i,-}))_{ij}| r_j \le |g_i(\check{x} - r_i e^{(i)})| \tag{6.5.5}$$

hold for $i = 1, \dots, n$, then f has a zero $x^* \in [x]$. The conditions (6.5.3)–(6.5.5) *are called the Moore–Kioustelidis test.*

Proof. Let $x \in [x]^{i,+}$. Then with some $\xi \in [x]^{i,+}$ and (6.5.4) we get

$$\begin{aligned} g_i(x) &= g_i(\check{x} + r_i e^{(i)}) + g_i'(\xi)(x_1 - \check{x}_1, \dots, x_{i-1} - \check{x}_{i-1}, 0, x_{i+1} - \check{x}_{i+1}, \dots, x_n - \check{x}_n)^T \\ &= g_i(\check{x} + r_i e^{(i)}) + \sum_{j\ne i} g_{ij}'(\xi)(x_j - \check{x}_j) \\ &\ge g_i(\check{x} + r_i e^{(i)}) - \sum_{j\ne i} |g_{ij}'([x]^{i,+})| r_j \ge 0, \quad \text{if } g_i(\check{x} + r_i e^{(i)}) \ge 0 \end{aligned}$$

and similarly

$$g_i(x) \le g_i(\check{x} + r_i e^{(i)}) + \sum_{j\ne i} |g_{ij}'([x]^{i,+})| r_j \le 0, \quad \text{if } g_i(\check{x} + r_i e^{(i)}) \le 0.$$

This shows that $g_i(x)$, $x \in [x]^{i,+}$, has the same sign as $g_i(\check{x} + r_i e^{(i)})$. In particular, it is zero if $g_i(\check{x} + r_i e^{(i)})$ is zero. A similar remark holds for $g(x)$, $x \in [x]^{i,-}$, and $g_i(\check{x} - r_i e^{(i)})$. This ensures $g_i(x) \cdot g_i(y) \le 0$ for all $x \in [x]^{i,+}$, $y \in [x]^{i,-}$, $i = 1, \dots, n$. Thus Miranda's theorem proves the assertion. □

In practical computation one has to use interval arithmetic evaluations even for (6.5.3) in order to be sure about the signs. Thus in (6.5.3) one starts with point intervals, computes with machine interval arithmetic and uses the upper bound for the interval resulting from the left-hand side. Similarly, one uses the upper bound for the left-hand side and the lower bound for the right-hand side in (6.5.4) and (6.5.5).

As a computational simplification one can replace (6.5.4) and (6.5.5) by

$$\sum_{j\neq i} |g'([x])_{ij}| r_j \le \min\{\, |g_i(\check{x} - r_i e^{(i)})|,\ |g_i(\check{x} + r_i e^{(i)})|\,\}. \tag{6.5.6}$$

This saves computational effort at the expense of restriction. It can be decreased even further as the following theorem in Frommer, Lang, Schnurr [108] shows.

Theorem 6.5.2. *Let $f\colon D \subseteq \mathbb{R}^n \to \mathbb{R}^n$, $f \in C^1(D)$, $0 \le r \in \mathbb{R}^n$, $\tilde{x} \in [x] \subseteq D$. Let $f([x])$ and $f'([x])$ exist and $C \in \mathbb{R}^{n\times n}$ be regular. With $g(x) = Cf(x)$ define*

$$[m]_C^{i,\pm} = g_i(\tilde{x}) + \left(Cf'([x])\right)_{i,*}([x]^{i,\pm} - \tilde{x}). \tag{6.5.7}$$

If

$$[m]_C^{i,-} \le 0 \le [m]_C^{i,+}, \quad i = 1, \dots, n, \tag{6.5.8}$$

then f has a zero $x^* \in [x]$. We call (6.5.8) the Miranda test.

The proof of this theorem is based on Miranda's theorem and on the mean value theorem applied to g_i, $x \in [x]^{i,\pm}$, and $\tilde{x}$, similarly to the proof of Theorem 6.5.1. In contrast to $\tilde{x} = \check{x}$ in Theorem 6.5.1, $\tilde{x} \in [x]$ is now arbitrary, and only one evaluation of g at $\tilde{x}$ is required whereas in (6.5.3)–(6.5.5) each component g_i is evaluated twice, at the midpoints of the faces $[x]^{i,\pm}$.

Notice that for some or all of the indices $i \in \{1, \dots, n\}$ the inequality (6.5.8) can be replaced by

$$[m]_C^{i,+} \le 0 \le [m]_C^{i,-}. \tag{6.5.9}$$

In order to enforce (6.5.8) in this case just multiply C by an appropriate signature matrix D. It can be easily seen that $[m]_C^{i,\pm}$ in (6.5.7) and (6.5.8) may be replaced by

$$[m]_N^{i,\pm} = g_i([x]^{i,\pm}) \tag{6.5.10}$$

or by

$$[m]_F^{i,\pm} = g_i(\tilde{x}^{i,\pm}) + \left(Cf'([x]^{i,\pm})\right)_{i,*}([x]^{i,\pm} - \tilde{x}^{i,\pm}) \quad \text{with } \tilde{x}^{i,\pm} \in [x]^{i,\pm}, \tag{6.5.11}$$

where the subscripts N, C, F indicate 'naive', 'centered', and 'face-centered' techniques, respectively, as introduced in Frommer, Lang [107]. The choice (6.5.11) together with the test (6.5.8) with $[m]_F^{i,\pm}$ instead of $[m]_C^{i,\pm}$ is a generalization of (6.5.3)–(6.5.5), where $\tilde{x}^{i,\pm}$ are the orthogonal projections of a single vector $\tilde{x} \in [x]$ onto the faces $[x]^{i,\pm}$. For this case we get the following result.

Theorem 6.5.3. *If (6.5.8) holds for $[m]_C^{i,\pm}$, then it also holds for $[m]_F^{i,\pm}$ if $\tilde{x}^{i,\pm}$ are chosen as orthogonal projections of the vector $\tilde{x} \in [x]$ onto the faces $[x]^{i,\pm}$.*

Proof. Since $[x]_i^{i,\pm} - \tilde{x}_i^{i,\pm} = 0$, $[x]_j^{i,\pm} - \tilde{x}_j^{i,\pm} = [x]_j^{i,\pm} - \tilde{x}_j$ for $j \neq i$, and $\tilde{x}^{i,\pm} - \tilde{x} = (0, \dots, 0, \tilde{x}_i^{i,\pm} - \tilde{x}_i, 0, \dots, 0)^T$, we get

$$\begin{aligned} g_i(\tilde{x}^{i,\pm}) &= g_i(\tilde{x}) + \left(CJ(\tilde{x}^{i,\pm}, \tilde{x})\right)_{i,*}(\tilde{x}^{i,\pm} - \tilde{x}) \\ &= g_i(\tilde{x}) + \left(CJ(\tilde{x}^{i,\pm}, \tilde{x})\right)_{ii}(\tilde{x}_i^{i,\pm} - \tilde{x}_i) \\ &\in g_i(\tilde{x}) + \left(Cf'([x])\right)_{ii}([x]_i^{i,\pm} - \tilde{x}_i) \end{aligned}$$

and

$$\begin{aligned}
[m]_F^{i,\pm} &= g_i(\tilde{x}^{i,\pm}) + \textstyle\sum_{j\neq i}(Cf'([x]^{i,\pm}))_{ij}([x]_j^{i,\pm} - \tilde{x}_j^{i,\pm}) \\
&\subseteq g_i(\tilde{x}) + (Cf'([x]))_{ii}([x]_i^{i,\pm} - \tilde{x}_i) + \textstyle\sum_{j\neq i}(Cf'([x]^{i,\pm}))_{ij}([x]_j^{i,\pm} - \tilde{x}_j^{i,\pm}) \\
&\subseteq g_i(\tilde{x}) + (Cf'([x]))_{i,*}([x]^{i,\pm} - \tilde{x}) = [m]_C^{i,\pm}.
\end{aligned}$$

□

The equivalence of the Miranda test (6.5.8), (6.5.9) for $[m]_F^{i,\pm}$ and the Moore–Kioustelidis test (6.5.3)–(6.5.5) is stated in the subsequent theorem.

Theorem 6.5.4. *Let the Miranda test consist of* (6.5.8) *or* (6.5.9) *depending on the index i. Then this test with the face-centered variant $[m]_F^{i,\pm}$ is equivalent to the Moore–Kioustelidis test* (6.5.3)–(6.5.5) *if $\tilde{x}^{i,\pm}$ in $[m]_F^{i,\pm}$ is chosen as orthogonal projection of the midpoint vector $\check{x}$ onto the faces $[x]^{i,\pm}$, i.e., as the midpoint $\check{x}^{i,\pm}$ of $[x]^{i,\pm}$.*

Proof. With $r = \mathrm{rad}([x])$,

$$[x]_j^{i,\pm} = \begin{cases} [x]_j, & \text{if } j \neq i \\ \check{x}_i \pm r_i e^{(i)}, & \text{if } j = i \end{cases}, \qquad \check{x}_j^{i,\pm} = \begin{cases} \check{x}_j, & \text{if } j \neq i \\ \check{x}_i \pm r_i e^{(i)}, & \text{if } j = i \end{cases}$$

we get

$$\begin{aligned}
[m]_F^{i,-} \le 0 \le [m]_F^{i,+} \Leftrightarrow\ & g_i(\check{x}^{i,-}) + \sum_{j\neq i}(Cf'([x]^{i,-}))_{ij}([x]_j^{i,-} - \check{x}_j^{i,-}) \\
=\ & g_i(\check{x} - r_i e^{(i)}) + \sum_{j\neq i}|Cf'([x]^{i,-})|_{ij}([x]_j - \check{x}_j) \le 0 \\
\le\ & g_i(\check{x}^{i,+}) + \sum_{j\neq i}(Cf'([x]^{i,+}))_{ij}([x]_j^{i,+} - \check{x}_j^{i,+}) \\
\Leftrightarrow\ & \text{(6.5.3)–(6.5.5) holds with} \\
& g_i(\check{x} - r_i e^{(i)}) \le 0 \le g_i(\check{x} + r_i e^{(i)}).
\end{aligned} \tag{6.5.12}$$

Here we used (6.5.12) and the fact that $[x]_j^{i,\pm} - \check{x}_j^{i,\pm}$ and $[x]_j - \check{x}_j$ are zero-symmetric intervals.

Similarly, we get the equivalence of $[m]_F^{i,+} \le 0 \le [m]_F^{i,-}$ and (6.5.3)–(6.5.5) with $g_i(\check{x} + r_i e^{(i)}) \le 0 \le g_i(\check{x} - r_i e^{(i)})$. □

As an additional refinement of (6.5.11) and in order to decrease the overestimation of the ranges $R_g([x]^{i,\pm})$, one can partition $[x]^{i,\pm}$ into subfaces $[x]^{i,\pm} = \bigcup_k [x]^{i,\pm,k}$ such that these subfaces are disjoint except for their relative boundaries. Opposite faces $[x]^{i,\pm}$ can be partitioned independently of each other. The expressions $[m]_C^{i,\pm}$ in (6.5.8) are then replaced successively by $[m]_F^{i,\pm,k}$ with

$$[m]_F^{i,\pm,k} = g_i(\tilde{x}^{i,\pm,k}) + (Cf'([x]^{i,\pm,k}))_{i,*}([x]^{i,\pm,k} - \tilde{x}^{i,\pm,k}),$$

where $\tilde{x}^{i,\pm,k} \in [x]^{i,\pm,k}$.

Next we prepare a test which is based on Theorem 1.5.10 of Newton–Kantorovich with $\|\cdot\| = \|\cdot\|_\infty$. A difficulty consists in proving the assumptions of this theorem by means of a computer; cf. the final part of Section 6.3. A bound $\beta > 0$ for $f'(\tilde{x})^{-1}$ and fixed $\tilde{x}$ can be found by enclosing the n solutions of the linear systems $f'(\tilde{x})z = e^{(i)}$, $i = 1, \ldots, n$, using interval analysis. If the resulting enclosure of $f'(\tilde{x})^{-1}$ is denoted by $[V]$, then the constant $\beta = \|\,|[V]|\,\|_\infty$ is certainly a bound as required in (ii) of Theorem 1.5.10. Similarly, a bound $\eta > 0$ of $\|f'(\tilde{x})^{-1}f(\tilde{x})\|_\infty$ can be derived enclosing the solution y^* of $f'(\tilde{x})y = f(\tilde{x})$. Remember that η is a bound for $\|x' - \tilde{x}\|_\infty$, where x' is the first Newton iterate when starting with $\tilde{x}$. In order to construct a Lipschitz constant $\gamma > 0$ for $f'(x)$ in D we remark first that the constant $\underline{t}$ in Theorem 1.5.10 satisfies

$$\underline{t} = \frac{1 - \sqrt{1 - 2\alpha}}{\beta\gamma} = \frac{2\eta}{1 + \sqrt{1 - 2\alpha}} < 2\eta \quad \text{if } \alpha < 1/2.$$

If $[x] = \tilde{x} + [-2\eta, 2\eta]e \subset D$, then $\overline{B}(\tilde{x}, \underline{t}) \subset \operatorname{int}([x]) \subset D$ is guaranteed as required in (v) of Theorem 1.5.10. Since usually we do not know a Lipschitz constant of f' on D, we will be content with one on $[x]$ and replace D by $\operatorname{int}([x])$ when applying the theorem of Newton–Kantorovich. On $[x]$ the Lipschitz constant γ can be obtained from

$$\begin{aligned}
\|f'(x) - f'(y)\|_\infty &= \max_{1\le i\le n} \sum_{j=1}^{n} |f'_{ij}(x) - f'_{ij}(y)| \\
&= \max_{1\le i\le n} \sum_{j=1}^{n} \sum_{s=1}^{n} \left| \frac{\partial^2}{\partial x_s \partial x_j} f_i(\xi_{js})(x_s - y_s) \right| \\
&\le \max_{1\le i\le n} \sum_{j=1}^{n} \sum_{s=1}^{n} \left| \frac{\partial^2}{\partial x_s \partial x_j} f_i([x]) \right| \cdot \|x - y\|_\infty \\
&= \gamma \|x - y\|_\infty
\end{aligned}$$

with $\xi_{js} \in [x]$ and

$$\gamma = \max_{1\le i\le n} \sum_{j=1}^{n} \sum_{s=1}^{n} \left| \frac{\partial^2}{\partial x_s \partial x_j} f_i([x]) \right|.$$

If

$$\alpha = \beta\gamma\eta \le 1/2 \tag{6.5.13}$$

holds with this γ, then f has a zero in $\tilde{x} + \underline{t}[-1, 1]e$ which is unique there, and in the case $\alpha < 1/2$ even in the larger set $(\tilde{x} + [-\overline{t}, \overline{t}]e) \cap [x] = \tilde{x} + \min\{\overline{t}, 2\eta\}[-1, 1]e$, where $\overline{t} = (1 + \sqrt{1 - 2\alpha})/(\beta\gamma)$ as in Theorem 1.5.10. We call the computation of α, β, γ and the test (6.5.13) the Newton–Kantorovich test.

When verifying this test on a computer we must take care in computing appropriate bounds of the constants being involved. For instance, one computes the expressions for α and $\underline{t}$ using interval arithmetic and chooses the resulting upper bound. Similarly, one proceeds with $\overline{t}$ and chooses the resulting lower bound. For details and improvements see Rall [279].

Based on Theorem 6.3.3 it is interesting to compare the Newton–Kantorovich test with the Krawczyk test in Section 6.3. As is indicated in Example 6.3.4, this latter test seems to be less powerful than the former. If, however, $f'([x])$ is replaced by the slope matrix

$$[f']([x], \check{x}) = \int_0^1 \square\{f'(\check{x} + t(y - \check{x})) \mid y \in [x]\}\, dt \subseteq f'([x]), \quad \check{x} \in [x], \tag{6.5.14}$$

for f, then Shen and Neumaier showed in [343] that the Newton–Kantorovich test implies the Krawczyk test if the remaining assumptions of Theorem 6.3.3 are kept with the exception of (6.3.12), which is replaced by the weaker inequality (6.5.13). The proof follows essentially the lines of that for Theorem 6.3.3 and is left to the reader as Exercise 6.5.2.

Applied to the Example 6.3.4 the Krawczyk operator with (6.5.14) yields

$$\begin{aligned} K([x]^0, \check{x}^0, C) &= K([\varepsilon, 2 - \varepsilon], 1, 1/2) \\ &= \frac{1}{2} + \frac{\varepsilon}{2} + \left|1 - \int_0^1 (1 + t[-1, 1](1 - \varepsilon))\, dt\right| (1 - \varepsilon)\,[-1, 1] \\ &= \frac{1}{2} + \frac{\varepsilon}{2} + \frac{(1 - \varepsilon)^2}{2}[-1, 1] \subseteq [x]^0 = [\varepsilon, 2 - \varepsilon] \end{aligned}$$

if and only if $0 \le \varepsilon \le 1$, i.e., $0 \le \alpha \le 1/2$ provided that one restricts ε to be positive. Thus the Newton–Kantorovich test and the Krawczyk test are equivalent here, in contrast to the following example of Shen, Neumaier [343].

Example 6.5.5. Let $f(x) = x^3 - \varepsilon$, $0 \le \varepsilon \le 1$, $\check{x}^0 = 1$, $C = (f'(\check{x}^0))^{-1} = 1/3$, $\beta = |C| = 1/3$, $\eta = (f'(\check{x}^0))^{-1} f(\check{x}^0) = (1 - \varepsilon)/3 \ge 0$, $[x]^0 = [\varepsilon, 2 - \varepsilon]$. Since $|f'(x) - f'(y)| \le 3|x + y| \cdot |x - y| \le 3 \max\{2|2 - \varepsilon|, 2|\varepsilon|\}|x - y| = 6(2 - \varepsilon)|x - y|$, $x, y \in [x]^0$ we can choose $\gamma = 6(2 - \varepsilon)$. Then $\alpha = \beta\gamma\eta = 2(2 - \varepsilon)(1 - \varepsilon)/3 \le 1/2$ holds if and only if $1/2 \le \varepsilon \le \min\{5/2, 1\} = 1$, i.e., the Newton–Kantorovich test is fulfilled if and only if $1/2 \le \varepsilon \le 1$. The Krawczyk operator applied with (6.5.14) yields

$$\begin{aligned} K([x]^0, \check{x}^0, C) &= K([\varepsilon, 2 - \varepsilon], 1, 1/3) \\ &= \frac{\varepsilon + 2}{3} + \left|1 - \int_0^1 (1 + t[-1, 1](1 - \varepsilon))^2\, dt\right| (1 - \varepsilon)\,[-1, 1] \\ &= \frac{\varepsilon + 2}{3} + \left((1 - \varepsilon)^2 + \frac{(1 - \varepsilon)^3}{3}\right)[-1, 1] \subseteq [x]^0 = [\varepsilon, 2 - \varepsilon] \end{aligned}$$

if and only if $(5 - \sqrt{17})/2 \le \varepsilon \le 1$. Therefore, the Krawczyk test holds with the modification (6.5.14) also for $(5 - \sqrt{17})/2 = 0.438447\ldots \le \varepsilon < 1/2$ while the Newton–Kantorovich test does not. □

Now we prove a result due to Rall [279] which shows that for $f'([x])$ the Newton–Kantorovich test implies the Krawczyk test (in its original form with $f'([x])$) at least after having carried out some steps of the classical Newton method.

Theorem 6.5.6. *Let the assumptions of Theorem 6.3.3 hold with the Lipschitz constant γ for $f'([x])$ on $\mathbb{I}([x]^0)$, $\tilde{x}^0 = \check{x}^0$, $\beta_0 \ge \|f'(\check{x}^0)^{-1}\|_\infty$, $\eta_0 \ge \|f'(\check{x}^0)^{-1}f(\check{x}^0)\|_\infty$,*

$$0 \le \alpha_0 = \beta_0\gamma\eta_0 < \frac{1}{2}. \tag{6.5.15}$$

Denote by (x^k) the Newton sequence starting with $x^0 = \check{x}^0$ and define $\eta_k, \beta_k, \alpha_k$ recursively by

$$\eta_{k+1} = \frac{1}{2}\frac{\alpha_k\eta_k}{1-\alpha_k}, \quad \beta_{k+1} = \frac{\beta_k}{1-\alpha_k}, \quad \alpha_{k+1} = \frac{1}{2}\left(\frac{\alpha_k}{1-\alpha_k}\right)^2, \quad k = 0, 1, \ldots .$$

Then

$$2^{-(k+1)}\eta_0 \ \ge\ \eta_{k+1} \ge \|x^{k+2} - x^{k+1}\|_\infty, \tag{6.5.16}$$

$$\beta_{k+1} \ge \|f'(x^{k+1})^{-1}\|_\infty, \tag{6.5.17}$$

$$0 \le \alpha_{k+1} = \beta_{k+1}\gamma\eta_{k+1} \ < \ 1/2. \tag{6.5.18}$$

In addition, there is a positive integer k_0 such that for $[x]^{k_0} = x^{k_0} + 2\eta_{k_0}[-e, e]$, $C_{k_0} = (f'(x^{k_0}))^{-1}$ one has

$$K([x]^{k_0}, x^{k_0}, C_{k_0}) \subseteq [x]^{k_0}.$$

Proof. We show by induction that (6.5.16)–(6.5.18) holds. For $k = -1$ these inequalities are valid by assumption. Let them hold for $k - 1$. Then

$$\begin{aligned}\|f'(x^{k+1})^{-1}\|_\infty &\le \|f'(x^{k+1})^{-1} - f'(x^k)^{-1}\|_\infty + \|f'(x^k)^{-1}\|_\infty \\ &\le \|f'(x^{k+1})^{-1}\|_\infty \|f'(x^k) - f'(x^{k+1})\|_\infty \|f'(x^k)^{-1}\|_\infty + \beta_k \\ &\le \|f'(x^{k+1})^{-1}\|_\infty \gamma\eta_k\beta_k + \beta_k = \|f'(x^{k+1})^{-1}\|_\infty \alpha_k + \beta_k,\end{aligned}$$

hence $\|(f'(x^{k+1}))^{-1}\|_\infty(1 - \alpha_k) \le \beta_k$ is valid and (6.5.17) follows from the definition of β_{k+1}.

Similarly, we have

$$\begin{aligned}\|x^{k+2} - x^{k+1}\|_\infty &= \|f'(x^{k+1})^{-1}f(x^{k+1})\|_\infty \\ &\le \beta_{k+1}\,\|f(x^k) + J(x^{k+1}, x^k)(x^{k+1} - x^k)\|_\infty.\end{aligned} \tag{6.5.19}$$

From $x^{k+1} = x^k - f'(x^k)^{-1}f(x^k)$ we get

$$f(x^k) = -f'(x^k)(x^{k+1} - x^k) = -\int_0^1 f'(x^k)\,dt\,(x^{k+1} - x^k).$$

Together with (6.5.17) and (6.5.19) this yields

$$\begin{aligned}
\|x^{k+2} - x^{k+1}\|_\infty &\le \beta_{k+1}\eta_k \int_0^1 \|-f'(x^k) + f'(x^k + t(x^{k+1} - x^k))\|_\infty \, dt \\
&\le \beta_{k+1}\eta_k^2\gamma \int_0^1 t\,dt = \frac{1}{2}\frac{\beta_k}{1-\alpha_k}\eta_k^2\gamma \\
&= \frac{1}{2}\frac{\alpha_k}{1-\alpha_k}\eta_k = \eta_{k+1} \le \frac{1}{2}\left(-1 + \frac{1}{1-\alpha_k}\right)2^{-k}\eta_0 \\
&\le \left(-1 + \frac{1}{1-1/2}\right)2^{-(k+1)}\eta_0 = 2^{-(k+1)}\eta_0.
\end{aligned} \tag{6.5.20}$$

This proves (6.5.16).

By the definition of α_{k+1}, β_{k+1}, and η_{k+1} and by the induction hypothesis for α_k we obtain

$$\alpha_{k+1} = \frac{1}{2}\left(\frac{\alpha_k}{1-\alpha_k}\right)^2 = \frac{1}{2}\frac{\alpha_k}{(1-\alpha_k)^2}\beta_k\gamma\eta_k = \eta_{k+1}\beta_{k+1}\gamma$$

as required in (6.5.18). Notice that $0 \le \alpha_k < 1/2$ implies $0 \le \alpha_k/(1-\alpha_k) < 1$ and therefore $0 \le \alpha_{k+1} < 1/2$.

Next we show that the difference equation

$$\alpha_{k+1} = \frac{1}{2}\left(\frac{\alpha_k}{1-\alpha_k}\right)^2, \quad k = 0, 1, \ldots \tag{6.5.21}$$

is solved by

$$\alpha_k = \frac{2\theta^{2^k}}{(1+\theta^{2^k})^2}, \quad k = 0, 1, \ldots, \tag{6.5.22}$$

where

$$\theta = \frac{1-\sqrt{1-2\alpha_0}}{1+\sqrt{1-2\alpha_0}} = \frac{2\alpha_0}{(1+\sqrt{1-2\alpha_0})^2} < 1$$

if $\alpha_0 \in [0, 1/2)$. We prove (6.5.22) by induction. For $k = 0$ we have

$$\frac{2\theta}{(1+\theta)^2} = \frac{2\,\frac{1-\sqrt{1-2\alpha_0}}{1+\sqrt{1-2\alpha_0}}}{\left(\frac{2}{(1+\sqrt{1-2\alpha_0})}\right)^2} = \alpha_0.$$

With (6.5.21) we get

$$\alpha_{k+1} = \frac{1}{2}\left(\frac{\alpha_k}{1-\alpha_k}\right)^2 = \frac{1}{2}\left(\frac{2\theta^{2^k}}{(1+\theta^{2^k})^2 - 2\theta^{2^k}}\right)^2 = \frac{2\theta^{2^{k+1}}}{(1+\theta^{2^{k+1}})^2}.$$

This proves (6.5.22).

By virtue of $\theta < 1$ we achieve $\alpha_k \le 1/4$ for some $k = k_0$, and Theorem 6.3.3 implies the assertion. □

Notice that the assumptions of Theorem 6.5.6 already allow the application of the theorem of Newton–Kantorovich which guarantees a unique zero x^* in $[x]^0$ and the convergence of the Newton sequence to x^*. The computation of the Krawczyk operator is therefore not necessary in this situation.

Our next test is based on the mapping degree for the function $g(x) = Cf(x)$ with a regular matrix $C \in \mathbb{R}^{n\times n}$. It originates from Frommer, Hoxha, Lang [105] and uses the homotopy

$$h(t, x) = tg(x) + (1 - t)(x - \tilde{x}), \quad 0 \le t \le 1,\ x \in [x],\ \tilde{x} \in \operatorname{int}([x]), \tag{6.5.23}$$

where $[x] \in \mathbb{IR}^n$ is given. If

$$h(t, x) \neq 0 \quad \text{for all } t \in [0, 1] \text{ and all } x \in \partial[x], \tag{6.5.24}$$

then by virtue of Theorem 1.5.3 (b) the mapping degree $\deg(\operatorname{int}([x]), h(t,\cdot), 0)$ does not depend on t, hence

$$\begin{aligned}\deg(\operatorname{int}([x]), g, 0) &= \deg(\operatorname{int}([x]), h(1,\cdot), 0) = \deg(\operatorname{int}([x]), h(0,\cdot), 0)\\ &= \deg(\operatorname{int}([x]), x - \tilde{x}, 0) = \operatorname{sign}(\det(I)) = 1.\end{aligned}$$

This guarantees a zero x^* of g and therefore of f in $[x]$ according to Theorem 1.5.3 (c).
In order to verify (6.5.24) we will use interval arithmetic tools.

Theorem 6.5.7. *Let $f\colon D \subseteq \mathbb{R}^n \to \mathbb{R}^n$, $f \in C(D)$, $[x] \in \mathbb{I}(D)$, $\tilde{x} \in \operatorname{int}([x])$, $C \in \mathbb{R}^{n\times n}$ regular, $g = Cf$. Assume that for some interval function $[g]\colon \mathbb{I}([x]) \to \mathbb{IR}^n$ the enclosure property $R_g([y]) \subseteq [g]([y])$ holds for the range $R_g([y])$ of g over any $[y] \in \mathbb{I}([x])$. Let $[y]^k \subseteq [x]$, $k = 1, \dots, m$, and $[t]^\ell \subseteq [0, 1]$, $\ell = 1, \dots, p$, satisfy*

$$\partial[x] = \bigcup_{k=1}^{m} [y]^k, \quad [0, 1] = \bigcup_{\ell=1}^{p} [t]^\ell. \tag{6.5.25}$$

If for all k and ℓ there exists an index $j = j(k, \ell) \in \{1, \dots, n\}$ such that the mapping degree test

$$0 \notin [t]^\ell \cdot [g]_j([y]^k) + (1 - [t]^\ell) \cdot ([y]_j^k - \tilde{x}_j) \tag{6.5.26}$$

holds, then f has a zero $x^* \in [x]$.

Proof. For all k and ℓ the condition (6.5.26) implies $0 \notin [t]^\ell \cdot [g]([y]^k) + (1 - [t]^\ell) \cdot ([y]^k - \tilde{x})$, whence

$$\begin{aligned} 0 &\notin \bigcup_{k=1}^{m} \bigcup_{\ell=1}^{p} ([t]^\ell \cdot [g]([y]^k) + (1 - [t]^\ell) \cdot ([y]^k - \tilde{x})) && (6.5.27)\\ &\supseteq \{ z \mid z = tg(x) + (1 - t)(x - \tilde{x}),\ t \in [0, 1],\ x \in \partial[x] \}. && (6.5.28)\end{aligned}$$

This guarantees (6.5.24) and proves the theorem. □

As an interval function $[g]([y])$ restricted to $\mathbb{I}([y]^k)$ one can choose for instance the interval arithmetic evaluation $g([y]) = Cf([y])$, the two mean value forms

$$[g]([y]) = g(\tilde{x}) + g'([x])([y] - \tilde{x}) \tag{6.5.29}$$

with $g'([x]) = Cf'([x])$, and

$$[g]([y]) = g(\tilde{y}^k) + g'([y]^k)([y] - \tilde{y}^k) \tag{6.5.30}$$

with $\tilde{y}^k \in [y]^k$, or the slope-based interval function

$$[g]([y]) = g(\tilde{y}^k) + [Y]^k([y] - \tilde{y}^k) \tag{6.5.31}$$

with $\tilde{y}^k \in [y]^k$ and $[Y]^k$ satisfying

$$g(y) - g(\tilde{y}^k) \in [Y]^k(y - \tilde{y}^k) \quad \text{for all } y \in [y]^k.$$

Here, we tacitly assume that f is sufficiently smooth and that $f([y])$, $f'([y])$, and $[Y]^k$ exist. Notice the differences between (6.5.29) and (6.5.30).

The mapping degree test (6.5.26) cannot be successful if $p = 1$, since then $[t]^1 = [0, 1] = 1 - [t]^1$, whence $0 \in [t]^1 \cdot [g]([y]^k) + (1 - [t]^1) \cdot ([y]^k - \tilde{x})$. A first choice for $[y]^k$ could be the faces $[x]^{i,\pm}$, $i = 1, \dots, n$. In this case the following corollary is a generalization of Theorem 6.5.2 (Miranda test) if there the inequalities are sharpened to strict inequalities.

Corollary 6.5.8. *Let the assumptions of Theorem 6.5.7 hold and choose*

$$[y]^{2i-1} = [x]^{i,-}, \quad [y]^{2i} = [x]^{i,+}, \quad i = 1, \dots, n,$$
$$[t]^1 = [0, 0.5], \quad [t]^2 = [0.5, 1]$$

(i.e., $m = 2n$, $p = 2$ in Theorem 6.5.7). Assume that for $i = 1, \dots, n$

$$[g]_i([x]^{i,+}) > 0, \quad [g]_i([x]^{i,-}) < 0 \tag{6.5.32}$$

holds. Then (6.5.26) is fulfilled with $j = j(k, \ell) = i$ for $k \in \{2i - 1, 2i\}$.

Proof. With $[x]_i^{i,-} - \tilde{x}_i < 0 < [x]_i^{i,+} - \tilde{x}_i$ the assumption (6.5.32) implies immediately $[t]^\ell \cdot [g]_i([x]^{i,+}) + (1 - [t]^\ell) \cdot ([x]_i^{i,+} - \tilde{x}_i) > 0$ and $[t]^\ell \cdot [g]_i([x]^{i,-}) + (1 - [t]^\ell) \cdot ([x]_i^{i,-} - \tilde{x}_i) < 0$ for $\ell = 1, 2$, i.e., (6.5.26) holds. □

The Moore–Kioustelidis test as well as the Miranda test can only be successful if g_i does not change its sign on $[x]^{i,+}$ and has opposite sign on $[x]^{i,-}$. The mapping degree test is more flexible as the following example from Frommer, Hoxha, Lang [105] and Frommer, Lang [107] shows.

Example 6.5.9. Define $f: \mathbb{R}^2 \to \mathbb{R}^2$ by

$$f(u, v) = \begin{pmatrix} 4 - 2(u - 1)^2 \\ (2 - (u + 1)^2)(2 - (v - 1)^2) \end{pmatrix}.$$

Then $x^* = (u^*, v^*)^T = (1 - \sqrt{2}, 1 - \sqrt{2})^T$ is a zero of f.

(a) Choose $[x] = ([-1,1],[-1,1])^T \in \mathbb{IR}^2$. In this interval vector x^* is the unique zero of f. We show that any test based on Miranda's theorem for $[x]$ must fail since $g_i(x) = (Cf(x))_i$ cannot have constant sign on $[x]^{i,+}$ or $[x]^{i,-}$ for any regular matrix

$$C = \begin{pmatrix} \alpha & \beta \\ \gamma & \delta \end{pmatrix}.$$

To this end we assume w.l.o.g. that $g_i(1,1) \in \{0,4\}$ holds for $i = 1,2$. Otherwise replace g by Dg with an appropriate regular diagonal matrix. Then $g_i(x)g_i(y) < 0$ implies $d_{ii}g_i(x) \cdot d_{ii}g_i(y) < 0$ and vice versa for $x \in [x]^{i,+}$, $y \in [x]^{i,-}$. If $g(1,1) = Cf(1,1) = C(4,-4)^T = 0$, then C is singular which is excluded. If $g(1,1) = C(4,-4)^T = (4,4)^T$, then $\beta = \alpha - 1$, $\delta = \gamma - 1$, hence $g(-1,1) = C(-4,4)^T = (-4,-4)^T$. Therefore, g_1 and g_2 change sign on $[x]^{2,+}$. If $g(1,1) = (4,0)^T$, then $\beta = \alpha - 1$, $\delta = \gamma$, and $g(-1,1) = (-4,0)^T$, $g(1,-1) = C(4,4)^T = (8\alpha - 4, 8\gamma)^T$, $g(-1,-1) = C(-4,-4)^T = (-8\alpha + 4, -8\gamma)^T$. Hence g_1 changes sign on $[x]^{2,+}$. In addition, since C is regular, we have $\gamma \neq 0$, hence $g_2(x)$ changes sign on $[x]^{2,-}$. If $g(1,1) = (0,4)$, then $\beta = \alpha$, $\delta = \gamma - 1$, $g(-1,1) = (0,-4)^T$, $g(1,-1) = (8\alpha, 8\gamma - 4)^T$, $g(-1,-1) = (-8\alpha, -8\gamma + 4)^T$. Hence g_2 changes sign on $[x]^{2,+}$ and g_1 on $[x]^{2,-}$. Therefore, the Miranda test cannot be successful.

(b) Now we choose $[x] = x^* + \alpha([-1,1],[-1,1])^T \in \mathbb{IR}^2$, $1 < \alpha < 2\sqrt{2}$, and $C = I$. Then $g = Cf = f$ satisfies

$$f_1(x_1^* \pm \alpha, \upsilon) = 4 - 2(-\sqrt{2} \pm \alpha)^2 = -2\alpha(\alpha \mp 2\sqrt{2}) \begin{cases} > 0 & \text{on } [x]^{1,+} \\ < 0 & \text{on } [x]^{1,-} \end{cases},$$

as required by the Miranda theorem. Similarly, for $f_2(u, x_2^* \pm \alpha)$ we have $f_2(u, x_2^* \pm \alpha) = h(u)(-\alpha)(\alpha \mp 2\sqrt{2})$ with $h(u) = 2 - (u+1)^2$, $u \in x_1^* + \alpha[-1,1]$. Since $f_2(u, x_2^* \pm \alpha)/h(u)$ differs from $f_1(x_1^* \pm \alpha, \upsilon)$ only by a factor 2 and since $h(x_1^*) = -4 + 4\sqrt{2} > 0$, $h(x_1^* + 1) = -9 + 6\sqrt{2} < 0$, the function f_2 changes sign on $[x]^{2,-}$ as well as on $[x]^{2,+}$. Thus the Miranda test cannot be successful. Numerical experiments in Frommer, Hoxha, Lang [105] (Fig. 2) show however that (6.5.26) is fulfilled if $\alpha = 2^{(q/8)-1}$, $q \in \{9, 10, \ldots, 14\}$, if $[t]^\ell = [(\ell-1)/p, \ell/p]$, $\ell = 1, \ldots, p$, $p \in \{2, 10\}$, and if $[y]^k$ is chosen according to the following strategy: One starts with the faces $[x]^{i,\pm}$ and carries on a bisection process along the actual longest edge until the test is positive or some maximal number of boxes in the bisection process is reached. For a positive mapping degree test between 4 and 16 subfaces per face are needed at most. □

Our last test stems from Frommer and Lang [107] and is based on Borsuk's Theorem 1.5.4 applied to $g(x) = Cf(x)$, C regular, and $[x] = \check{x} + \mathrm{rad}([x])[-1,1]$. Obviously, $\check{x} + \delta \in \partial[x]$ if and only if $\check{x} - \delta \in \partial[x]$. Therefore, the topological prerequisite of Borsuk's theorem is certainly fulfilled. In order to verify the computational one, i.e.,

$$g(\check{x} + \delta) \neq \lambda g(\check{x} - \delta) \quad \text{for all } \lambda > 0 \text{ and all } \check{x} + \delta \in \partial[x], \tag{6.5.33}$$

we are going to derive three possibilities:

(1) From the Cauchy–Schwarz inequality we know that for $g(\check{x} \pm \delta) \neq 0$ we have

$$s(\check{x}, \delta) = \frac{g(\check{x} + \delta)^T g(\check{x} - \delta)}{\|g(\check{x} + \delta)\|_2 \, \|g(\check{x} - \delta)\|_2} \in [-1, 1] \tag{6.5.34}$$

with $s(\tilde{x}, \delta) = 1$ if and only if

$$g(\check{x} + \delta) = \lambda g(\check{x} - \delta) \quad \text{for some } \lambda > 0. \tag{6.5.35}$$

Thus, for the range of s over $[x]^{i,+}$ the condition

$$\sup R_s([x]^{i,+}) < 1, \quad i = 1, \dots, n, \tag{6.5.36}$$

is equivalent to (6.5.33). It is our first criterion. The condition (6.5.36) is certainly fulfilled if we can find intervals $[b]_S^i$, $i = 1, \dots, n$, which satisfy $[b]_S^i \supseteq R_s([x]^{i,+})$ and $\overline{b}_S^i < 1$. An example could be

$$[b]_S^i = \frac{g([x]^{i,+})^T \cdot g([x]^{i,-})}{\sqrt{\sum_{j=1}^n (g_j([x]^{i,+}))^2} \cdot \sqrt{\sum_{j=1}^n (g_j([x]^{i,-}))^2}}.$$

It is obvious that this interval expression lets us expect a tremendous overestimation of $R_S([x]^{i,+})$ due to the dependency of the variables in (6.5.34). Therefore, a higher order form and/or decompositions of $[x]^{i,\pm}$ should be used to enclose $g([x]^{i,\pm})$ more tightly; cf. Theorem 4.1.22.

(2) The second possibility to apply Borsuk's theorem starts with

$$g_j(\check{x} + \delta) = \lambda g_j(\check{x} - \delta), \quad \check{x} + \delta \in \partial[x], \ \lambda > 0, \tag{6.5.37}$$

which we do *not* want to be fulfilled. From (6.5.37) we obtain

$$\lambda = \frac{g_j(\check{x} + \delta)}{g_j(\check{x} - \delta)} > 0,$$

if the denominator differs from zero. Therefore, (6.5.33) holds if

$$\bigcap_{j=1}^n \frac{R_{g_j}([x]^{i,+})}{R_{g_j}([x]^{i,-})} \cap (0, \infty) = \emptyset \quad \text{for } i = 1, \dots, n, \tag{6.5.38}$$

which is our second criterion. Notice that the ranges in the numerator and the denominator are intervals. If $0 \in R_{g_j}([x]^{i,-})$, we have to interpret the quotient along the lines of Section 6.1, cf. particularly Remark 6.1.2 (b). As for the Miranda test, we use three techniques to enclose the intersection

$$\bigcap_{j=1}^n \frac{R_{g_j}([x]^{i,+})}{R_{g_j}([x]^{i,-})}$$

in (6.5.38), a naive, a centered and a face-centered one, which lead to the intervals

$$[b]_N^i := \bigcap_{j=1}^n \frac{g_j([x]^{i,+})}{g_j([x]^{i,-})},$$

$$[b]_C^i := \bigcap_{j=1}^n \frac{g_j(\tilde{x}) + (Cf'([x]))_{j,*}([x]^{i,+} - \tilde{x})}{g_j(\tilde{x}) + (Cf'([x]))_{j,*}([x]^{i,-} - \tilde{x})},$$

$$[b]_F^i := \bigcap_{j=1}^n \frac{g_j(\tilde{x}^{i,+}) + (Cf'([x]^{i,+}))_{j,*}([x]^{i,+} - \tilde{x}^{i,+})}{g_j(\tilde{x}^{i,-}) + (Cf'([x]^{i,-}))_{j,*}([x]^{i,-} - \tilde{x}^{i,-})}.$$

(3) The third possibility for applying Borsuk's theorem starts again with (6.5.37) and evaluates g_j around $\check{x}$. Then one gets

$$g_j(\check{x}) + (CJ(\check{x} + \delta, \check{x}))_{j,*}\delta = \lambda\left(g_j(\check{x}) - (CJ(\check{x} - \delta, \check{x}))_{j,*}\delta\right),$$

whence

$$\begin{aligned}(1-\lambda)g_j(\check{x}) &= -\left(C(\lambda J(\check{x} - \delta, \check{x}) + J(\check{x} + \delta, \check{x}))\right)_{j,*}\delta \\ &\in -\left(C(\lambda f'([x]) + f'([x]))\right)_{j,*}([x]^{i,+} - \check{x}) \\ &= -(\lambda + 1)(Cf'([x]))_{j,*}([x]^{i,+} - \check{x}).\end{aligned}$$

In the last equality we exploited $\lambda > 0$. This yields

$$\frac{\lambda - 1}{\lambda + 1} g_j(\check{x}) \in (Cf'([x]))_{j,*}([x]^{i,+} - \check{x})$$

and finally

$$\frac{\lambda - 1}{\lambda + 1} \in \frac{(Cf'([x]))_{j,*}([x]^{i,+} - \check{x})}{g_j(\check{x})}$$

with the convention

$$\frac{[a]}{0} = \begin{cases} \mathbb{R} & \text{if } 0 \in [a], \\ \emptyset & \text{otherwise.} \end{cases}$$

Since $\lambda > 0$ implies $(\lambda - 1)/(\lambda + 1) \in (-1, 1)$ and since we do *not* want λ to be positive we arrive at

$$[b]_M^i \cap (-1, 1) = \emptyset, \quad i = 1, \ldots, n, \tag{6.5.39}$$

with

$$[b]_M^i := \bigcap_{j=1}^n \frac{(Cf'([x]))_{j,*}([x]^{i,+} - \check{x})}{g_j(\check{x})}. \tag{6.5.40}$$

Summarizing the computational tests (6.5.36), (6.5.38), and (6.5.39) leads to the following terminology.

Definition 6.5.10. We say that the Borsuk test is successful if for each $i \in \{1, \dots, n\}$ at least one of the following conditions holds:

$$[b]_M^i \cap (-1, 1) = \emptyset, \tag{6.5.41}$$

$$[b]_N^i \cap (0, \infty) = \emptyset, \tag{6.5.42}$$

$$[b]_C^i \cap (0, \infty) = \emptyset, \tag{6.5.43}$$

$$[b]_F^i \cap (0, \infty) = \emptyset, \tag{6.5.44}$$

$$\sup [b]_S^i < 1. \tag{6.5.45}$$

Notice that we do not require the same condition to be fulfilled for all i simultaneously.

Similar to Miranda's test, we can replace $[b]_F^i$ by using subfaces. But this time the subfaces $[x]^{i,+,k}$, $[x]^{i,-,k}$ cannot be chosen independently of each other, but must be symmetric with respect to the midpoint $\check{x}$ of the box $[x]$; cf. Frommer, Lang [106] for details. While the preconditioned matrix C cannot be changed during the Miranda test, we have more flexibility in the Borsuk test. In Frommer, Lang [106] the choice $C = C^{i,k}$ was allowed to depend on each pair of opposite subfaces $[x]^{i,\pm,k}$ such that the function $g = C^{i,k} \cdot f$ points in almost opposite directions on $[x]^{i,+,k}$ and $[x]^{i,-,k}$. This is approximately achieved if

$$g^+ = C^{i,k} \cdot f^+ = (100, 1, 0, \dots, 0)^T \quad \text{and}$$
$$g^- = C^{i,k} \cdot f^- = (-100, 1, 0, \dots, 0)^T,$$

where $f^\pm = f(\text{mid}([x]^{i,\pm,k}))$. Unless already $f^+ = -f^-$ we cannot obtain $g^+ = -g^-$ exactly with a regular matrix $C^{i,k}$. In order to construct $C^{i,k}$, one first computes the QR decompositions $(f^+, f^-) = Q_f \cdot R_f$ and $(g^+, g^-) = Q_g \cdot R_g$ and defines (in the usual MATLAB notation)

$$C^{i,k} = (g^+, g^-, Q_g(:, 3{:}n)) \cdot (f^+, f^-, Q_f(:, 3{:}n))^{-1}$$

which implies $C^{i,k} f^+ = (g^+, g^-, Q_g(:, 3{:}n))e^{(1)} = g^+$, and similarly, $C^{i,k} f^- = g^-$. If this matrix is very ill-conditioned, choose $C^{i,k} = I \in \mathbb{R}^{n\times n}$. Frommer and Lang report in [106] on very good results if such varying preconditioners are applied to Example 6.5.9.

We conclude this section with some comparisons of our tests. In the Sections 6.3 and 6.4 we have already remarked that the Hansen test is more general than the Moore–Qi test and the latter is more general than the Krawczyk test. By virtue of Theorem 6.5.3 we have seen that if the Miranda test for $[m]_C^{i,\pm}$ is successful, it also holds for $[m]_F^{i,\pm}$. In Corollary 6.5.8 we have shown that under mild restrictions a positive Miranda test (6.5.8) guarantees that the mapping degree test is fulfilled, if applied appropriately. Now we prove that a positive Krawczyk test implies a positive Miranda test (6.5.8).

Theorem 6.5.11. *Let the notation and the assumptions of the Theorems 6.3.1 and 6.5.1 hold. If the Krawczyk test (6.3.4) is successful, then the centered Miranda test (6.5.8) (with $[m]_C^{i,\pm}$) is satisfied, too.*

If the conditions

$$\sup\{\,(I - Cf'([x]))_{ii}(\overline{x}_i - \tilde{x}_i)\,\} = \sup\{\,(I - Cf'([x]))_{ii}([x]_i - \tilde{x}_i)\,\} \tag{6.5.46}$$

$$\inf\{\,(I - Cf'([x]))_{ii}(\underline{x}_i - \tilde{x}_i)\,\} = \inf\{\,(I - Cf'([x]))_{ii}([x]_i - \tilde{x}_i)\,\} \tag{6.5.47}$$

hold for $i = 1, \ldots, n$, *then both tests are equivalent.*

Proof. Let the Krawczyk test be satisfied. Then by virtue of $[x]_j^{i,-} = [x]_j$ for $j \neq i$ and $[x]_i^{i,-} = \underline{x}_i$ we have

$$\begin{aligned}
\underline{x}_i &\le \inf K([x], \tilde{x}, C)_i && (6.5.48)\\
&= \inf\Big\{\, \tilde{x}_i - g_i(\tilde{x}) + (I - Cf'([x]))_{ii}([x]_i - \tilde{x}_i) + \sum_{j\neq i}^{n}(I - Cf'([x]))_{ij}([x]_j - \tilde{x}_j)\Big\}\\
&\le \inf\Big\{\, \tilde{x}_i - g_i(\tilde{x}) + (I - Cf'([x]))_{ii}(\underline{x}_i - \tilde{x}_i) + \sum_{j\neq i}^{n}(I - Cf'([x]))_{ij}([x]_j - \tilde{x}_j)\Big\} && (6.5.49)\\
&= \inf\{\, \tilde{x}_i - g_i(\tilde{x}) + (\underline{x}_i - \tilde{x}_i) - Cf'([x])([x]^{i,-} - \tilde{x})\,\}\\
&= \inf\{\, \underline{x}_i - [m]_C^{i,-}\,\},
\end{aligned}$$

whence

$$0 \ge [m]_C^{i,-}. \tag{6.5.50}$$

If (6.5.47) holds, then the inequality in (6.5.49) can be replaced by equality, hence (6.5.48) and (6.5.50) are equivalent. Similarly, $\sup K([x], \tilde{x}, C)_i \le \overline{x}_i$ implies $\inf [m]_C^{i,+} \ge 0$ with equality if (6.5.46) is valid. □

Corollary 6.5.12. *In each of the following three cases the Krawczyk test* (6.3.4) *and the centered Miranda test* (6.5.8) *are equivalent.*

(i) C *and* $[x]$ *are such that* $\inf\,(I - Cf'([x]))_{ii} \ge 0$, $i = 1, \ldots, n$.
(ii) $\tilde{x} = \check{x}$ *and* $\operatorname{mid}(I - Cf'([x]))_{ii} \ge 0$, $i = 1, \ldots, n$.
(iii) $\tilde{x} = \check{x}$ *and* $C = (\operatorname{mid}(f'([x])))^{-1}$.

Proof. (i) implies (6.5.46), (6.5.47).

(ii) yields

$$\begin{aligned}
\sup\{\,(I - Cf'([x]))_{ii}([x]_i - \check{x}_i)\,\} &= |I - Cf'([x])|_{ii}\operatorname{rad}([x]_i)\\
&= \sup\{\,(I - Cf'([x]))_{ii}\operatorname{rad}([x]_i)\,\}\\
&= \sup\{\,(I - Cf'([x]))_{ii}(\overline{x}_i - \check{x}_i)\,\}
\end{aligned}$$

and

$$\begin{aligned}
\inf\{\,(I - Cf'([x]))_{ii}([x]_i - \check{x}_i)\,\} &= |I - Cf'([x])|_{ii}(-\operatorname{rad}([x]_i))\\
&= \inf\{\,(I - Cf'([x]))_{ii}(\underline{x}_i - \check{x}_i)\,\}.
\end{aligned}$$

Hence (6.5.46) and (6.5.47) hold.

(iii) follows from (ii) since with $f'([x]) =: \check{f}' + [-r_{f'}, r_{f'}]$ and $C = (\check{f}')^{-1}$ we have $I - Cf'([x]) = I - (I + |C|[-r_{f'}, r_{f'}])$, whence $\operatorname{mid}(I - Cf'([x])) = 0$. □

Next we will show that the modified Hansen–Sengupta test (6.4.19) and the Miranda test (6.5.8) are equivalent if we supplement this test by its second possibility (6.5.9) and if we replace the inequality signs by strict inequality. That means, we now require

$$[m]_C^{i,-} < 0 < [m]_C^{i,+} \quad \text{or} \quad [m]_C^{i,+} < 0 < [m]_C^{i,-}, \quad i = 1, \dots, n. \tag{6.5.51}$$

We call (6.5.51) the strict Miranda test. At first glance (6.5.51) seems more flexible than the modified Hansen–Sengupta test since there

$$0 \notin (Cf'([x]))_{ii}, \quad i = 1, \dots, n, \tag{6.5.52}$$

is necessary in order to guarantee the applicability of the method while such a condition is not needed for the Miranda test. The following lemma shows however, that the strict version of this test will fail if (6.5.52) is not fulfilled.

Lemma 6.5.13. *For $[m]_C^{i,\pm}$ as in (6.5.7) we have the following properties.*

$$\begin{aligned} &\textit{If } [m]_C^{i,-} < 0 < [m]_C^{i,+}, \textit{ then } (Cf'([x]))_{ii} > 0, \\ &\textit{if } [m]_C^{i,+} < 0 < [m]_C^{i,-}, \textit{ then } (Cf'([x]))_{ii} < 0, \end{aligned} \tag{6.5.53}$$

hence $0 \in (Cf'([x]))_{ii}$ for some $i \in \{1, \dots, n\}$ causes the strict Miranda test (6.5.51) to fail.

Proof. Assume that (6.5.53) is false. Then there is some index i_0 and some element $\tilde{d}_{i_0} \in (Cf'([x]))_{i_0 i_0}$ such that $\tilde{d}_{i_0} \leq 0$. Since $\tilde{x}_j \in [x]_j$ we have $0 \in [x]_j - \tilde{x}_j$ and $0 \in \sum_{j \neq i_0} (Cf'([x]))_{i_0 j}([x]_j - \tilde{x}_j)$, whence $(Cf(\tilde{x}))_{i_0} + \tilde{d}_{i_0}(x_{i_0}^{\pm} - \tilde{x}_{i_0}) + 0 \in [m]_C^{i_0,\pm}$ for $x_{i_0}^{+} := \overline{x}_{i_0}$, $x_{i_0}^{-} := \underline{x}_{i_0}$. This implies the contradiction

$$\begin{aligned} (Cf(\tilde{x}))_{i_0} &\leq (Cf(\tilde{x}))_{i_0} + \tilde{d}_{i_0}(x_{i_0}^{-} - \tilde{x}_{i_0}) < 0 \\ &< (Cf(\tilde{x}))_{i_0} + \tilde{d}_{i_0}(x_{i_0}^{+} - \tilde{x}_{i_0}) \leq (Cf(\tilde{x}))_{i_0}. \end{aligned}$$

□

In order to prove the indicated equivalence of the two tests we need another auxiliary result.

Lemma 6.5.14. *Let $[a], [b] \in \mathbb{IR}$, and $t \in \mathbb{R}$. Then*

$$[a] > 0 \textit{ and } 0 < t[a] + [b] \textit{ holds if and only if } [a] > 0 \textit{ and } -[b]/[a] < t, \tag{6.5.54}$$

$$[a] < 0 \textit{ and } 0 > t[a] + [b] \textit{ holds if and only if } [a] < 0 \textit{ and } -[b]/[a] < t. \tag{6.5.55}$$

Proof. Since $\tilde{a} > 0$ and $0 < t\tilde{a} + \tilde{b}$ is equivalent to $\tilde{a} > 0$ and $-\tilde{b}/\tilde{a} < t$, the equivalence (6.5.54) follows by varying $\tilde{a} \in [a]$ and $\tilde{b} \in [b]$. The second equivalence is proved analogously. □

Theorem 6.5.15. *The strict Miranda test (6.5.51) is successful if and only if the modified Hansen–Sengupta test (6.4.19) is successful.*

Proof. By virtue of Lemma 6.5.13 we can assume that $0 \notin (Cf'([x]))_{ii}$ holds for $i = 1, \ldots, n$. As a working hypothesis we suppose in addition that $0 < (Cf'([x]))_{ii}$ holds so that Lemma 6.5.13 implies

$$[m]_C^{i,-} < 0 < [m]_C^{i,+} \tag{6.5.56}$$

as the relevant part of the strict Miranda test in (6.5.51). Define $t^+ = \overline{x}_i - \tilde{x}_i$, $t^- = \underline{x}_i - \tilde{x}_i$,

$$[a] = (Cf'([x]))_{ii}, \quad [b] = (Cf(\tilde{x}))_i + \sum\nolimits_{j \neq i} (Cf'([x]))_{ij}([x]_j - \tilde{x}_j).$$

Obviously, $t^+ = [x]_i^{i,+} - \tilde{x}_i \geq 0$, $t^- = [x]_i^{i,-} - \tilde{x}_i \leq 0$, and $[a] > 0$. Therefore, we have $[m]_C^{i,\pm} = t^\pm [a] + [b]$, and (6.5.56) is equivalent to

$$[m]_C^{i,-} = (-t^-)(-[a]) + [b] < 0 < t^+[a] + [b] = [m]_C^{i,+}.$$

By Lemma 6.5.14 this is equivalent to $-[b]/(-[a]) < -t^-$ and $-[b]/[a] < t^+$ which can be rewritten as

$$\underline{x}_i < \tilde{x}_i - [b]/[a] = H_i^{\mathrm{mod}}([x], \tilde{x}, C),$$
$$\overline{x}_i > \tilde{x}_i - [b]/[a] = H_i^{\mathrm{mod}}([x], \tilde{x}, C),$$

i.e., $H_i^{\mathrm{mod}}([x], \tilde{x}, C) \subseteq [x]_i$. This also holds if our working hypothesis is replaced by $0 > (Cf'([x]))_{ii}$. The inequality signs in (6.5.56) then have to be reversed. The details are left to the reader; cf. also Goldstejn [120]. □

Next we prove that each variant (6.5.42)–(6.5.44) of the Borsuk test holds if the corresponding variant of the Miranda test is satisfied.

Theorem 6.5.16. *Let the Miranda test (6.5.8) or its variant (6.5.9) be fulfilled with $[m]_N^{i,\pm}$ as in (6.5.10) (resp. $[m]_C^{i,\pm}$ as in (6.5.7), resp. $[m]_F^{i,\pm}$ as in (6.5.11)).*

Then the corresponding variant (6.5.42) (resp. (6.5.43), resp. (6.5.44)) of the Borsuk test holds with the same i, $[x]$, $\tilde{x}$, $\tilde{x}^{i,\pm}$, C.

Proof. Assume that for the Miranda test the naive variant (6.5.10) holds. Then $g_i([x]^{i,+})/g_i([x]^{i,-}) \subseteq (-\infty, 0]$ follows, which immediately implies (6.5.42). The other variants are proved similarly. □

Let $\boxed{T_1} \to \boxed{T_2}$ mean that test T_1 implies test T_2 (eventually by adapting parameters appropriately) so that test T_2 is more powerful than test T_1. Then we have proved the scheme in Figure 6.5.1.

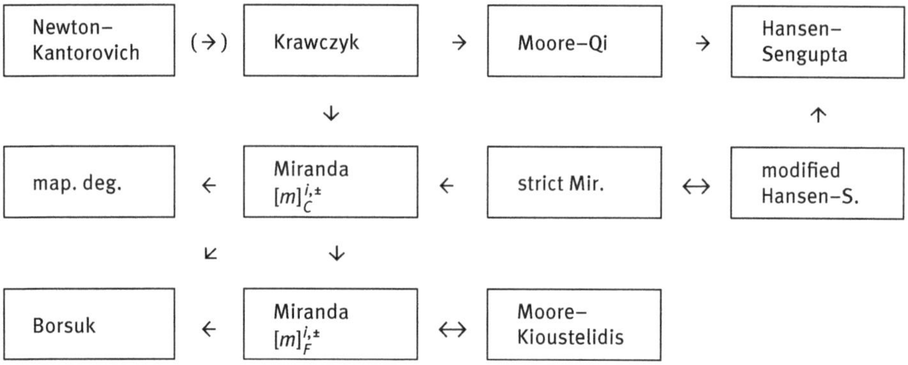

Fig. 6.5.1: Implications of tests.

Exercises

Ex. 6.5.1 (Shen, Neumaier [343]). Show by the example

$$f(x) = \begin{pmatrix} x_1 + x_2 \\ x_1 - x_2 \end{pmatrix}, \quad [x] = \begin{pmatrix} [-\varepsilon, \varepsilon] \\ [-2\varepsilon, 2\varepsilon] \end{pmatrix}, \quad \varepsilon > 0,$$

that the assumptions of Miranda's theorem 1.5.9 can be violated in arbitrary small neighborhoods of a zero of f.

Ex. 6.5.2 (Shen, Neumaier [343]). Show that the Newton–Kantorovich test implies the Krawczyk test if the Krawczyk operator is formulated with $[f'](\,[x], \tilde{x})$ from (6.5.14) instead of $f'([x])$, and if (6.3.12) is weakened to (6.5.13) while the other assumptions in Theorem 6.3.3 are kept.

6.6 Bisection method

Another very simple, but slow and in practice not recommendable method to verify zeros $x^* \in [x]^0$ of some given function f is based on a continued bisection of the (one- or multidimensional) interval $[x]^0$, evaluating f at the resulting subintervals, discarding those intervals $[x]$ for which $0 \notin f([x])$ can be shown and putting the remaining ones back into a list. From there they are fetched and bisected anew up to a stage where their diameter goes below some limit. Then they are tested by means of some of the criteria in the previous sections whether they contain a zero of f. The algorithm is similar to that of the modified Newton method in Section 6.1. We list here only its basic form and restrict ourselves to functions $f\colon D \subseteq \mathbb{R} \to \mathbb{R}$. Details for practical computation and generalizations to functions $f\colon D \subseteq \mathbb{R}^m \to \mathbb{R}^n$ are left to the reader.

Bisection method – one-dimensional case

Let $f: D \subseteq \mathbb{R} \to \mathbb{R}$, $f \in C(D)$, $[x]^0 \subseteq D$, $\varepsilon > 0$ (small). Let $[v]$ be a vector of dynamic length n called a stack.

(1) $n := 1$, $[v]_1 := [x]^0$

(2) $[x] := [v]_n$, $n := n - 1$

(3) compute $f([x])$

 (a) if $0 \notin f([x])$ then $[x]$ does not contain any zero of f and is discarded

 (b) if $0 \in f([x])$ then $[x]$ is a candidate for a zero of f

  (i) if $d([x]) \le \varepsilon$ then try to verify a zero by means of an existence test; print $[x]$, $f([x])$ and the result of the verification (zero, candidate)

  (ii) if $d([x]) > \varepsilon$ then bisect $[x]$ into $[x] = [y]^1 \cup [y]^2$ with $[y]^1 \le [y]^2$
  $n := n + 2$, $[v]_{n-1} := [y]^2$, $[v]_n := [y]^1$

(4) if $n > 0$ then goto (2).

We emphasize that the algorithm delivers only candidates for zeros x^* of f if the existence test in step 3 (b) (i) fails. Another problem can occur if x^* is a common endpoint of two consecutive intervals. This pretends that there are two zeros where only one exists. Such a situation can be avoided if final consecutive intervals are joined. We illustrate this phenomenon in Example 6.6.1.

Example 6.6.1. Let $f(x) = x^3 - 5x^2 - 4x + 20 = (x + 2)(x - 2)(x - 5)$. We look for the zeros of f. To this end we consider the Frobenius companion matrix

$$F = \begin{pmatrix} 0 & 0 & -20 \\ 1 & 0 & 4 \\ 0 & 1 & 5 \end{pmatrix}$$

which has $f(x)$ as characteristic polynomial and therefore the zeros of f as eigenvalues. From Gershgorin's Theorem 1.8.19 or Theorem 1.3.10 we know that the *real* eigenvalues of F are contained in $[x]^0 = [-20, 20]$. With $\varepsilon = 10^{-5}$ we see at once that a depth of at most 22 bisections will occur since $2^{22} = 4\,194\,304 \ge d([x]^0)/\varepsilon = 4\,000\,000$. The algorithm computes 10 candidates which partly overlap. After joining overlapping intervals one obtains the candidates

$$\begin{aligned}
&[-2.00000762939454, -1.99999809265136],\\
&[\ \ 1.99998855590820,\ \ 2.00001716613770],\\
&[\ \ 4.99997138977050,\ \ 5.00002861022950],
\end{aligned}$$

for which the zeros must still be verified. During the bisection, the stack length n was at most 21. The algorithm cycled 183 times through the steps (2)–(4).

If $\varepsilon = 10\,\texttt{eps}$ with MATLAB's machine precision eps, we get

$$\begin{aligned}&[-2.00000000000001, -1.99999999999999\,],\\&[\ \ 1.99999999999999,\ \ 2.00000000000001\,],\\&[\ \ 4.99999999999999,\ \ 5.00000000000001\,],\end{aligned}$$

with 508 cycles and a stack length of at most 53. This time the algorithm delivers 14 partly overlapping intervals which can be joined to the ones being listed above.

Notes to Chapter 6

To 6.1: On the variety of enclosure methods for nonlinear equations we mention here only Alefeld [6], Alefeld, Potra [43, 44], Alefeld, Potra, Shi [46]. The one-dimensional interval Newton method is already roughly described (without intersection) in Sunaga [351]. It is also studied (with intersection) in Moore [232] and Nickel [260]. For its convergence order see Alefeld [12]. The modified interval Newton method can be found in Alefeld [5]. Neumaier considered parameter dependent systems in Neumaier [256]. Heindl provides computable hypernorm bounds for the error of an approximate solution to a system of nonlinear equations in Heindl [139].

To 6.2: The interval Newton method in a slightly different form than in this book seems to go back to Moore [231, 232]. Convergence was proved in Alefeld, Herzberger [24]. Theorem 6.2.4 and Corollary 6.2.5 can be found in Alefeld [8]. See also Mayer [206].

To 6.3: The Krawczyk operator was introduced in Krawczyk [172]. The Krawczyk test (6.3.4) originates from Moore [233]; see also Moore [234]. An analogue of the quadratic convergence theorem 6.3.5 for the Alefeld method was proved in Alefeld, Platzöder [42]. A generalization of the Krawczyk algorithm is contained in Qi [277].

To 6.4: The Hansen–Sengupta method goes back to Hansen, Sengupta [132]. Lemma 6.4.2 and a great number of the proofs in Section 6.4 follow the lines of Neumaier [257].

To 6.5: The contents of this section are mainly contained in Moore, Kioustelidis [238], Rall [279], Frommer, Hoxha, Lang [105], Frommer, Lang [106, 107], Frommer, Lang, Schnurr [108]. See also Kioustelidis [163].

The comparisons on existence theorems of Kantorovich, Moore, Miranda, and Borsuk can be found in Alefeld, Frommer, Heindl, J. Mayer [21] and Alefeld, Potra, Shen [45]. See also Shen, Wolfe [344]. Relations between some of these theorems can also be found in Neumaier, Shen [259], Schäfer, Schnurr [329], and in Schnurr [330]. A second-derivative test and its comparison with the Kantorovich theorem is presented in Qi [278]

7 Eigenvalue problems and related ones

In the present chapter we deal with the verification and enclosure of eigenpairs (x^*, λ^*) – primarily for a single real matrix $A \in \mathbb{R}^{n\times n}$ but on second glance also for an interval matrix $[A] \in \mathbb{IR}^{n\times n}$. In the latter case 'verification and enclosure' means that we look for an interval vector $[x]$ and an interval $[\lambda]$ such that for each matrix $\tilde{A} \in [A]$ there is an eigenpair $(\tilde{x}^*, \tilde{\lambda}^*) \in [x] \times [\lambda]$. We will denote this as 'interval eigenvalue problem for $[A]$'.

Since eigenvectors x^* are not unique, we will normalize them in many cases by fixing the n-th component to be equal to one, i.e.,

$$x_n^* = (e^{(n)})^T x^* = 1.$$

In this case we will speak of a normalized eigenpair (x^*, λ^*). The value one can be replaced by any nonzero value; cf. Mayer [208] for details. The normalization of the n-th component of x^* instead of another one will simplify the notation. It is, of course, impossible if originally $x_n^* = 0$. But if so, we can rewrite the eigenvalue-eigenvector equation $Ax = \lambda x$ as $PAP^T(Px) = \lambda(Px)$ using an appropriate permutation matrix P such that $(Px^*)_n \neq 0$. This does not change the eigenvalues but influences the original order of components of x^*. Therefore, our normalization can be thought of 'without loss of generality'.

We will restrict ourselves to real eigenpairs of real matrices. Such eigenpairs certainly exist if we consider *simple* eigenvalues of $A \in \mathbb{R}^{n\times n}$ (Section 7.2) or real *symmetric* matrices (Section 7.3 and 7.6). For simple eigenvalues we will present a Krawczyk-like method. For symmetric matrices we will describe a method based on Jacobi transformations followed by estimations which grow out of Gershgorin's Theorem 1.8.19 and Theorem 1.8.25 due to Wilkinson.

Complex eigenvalues of real matrices and eigenvalues of complex matrices do not cause any severe additional problems but assume a complex interval arithmetic which we defer to Chapter 9. Multiple eigenvalues of *nonsymmetric* real matrices are more difficult to handle. We consider here only double eigenvalues (Section 7.4), which in practical computation cannot easily be separated from eigenvalues which lie closely together. We will call this situation 'nearly double eigenvalues'. It is a special case of a cluster of eigenvalues defined as a set of eigenvalues which hardly differ in value up to multiple eigenvalues. It is obvious that all methods for verifying zeros of nonlinear functions (cf. Chapter 6) can be applied to the function

$$f(x, \lambda) = \begin{pmatrix} Ax - \lambda x \\ (e^{(n)})^T x - 1 \end{pmatrix}$$

in order to verify normalized eigenpairs. But as in traditional numerics, the simple form of f suggests methods of their own. As a small extension we will consider the generalized eigenvalue problem $Ax = \lambda Bx$ with a regular matrix $B \in \mathbb{R}^{n\times n}$ for simple eigenvalues in Section 7.5. Moreover we will introduce into the ideas of Behnke

DOI 10.1515/9783110499469-008

(Section 7.6) which can also be used for the nonalgebraic eigenvalue problem. We will not further pursue this interesting more general topic in our book, but will restrict ourselves only to clusters of eigenvalues of the generalized algebraic eigenvalue problem.

Enclosures of singular values of real $m \times n$ matrices are handled in Section 7.7. Similar to previous sections, the method starts with an ansatz which consists of an approximation of the singular values and some error enclosing zero-symmetric interval vector.

Up to now all our problems are particular quadratic systems which we will handle essentially with the same idea. Therefore, although not practically relevant, we start this chapter with quadratic systems in Section 7.1, proving a crucial theorem which is basic for many of the subsequent sections.

We close the chapter with an applied problem in Section 7.8. There, we introduce an inverse eigenvalue problem and present a method for verifying a solution.

7.1 Quadratic systems

Let

$$g \colon \begin{cases} \mathbb{R}^n \to \mathbb{R}^n \\ x \mapsto g(x) = r + Sx + T(x, x) \end{cases} \tag{7.1.1}$$

with $r \in \mathbb{R}^n$, $S \in \mathbb{R}^{n\times n}$, and

$$T \colon \begin{cases} \mathbb{R}^n \times \mathbb{R}^n \to \mathbb{R}^n \\ (x, y) \mapsto \left(\sum_{j=1}^{n} \sum_{k=1}^{n} t_{ijk} x_k y_j \right) \end{cases}.$$

We want to construct an interval vector which, under certain conditions, contains a fixed point of g. Obviously at most quadratic terms in x_i enter into the equation $x = g(x)$. This justifies the title of this section.

In the following theorem we will use the interval arithmetic evaluation of $g(x)$ with

$$\big(T([x], [y])\big)_i = \sum_{j=1}^{n} \sum_{k=1}^{n} t_{ijk} [x]_k [y]_j \quad \text{for } i = 1, \dots, n.$$

Theorem 7.1.1. *Let g be defined as in* (7.1.1), *and let*

$$\rho = \|r\|_\infty, \quad \sigma = \|S\|_\infty, \quad \tau = \|T\|_\infty = \max_{1\le i\le n} \left\{ \sum_{j=1}^{n} \sum_{k=1}^{n} |t_{ijk}| \right\}, \tag{7.1.2}$$

$$\sigma \le 1, \quad \Delta = (1-\sigma)^2 - 4\rho\tau \ge 0, \tag{7.1.3}$$

$$\left.\begin{aligned} \beta^- &= (1-\sigma-\sqrt{\Delta})/(2\tau) = 2\rho/(1-\sigma+\sqrt{\Delta}), \\ \beta^+ &= (1-\sigma+\sqrt{\Delta})/(2\tau). \end{aligned}\right\} \tag{7.1.4}$$

(a) *For any* $\beta \in [\beta^-, \beta^+]$ *the function* g *has at least one fixed point* x^* *in* $[x]^0 = [-\beta, \beta]e$*, and the iteration*

$$[x]^{(k+1)} = g([x]^{(k)}), \quad k = 0, 1, \dots, \tag{7.1.5}$$

converges to some interval vector $[x]^*$ *with*

$$x^* \in [x]^* \subseteq [x]^{(k)} \subseteq [x]^{(k-1)} \subseteq \dots \subseteq [x]^{(0)}, \quad k \in \mathbb{N}. \tag{7.1.6}$$

(b) *For any* $\beta \in [\beta^-, (\beta^- + \beta^+)/2)$ *the function* g *has a unique fixed point* x^* *in* $[x]^{(0)} = [-\beta, \beta]e$*, and (7.1.6) holds with* $[x]^* = [x^*, x^*]$*, i.e., (7.1.5) converges to* x^**.*

(c) *For any* $\beta \in (\beta^-, \beta^+)$ *and* $[x] = [-\beta, \beta]e$ *the function* g *satisfies*

$$g([x]) \subseteq \operatorname{int}([x]). \tag{7.1.7}$$

Proof. (a) We show that $[x] = [-\beta, \beta]e$ satisfies

$$g([x]) = r + S[x] + T([x], [x]) \subseteq [x] \tag{7.1.8}$$

whenever $\beta \in [\beta^-, \beta^+]$. Then Brouwer's fixed point theorem 1.5.8 implies the first part of the assertion while the second is trivial by virtue of the Theorems 2.5.5 and 4.1.4. The inclusion (7.1.8) is equivalent to

$$\begin{aligned} [-\beta, \beta]e &\supseteq r + \left(\sum_{j=1}^{n} s_{ij}[-\beta, \beta]\right) + \left(\sum_{j=1}^{n}\sum_{k=1}^{n} t_{ijk}[-\beta, \beta][-\beta, \beta]\right) \\ &= r + [-\beta, \beta]\,|S|e + [-\beta^2, \beta^2]\,|T|(e, e) \end{aligned} \tag{7.1.9}$$

where $|T|(x, y) = (\sum_{k,j} |t_{ijk}| x_j y_k) \in \mathbb{R}^n$. This is equivalent to

$$|r| + \beta|S|e + \beta^2|T|(e, e) \le \beta e. \tag{7.1.10}$$

The inequality (7.1.10) certainly holds if

$$\rho + \beta\sigma + \beta^2\tau \le \beta,$$

whence $\beta \in [\beta^-, \beta^+]$.

(b) Let $[x]^*$ be the limit of (7.1.5). For the diameter we get

$$\begin{aligned} d([x]^*) &= d(g([x]^*)) \\ &= |S|d([x]^*) + \left(\sum_{j=1}^{n}\sum_{k=1}^{n} |t_{ijk}| d([x]_k^* [x]_j^*)\right), \end{aligned}$$

and with

$$d_\infty = \|d([x]^*)\|_\infty$$

and

$$\begin{aligned} d([x]_k^* [x]_j^*) &\le d([x]_k^*)\,|[x]_j^0| + |[x]_k^0|\,d([x]_j^*) \\ &\le \beta\{d([x]_k^*) + d([x]_j^*)\} \end{aligned}$$

we obtain

$$d_\infty \le \sigma d_\infty + 2\beta\tau d_\infty. \tag{7.1.11}$$

If $d_\infty > 0$, then (7.1.11) implies $1 \le \sigma + 2\beta\tau$, hence

$$\beta \ge \frac{1-\sigma}{2\tau} = \frac{\beta_1+\beta_2}{2}$$

which contradicts our assumption. Thus, $d_\infty = 0$, and $x^* \in [x]^*$ yields $[x]^* = [x^*, x^*]$.

Since each fixed point of g, which is contained in $[x]^{(0)}$, remains in $[x]^{(k)}$, $k \in \mathbb{N}$, the uniqueness of x^* follows from $d_\infty = 0$.

(c) One can proceed as in (a) ending up with

$$\rho + \beta\sigma + \beta^2\tau < \beta,$$

which means $\beta \in (\beta^-, \beta^+)$. □

The assumptions (7.1.3) are needed for the existence of β^-, β^+, and for $\beta^- \ge 0$. In our applications we shall have $\rho \approx 0$, $\sigma \approx 0$ so that these assumptions are fulfilled.

Instead of a single real quadratic system, one sometimes has to handle several simultaneously. To this end one considers the interval arithmetic evaluation of g with respect to x, r, S and T, i.e., interval functions defined by

$$g([x],[r],[S],[T]) = [r] + [S][x] + [T]([x],[x])$$

with $[r] \in \mathbb{IR}^n$, $[S] \in \mathbb{IR}^{n\times n}$, and $[T] = ([t]_{ijk}) \in \mathbb{IR}^{n\times n\times n}$. It is clear that an interval vector $[x]$, which satisfies

$$g([x],[r],[S],[T]) \subseteq [x]$$

contains for *each* function $g(x,r,S,T) = r + Sx + T(x,x)$ with $r \in [r]$, $S \in [S]$, $T \in [T]$ at least one fixed point. We leave it to the reader as an easy exercise (Exercise 7.1.1) to reformulate Theorem 7.1.1 for interval data. We only mention that ρ, σ, τ have to be replaced by

$$\rho = \|\,|[r]|\,\|_\infty, \quad \sigma = \|\,|[S]|\,\|_\infty, \quad \tau = \max_{1\le i\le n}\left\{\sum_{j=1}^n \sum_{k=1}^n |[t_{ijk}]|\right\}.$$

Refer to Section 7.8 for an application of this case.

Exercises

Ex. 7.1.1. Reformulate Theorem 7.1.1 for interval data.

7.2 A Krawczyk-like method

We start with a Krawczyk-like method tailored for simple eigenvalues. Let (x^*, λ^*) be a normalized real eigenpair of $A \in \mathbb{R}^{n\times n}$. Let $(\tilde{x}, \tilde{\lambda})$ with $\tilde{x}_n = 1$ be a real approximation of this eigenpair which can be obtained by any traditional method such as the inverse power method with shift, or by the MATLAB function `eig`, for example. Obviously, (x^*, λ^*) solves the equation

$$f(x,\lambda) = \begin{pmatrix} Ax - \lambda x \\ (e^{(n)})^T x - 1 \end{pmatrix} = 0. \tag{7.2.1}$$

Replacing (x, λ) by $(\tilde{x} + \Delta x, \tilde{\lambda} + \Delta\lambda)$ with

$$\Delta x := x - \tilde{x}, \quad \Delta\lambda := \lambda - \tilde{\lambda}. \tag{7.2.2}$$

yields

$$f(x,\lambda) = f(\tilde{x},\tilde{\lambda}) + \begin{pmatrix} A - \tilde{\lambda} I_n & -\tilde{x} \\ (e^{(n)})^T & 0 \end{pmatrix} \begin{pmatrix} \Delta x \\ \Delta\lambda \end{pmatrix} - \begin{pmatrix} \Delta\lambda\, \Delta x \\ 0 \end{pmatrix} = 0. \tag{7.2.3}$$

We multiply (7.2.3) by some preconditioning matrix $-C$, which will be determined later on, and we add $\binom{\Delta x}{\Delta\lambda}$ on both sides in order to end up with the fixed point equation

$$\begin{pmatrix} \Delta x \\ \Delta\lambda \end{pmatrix} = g(\Delta x, \Delta\lambda), \tag{7.2.4}$$

where

$$g(\Delta x, \Delta\lambda) = -Cf(\tilde{x},\tilde{\lambda}) + \left\{ I_{n+1} - C \begin{pmatrix} A - \tilde{\lambda} I_n & -(\tilde{x} + \Delta x) \\ (e^{(n)})^T & 0 \end{pmatrix} \right\} \begin{pmatrix} \Delta x \\ \Delta\lambda \end{pmatrix}. \tag{7.2.5}$$

The construction of g from (7.2.3) shows that

$$g(\Delta x, \Delta\lambda) = s(x, \lambda) \tag{7.2.6}$$

holds for

$$s(x,\lambda) = \begin{pmatrix} x - \tilde{x} \\ \lambda - \tilde{\lambda} \end{pmatrix} - Cf(x,\lambda) = \begin{pmatrix} \Delta x \\ \Delta\lambda \end{pmatrix} - Cf(x,\lambda). \tag{7.2.7}$$

The error $(\Delta x^*, \Delta\lambda^*) := (x^* - \tilde{x}, \lambda^* - \tilde{\lambda})$ is a solution of (7.2.4).

Now we forget the meaning of (x^*, λ^*) and $(\tilde{x}, \tilde{\lambda})$, and we consider only the function g and its interval arithmetic evaluation at

$$[\Delta x] = [x] - \tilde{x}, \quad [\Delta\lambda] = [\lambda] - \tilde{\lambda},$$

where $[x] \in \mathbb{IR}^n$, $[\lambda] \in \mathbb{IR}$. Then we end up with the following main result of this section.

Theorem 7.2.1. *Let $A \in \mathbb{R}^{n\times n}$, $\tilde{\lambda} \in \mathbb{R}$, $\tilde{x} \in \mathbb{R}^n$, $C \in \mathbb{R}^{(n+1)\times(n+1)}$, and define g by (7.2.5). Let $\tilde{x}$ be normalized by $\tilde{x}_n = 1$. If g fulfills the inclusion*

$$g([\Delta x], [\Delta\lambda]) \subseteq \operatorname{int}([\Delta x]^T, [\Delta\lambda])^T, \tag{7.2.8}$$

then the following assertions hold.

(a) *C is nonsingular.*

(b) *There exists exactly one eigenvector $x^* \in \tilde{x} + [\Delta x]$ of A which is normalized by $x_n^* = 1$.*

(c) *There exists exactly one eigenvalue $\lambda^* \in \tilde{\lambda} + [\Delta\lambda]$ of A.*

The following items refer to x^* from (b) *and λ^* from* (c).

(d) *The equation $Ax^* = \lambda^* x^*$ holds.*

(e) *The number λ^* is a geometrically simple eigenvalue.*

(f) *If $(\tilde{x}, \tilde{\lambda})$ are sufficiently good approximations of (x^*, λ^*), then it can be guaranteed that λ^* is algebraically simple.*

(g) *If one starts the iteration*

$$\begin{pmatrix} [\Delta x]^{(k+1)} \\ [\Delta\lambda]^{(k+1)} \end{pmatrix} = g([\Delta x]^{(k)}, [\Delta\lambda]^{(k)}), \quad k = 0, 1, \ldots, \tag{7.2.9}$$

with $([\Delta x]^{(0)}, [\Delta\lambda]^{(0)}) = ([\Delta x], [\Delta\lambda])$ from (7.2.8), then the iterates converge satisfying

$$([\Delta x]^{(k+1)}, [\Delta\lambda]^{(k+1)}) \subseteq ([\Delta x]^{(k)}, [\Delta\lambda]^{(k)}), \quad k = 0, 1, \ldots,$$

and

$$(x^*, \lambda^*) \in (\tilde{x}, \tilde{\lambda}) + ([\Delta x]^{(k)}, [\Delta\lambda]^{(k)}), \quad k = 0, 1, \ldots.$$

In order to prove Theorem 7.2.1 we need the subsequent auxiliary results.

Our first lemma is an immediate consequence of Theorem 3.1.6 (b). It shows that the last two matrices in (7.2.5) are nonsingular if the crucial subset property (7.2.8) of Theorem 7.2.1 holds. In particular, it proves part (a) of the theorem.

Lemma 7.2.2. *With the notation of Theorem 7.2.1 the assumption (7.2.8) implies that C and*

$$\begin{pmatrix} A - \tilde{\lambda} I_n & -(\tilde{x} + \Delta x) \\ (e^{(n)})^T & 0 \end{pmatrix}$$

are nonsingular for all $\Delta x \in [\Delta x]$.

Our next lemma guarantees that the assumption of Theorem 1.8.3 is fulfilled. Recall that this theorem states properties which guarantee simplicity of an eigenvalue.

Lemma 7.2.3. *Let (x^*, λ^*) be an eigenpair of $A \in \mathbb{R}^{n\times n}$ with $\Delta x^* \in [\Delta x] \in \mathbb{IR}^n$. Then (7.2.8) implies $x_n^* \neq 0$.*

Proof. Assume that $x_n^* = 0$ holds. Then

$$\begin{pmatrix} A - \tilde{\lambda} I_n & -x^* \\ \left(e^{(n)}\right)^T & 0 \end{pmatrix} \begin{pmatrix} x^* \\ \Delta\lambda^* \end{pmatrix} = \begin{pmatrix} Ax^* - \lambda^* x^* \\ 0 \end{pmatrix} = 0,$$

and Lemma 7.2.2 implies the contradiction $x^* = 0$. □

In the next lemma we relate the errors Δx^*, $\Delta\lambda^*$ to the interval vector which satisfies (7.2.8).

Lemma 7.2.4. *Let (x^*, λ^*) be an eigenpair of $A \in \mathbb{R}^{n\times n}$ with $\Delta x^* \in [\Delta x] \in \mathbb{IR}^n$, $x_n^* = 1$. Then* (7.2.8) *implies $\Delta\lambda^* \in [\Delta\lambda]$.*

Proof. With s, g from (7.2.7), (7.2.5) define p as the projection

$$p(\Delta\lambda) = g_{n+1}(\Delta x^*, \Delta\lambda) = s_{n+1}(x^*, \lambda) = \left\{\begin{pmatrix} \Delta x^* \\ \Delta\lambda \end{pmatrix} - Cf(x^*, \lambda)\right\}_{n+1}.$$

Then the inclusion (7.2.8) implies $p(\Delta\lambda) \in \operatorname{int}([\Delta\lambda])$ for all $\Delta\lambda \in [\Delta\lambda]$. Thus by virtue of Brouwer's fixed point theorem there is a fixed point $\Delta\hat{\lambda} \in \operatorname{int}([\Delta\lambda])$ of p. We will show that $\Delta\hat{\lambda} = \Delta\lambda^*$ holds. Let $\hat{\lambda} = \tilde{\lambda} + \Delta\hat{\lambda}$ and define $\Delta\hat{y} \in \mathbb{R}^n$ as the leading part of the vector $g(\Delta x^*, \Delta\hat{\lambda})$, i.e.,

$$\begin{pmatrix} \Delta\hat{y} \\ \Delta\hat{\lambda} \end{pmatrix} = g(\Delta x^*, \Delta\hat{\lambda}).$$

Let $\hat{y} = \tilde{x} + \Delta\hat{y}$ and assume for the moment $\Delta\hat{\lambda} \neq \Delta\lambda^*$, or, equivalently, $\hat{\lambda} \neq \lambda^*$. With (7.2.6) we obtain

$$\begin{aligned}\begin{pmatrix} \Delta\hat{y} \\ \Delta\hat{\lambda} \end{pmatrix} &= g(\Delta x^*, \Delta\hat{\lambda}) = s(x^*, \hat{\lambda}) = \begin{pmatrix} \Delta x^* \\ \Delta\hat{\lambda} \end{pmatrix} - C\begin{pmatrix} Ax^* - \hat{\lambda}x^* \\ 0 \end{pmatrix} \\ &= \begin{pmatrix} \Delta x^* \\ \Delta\hat{\lambda} \end{pmatrix} - (\lambda^* - \hat{\lambda})C\begin{pmatrix} x^* \\ 0 \end{pmatrix}.\end{aligned}$$

Therefore,

$$C\begin{pmatrix} x^* \\ 0 \end{pmatrix} = \frac{1}{\lambda^* - \hat{\lambda}}\begin{pmatrix} x^* - \hat{y} \\ 0 \end{pmatrix},$$

and with $\underline{\lambda} = \tilde{\lambda} + \underline{\Delta\lambda}$ we have

$$\begin{aligned} g(\Delta x^*, \underline{\Delta\lambda}) = s(x^*, \underline{\lambda}) &= \begin{pmatrix} \Delta x^* \\ \underline{\Delta\lambda} \end{pmatrix} - C\begin{pmatrix} Ax^* - \underline{\lambda}x^* \\ 0 \end{pmatrix} \\ &= \begin{pmatrix} \Delta x^* \\ \underline{\Delta\lambda} \end{pmatrix} - \frac{\lambda^* - \underline{\lambda}}{\lambda^* - \hat{\lambda}}\begin{pmatrix} x^* - \hat{y} \\ 0 \end{pmatrix}.\end{aligned}$$

This implies the contradiction

$$p(\underline{\Delta\lambda}) = \underline{\Delta\lambda} = g_{n+1}(\Delta x^*, \underline{\Delta\lambda}) \in \operatorname{int}([\Delta\lambda]).$$ □

The following lemma is a kind of counterpart to the previous one where the roles of Δx^* and $\Delta\lambda^*$ are exchanged.

Lemma 7.2.5. *Let (x^*, λ^*) be an eigenpair of $A \in \mathbb{R}^{n\times n}$ with $\Delta\lambda^* \in [\Delta\lambda]$ and no restriction on x_n^*. Then* (7.2.8) *implies the existence of some vector $\Delta y^* \in [\Delta x]$ such that $(y^*, \lambda^*) = (\tilde{x} + \Delta y^*, \lambda^*)$ is an eigenpair of A satisfying $\Delta y^* = y^* - \tilde{x} \in [\Delta x]$ and $(\Delta y^*)_n = 0$.*

Proof. Let p be the projection defined by

$$p(\Delta x) = \begin{pmatrix} g_1(\Delta x, \Delta\lambda^*) \\ \vdots \\ g_n(\Delta x, \Delta\lambda^*) \end{pmatrix} = \begin{pmatrix} s_1(x, \lambda^*) \\ \vdots \\ s_n(x, \lambda^*) \end{pmatrix}.$$

For all $\Delta x \in [\Delta x]$ the inclusion (7.2.8) implies

$$p(\Delta x) \in \operatorname{int}([\Delta x]), \tag{7.2.10}$$

hence Brouwer's fixed point theorem guarantees a fixed point $\Delta\hat{y} \in \operatorname{int}([\Delta x])$ of p. Let $\hat{y} = \tilde{x} + \Delta\hat{y}$, define $\Delta\hat{\lambda}$ by means of the equation

$$s(\hat{y}, \lambda^*) = g(\Delta\hat{y}, \Delta\lambda^*) = \begin{pmatrix} \Delta\hat{y} \\ \Delta\hat{\lambda} \end{pmatrix}, \tag{7.2.11}$$

and assume for the moment

$$\Delta\hat{\lambda} \neq \Delta\lambda^*, \tag{7.2.12}$$

or, equivalently, $\hat{\lambda} \neq \lambda^*$. Let

$$u(t) = (1 - x_n^* t)\hat{y} + tx^* = \hat{y} + t(x^* - x_n^*\hat{y})$$

and $\Delta u(t) = u(t) - \tilde{x}$.

If $x^* - x_n^*\hat{y} = 0$, then $x_n^* \neq 0$ because of $x^* \neq 0$. Hence $\hat{y} = x^*/x_n^*$. This implies $\hat{y}_n = 1$ and, by (7.2.11),

$$\begin{pmatrix} \Delta\hat{y} \\ \Delta\lambda^* \end{pmatrix} = s(\hat{y}, \lambda^*) = g(\Delta\hat{y}, \Delta\lambda^*) = \begin{pmatrix} \Delta\hat{y} \\ \Delta\hat{\lambda} \end{pmatrix},$$

contradicting (7.2.12). Therefore, let $x^* - x_n^*\hat{y} \neq 0$. Then $\lim_{t\to\infty}\|\Delta u(t)\|_\infty = \infty$ and $\Delta u(0) = \Delta\hat{y} \in [\Delta x]$. Choose $\tilde{t} \in \mathbb{R}$ such that $\Delta u(\tilde{t})$ lies on the boundary $\partial[\Delta x]$ of $[\Delta x]$, hence

$$p(\Delta u(\tilde{t})) \in \operatorname{int}([\Delta x]) \tag{7.2.13}$$

by virtue of (7.2.10). On the other hand we get

$$\begin{aligned} s(u(\tilde{t}), \lambda^*) &= \begin{pmatrix} \Delta u(\tilde{t}) \\ \Delta\lambda^* \end{pmatrix} - C \begin{pmatrix} (1 - x_n^*\tilde{t})(A\hat{y} - \lambda^*\hat{y}) \\ (1 - x_n^*\tilde{t})\hat{y}_n + \tilde{t}x_n^* - 1 \end{pmatrix} \\ &= \begin{pmatrix} \Delta u(\tilde{t}) \\ \Delta\lambda^* \end{pmatrix} + (1 - x_n^*\tilde{t}) \left\{ s(\hat{y}, \lambda^*) - \begin{pmatrix} \Delta\hat{y} \\ \Delta\lambda^* \end{pmatrix} \right\} \\ &= \begin{pmatrix} \Delta u(\tilde{t}) \\ \Delta\lambda^* \end{pmatrix} + \begin{pmatrix} 0 \\ (1 - x_n^*\tilde{t})(\hat{\lambda} - \lambda^*) \end{pmatrix}. \end{aligned}$$

This implies

$$p(\Delta u(\tilde t)) = \Delta u(\tilde t) \in \partial[\Delta x],$$

contradicting (7.2.13). Hence $\hat\lambda = \lambda^*$, and (7.2.11) implies $(\Delta\hat y)_n = 0$. The assertion follows with $y^* := \hat y$. □

Our final auxiliary result guarantees existence and uniqueness of an eigenpair of A within some appropriate set.

Lemma 7.2.6. *If* (7.2.8) *is valid, then there is exactly one eigenpair* $(x^*, \lambda^*) \in (\tilde x, \tilde\lambda) + ([\Delta x], [\Delta\lambda])$ *of* A *with* x^* *being normalized by* $x_n^* = 1$.

Proof. The existence of (x^*, λ^*) follows at once from Lemma 7.2.2 and from Brouwer's fixed point theorem applied to g or s from (7.2.5) and (7.2.7), respectively.

To prove the uniqueness, assume that there is a second eigenpair (y^*, μ^*) of A satisfying $y_n^* = 1$, $y^* \neq x^*$ and

$$(\Delta y^*, \Delta\mu^*) := (y^* - \tilde x, \mu^* - \tilde\lambda) \in ([\Delta x], [\Delta\lambda]).$$

Case 1, $\lambda^* \neq \mu^*$: Assume that $\lambda^* = \tilde\lambda$ holds. Then

$$\begin{pmatrix} A - \tilde\lambda I_n & -y^* \\ (e^{(n)})^T & 0 \end{pmatrix} \begin{pmatrix} y^* - x^* \\ \mu^* - \tilde\lambda \end{pmatrix} = \begin{pmatrix} (\mu^* - \tilde\lambda)\, y^* - y^*(\mu^* - \tilde\lambda) \\ 0 \end{pmatrix} = 0,$$

and Lemma 7.2.2 implies the contradiction $y^* = x^*$. Hence $\lambda^* \neq \tilde\lambda$, and analogously one shows $\mu^* \neq \tilde\lambda$.

Let

$$h(t) = (1-t)(\mu^* - \tilde\lambda) + t(\lambda^* - \tilde\lambda) = \mu^* - \tilde\lambda + t(\lambda^* - \mu^*), \quad t \in \mathbb{R},$$

and notice that

$$h(t) = 0 \quad \text{is equivalent to} \quad t = \frac{\mu^* - \tilde\lambda}{\mu^* - \lambda^*} \tag{7.2.14}$$

with t differing from zero by the previous conclusions. For $z \in \mathbb{R}$ and $t \in \mathbb{R}$ we define the expressions

$$g_z(\Delta x, \Delta, \lambda) = s(\tilde x, \tilde\lambda) + \left\{ I_{n+1} - C \begin{pmatrix} A - \tilde\lambda I_n & -z \\ \left(e^{(n)}\right)^T & 0 \end{pmatrix} \right\} \begin{pmatrix} \Delta x \\ \Delta\lambda \end{pmatrix},$$

$$u(t) = x^* + t(y^* - x^*), \quad v(t) = x^* + t\frac{\lambda^* - \tilde\lambda}{h(t)}(y^* - x^*), \quad \sigma(t) = \tilde\lambda + \frac{\lambda^* - \tilde\lambda}{h(t)}(\mu^* - \tilde\lambda),$$

$$\Delta u(t) = u(t) - \tilde x, \qquad \Delta v(t) = v(t) - \tilde x, \qquad \Delta\sigma(t) = \sigma(t) - \tilde\lambda,$$

where we exclude $t = \frac{\mu^* - \tilde\lambda}{\mu^* - \lambda^*}$ in the definition of $v(t), \sigma(t), \Delta v(t), \Delta\sigma(t)$.

Then $A\upsilon(t) - \tilde{\lambda}\upsilon(t) - (\sigma - \tilde{\lambda})u(t) = 0$, $\upsilon_n(t) - 1 = 0$, and finally

$$\begin{aligned} g_{u(t)}(\Delta\upsilon(t), \Delta\sigma(t)) &= s(\tilde{x}, \tilde{\lambda}) + \begin{pmatrix} \Delta\upsilon(t) \\ \Delta\sigma(t) \end{pmatrix} - C \begin{pmatrix} A\upsilon(t) - \tilde{\lambda}\upsilon(t) - (A\tilde{x} - \tilde{\lambda}\tilde{x}) - (\sigma - \tilde{\lambda})u(t) \\ \upsilon_n(t) - \tilde{x}_n \end{pmatrix} \\ &= -C \begin{pmatrix} A\tilde{x} - \tilde{\lambda}\tilde{x} \\ \tilde{x}_n - 1 \end{pmatrix} + \begin{pmatrix} \Delta\upsilon(t) \\ \Delta\sigma(t) \end{pmatrix} + C \begin{pmatrix} A\tilde{x} - \tilde{\lambda}\tilde{x} \\ 0 \end{pmatrix} \\ &= \begin{pmatrix} \Delta\upsilon(t) \\ \Delta\sigma(t) \end{pmatrix}. \end{aligned}$$

By virtue of (7.2.8) we get

$$\begin{pmatrix} \Delta x^* \\ \Delta\lambda^* \end{pmatrix} = s(x^*, \lambda^*) = g(\Delta x^*, \Delta\lambda^*) \in \operatorname{int} \begin{pmatrix} [\Delta x] \\ [\Delta\lambda] \end{pmatrix}, \tag{7.2.15}$$

whence $\Delta u(0) = \Delta x^* \in \operatorname{int}([\Delta x])$. Analogously, $\Delta u(1) = \Delta y^* \in \operatorname{int}([\Delta x])$.

Therefore, there are real numbers $\underline{t} < 0$, $\bar{t} > 1$ such that $\Delta u(\underline{t})$, $\Delta u(\bar{t})$ lie on the boundary $\partial[\Delta x]$ of $[\Delta x]$ with $\Delta u(t) \in \operatorname{int}(\Delta x)$ for $\underline{t} < t < \bar{t}$.

Now we will show

$$\begin{pmatrix} \Delta\upsilon(t) \\ \Delta\sigma(t) \end{pmatrix} \in \operatorname{int} \begin{pmatrix} [\Delta x] \\ [\Delta\lambda] \end{pmatrix} \quad \text{for all } t \in [\underline{t}, \bar{t}]. \tag{7.2.16}$$

By virtue of $(\Delta\upsilon(0), \Delta\sigma(0)) = (\Delta x^*, \Delta\lambda^*)$ and (7.2.15) this relation is certainly true for $t = 0$. Assuming

$$\begin{pmatrix} \Delta\upsilon(\tilde{t}) \\ \Delta\sigma(\tilde{t}) \end{pmatrix} \in \partial \begin{pmatrix} [\Delta x] \\ [\Delta\lambda] \end{pmatrix} \quad \text{for some } \tilde{t} \in [\underline{t}, \bar{t}] \tag{7.2.17}$$

yields the contradiction

$$\begin{pmatrix} \Delta\upsilon(\tilde{t}) \\ \Delta\sigma(\tilde{t}) \end{pmatrix} = g_{u(\tilde{t})}(\Delta\upsilon(\tilde{t}), \Delta\sigma(\tilde{t})) \in g([\Delta x], [\Delta\lambda]) \subseteq \operatorname{int} \begin{pmatrix} [\Delta x] \\ [\Delta\lambda] \end{pmatrix}.$$

Thus, (7.2.16) is valid.

Next notice that

$$\frac{\mu^* - \tilde{\lambda}}{\mu^* - \lambda^*} \notin [\underline{t}, \bar{t}] \tag{7.2.18}$$

holds. Otherwise let t tend to $\frac{\mu^* - \tilde{\lambda}}{\mu^* - \lambda^*} \neq 0$. Then $h(t)$ tends to zero and $\|\Delta\upsilon(t)\|_\infty$ tends to infinity contradicting (7.2.16). Therefore, (7.2.18) is valid. Together with (7.2.14) it implies $h(t) \neq 0$ for $\underline{t} \le t \le \bar{t}$. W.l.o.g. let $h(t) > 0$ for $\underline{t} \le t \le \bar{t}$.

We will prove the inequality

$$\underline{t} < t\frac{\lambda^* - \tilde{\lambda}}{h(t)} < \bar{t} \quad \text{for } \underline{t} \le t \le \bar{t}. \tag{7.2.19}$$

Assume that there is a real number $t_1 \in [\underline{t}, \bar{t}]$ such that $t_1 \frac{\lambda^* - \tilde{\lambda}}{h(t_1)} \le \underline{t}$ is fulfilled. Since (7.2.19) holds for $t = 0$, there is some $t' \in [\underline{t}, \bar{t}]$ satisfying

$$t' \frac{\lambda^* - \tilde{\lambda}}{h(t')} = \underline{t}.$$

This yields

$$\Delta v(t') = \Delta x^* + \underline{t}(y^* - x^*) = \Delta u(\underline{t}) \in \partial([\Delta x])$$

contradicting (7.2.16). Thus the left inequality of (7.2.19) holds, and the right inequality can be proven analogously.

Choose $t = \underline{t}$ in (7.2.19). Taking into account $\underline{t} < 0$, $h(\underline{t}) > 0$, this implies the equivalence

$$\begin{aligned}\underline{t} < \underline{t}\,\frac{\lambda^* - \tilde{\lambda}}{h(\underline{t})} &\iff (1-\underline{t})(\mu^* - \tilde{\lambda}) + \underline{t}(\lambda^* - \tilde{\lambda}) > \lambda^* - \tilde{\lambda} \\ &\iff (1-\underline{t})(\mu^* - \tilde{\lambda}) > (1-\underline{t})(\lambda^* - \tilde{\lambda}) \\ &\iff \mu^* > \lambda^* .\end{aligned}$$

This contradicts $\mu^* < \lambda^*$, which is equivalent to the right inequality $\bar{t}\frac{\lambda^* - \tilde{\lambda}}{h(\bar{t})} < \bar{t}$ of (7.2.19) when choosing there $t = \bar{t}$.

Case 2, $\lambda^* = \mu^*$: Analogously to Case 1 one proves $\lambda^* = \mu^* \neq \tilde{\lambda}$, and with

$$\begin{aligned} &h(t) = \lambda^* - \tilde{\lambda} \neq 0, \\ u(t) = v(t) = x^* + t(y^* - x^*), \qquad &\sigma(t) = \lambda^*, \\ \Delta u(t) = \Delta v(t) = v(t) - \tilde{x}, \qquad &\Delta\sigma(t) = \sigma(t) - \tilde{\lambda}\end{aligned}$$

one can repeat all the steps of the previous case up to (7.2.17) which yields

$$\Delta u(\tilde{t}) \in \partial([\Delta x]),$$

contradicting

$$\begin{aligned}\begin{pmatrix}\Delta u(\tilde{t}) \\ \Delta\sigma(\tilde{t})\end{pmatrix} = \begin{pmatrix}\Delta v(\tilde{t}) \\ \Delta\sigma(\tilde{t})\end{pmatrix} &= g_{u(\tilde{t})}\left(\Delta v(\tilde{t}), \Delta\sigma(\tilde{t})\right) \\ &\in g([\Delta x], [\Delta\lambda]) \subseteq \operatorname{int}\begin{pmatrix}[\Delta x] \\ [\Delta\lambda]\end{pmatrix}.\end{aligned}$$ □

Now we can prove Theorem 7.2.1.

Proof of Theorem 7.2.1. (a) is contained in Lemma 7.2.2.

(b) The existence of a normalized eigenvector $x^* \in \tilde{x} + [\Delta x]$, $x_n^* = 1$, follows from Lemma 7.2.6.

If there are two linearly independent normalized eigenvectors $x^*, y^* \in \tilde{x} + [\Delta x]$ which are associated with two eigenvalues λ^* and μ^* of A. Then these eigenvalues are contained in $\tilde{\lambda} + [\Delta\lambda]$ by virtue of Lemma 7.2.4. Hence Lemma 7.2.6 implies $\lambda^* = \mu^*$ and $x^* = y^*$, contradicting the linear independence of x^*, y^*.

(c) The existence of $\lambda^* \in \tilde{\lambda} + [\Delta\lambda]$ follows again from Lemma 7.2.6. If there is a second eigenvalue μ^* with $\lambda^* \neq \mu^* \in \tilde{\lambda} + [\Delta\lambda]$, then the Lemmas 7.2.5 and 7.2.6 yield the contradiction $\lambda^* = \mu^*$.

(d) follows from Lemma 7.2.6 and (b), (c).

(e) If λ^* is not geometrically simple, then there are two linearly independent eigenvectors x^*, y^* associated with λ^*, with $x^* \in \tilde{x} + [\Delta x]$ being normalized by $x_n^* = 1$. Since $((\Delta x^*)^T, \Delta\lambda^*)^T$ is a solution of (7.2.4), the assumption (7.2.8) guarantees $\Delta x^* \in \operatorname{int}([\Delta x])$. Therefore, for $\varepsilon > 0$ sufficiently small, the vector

$$z^* = \begin{cases} x^* + \varepsilon\left(\frac{y^*}{y_n^*} - x^*\right), & \text{if } y_n^* \neq 0 \\ x^* + \varepsilon y^*, & \text{if } y_n^* = 0 \end{cases}$$

is a normalized eigenvector of A associated with λ^* which is linearly independent of x^* and which satisfies $z^* \in \tilde{x} + [\Delta x]$. This contradicts (b).

(f) Since, by virtue of Lemma 7.2.2, the matrix

$$\begin{pmatrix} A - \tilde{\lambda} I_n & -x^* \\ (e^{(n)})^T & 0 \end{pmatrix}$$

is nonsingular, the same holds by continuity for

$$\begin{pmatrix} A - \lambda^* I_n & -x^* \\ (e^{(n)})^T & 0 \end{pmatrix}$$

if $\tilde{\lambda}$ approximates λ^* sufficiently well. Theorem 1.8.3 proves the assertion.

(g) follows from (7.2.8) and from Theorem 4.3.1. □

In Theorem 7.2.1 we assumed A to be a real point matrix. It is easy to see that the assertions of this theorem remain true if A is replaced by an interval matrix $[A]$. The statements then hold for all $A \in [A]$ and must be reformulated appropriately.

Unfortunately, Theorem 7.2.1 does not give any hint as to how to choose C and $[\Delta x]$, $[\Delta\lambda]$ such that the crucial condition (7.2.8) is fulfilled, nor does it guarantee that the iterates from (7.2.9) contract to the exact error $(\Delta x^*, \Delta\lambda^*)$. One possibility for C could be

$$C = B^{-1}, \quad \text{where } B = \begin{pmatrix} A - \tilde{\lambda} I_n & -\tilde{x} \\ (e^{(n)})^T & 0 \end{pmatrix}, \tag{7.2.20}$$

or at least $C \approx B^{-1}$, so that the zero matrix is contained in the matrix within braces of (7.2.5) if Δx is replaced there by an interval vector which contains the zero vector. Inspecting the cofactors of B reveals that in the case (7.2.20) C has the block form

$$C = \begin{pmatrix} C_{11} & C_{12} \\ 0 & 1 \\ C_{21} & C_{22} \end{pmatrix} \tag{7.2.21}$$

with $C_{11} \in \mathbb{R}^{(n-1)\times n}$, $C_{12} \in \mathbb{R}^{n-1}$, $C_{21} \in \mathbb{R}^{1\times n}$, $C_{22} \in \mathbb{R}$. Any preconditioning matrix C of the form (7.2.21) effects

$$g_n([\Delta x], [\Delta\lambda]) = 0, \tag{7.2.22}$$

if $\tilde{x}$ is normalized by $\tilde{x}_n = 1$. In particular, this means

$$[\Delta x]_n^{(k+1)} = 0, \quad k = 0, 1, \ldots,$$

in (7.2.9). Therefore, it seems reasonable to allow only starting vectors $[\Delta x]^{(0)}$ with $[\Delta x]_n^{(0)} = 0$. In this case, the blocks C_{12}, C_{22} do not influence the values of g since they are multiplied by zero. One can shrink the whole problem to an n-dimensional one, deleting the n-th component of g and of $[\Delta x]$ and using $\hat{C} = \binom{C_{11}}{C_{21}} \in \mathbb{R}^{n\times n}$ instead of C.

This results in a new function $\hat{g}$, defined by

$$\hat{g}(\Delta\hat{x}, \Delta\lambda) = \hat{C}(\tilde{\lambda}\tilde{x} - A\tilde{x}) + \left\{I - \hat{C}\left(\hat{B} + \begin{pmatrix} 0 & -\Delta\hat{x} \\ 0 & 0 \end{pmatrix}\right)\right\}\begin{pmatrix} \Delta\hat{x} \\ \Delta\lambda \end{pmatrix} \tag{7.2.23}$$

with

$$\Delta\hat{x} = (\Delta x_1, \ldots, \Delta x_{n-1})^T \tag{7.2.24}$$

and

$$\hat{B} = ((A - \tilde{\lambda}I)_{*,1}, \ldots, (A - \tilde{\lambda}I)_{*,n-1}, -\tilde{x}). \tag{7.2.25}$$

Because of $\det(\hat{C}) = -\det(C)$ it is clear that $\hat{C}$ is nonsingular if and only if C has this property. In addition, $\hat{C} = \hat{B}^{-1}$ holds if $C = B^{-1}$; cf. Exercise 7.2.1.

Theorem 7.2.1 remains valid for the new situation, the proof can be reduced to the former one: reconstruct g from $\hat{g}$ via (7.2.21), (7.2.22) with $C_{12} = C_{22} = 0$; let

$$\hat{g}([\Delta x]', [\Delta\lambda]) \subseteq \operatorname{int}\begin{pmatrix} [\Delta x]' \\ [\Delta\lambda] \end{pmatrix}$$

hold and define

$$[\Delta x] = \begin{pmatrix} [\Delta x]' \\ [-\varepsilon, \varepsilon] \end{pmatrix}, \quad \varepsilon > 0.$$

For sufficiently small ε we get

$$\begin{aligned} g_i([\Delta x], [\Delta\lambda]) &\subseteq \operatorname{int}([\Delta x]_i), \quad i = 1, \ldots, n-1, \\ g_{n+1}([\Delta x], [\Delta\lambda]) &\subseteq \operatorname{int}([\Delta\lambda]), \end{aligned}$$

and, by (7.2.22),

$$0 = g_n([\Delta x], [\Delta\lambda]) \subseteq \operatorname{int}([\Delta x]_n) = (-\varepsilon, \varepsilon).$$

Hence (7.2.8) is fulfilled for $([\Delta x], [\Delta\lambda])$. Taking into account $x_n^* = \tilde{x}_n = 1$ and $\Delta x_n^* = 0$ one can replace $[\Delta x]_n$ by $[\Delta x]_n = 0$ in Theorem 7.2.1 (b). With these remarks the analogue of Theorem 7.2.1 is easily seen.

Having chosen C, one can combine the iteration (7.2.9) with ε-inflation starting with $[\Delta x]^{(0)} = 0$ and $[\Delta\lambda]^{(0)} = 0$. If λ^* is a simple eigenvalue and if the approximations $\tilde{x}$, $\tilde{\lambda}$ are sufficiently good, then Example 4.2.10 reveals that $g([x], [\lambda])$ is a local P-contraction. This lets us expect that the assumptions of Theorem 4.3.5 hold, hence

ε-inflation is successful, resulting in (7.2.8) after finitely many steps of iterations. In addition, in this case the iterates $(\tilde{x} + [\Delta x]^k, \tilde{\lambda} + [\Delta\lambda]^k)$ contract to (x^*, λ^*).

Our next theorem provides an expression for $[x]^0$, $[\lambda]^0$ such that (7.2.8) is fulfilled in advance and the convergence to (x^*, λ^*) is guaranteed.

Theorem 7.2.7. *With the notation from Theorem 7.2.1, with f from (7.2.1), B from (7.2.20), and $\tilde{x}_n = 1$ define*

$$\rho = \left\| C \begin{pmatrix} A\tilde{x} - \tilde{\lambda}\tilde{x} \\ 0 \end{pmatrix} \right\|_\infty = \|Cf(\tilde{x}, \tilde{\lambda})\|_\infty,$$

$$\sigma = \left\| I_{n+1} - C \begin{pmatrix} A - \tilde{\lambda} I_n & -\tilde{x} \\ (e^{(n)})^T & 0 \end{pmatrix} \right\|_\infty = \|I_{n+1} - CB\|_\infty,$$

$$\tau = \|C\|_\infty \tag{7.2.26}$$

and assume

$$\sigma < 1, \quad \Delta = (1 - \sigma)^2 - 4\rho\tau \ge 0. \tag{7.2.27}$$

Then the numbers

$$\beta^- = (1 - \sigma - \sqrt{\Delta})/(2\tau) = \frac{2\rho}{1 - \sigma + \sqrt{\Delta}},$$

$$\beta^+ = (1 - \sigma + \sqrt{\Delta})/(2\tau)$$

are nonnegative, and the condition (7.2.8) of Theorem 7.2.1 is fulfilled for $([\Delta x]^T, [\Delta\lambda])^T = [-\beta, \beta]e \in \mathbb{IR}^{n+1}$ with arbitrary $\beta \in (\beta^-, \beta^+)$. In particular, all the assertions of that theorem hold.

If β is restricted to $[\beta^-, (\beta^- + \beta^+)/2)$, then the iterates of (7.2.9) converge to the error $\binom{\Delta x^*}{\Delta\lambda^*}$.

Proof. Notice that $\sigma < 1$ implies that C is nonsingular. Now apply Theorem 7.1.1 with

$$r = -C \begin{pmatrix} A\tilde{x} - \lambda\tilde{x} \\ 0 \end{pmatrix}, \quad S = I_{n+1} - C \begin{pmatrix} A - \tilde{\lambda} I_n & -\tilde{x} \\ (e^{(n)})^T & 0 \end{pmatrix},$$

$$t_{ijk} = \begin{cases} c_{ij}, & \text{if } j \in \{1, \ldots, n\} \text{ and } k = n+1 \\ 0 & \text{otherwise} \end{cases}, \quad i = 1, \ldots, n+1,$$

and notice that

$$\begin{aligned} g([\Delta x], [\Delta\lambda]) &\subseteq r + S \begin{pmatrix} [\Delta x] \\ [\Delta\lambda] \end{pmatrix} + C \left\{ \begin{pmatrix} 0 & [\Delta x] \\ 0 & 0 \end{pmatrix} \begin{pmatrix} [\Delta x] \\ [\Delta\lambda] \end{pmatrix} \right\} \\ &= r + S \begin{pmatrix} [\Delta x] \\ [\Delta\lambda] \end{pmatrix} + C \left\{ \begin{pmatrix} [\Delta x] \\ 0 \end{pmatrix} [\Delta\lambda] \right\} \end{aligned} \tag{7.2.28}$$

holds because of Theorem 3.1.2 (f). Denote the quantities in (7.1.2) and (7.1.4) by $\hat{\rho}, \hat{\sigma}, \hat{\tau}, \hat{\beta}^-, \hat{\beta}^+$ in order to distinguish them from those in Theorem 7.2.7. Then $\rho = \hat{\rho}$, $\sigma = \hat{\sigma}$, $\tau \ge \hat{\tau}$, hence $\hat{\beta}^- \le \beta^-$, $\beta^+ \le \hat{\beta}^+$, and Theorem 7.1.1 proves the assertion. □

The subset property in (7.2.28) led us to the representation (7.2.5) for g instead of using the last expression in (7.2.28) although equality holds there for point intervals and although the quadratic small term of g is better visible by the last expression of (7.2.28). In practical computation, one can store the major part of the matrix within braces in (7.2.5) and needs to update only its last column

$$e^{(n+1)} + C\begin{pmatrix} \tilde{x} + [\Delta x] \\ 0 \end{pmatrix}$$

in each step of the iteration (7.2.9). Apart from the initialization, this effects essentially two matrix vector multiplications per step and is not more work than iterating with the interval arithmetic evaluation of the last expression in (7.2.28).

Notice that good approximations $(\tilde{x}, \tilde{\lambda}) \approx (x^*, \lambda^*)$ and $C \approx B^{-1}$ yield $\rho \approx 0$, $\sigma \approx 0 < 1$ and $\Delta \approx 1 \geq 0$ such that the assumptions (7.2.27) are fulfilled. In order to satisfy the condition (7.2.8), we need β from the open interval (β^-, β^+) as the proof of Theorem 7.1.1 shows. Uniqueness of (x^*, λ^*) follows then from Theorem 7.2.1. Theorem 7.1.1 guarantees convergence of the iterates (7.2.9) for β even from the closed interval $[\beta^-, \beta^+]$, but uniqueness of (x^*, λ^*) only for $\beta \in [\beta^-, (\beta^- + \beta^+)/2)$. Experience shows that in practical computation rounding errors often prevent verification when choosing $\beta = \beta^-$, in contrast to the theory. Therefore, in our examples we often used the geometric mean

$$\beta = g_m = \sqrt{\beta^- \cdot (\beta^- + \beta^+)/2} \tag{7.2.29}$$

which lies between β^- and $(\beta^- + \beta^+)/2$.

Theorem 7.2.7 holds similarly for $\hat{g}$ and $\hat{B}$ from (7.2.23) and (7.2.25). One only has to replace ρ, σ, τ by

$$\rho = \|\hat{C}(\tilde{\lambda}\tilde{x} - A\tilde{x})\|_\infty, \qquad \sigma = \|I - \hat{C}\hat{B}\|_\infty, \qquad \tau = \|\hat{C}\|_\infty,$$

and $[\Delta x]$ by $[\Delta\hat{x}] \in \mathbb{IR}^{n-1}$. Refer to Alefeld [10] or Mayer [208] for details.

The advantage of $\hat{g}$ and $\hat{B}$ is that the dimension is now one less than with g and B and that $\tilde{x}_n = 1 = x_n^*$ and $\Delta x_n^* = 0$ hold rigourously since the component $[\Delta x]_n^{(k)}$ no longer occurs in the iteration explicitly and is defined to be zero.

We close this subsection with two examples in which we computed the approximations of the eigenvalues and eigenvectors by using the MATLAB function `eig` and a normalization of the last component of the eigenvectors to be one. The approximate inverse C of the matrix B in (7.2.20) is obtained by the MATLAB instruction `B\eye(n)`, where n is the dimension of B. We verify the eigenpairs by means of INTLAB via (7.2.8) and Theorem 7.2.7 using the geometric mean $\beta = g_m$ in (7.2.29) and the enclosure `AccDot(·,·,[])` of the accurate dot product of INTLAB for $f(\tilde{x}, \tilde{\lambda})$ in (7.2.5). We start the iterative process (7.2.9) with $[-g_m, g_m]e$, and we stop it similarly as in Remark 5.5.1 whenever two successive iterates coincide completely for the first time.

Instead of listing the computed enclosures for the eigenvectors we mention only upper bounds of their modified relative width $q([x])$ which we define for $[x] \in \mathbb{IR}^n$ by

$$q = q([x]) = \max_{i=1,\ldots,n} q_i \tag{7.2.30}$$

with

$$q_i = \begin{cases} d([x]_i)/|[x]_i| < 1, & \text{if } 0 \notin [x]_i, \\ d([x]_i), & \text{if } 0 \in [x]_i, \end{cases} \quad i = 1, \ldots, n.$$

Obviously,

$$[x]_i \subseteq \begin{cases} \operatorname{sign}(\check{x}_i)|[x]_i|[1-q, 1], & \text{if } 0 \notin [x]_i, \\ [-q, q], & \text{if } 0 \in [x]_i, \end{cases} \quad i = 1, \ldots, n$$

holds.

Example 7.2.8. The symmetric matrix

$$A_n = (a_{ij}) = \begin{pmatrix} n & n-1 & n-2 & \ldots & 2 & 1 \\ n-1 & n-1 & n-2 & \ldots & 2 & 1 \\ n-2 & n-2 & n-2 & \ldots & 2 & 1 \\ \vdots & \vdots & \vdots & \ddots & \vdots & \vdots \\ 2 & 2 & 2 & \ldots & 2 & 1 \\ 1 & 1 & 1 & \ldots & 1 & 1 \end{pmatrix} \in \mathbb{R}^{n\times n}$$

is defined by $a_{ij} = n + 1 - \max\{i, j\}$. It is a test matrix from Gregory, Karney [122] whose eigenvalues are known to be

$$\lambda_i^* = \frac{1}{2}\left(1 - \cos\frac{(2i-1)\pi}{2n+1}\right)^{-1}, \quad i = 1, \ldots, n;$$

cf. also Mayer [210]. We chose $n = 100$ and verified eigenpairs for λ_1, $\lambda_{n/2}$ and λ_n after having ordered the eigenvalues decreasingly. We list our results in Table 7.2.1 where $[\lambda]_i$ denotes the enclosure of the eigenvalue λ_i^*. Only $k + 1 = 4$ iterates had to be computed in order to fulfill the stopping criterion for $i = 1$ and $i = 50$, and five iterates for $i = 100$. □

Tab. 7.2.1: Results of Example 7.2.8.

i	$[\lambda]_i$	$q([x])$
1	$4.093\,560\,474\,685\,3_1^2 \cdot 10^3$	$8.7 \cdot 10^{-16}$
50	$0.512\,002\,660\,043\,1_2^3$	$3.1 \cdot 10^{-13}$
100	$0.250\,061\,082\,720_{69}^{70}$	$1.3 \cdot 10^{-12}$

Our second example illustrates the verification process for an unsymmetric matrix which occurs in connection with the Riemann hypothesis (cf. Roesler [296]).

Example 7.2.9. Let the $(n-1)\times(n-1)$ matrix A_n be defined by

$$A_n = (a_{ij}) = \begin{pmatrix} 1 & -1 & 1 & -1 & 1 & -1 & 1 & \dots & \vdots \\ & 2 & -1 & -1 & 2 & -1 & -1 & \dots & \vdots \\ & & 3 & -1 & -1 & -1 & 3 & \dots & \vdots \\ & & & 4 & -1 & -1 & -1 & \dots & \vdots \\ & & & & 5 & -1 & -1 & \dots & \vdots \\ & & & & & 6 & -1 & \dots & \vdots \\ & & -1 & & & & 7 & \dots & \vdots \\ & & & & & & & \ddots & \vdots \\ & & & & & & & & n-1 \end{pmatrix}$$

with

$$a_{ij} = (i+1)\delta_{(i+1)|(j+1)} - 1$$

and

$$\delta_{i|j} = \begin{cases} 1, & \text{if } i \text{ is a divisor of } j \\ 0 & \text{otherwise} \end{cases} .$$

Roesler showed in [296] that the Riemann hypothesis is valid if and only if the determinant $\det(A_n)$ of A_n increases for any $\varepsilon > 0$ at most like

$$\det(A_n) = O(n!\, n^{-1/2+\varepsilon}), \qquad n \to \infty.$$

Since $\det(A_n)$ can be represented as the product of the eigenvalues of A_n (cf. Theorem 1.8.1), we get a relationship to the subject of this section. For $n = 11$, which means $A_n \in \mathbb{R}^{10\times 10}$, we listed in Table 7.2.2 the enclosures $[\lambda]_i$ for the eigenvalues λ_i^* and upper bounds $q([x])$ for the modified relative widths (7.2.30) of the corresponding eigenvectors. Only four iterates had to be computed in order to fulfill the stopping criterion.

As one might guess, the exact values of λ_4^* and λ_5^* are 4 and 5, respectively; cf. Roesler [296]. □

Tab. 7.2.2: Results of Example 7.2.9.

i	$[\lambda]_i$	$q([x])$
1	$-0.019\,702\,143\,297\,5_5^4$	$3.4 \cdot 10^{-16}$
2	$0.375\,851\,705\,484\,4_6^7$	$1.6 \cdot 10^{-15}$
3	$2.714\,315\,143\,311\,9_3^4$	$2.8 \cdot 10^{-15}$
4	$4.0 + [-1, 1] \cdot 10^{-14}$	$3.1 \cdot 10^{-15}$
5	$5.0 + [-1, 1] \cdot 10^{-14}$	$4.8 \cdot 10^{-15}$
6	$6.534\,132\,065\,892\,6_3^5$	$9.6 \cdot 10^{-15}$
7	$7.314\,390\,058\,013\,4_1^3$	$1.1 \cdot 10^{-14}$
8	$8.655\,903\,539\,939\,0_0^1$	$7.9 \cdot 10^{-15}$
9	$9.588\,680\,211\,084\,1_4^6$	$3.5 \cdot 10^{-14}$
10	$10.836\,429\,419\,571\,9_3^4$	$4.7 \cdot 10^{-15}$

In passing, we note that Riemann's ζ function is defined in the complex half-plane $\operatorname{Re} z > 1$ by

$$\zeta(z) = \sum_{m=1}^{\infty} \frac{1}{m^z}. \tag{7.2.31}$$

For $\operatorname{Re} z \leq 1$, $z \neq 1$, it is defined as an analytical continuation of (7.2.31). This function has a simple pole in $z = 1$ and the only *real* zeros in $-2, -4, -6, \ldots$, which are usually called the *trivial* zeros. Riemann's hypothesis says that all nontrivial zeros of ζ are on the straight line $\operatorname{Re} z = \frac{1}{2}$. If this hypothesis is true, one could improve some error estimates on the prime number theorem

$$\pi(x) = \frac{x}{\ln x} + o\left(\frac{x}{\ln x}\right), \quad x \to \infty,$$

where $\pi(x)$ denotes the prime-counting function which counts the number of primes less than or equal to x. In 1914 Hardy showed in [134] that there are infinite zeros of ζ on $\operatorname{Re} z = \frac{1}{2}$. In 1986 Van de Lune, te Riele, and Winter showed in [196] by means of a vector computer running approximately 1 500 hours that in the rectangle

$$\{z \mid 0 < \operatorname{Re} z < 1\} \cap \{z \mid 0 < \operatorname{Im} z \leq 545\,439\,823.215\}$$

there are exactly 1 500 000 001 zeros of ζ, and that these zeros are all on the line $\operatorname{Re} z = \frac{1}{2}$.

For some other matrices we realized that the iterative process (7.2.9) as described above does not work satisfactory. We found out that a crucial obstacle was the approximate inverse C of the matrix B in (7.2.20), which was often computed too roughly. This was also the reason why we preferred $C = B \setminus$ `eye(n)` instead of applying MATLAB's function $C =$ `inv(B)`. An even higher precision for the inverse C can be achieved when using some kind of staggered correction as described in Rump [321], where among others INTLAB's exact dot product `AccDot(·,·,·)` is used based on MATLAB's cell concept. We leave it to the reader to experiment with that; cf. Exercise G.3.

Exercises

Ex. 7.2.1. Let $C = B^{-1}$ for $B \in \mathbb{R}^{(n+1)\times(n+1)}$ from (7.2.20). Prove that C has the block form (7.2.21) and that

$$\hat{C} = \begin{pmatrix} C_{11} \\ C_{21} \end{pmatrix} \in \mathbb{R}^{n\times n}$$

with blocks as there satisfies $\hat{C} = \hat{B}^{-1}$ for $\hat{B}$ in (7.2.25).

Ex. 7.2.2. Show that Theorem 7.1.1 can also be applied to the function $\hat{g}$ of Section 7.2 with

$$\rho = \|\hat{C}(A\tilde{x} - \tilde{\lambda}\tilde{x})\|_\infty, \quad \sigma = \|I - \hat{C}\hat{B}\|_\infty, \quad \tau = \|\hat{C}\|_\infty.$$

Ex. 7.2.3. Which statements of Section 7.2 remain true or can be reformulated correspondingly if the normalization $\tilde{x}_n = 1$ is replaced by the Euclidean normalization $\tilde{x}^T\tilde{x} = 1$?

7.3 Lohner method

The Lohner method is tailored for verifying and enclosing eigenvalues and – to a lesser extent – eigenvectors of *symmetric* matrices A. It is presented in Lohner [193] and essentially based on a similarity transformation $T^{-1}AT$ and on Gershgorin's theorem 1.8.19 combined with a tricky switching between single and multiple precision computation and an appropriate application of interval analysis in the later stage of the algorithm. Eigenvectors are enclosed using Wilkinson's estimation in Theorem 1.8.25. Since the transformation matrix T results from at least one application of the Jacobi method for the algebraic eigenvalue problem, we recall this method first. Its idea is to zero some off-diagonal entry $a_{pq} = a_{qp}$, $p < q$, of A by means of an appropriate orthogonal transformation $J_{pq}^T A J_{pq} = A'$ and to repeat this step with A' and its succeeding transformed matrices in a prescribed way. The zeros generated in one step are normally destroyed in the next one. The orthogonal matrices $J_{pq} \in \mathbb{R}^{n\times n}$ are Jacobi (or Givens) rotations (cf. Golub, van Loan [121]) which are defined by

$$J_{pq} = \begin{pmatrix} 1 & & & & & & & & & & \\ & \ddots & & & & & & & & O & \\ & & 1 & & & & & & & & \\ & & & c & 0 & \cdots & 0 & -s & & & \\ & & & 0 & 1 & & & 0 & & & \\ & & & \vdots & & \ddots & & \vdots & & & \\ & & & 0 & & & 1 & 0 & & & \\ & & & s & 0 & \cdots & 0 & c & & & \\ & & & & & & & & 1 & & \\ & O & & & & & & & & \ddots & \\ & & & \uparrow & & & & \uparrow & & & 1 \\ & & & p & & & & q & & & \end{pmatrix} \begin{matrix} \\ \\ \\ \leftarrow p \\ \\ \\ \\ \leftarrow q \\ \\ \\ \\ \\ \end{matrix} \tag{7.3.1}$$

with $c = \cos\varphi$, $s = \sin\varphi$. The angle φ in (7.3.1) is determined such that $\varphi \in [-\pi/4, \pi/4]$ and $a'_{pq} = 0$ holds. In our programs we used the so-called row cyclic variant of the Jacobi method in which (p, q) is chosen in a row cyclic way, i.e., (p, q) runs through the scheme

$$\begin{array}{ccccccc} (1,2) & \to & (1,3) & \to & \dots & \to & (1,n) \\ & & & & & & \hookleftarrow \\ & & (2,3) & \to & \dots & \to & (2,n) \\ & & & & \ddots & & \vdots \\ & & & & & & (n-1,n) \end{array}$$

row-wise from left to right starting with $(1, 2)$, passing through $(1, n)$ in the first row, and continuing with $(2, 3)$ in the second row, etc.; one restarts with the index pair $(1, 2)$ if one has arrived at $(n - 1, n)$. The entries of A' are computed via the following procedure, provided that $a_{pq} \neq 0$.

Algorithm 7.3.1 (Jacobi method; one step).

$$\theta = \frac{a_{pp} - a_{qq}}{2a_{pq}} \quad (= \cot(2\varphi))$$

$$t = \tan\varphi = \begin{cases} \dfrac{1}{\theta + (\operatorname{sign}\theta)\sqrt{\theta^2 + 1}} & \text{if } \theta \neq 0 \\ 1 & \text{if } \theta = 0 \end{cases}$$

$$c = \frac{1}{\sqrt{1 + t^2}}; \quad s = ct; \quad \tau = \tan\left(\frac{\varphi}{2}\right) = \frac{s}{1 + c}$$

$$a'_{pp} = a_{pp} + ta_{pq}$$
$$a'_{qq} = a_{qq} - ta_{pq}$$
$$a'_{pq} = a'_{qp} = 0$$
$$a'_{pj} = a'_{jp} = a_{pj} + s(a_{qj} - \tau a_{pj}), \quad j \notin \{p, q\}$$
$$a'_{qj} = a'_{jq} = a_{qj} - s(a_{pj} + \tau a_{qj}), \quad j \notin \{p, q\}$$
$$a'_{ij} = a_{ij}, \quad i, j \notin \{p, q\}. \qquad \square$$

Notice that sign(θ) in the definition of t prevents cancellation. In addition, it is responsible for $\varphi \in [-\pi/4, \pi/4]$, which is necessary for the convergence of the method; cf. Excercise 7.3.1 otherwise. But an explicit computation of φ is not necessary.

Instead of assuming $a_{pq} \neq 0$ for Algorithm 7.3.1, one often uses a threshold variant of the Jacobi method, skipping the computational process for an index pair (p, q) if $|a_{pq}| \leq \varepsilon_k$ with $\varepsilon_k > 0$ being some given small number acting as threshold. If $|a_{ij}| < \varepsilon_k$ for all $i < j$, then ε_k is replaced by some $\varepsilon_{k+1} \in (0, \varepsilon_k)$ such that the sequence $\{\varepsilon_k\}_{k=1}^{\infty}$ decreases strictly to zero. Taking into account $J_{pq}^T = J_{pq}^{-1}$, it is known (Appendix D, Forsythe, Henrici [99], Henrici [141]) that the (infinite) row cyclic variant of the Jacobi method converges to the Jordan normal form of A, which, by the symmetry of A, is a diagonal matrix with the eigenvalues of A as diagonal entries. The product of the transformation matrices J_{pq} tends to an orthogonal transformation matrix whose columns are pairwise orthonormal eigenvectors of A.

In the sequel we will also need the following lemma which provides an enclosure of T^{-1}.

Lemma 7.3.2. *Let*

$$\|I - CT\|_\infty < 1 \text{ hold for } C, T \in \mathbb{R}^{n\times n}. \tag{7.3.2}$$

Then C, T are nonsingular, and

$$T^{-1} \in C + [-\alpha, \alpha]ee^T \quad \text{with } \alpha = \|C\|_\infty \frac{\|I - CT\|_\infty}{1 - \|I - CT\|_\infty}.$$

Proof. The existence of C^{-1}, T^{-1} follows immediately from (7.3.2). By Neumann's series, we obtain

$$\begin{aligned} T^{-1} - C &= (I - CT)\{I - (I - CT)\}^{-1}C \\ &= (I - CT)\sum_{k=0}^{\infty}(I - CT)^k C, \end{aligned}$$

hence $\|T^{-1} - C\|_\infty \leq \|I - CT\|_\infty \sum_{k=0}^{\infty}\|I - CT\|_\infty^k \|C\|_\infty = \alpha$. $\square$

Lohner's method proceeds in several steps.

Step 1. In the first step A is transformed approximately to diagonal form by means of a nearly orthogonal transformation matrix S. This can be achieved by using any traditional algorithm in single precision which aims at such a transformation. Lohner uses the row cyclic variant of the Jacobi method which he stops with the transformed matrix $\tilde{D}$ and the transformation matrix S as product of the Jacobi rotations J_{pq}. The diagonal entries of $\tilde{D}$ are obviously approximations of the eigenvalues of A, the columns of S are pairwise (nearly) orthonormal approximations of corresponding eigenvectors. With $\tilde{d}_{ii}$ we construct the diagonal matrix D by $d_{ii} = \tilde{d}_{ii}$. In our programs we applied MATLAB's function `eig` in order to obtain D and S, where D contains the computed eigenvalues in its diagonal and where S is the computed matrix of eigenvectors. This matrix may not be orthogonal; it will be orthogonalized in Step 2.

Since we cannot expect that the computed columns of S are 'perfect' eigenvectors, we try to improve them in Steps 2–5. Our final transformation matrix T will be computed in Step 5.

Step 2. Reorthogonalize the columns of S using the Gram–Schmidt method in double precision. 'Double precision' can be realized using a staggered correction format as described in Section 2.7. In our programs we applied MATLAB's cell concept and INTLAB's command `AccDot`; cf. Appendix G. If S is regular and if s_j denotes the j-th column of S, the formulae for the orthogonalization process are the following:

$$\tilde{t}_1 = s_1/\sqrt{s_1^T s_1}, \qquad \left.\begin{aligned} r_j &= s_j - \sum_{k=1}^{j-1} (\tilde{t}_k^T s_j)\,\tilde{t}_k \\ \tilde{t}_j &= r_j/\sqrt{r_j^T r_j} \end{aligned}\right\}, \quad j = 2, \ldots, n.$$

Assume that the output in Step 2 is the matrices $\tilde{T}_1$, $\tilde{T}_2$ which form the components of the staggered correction format such that $\tilde{T} = \tilde{T}_1 + \tilde{T}_2$ is now a better 'nearly orthogonal' matrix.

Step 3. Compute $\tilde{T}^T A \tilde{T}$ and split the result into

$$\tilde{T}^T A \tilde{T} = D + \tilde{A}_1 \tag{7.3.3}$$

with D from Step 1. We get

$$\tilde{A}_1 = \tilde{T}^T A \tilde{T} - D \approx \tilde{A}_2 := \tilde{T}^T (A\tilde{T} - \tilde{T}D), \tag{7.3.4}$$

using

$$\tilde{T}^T \tilde{T} \approx I \tag{7.3.5}$$

by virtue of Step 2. The expression within brackets in (7.3.4) is evaluated as

$$A\tilde{T}_1 + A\tilde{T}_2 - \tilde{T}_1 D - \tilde{T}_2 D.$$

Here the computation must be done with an exact dot product as provided in programming languages like PASCAL-XSC, C-XSC, and by the command `AccDot` in INTLAB; cf. Klatte et al. [164, 165], Rump [320]. The final computation of $\tilde{A}_2$ can be done in simple precision since we need only an approximation of $\tilde{A}_1$. By virtue of the choice of $\tilde{T}$ we can expect that the entries of $\tilde{A}_1$ and $\tilde{A}_2$ are small. By (7.3.4) and (7.3.5) we can assume that, in addition, $\tilde{A}_2$ is symmetric – at least when computed in single precision. In fact, a small deviation from symmetry does not matter since we still try to improve S from Step 1. If necessary, just reflect the entries in the strict upper triangle of $\tilde{A}_2$ at its diagonal and denote the new matrix again by $\tilde{A}_2$.

Step 4. With D from Step 1 we consider

$$\tilde{A} = D + \tilde{A}_2 \tag{7.3.6}$$

instead of $D + \tilde{A}_1$ and apply the row cyclic Jacobi method to $D + \tilde{A}_2$. The splitting in (7.3.6) has the advantage that the entries of $\tilde{A}_2$ are all small and of approximately the same magnitude. The formulae of the Jacobi method can now be applied as in Algorithm 7.3.1 – essentially in single precision. Only for θ – particularly in the presence of an eigenvalue cluster – one should compute

$$d_{pp} + (\tilde{A}_2)_{pp} - d_{qq} - (\tilde{A}_2)_{qq}$$

'exactly', for instance via the INTLAB command

$$\texttt{AccSum}(\,[d_{pp}, (\tilde{A}_2)_{pp}, -d_{qq}, -(\tilde{A}_2)_{qq}]\,)$$

or via

$$\texttt{AccDot}(\,[d_{pp}, (\tilde{A}_2)_{pp}, -d_{qq}, -(\tilde{A}_2)_{qq}]\,,\texttt{ones(4,1)}).$$

Denote by $\tilde{A}^{\text{new}} = (\tilde{a}_{ij}^{\text{new}})$ the matrix resulting from $\tilde{A}$ by a single Jacobi rotation which zeros $\tilde{a}_{pq}$. Then for the new entries $\tilde{a}_{pj}^{\text{new}} = \tilde{a}_{jp}^{\text{new}}$ and $\tilde{a}_{qj}^{\text{new}} = \tilde{a}_{jq}^{\text{new}}$, $j = 1, \ldots, n$, one can skip the matrix D: For $j \notin \{p, q\}$ its entries do not occur in the formulae, and for $\tilde{a}_{pp}^{\text{new}}$ one gets $\tilde{a}_{pp}^{\text{new}} = d_{pp} + (\tilde{A}_2)_{pp} + t(\tilde{A}_2)_{pq}$. Therefore, if one splits $\tilde{A}^{\text{new}}$ into $\tilde{A}^{\text{new}} = D + \tilde{A}_2^{\text{new}}$ similarly as in (7.3.6), then one only has to update $\tilde{A}_2$ in (7.3.6) which leads to $(\tilde{A}_2^{\text{new}})_{pp} = (\tilde{A}_2)_{pp} + t(\tilde{A}_2)_{pq}$ without using D. This holds similarly for $(\tilde{A}_2^{\text{new}})_{qq}$.

Since $\tilde{A}$ was nearly diagonal, one can expect that $\tilde{a}_{pq}$ is small. This implies θ in Algorithm 7.3.1 to be large (unless $\tilde{a}_{pp} \approx \tilde{a}_{qq}$) whence $t \approx 0$ and $s \approx 0$. Thus the transformation matrix J_{pq} is nearly the identity matrix. Therefore, it makes sense to consider the deviation from the identity $W_{pq} := J_{pq} - I$ instead of J_{pq}. Notice that aside from $(W_{pq})_{pp} = (W_{pq})_{qq} = c - 1$, $(W_{pq})_{pq} = -(W_{pq})_{qp} = -s$ the entries of W_{pq} are zero. The transformation update $S^{\text{new}} = I + W^{\text{new}}$ of the previous transformation matrix $S = I + W$ can then be written as $S^{\text{new}} = S \cdot J_{pq} = (I + W)(I + W_{pq})$, whence $W^{\text{new}} = W + W_{pq} + W \cdot W_{pq}$ (still computed in single precision). One can stop if $\max_{i,j} |\tilde{a}_{ij}^{\text{new}}| \le \texttt{eps}$ (machine precision) holds after having run through full Jacobi cycles.

Step 5. Now we combine $\tilde{T}$ from Step 2 with the final transformation matrix $\tilde{S} = I + \tilde{W}$ of the Jacobi method in Step 4. Since $\tilde{T} = \tilde{T}_1 + \tilde{T}_2$ was computed in double precision we do the same for

$$\tilde{T}\tilde{S} = (\tilde{T}_1 + \tilde{T}_2)(I + \tilde{W}) = \tilde{T}_1 + \tilde{T}_2 + \tilde{T}_1\tilde{W} + \tilde{T}_2\tilde{W},$$

in INTLAB for instance via `AccDot([`$\tilde{T}_1, \tilde{T}_2, \tilde{T}_1, \tilde{T}_2$`],[`$I; I; \tilde{W}; \tilde{W}$`],2)`. In staggered correction format (or as corresponding cells in MATLAB) one gets T_1, T_2 with $T := T_1 + T_2$. This *computed* matrix T is our final transformation matrix which we will apply to the original matrix A. Its columns are eigenvector approximations which often have double accuracy.

Step 6. Now we consider the similarity transformation $T^{-1}AT$. Although we computed precisely, we cannot expect that $T^{-1} = T^T$ holds. Even so, $T^{-1}AT$ is normally not exactly available on a computer. Therefore, we need a tight enclosure of $A^{\text{new}} = T^{-1}AT$ by an interval matrix. To this end we split this matrix into $A^{\text{new}} = D + A_1$, similarly as in (7.3.6). Then

$$A_1 = T^{-1}(AT - TD) = T^{-1}(AT_1 + AT_2 - T_1D - T_2D), \tag{7.3.7}$$

where the expression in brackets must be enclosed tightly by an interval matrix, for instance, via INTLAB's command `AccDot`, this time with the option `[ ]`. The inverse T^{-1} is enclosed by an interval matrix $[T^{-1}]$ using Lemma 7.3.2 with $C = T^T$ and single precision. The assumption (7.3.2) can be expected to be fulfilled because of $T^TT \approx I$. Then

$$T^{-1} \in [T^{-1}] := T^T + [-\alpha, \alpha]ee^T \tag{7.3.8}$$

holds with

$$\alpha = \|T^T\|_\infty \frac{\eta}{1-\eta}, \quad \eta = \|I - T^TT\|_\infty .$$

In order to guarantee $T^{-1} \in [T^{-1}]$ we must compute α , $[T^{-1}]$, and all norms by means of interval arithmetic. To this end we notice that the factor $\eta/(1-\eta) = -1 + 1/(1-\eta)$ increases if η does, and the INTLAB command `sup(norm([A],inf))` delivers an upper bound of $\| |[A]| \|_\infty$ for $[A] \in \mathbb{IR}^{n\times n}$.

With $[T^{-1}]$ we compute an enclosure $[A_1]$ of A_1 in (7.3.7). Then $A^{\text{new}} = T^{-1}AT \in D + [A_1]$ follows so that we can verify and enclose the eigenvalues of A in our next step.

Step 7. We apply Gershgorin's theorem 1.8.19 to $D + [A_1]$. Then the eigenvalues of A are contained in the union $[g]$ of the Gershgorin intervals

$$[g]_i = \left[\underline{d}_{ii} + (\underline{A}_1)_{ii} - \sum_{\substack{j=1\\ j\neq i}}^{n} |[A_1]_{ij}|,\ \overline{d}_{ii} + (\overline{A}_1)_{ii} + \sum_{\substack{j=1\\ j\neq i}}^{n} |[A_1]_{ij}| \right] \quad i = 1, \ldots, n,$$

where the bounds of $[g]_i$ must be computed using interval arithmetic or directed rounding.

If

$$[g]_i \cap \bigcup_{\substack{j=1\\ j\neq i}}^{n} [g]_j = \emptyset \qquad (7.3.9)$$

then we call a Gershgorin interval $[g]_i$ isolated.

Such an isolated interval $[g]_i$ contains exactly one eigenvalue λ^* of A by Theorem 1.8.19 (b). It is algebraically simple. This remark holds trivially for *all* eigenvalues of A if the intervals $[g]_i$, $i = 1, \ldots, n$, are pairwise disjoint. Such pairwise disjoint Gershgorin intervals are expected when dealing with the inverse eigenvalue problem described in Section 7.8. Clustering can also be verified by the method above by applying Theorem 1.8.19 (b) in a straightforward way.

The bounds $\underline{g}_i, \overline{g}_i$ can be improved if one intersects $[g]_i$ with additional enclosures based on estimates like those of Theorem 1.8.16.

Step 8: Let λ^* be a simple eigenvalue of A which is enclosed by an isolated Gershgorin interval $[g]_i$. In order to obtain an enclosure for a corresponding eigenvector x^* which is normalized by $(x^*)^T x^* = 1$, apply Theorem 1.8.25 with

$$\begin{aligned} \tilde{x} &= \frac{\tilde{T}_{*,i}}{\|\tilde{T}_{*,i}\|_2} \\ \tilde{\lambda} &= \operatorname{mid}([g]_i), \qquad \delta = \|A\tilde{x} - \tilde{\lambda}\tilde{x}\|_2, \\ a &= \min_{\substack{1\le j\le n\\ j\neq i}} \left\{\min\{|c| \mid c \in [g]_j - \tilde{\lambda}\}\right\} - \eta \quad (\eta \text{ any small number}). \end{aligned}$$

Define β as square root of the right-hand side in (1.8.43). If $a \ge \delta$ holds, then $x^* \in \tilde{x} + [-\beta, \beta]e$.

If $[g]_i$ is not isolated, then consider the connected component $[g]^c$ of $[g]$ to which $[g]_i$ belongs. Here one cannot decide whether $[g]^c$ contains multiple eigenvalues and/or two or more separate eigenvalues. We can only enclose a basis of eigenvectors corresponding to the eigenvalues in $[g]^c$. For details we refer to Kreß [176]. □

We close this section with two examples.

Example 7.3.3. The $n \times n$ Hilbert matrix $H_n = (h_{ij})$ is defined by $h_{ij} = 1/(i + j - 1)$. Unfortunately, H_n is not representable by machine numbers if $n > 1$. Therefore, we multiply H_n by the least common multiple $c_n = \operatorname{lcm}(1, \ldots, 2n - 1)$ such that the entries of the resulting so-called preconditioned Hilbert matrix $H_n^{\text{prec}} = c_n H_n$ are integers only and therefore exactly representable in MATLAB (if necessary as floating point numbers) on a computer for $n \le 18$. Typical values are $c_{11} = 232\,792\,560$ and $c_{18} = 144\,403\,552\,893\,600$. It is well known that H_n is symmetric, positive definite, and very ill-conditioned for linear systems $H_n x = b$. Eigenpairs for H_n with smaller dimension n are listed in Gregory, Karney [122]. The eigenvalues of H_n are smaller by the factor c_n than those of H_n^{prec}, the eigenvectors remain the same.

Using INTLAB for the Lohner method yields the following enclosures $[\lambda]_i$ for selected eigenvalues λ_i^*. All Gershgorin intervals are isolated. The results are contained in the Tables 7.3.1 and 7.3.2.

Case $n = 11$: Here the method needs $k = 2$ Jacobi cycles until the off-diagonal entries are smaller than MATLAB's machine precision `eps`.

Case $n = 18$: This time $k = 4$ Jacobi cycles are necessary. The smallest eigenvalue λ_1 is particularly difficult to verify. This is already stated in Lohner [193]. □

Tab. 7.3.1: Results of Example 7.3.3 in the case $n = 11$.

i	$[\lambda]_i$	i	$[\lambda]_i$
1	$7.899\,160\,434\,834_{179}^{203} \cdot 10^{-7}$	7	$4.130\,403\,774\,512\,8_{68}^{74} \cdot 10^{4}$
2	$1.817\,429\,864\,965\,0_{58}^{61} \cdot 10^{-4}$	8	$7.250\,171\,400\,862\,72_{0}^{4} \cdot 10^{5}$
3	$1.928\,297\,327\,418\,4_{18}^{21} \cdot 10^{-2}$	9	$9.38\,780\,029\,417\,18_{3}^{9} \cdot 10^{6}$
4	$1.249\,718\,543\,026\,62_{6}^{8} \cdot 10^{0}$	10	$8.435\,986\,343\,818_{797}^{805} \cdot 10^{7}$
5	$5.521\,279\,596\,429_{197}^{201} \cdot 10^{1}$	11	$4.131\,795\,990\,566\,00_{3}^{8} \cdot 10^{8}$
6	$1.755\,815\,982\,827\,37_{0}^{1} \cdot 10^{3}$		

Tab. 7.3.2: Results of Example 7.3.3 in the case $n = 18$.

i	$[\lambda]_i$
1	$1.22_{520}^{602} \cdot 10^{-11}$
2	$4.765\,3_{34}^{43} \cdot 10^{-9}$
5	$8.722\,494\,855\,3_{87}^{96} \cdot 10^{-3}$
10	$1.020\,114\,001\,053\,09_{1}^{2} \cdot 10^{6}$
18	$2.722\,318\,555\,347\,6_{58}^{63} \cdot 10^{14}$

Example 7.3.4. The Wilkinson matrix $W_{2n+1}^+ \in \mathbb{R}^{(2n+1)\times(2n+1)}$ is defined by

$$W_{2n+1}^+ = (w_{ij}^{(2n+1)}) = \begin{pmatrix} n & 1 & & & & & & & O \\ 1 & n-1 & 1 & & & & & & \\ & 1 & n-2 & 1 & & & & & \\ & & \ddots & \ddots & \ddots & & & & \\ & & & 1 & 0 & 1 & & & \\ & & & & \ddots & \ddots & \ddots & & \\ & & & & & 1 & n-2 & 1 & \\ & O & & & & & 1 & n-1 & 1 \\ & & & & & & & 1 & n \end{pmatrix};$$

see Wilkinson [362]. Let $n = 10$. We want to verify the two largest eigenvalues λ_{20}^* and λ_{21}^*, which are known to differ only in the sixteenth significant digit. Lohner's method delivers the enclosures

$$\lambda_{20}^* \in [10.746\,194\,182\,903\,3_1^3]$$

and

$$\lambda_{21}^* \in [10.746\,194\,182\,903_{39}^{40}].$$

Already one Jacobi cycle transforms the off-diagonal entries below eps, all Gershgorin intervals are isolated. □

Exercises

Ex. 7.3.1. Show that the row cyclic Jacobi method does not need to converge if the restriction $\varphi \in [-\pi/4, \pi/4]$ is replaced by $\varphi \in [-\pi/2, \pi/2]$. To this end use the example

$$A = A_0 = \begin{pmatrix} 2 & 0 & 1 \\ 0 & 3 & 0 \\ 1 & 0 & 4 \end{pmatrix}$$

from Forsythe, Henrici [99] and choose $\varphi = \pi/2$ for each Jacobi matrix J_{pq} in order to zero the nonzero entries in the strict upper triangle. Compute $A_1, A_2, \dots, A_6$ and show that $A_6 = A_0$ holds.

Ex. 7.3.2. Let $f \in C([0,1])$. Find the coefficients a_i of the polynomial $p(x) = a_0 + a_1 x + \dots + a_{n-1}x^{n-1}$ such that $\|f - p\|_2 := \left\{\int_0^1 [f(x) - p(x)]^2\,dx\right\}^{1/2}$ is minimal. To this end consider the square $\|f - p\|_2^2$ and show that one has to solve the linear system $H_n\,a = b$, where H_n is the $n \times n$ Hilbert matrix and $a = (a_0, a_1, \dots, a_{n-1})^T$. What does $b \in \mathbb{R}^n$ look like?

7.4 Double or nearly double eigenvalues

While the method in Section 7.2 is tailored for simple eigenvalues, we now address a real double eigenvalue or two different real eigenvalues with clustering being allowed. That means, we consider the following three cases:

(i) $\lambda_i^* \neq \lambda_j^*$ are two algebraically simple eigenvalues of A;
(ii) $\lambda_i^* = \lambda_j^*$ is a geometrically and algebraically double eigenvalue of A;
(iii) $\lambda_i^* = \lambda_j^*$ is an algebraically double but geometrically simple eigenvalue of A.

All three cases share the same property: The zero vector together with the real eigenvectors of A associated with $\lambda_i^*, \lambda_j^* \in \mathbb{R}$, and (in case (iii)) the corresponding real principal vectors span a two-dimensional subspace V of $\mathbb{R}^n$ which is invariant with respect to the mapping represented by A; i.e., $w \in V$ implies $Aw \in V$. Any pair u^*, v^* of linearly independent vectors from V are called generators of V since each element w of V can be represented as a linear combination $w = \mu u^* + \nu v^*$, $\mu, \nu \in \mathbb{R}$, of u^* and v^*. For each pair of generators u^*, v^* there is a matrix $M \in \mathbb{R}^{2\times 2}$ such that

$$A\,U^* = U^*\,M \tag{7.4.1}$$

holds with $U^* = (u^*, v^*) \in \mathbb{R}^{n\times 2}$.

In the cases (i) and (ii), u^*, v^* can be chosen to be eigenvectors x^*, y^*, in the case (iii) it is possible to choose u^* as eigenvector x^* and v^* as an associated principal vector y^* of degree two. Then M has the form

$$M = J = \begin{pmatrix} \lambda_i^* & \kappa \\ 0 & \lambda_j^* \end{pmatrix} \tag{7.4.2}$$

with $\kappa = 0$ in the cases (i), (ii) and $\kappa = 1$ for (iii). If u^*, v^* are linear combinations of x^*, y^*, one can obtain the underlying vectors x^*, y^* by essentially finding the eigenvalues of the 2×2 matrix M and by reducing it to the Jordan normal form (7.4.2). If $M = S^{-1}JS$ holds with some 2×2 matrix S and with J from (7.4.2), then (7.4.1) is equivalent to

$$A(U^*S) = (U^*S)(S^{-1}MS) = (U^*S)J$$

where the columns of U^*S are x^* and y^*, respectively. Although our aim consists in computing enclosures of eigenvectors and principal vectors, we will start with the general case (7.4.1) where u^*, v^* are any generators of V. First we remark that it is always possible to find two components i_1, i_2 out of $n \geq 2$ and two generators u^*, v^* of V for which

$$u_{i_1}^* = \alpha, \quad u_{i_2}^* = \beta, \quad v_{i_1}^* = \gamma, \quad v_{i_2}^* = \delta \tag{7.4.3}$$

holds with prescribed values $\alpha, \beta, \gamma, \delta \in \mathbb{R}$ provided that $\binom{\alpha}{\beta}$, $\binom{\gamma}{\delta}$ are linearly independent. This makes the generators unique. If u, v are generators which do not yet satisfy the normalizations (7.4.3), we make the ansatz

$$u^* = d_{11}u + d_{12}v, \quad v^* = d_{21}u + d_{22}v$$

requiring

$$\left.\begin{aligned} u_{i_1}^* &= d_{11}u_{i_1} + d_{12}v_{i_1} = \alpha \\ u_{i_2}^* &= d_{11}u_{i_2} + d_{12}v_{i_2} = \beta \end{aligned}\right\} \tag{7.4.4}$$

and

$$\left.\begin{aligned} v_{i_1}^* &= d_{21}u_{i_1} + d_{22}v_{i_1} = \gamma \\ v_{i_2}^* &= d_{21}u_{i_2} + d_{22}v_{i_2} = \delta \end{aligned}\right\}. \tag{7.4.5}$$

The two 2×2 systems (7.4.4), (7.4.5) have the same coefficient matrix

$$K = \begin{pmatrix} u_{i_1} & v_{i_1} \\ u_{i_2} & v_{i_2} \end{pmatrix}.$$

Therefore, both are uniquely solvable if and only if $\det K \neq 0$, i.e., $u_{i_1} v_{i_2} - u_{i_2} v_{i_1} \neq 0$. This is possible by using the same strategy as in Alefeld, Spreuer [48] and Dongarra, Moler, Wilkinson [85]: First define i_1 by

$$u_{i_1} = \max_{1 \le i \le n} |u_i|. \tag{7.4.6}$$

The value u_{i_1} certainly differs from zero because u, v are generators of V. In particular, they are linearly independent. With i_1 define i_2 via

$$u_{i_1} v_{i_2} - u_{i_2} v_{i_1} = \max_{1 \le j \le n} |u_{i_1} v_j - u_j v_{i_1}|. \tag{7.4.7}$$

If the left-hand side of (7.4.7) were zero, then $v = \frac{v_{i_1}}{u_{i_1}} u$, hence u, v are linearly dependent contradicting the generator property of u, v. Therefore, (7.4.3) is possible. Since $\binom{\alpha}{\beta}$, $\binom{\gamma}{\delta}$ are assumed to be linearly independent, the same holds for u^*, v^*. Hence they are generators of V. For ease of notation we assume $i_1 = n - 1$ and $i_2 = n$ in the sequel.

Let

$$f\colon \begin{cases} \mathbb{R}^{2n+4} \to \mathbb{R}^{2n+4} \\ \qquad z \mapsto f(z) \end{cases}$$

be defined by

$$z = (u^T, m_{11}, m_{21}, v^T, m_{12}, m_{22})^T, \quad f(z) = \begin{pmatrix} Au - m_{11}u - m_{21}v \\ u_{n-1} - \alpha \\ u_n - \beta \\ Av - m_{12}u - m_{22}v \\ v_{n-1} - \gamma \\ v_n - \delta \end{pmatrix}.$$

It is obvious that the vectors u^*, v^* in the representation

$$z^* = ((u^*)^T, m_{11}^*, m_{21}^*, (v^*)^T, m_{12}^*, m_{22}^*)^T$$

for a zero z^* of f are the generators of V satisfying (7.4.3).

Assume now that we are given real approximations $\tilde u, \tilde v, \tilde M$ of u^*, v^*, M, which are gathered in the vector $\tilde z = (\tilde u^T, \tilde m_{11}, \tilde m_{21}, \tilde v^T, \tilde m_{12}, \tilde m_{22})^T$. Let $\tilde u, \tilde v$ be linearly independent. Dongarra, Moler, Wilkinson [85] describe how to obtain $\tilde u, \tilde v, \tilde M$ using the QR algorithm. We proceed now as in Section 7.2, expanding the function $z - \tilde z - Cf(z)$ at $\tilde z$, where $C \in \mathbb{R}^{(2n+4)\times(2n+4)}$ is some nonsingular preconditioning matrix. Introducing

$$\Delta z = (\Delta u^T, \Delta m_{11}, \Delta m_{21}, \Delta v^T, \Delta m_{12}, \Delta m_{22})^T = z - \tilde z$$

we get

$$f(z) = 0 \Longleftrightarrow \Delta z = g(\Delta z) = -Cf(\tilde{z}) + (I_{2n+4} - CB)\Delta z + \tilde{T}\Delta z \tag{7.4.8}$$

with the $(2n+4) \times (2n+4)$ matrices

$$B = \begin{pmatrix} A - \tilde{m}_{11}I_n & -\tilde{u} & -\tilde{v} & -\tilde{m}_{21}I_n & 0 & 0 \\ (e^{(n-1)})^T & 0 & 0 & 0 & 0 & 0 \\ (e^{(n)})^T & 0 & 0 & 0 & 0 & 0 \\ -\tilde{m}_{12}I_n & 0 & 0 & A - \tilde{m}_{22}I_n & -\tilde{u} & -\tilde{v} \\ 0 & 0 & 0 & (e^{(n-1)})^T & 0 & 0 \\ 0 & 0 & 0 & (e^{(n)})^T & 0 & 0 \end{pmatrix} \tag{7.4.9}$$

and

$$\tilde{T} = C \begin{pmatrix} 0 & \Delta u & \Delta v & 0 & 0 & 0 \\ 0 & 0 & 0 & 0 & 0 & 0 \\ 0 & 0 & 0 & 0 & 0 & 0 \\ 0 & 0 & 0 & 0 & \Delta u & \Delta v \\ 0 & 0 & 0 & 0 & 0 & 0 \\ 0 & 0 & 0 & 0 & 0 & 0 \end{pmatrix} \tag{7.4.10}$$

which have the same block partitioning. Thus we arrive again at a quadratic function g, for which Theorem 7.1.1 applies with

$$r = -Cf(\tilde{z}), \quad S = I - CB, \tag{7.4.11}$$

and

$$t_{ijk} = \begin{cases} c_{i,j+k-(n+1)}, & \text{if } k \in \{n+1, 2n+3\} \text{ and } j \in \{1, \dots, n\}, \\ c_{i,j+k-(2n+4)}, & \text{if } k \in \{n+2, 2n+4\} \text{ and } j \in \{n+3, \dots, 2n+2\}, \\ 0 & \text{otherwise}, \end{cases} \tag{7.4.12}$$

$$i = 1, \dots, 2n+4.$$

Hence

$$\left.\begin{aligned} &\rho = \|Cf(\tilde{z})\|_\infty, \quad \sigma = \|I - CB\|_\infty, \\ &\tau = \|T\|_\infty = \left\|2|C|\,(\hat{e}^T, 0, 0, \hat{e}^T, 0, 0)^T\right\|_\infty \le 2\|C\|_\infty \end{aligned}\right\} \tag{7.4.13}$$

where $\hat{e} = (1, \dots, 1)^T \in \mathbb{R}^n$.

Then we get the following result in which we require the normalization (7.4.3) with

$$\alpha = \tilde{u}_{n-1}, \quad \beta = \tilde{u}_n, \quad \gamma = \tilde{v}_{n-1}, \quad \delta = \tilde{v}_n. \tag{7.4.14}$$

Theorem 7.4.1. *Let g, ρ, σ, τ be defined as in (7.4.8), (7.4.13), and let (7.4.14) hold. Assume*

$$\sigma < 1 \quad \textit{and} \quad \Delta = (1-\sigma)^2 - 4\rho\tau \ge 0, \tag{7.4.15}$$

and let

$$\beta^- = (1 - \sigma - \sqrt{\Delta})/(2\tau),$$

$$\beta^+ = (1 - \sigma + \sqrt{\Delta})/(2\tau).$$

(a) *If $\beta \in [\beta^-, \beta^+]$, then g has in $[-\beta, \beta]e \in \mathbb{R}^{2n+4}$ at least one fixed point Δz^*. With*

$$U^* = (u^*, v^*) \quad \text{and} \quad M = \begin{pmatrix} m_{11}^* & m_{12}^* \\ m_{21}^* & m_{22}^* \end{pmatrix},$$

and

$$z^* = \tilde{z} + \Delta z^* = ((u^*)^T, m_{11}^*, m_{21}^*, (v^*)^T, m_{12}^*, m_{22}^*)^T,$$

the equation (7.4.1) holds. In particular, u^*, v^* are linearly independent vectors which are normalized by (7.4.14) and which are generators of an invariant two-dimensional subspace of $\mathbb{R}^n$.

The iteration

$$[\Delta z]^{(k+1)} = g([\Delta z]^{(k)}), \quad k = 0, 1, \dots \tag{7.4.16}$$

converges to some interval vector $[\Delta z]^*$ with

$$\Delta z^* \in [\Delta z]^* \subseteq [\Delta z]^{(k)} \subseteq \dots \subseteq [\Delta z]^{(0)}, \quad k = 0, 1, \dots. \tag{7.4.17}$$

(b) *If $\beta \in [\beta^-, (\beta^- + \beta^+)/2)$, then g has a unique fixed point Δz^* in $[\Delta z]^{(0)} = [-\beta, \beta]e \in \mathbb{R}^{2n+4}$, and (7.4.17) holds with $[\Delta z]^* = \Delta z^*$, i.e., (7.4.16) converges by contracting to Δz^*.* □

We notice that C^{-1}, B^{-1} exist since $\sigma < 1$. Choosing $C = B^{-1}$ leads to the block form

$$\begin{array}{c} \begin{pmatrix} C_{11} & C_{12} & C_{13} & C_{14} \\ 0 & 1\,0 & 0 & 0 \\ 0 & 0\,1 & 0 & 0 \\ C_{41} & C_{42} & C_{43} & C_{44} \\ 0 & 0 & 0 & 1\,0 \\ 0 & 0 & 0 & 0\,1 \\ C_{71} & C_{72} & C_{73} & C_{74} \end{pmatrix} \begin{array}{l} \\ \leftarrow n-1 \quad \text{(row index)} \\ \leftarrow n \\ \\ \leftarrow 2n+1 \\ \leftarrow 2n+2 \\ \\ \end{array} \\ \begin{array}{cccc} \underbrace{}_{n} & \underbrace{}_{2} & \underbrace{}_{n} & \underbrace{}_{2} \end{array} \quad \leftarrow \text{number of columns} \end{array}$$

whence $g_i(\Delta z) = 0$ for $i \in \{n-1, n, 2n+1, 2n+2\}$ and for any $\Delta z \in \mathbb{R}^{2n+4}$ with $\tilde{z}$ satisfying (7.4.14). Thus, as in the previous subsection, it makes sense to start with interval vectors $[\Delta z]$ for which $[\Delta z]_i = 0$, $i \in \{n-1, n, 2n+1, 2n+2\}$ holds, shrinking the matrices and vectors in g to get a new function $\hat{g}\colon \mathbb{R}^{2n} \to \mathbb{R}^{2n}$ with modifications analogous to those in Section 7.2. Essentially, the method (7.4.16) has been presented in Alefeld, Spreuer [48] in this form.

We want to consider another specialization of (7.4.16). To this end, let $\tilde{m}_{21} = 0$ and $\tilde{m}_{12} \in \{0, 1\}$. Then $\tilde{M}$ has Jordan normal form, hence $\tilde{m}_{11}, \tilde{m}_{22}$ can be thought to approximate eigenvalues λ^*, μ^*, and $\tilde{u}, \tilde{v}$ can be considered to approximate corresponding eigenvectors and/or principal vectors according to the cases (i)–(iii). The matrix B has the block form

$$B = \begin{pmatrix} B_{11} & 0 \\ B_{21} & B_{22} \end{pmatrix} \quad \text{with } B_{ii} = \begin{pmatrix} A - \tilde{m}_{ii} I_n & -\tilde{u} & -\tilde{v} \\ (e^{(n-1)})^T & 0 & 0 \\ (e^{(n)})^T & 0 & 0 \end{pmatrix}, \quad i = 1, 2.$$

Hence B is nonsingular if and only if B_{ii}^{-1} exists for $i = 1, 2$. This certainly holds if one of the cases (i)–(iii) is handled and if the approximation $\tilde{z}$ is sufficiently good, as can be immediately seen from the subsequent result which reminds us of Theorem 1.8.3.

Theorem 7.4.2. *Let*

$$B_i^* = \begin{pmatrix} A - \lambda_i^* I_n & -u^* & -v^* \\ (e^{(n-1)})^T & 0 & 0 \\ (e^{(n)})^T & 0 & 0 \end{pmatrix} \in \mathbb{R}^{(n+2)\times(n+2)}, \quad i = 1, 2,$$

with λ_1^*, λ_2^* being two real eigenvalues of A ($\lambda_1^* = \lambda_2^*$ being allowed) and with u^*, v^* being two real linearly independent vectors from the largest invariant real subspace V belonging to λ_1^*, λ_2^* which satisfy

$$u_{n-1}^* \neq 0, \quad u_{n-1}^* v_n^* - u_n^* v_{n-1}^* \neq 0. \tag{7.4.18}$$

Let (7.4.1) hold with $U^* = (u^*, v^*)$. Then the following statements are equivalent.
(a) *Either λ_1^*, λ_2^* are two different algebraically simple eigenvalues of A or $\lambda_1^* = \lambda_2^*$ is an algebraically double eigenvalue of A.*
(b) *B_1 and B_2 are both nonsingular.*

Proof. Let (a) hold and let u^*, v^* be generators of the invariant two-dimensional subspace V associated with λ_1^*, λ_2^*. Assume that u^*, v^* satisfy (7.4.18) and that B_1^* is singular. (If B_2^* is singular the proof proceeds analogously.) Then there is a vector $w \in \mathbb{R}^{n+2} \setminus \{0\}$ such that

$$B_1^* w = 0. \tag{7.4.19}$$

Decomposing w into $w = ((w^*)^T, w_{n+1}, w_{n+2})^T$ with $w^* \in \mathbb{R}^n$ yields

$$(A - \lambda_1^* I_n) w^* = w_{n+1} u^* + w_{n+2} v^*, \tag{7.4.20}$$

$$w_{n-1}^* = 0, \tag{7.4.21}$$

$$w_n^* = 0. \tag{7.4.22}$$

If $w^* = 0$, then $w_{n+1} u^* + w_{n+2} v^* = 0$. Since u^*, v^* are linearly independent we get $w_{n+1} = w_{n+2} = 0$ contradicting $w \neq 0$. Therefore, $w^* \neq 0$ with $w_{n-2}^* \neq 0$ w.l.o.g., whence by (7.4.18), (7.4.21), (7.4.22)

$$\det \begin{pmatrix} w_{n-2}^* & u_{n-2}^* & v_{n-2}^* \\ w_{n-1}^* & u_{n-1}^* & v_{n-1}^* \\ w_n^* & u_n^* & v_n^* \end{pmatrix} = w_{n-2}^* (u_{n-1}^* v_n^* - u_n^* v_{n-1}^*) \neq 0.$$

Hence u^*, v^*, w^* are linearly independent.

If $w_{n+1} = w_{n+2} = 0$, then w^* is an eigenvector associated with λ_1^*. Hence the dimension $\dim V$ of the invariant subspace V exceeds two, contradicting (a). If $w_{n+1} \neq 0$ or $w_{n+2} \neq 0$, represent u^*, v^* as linear combinations of corresponding eigenvectors / principal vectors x^*, y^* contained in V. Then (7.4.20) implies

$$(A - \lambda_1^* I_n) w^* = \alpha x^* + \beta y^* \quad \text{with } \alpha^2 + \beta^2 \neq 0. \tag{7.4.23}$$

If $\lambda_1^* \neq \lambda_2^*$ and $\alpha \cdot \beta = 0$, then w^* is a principal vector associated with λ_1^* or $(A - \lambda_1^* I_n)w^*$ is a principal vector associated with λ_2^* although λ_1^*, λ_2^* are simple eigenvalues. If $\lambda_1^* \neq \lambda_2^*$ and $\alpha \cdot \beta \neq 0$, then multiply (7.4.23) by $(A - \lambda_2^* I_n)$ and commute both matrix factors to see that $(A - \lambda_2^* I_n)w^*$ is a principal vector associated with λ_1^*. This again contradicts the simplicity of λ_1^*. If $\lambda_1^* = \lambda_2^*$ is a geometric double eigenvalue, then w^* is a corresponding principal vector of degree two. If $\lambda_1^* = \lambda_2^*$ is a geometrically simple but algebraically double eigenvalue, then w^* is a corresponding principal vector of degree three. Hence in all cases we get $\dim V \geq 3$ contradicting (a). Therefore, B_1^* cannot be singular, and (b) is valid.

In order to prove the converse, let (b) hold and assume (a) to be false. Since (7.4.1) holds for U^*, the subspace $\hat{V}$ spanned by u^*, v^* is invariant with respect to A. Due to this fact, u^*, v^* can be written as linear combinations of two eigenvectors x^*, y^* of A or of an eigenvector x^* and a corresponding principal vector y^* of degree 2. Since we assumed (a) to be false we must have $\dim V \geq 3$, hence the Jordan normal form of A shows that there is a left eigenvector w^* of A associated with λ_1^* or λ_2^* which is orthogonal to x^* and y^*. Therefore $(w^*)^T u^* = (w^*)^T v^* = 0$, whence $((w^*)^T, 0, 0)B_i^* = 0$ for $i = 1$ or $i = 2$. This contradicts (b). □

We notice that an analogous theorem also holds in the modified case dealing with $\hat{g}$ mentioned above. This can be seen as in the proof of Theorem 1.8.3. For this modification, and with $\tilde{m}_{21} = 0$, B has the form

$$B = \hat{B} = \begin{pmatrix} \hat{B}_{11} & 0 \\ -\tilde{m}_{12} I' & \hat{B}_{22} \end{pmatrix}$$

where $\hat{B}_{ii}$, $i = 1, 2$, is $A - \tilde{m}_{ii} I_n$ with the columns $n - 1$ and n being replaced by $-\tilde{u}, -\tilde{v}$, respectively, and where $I' = I_n - e^{(n-1)}(e^{(n-1)})^T - e^{(n)}(e^{(n)})^T$. Its inverse reads

$$\hat{B}^{-1} = \begin{pmatrix} \hat{B}_{11}^{-1} & 0 \\ \hat{B}_{22}^{-1} \tilde{m}_{12} I' \hat{B}_{11}^{-1} & \hat{B}_{22}^{-1} \end{pmatrix}.$$

Applying Theorem 7.4.2 with C having the block form

$$C = \begin{pmatrix} C_1 & 0 \\ C_2 \tilde{m}_{12} I' C_1 & C_2 \end{pmatrix} \tag{7.4.24}$$

yields expressions for ρ, σ, τ which are essentially identical with the corresponding quantities in Alefeld, Spreuer [48], (2.11)–(2.13).

For numerical examples we refer to Alefeld, Spreuer [48], where among others the following one, computed in PASCAL-XSC (cf. Klatte et al. [164]), was presented.

Example 7.4.3. Consider the 7×7 matrix

$$A = \begin{pmatrix} -6 & 0 & 0 & -1 & -4 & -4 & 0 \\ 0 & 4 & 1 & 0 & 0 & 0 & 2 \\ 0 & 1 & 4 & 0 & 0 & 0 & 0 \\ -1 & 0 & 0 & -6 & -4 & -4 & 0 \\ -4 & 0 & 0 & -4 & -6 & -1 & 0 \\ -4 & 0 & 0 & -4 & -1 & -6 & 0 \\ 0 & 1 & 1 & 0 & 0 & 0 & 4 \end{pmatrix}$$

with the eigenvalues

$$\lambda_1 = 6, \quad \lambda_2 = \lambda_3 = 3, \quad \lambda_4 = 1, \quad \lambda_5 = \lambda_6 = -5, \quad \lambda_7 = -15$$

and with the corresponding eigenvectors/principal vectors

$$x^1 = \frac{1}{4}\begin{pmatrix} 0 \\ 4 \\ 2 \\ 0 \\ 0 \\ 0 \\ 3 \end{pmatrix}, \; x^2 = \begin{pmatrix} 0 \\ 1 \\ -1 \\ 0 \\ 0 \\ 0 \\ 0 \end{pmatrix}, \; x^3 = \begin{pmatrix} 0 \\ 0 \\ -1 \\ 0 \\ 0 \\ 0 \\ 1 \end{pmatrix}, \; x^4 = \begin{pmatrix} 1 \\ 0 \\ 0 \\ 1 \\ -1 \\ -1 \\ 0 \end{pmatrix},$$

$$x^5 = \begin{pmatrix} 1 \\ 0 \\ 0 \\ -1 \\ -1 \\ 1 \\ 0 \end{pmatrix}, x^6 = \begin{pmatrix} 1 \\ 0 \\ 0 \\ -1 \\ 1 \\ -1 \\ 0 \end{pmatrix}, \; x^7 = \begin{pmatrix} 1 \\ 0 \\ 0 \\ 1 \\ 1 \\ 1 \\ 0 \end{pmatrix}.$$

Here x^3 is a principal vector associated with $\lambda_2 = \lambda_3 = 3$ and x^5, x^6 are two linearly independent eigenvectors belonging to $\lambda_5 = \lambda_6 = -5$. The approximations $\tilde{M}, \tilde{u}, \tilde{v}$ were chosen to be

$$\tilde{M} = \begin{pmatrix} -4.999\,999\,99 & 0.000\,000\,01 \\ 0 & -5.000\,000\,01 \end{pmatrix},$$

$$\tilde{u} = \begin{pmatrix} 0.999\,999\,99 \\ 0.000\,000\,01 \\ -0.000\,000\,01 \\ -1 \\ -0.999\,999\,99 \\ 0.999\,999\,99 \\ 0.000\,000\,01 \end{pmatrix} \approx x^5, \qquad \tilde{v} = \begin{pmatrix} 1 \\ 0.000\,000\,01 \\ -0.000\,000\,01 \\ -0.999\,999\,99 \\ 0.999\,999\,99 \\ -0.999\,999\,99 \\ 0.000\,000\,01 \end{pmatrix} \approx x^6.$$

The modified method with $i_1 = 4$, $i_2 = 5$ (instead of $i_1 = n - 1 = 6$, $i_2 = n = 7$) and with $C \approx B^{-1}$ from (7.4.24) (adapted) yields

$$\begin{pmatrix} -5 & 0 \\ 0 & -5 \end{pmatrix} + \begin{pmatrix} [-1,1]\cdot 10^{-12} & [-1,0.3]\cdot 10^{-19} \\ [-2.6,2.6]\cdot 10^{-20} & [-1,1]\cdot 10^{-12} \end{pmatrix},$$

$$x^5 + \begin{pmatrix} [-0.1,1]\cdot 10^{-11} \\ [-1,0.2]\cdot 10^{-19} \\ [-0.2,1]\cdot 10^{-19} \\ 0 \\ 1.0\cdot 10^{-8} \\ [-1.0001 - 0.0099]\cdot 10^{-8} \\ [-1,0.2]\cdot 10^{-19} \end{pmatrix}, \quad x^6 + \begin{pmatrix} [-1.0001 - 0.9999]\cdot 10^{-8} \\ [-2,1.2]\cdot 10^{-19} \\ [-0.3,1]\cdot 10^{-19} \\ -0.99999999 \\ 0.99999999 \\ [0.9999, 1.0001]\cdot 10^{-8} \\ [-1,0.3]\cdot 10^{-19} \end{pmatrix}$$

as verified enclosures for

$$M = \begin{pmatrix} -5 & 0 \\ 0 & -5 \end{pmatrix}$$

from (7.4.1) and for the generators $u^* = (1 - 0.5\cdot 10^{-8})\,x^5 + 0.5\cdot 10^{-8}\,x^6$ and $v^* = 0.99999999\cdot x^6$, respectively, which are trivially linear combinations of x^5, x^6 and satisfy $u_i^* = \tilde{u}_i$, $v_i^* = \tilde{v}_i$, $i \in \{4,5\}$. □

7.5 The generalized eigenvalue problem

As is well known, the generalized eigenvalue problem

$$Ax = \lambda Bx, \quad A, B \in \mathbb{R}^{n\times n}, \quad B \text{ nonsingular}, \tag{7.5.1}$$

can be reduced at once to the standard eigenvalue problem by multiplying (7.5.1) with B^{-1}. In contrast to Sections 1.8 and 7.6 we do not assume here that A, B are symmetric and that B is positive definite. In practice, because of rounding errors, $B^{-1}A$ normally cannot be computed exactly, but it can be enclosed 'solving', e. g., the n linear systems

$$Bz^j = A_{*,j}, \quad j = 1, \dots, n, \quad (A_{*,j}\colon j\text{-th column of } A)$$

by the verification methods mentioned in Chapter 5. The resulting inclusions $[z]^j$ of z^j then yield an enclosure for the columns of $B^{-1}A$, and the eigenvalue/eigenvector methods of Section 7.2 or 7.3 can be applied to the interval matrix $([z]^1, \dots, [z]^n)$. There are also other possibilities to handle (7.5.1). The first one can be used for simple eigenvalues $\lambda = \lambda^*$ of (7.5.1), i.e., for simple eigenvalues of

$$B^{-1}A\,x = \lambda x, \tag{7.5.2}$$

although there is no need to require simplicity for λ^* from the start. But it will turn out as in Section 7.2 that some matrix that is involved is nonsingular if λ^* is simple.

The method proceeds analogously to that in Section 7.2 for the standard eigenvalue problem. Start with

$$f(x,\lambda)=\begin{pmatrix}Ax-\lambda Bx\\ x_n-1\end{pmatrix},$$

precondition with $-C\in\mathbb{R}^{(n+1)\times(n+1)}$, choose an approximation $(\tilde{x},\tilde{\lambda})$, $\tilde{x}_n=1$, for an eigenpair (x^*,λ^*) of (7.5.1) and expand

$$s(x,\lambda)=\begin{pmatrix}\Delta x\\ \Delta\lambda\end{pmatrix}-Cf(x,\lambda)\quad\text{at}\quad\begin{pmatrix}\tilde{x}\\ \tilde{\lambda}\end{pmatrix}.$$

Then the exact error

$$\begin{pmatrix}\Delta x^*\\ \Delta\lambda^*\end{pmatrix}=\begin{pmatrix}x^*-\tilde{x}\\ \lambda^*-\tilde{\lambda}\end{pmatrix}$$

is a solution of the equation

$$\begin{pmatrix}\Delta x\\ \Delta\lambda\end{pmatrix}=g(\Delta x,\Delta\lambda),$$

where g is defined by

$$\begin{aligned}g(\Delta x,\Delta\lambda)&=-Cf(\tilde{x},\tilde{\lambda})+\left\{I_{n+1}-C\begin{pmatrix}A-\tilde{\lambda}B & -B(\tilde{x}+\Delta x)\\ (e^{(n)})^T & 0\end{pmatrix}\right\}\begin{pmatrix}\Delta x\\ \Delta\lambda\end{pmatrix}\\ &=s(x,\lambda).\end{aligned}$$

Checking the proofs for the lemmas and theorems in Section 7.2 shows that with minor modifications the analogue of Theorem 7.2.7 holds with

$$\begin{aligned}B'&=\begin{pmatrix}B\\ 0\end{pmatrix}\in\mathbb{R}^{(n+1)\times n}\\ \rho&=\left\|C\begin{pmatrix}A\tilde{x}-\tilde{\lambda}B\tilde{x}\\ 0\end{pmatrix}\right\|_\infty,\quad \sigma=\left\|I_{n+1}-C\begin{pmatrix}A-\tilde{\lambda}B & -B\tilde{x}\\ (e^{(n)})^T & 0\end{pmatrix}\right\|_\infty,\\ \tau&=\|\,|C|\cdot|B'|\,\|_\infty,\quad\text{and}\\ [t]_{ijk}&=\begin{cases}(|C|\cdot|B'|)_{ij}\cdot[-1,1] & \text{if } j\in\{1,\dots,n\}\text{ and if } k=n+1,\\ 0 & \text{otherwise,}\end{cases}\\ &\quad i=1,\dots,n+1.\end{aligned}$$

The absolute values arise because of

$$\begin{aligned}\left(-C\begin{pmatrix}-B[\Delta x]\\ 0\end{pmatrix}\right)_i[\Delta\lambda]&=\left(\sum_{s=1}^n\sum_{j=1}^n c_{is}b_{sj}[\Delta x]_j\right)[\Delta\lambda]\\ &\subseteq\sum_{j=1}^n\sum_{s=1}^n c_{is}b_{sj}[-1,1][\Delta x]_j[\Delta\lambda]\\ &=\sum_{j=1}^n\left(\sum_{s=1}^n|c_{is}|\cdot|b_{sj}|[-1,1]\right)[\Delta x]_j[\Delta\lambda].\end{aligned}$$

In particular, the interval version of Theorem 7.1.1 has to be applied.

Notice that Theorem 1.8.3, which is needed for the proofs, still holds if its matrices are adapted according to (7.5.1). The modification of g in Section 7.2 also holds; simply replace $\hat{B}$ in (7.2.25) by

$$\hat{B} = ((A - \tilde{\lambda}B)_{*,1}, \ldots, (A - \tilde{\lambda}B)_{*,n-1}, -B\tilde{x}) \in \mathbb{R}^{n\times n}. \tag{7.5.3}$$

We also remark that A and B can be replaced by interval matrices $[A]$, $[B]$.

Another procedure for (7.5.1) is presented in Section 7.6 provided that A and B are symmetric and B is positive definite.

Example 7.5.1 (Alefeld [14], modified). Let

$$A = \begin{pmatrix} 10 & 2 & 3 & 1 & 1 \\ 2 & 12 & 1 & 2 & 1 \\ 3 & 1 & 11 & 1 & -1 \\ 1 & 2 & 1 & 9 & 1 \\ 1 & 1 & -1 & 1 & 15 \end{pmatrix}, \qquad B = \begin{pmatrix} 12 & 1 & -1 & 2 & 1 \\ 1 & 14 & 1 & -1 & 1 \\ -1 & 1 & 16 & -1 & 1 \\ 2 & -1 & -1 & 12 & -1 \\ 1 & 1 & 1 & -1 & 11 \end{pmatrix}$$

be the matrices from Wilkinson, Reinsch [363], p. 312. To enclose the smallest eigenvalue λ^* of the generalized eigenvalue problem $Ax = \lambda Bx$ we chose

$$\tilde{\lambda} = 0.432\,787 \quad \text{and} \quad \tilde{x} = \begin{pmatrix} -0.852\,365 \\ 0.388\,181 \\ 1.0 \\ -0.693\,249 \\ 0.262\,649 \end{pmatrix}.$$

With the exception of $\tilde{x}_3$ these values coincide with the first six significant digits of the vector v in Table 3 of Wilkinson, Reinsch [363], p. 313, when being divided by v_3. Prescribing the third component of x^* by $x_3^* = 1$ instead of the n-th one, one has to move the last column $-B\tilde{x}$ in (7.5.3) to the third one and to use $-B(\tilde{x} + (\Delta x_1, \Delta x_2, 0, \Delta x_4, \Delta x_5)^T)$ instead of $-B(\tilde{x} + \Delta x)$ in the analogue of $\hat{g}$ of (7.2.23). The last factor there has to be replaced by $(\Delta x_1, \Delta x_2, \Delta\lambda, \Delta x_4, \Delta x_5)^T$. The matrix $\hat{C} \approx \hat{B}^{-1}$ is computed in MATLAB via $\hat{B}$ `\eye(5)`. Then the modified method described above delivers the verified enclosures

$$\lambda^* \in [\,0.432\,787\,211\,016\,9_6^7\,] \quad \text{and} \quad x^* \in \begin{pmatrix} [-0.852\,364\,724\,653\,9_5^4\,] \\ [\ \ 0.388\,180\,670\,492\,1_0^1\,] \\ [\ \ 1.0_0^0 \qquad\qquad\qquad\quad] \\ [-0.693\,248\,964\,367\,9_6^5\,] \\ [\ \ 0.262\,649\,390\,965\,3_2^1\,] \end{pmatrix}.$$

Here we again used INTLAB and iterated as in Section 7.2. The analogous stopping criterion was fulfilled for $k = 4$.

Another example and more details can be found in Alefeld [14].

Exercises

Ex. 7.5.1. Show that the generalized eigenvalue problem 7.5.1 has exactly n eigenvalues if B is regular and if the eigenvalues are counted according to their algebraic multiplicity.

Show that this is no longer true if B is singular.

Hint: Combine each of the following matrices A_1, A_2 with each of the two singular matrices B_1, B_2. Which cases do occur?

$$A_1 = \begin{pmatrix} 1 & 2 \\ 0 & 3 \end{pmatrix}, \quad A_2 = \begin{pmatrix} 1 & 2 \\ 0 & 0 \end{pmatrix}; \qquad B_1 = \begin{pmatrix} 1 & 0 \\ 0 & 0 \end{pmatrix}, \quad B_2 = \begin{pmatrix} 0 & 1 \\ 0 & 0 \end{pmatrix}.$$

7.6 A method due to Behnke

In this section we construct enclosures $[\lambda]_i$, $i = r, \dots, s$, for a cluster

$$\{\lambda_r, \lambda_{r+1}, \dots, \lambda_s\} \tag{7.6.1}$$

of eigenvalues of the generalized eigenvalue problem

$$Ax = \lambda Bx, \quad A, B \in \mathbb{R}^{n\times n} \text{ symmetric, } B \text{ positive definite, } x \in \mathbb{R}^n \setminus \{0\}, \tag{7.6.2}$$

where multiple eigenvalues are admitted.

Denote by λ_i, $i = 1, \dots, n$, the eigenvalues of (7.6.2) and assume

$$\lambda_1 \le \lambda_2 \le \dots \le \lambda_{r-1} \ll \lambda_r \le \dots \le \lambda_s \ll \lambda_{s+1} \le \lambda_n, \tag{7.6.3}$$

i.e., the eigenvalues in (7.6.1) are well separated from the other ones. We are going to present an algorithm due to Behnke [65, 66, 67], which is essentially based on Algorithm 1.8.8 of Bunch, Kaufman, Parlett, on Theorem 1.8.21, and on Corollary 1.8.23. The Algorithm 1.8.8 is modified in such a way that it can start with a *regular* interval matrix $[A] = [A]^T$ – preferably of small width – and outputs an interval block diagonal matrix $[D]$, which for all point matrices $A \in [A]$ contains a corresponding block diagonal matrix D similar to that in Theorem 1.8.9, which satisfies $\operatorname{ind}(\tilde{D})$ = constant for all $\tilde{D} \in [D]$. The transition to interval matrices is necessary in order to take rounding errors into account. With the notation as in Algorithm 1.8.8, in particular with

$$\tilde{P}[A]\tilde{P}^T = \begin{pmatrix} [E] & [C]^T \\ [C] & [B] \end{pmatrix}, \quad \tilde{P} \text{ permutation matrix, } [E] \in \mathbb{IR}^{s\times s},$$

the Modified Algorithm of Bunch, Kaufman, Parlett, proceeds as follows, where the meaning of r, s is different from that in (7.6.1).

Algorithm 7.6.1. Choose $\alpha \in (0,1)$ and let $m = n$;
if $m = 1$ then $s = 1$, $\tilde{P} = I$, $[e]_{11} = [a]_{11}$;
if $m > 1$
choose $r \in \{2, \dots, m\}$ such that $|[a]_{r1}| = \max_{2 \le i \le m} |[a]_{i1}|$;
if

$$|[a]_{11}| \ge \alpha |[a]_{r1}| \tag{7.6.4}$$

then $s = 1$, $\tilde{P} = I$, $[e]_{11} = [a]_{11}$
else
choose $p \in \{1, \dots, m\} \setminus \{r\}$ such that $|[a]_{pr}| = \max_{1 \le i \le m,\, i \ne r} |[a]_{ir}|$;
if

$$|[a]_{11}| \cdot |[a]_{pr}| \ge \alpha |[a]_{r1}|^2 \tag{7.6.5}$$

then $s = 1$, $\tilde{P} = I$, $[e]_{11} = [a]_{11}$
else
if

$$|[a]_{rr}| \ge \alpha |[a]_{pr}| \tag{7.6.6}$$

then $s = 1$ and $\tilde{P}$ such that $[e]_{11} = [a]_{rr}$
else
choose $s = 2$ and $\tilde{P}$ such that

$$[E] = \begin{pmatrix} [a]_{11} & [a]_{r1} \\ [a]_{r1} & [a]_{rr} \end{pmatrix}; \tag{7.6.7}$$

(notice $[a]_{r1} = [a]_{1r}$)
if

$$(s = 1 \text{ and } 0 \in [e]_{11}) \text{ or } (s = 2 \text{ and } 0 \in [e]_{11}[e]_{22} - [e]_{12}^2 = [a]_{11}[a]_{rr} - [a]_{r1}^2) \tag{7.6.8}$$

then stop with failure since $[E]$ contains a singular symmetric element; if necessary change α and restart the algorithm; otherwise choose $[E]$ as diagonal block of $[D]$ in its rows $n - m + i$, $i \in \{1, s\}$;
redefine m by $m - s$;
if $m > 0$ then repeat the steps for

$$[B]' = \begin{cases} [B], & \text{if } [C] = O \\ [B] - [C][E_{\text{sym}}^{-1}][C]^T, & \text{if } [C] \ne O \end{cases}, \tag{7.6.9}$$

where

$$[E_{\text{sym}}^{-1}] := \begin{cases} \dfrac{1}{[e]_{11}}, & \text{if } s = 1, \\ \dfrac{1}{[a]_{11}[a]_{rr} - [a]_{1r}^2} \begin{pmatrix} [a]_{rr} & -[a]_{r1} \\ -[a]_{r1} & [a]_{11} \end{pmatrix}, & \text{if } s = 2, \end{cases} \tag{7.6.10}$$

else terminate the algorithm. □

Algorithm 7.6.1 provides a decomposition $PAP^T = LDL^T$ as in Theorem 1.8.9 for every matrix $A \in [A]$, but not always according to the strategy in the original Bunch–Kaufman–Parlett algorithm 1.8.8. Thus the choice of r in Algorithm 7.6.1 cannot guarantee $|a_{r1}| = \max_{2 \le i \le n} |a_{i1}|$ for each matrix $A \in [A]$. This does not matter since we are mainly interested in the number of positive and negative eigenvalues of $A \in [A]$.

Let us comment on this algorithm. If it terminates regularly with some block diagonal interval matrix $[D]$, then the failure of condition (7.6.8) implies that the elements of a 1×1 diagonal block are nonzero and do not change their sign, and the same holds for the determinants $\det \tilde{E}$ of the *symmetric* elements $\tilde{E}$ of a 2×2 diagonal block $[E]$. In the latter case, and in the notation of the algorithm, we get

$$\det \tilde{E} \le |[a]_{11}| \cdot |[a]_{rr}| - |[a]_{r1}|^2 < 0 \tag{7.6.11}$$

analogously to (1.8.17) if $\tilde{e}_{12} = \tilde{e}_{21}$ satisfies $|\tilde{e}_{12}| = |[a]_{r1}|$. Since $\det \tilde{E}$ depends continuously on the entries of $\tilde{E}$ and differs from zero at least for symmetric matrices $\tilde{E} \in [E]$, the inequality (7.6.11) holds for arbitrary symmetric elements $\tilde{E}$ of $[E]$. In particular, the inverse $\tilde{E}^{-1}$ exists for such elements and so does $[E^{-1}_{\text{sym}}]$ which encloses it.

Notice that by virtue of the square the failure of the second condition in (7.6.8) does not imply regularity of $[E] \in \mathbb{IR}^{2\times 2}$ as the example

$$[E] = \begin{pmatrix} 1/4 & [-1, 1] \\ [-1, 1] & -1/4 \end{pmatrix}$$

shows.

Based on these remarks $\operatorname{ind}(\tilde{D})$ is constant for all symmetric matrices $\tilde{D} \in [D]$, where in the modified algorithm no eigenvalue zero occurs, and each 2×2 block contributes one positive and one negative eigenvalue such that $\operatorname{ind}(\tilde{D}) = (n_1, 0, n - n_1)$ can be easily computed from $\underline{D}$ or $\overline{D}$. For simplicity we will say that in this case the *interval* matrix $[D]$ has n_1 negative eigenvalues and $n - n_1$ positive ones keeping in mind that we only consider the symmetric matrices in $[D]$.

For regular point matrices $[A] \equiv A$ a failure due to (7.6.8) cannot occur in exact arithmetic since then zero is an eigenvalue of $[D] \equiv D$ contradicting $\operatorname{ind}(A) = \operatorname{ind}(D)$. Therefore, for regular interval matrices $[A]$ of small width one can expect that (7.6.8) is never fulfilled.

Now we want to outline the algorithm of Behnke for computing bounds of (7.6.1). It starts with approximations of eigenpairs, computes from them rough bounds for the cluster, associates with the problem a smaller generalized eigenvalue problem, encloses the eigenvalues of the new problem by means of bisection and the Modified Algorithm of Bunch, Kaufman, Parlett and derives bounds from them for the individual eigenvalues of the cluster. The details read as follows; they will be commented afterwards.

(1) Approximations of eigenpairs.
Compute approximate eigenpairs $(\tilde{x}_{r-1}, \tilde{\lambda}_{r-1}), \ldots, (\tilde{x}_{s+1}, \tilde{\lambda}_{s+1})$ of the eigenvalue problem (7.6.2) by your favorite software (for instance MATLAB) assuming (7.6.1) and (7.6.3) for the eigenvalues and approximate B-orthonormality for the eigenvectors. W.l.o.g. assume

$$\tilde{\lambda}_{r-1} + 0.01|\tilde{\lambda}_{r-1}| < \tilde{\lambda}_r \quad \text{if } r > 1, \tag{7.6.12}$$

$$\tilde{\lambda}_s < \tilde{\lambda}_{s+1} - 0.01|\tilde{\lambda}_{s+1}| \quad \text{if } s < n, \tag{7.6.13}$$

with obvious slight modifications here and in the sequel if $\tilde{\lambda}_{r-1} = 0$ or $\tilde{\lambda}_{s+1} = 0$.

(2) Rough lower bound $\underline{\alpha}$ of the cluster.
If $r = 1$ then choose $\underline{\alpha} = \tilde{\lambda}_1 - 0.005|\tilde{\lambda}_1|$
else choose

$$\underline{\alpha} = \tilde{\lambda}_{r-1} + 0.005|\tilde{\lambda}_{r-1}| \tag{7.6.14}$$

(again with an obvious modification if $\tilde{\lambda}_{r-1} = 0$) and assume that $\underline{\alpha}$ is not an eigenvalue of (7.6.2). Notice that (7.6.13) and (7.6.14) imply $\tilde{\lambda}_{r-1} < \underline{\alpha} < \tilde{\lambda}_r$. But this does not mean that $\lambda_{r-1} < \underline{\alpha} < \lambda_r$ is fulfilled automatically, since $\tilde{\lambda}_{r-1}, \tilde{\lambda}_r$ in (7.6.13), (7.6.14) are only approximations of eigenvalues. The same holds for $\underline{\alpha}$ in the case $r = 1$. Apply the Modified Algorithm 7.6.1 of Bunch, Kaufman, Parlett to the matrix $[C] = A - \underline{\alpha}B$ using machine interval arithmetic. Check whether the resulting block diagonal matrix $[D]$ has $r - 1$ negative and $n - (r - 1)$ positive eigenvalues; if not, change $\underline{\alpha}$ appropriately and redo the test, or stop with failure.
Sylvester's law of inertia and Theorem 1.8.21 guarantee $\underline{\alpha} < \lambda_1$, respectively $\underline{\alpha} \in (\lambda_{r-1}, \lambda_r)$.

(3) Rough upper bound $\overline{\alpha}$ of the cluster.
If $s = n$ then choose $\overline{\alpha} = \tilde{\lambda}_n + 0.005|\tilde{\lambda}_n|$
else choose

$$\overline{\alpha} = \tilde{\lambda}_{s+1} - 0.005|\tilde{\lambda}_{s+1}|$$

(again with an obvious modification if $\tilde{\lambda}_n = 0$, respectively $\tilde{\lambda}_{s+1} = 0$) and assume that $\overline{\alpha}$ is not an eigenvalue of (7.6.2). Apply the Modified Algorithm 7.6.1 of Bunch, Kaufman, Parlett to the matrix $[C] = A - \overline{\alpha}B$ using machine interval arithmetic. Check whether the resulting block diagonal matrix $[D]$ has s negative and $n - s$ positive eigenvalues; if not, change $\overline{\alpha}$ appropriately and redo the test, or stop with failure.
By the same reason as in step (2), we obtain $\lambda_n < \overline{\alpha}$, respectively $\overline{\alpha} \in (\lambda_s, \lambda_{s+1})$.

(4) Spectral shift and smaller eigenvalue problem.
Let $k = s - r + 1$ and choose $j \in \{r, r + 1, \ldots, s\}$. Define

$$\hat{\underline{\alpha}} = \underline{\alpha} - \tilde{\lambda}_j, \quad \hat{\overline{\alpha}} = \overline{\alpha} - \tilde{\lambda}_j,$$

$$u^1 = \tilde{x}^r, \ \ldots, \ u^k = \tilde{x}^s, \quad U = (u^1, \ldots, u^k).$$

Compute $\underline{\hat{\alpha}}$ using upward rounding and $\overline{\hat{\alpha}}$ using downward rounding.
Define $[B] \equiv B$ and use machine interval arithmetic to compute

$$[\hat{A}] = A - \tilde{\lambda}_j[B].$$

For $i = 1, \dots, k$ compute $[y]^i = [\hat{A}]u^i$, solve the interval linear system $Bv^i = [y]^i$ by your favorite interval method, for instance by means of the interval Gaussian algorithm, and denote the solution by $[v]^i$.
Compute the $k \times k$ interval matrices

$$\begin{aligned}[A]_0 = ((u^i)^T[B]u^j) = U^T[B]U, \quad [A]_1 = ((u^i)^T[y]^j) = U^T[\hat{A}]U,\\ [A]_2 = (([v]^i)^T[y]^j) \ni U^T(\hat{A}B^{-1}\hat{A})U, \quad \hat{A} = \hat{A}^T \in [\hat{A}],\end{aligned} \tag{7.6.15}$$

and consider the smaller generalized interval eigenvalue problem

$$[A]_1 x = \tau[A]_0 x \tag{7.6.16}$$

in the case $r = 1$ or $s = n$, and

$$([A]_2 - \alpha[A]_1)x = \tau([A]_1 - \alpha[A]_0)x \tag{7.6.17}$$

in the case $1 < r \le s < n$. Both include the corresponding point problems, i.e.,

$$U^T\hat{A}Ux = \tau U^T BUx, \quad \hat{A} = \hat{A}^T \in [\hat{A}], \tag{7.6.18}$$

in the first case and

$$\{U^T(\hat{A}B^{-1}\hat{A} - \alpha\hat{A})U\}x = \tau\{U^T(\hat{A} - \alpha B)U\}x, \quad \hat{A} = \hat{A}^T \in [\hat{A}], \tag{7.6.19}$$

in the second one.

(5) Upper bounds for λ_i, $i = r, \dots, s$.
Case $r = 1$: Compute enclosures $[\tau]_1, \dots [\tau]_k$ for the eigenvalues of all generalized eigenvalue problems contained in

$$[A]_1 x = \tau[A]_0 x. \tag{7.6.20}$$

To this end choose a starting interval which encloses all eigenvalues of (7.6.20). Bisect it repeatedly and apply the Modified Algorithm 7.6.1 of Bunch, Kaufman, Parlett to $[A]_1 - \tau[A]_0$, where τ is the parameter to be replaced by the bisection values.
Case $r > 1$: Choose $\alpha = \underline{\hat{\alpha}}$ in (7.6.17).
(a) Test whether each symmetric matrix in $[C] = [A]_1 - \alpha[A]_0$ has only positive eigenvalues. To this end check, for instance, whether

$$\underline{c}_{ii} > \sum_{\substack{j=1\\ j\ne i}}^{k} |[c]_{ij}| \tag{7.6.21}$$

holds for $i = 1, \dots, k$, and use Gershgorin's theorem. If this is not the case then stop the algorithm with failure.

(b) For each fixed index $i \in \{1, \ldots, k\}$, compute a common enclosure $[\tau]_i$ for the eigenvalue τ_i of each generalized eigenvalue problem contained in (7.6.17). Do this as in the case $r = 1$ via bisections and the Modified Algorithm 7.6.1 of Bunch, Kaufman, Parlett, applied to $([A]_2 - \alpha[A]_1) - \tau([A]_1 - \alpha[A]_0)$, where τ is again the parameter which is replaced by the bisection values.
Test whether $\underline{\hat{\alpha}} < \underline{\tau}_1$; if not, stop with failure.

In both cases define $\overline{\lambda}_r = \tilde{\lambda}_j + \overline{\tau}_1, \ldots, \overline{\lambda}_s = \tilde{\lambda}_j + \overline{\tau}_k$ using upward rounding.

(6) Lower bounds for λ_i, $i = r, \ldots, s$.
Let $\tau' = -\tau$.
Case $s = n$: Compute enclosures $[\tau]'_1, \ldots [\tau]'_k$ for the eigenvalues of all generalized eigenvalue problems contained in

$$-[A]_1 x = \tau' [A]_0 x$$

analogously to step (5) and let $[\tau]_i = -[\tau]'_{k+1-i}$, $i = 1, \ldots, k$.

Case $s < n$: Choose $\alpha = \overline{\hat{\alpha}}$ in (7.6.17).

(a) Test similarly as in step (5) whether each symmetric matrix in $[C] = -([A]_1 - \alpha[A]_0)$ has only positive eigenvalues. If this is not the case, then stop the algorithm with failure.

(b) For each fixed index $i \in \{1, \ldots, k\}$ compute a common enclosure $[\tau]'_i$ for the eigenvalue τ'_i of each generalized eigenvalue problem contained in

$$([A]_2 - \alpha[A]_1)x = \tau'\{-([A]_1 - \alpha[A]_0)\}x.$$

Do this via bisections and the Modified Algorithm 7.6.1 of Bunch, Kaufman, Parlett, applied to $([A]_2 - \alpha[A]_1) + \tau'([A]_1 - \alpha[A]_0)$, where τ' is again the parameter which is replaced by the bisection values. Let $[\tau]_i = -[\tau]'_{k+1-i}$, $i = 1, \ldots, k$, as in the case $s = n$.
Test whether $\overline{\tau}_k < \overline{\hat{\alpha}}$; if not, then stop with failure.

In both cases define $\underline{\lambda}_r = \tilde{\lambda}_j + \underline{\tau}_1, \ldots, \underline{\lambda}_s = \tilde{\lambda}_j + \underline{\tau}_k$ using downward rounding.

The intervals $[\lambda]_i$, $i = r, \ldots, s$, constructed with the bounds in steps (5) and (6) enclose the eigenvalues (7.6.1) of the cluster. □

We will comment on the individual steps of the algorithm.

Steps (2), (3). Machine interval arithmetic is applied in order to guarantee $\underline{\alpha} \in (\lambda_{r-1}, \lambda_r)$ and $\overline{\alpha} \in (\lambda_s, \lambda_{s+1})$.

Step (4). Notice that $\underline{\hat{\alpha}}$ is computed with upward rounding. It can also be computed with machine interval arithmetic using the upper bound of the resulting interval. A similar remark holds for the computation of $\overline{\alpha}$ using downward rounding, and the lower bound, respectively. Based on machine interval arithmetic the matrix $[\hat{A}]$ will normally not be a point matrix but only an enclosure of $A - \tilde{\lambda}_j B$, and a similar remark holds for $[y]^i$ and $[v]^i$.

Step (5). We first start with some general remarks. Since the eigenvector approximations $\tilde{x}_i$ are nearly B-orthonormal, the matrix $A_0 = U^T BU$ is approximately the $k \times k$ unit matrix I_k. Moreover, with (7.6.2) and with the diagonal matrix $D_\lambda = \operatorname{diag}(\lambda_r, \ldots, \lambda_s) \approx \tilde{\lambda}_j I_k$ we get

$$\begin{aligned}
A_1 &= U^T(A - \tilde{\lambda}_j B)U \approx U^T(BUD_\lambda - BU\tilde{\lambda}_j) \\
&\approx D_\lambda - \tilde{\lambda}_j I_k, \\
A_2 &= U^T(A - \tilde{\lambda}_j B)B^{-1}(A - \tilde{\lambda}_j B)U \\
&\approx ((A - \tilde{\lambda}_j B)U)^T B^{-1} BU(D_\lambda - \tilde{\lambda}_j I_k) \\
&\approx (BU(D_\lambda - \tilde{\lambda}_j I_k))^T U(D_\lambda - \tilde{\lambda}_j I_k) \approx (D_\lambda - \tilde{\lambda}_j I_k)^2, \\
C &= A_1 - \underline{\hat{\alpha}} A_0 \approx D_\lambda - \tilde{\lambda}_j I_k - (\underline{\alpha} - \tilde{\lambda}_j)I_k = D_\lambda - \underline{\alpha} I_k \geq O.
\end{aligned}$$

Therefore, the corresponding interval matrices in step (5) are approximately of the same type. In particular, they can be expected to be nearly diagonal and $[C]$ to be positive definite. This facilitates the application of Gershgorin's theorem. In this connection the right-hand side of the Gershgorin test (7.6.21) must be computed carefully using for example the `abs`-function of INTLAB, summing up the intervals $\operatorname{abs}([c]_{ij})$ and using the upper bound of this sum.

Case $r = 1$: We have to compute an initial enclosure $[\tau]$ for all eigenvalues τ of all generalized eigenvalue problems (7.6.18) with $\hat{A} = \hat{A}^T \in [\hat{A}]$. To this end we transform (7.6.18) equivalently into the standard eigenvalue problem

$$(U^T BU)^{-1}(U^T \hat{A} U)x = \tau x \tag{7.6.22}$$

and notice that τ is real according to Theorem 1.8.20. We solve the k interval linear systems $[A]_0 z = ([A]_1)_{*,j}$, $j = 1, \ldots, k$, for instance with the interval Gaussian algorithm, and denote the corresponding solutions by $[z]_j$. The matrix $[H] = ([z]_1, \ldots, [z]_k) \in \mathbb{IR}^{k\times k}$ encloses the matrix in (7.6.22) for all $\hat{A} \in [\hat{A}]$. Hence

$$[\tau] = [-\| |[H]| \|,\ \| |[H]| \|]$$

is an enclosure for τ in (7.6.18) and (7.6.22) for an arbitrary matrix norm $\|\cdot\|$ and any $\hat{A} = \hat{A}^T \in [\hat{A}]$.

Another enclosure is given by

$$[\tau] = \left[\min_i\left(\underline{h}_{ii} - \sum_{\substack{j=1\\ j\neq i}}^{k} |[h]_{ij}|\right),\ \max_i\left(\overline{h}_{ii} + \sum_{\substack{j=1\\ j\neq i}}^{k} |[h]_{ij}|\right)\right]$$

as Gershgorin's theorem shows, applied to the matrix in (7.6.22).

Now inflate the enclosure $[\tau]$ slightly to some interval $[\tau_0^*, \tau_k^*]$ which contains $[\tau]$ in its interior so that τ_0^*, τ_k^* are certainly not eigenvalues of (7.6.18). Put $[\tau_0^*, \tau_k^*]$ on a list L and start the bisection process.

Assume that at some stage of the bisection process the list L contains at most k intervals $[\tau_{i_j}^*, \tau_{i_{j+1}}^*]$, $0 \le i_j < i_{j+1} \le k$, with the following three properties which are certainly satisfied for the initial interval $[\tau_0^*, \tau_k^*]$.

(1) The bounds $\tau_{i_j}^*$, $\tau_{i_{j+1}}^*$ are not eigenvalues of (7.6.18).
(2) The interval $[\tau_0^*, \tau_{i_j}^*]$ contains exactly i_j eigenvalues of (7.6.18).
(3) The interval $[\tau_0^*, \tau_{i_{j+1}}^*]$ contains exactly i_{j+1} eigenvalues of (7.6.18).

Then in $[\tau_{i_j}^*, \tau_{i_{j+1}}^*]$ there are exactly $i_{j+1} - i_j$ eigenvalues of (7.6.18).

The bisection process proceeds as follows:

Fetch some interval $[\tau_{i_j}^*, \tau_{i_{j+1}}^*]$ from the list L and compute the midpoint $\tau^* = (\tau_{i_j}^* + \tau_{i_{j+1}}^*)/2$. Apply the Modified Algorithm 7.6.1 of Bunch, Kaufman, Parlett to $[A]_1 - \tau^*[A]_0$ and assume that it terminates regularly. Then zero is not an eigenvalue of $U^T\hat{A}U - \tau^* U^T BU$ for $\hat{A} = \hat{A}^T \in [\hat{A}]$. Let n^* be the number of negative eigenvalues. By virtue of Theorem 1.8.21, the interval $[\tau_0^*, \tau^*]$ contains exactly n^* eigenvalues of (7.6.18).

If $n^* = i_j$, then the interval $[\tau_{i_j}^*, \tau^*]$ does not contain an eigenvalue of (7.6.18) and can be deleted. In this case redefine $\tau_{i_j}^* = \tau^*$ and put $[\tau_{i_j}^*, \tau_{i_{j+1}}^*]$ back into the list.

Similarly, if $n^* = i_{j+1}$, redefine $\tau_{i_{j+1}}^* = \tau^*$ and put $[\tau_{i_j}^*, \tau_{i_{j+1}}^*]$ back into the list.

If $i_j < n^* < i_{j+1}$, the list contained less than k intervals before the present bisection took place. Define $\tau_{n^*}^* = \tau^*$ and put the two intervals $[\tau_{i_j}^*, \tau_{n^*}^*]$, $[\tau_{n^*}^*, \tau_{i_{j+1}}^*]$ back into the list.

In each of the three cases the new interval bound τ^* is not an eigenvalue of (7.6.18), and all three properties are fulfilled for each interval which is put into the list.

Run cyclically through the list in order to bisect all intervals before bisecting some interval again. Stop the bisection process if the length of the intervals are sufficiently small. If $i_{j+1} = i_j + 1$, then the interval $[\tau_{i_j}^*, \tau_{i_{j+1}}^*]$ contains exactly one eigenvalue which, in addition, is simple. Notice that multiple and nearly multiple eigenvalues induce $i_{j+1} > i_j + 1$.

Having finally enclosed the eigenvalues τ_i by intervals $[\tau]_i$, $i = 1, \dots, k$, Corollary 1.8.24 guarantees $\lambda_i - \tilde{\lambda}_j \le \overline{\tau}_i$, hence $\lambda_i \le \tilde{\lambda}_j + \overline{\tau}_i$, for the eigenvalues λ_i of the original eigenvalue problem (7.6.2).

Notice that the case $r = 1$ can also alternatively be treated like the succeeding case $r > 1$.

Case $r > 1$: First we remark that the test in (a) ensures the applicability of Theorem 1.8.21 to the interval eigenvalue problem (7.6.17), necessary for the bisection process in part (b). In fact, by virtue of the positive eigenvalues it guarantees that each symmetric matrix out of $[A]_1 - \underline{\hat{\alpha}}[A]_0$ is positive definite.

Since we need an initial enclosure of all eigenvalues of (7.6.19), we choose $[\tau] = [\underline{\hat{\alpha}}, \overline{\hat{\alpha}}]$ as a trial and check similarly as above by means of the Modified Algorithm 7.6.1

of Bunch, Kaufman, Parlett whether the the symmetric matrices in $([A]_2 - \underline{\hat{\alpha}}[A]_1) - \underline{\hat{\alpha}}([A]_1 - \underline{\hat{\alpha}}[A]_0)$ have only positive eigenvalues and those in $([A]_2 - \underline{\hat{\alpha}}[A]_1) - \overline{\hat{\alpha}}([A]_1 - \underline{\hat{\alpha}}[A]_0)$ have only negative ones. Then Theorem 1.8.21 guarantees that all eigenvalues τ of (7.6.19) are greater than $\underline{\hat{\alpha}}$ and less than $\overline{\hat{\alpha}}$. Success for the first check is expected since

$$U^T(\hat{A}B^{-1}\hat{A} - \alpha\hat{A})U - \alpha U^T(\hat{A} - \alpha B)U = U^T(\hat{A} - \alpha B)B^{-1}(\hat{A} - \alpha B)U$$

is a symmetric positive definite matrix if $\hat{A} = \hat{A}^T$, if U has full rank, and if α is not an eigenvalue of $\hat{A} - \alpha B$. The second check *could* succeed based on the following heuristics. Corollary 1.8.23 (b) with $\alpha = \underline{\hat{\alpha}}$, $q = k$, $i = k$ guarantees that $(\underline{\hat{\alpha}}, \overline{\tau}_k]$ contains at least k eigenvalues of $\hat{A}x = \tau Bx$. Therefore, $\lambda_{r-1+i} - \tilde{\lambda}_j \le \overline{\tau}_k$, $i = 1, \ldots, k$, must hold. In particular, $i = k$ implies $\lambda_s - \tilde{\lambda}_j \le \overline{\tau}_k$. Since $\lambda_s - \tilde{\lambda}_j < \overline{\hat{\alpha}}$ also holds the trial $\overline{\tau} = \overline{\hat{\alpha}}$ seems to be appropriate. If the check fails, increase $\overline{\tau}$ or stop.

For an initial enclosure $[\tau]$ one can, of course, also proceed as in the case $r = 1$ with $[H]$ being modified correspondingly.

The bisection process is realized as in the case $r = 1$. Corollary 1.8.23 implies

$$\underline{\hat{\alpha}} < \lambda_{r-1+i} - \tilde{\lambda}_j \le \overline{\tau}_i. \tag{7.6.23}$$

In practice we computed $\underline{\hat{\alpha}}$ using upward rounding so that, for instance, $\underline{\hat{\alpha}} < \lambda_r - \tilde{\lambda}_j$ cannot be guaranteed in machine arithmetic. If this inequality does not hold, then the index $r - 1 + i$ must be increased in (7.6.23), but by virtue of the increasing order of the eigenvalues λ_i the original right inequality in (7.6.23) still remains true. In addition,

$$\underline{\alpha} < \lambda_{r-1+i} \le \overline{\tau}_i + \tilde{\lambda}_j, \quad i = 1, \ldots, k,$$

is guaranteed because we used upward rounding when computing $\overline{\tau}_i + \tilde{\lambda}_j$. This rounding can be realized in practice as in step (4) using interval arithmetic.

The test $\underline{\hat{\alpha}} < \underline{\tau}_1$ should work if λ_{r-1} and λ_r are well separated. It is necessary if one uses an initial enclosure $[\tau] \ne [\underline{\hat{\alpha}}, \overline{\hat{\alpha}}]$.

Step (6). The remarks for step (5) are similarly valid for step (6).

Case $s = n$: The algorithm for the case $r = 1$ in step (5) can be applied to the interval eigenvalue problem $-[A]_1x = \tau'[A]_0x$ with $\tau' = -\tau$, hence $\tau_i = -\tau'_{k+1-i}$ and $[\tau]_i = -[\tau]'_{k+1-i}$. At the end, Corollary 1.8.24 guarantees that there are at least $k + 1 - i$ eigenvalues $-(\lambda_\ell - \tilde{\lambda}_j) < \tau'_{k+1-i} = -\tau_i$ or, equivalently, $\lambda_\ell - \tilde{\lambda}_j > \tau_i$. Therefore, $\tilde{\lambda}_j + \underline{\tau}_i$ is a lower bound for the eigenvalues $\lambda_n = \lambda_s, \lambda_{n-1} = \lambda_{s-1}, \ldots, \lambda_{n+1-(k+1-i)} = \lambda_{s-k+i} = \lambda_{r-1+i}$, in particular, for λ_{r-1+i}.

The case $s = n$ can alternatively be treated in the same way as the case $s < n$.

Case $s < n$: In order to apply Theorem 1.8.21 and Corollary 1.8.23 we must rewrite (7.6.17) as

$$([A]_2 - \alpha[A]_1)x = (-\tau)\{-([A]_1 - \alpha[A]_0)\}x. \tag{7.6.24}$$

By conclusions similar to those in step (5) we get $C = -(A_1 - \overline{\hat{\alpha}}A_0) \approx -(D_\lambda - \overline{\hat{\alpha}}I) \ge O$, i.e. C can be expected to be symmetric and positive definite. The latter is guaranteed for each matrix $C \in -([A]_1 - \overline{\hat{\alpha}}[A]_0)$ by the test in (a) and led us to (7.6.24).

With $\tau' = -\tau$ we can apply the algorithm for $r > 1$ to (7.6.24), where now $-\bar{\alpha}$ plays the same role as $\hat{\underline{\alpha}}$ previously. This implies $[\tau]_i = -[\tau]'_{k+1-i}$ as in the case $s = n$. □

We comment on the algorithm for the particular case of a simple eigenvalue $\lambda_r = \lambda_s$. To this end, let $\tilde{x}$ be an approximation of an eigenvector x^r associated with the eigenvalue λ_r of $Ax = \lambda Bx$. Assume that $\tilde{x}$ is normalized by $\tilde{x}^T\tilde{x} = 1$. Let $\tilde{\lambda}_r$ be an approximation of λ_r and compute the matrices in (7.6.18), (7.6.19) with $U \equiv \tilde{x} \in \mathbb{R}^n$. Then $[A]_0, [A]_1, [A]_2 \in \mathbb{IR}^{1\times 1}$, hence x in (7.6.18), (7.6.19) is in $\mathbb{R}^1 \setminus \{0\}$ and can be deleted by division. Thus

$$\tau \in \frac{[A]_1}{[A]_0} = [\tau] \qquad \text{in the case } r = s = 1 \text{ or } r = s = n, \tag{7.6.25}$$

and

$$\tau \in \frac{[A]_2 - \alpha[A]_1}{[A]_1 - \alpha[A]_0} = [\tau] \qquad \text{in the case } 1 < r = s < n, \tag{7.6.26}$$

which are initial enclosures of the eigenvalue τ in (7.6.18), and (7.6.19), respectively, as required above. After a slight inflation of $[\tau]$ to $[\tau_0^*, \tau_1^*]$ the bisection process in step (5) of the algorithm is very much simplified: if in the case $r = s = 1$ the interval $[A]_1 - \tau^*[A]_0$ contains zero, the algorithm must stop with lack of decision. If it lies left of zero, then redefine $\tau_1^* = \tau^*$ otherwise $\tau_0^* = \tau^*$, and put the smaller new interval $[\tau_0^*, \tau_1^*]$ back into the list. Proceed analogously with $([A]_2 - \hat{\underline{\alpha}}[A]_1) - \tau^*([A]_1 - \hat{\underline{\alpha}}[A]_0)$ in the case $1 < r < n$. The bisection process in Step 6 is done analogously.

We illustrate Behnke's algorithm by his example in Behnke [67]; cf. also Gregory, Karney [122].

Example 7.6.2. Define the 18×18 matrices A, B by

$$a_{ij} = \begin{cases} 0 & \text{if } (i+j) \equiv 0 \mod 19, \\ 1 & \text{if } (i+j) \not\equiv 0 \mod 19 \text{ and } (i+j) \equiv k^2 \mod 19 \text{ for a } k \in \mathbb{N}, \\ -1 & \text{otherwise,} \end{cases}$$

and

$$b_{ij} = \begin{cases} 1 & \text{if } i = j, \\ 10^{-6} & \text{if } |i-j| = 3, \\ 0 & \text{otherwise.} \end{cases}$$

The matrix A has only the entries $-1, 0, 1$. The value 1 occurs for

$$i + j \in \{4, 5, 6, 7, 9, 11, 16, 17, 20, 23, 24, 25, 26, 28, 30, 35, 36\}; \tag{7.6.27}$$

see Exercise 7.6.1. It can be shown that A has the eightfold eigenvalues $\pm\sqrt{19}$ and the simple eigenvalues ± 1. The matrix B is an approximation of the identity matrix.

Therefore, it is not surprising that the eigenvalue problem $Ax = \lambda Bx$ has two clusters of eigenvalues:

$$\lambda_1 \le \cdots \le \lambda_8 \ll \lambda_9 \ll \lambda_{10} \ll \lambda_{11} \le \cdots \le \lambda_{18}.$$

We programmed in INTLAB using MATLAB's function `eig` in order to get approximations of eigenpairs of (7.6.2). We computed the cluster indices r, s according to (7.6.12), (7.6.13), i.e., we started the cluster with r if (7.6.12) was fulfilled, decided that $\tilde{\lambda}_i$, $i = r+1, r+2, \ldots$ belonged to this cluster as long as $\tilde{\lambda}_{i-1} + 0.01\,|\tilde{\lambda}_{i-1}| \ge \tilde{\lambda}_i$ held and stopped the cluster with $i = s$ if (7.6.13) was fulfilled for the first time. We used a mask vector m of length n in order to characterize the cluster points by $m(r+j) = j+1$, $j = 0, \ldots, s-r$ and 'isolated' approximations $\tilde{\lambda}_i$ by $m(i) = 0$. We followed the algorithm through the steps (2)–(4) using the interval Gaussian algorithm for the enclosures $[v]^i$. For the initial enclosure $[\tau]$ of the steps (5) and (6), we used Gershgorin intervals in the cases $r = 1$ and $s = n$, and $[\tau] = [\underline{\hat{\alpha}}, \overline{\hat{\alpha}}]$ otherwise. After 28 cycles of bisections we stopped the algorithm since for further cycles the stopping criterion (7.6.8) held partially. The diameter of the enclosures of the eigenvalues were then about 10^{-6}, the enclosures $[\lambda]_i$ themselves were pairwise disjoint so that we could improve τ by (7.6.25) and (7.6.26) with a shift of $\mathrm{mid}([\lambda]_i)$ for $[A]_1$, $[A]_2$, with the eigenvector approximations from step (1) of the algorithm, and with $\underline{\alpha}$, $\overline{\alpha}$, $\underline{\hat{\alpha}}$, $\overline{\hat{\alpha}}$ as in the steps (2)–(4) there. The result is listed below. In view of MATLAB's machine precision it is quite satisfactory.

$$\begin{aligned}
\lambda_1, -\lambda_{18} &\in [-4.358\,906\,293\,914\,3_3^1\,]\\
\lambda_2, -\lambda_{17} &\in [-4.358\,904\,905\,898\,5_8^6\,]\\
\lambda_3, -\lambda_{16} &\in [-4.358\,903\,409\,616\,4_6^5\,]\\
\lambda_4, -\lambda_{15} &\in [-4.358\,900\,453\,991\,2_5^2\,]\\
\lambda_5, -\lambda_{14} &\in [-4.358\,898\,673\,162\,8_3^1\,]\\
\lambda_6, -\lambda_{13} &\in [-4.358\,895\,813\,100\,5_4^2\,]\\
\lambda_7, -\lambda_{12} &\in [-4.358\,893\,709\,563_{10}^{09}\,]\\
\lambda_8, -\lambda_{11} &\in [-4.358\,891\,195\,065\,1_2^0\,]\\
\lambda_9, -\lambda_{10} &\in [-0.999\,999\,333\,334\,1_5^4\,].
\end{aligned}$$

Exercises

Ex. 7.6.1. Let $p > 2$ be a prime number. Show that the squares of the numbers $0, 1, \ldots, (p-1)/2$ form – modulo p – already *all* squares $k^2 \bmod p$, $k \in \mathbb{N}$. Use this result in order to prove (7.6.27).

7.7 Verification of singular values

In this section we want to verify singular values σ_i of a rectangular matrix $A \in \mathbb{R}^{m\times n}$ and the corresponding left and right singular vectors u^i, v^i which are columns of the orthogonal matrices $U \in \mathbb{R}^{n\times n}$, $V \in \mathbb{R}^{m\times m}$ of the singular value decomposition

$$A = V\Sigma U^T \tag{7.7.1}$$

as described in Theorem 1.7.10. Since (7.7.1) yields

$$\left.\begin{aligned} Au^i &= \sigma_i v^i \\ A^T v^i &= \sigma_i u^i \end{aligned}\right\} \quad i = 1, \dots, \min\{m, n\} \tag{7.7.2}$$

the vectors u^i, v^i are eigenvectors of the symmetric matrices $A^TA \in \mathbb{R}^{n\times n}$ and $AA^T \in \mathbb{R}^{m\times m}$, respectively, associated with the eigenvalues σ_i^2 as was already mentioned in Section 1.7. In this respect one could use the methods in the Sections 7.2 and 7.3 to enclose u^i, v^i and σ_i^2. Another way consists of considering the function

$$f\colon \mathbb{R}^{n+m+1} \to \mathbb{R}^{n+m+1}$$

defined by

$$f(u, v, \sigma) = \begin{pmatrix} Au - \sigma v \\ A^T v - \sigma u \\ u^T u - 1 \end{pmatrix} \tag{7.7.3}$$

whose zeros (u^*, v^*, σ^*), $\sigma^* \neq 0$, contain a pair of singular vectors u^*, v^* associated with the singular value σ^*. The singular vectors are normalized according to

$$(u^*)^T u^* = 1, \quad (v^*)^T v^* = 1 \tag{7.7.4}$$

which follows from the definition of f and from

$$Au^* = \sigma^* v^*, \quad A^T v^* = \sigma^* u^*,$$

whence

$$(v^*)^T v^* = (v^*)^T \frac{1}{\sigma^*} Au^* = \frac{1}{\sigma^*}(A^T v^*)^T u^* = (u^*)^T u^* = 1.$$

For $\sigma^* = 0$ the zeros (u^*, v^*, σ^*) of f do not necessarily fulfill $(v^*)^T v^* = 1$, since obviously $f(u^*, 0, 0) = 0$ for each right singular vector u^* of A which is associated with $\sigma^* = 0$ and normalized by $(u^*)^T u^* = 1$. In order to force v^* to be normalized by (7.7.4) even in this case, one can start with the function

$$f(u, v, \sigma, \sigma') = \begin{pmatrix} Au - \sigma v \\ A^T v - \sigma' u \\ u^T u - 1 \\ v^T v - 1 \end{pmatrix} \tag{7.7.5}$$

repeating the steps to follow. The zeros of (7.7.5) trivially fulfill (7.7.4), and they also satisfy $\sigma = \sigma'$. This latter equality can be seen from (7.7.4) and from

$$\begin{aligned} Au = \sigma v \quad &\Rightarrow \quad v^T A u = \sigma v^T v = \sigma, \\ A^T v = \sigma' u \quad &\Rightarrow \quad u^T A^T v = \sigma' u^T u = \sigma'. \end{aligned}$$

Refer to Alefeld [11] for details.

Introducing approximations $\tilde{u}, \tilde{v}, \tilde{\sigma}$ and the differences

$$\Delta u = u - \tilde{u}, \quad \Delta v = v - \tilde{v}, \quad \Delta\sigma = \sigma - \tilde{\sigma}$$

yields the fixed point problem

$$\begin{pmatrix} \Delta u \\ \Delta v \\ \Delta\sigma \end{pmatrix} = g(\Delta u, \Delta v, \Delta\sigma) = -Cf(\tilde{u}, \tilde{v}, \tilde{\sigma}) + (I - CB)\begin{pmatrix} \Delta u \\ \Delta v \\ \Delta\sigma \end{pmatrix} + \tilde{T}\begin{pmatrix} \Delta u \\ \Delta v \\ \Delta\sigma \end{pmatrix} \tag{7.7.6}$$

with the $(m + n + 1) \times (n + m + 1)$ matrices

$$B = \begin{pmatrix} A & -\tilde{\sigma} I_m & -\tilde{v} \\ -\tilde{\sigma} I_n & A^T & -\tilde{u} \\ 2\tilde{u}^T & 0 & 0 \end{pmatrix}, \quad \tilde{T} = C\begin{pmatrix} 0 & 0 & \Delta v \\ 0 & 0 & \Delta u \\ -(\Delta u)^T & 0 & 0 \end{pmatrix}. \tag{7.7.7}$$

This follows analogously to Section 7.2 by preconditioning f with

$$-C \in \mathbb{R}^{(m+n+1)\times(m+n+1)},$$

adding $((\Delta u)^T, (\Delta v)^T, \Delta\sigma)^T$ on both sides of the equation $0 = -Cf(u, v, \sigma)$, and evaluating the right-hand side at the approximation $(\tilde{u}^T, \tilde{v}^T\ \tilde{\sigma})^T$. The series terminates after the third summand of the expansion, hence g is a quadratic function as defined in (7.1.1) with $r = -Cf(\tilde{u}, \tilde{v}, \tilde{\sigma})$, $S = I - CB$, and with $T = (t_{ijk})$ given by

$$t_{ijk} = \begin{cases} c_{i,m+j} & \text{if } j \in \{1, \ldots, n\} \text{ and } k = n + m + 1, \\ c_{i,j-n} & \text{if } j \in \{n + 1, \ldots, n + m\} \text{ and } k = n + m + 1, \\ -c_{i,m+n+1} & \text{if } j = k \in \{1, \ldots, n\}, \\ 0 & \text{otherwise,} \end{cases}$$

$i = 1, \ldots, n + m + 1$. Therefore, Theorem 7.1.1 yields a method and at the same time a criterion for verifying solutions (u^*, v^*, σ^*) of the singular value problem. We replace there the notation σ by $\hat{\sigma}$ in order to distinguish it from a singular value. Then with

$$\rho = \|Cf(\tilde{u}, \tilde{v}, \tilde{\sigma})\|_\infty, \quad \hat{\sigma} = \|I - CB\|_\infty, \quad \text{and } \tau = \|\,|C| \cdot (1, \ldots, 1, n)^T\,\|_\infty$$

we get at once the following result.

Theorem 7.7.1. *With the notation above let*

$$\begin{aligned} \hat\sigma < 1, \quad \Delta &= (1-\hat\sigma)^2 - 4\rho\tau \ge 0, \\ \beta^- &= (1-\hat\sigma-\sqrt{\Delta})/(2\tau), \\ \beta^+ &= (1-\hat\sigma+\sqrt{\Delta})/(2\tau). \end{aligned} \tag{7.7.8}$$

(a) *If $\beta \in [\beta^-, \beta^+]$, then g has at least one fixed point*

$$\begin{pmatrix} \Delta u^* \\ \Delta v^* \\ \Delta\sigma^* \end{pmatrix} \in \mathbb{R}^{n+m+1} \quad \textit{in} \begin{pmatrix} [\Delta u]^0 \\ [\Delta v]^0 \\ [\Delta\sigma]^0 \end{pmatrix} = [-\beta, \beta]e \in \mathbb{IR}^{n+m+1}.$$

If, in addition, $\sigma^* = \tilde\sigma + \Delta\sigma^* \ne 0$, then $u^* = \tilde u + \Delta u^* \in \mathbb{R}^n$, $v^* = \tilde v + \Delta v^* \in \mathbb{R}^m$, $\sigma^* \in \mathbb{R}$ form a triple of a right singular vector, a left singular vector, and a corresponding singular value such that the normalizations (7.7.4) are satisfied. The iteration

$$\begin{pmatrix} [\Delta u]^{k+1} \\ [\Delta v]^{k+1} \\ [\Delta\sigma]^{k+1} \end{pmatrix} = g\left([\Delta u]^k, [\Delta v]^k, [\Delta\sigma]^k\right), \quad k = 0, 1, \ldots, \tag{7.7.9}$$

converges to some interval vector $\begin{pmatrix} [\Delta u]^* \\ [\Delta v]^* \\ [\Delta\sigma]^* \end{pmatrix}$ *with*

$$\begin{pmatrix} \Delta u^* \\ \Delta v^* \\ \Delta\sigma^* \end{pmatrix} \in \begin{pmatrix} [\Delta u]^* \\ [\Delta v]^* \\ [\Delta\sigma]^* \end{pmatrix} \subseteq \begin{pmatrix} [\Delta u]^k \\ [\Delta v]^k \\ [\Delta\sigma]^k \end{pmatrix} \subseteq \begin{pmatrix} [\Delta u]^{k-1} \\ [\Delta v]^{k-1} \\ [\Delta\sigma]^{k-1} \end{pmatrix} \subseteq \cdots \subseteq \begin{pmatrix} [\Delta u]^0 \\ [\Delta v]^0 \\ [\Delta\sigma]^0 \end{pmatrix}, \quad k \in \mathbb{N}. \tag{7.7.10}$$

(b) *If $\beta \in [\beta^-, (\beta^- + \beta^+)/2)$, then g has a unique fixed point*

$$\begin{pmatrix} \Delta u^* \\ \Delta v^* \\ \Delta\lambda^* \end{pmatrix} \textit{ in } \begin{pmatrix} [\Delta u]^0 \\ [\Delta v]^0 \\ [\Delta\sigma]^0 \end{pmatrix} = [-\beta, \beta]e \in \mathbb{IR}^{n+m+1},$$

and (7.7.10) holds with

$$\begin{pmatrix} [\Delta u]^* \\ [\Delta v]^* \\ [\Delta\sigma]^* \end{pmatrix} = \begin{pmatrix} \Delta u^* \\ \Delta v^* \\ \Delta\sigma^* \end{pmatrix},$$

i.e., the iteration (7.7.9) converges to

$$\begin{pmatrix} \Delta u^* \\ \Delta v^* \\ \Delta\sigma^* \end{pmatrix}.$$

□

As in Section 7.2, one normally chooses $C \approx B^{-1}$ so that $\hat{\sigma} \approx 0$. If the approximations $\tilde{u}, \tilde{v}, \tilde{\sigma}$ are sufficiently close to a solution $u^*, v^*, \sigma^* \neq 0$, the weighted residual r will be small, hence the assumptions (7.7.8) of Theorem 7.7.1 will be fulfilled. Notice that $\sigma \neq 0$ certainly holds if $0 \notin \tilde{\sigma} + [\Delta\sigma]^0$ and that a similar theorem holds if g is based on (7.7.5).

As we are going to see by the subsequent Theorem 7.7.2 and by continuity arguments, the inverse B^{-1} certainly exists if the approximations $\tilde{u}, \tilde{v}$ of the singular values u^*, v^* are not too bad and if the corresponding singular value σ^* is simple, i.e., if it is a simple eigenvalue of the matrix A^TA. We already noticed that the columns u^i, $i = 1, \ldots, \min\{m, n\}$, of the matrix U from (7.7.1) fulfill

$$A^TA\,u^i = (\sigma_i^*)^2\,u^i, \tag{7.7.11}$$

with σ_i^* being the i-th singular value. Therefore, because U is nonsingular, a singular value $\sigma^* \neq 0$ is simple if and only if it occurs only once in the matrix Σ whence, by virtue of

$$AA^Tv^i = (\sigma_i^*)^2 v^i, \quad v^i\text{: } i\text{-th column of } V,$$

a simple nonzero singular value is also a simple nonzero eigenvalue of AA^T and vice versa.

We are now ready to prove the following theorem on the simplicity of singular values.

Theorem 7.7.2. *Let $\sigma^* \neq 0$ be a singular value of $A \in \mathbb{R}^{m\times n}$. Then σ^* is simple if and only if the matrix*

$$B^* = \begin{pmatrix} A & -\sigma^* I_m & -v^* \\ -\sigma^* I_n & A^T & -u^* \\ 2(u^*)^T & 0 & 0 \end{pmatrix} \in \mathbb{R}^{(m+n+1)\times(n+m+1)}$$

is nonsingular with u^*, v^* denoting a right and a left singular vector of A, respectively, associated with $\sigma^* \neq 0$.

Proof. The idea of the proof is already contained in the proof of Theorem 1.8.3. Let $\sigma^* \neq 0$ be simple and assume B^* to be singular. Then there is a vector

$$z = \begin{pmatrix} z^1 \\ z^2 \\ z' \end{pmatrix} \quad \text{with } z^1 \in \mathbb{R}^n,\ z^2 \in \mathbb{R}^m \text{ and } z' \in \mathbb{R}$$

such that $z \neq 0$ and $B^*z = 0$. Hence

$$Az^1 - \sigma^* z^2 = z'v^*, \tag{7.7.12}$$
$$A^Tz^2 - \sigma^* z^1 = z'u^*, \tag{7.7.13}$$
$$(u^*)^Tz^1 = 0. \tag{7.7.14}$$

If $z' = 0$, then

$$A^T A z^1 = \sigma^* A^T z^2 = (\sigma^*)^2 z^1$$

whence, by the simplicity of σ^*, we have $z^1 = \alpha u^*$. Therefore, (7.7.14) implies $z^1 = 0$, and (7.7.12) together with $\sigma^* \neq 0$ yields $z^2 = 0$, contradicting $z \neq 0$.

If $z' \neq 0$, then we get from (7.7.12), (7.7.13)

$$A^T A z^1 - (\sigma^*)^2 z^1 - z'\sigma^* u^* = z' A^T v^* = z'\sigma^* u^*$$

hence

$$(A^T A - (\sigma^*)^2 I_n) z^1 = 2z'\sigma^* u^*.$$

Multiplying this equation with $A^T A - (\sigma^*)^2 I_n$ and taking into account (7.7.11) shows that z^1 is a principal vector of degree two for the symmetric matrix $A^T A$ which is impossible since $A^T A$ is diagonalizable. Therefore, B^* is nonsingular.

Assume now, conversely, that B^* is nonsingular and $\sigma^* \neq 0$ is a multiple singular value. According to the discussion preceding Theorem 7.7.2, σ^* occurs multiply in Σ, whence there is a second pair $\hat{u} \in \mathbb{R}^n$, $\hat{v} \in \mathbb{R}^m$ of singular vectors, associated with σ^*. With (7.7.2) and by the orthonormality of U, V this yields the contradiction

$$\begin{aligned} &((\hat{v}^*)^T, (\hat{u}^*)^T, 0)B^* \\ &= ((\hat{v}^*)^T A - \sigma^*(\hat{u}^*)^T, (\hat{u}^*)^T A^T - \sigma^*(\hat{v}^*)^T, -(\hat{u}^*)^T u^* - (\hat{v}^*)^T v^*) = 0. \qquad \square \end{aligned}$$

We finally remark that A can again be replaced by interval matrices $[A] \in \mathbb{IR}^{m\times n}$ and that the complex case (cf. Chapter 9) can be handled analogously.

The subsequent example which we borrowed from Alefeld [11] illustrates the efficiency of the method described in Theorem 7.7.1.

Example 7.7.3. We want to enclose the largest singular value σ_1 of the 5×3 matrix

$$A = \begin{pmatrix} 1 & 6 & 11 \\ 2 & 7 & 12 \\ 3 & 8 & 13 \\ 4 & 9 & 14 \\ 5 & 10 & 15 \end{pmatrix}$$

with the approximations $\tilde{\sigma}_1 = 0.351\,272\,233 \cdot 10^2$,

$$\tilde{u} = \begin{pmatrix} 0.201\,664\,911 \\ 0.516\,830\,501 \\ 0.831\,996\,092 \end{pmatrix}, \quad \tilde{v} = \begin{pmatrix} 0.354\,557\,057 \\ 0.398\,696\,370 \\ 0.442\,835\,683 \\ 0.486\,974\,996 \\ 0.531\,114\,309 \end{pmatrix}.$$

With $C = B^{-1}$ (cf. the paragraphs following the proof of Theorem 7.2.1 for a discussion of this choice) and a computation in INTLAB similar to Section 7.2 we got

$$\sigma_1 \in [\,35.127\,223\,333\,574_{66}^{70}\,],$$

$$u^* \in \begin{pmatrix} [\,0.201\,664\,911\,192_{69}^{70}\,] \\ [\,0.516\,830\,501\,392\,3_0^1\,] \\ [\,0.831\,996\,091\,591\,9_1^2\,] \end{pmatrix}, \quad \text{and} \quad v^* \in \begin{pmatrix} [\,0.354\,557\,057\,037\,6_8^9\,] \\ [\,0.398\,696\,369\,998\,8_3^4\,] \\ [\,0.442\,835\,682\,959\,9_8^9\,] \\ [\,0.486\,974\,995\,921\,1_3^4\,] \\ [\,0.531\,114\,308\,882\,2_8^9\,] \end{pmatrix}.$$

The stopping criterion (5.5.27) according to Remark 5.5.2 was fulfilled for $k = 4$.

7.8 An inverse eigenvalue problem

The inverse eigenvalue problem which we want to consider in this section consists in finding n real numbers c_i^*, $i = 1, \ldots, n$, such that the matrix

$$A(c) = A_0 + \sum_{i=1}^{n} c_i A_i, \quad c = (c_i) \in \mathbb{R}^n, \tag{7.8.1}$$

has prescribed eigenvalues

$$\lambda_1^* < \lambda_2^* < \cdots < \lambda_n^* \tag{7.8.2}$$

for $c = c^* = (c_i^*)$. Here, A_i, $i = 0, \ldots, n$, are given symmetric $n \times n$ matrices such that $A(c) = (A(c))^T$ holds. The problem arises – among others – if one wants to find the spring constants c_i in a spring-mass-wall device sketched in Figure 7.8.1, where the eigenfrequencies of the system and one single spring constant, for instance c_0, are given.

Fig. 7.8.1: A spring-mass-wall device.

The situation is made clearer in the following example.

Example 7.8.1. Let n masses m_i be coupled by springs on a straight line and by two rigid walls standing in parallel as is indicated in Figure 7.8.1.

By Hooke's law and by Newton's second law, the system is described by

$$\left.\begin{aligned} &B\ddot{x}(t) = -Ax(t), \quad x(t) = (x_i(t)) \in \mathbb{R}^n \\ &x(t_0) = x_0 \\ &\dot{x}(t_0) = \dot{x}_0 \end{aligned}\right\} \tag{7.8.3}$$

with $x_i(t)$ denoting the distance of the i-th mass from its equilibrium position at the time t, with the mass matrix

$$B = \operatorname{diag}(m_1, \ldots, m_n) \in \mathbb{R}^{n\times n}$$

and with the symmetric $n \times n$ matrix

$$A = \begin{pmatrix} c_0 + c_1 & -c_1 & & & & & \\ -c_1 & c_1 + c_2 & -c_2 & & & 0 & \\ & -c_2 & c_2 + c_3 & -c_3 & & & \\ & & \ddots & \ddots & \ddots & & \\ & 0 & & -c_{n-2} & c_{n-2} + c_{n-1} & -c_{n-1} \\ & & & & -c_{n-1} & c_{n-1} + c_n \end{pmatrix}$$

in which c_i are the above-mentioned spring constants. The vectors x_0 and $\dot{x}_0$ are the initial values for the displacement x and for the velocity $\dot{x}$. A fundamental system of (7.8.3) is given by

$$\{\, x(t) \mid x(t) = v^i \cos(\sqrt{\lambda_i} t) \text{ or } x(t) = v^i \sin(\sqrt{\lambda_i} t),\ i = 1, \ldots, n \,\}.$$

It satisfies

$$\lambda_i B v^i = A v^i. \tag{7.8.4}$$

As is well known, $\sqrt{\lambda_i}$ are the eigenfrequencies of the oscillating system. We will introduce a matrix of the form (7.8.1) into (7.8.4). To this end let $B^{\frac{1}{2}} = \operatorname{diag}(\sqrt{m_1}, \ldots, \sqrt{m_n})$, $w^i = B^{\frac{1}{2}} v^i$, and define the symmetric $n \times n$ matrices $H^{(l)}$, $l = 1, \ldots, n$, by

$$h_{ij}^{(0)} = \begin{cases} c_0 & \text{if } i = j = 1 \\ 0 & \text{otherwise} \end{cases},$$

$$h_{ij}^{(l)} = \begin{cases} 1 & \text{if } i = j \in \{l, l+1\} \\ -1 & \text{if } (i,j) \in \{(l, l+1),\ (l+1, l)\} \\ 0 & \text{otherwise} \end{cases}, \qquad l = 1, \ldots, n-1,$$

$$h_{ij}^{(n)} = \begin{cases} 1 & \text{if } i = j = n \\ 0 & \text{otherwise} \end{cases}.$$

With $B^{-\frac{1}{2}} = (B^{\frac{1}{2}})^{-1}$, equation (7.8.4) can be written equivalently as

$$\lambda_i w^i = B^{-\frac{1}{2}} A B^{-\frac{1}{2}} w^i = \left(B^{-\frac{1}{2}} H^{(0)} B^{-\frac{1}{2}} + \sum_{l=1}^{n} c_l B^{-\frac{1}{2}} H^{(l)} B^{-\frac{1}{2}} \right) w^i$$

which is (7.8.1) with the symmetric matrices $A_i = B^{-\frac{1}{2}} H^{(i)} B^{-\frac{1}{2}}$. □

In order to find a verified solution $c^* \in \mathbb{R}^n$ of the inverse eigenvalue problem we start with the function

$$f: \begin{cases} U^* \to \mathbb{R}^n \\ c \mapsto \lambda(c) - \lambda^* \end{cases} \tag{7.8.5}$$

where $\lambda(c) = (\lambda_i(c))$ is the vector whose the components are the eigenvalues of $A(c)$, ordered increasingly like the components of $\lambda^* = (\lambda_i^*) = \lambda(c^*)$; cf. (7.8.2). If we assume for the moment that c^* exists, then, by Theorem 1.8.2 (b), there is certainly a neighborhood U^* of c^* such that for $c \in U^*$ the eigenvalues of $A(c)$ are simple and can therefore be thought to be ordered as required. This neighborhood has been chosen in (7.8.5) to define f. Trivially, the zeros of f are just the solutions of the inverse eigenvalue problem.

We want to apply Newton's method to obtain an approximation of c^*. To this end we express the Jacobian f' of f by means of the given matrices A_i and by the eigenvectors $x^i(c)$ of $A(c)$, associated with $\lambda_i(c)$ and normalized by

$$(x^i(c))^T x^i(c) = 1. \tag{7.8.6}$$

In order to avoid ambiguity, we think of $\operatorname{sign}(x^i_{i_0}(c))$ being fixed for some component $x^i_{i_0}(c)$ of $x^i(c)$, for instance, $i_0 = n$. Although Theorem 7.2.1 (b) was proved assuming the normalization $x_n = 1$, it is easy to see that its assertion also holds in the present situation. By virtue of the simplicity of $\lambda_i(c)$ one only has to consider

$$\frac{x^*}{\sqrt{(x^*)^T x^*}} \quad \text{with } x_n^* = 1,$$

where x^* is the eigenvector from Theorem 7.2.1 (b); cf. also Exercise 1.8.12 and Exercise 7.2.3. Multiplying

$$A(c)x^i(c) = \lambda_i(c)x^i(c)$$

by $(x^i(c))^T$ from the left and taking into account (7.8.6) yields

$$(x^i(c))^T A(c)x^i(c) = \lambda_i(c).$$

Differentiating both sides with respect to c_j results in

$$\begin{aligned}\frac{\partial \lambda_i(c)}{\partial c_j} &= \frac{\partial (x^i(c))^T}{\partial c_j} A(c)x^i(c) + (x^i(c))^T A_j x^i(c) \\ &\quad + (x^i(c))^T A(c) \frac{\partial x^i(c)}{\partial c_j} \\ &= (x^i(c))^T A_j x^i(c) + 2\lambda_i(c)(x^i(c))^T \frac{\partial x^i(c)}{\partial c_j}.\end{aligned}$$

Differentiating (7.8.6) with respect to c_j yields

$$2(x^i(c))^T \frac{\partial x^i(c)}{\partial c_j} = 0,$$

thus

$$\frac{\lambda_i(c)}{\partial c_j} = (x^i(c))^T \cdot A_j \cdot x^i(c). \tag{7.8.7}$$

Given c, the vectors $\lambda(c)$ and $x^i(c)$, $i = 1, \dots, n$, can be computed approximately using for instance MATLAB. Therefore, the matrix and the right-hand side of the Newton equation

$$\left((x^i(c^k))^T \cdot A_j \cdot x^i(c^k)\right)(c^{k+1} - c^k) = -(\lambda(c^k) - \lambda^*) \tag{7.8.8}$$

are known at least approximately; hence the Newton iterate c^{k+1} can be computed from (7.8.8). Stopping the iteration and verifying c^* can now be done following the lines of Alefeld [18] or the final part of Section 6.3. In the verification step, $f(\check{c})$ has to be enclosed by a tight interval vector, i.e., the eigenvalues of $A(\check{c})$ have to be enclosed by one of the methods described in Sections 7.2, 7.3, or 7.6. In addition, an enclosure of $f'(c)$ is needed with c varying in an interval vector $[c]$. This means that the normalized eigenvectors $x^i(c)$, $c \in [c]$ have to be enclosed. Again, the methods in the Sections 7.2, 7.3, or 7.6 can be applied, this time on the interval matrix $A([c]) = A_0 + \sum_{i=1}^n [c]_i A_i$ showing that enclosure methods are also needed for a whole set of point problems and not only for a single one.

The following numerical examples are taken from Alefeld, Gienger, Mayer [22]. There it is proved that the inverse eigenvalue problem can have several solutions, a phenomenon which has already been remarked on in Friedland [101].

We reprogrammed both examples in INTLAB. To this end we store the matrices A_i, $i = 1, \dots, n$, in a cell array and proceed similarly as in Example 6.3.9 using the Newton method (7.8.8) and its stopping criterion (6.3.47). The eigenpairs in (7.8.8) are computed approximately using MATLAB's function `eig(.)`. At the end we get an initial enclosure $[x] = [c]$ for c^* via (6.3.40) which we verify by means of (6.3.42). For the computation of the Krawczyk interval vector $[x]_K = [c]_K$ we make the following remarks:

We use $\tilde{x} = \check{c}$ and the approximate inverse $C = C_K$ of $f'(\check{c})$ instead of $f'(c_{\text{old}})$, where c_{old} denotes the next to last iterate of the Newton method. We compute C_K in MATLAB via (7.8.7) as $C_K = ((x^i(\check{c}))^T A_j x^i(\check{c})) \backslash$ `eye(n)`. Since c^* is unknown, we have to replace $A(c)$ in (7.8.1) by the interval matrix $[A]_c = A([c])$ and, correspondingly, we need enclosures $(x^i([c]), \lambda_i([c]))$, $i = 1, \dots, n$, for the i-th eigenpair of *all* matrices $A \in [A]_c$. To this end we use the Krawczyk-like method in Section 7.2 and proceed similarly as for the examples there. But notice that we have to replace the point matrix A now by the *interval* matrix $[A]_c$ and we have to take into account the normalization

$$x^T x = 1 \tag{7.8.9}$$

instead of $x_n = 1$. We choose the appropriate eigenpair $(\tilde{x}^i, \tilde{\lambda}_i)$, $i = 1, \dots, n$, from the MATLAB result `eig(`$\check{A}_C$`)`, sorting the eigenvalues $\tilde{\lambda} = \tilde{\lambda}_i$ ascendingly and normalizing the eigenvectors $\tilde{x} = \tilde{x}^i$ by (7.8.9) and $\tilde{x}_n^i \geq 0$. For each $i \in \{1, \dots, n\}$ the following steps have to be done:

For the matrix B in (7.2.20) (used for the approximate midpoint inverse $C = C_{\text{ewp}}$ and the number σ in (7.2.26)) we have to substitute the row vector $(e^{(n)})^T$ by $2\tilde{x}^T$, and the same vector in (7.2.5) by $(2\tilde{x} + \Delta x)^T$. The value $f(\tilde{x}, \tilde{\lambda})$ is now the interval vector

$$\begin{pmatrix} [A]_c\tilde{x} - \tilde{\lambda}\tilde{x} \\ 0 \end{pmatrix} \in \mathbb{IR}^{n+1}$$

which also replaces

$$\begin{pmatrix} A\tilde{x} - \tilde{\lambda}\tilde{x} \\ 0 \end{pmatrix}$$

for ρ in (7.2.26). The number τ there is redefined by $\tau = \|C_{\text{ewp}}(1, \ldots, 1, n)^T\|_\infty$. We verify the enclosure of eigenpairs via (7.2.8) and stop the iteration (7.2.9) when two successive iterates coincide on the computer for the first time. With these enclosures of eigenpairs we compute an enclosure of $f'([c])$ in (6.3.1) via $([x]^i([c])^T A_j x^i([c])) \in \mathbb{IR}^{n\times n}$.

Having verified c^* by $[c]_K \subseteq [c]$ we improve $[c]$ using the Krawczyk method (6.3.2) with the preconditioning matrix $C = C_K$ as previously. We choose $\tilde{x}^k = \tilde{c}^k = \check{c}^k$ and $A([c]^k)$ similarly as above. The remaining steps of the iteration are a copy of the computation of $[c]_K$ with an additional final intersection as in (6.3.2).

Example 7.8.2. The matrices $A_i \in \mathbb{R}^{5\times 5}$, $i = 0, \ldots, 5$, are given by

$$A_0 = \begin{pmatrix} 6 & 1 & 3 & -2 & 0 \\ 1 & 2 & 2 & 0 & 4 \\ 3 & 2 & 1 & 2 & 0 \\ -2 & 0 & 2 & -2 & 0 \\ 0 & 4 & 0 & 0 & -3 \end{pmatrix}, \quad A_1 = \begin{pmatrix} 2 & 1 & 0 & -1 & 1 \\ 1 & 0 & -4 & -1 & 0 \\ 0 & -4 & -2 & 1 & 3 \\ -1 & -1 & 1 & 0 & 5 \\ 1 & 0 & 3 & 5 & -1 \end{pmatrix},$$

$$A_2 = \begin{pmatrix} 1 & 2 & -3 & 0 & -1 \\ 2 & -1 & -3 & 1 & 0 \\ -3 & -3 & 0 & -2 & 2 \\ 0 & 1 & -2 & 0 & 6 \\ -1 & 0 & 2 & 6 & 1 \end{pmatrix}, \quad A_3 = \begin{pmatrix} 2 & -1 & 0 & 2 & 1 \\ -1 & 2 & 1 & 0 & -6 \\ 0 & 1 & -3 & 8 & -3 \\ 2 & 0 & 8 & 6 & -3 \\ 1 & -6 & -3 & -3 & 4 \end{pmatrix},$$

$$A_4 = \begin{pmatrix} -3 & -2 & 2 & 0 & 4 \\ -2 & 1 & 2 & -4 & 0 \\ 2 & 2 & -2 & -1 & 2 \\ 0 & -4 & -1 & -5 & 0 \\ 4 & 0 & 2 & 0 & 1 \end{pmatrix}, \quad A_5 = \begin{pmatrix} -3 & -1 & -5 & 3 & 2 \\ -1 & 2 & 7 & -1 & -2 \\ -5 & 7 & 5 & -3 & 0 \\ 3 & -1 & -3 & 0 & -2 \\ 2 & -2 & 0 & -2 & 4 \end{pmatrix}.$$

The eigenvalues λ_i^* are prescribed by

$$\lambda^* = (-10, -5, -1, 4, 10)^T.$$

As can be seen by a direct computation, the vector $c^* = (-3, 4, 1, 2, -1)^T$ is a solution of the problem. It is verified when starting the Newton iteration with $c^0 =$

$(-2.9, 4.1, 0.9, 2.01, -1.01)^T$. With this iterative process we got the enclosure

$$[c] = (-3, 4, 1, 2, -1)^T + 10^{-15} \cdot [-1, 1] \cdot e.$$

For the Newton iteration we needed $k_{\text{newton}} + 1 = 5$ iterates, for the enclosure of eigenpairs $k_{\text{ewp}} + 1 = 4$ iterates, for the enclosure of c^* finally $k_{c^*} + 1 = 3$ iterates. Starting with $c^0 = (10, 10, 10, 10, 10)^T$ yields

$$[c] = \begin{pmatrix} [-3.879\,049\,564\,183\,7^2_5] \\ [\ 4.305\,375\,937\,429^{11}_{08}] \\ [\ 0.729\,062\,953\,735\,3^7_5] \\ [\ 1.682\,982\,632\,583^{81}_{78}] \\ [-1.092\,532\,116\,503\,9^2_4] \end{pmatrix}$$

which means that

$$c^{**} = \begin{pmatrix} -3.879\,049\,564\,183\,7\dots \\ 4.305\,375\,937\,429\dots \\ 0.729\,062\,953\,735\,3\dots \\ 1.682\,982\,632\,583\dots \\ -1.092\,532\,116\,503\,9\dots \end{pmatrix}$$

is another solution of the problem. Here we got $k_{\text{newton}} = 11$, $k_{\text{ewp}} = 3$, $k_{c^*} = 2$. □

Our second example originates from Friedland, Nocedal, Overton [102], where an approximation of c^* has been derived.

Example 7.8.3.

$$A_0 = \begin{pmatrix} 0 & 4 & -1 & 1 & 1 & 5 & -1 & 1 \\ 4 & 0 & -1 & 2 & 1 & 4 & -1 & 2 \\ -1 & -1 & 0 & 3 & 1 & 3 & -1 & 3 \\ 1 & 2 & 3 & 0 & 1 & 2 & -1 & 4 \\ 1 & 1 & 1 & 1 & 0 & 1 & -1 & 5 \\ 5 & 4 & 3 & 2 & 1 & 0 & -1 & 6 \\ -1 & -1 & -1 & -1 & -1 & -1 & 0 & 7 \\ 1 & 2 & 3 & 4 & 5 & 6 & 7 & 0 \end{pmatrix}$$

$$A_i = e^{(i)}(e^{(i)})^T, \quad i = 1, \dots, 8, \quad \lambda^* = (10, 20, 30, 40, 50, 60, 70, 80)^T.$$

Notice that $A(c)_{ii} = c_i$, $i = 1, \dots, n$, while $A(c)_{ij} = (A_0)_{ij}$ for $i \neq j$.

Starting with $c^0 = (10, 20, 30, 40, 50, 60, 70, 80)^T$ yields

$$[c] = \begin{pmatrix} [11.907\,876\,102\,472^{73}_{68}] \\ [19.705\,521\,508\,08^{701}_{695}] \\ [30.545\,498\,186\,977\,0^9_2] \\ [40.062\,657\,488\,448\,0^8_0] \\ [51.587\,140\,290\,725^{53}_{44}] \\ [64.702\,131\,432\,179^{63}_{41}] \\ [70.170\,675\,820\,891^{28}_{03}] \\ [71.318\,499\,170\,21^{916}_{897}] \end{pmatrix}$$

with $k_{\text{newton}} = 5$, $k_{\text{ewp}} = 3$, $k_{c^*} = 4$, similar to Example 7.8.2.

Starting with $c^0 = (-10, -10, -30, -30, -50, -50, -70, -70)^T$ results in

$$[c] = \begin{pmatrix} [11.461\,354\,297\,738\,6^{6}_{1}] \\ [78.880\,829\,360\,854^{38}_{19}] \\ [68.353\,399\,602\,851^{43}_{24}] \\ [49.878\,330\,411\,746\,6^{9}_{1}] \\ [59.168\,917\,833\,392^{34}_{24}] \\ [30.410\,470\,147\,540^{42}_{36}] \\ [24.834\,324\,014\,386^{23}_{16}] \\ [37.012\,374\,331\,490^{26}_{15}] \end{pmatrix},$$

with $k_{\text{newton}} = 5$, $k_{\text{ewp}} \le 4$, $k_{c^*} = 7$. □

Exercises

Ex. 7.8.1. Find examples like Example 7.8.2 for yourself and experiment with them. Start for instance with an $n \times n$ symmetric diagonal block matrix A with 1×1 and 2×2 blocks such that A has n prescribed different eigenvalues λ_i^*. Then choose symmetric matrices $A_i \in \mathbb{R}^{n \times n}$ and real numbers c_i, $i = 1, \ldots, n$, arbitrarily and finally adapt A_0 such that $A = A_0 + \sum_{i=1}^{n} c_i A_i$ holds.

Notes to Chapter 7

This chapter is a revised and extended version of the author's contribution 'Result verification for eigenvectors and eigenvalues' to the book Herzberger [144]; cf. pp. 209–276. The reduction of many proofs to the crucial Theorem 7.1.1 was published by the author in that book for the first time. But we emphasize that the underlying basic results have different authorship – cf. the list below. In the present version of our contribution we made some corrections and extensions, prepared some exercises, enlarged Section 7.3 considerably, and added Section 7.6. Another overview on enclosure methods for eigenpairs is presented in Mayer [207].

To 7.1: The results of this chapter on quadratic systems are completely contained in Alefeld [13].

To 7.2: Most results of this section originate from Rump [311, 312]. Variants of the method (partly with preconditioning from the right) and first verification and convergence results are already contained in Krawczyk [171, 173] based on a paper by Unger [354]. Theorem 7.2.7 goes back to Alefeld [10]. The examples are contained in Mayer [210]. The eigenvalue problem for an interval matrix $[A]$ as described in the beginning of Chapter 7 is mentioned (certainly not for the first time) in Alefeld, Kreinovich, Mayer [32]. Predecessors are for instance Krawczyk [173] and Deif [81]. Complex eigenpairs are considered in Krawczyk [174] and Grüner [124]. A unified approach to enclosure methods for eigenpairs was given in Mayer [209].

To 7.3: The idea of applying Jacobi matrices and the details of the method are due to Lohner [193]. The application of Gershgorin's theorem in combination with interval analysis can already be found in Kreß [176]. Using a Jacobi method in order to enclose eigenpairs of Hermitian matrices is presented by Fernando and Pont in [94].

To 7.4: Double and nearly double eigenvalues are handled by Alefeld, Spreuer in [48] whose results are presented in this section. For real symmetric matrices they can also be handled by Lohner's method described in Section 7.3; see also Kreß [176].

To 7.5: The method for the generalized eigenvalue problem and the results on it are introduced in Alefeld [14]. Another method in this respect is studied in Miyajima, Ogita, Rump, Oishi [229].

To 7.6: Behnke's method was presented in several papers, among them Behnke [65, 66, 67].

To 7.7: Enclosures for singular values are given in Alefeld [11], for singular values of *interval* matrices in Deif [82]. Enclosures for generalized singular values are studied in Alefeld, Hoffmann, Mayer [27].

To 7.8: Verification and enclosure in connection with the inverse eigenvalue problem is considered in Alefeld, Mayer [36] and Alefeld, Gienger, Mayer [22].

8 Automatic differentiation

Automatic differentiation is a technique to compute the *value* $f'(x_0)$ of the derivative function f' without knowing an explicit expression $f'(x)$. The value is computed directly together with the function value $f(x_0)$ running through the formula tree of $f(x)$. Thus one starts with x_0 and constants c together with the derivatives 1 and 0 of x and c, respectively, and builds up $f(x_0)$ together with $f'(x_0)$, applying the corresponding rules for the differentiation. This technique should not be mixed up with symbolic differentiation, which can be found in algebra systems like MAPLE and which computes an expression $f'(x)$ of f' from that of f first and evaluates $f'(x)$ at x_0 afterwards. It also differs from numerical differentiation that only delivers an approximation of $f'(x_0)$ using appropriate values of f.

There are essentially two modes to apply automatic differentiation: the forward mode and the backward mode. The first can be easily understood while the second needs fewer operations in some situations. Both can also be applied to functions $f: D = D_f \subseteq \mathbb{R}^m \to \mathbb{R}^n$ by computing $\frac{\partial f_i}{\partial x_j}(x^0)$, $x^0 \in D$, for $i \in \{1, \dots, n\}$, $j \in \{1, \dots, m\}$ separately in a similar way. It is easy to see that the interval arithmetical evaluation $\frac{\partial f_i}{\partial x_j}([x]^0)$ at an interval vector $[x]^0$ can be computed in the same way providing a method to obtain the interval arithmetic evaluation $f'([x]^0)$ of the Jacobian f' which is used in Chapter 6. In addition, the technique can be generalized to compute the Taylor coefficients $(f)_k := f^{(k)}(x_0)/k!$, $k = 0, 1, \dots, p$, of a given function $f: D \subseteq \mathbb{R} \to \mathbb{R}$, which are used in particular methods for enclosing solutions of initial value problems or boundary value problems. Here, one starts as above but supplements the first derivative of x and c by 0 for the higher order ones. We tacitly assume in the sequel that all function values and derivatives exist without mentioning this explicitly. In addition, we suppose in the whole chapter that $f(x)$ is an expression from the set $\mathbb{E}$ in Definition 4.1.2, or, equivalently, that f is factorable with a fixed factor sequence $f^1, \dots, f^n, \dots, f^m, \dots, f^s$ as in Definition 4.1.1.

8.1 Forward mode

In this section we describe the computation of the Taylor coefficients

$$(f)_k := \frac{f^{(k)}(x_0)}{k!}, \quad k = 0, 1, \dots, p \quad (x_0 \in D_f \subseteq \mathbb{R})$$

in the so-called forward mode. To this end we start with a trivial but fundamental lemma which can be proved with elementary knowledge of calculus. Here and in the sequel we interpret $((f)_i)_j$ always as

$$((f)_i)_j = \frac{1}{i!\,j!}\frac{df^{(i)}(x_0)}{dx^j} := \frac{1}{i!\,j!}\frac{df^{(i)}(x)}{dx^j}\bigg|_{x=x_0} = \frac{1}{i!\,j!}f^{(i+j)}(x_0).$$

DOI 10.1515/9783110499469-009

Lemma 8.1.1. *For $f(x), g(x) \in \mathbb{E}$ and fixed $x_0 \in D_f \cap D_g \subseteq \mathbb{R}$ we have*
(a) $(f)_1 = f'(x_0)$.
(b) $(f)_k = \frac{1}{k}((f)_1)_{k-1}$.
(c) $(f \circ g)_1 = (f \circ g)'(x_0) = ((f' \circ g) \cdot g')(x_0) = ((f' \circ g)_0 \cdot (g)_1$, *where $\circ$ denotes the usual composition of functions.*

Based on Lemma 8.1.1 our next result shows how $(f)_k$ can be computed recursively for the operations und functions occurring in $\mathbb{E}$.

Theorem 8.1.2. *For $k = 0, 1, \dots, p$ and $\alpha \in \mathbb{R}$ the following properties hold for functions $f, g\colon D \to \mathbb{R}$, $D \subseteq \mathbb{R}$.*

(a) $$(\pm f)_k = \pm(f)_k.$$

(b) $$(f + g)_k = (f)_k + (g)_k.$$

(c) $$(f - g)_k = (f)_k - (g)_k.$$

(d) $$(f \cdot g)_k = \sum_{j=0}^{k} (f)_j \cdot (g)_{k-j}.$$

(e) $$\begin{aligned}(f/g)_k &= \left\{(f)_k - \sum_{j=1}^{k} (g)_j (f/g)_{k-j}\right\} \Big/ (g)_0 \\ &= \left\{(f)_k - \sum_{j=0}^{k-1} (g)_{k-j} (f/g)_j\right\} \Big/ (g)_0, \quad \text{if } (g)_0 \neq 0.\end{aligned}$$

(f) $$\begin{aligned}(\exp f)_0 &= \exp (f)_0, \\ (\exp f)_k &= \sum_{j=0}^{k-1} \left(1 - \frac{j}{k}\right) (\exp f)_j (f)_{k-j} \\ &= \sum_{j=1}^{k} \frac{j}{k} (\exp f)_{k-j} (f)_j, \quad k \geq 1.\end{aligned}$$

(g) $$\begin{aligned}(\ln f)_0 &= \ln (f)_0, \quad (\ln f)_1 = (f)_1/(f)_0, \\ (\ln f)_k &= \left\{(f)_k - \sum_{j=1}^{k-1} \frac{j}{k} (\ln f)_j (f)_{k-j}\right\} \Big/ (f)_0, \quad k \geq 2.\end{aligned}$$

(h) $$\begin{aligned}(\sin f)_0 &= \sin (f)_0, \quad (\cos f)_0 = \cos (f)_0, \\ (\sin f)_k &= \sum_{j=0}^{k-1} \left(1 - \frac{j}{k}\right) (\cos f)_j (f)_{k-j}, \\ (\cos f)_k &= -\sum_{j=0}^{k-1} \left(1 - \frac{j}{k}\right) (\sin f)_j (f)_{k-j}.\end{aligned}$$

(i) $$(\arctan f)_0 = \arctan(f)_0,$$

$$(\arctan f)_k = \frac{1}{1+((f)_0)^2}\left\{(f)_k - \frac{1}{k}\sum_{j=1}^{k-1} j\,(\arctan f)_j(1+((f)_0)^2)_{k-j}\right\}, \quad k \ge 1.$$

(j) $$(f^g)_k = (\exp(g\cdot \ln f))_k, \quad \textit{if } (f)_0 > 0.$$

(k) $$(f^\alpha)_k = \left\{\sum_{j=0}^{k-1}\left(\alpha - \frac{j(\alpha+1)}{k}\right)(f^\alpha)_j(f)_{k-j}\right\} \Big/ (f)_0, \quad \textit{if } (f)_0 > 0.$$

Proof. (a), (b), and (c) are obvious.

(d) The formula is trivially true for $k = 0$. Let it hold for some k. Then

$$\begin{aligned}
(f\cdot g)_{k+1} &= \frac{1}{k+1}((f\cdot g)_1)_k = \frac{1}{k+1}((f)_1(g)_0 + (f)_0(g)_1)_k\\
&= \frac{1}{k+1}\left\{\sum_{j=0}^{k}((f)_1)_j((g)_0)_{k-j} + \sum_{j=0}^{k}((f)_0)_j((g)_1)_{k-j}\right\}\\
&= \frac{1}{k+1}\left\{\sum_{j=0}^{k}(j+1)(f)_{j+1}(g)_{k-j} + \sum_{j=0}^{k}(k+1-j)(f)_j(g)_{k+1-j}\right\}\\
&= \frac{1}{k+1}\left\{\sum_{j=1}^{k+1} j(f)_j(g)_{k+1-j} + \sum_{j=0}^{k}(k+1-j)(f)_j(g)_{k+1-j}\right\}\\
&= \frac{1}{k+1}\left\{(k+1)(f)_{k+1}(g)_0 + \sum_{j=0}^{k}(k+1)(f)_j(g)_{k+1-j}\right\}\\
&= \sum_{j=0}^{k+1}(f)_j\cdot(g)_{k+1-j}.
\end{aligned}$$

(e) With (d) we get

$$\begin{aligned}
(f)_k &= \left(g\cdot\frac{f}{g}\right)_k = \sum_{j=0}^{k}(g)_j\left(\frac{f}{g}\right)_{k-j} = (g)_0\left(\frac{f}{g}\right)_k + \sum_{j=1}^{k}(g)_j\left(\frac{f}{g}\right)_{k-j}\\
&= (g)_0\left(\frac{f}{g}\right)_k + \sum_{j=0}^{k-1}(g)_{k-j}\left(\frac{f}{g}\right)_j,
\end{aligned}$$

hence (e) follows.

(f) From $(\exp f)' = (\exp f)\cdot f'$ we get $(\exp f)_1 = (\exp f)_0(f)_1$ which together with (d) implies

$$\begin{aligned}
(\exp f)_k &= \frac{1}{k}((\exp f)_1)_{k-1} = \frac{1}{k}((\exp f)_0(f)_1)_{k-1}\\
&= \frac{1}{k}\sum_{j=0}^{k-1}(\exp f)_j((f)_1)_{k-1-j} = \frac{1}{k}\sum_{j=0}^{k-1}(\exp f)_j(k-j)(f)_{k-j}\\
&= \sum_{j=0}^{k-1}\left(1-\frac{j}{k}\right)(\exp f)_j(f)_{k-j} = \sum_{j=1}^{k}\frac{j}{k}(\exp f)_{k-j}(f)_j, \quad k \ge 1.
\end{aligned}$$

(g) From $(\ln f)' = f'/f$ we get $(\ln f)_1 = (f)_1/(f)_0$ and furthermore

$$\begin{aligned}(\ln f)_k &= \frac{1}{k}\big((\ln f)_1\big)_{k-1} = \frac{1}{k}\big((f)_1/(f)_0\big)_{k-1} \\ &= \frac{1}{k}\left\{\big((f)_1\big)_{k-1} - \sum_{j=1}^{k-1}(f)_j\Big(\frac{(f)_1}{(f)_0}\Big)_{k-1-j}\right\} \Big/ (f)_0 \\ &= \left\{(f)_k - \frac{1}{k}\sum_{j=1}^{k-1}(f)_j\big((\ln f)_1\big)_{k-1-j}\right\} \Big/ (f)_0 \\ &= \left\{(f)_k - \frac{1}{k}\sum_{j=1}^{k-1}(f)_j(k-j)(\ln f)_{k-j}\right\} \Big/ (f)_0 \\ &= \left\{(f)_k - \sum_{j=1}^{k-1}\frac{j}{k}(\ln f)_j(f)_{k-j}\right\} \Big/ (f)_0, \quad k \geq 2.\end{aligned}$$

(h) From $(\sin f)' = (\cos f)f'$ we get $(\sin f)_1 = (\cos f)_0(f)_1$ and with (d)

$$\begin{aligned}(\sin f)_k &= \frac{1}{k}\big((\sin f)_1\big)_{k-1} = \frac{1}{k}\big((\cos f)_0(f)_1\big)_{k-1} \\ &= \frac{1}{k}\sum_{j=0}^{k-1}(\cos f)_j\big((f)_1\big)_{k-1-j} = \frac{1}{k}\sum_{j=0}^{k-1}(k-j)(\cos f)_j(f)_{k-j}\end{aligned}$$

follows. The remaining part can be shown similarly.

(i) With (e) we get

$$\begin{aligned}(\arctan f)_k &= \frac{1}{k}\big((\arctan f)_1\big)_{k-1} = \frac{1}{k}\Big((f)_1/\big(1+((f)_0)^2\big)\Big)_{k-1} \\ &= \frac{1}{1+((f)_0)^2}\left\{\frac{1}{k}\big((f)_1\big)_{k-1} - \frac{1}{k}\sum_{j=0}^{k-2}\Big((f)_1/\big(1+((f)_0)^2\big)\Big)_j\big(1+((f)_0)^2\big)_{k-1-j}\right\} \\ &= \frac{1}{1+((f)_0)^2}\left\{(f)_k - \frac{1}{k}\sum_{j=0}^{k-2}(j+1)(\arctan f)_{j+1}\big(1+((f)_0)^2\big)_{k-(j+1)}\right\} \\ &= \frac{1}{1+((f)_0)^2}\left\{(f)_k - \frac{1}{k}\sum_{j=1}^{k-1}j(\arctan f)_j\big(1+((f)_0)^2\big)_{k-j}\right\}, \quad k \geq 1.\end{aligned}$$

(j) is obvious.

(k) Equating the two representations

$$\begin{aligned}(f^{\alpha+1})_k &= \frac{1}{k}\big((f^{\alpha+1})_1\big)_{k-1} = \frac{\alpha+1}{k}\big((f^\alpha)_0(f)_1\big)_{k-1} \\ &= \frac{\alpha+1}{k}\sum_{j=0}^{k-1}(f^\alpha)_j\big((f)_1\big)_{k-1-j} \\ &= \frac{\alpha+1}{k}\sum_{j=0}^{k-1}(k-j)(f^\alpha)_j(f)_{k-j}\end{aligned}$$

and

$$(f^{\alpha+1})_k = (f^\alpha f)_k = \sum_{j=0}^{k} (f^\alpha)_j (f)_{k-j}$$

yields

$$(f^\alpha)_k (f)_0 = \sum_{j=0}^{k-1} \left\{ -1 + \frac{\alpha+1}{k}(k-j) \right\} (f^\alpha)_j (f)_{k-j}$$

and finally the assertion. □

Notice that we did not mention the functions $f(x) = x^2$ and $f(x) = \sqrt{x}$ explicitly in our list above as is done, for instance, in Fischer [97]. These functions are integrated in $f^\alpha(x)$, although it is recommendable to handle them separately when writing software which uses automatic differentiation. For details we refer to Fischer [97].

In a practical realization of automatic differentiation and of automatic generation of Taylor coefficients one can work with vectors $v \in \mathbb{R}^{p+1}$ with $v_{k+1} = (f)_k$, $k = 0, \ldots, p$. An explicit formula tree or an explicit factor sequence of f need not be known by the user. A separate data type (`taylor`, e.g.) and the possibility of operator overloading are useful unless the operations for expressions in $\mathbb{E}$ are renamed and applied as functions or procedures such as `plus`, `minus`, etc. MATLAB provides such a possibility. One can also work with structures, in particular, if only $f(x_0)$ and $f'(x_0)$ are required. Thus if the components of a structure are denoted by $f.x$ for the function value $f(x_0) = (f)_0$ and $f.dx$ for the derivative $f'(x_0) = (f)_1$ one starts with $x.x = x_0$, $x.dx = 1$ and $c.x = c$, $c.dx = 0$ for x and a constant c in an expression $f(x) \in \mathbb{E}$, and one builds up $f(x_0)$ and $f'(x_0)$ according to the rules in Theorem 8.1.2. As an illustration consider the following example.

Example 8.1.3. Let $f(x) = (x+5)\sin(3x)$ be evaluated at $x = x_0 = 7$ using the factor sequence $f^1(x) = x$, $f^2(x) = 3$, $f^3(x) = f^m(x) = 5$, $f^4(x) = f^1(x) + f^3(x) = x + 5$, $f^5(x) = f^2(x) \cdot f^1(x) = 3 \cdot x$, $f^6(x) = \sin(f^5(x)) = \sin(3 \cdot x)$, $f^7(x) = f^s(x) = f^4(x) \cdot f^6(x) = f(x)$. Then

$$\begin{aligned}
f^1.x &= x_0 = 7, & f^1.dx &= df^1(x_0)/dx = 1,\\
f^2.x &= 3, & f^2.dx &= df^2(x_0)/dx = 0,\\
f^3.x &= 5, & f^3.dx &= df^3(x_0)/dx = 0,\\
f^4.x &= f^1.x + f^3.x, & f^4.dx &= f^1.dx + f^3.dx,\\
f^5.x &= f^2.x \cdot f^1.x, & f^5.dx &= f^2.dx \cdot f^1.x + f^2.x \cdot f^1.dx,\\
f^6.x &= \sin(f^5.x), & f^6.dx &= \cos(f^5.x) \cdot f^5.dx,\\
f^7.x &= f^4.x \cdot f^6.x, & f^7.dx &= f^4.dx \cdot f^6.x + f^4.x \cdot f^6.dx.
\end{aligned}$$

Inspecting these formulae one recognizes at once that the values of $f^i.dx$ are constant for $i = 1, \ldots, m = 3$, and for $i > m$ they depend linearly on previous values $f^j.dx$, $j < i$. In such a linear combination, at most two coefficients are nonzero and then of the form $f^j.x$ or $\varphi(f^j.x)$, $j < i$, depending on the underlying operation: one coefficient for a unary operation, two for a binary one. These coefficients can be expressed as $\partial f^i/\partial f^j$. Therefore, if one defines the strict lower triangular matrix $L \in \mathbb{R}^{7\times 7}$ and the vector $b \in \mathbb{R}^7$ by

$$l_{ij} := \begin{cases} \frac{\partial f^i}{\partial f^j}, & \text{if } j < i \\ 0 & \text{otherwise} \end{cases}, \qquad b_i := \begin{cases} 1, & \text{if } i = 1 \\ 0 & \text{otherwise} \end{cases}$$

then the vector $h = (f^1.dx, \ldots, f^7.dx)^T$ satisfies the equation

$$h = Lh + b. \tag{8.1.1}$$

Notice that the vector b contains the 'starting values' $f^i.dx$, $i = 1, \ldots, m$, while the corresponding rows of L are zero. For $i = m+1, \ldots, s = 7$ the components b_i are zero, while the corresponding rows of L are zero with the exception of at most two entries. The form of L indicates that the components of h can be computed in a forward way. This justifies the terminology 'forward mode'.

From (8.1.1) one sees immediately that h can also be computed as

$$h = (I - L)^{-1} b, \tag{8.1.2}$$

which will be the starting point for the backward mode in Section 8.2. □

For functions $f\colon D \subseteq \mathbb{R}^n \to \mathbb{R}$ a partial derivative $\frac{\partial f}{\partial x_j}(x^0)$ can be computed using the formulae for the Taylor coefficients with $p = 1$ and

$$(x_i)_0 = x_i^0, \quad (x_i)_1 = \begin{cases} 1, & \text{if } i = j \\ 0 & \text{otherwise} \end{cases}, \qquad i = 1, \ldots, n. \tag{8.1.3}$$

Here again one can use vectors v to represent $f(x^0)$ together with $\nabla f(x^0) = \operatorname{grad}(f(x^0))$, choosing $v \in \mathbb{R}^{n+1}$ with $v_1 = f(x^0)$ and $v_{j+1} = \partial f(x^0)/\partial x_j$. A datatype like `gradient` and the formulae

$$\left.\begin{aligned}
\nabla x_i &= e^{(i)}, \quad i = 1, \ldots, n, \\
\nabla c &= 0 \in \mathbb{R}^n, \\
\nabla(\pm f) &= \pm \nabla f, \\
\nabla(f \pm g) &= \nabla f \pm \nabla g, \\
\nabla(f \cdot g) &= (\nabla f) g + f \nabla g, \\
\nabla(f/g) &= (\nabla f)/g - (f/g^2)\nabla g = (\nabla f - (f/g)\nabla g)/g, \\
\nabla \exp(f) &= (\exp f)\nabla f, \\
\nabla \ln(f) &= (\nabla f)/f, \\
\nabla \sin(f) &= (\cos f)\nabla f, \\
\nabla \cos(f) &= (-\sin f)\nabla f, \\
\nabla \arctan(f) &= (\nabla f)/(1 + f^2), \\
\nabla f^\alpha &= \alpha f^{\alpha-1}\nabla f,
\end{aligned}\right\} \tag{8.1.4}$$

$(f, g\colon D \subseteq \mathbb{R}^n \to \mathbb{R})$ form the basis of a gradient arithmetic.

The Jacobian $f'(x^0)$ of a function $f\colon D \subseteq \mathbb{R}^n \to \mathbb{R}^p$ can be realized as a vector of gradients using gradient arithmetic for the individual components f_i of f and storing the result $f_i(x^0)$, $(\nabla f_i(x^0))^T$, $i = 1, \ldots, n$, as i-th row of a $p \times (n+1)$ matrix.

Exercises

Ex. 8.1.1. Extend Theorem 8.1.2 by the functions $\sinh x$ and $\cosh x$.

8.2 Backward mode

Let $f\colon D \subseteq \mathbb{R} \to \mathbb{R}$ be a factorable function with factor sequence $f^1, \ldots, f^m, \ldots, f^s$ as in Definition 4.1.2. The idea of representing the derivatives $h_j = (f^j)'(x^0)$, $j = 1, \ldots, s$, in the form

$$h = Lh + b, \quad h = (h_i) \in \mathbb{R}^s, \tag{8.2.1}$$

as in Example 8.1.3 holds generally. Thus if one defines

$$l_{ij} = \begin{cases} \frac{\partial f^i}{\partial f^j} & \text{for } i > m \text{ and } j < i, \\ 0 & \text{otherwise,} \end{cases}$$

one obtains

$$\begin{array}{ll}
l_{i\ell} = \pm 1, & \text{if } f^i = \pm f^\ell, \\
l_{i\ell} = 1,\ l_{ir} = 1, & \text{if } f^i = f^\ell + f^r, \\
l_{i\ell} = 1,\ l_{ir} = -1, & \text{if } f^i = f^\ell - f^r, \\
l_{i\ell} = f^r,\ l_{ir} = f^\ell, & \text{if } f^i = f^\ell \cdot f^r, \\
l_{i\ell} = 1/f^r,\ l_{ir} = -f^i/f^r, & \text{if } f^i = f^\ell/f^r, \\
l_{i\ell} = \varphi'(f^\ell), & \text{if } f^i = \varphi(f^\ell),\ \varphi \in \mathbb{F},
\end{array}$$

and $l_{ij} = 0$ otherwise. The indices $\ell, r < i$ remind us of 'left' and 'right' again, according to the position of the operands involved. The vector $b \in \mathbb{R}^s$ contains the 'starting values' $b_1 = 1$, $b_i = 0$ otherwise.

For a function $f\colon D \subseteq \mathbb{R}^n \to \mathbb{R}$ with a factor sequence (f^k) similar to that above, the formulae in (8.1.4) show that in the computation of a *fixed* partial derivative $\partial f(x^0)/\partial x_\nu$ the index ν enters *directly* only into the starting values (8.1.3) while the calculation rules (8.1.4) coincide for *all* components of $\operatorname{grad}(f^k)$ and any fixed $k \in \{n+1, \ldots, s\}$. Therefore, the matrix L in (8.2.1) is independent of ν while $b_i = b_i^\nu = \partial x_i/\partial x_\nu = \partial f^i/\partial x_\nu$ holds for $i = 1, \ldots, n$, i.e., $b = b^\nu = e^{(\nu)} \in \mathbb{R}^s$.

In the forward mode the right-hand side $Lh + b^\nu$ in (8.2.1) must be evaluated *for each* $b = b^\nu$, $\nu = 1, \ldots, n$, if $\operatorname{grad} f(x^0)$ is required. In the so-called *backward mode* one transforms (8.2.1) equivalently to

$$h = (I - L)^{-1} b^\nu. \tag{8.2.2}$$

Since b^ν equals $e^{(\nu)}$ no explicit matrix-vector multiplication must be done in (8.2.2) – at the expense of computing the inverse of a sparse triangular matrix with unit diagonal. Thus, h with $h_i := \partial f^i(x^0)/\partial x_\nu$, $i = 1, \dots, s$, is just the ν-th column of $(I - L)^{-1}$.

Lemma 8.2.1. *Let $C = (I - L)^{-1} \in \mathbb{R}^{n\times n}$ with a strictly lower triangular matrix L. Then the entries c_{ij} of C can be computed as*

$$c_{ij} = \begin{cases} 1, & \text{if } j = i, \\ \sum_{k=j+1}^{i} c_{ik} l_{kj}, & \text{if } j = i-1, i-2, \dots, 1, \\ 0, & \text{if } j > i, \end{cases} \tag{8.2.3}$$

$$= \delta_{ij} + \sum_{k=j+1}^{i} c_{ik} l_{kj}. \tag{8.2.4}$$

The proof is immediate using $C(I - L) = I$.

The formula (8.2.4) shows that for the computation of c_{ij}, $j < i$, the succeeding entries $c_{i,j+1}, c_{i,j+2}, \dots, c_{ii}$ must be known. This is the reason why one proceeds backward with respect to j in the middle expression of (8.2.3). This backward procedure is responsible for the name ‘backward mode’ in automatic differentiation in which we will compute c_{sj}, $j = s, s-1, \dots, 1$, after having computed and stored the factor sequence $f^1(x^0), \dots, f^s(x^0)$. By our preceding remarks (with $\nu = j$) the relation

$$c_{sj} = \frac{\partial f^s(x^0)}{\partial x_j} = \frac{\partial f(x^0)}{\partial x_j}, \quad j = 1, \dots, n,$$

follows. The components $d^j := c_{sj}$ can be obtained in the following recursive way.

(1) Compute and store $f^1(x^0), \dots, f^s(x^0)$.

(2) Initialize

$$d^j = \begin{cases} 1, & \text{if } j = s \\ 0, & \text{if } j < s \end{cases}.$$

(3) Backward step:
For $k = s$ down to $m + 1$ compute

$$\begin{array}{ll} d^\ell = d^\ell \pm d^k, & \text{if } f^k = \pm f^\ell, \\ d^\ell = d^\ell + d^k,\ d^r = d^r \pm d^k, & \text{if } f^k = f^\ell \pm f^r, \\ d^\ell = d^\ell + f^r \cdot d^k,\ d^r = d^r + f^\ell \cdot d^k, & \text{if } f^k = f^\ell \cdot f^r, \\ d^\ell = d^\ell + (d^k/f^r),\ d^r = d^r - f^k \cdot (d^k/f^r), & \text{if } f^k = f^\ell/f^r, \\ d^\ell = d^\ell + \varphi'(f^\ell) \cdot d^k, & \text{if } f^k = \varphi(f^\ell),\ \varphi \in \mathbb{F}. \end{array}$$

At the end of the k-loop one gets $c_{sj} = d^j$, $j = 1, \dots, n$.

In order to understand the backward step look at (8.2.4). If, for instance, $f^k = f^\ell - f^r$, then $l_{k\ell} = 1$, $l_{kr} = -1$. These entries of L are responsible for the contributions $c_{sk} l_{k\ell} = c_{sk} = d^k$ and $c_{sk} l_{kr} = -c_{sk} = -d^k$ to the *final* values $d^\ell = c_{s\ell}$ and $d^r = c_{sr}$, respectively. Therefore, the *actual* partial sums d^ℓ and d^r are updated to $d^\ell + d^k$, and $d^r - d^k$, respectively.

A possible disadvantage of the backward mode is the storage of all values of the factor sequence $(f^k(x^0))$ that are needed for computing $d = (d^i) \in \mathbb{R}^n$.

We want to compare now the amount of work $W(f, \nabla f)$ to compute f and ∇f with $W(f)$, the amount of work for f alone. To this end we consider one step of (8.1.4), i.e., the work from f^{k-1} to f^k, if (f^k) is the factor sequence of f. Let 'A' denote addition/subtraction, 'M' multiplication/division, 'F' the evaluation of an elementary function $\varphi \in \mathbb{F}$. Then Table 8.2.1 is immediate.

Tab. 8.2.1: Amount of work.

f^k	$W(f^k)$	$W(\nabla f^k)_{\text{forward}}$	$W(\nabla f^k)_{\text{backward}}$
variable, constant	—	—	—
± (unary)	1 A	n A	1 A
± (binary)	1 A	n A	2 A
·	1 M	n A + 2n M	2 A + 2 M
/	1 M	n A + 2n M	2 A + 2 M
φ	1 F	n M + 1 F	1 A + 1 M + 1 F

If all operations are weighted equally, one sees at once that for the forward mode we get

$$\frac{W(f, \nabla f)}{W(f)} \le 3n + 1, \tag{8.2.5}$$

while for the backward mode we obtain

$$\frac{W(f, \nabla f)}{W(f)} \le 5 \tag{8.2.6}$$

for functions $f: D \subseteq \mathbb{R}^n \to \mathbb{R}$. Thus the backward mode is much less expensive than the forward mode provided that $n > 1$. The inequality (8.2.6) illustrates a result by Baur and Strassen who showed in [61] that for rational functions f in n variables the complexity for the computation of a function value $f(x^0)$ differs from that of the corresponding gradient $\nabla f(x^0)$ only by a constant factor which is independent of n.

The example $f(x) = \prod_{i=1}^n x_i$ shows that in the forward mode $W(f, \nabla f)/W(f) = O(n)$ cannot be improved, the example $f(x_1, x_2) = x_1/x_2$ implies $3 \le W(f, \nabla f)/W(f)$.

We recommend to test both modes using the Helmholtz energy function

$$f(x) = \sum_{i=1}^n x_i \ln \frac{x_i}{1 - b^T x} - \frac{x^T A x}{\sqrt{8}\, b^T x} \ln \frac{1 + (1 + \sqrt{2})\, b^T x}{1 + (1 - \sqrt{2})\, b^T x},$$

$0 \le x, b \in \mathbb{R}^n$, $A = A^T \in \mathbb{R}^{n \times n}$; cf. Fischer [97] or Griewank [123].

We leave it to the reader to compute the Hessian H of f and the products $(\nabla f)^T v$, Hv for some vector v in both modes with the corresponding work saving in the backward mode. Furthermore, we mention that both methods can be applied to compute slopes. For details we refer again to Fischer [97].

Notes to Chapter 8

To 8.1: The forward mode of automatic differentiation and, more generally, the automatic generation of Taylor coefficients, can already be found in Moore [232]. Later on, it was discussed in various works; cf. Rall [280], Lohner [192], Griewank [123], Tucker [353], e.g. An algorithmic description is given in Hammer, Hocks, Kulisch, Ratz [129, 130]. Automatic generation of slopes can be found in Neumaier [257] and of slopes of higher order in Schnurr [331]. A large bibliography on automatic differentiation can be found in Corliss [79].

To 8.2: Our way to describe the backward mode of automatic differentiation is taken from Fischer [97]. See also Fischer [98].

9 Complex intervals

The set $\mathbb{C}$ of complex numbers together with an arithmetic on $\mathbb{C}$ can be introduced in many ways. For engineers one usually defines $\mathbb{C}$ as the set

$$\mathbb{C} = \{ z = a + ib \mid a, b \in \mathbb{R},\ i^2 = -1 \}$$

and an arithmetic as for algebraic terms in $\mathbb{R}$. A more rigorous definition starts with pairs $(a, b) \in \mathbb{R}^2$ and an arithmetic which adds and subtracts entrywise but multiplies by

$$(a, b) \cdot (c, d) = (ac - bd, ad + bc)$$

and divides by

$$(a, b)/(c, d) = \left(\frac{ac + bd}{c^2 + d^2}, \frac{bc - ad}{c^2 + d^2} \right)$$

if $(c, d) \neq (0, 0)$, cf. for instance Hille [148]. It is obvious that the resulting structures $(\mathbb{C}, +, -, \cdot, /)$ and $(\mathbb{R}^2, +, -, \cdot, /)$ are isomorphic and can be identified. An algebraic access consists (up to isomorphism) of the smallest field extension of $\mathbb{R}$ such that $x^2 + 1 = 0$ is solvable; cf. Ahlfors [4], for example. It is left to the reader to show that the set of all 2×2 matrices of the particular form

$$A = \begin{pmatrix} a & b \\ -b & a \end{pmatrix}, \qquad a, b \in \mathbb{R}, \tag{9.0.1}$$

together with the usual four matrix operations $+, -, \cdot, (.)^{-1}$ is another realization of $(\mathbb{C}, +, -, \cdot, /)$. It will be (9.0.1) that facilitates the handling of iterative methods for complex linear systems of equations.

It is standard knowledge in complex calculus that a complex number $z = a + ib$ can be illustrated geometrically as a point in $\mathbb{R}^2$ with the x-axis as the real axis and with the y-axis as the imaginary one. Thus every complex number z can be written in polar coordinate form $z = re^{i\varphi}$ with $e^{i\varphi} := \cos\varphi + i \sin\varphi$, where $r = |z| := \sqrt{a^2 + b^2}$, $\varphi \in [0, 2\pi)$ with φ to be defined arbitrarily if $r = 0$, and $\cos\varphi = a/r$, $\sin\varphi = b/r$ if $r > 0$. Both representations, $z = a + ib$ and $z = re^{i\varphi}$, form the starting point of complex intervals and of a corresponding arithmetic which we will consider in separate sections. The first representation leads to rectangles in the complex plane, the second one ends with discs there.

DOI 10.1515/9783110499469-010

9.1 Rectangular complex intervals

We introduce rectangular complex intervals in the following way.

Definition 9.1.1. Let $[a], [b] \in \mathbb{IR}$. The set

$$[z] = [a] + i[b] = \{\, \tilde{a} + i\tilde{b} \mid \tilde{a} \in [a],\ \tilde{b} \in [b] \,\} \subseteq \mathbb{C}$$

is called a rectangular complex interval. The set of all such intervals is denoted by $\mathbb{IC}_R$. In $\mathbb{IC}_R$ the relations $[z] = [z]'$, $[z] \subseteq [z]'$, and $[z] \cap [z]'$ are defined in the set theoretic way.

As the terminology indicates, a rectangular complex interval can be represented as a rectangle in the complex plane. Equality $[z] = [a] + i[b] = [z]' := [a]' + i[b]'$ holds if and only if $[a] = [a]'$ and $[b] = [b]'$, and $[z] \subseteq [z]'$ is valid if and only $[a] \subseteq [a]'$ and $[b] \subseteq [b]'$. Unless $[z] \cap [z]' = \emptyset$ this intersection is again an element of $\mathbb{IC}_R$.

On $\mathbb{IC}_R$ we introduce the subsequent operations.

Definition 9.1.2. Let $[z] = [a] + i[b]$, $[w] = [c] + i[d] \in \mathbb{IC}_R$. Then we define

$$\begin{aligned}
[z] + [w] &= ([a] + [c]) + i([b] + [d]),\\
[z] - [w] &= ([a] - [c]) + i([b] - [d]),\\
[z] \cdot [w] &= ([a][c] - [b][d]) + i([a][d] + [b][c]),\\
[z]/[w] &= \frac{[a][c] + [b][d]}{[c]^2 + [d]^2} + i\frac{[b][c] - [a][d]}{[c]^2 + [d]^2},
\end{aligned}$$

where for the denominator of the division $0 \notin [w]$ is assumed and the square function of $\mathbb{IR}$ is used.

This arithmetic in $\mathbb{IC}_R$ is called rectangular arithmetic. If we write $[z], [w] \in \mathbb{IC}_R$, we always assume it to hold. It is easy to see that $(\mathbb{C}, +, -, \cdot, /)$ is isomorphically embedded in $(\mathbb{IC}_R, +, -, \cdot, /)$ while $(\mathbb{IR}, +, -, \cdot, /)$ is not, although $\mathbb{IR}$ is a subset of $\mathbb{IC}_R$, of course. As a counterexample consider for instance $1/[1, 2]$, which is $[0.5, 1]$ in $\mathbb{IR}$ and $[0.25, 2]$ in $\mathbb{IC}_R$.

We will now list some properties of this arithmetic which can be seen directly from its definition and Theorem 4.1.5.

Theorem 9.1.3. *Let* $\circ \in \{+, -, \cdot, /\}$, $[w], [z] \in \mathbb{IC}_R$.

(a) $\{\tilde{z} \circ \tilde{w} \mid \tilde{z} \in [z], \tilde{w} \in [w]\} \subseteq [z] \circ [w]$ *with equality for* $\circ \in \{+, -\}$.

(b) $\square\{\tilde{z} \cdot \tilde{w} \mid \tilde{z} \in [z], \tilde{w} \in [w]\} = [z] \cdot [w]$, *where the interval hull* $\square S$ *of* $S \subseteq \mathbb{C}$ *is that superset of* S *in* $\mathbb{IC}_R$ *which has the smallest area.*

Theorem 9.1.3 becomes wrong if the hull symbol $\square$ is dropped on the left-hand side in (b). This can be seen from $[z] = [1, 2]$, $[w] = 1 + i$. Here $[z][w] = [1, 2] + i[1, 2]$ while $\{\, \tilde{z}\tilde{w} \mid \tilde{z} \in [1, 2],\ \tilde{w} = 1 + i \,\} = \{\, t(1 + i) \mid 1 \le t \le 2 \,\}$ is a diagonal of the rectangle $[z][w]$. The counterexample above, Theorem 9.1.3 shows that the (b)-part of this theorem becomes wrong for division. The division $[z]/[w]$ can be improved with respect to enclosure if one defines $[z]/[w] = [z] \cdot (1/[w])$ and if one computes $1/[w]$ as $\square\{\,1/\tilde{w} \mid \tilde{w} \in [w]\,\}$. Unfortunately, the computation of this hull is very costly. For details see Rokne/Lancaster [310].

Our next theorem can be proved directly from the definitions in $\mathbb{IC}_R$ or from inclusion monotony in $\mathbb{IR}$.

Theorem 9.1.4. *Let* $[u], [v], [w], [z] \in \mathbb{IC}_R$.
(a) $(\mathbb{IC}_R, +)$ *is a commutative semigroup with neutral elements.*
(b) $(\mathbb{IC}_R, +, \cdot)$ *has no zero divisors.*
(c) $[z]$ *is invertible with respect to* $+$ *and* $\cdot$, *respectively, if and only if* $[z] \equiv z \in \mathbb{C}$ *and* $[z] \equiv z \in \mathbb{C} \setminus \{0\}$, *respectively.*
(d) $0 \in [z] - [z]$, $1 \in [z]/[z]$ *(where* $0 \notin [z]$ *in case of division).*
(e) $[z]([v] + [w]) \subseteq [z][v] + [z][w]$ *(subdistributivity)*
$z([v] + [w]) = z[v] + z[w]$ *for* $z \in \mathbb{C}$.
(f) $[u] \subseteq [v]$, $[w] \subseteq [z] \Rightarrow [u] \circ [w] \subseteq [v] \circ [z]$ *for* $\circ \in \{+, -, \cdot, /\}$.

Notice that associativity does not hold for $(\mathbb{IC}_R, \cdot)$ as the example $[v] = [1, 2]$, $[w] = 1 + i$, $[z] = 1 + i$ shows. Here,

$$\begin{aligned}([v][w])[z] &= ([1, 2] + [1, 2]i)(1 + i) = [-1, 1] + [2, 4]i \\ &\neq [v]([w][z]) = [1, 2] \cdot 2i = [2, 4]i.\end{aligned}$$

But $(\mathbb{IC}_R, \cdot)$ is commutative and has a neutral element.

Now we introduce several auxiliary functions for complex rectangular intervals.

Definition 9.1.5.
(a) Let $[z] = [a] + i[b]$, $[w] = [c] + i[d] \in \mathbb{IC}_R$. Then we define the distance q between $[z]$ and $[w]$ by

$$q([z], [w]) = q([a], [c]) + q([b], [d]).$$

(b) If $([z]_k)$ is a sequence in $\mathbb{IC}_R$, then we define $\lim_{k\to\infty}[z]_k = [z]$ if and only if $\lim_{k\to\infty} q([z]_k, [z]) = 0$.

It is easy to see that for $[z]_k = [a]_k + i[b]_k$, $[z] = [a] + i[b]$ we have $\lim_{k\to\infty}[z]_k = [z]$ if and only if $\lim_{k\to\infty}[a]_k = [a]$ and $\lim_{k\to\infty}[b]_k = [b]$. In addition, $(\mathbb{IC}_R, q)$ is a complete metric space.

Definition 9.1.6. Let $[z] = [a] + i[b] \in \mathbb{IC}_R$.
(a) The midpoint $\check{z} = \mathrm{mid}([z])$ of $[z]$ is defined by

$$\check{z} = (\underline{z} + \overline{z})/2 = \check{a} + i\check{b}.$$

(b) The radius $r_z = \mathrm{rad}([z])$ of $[z]$ is defined by

$$\mathrm{rad}([z]) = \mathrm{rad}([a]) + \mathrm{rad}([b]),$$

its diameter or width $d([z])$ by $d([z]) = 2\,\mathrm{rad}([z]) = d([a]) + d([b])$.

(c) The absolute value $|[z]|$ of $[z]$ is defined by

$$|[z]| = q([z], 0) = |[a]| + |[b]|.$$

Notice that for $[z] \equiv z \in \mathbb{C}$ the absolute value in $\mathbb{IC}_R$ does not reduce to the usual absolute value for complex numbers, i.e., to the Euclidean distance of z and 0 in the complex plane.

For q, mid, rad, $|\cdot|$ similar properties hold as in Chapter 2.

Theorem 9.1.7. *Let* $[u], [v], [w], [z] \in \mathbb{IC}_R$. *Then we have*

$$q([u] + [z], [u] + [w]) = q([z], [w]),$$
$$q([u] + [z], [v] + [w]) \le q([u], [v]) + q([z], [w]),$$
$$q([u][z], [u][w]) \le |[u]|q([z], [w]), \tag{9.1.1}$$
$$|[z] + [w]| \le |[z]| + |[w]|$$
$$|[u][z]| \le |[u]| \cdot |[z]| \tag{9.1.2}$$
$$\mathrm{rad}([z] \pm [w]) = \mathrm{rad}([u]) + \mathrm{rad}([w])$$
$$|[u]|\,\mathrm{rad}([z]) \le \mathrm{rad}([u][z]) \le |[u]|\,\mathrm{rad}([z]) + \mathrm{rad}([u])|[z]|$$
$$\mathrm{rad}(u[z]) = |u|\,\mathrm{rad}([z]), \quad u \in \mathbb{C}$$
$$[z] \subseteq [w] \Rightarrow \mathrm{rad}([w]) - \mathrm{rad}([z]) \le q([z], [w]) \le d([w]) - d([z]).$$

The proof uses the definition of q, rad, $|\cdot|$ and is left to the reader. Notice that in (9.1.1) and (9.1.2) strict inequality can hold even if $[u] \equiv u \in \mathbb{C}$. Thus these results differ from those in Chapter 2.

Rectangular complex interval vectors with n components and rectangular complex $m \times n$ matrices are defined as vectors resp. matrices of the corresponding size with rectangular complex intervals as entries. The set of such vectors resp. matrices is denoted by $\mathbb{IC}_R^n$ resp. $\mathbb{IC}_R^{m\times n}$. Addition, subtraction, and multiplication are defined in a straightforward way with properties literally as in Theorem 3.1.2. Absolute value, distance, midpoint, radius, diameter are defined entrywise, again with properties similar to those in Theorem 3.1.5.

Now we introduce a bijective mapping κ from $\mathbb{IC}_R$ resp. $\mathbb{IC}_R^{m\times n}$ onto some particular classes of real matrices. This mapping was defined in Arndt [52] and can help to reformulate complex problems as real ones.

Definition 9.1.8. Let $[z] = [a] + i[b] \in \mathbb{IC}_R$, $[a], [b] \in \mathbb{IR}$.
(a) Then $\kappa([z])$ is defined as

$$\kappa([z]) = \begin{pmatrix} [a] & [b] \\ -[b] & [a] \end{pmatrix} \in \mathbb{IR}^{2\times 2}.$$

(b) If $[v] = [a] + i[b] \in \mathbb{IC}_R^n$ with $[a], [b] \in \mathbb{IR}^n$, then we define

$$\kappa([v]) = \begin{pmatrix} [a] & [b] \\ -[b] & [a] \end{pmatrix} \in \mathbb{IR}^{2n\times 2}.$$

(c) If $[C] = [A] + i[B] \in \mathbb{IC}_R^{m\times n}$ with $[A], [B] \in \mathbb{IR}^{m\times n}$, then

$$\kappa([C]) = \begin{pmatrix} [A] & [B] \\ -[B] & [A] \end{pmatrix} \in \mathbb{IR}^{2m\times 2n}.$$

Obviously, (a) of Definition 9.1.8 is derived from (9.0.1), and (b) is contained in (c) if $\mathbb{IC}_R^n$ is identified with $\mathbb{IC}_R^{n\times 1}$.

Theorem 9.1.9. *Let $[C], [C]' \in \mathbb{IC}_R^{m\times n}$, $[C]'' \in \mathbb{IC}_R^{n\times r}$. Define the permutation matrix $P_s \in \mathbb{R}^{2s\times 2s}$ by $P_s = (e^{(1)}, e^{(3)}, e^{(5)}, \ldots, e^{(2s-1)} | e^{(2)}, e^{(4)}, e^{(6)}, \ldots, e^{(2s)})$. Then we have*
(a) $\kappa([C]) = P_m^T(\kappa([c]_{ij}))P_n$,
(b) $\kappa([C] \pm [C]') = \kappa([C]) \pm \kappa([C]')$,
(c) $\kappa([C] \cdot [C]'') = \kappa([C]) \cdot \kappa([C]'')$.

Proof. (a) and (b) are obvious by the definition of P_s and of $\kappa(\cdot)$.

With $[C] = [A] + i[B]$, $[C]'' = [A]'' + i[B]''$, $([A], [B] \in \mathbb{IR}^{m\times n}$, $[A]'', [B]'' \in \mathbb{IR}^{n\times r})$ the equality in (c) follows directly from $[C] \cdot [C]'' = [A][A]'' - [B][B]'' + i([A][B]'' + [B][A]'')$ and the definition of $\kappa(\cdot)$. □

Theorem 9.1.10. *Let $[C] = [A] + i[B] \in \mathbb{IC}_R^{n\times n}$ with $[A], [B] \in \mathbb{IR}^{n\times n}$ and let*

$$M = |\kappa([C])| = \begin{pmatrix} |[A]| & |[B]| \\ |[B]| & |[A]| \end{pmatrix}.$$

(a) *If $[a]_{i_0 j_0} \cdot [b]_{i_0 j_0} \neq 0$ for some index pair (i_0, j_0) with $i_0, j_0 \leq n$, then $|[C]|$ and M are irreducible simultaneously.*
(b) *Let $\rho = \rho(|[C]|) = \rho(|[A]| + |[B]|)$. Then $\rho = \rho(M)$ holds. The eigenvalues of $|[C]| = |[A]| + |[B]|$ and those of $C^- = |[A]| - |[B]|$ are eigenvalues of M.*

Proof. (a) Define $J_1 = \{1, \ldots, n\}$, $J_2 = \{n+1, \ldots, 2n\}$ and let $|[C]|$ be irreducible. Choose $i_1', j_1' \in J_1 \cup J_2$ and define

$$i_1 = \begin{cases} i_1', & \text{if } i_1' \in J_1 \\ i_1' - n, & \text{if } i_1' \in J_2 \end{cases} \qquad j_1 = \begin{cases} j_1', & \text{if } j_1' \in J_1 \\ j_1' - n, & \text{if } j_1' \in J_2 \end{cases}.$$

By virtue of the assumption on $|[C]|$ and $[a]_{i_0 j_0}$, $[b]_{i_0 j_0}$ there is a path

$$i_1 \to \cdots \to i_s = i_0 \to j_0 = j_t \to j_{t-1} \to \cdots \to j_1$$

in the directed graph $G(|[C]|)$ of $|[C]|$. From this path we will construct a path in the directed graph $G(M)$ of M which connects i_1' with j_1'. To this end define

$$i_\ell' = \begin{cases} i_\ell & \text{if } i_{\ell-1}' \leq n \text{ and } [a]_{i_{\ell-1}, i_\ell} \neq 0 \\ i_\ell + n & \text{if } i_{\ell-1}' \leq n \text{ and } [a]_{i_{\ell-1}, i_\ell} = 0 \\ i_\ell + n & \text{if } i_{\ell-1}' > n \text{ and } [a]_{i_{\ell-1}, i_\ell} \neq 0 \\ i_\ell & \text{if } i_{\ell-1}' > n \text{ and } [a]_{i_{\ell-1}, i_\ell} = 0 \end{cases}$$

for $\ell = 2, 3, \dots, s$. Similarly, for $t = t, t-1, \dots, 2$ define

$$j'_\ell = \begin{cases} j_\ell & \text{if } j'_{\ell-1} \le n \quad \text{and} \quad [a]_{j_\ell, j_{\ell-1}} \ne 0 \\ j_\ell + n & \text{if } j'_{\ell-1} \le n \quad \text{and} \quad [a]_{j_\ell, j_{\ell-1}} = 0 \\ j_\ell + n & \text{if } j'_{\ell-1} > n \quad \text{and} \quad [a]_{j_\ell, j_{\ell-1}} \ne 0 \\ j_\ell & \text{if } j'_{\ell-1} > n \quad \text{and} \quad [a]_{j_\ell, j_{\ell-1}} = 0 \end{cases}.$$

Hence $i'_s \in \{i_0, i_0 + n\}$, $j'_t \in \{j_0, j_0 + n\}$. Since $[a]_{i_0 j_0} \cdot [b]_{i_0 j_0} \ne 0$, by assumption $i_0 \to j_0$, $i_0 \to j_0 + n$, $i_0 + n \to j_0$, $i_0 + n \to j_0 + n$ are edges in $G(M)$, hence

$$i'_1 \to \dots \to i'_s \to j'_t \to j'_{t-1} \to \dots \to j'_1$$

is a path in $G(M)$ which connects i'_1 with j'_1. Thus M is irreducible.

Let now, conversely, M be irreducible and consider an edge $i_1 \to i_2$ in $G(M)$. Then

$$0 < m_{i_1, i_2} = \begin{cases} |[a]_{i_1, i_2}| \le |[c]_{i_1, i_2}|, & \text{if } i_1, i_2 \in J_1 \\ |[b]_{i_1, i_2 - n}| \le |[c]_{i_1, i_2 - n}|, & \text{if } i_1 \in J_1,\ i_2 \in J_2 \\ |[b]_{i_1 - n, i_2}| \le |[c]_{i_1 - n, i_2}|, & \text{if } i_1 \in J_2,\ i_2 \in J_1 \\ |[a]_{i_1 - n, i_2 - n}| \le |[c]_{i_1 - n, i_2 - n}|, & \text{if } i_1, i_2 \in J_2 \end{cases}$$

holds. Let $i, j \in J_1$. There is a path in $G(M)$ which connects i with j. Whenever in this path there is an edge $i_1 \to i_2$ with $i_1 \in J_2$ or $i_2 \in J_2$ replace i_1 by $i_1 - n$ in the first case and i_2 by $i_2 - n$ in the second. The result is a path in $G(|[C]|)$ which connects i with j. Hence $|[C]|$ is irreducible.

(b) Let $\varepsilon > 0$. Then $(|[A]| + \varepsilon ee^T) + (|[B]| + \varepsilon ee^T)$ is irreducible and has a Perron vector x_ε associated with the spectral radius ρ_ε as eigenvalue. The matrix

$$M_\varepsilon = \begin{pmatrix} |[A]| + \varepsilon ee^T & |[B]| + \varepsilon ee^T \\ |[B]| + \varepsilon ee^T & |[A]| + \varepsilon ee^T \end{pmatrix}$$

is irreducible, too, and has the positive eigenvector $(x_\varepsilon^T, x_\varepsilon^T)^T$ associated with the same eigenvalue ρ_ε. According to the Theorems 1.9.4 and 1.9.5 this eigenvalue must be the spectral radius of M_ε. Letting ε tend to zero proves $\rho = \rho(M)$. If (x, λ) is an eigenpair of $|[C]|$ then

$$M \begin{pmatrix} x \\ x \end{pmatrix} = \lambda \begin{pmatrix} x \\ x \end{pmatrix}.$$

Similarly, if (x, λ) is an eigenpair of C^- then

$$M \begin{pmatrix} x \\ -x \end{pmatrix} = \lambda \begin{pmatrix} x \\ -x \end{pmatrix}.$$

This concludes the proof. □

As can be seen from the proof of (a), the assumption $[a]_{i_0 j_0} \cdot [b]_{i_0 j_0} \ne 0$ is not needed for the direction 'M irreducible $\Rightarrow$ $|[C]|$ irreducible'. The 1×1 matrix $[C] = (1)$ with $M = I_2 \in \mathbb{R}^{2\times 2}$ shows that it is needed for the converse direction.

Exercises

Ex. 9.1.1. Compute the expressions $[z] + [z]'$, $[z] - [z]'$, $[z] \cdot [z]'$, $[z]/[z]'$ for $[z] = [0, 2] + i[-1, 3]$, $[z]' = [-1, 1] + i[1, 2]$.

Ex. 9.1.2. Show that for $[z] = [2, 4]$ and $[z]' = 1 + i$ the strict subset property $\{z \cdot z' \mid z \in [z],\ z' \in [z]'\} \subset [z][z]'$ holds.

Ex. 9.1.3. Show by means of the example $[z] = [-2, -1] + i[-1, 1]$, $[z]' = [1, 2] + i[-1, 1]$ that $0 \notin [z]$ and $0 \notin [z]'$ does not imply $0 \notin [z] \cdot [z]'$.

9.2 Circular complex intervals

Definition 9.2.1. Let $\check{z} \in \mathbb{C}$, $0 \le r_z \in \mathbb{R}$. We call the disc

$$[z] = \langle \check{z}, r_z \rangle = \{ z \in \mathbb{C} \mid |z - \check{z}| \le r_z \} \tag{9.2.1}$$

a circular complex interval. It is determined by its midpoint $\mathrm{mid}([z]) := \check{z}$ and its radius $\mathrm{rad}([z]) := r_z$. The set of all such intervals is denoted by $\mathbb{IC}_C$. In $\mathbb{IC}_C$ the relations $[z] = [z]'$ and $[z] \subseteq [z]'$ are defined in the set theoretic way.

The notation in (9.2.1) should not be mixed up with the mignitude in Chapter 2.

Notice that $[z] := \langle \check{z}, r_z \rangle = [z]' := \langle \check{z}', r_{z'} \rangle$ holds if and only if $\check{z} = \check{z}'$ and $r_z = r_{z'}$, and that $[z] \subseteq [z]'$ is valid if and only if $|\check{z} - \check{z}'| \le r_{z'} - r_z$. The latter can be seen as in Theorem 2.4.8 (a). A nonempty intersection $[z] \cap [z]'$ is generally not an element of $\mathbb{IC}_C$ if defined in a set theoretic way.

On $\mathbb{IC}_C$ we introduce the following circular arithmetic.

Definition 9.2.2. Let $[z] = \langle \check{z}, r_z \rangle$, $[w] = \langle \check{w}, r_w \rangle \in \mathbb{IC}_C$. Then we define

$$\begin{aligned}
[z] + [w] &= \langle \check{z} + \check{w}, r_z + r_w \rangle, \\
[z] - [w] &= \langle \check{z} - \check{w}, r_z + r_w \rangle, \\
[z] \cdot [w] &= \langle \check{z} \cdot \check{w}, |\check{z}| r_w + |\check{w}| r_z + r_z r_w \rangle, \\
1/[z] &= \left\langle \frac{\bar{\check{z}}}{|\check{z}|^2 - r_z^2}, \frac{r_z}{|\check{z}|^2 - r_z^2} \right\rangle, \\
[w]/[z] &= [w] \cdot \frac{1}{[z]},
\end{aligned}$$

where for the denominator of the division $0 \notin [z]$, i.e., $r_z < |\check{z}|$ is assumed.

Notice that the bar in $\bar{\check{z}}$ denotes the conjugate complex number of $\check{z}$ and not the upper bound of some interval; similarly for the rest of this section.

On a computer the midpoints in the right-hand sides of Definition 9.2.2 are rarely machine numbers but must be rounded to them. Therefore, in a practical realization of the circular arithmetic the radius must take into account this rounding. How this can be done is implemented in INTLAB (see Appendix G and Rump [320]).

We leave it to the reader to prove results literally as in Theorem 9.1.3 (a) and Theorem 9.1.4 for $(\mathbb{IC}_C, +, -, \cdot, /)$. See also Chapter 5 in Alefeld, Herzberger [26] and consider the circle

$$(\check{z} + r_z e^{i\varphi}) \pm (\check{w} \pm r_w e^{i\varphi}) = (\check{z} + r_z e^{i\varphi} \pm \check{w}) + r_w e^{i\varphi} = \check{z} \pm \check{w} + (r_z + r_w) e^{i\varphi},$$

$0 \le \varphi < 2\pi$, (notice the same angle φ!) which traces the envelope of all discs

$$\langle \check{z} + r_z e^{i\varphi} + \check{w}, r_w \rangle, \quad \varphi \in [0, 2\pi) \text{ fixed},$$

when proving

$$[z] \pm [w] = \{\tilde{z} \pm \tilde{w} \mid \tilde{z} \in [z],\ \tilde{w} \in [w]\}$$

for the circular arithmetic.

In contrast to $(\mathbb{IC}_R, \cdot)$, associativity holds in $(\mathbb{IC}_C, \cdot)$ so that this structure is a commutative semigroup with a neutral element. But Theorem 9.1.3 (b) does not hold for $\mathbb{IC}_C$ if one defines $\square S$ to be that disc which encloses a bounded set $S \subseteq \mathbb{C}$ optimally in the sense of smallest radius. This can be seen from the following example.

Example 9.2.3. Let $[z] = [w] = \langle 1, 1\rangle$. Then $[u] = [z][w] = \langle 1, 3\rangle$ encloses $S = \{\tilde{z}\tilde{w} \mid \tilde{z}, \tilde{w} \in \langle 1, 1\rangle\}$. But $[u]' = \langle 2, \sqrt{8}\rangle$ also does so with a radius that is smaller than that of $[u]$. Hence $\square S \neq [u]$. In order to prove $S \subseteq [u]'$ let $\tilde{u} \in S$. Then $\tilde{u} = (1 + re^{i\varphi})(1 + \rho e^{i\theta})$ for some $r, \rho \in [0, 1]$, $\varphi, \theta \in [0, 2\pi)$. Hence

$$\begin{aligned}
|\tilde{u} - 2|^2 &= |-1 + re^{i\varphi} + \rho e^{i\theta} + r\rho e^{i(\varphi+\theta)}|^2 \\
&= 1 + r^2 + \rho^2 + r^2\rho^2 - 2r(1 - \rho^2)\cos\varphi \\
&\quad - 2\rho(1 - r^2)\cos\theta - 2r\rho\cos(\varphi + \theta) + 2r\rho\cos(\varphi - \theta) \\
&\le 4 + 2r(1 - \rho^2) + 2\rho(1 - r^2) + 4r\rho =: f(r, \rho).
\end{aligned}$$

On $[0, 1] \times [0, 1]$ we have

$$\frac{\partial f(r, \rho)}{\partial r} = 2(1 - \rho^2) - 4\rho r + 4\rho = 2(1 - \rho^2) + 4\rho(1 - r) \ge 0.$$

Hence f increases in r for fixed ρ and we get

$$4 \le 4 + 2\rho = f(0, \rho) \le f(r, \rho) \le f(1, \rho) = 4 + 2(1 - \rho^2) + 4\rho = 8 - 2(1 - \rho)^2 \le 8.$$

This proves $|\tilde{u} - 2|^2 \le 8$, whence $\tilde{u} \in [u]'$. □

Notice that we do not claim $[u]' = \square S$ in our example. In fact, $\square S = \langle 3/2, 3\sqrt{3}/2\rangle$ since according to the *optimal enclosing circular arithmetic* in Krier [177], p. 42, we generally get

$$\begin{aligned}
[v] &:= \square\{\tilde{z}\tilde{w} \mid \tilde{z} \in \langle \check{z}, r_z\rangle,\ \tilde{w} \in \langle \check{w}, r_w\rangle\} \\
&= \langle \check{z}\check{w}(1 + x_0),\ (3|\check{z}\check{w}|^2 x_0^2 + 2(|\check{z}\check{w}|^2 + |r_z\check{w}|^2 + |\check{z}r_w|^2)x_0 \\
&\quad + |r_z\check{w}|^2 + |\check{z}r_w|^2 + (r_z r_w)^2)^{1/2}\rangle, \qquad (9.2.2)
\end{aligned}$$

where x_0 is the (only) nonnegative zero of the polynomial

$$p(x) = 2|\check z\check w|^2 x^3 + (|\check z\check w|^2 + |r_z\check w|^2 + |\check z r_w|^2)\, x^2 - (r_z r_w)^2 \tag{9.2.3}$$

if the degree of p is at least 2, and $x_0 = 0$ otherwise. In our example x_0 is $1/2$. The formulae for this optimal arithmetic differ from that in Definition 9.2.2 by the multiplication, which is given by (9.2.2) and which is used when computing $[z]/[w] := [z] \cdot (1/[w])$. The drawback of this arithmetic is the computation of x_0, of course.

We emphasize that the midpoint equation $\check v = \check z \cdot \check w$ no longer holds in formula (9.2.2) and thus for $[z]$ and $[w]$ in Example 9.2.3. If one redefines $\square$ in Theorem 9.1.3 (b) to be the smallest disc with midpoint $\check z \cdot \check w$ such that $S \subseteq \langle \check z\check w, r\rangle$, then Theorem 9.1.3 (b) holds true for $\mathbb{IC}_C$ instead of $\mathbb{IC}_R$. This can be seen by choosing $\tilde z = \check z + r_z e^{i\arg(\check z)} = e^{i\arg(\check z)}(|\check z| + r_z)$, $\tilde w = \check w + r_w e^{i\arg(\check w)} = e^{i\arg(\check w)}(|\check w| + r_w)$, whence $\tilde z\tilde w = \check z\check w + e^{i(\arg\check z + \arg\check w)}(|\check z| r_w + |\check w| r_z + r_z r_w)$ and $|\tilde z\tilde w - \check z\check w| = |\check z| r_w + r_z|\check w| + r_z r_w = \mathrm{rad}([z][w])$.

Now we will show that

$$\frac{1}{[z]} = \left\{ \frac{1}{\tilde z} \;\middle|\; \tilde z \in [z] \right\} \tag{9.2.4}$$

holds provided that $0 \notin [z]$. In order to prove (9.2.4) we first note that the mapping $f\colon \mathbb{C}\setminus\{0\} \to \mathbb{C}\setminus\{0\}$ with $f(z) := 1/z$ is obviously bijective. Next we show that for a given disc $\langle \check z, r_z\rangle$ with $r_z < |\check z|$ this function f maps the circle $C_z = \{z \mid |z - \check z| = r_z\}$ onto the circle $C_w = \{w \in \mathbb{C} \mid |w - \check w| = r_w\}$ with $\check w = \bar{\check z}/(|\check z|^2 - r_z^2)$ and $r_w = r_z/(|\check z^2| - r_z^2)$. To this end choose $z \in C_z$ arbitrarily, say $z = \check z + r_z e^{i\theta}$ with some $\theta \in [0, 2\pi)$. Then

$$\begin{aligned}\left|\frac{1}{z} - \check w\right| &= \left|\frac{1}{\check z + r_z e^{i\theta}} - \frac{\bar{\check z}}{|\check z|^2 - r_z^2}\right| = \frac{|-r_z^2 - \bar{\check z} r_z e^{i\theta}|}{|(\check z + r_z e^{i\theta})(|\check z|^2 - r_z^2)|}\\ &= \frac{|\bar{\check z} + r_z e^{-i\theta}|}{|\check z + r_z e^{i\theta}|} r_w = \frac{|\bar z|}{|z|} r_w = r_w.\end{aligned}$$

This shows $1/z \in C_w$, and since f is bijective and continuous it maps the closed curve $z = \check z + r_z e^{i\theta}$, $0 \le \theta \le 2\pi$ onto a closed subcurve of C_w which cannot have any double point for $0 < \theta < 2\pi$. This implies $f(C_z) = C_w$. If $\check z$ is represented as $\check z = |\check z| e^{i\varphi}$, then $z_0 := \check z - (r_z/|\check z|) r_z e^{i\varphi}$ lies in the interior of C_z because of $r_z < |\check z|$. Since

$$\frac{1}{z_0} = \frac{1}{(|\check z| - r_z^2/|\check z|)e^{i\varphi}} = \frac{\bar{\check z}}{|\check z|^2 - r_z^2} = \check w$$

lies in the interior of C_w and since f is bijective, each path in the interior of C_w with endpoint $\check w$ is the image of a path p_z in the interior of C_z with endpoint z_0. Otherwise p_z has to cross the boundary C_z in contrast to f being bijective and $f(C_z) = C_w$. Therefore, $|f^{-1}(w) - \check z| < r_z$ for each w with $|w - \check w| < r_w$, whence $f(\langle \check z, r_z\rangle) = \langle \check w, r_w\rangle$.

Notice that (9.2.4) also follows directly using basic knowledge on Möbius transformations, i.e., on particular conformal mappings in the complex plane. See Ahlfors [4] and Alefeld, Herzberger [26], for example.

Since $[z] \cdot [w]$ is not enclosed optimally, we cannot expect that $[w]/[z]$ behaves differently. Therefore, it makes sense to look for bounds of the quotient

$$\frac{\mathrm{rad}([z][w])}{\mathrm{rad}(S)} \tag{9.2.5}$$

with

$$S = \{\check{z}\check{w} \mid \check{z} \in [z],\ \check{w} \in [w]\} \quad \text{and} \quad \mathrm{rad}(S) = \max\{|z' - z''|/2 \mid z', z'' \in S\}. \tag{9.2.6}$$

The following nice result in this respect originates from Krier [177] and Rump [319]. It bounds the quotient in (9.2.5) by the constant 1.5.

Theorem 9.2.4. *With S and* rad(S) *as in* (9.2.6) *we have*

$$\mathrm{rad}([z][w]) \le \frac{3}{2}\,\mathrm{rad}(S). \tag{9.2.7}$$

Proof. If $\check{z} = 0$, then

$$S = \{\tilde{u} \mid \tilde{u} = \check{w}re^{i\theta} + rRe^{i(\theta+\varphi)},\ 0 \le r \le r_z,\ 0 \le R \le r_w,\ 0 \le \theta,\ \varphi < 2\pi\}$$

is a zero-centered disc with radius $|\check{w}|r_z + r_z r_w$ which coincides with $[z][w]$. Hence (9.2.7) holds. Similarly, $\check{w} = 0$ ends up with (9.2.7). If $r_z = 0$, then

$$S = \{\tilde{u} \mid \tilde{u} = \check{w}\check{z} + |\check{z}|re^{i(\varphi+\arg\check{z})},\ 0 \le r \le r_w,\ 0 \le \varphi < 2\pi\}$$

is again a disc, this time with midpoint $\check{z}\check{w}$ and radius $|\check{z}|r_w$. It coincides again with $[z][w]$; similarly for $r_w = 0$. Therefore, without loss of generality we may assume $\check{z} \ne 0$, $\check{w} \ne 0$, $r_z > 0$, $r_w > 0$ from now on. Next we remark that (9.2.7) does not depend on a fixed nonzero scaling factor for $[z]$ and for $[w]$. Therefore, we can consider $([z]/\check{z}) \cdot ([w]/\check{w})$, i.e., without loss of generality we may assume $\check{z} = \check{w} = 1$. Since $(1 \pm r_z)(1 \pm r_w) \in S$ we get

$$\mathrm{rad}(S) \ge \max\{(1 + r_z)(1 + r_w) - (1 \pm r_z)(1 \pm r_w)\}, \tag{9.2.8}$$

where all possible combinations + and − are allowed for the maximum. This implies

$$\mathrm{rad}(S) \ge \frac{1}{2}\begin{cases} (1 + r_z)(1 + r_w) - (1 - r_z)(1 - r_w), & \text{if } r_z,\ r_w \in (0, 1) \\ (1 + r_z)(1 + r_w) - (1 - r_z)(1 + r_w), & \text{if } r_z \ge r_w \text{ and } r_z \ge 1 \\ (1 + r_z)(1 + r_w) - (1 + r_z)(1 - r_w), & \text{if } r_w \ge r_z \text{ and } r_w \ge 1 \end{cases}$$

$$= \left.\begin{cases} r_z + r_w, & \text{if } r_z,\ r_w \in (0, 1) \\ r_z + r_z r_w, & \text{if } r_z \ge r_w, \text{ and } r_z \ge 1 \\ r_w + r_z r_w, & \text{if } r_w \ge r_z, \text{ and } r_w \ge 1 \end{cases}\right\} > 0.$$

With $\operatorname{rad}([z][w]) = r_z + r_w + r_z r_w$ we finally obtain

$$\frac{\operatorname{rad}([z][w])}{\operatorname{rad}(S)} \le 1 + \left\{\begin{array}{ll} \frac{1}{(1/r_z)+(1/r_w)}, & \text{if } r_z,\ r_w \in (0,1) \\ \frac{1}{(r_z/r_w)+r_z}, & \text{if } r_z \ge r_w, \text{ and } r_z \ge 1 \\ \frac{1}{(r_w/r_z)+r_w}, & \text{if } r_w \ge r_z, \text{ and } r_w \ge 1 \end{array}\right\}$$

$$\le 1 + \frac{1}{1+1} = \frac{3}{2},$$

which proves (9.2.7). □

Notice that Theorem 9.2.4 also holds if one transfers the circular arithmetic to real intervals in midpoint-radius form as was done in the MATLAB package INTLAB.

As our remark following Example 9.2.3 shows, strict inequality can hold in (9.2.8). In our subsequent example we will prove that in particular cases S in (9.2.6) can be bounded by a cardioid.

Example 9.2.5. Let $[z] = [w] = \langle 1, 1\rangle$ as in Example 9.2.3. Then the boundary ∂S of S in (9.2.6) is a cardioid.

In order to prove this statement we start with

$$\tilde{z}\tilde{w} = (1 + re^{i\theta}) \cdot (1 + Re^{i\varphi}),$$
$$0 \le r \le r_z = 1,\ 0 \le R \le r_w = 1,\ 0 \le \theta,\ \varphi \le 2\pi. \tag{9.2.9}$$

It is easy to see that ∂S is a subset of all products in (9.2.9) for which $r = r_z = 1$ and $R = r_w = 1$. In fact ∂S is the envelope of the family of curves C_φ defined by

$$\begin{aligned} x = u(\theta, \varphi) &= (1 + \cos\theta)(1 + \cos\varphi) - (\sin\theta)\sin\varphi \\ &= 1 + \cos\theta + \cos\varphi + \cos(\theta + \varphi), \\ y = v(\theta, \varphi) &= (1 + \cos\theta)\sin\varphi + (\sin\theta)(1 + \cos\varphi) \\ &= \sin\varphi + \sin\theta + \sin(\theta + \varphi) \end{aligned}$$

with $0 \le \theta,\ \varphi \le 2\pi$. These parametric equations result from the real and the imaginary part in (9.2.9). As the parameter of a single curve C_{φ_0} we choose θ, while φ is the family parameter. (We can interchange these roles, of course.) The envelope E is that curve $x = p(\varphi)$, $y = q(\varphi)$ for which there is a function $\theta = f(\varphi)$ such that

$$p(\varphi) = u(f(\varphi), \varphi), \quad q(\varphi) = v(f(\varphi), \varphi)$$

with parallel tangential vectors $(p'(\varphi), q'(\varphi))^T \ne 0$ and $(u_\theta, v_\theta)^T$. Thus both vectors must be linearly dependent, i.e.,

$$\det\begin{pmatrix} p'(\varphi) & u_\theta \\ q'(\varphi) & v_\theta \end{pmatrix} = \det\begin{pmatrix} u_\theta f'(\varphi) + u_\varphi & u_\theta \\ v_\theta f'(\varphi) + v_\varphi & v_\theta \end{pmatrix} = \det\begin{pmatrix} u_\varphi & u_\theta \\ v_\varphi & v_\theta \end{pmatrix}$$
$$= u_\varphi v_\theta - u_\theta v_\varphi = 0.$$

After some computations this results in

$$\sin(\theta - \varphi) - \sin\varphi + \sin\theta = 0,$$

or, equivalently, in

$$(1 + \cos\varphi)\sin\theta = (1 + \cos\theta)\sin\varphi. \tag{9.2.10}$$

By squaring both sides one gets

$$(1 + \cos\varphi)^2(1 - \cos^2\theta) = (1 + \cos\theta)^2(1 - \cos^2\varphi). \tag{9.2.11}$$

For $\theta \neq \pi$ and $\varphi \neq \pi$ the equation (9.2.11) is equivalent to

$$(1 + \cos\varphi)(1 - \cos\theta) = (1 + \cos\theta)(1 - \cos\varphi),$$

i.e., to

$$\cos\theta = \cos\varphi. \tag{9.2.12}$$

It is obvious that (9.2.12) implies (9.2.11) also for $\theta = \pi$ or $\varphi = \pi$. From (9.2.12) one gets $\sin\theta = \pm\sin\varphi$. With $\sin\theta = -\sin\varphi$ and (9.2.12), equation (9.2.10) becomes $(1 + \cos\varphi)\sin\varphi = 0$ which is only true for particular values of φ. Therefore, $\sin\theta = \sin\varphi$ is the only useful solution. It yields the envelope E:

$$x = u(f(\varphi), \varphi) = (1 + \cos\varphi)^2 - \sin^2\varphi = 2(1 + \cos\varphi)\cos\varphi,$$
$$y = v(f(\varphi), \varphi) = 2(1 + \cos\varphi)\sin\varphi.$$

In the complex plane this curve can be written as $z(\varphi) = 2(1 + \cos\varphi)e^{i\varphi} = 2(\cos\varphi)e^{i\varphi} + 2e^{i\varphi}$. A simple calculation shows that the first summand $2(\cos\varphi)e^{i\varphi}$ describes a point P on the circle C: $|z - 1| = 1$. The point Z described by $z(\varphi)$ lies on the ray OP such that the line segment $\overline{PZ}$ has constant length 2. Thus E is the conchoid of the circle C with respect to zero and with a distance $\overline{PZ}$ which is the diameter of C. Therefore, it is a cardioid, cf. Strubecker [349] or Bronstein et al. [72]. □

Now we introduce distance, diameter and absolute value in $\mathbb{IC}_C$.

Definition 9.2.6.

(a) Let $[z] = \langle \check{z}, r_z\rangle$, $[w] = \langle \check{w}, r_w\rangle \in \mathbb{IC}_C$. Then we define the distance q between $[z]$ and $[w]$ by

$$q([z], [w]) = |\check{z} - \check{w}| + |r_z - r_w|.$$

(b) If $([z]_k)$ is a sequence in $\mathbb{IC}_R$, then we define $\lim_{k\to\infty}[z]_k = [z]$ if and only if $\lim_{k\to\infty} q([z]_k, [z]) = 0$.

Obviously, $\lim_{k\to\infty}[z]_k = [z]$ if and only if $\lim_{k\to\infty}\check{z}_k = \check{z}$ and $\lim_{k\to\infty}\operatorname{rad}([z]_k) = r_z$. In addition, $(\mathbb{IC}_C, q)$ is a complete metric space.

Definition 9.2.7. Let $[z] = \langle \check{z}, r_z\rangle \in \mathbb{IC}_C$. Then we define the diameter $d([a])$ of $[z]$ by $d([a]) = 2r_z$ and the absolute value $|[z]|$ by $|[z]| = |\check{z}| + r_z$.

Theorem 9.1.7 holds literally for $\mathbb{IC}_C$ instead of $\mathbb{IC}_R$. In (9.1.1) and (9.1.2) equality holds if $[u] \equiv u \in \mathbb{C}$. We leave the proofs to the reader as Exercise 9.2.8.

Exercises

Ex. 9.2.1. Compute the expressions $[z] + [z]'$, $[z] - [z]'$, $[z] \cdot [z]'$, $[z]/[z]'$ for $[z] = \langle 1 + i, 2 \rangle$, $[z]' = \langle 2i, 1 \rangle$.

Ex. 9.2.2. Compute the smallest circular complex interval $[z]'$ which encloses the rectangular complex interval $[z] = [-1, 3] + i[0, 2]$, i.e., compute the hull in $\mathbb{IC}_C$ of $[z] \in \mathbb{IC}_R$. Do the same for a general rectangular complex interval $[z]$.

Ex. 9.2.3. Compute the smallest circular complex interval $[z]'$ which encloses the intersection of the two intervals $[z], [w] \in \mathbb{IC}_C$ if $[z] \cap [w] \neq \emptyset$.

Ex. 9.2.4. Show by means of the example $[z] = \langle -2, 1 \rangle$, $[w] = \langle 2, 1 \rangle$ that $0 \notin [z]$ and $0 \notin [w]$ does not imply $0 \notin [z] \cdot [w]$. This holds, in particular, if one considers only the projections of $[z], [w], [z] \cdot [w]$ on the real axis if the multiplication is done first in the complex plane according to Definition 9.2.2 before projecting. What about the multiplication $[a] \cdot [b]$ of $[a], [b] \in \mathbb{IR}$ in this respect if carried out by the real arithmetic in Definition 2.2.1?

Ex. 9.2.5. Show by induction for $[z]_k = \langle \check{z}_k, r_k \rangle \in \mathbb{IC}_C$, $k = 1, \ldots, m$:

$$\prod_{k=1}^{m} [z]_k = \prod_{k=1}^{m} \langle \check{z}_k, r_k \rangle = \left\langle \prod_{k=1}^{m} \check{z}_k, \prod_{k=1}^{m} (|\check{z}_k| + r_k) - \prod_{k=1}^{m} |\check{z}_k| \right\rangle.$$

Ex. 9.2.6. Show that the polynomial p from (9.2.3) has exactly one nonnegative zero if the degree of p is at least 2. To this end consider for instance the signs of $p(0)$ and $p(x)$ for $x > 0$ sufficiently large, and show that p is strictly monotonously increasing for $x \geq 0$.

Ex. 9.2.7. Show that in complex analysis the image of any circle through the origin is a cardioid under the mapping $z \to z^2$. Let $[z]$ be any disc with zero at its boundary. Describe the boundary of the set $S = \{\tilde{z}\tilde{w} \mid \tilde{z}, \tilde{w} \in [z]\}$.

Ex. 9.2.8. Prove the Theorems 9.1.4 and 9.1.7 for $\mathbb{IC}_R$ and $\mathbb{IC}_C$.

9.3 Applications of complex intervals

We start this section with the fixed point iteration

$$[z]^{k+1} = [A][z]^k + [b], \quad k = 0, 1 \ldots, \tag{9.3.1}$$

where the entries of $[A]$ and $[b]$ are from $\mathbb{IC}_R$.

Theorem 9.3.1. *In $\mathbb{IC}_R$ the fixed point iteration (9.3.1) is globally convergent if and only if $\rho(|[A]|) < 1$.*

Proof. One possibility to prove Theorem 9.3.1 consists in repeating the steps in the proof of Theorem 5.5.10.

A second one transfers (9.3.1) to the real problem

$$\begin{pmatrix} [y]^{k+1} \\ [x]^{k+1} \end{pmatrix} = \kappa([A]) \begin{pmatrix} [y]^k \\ [x]^k \end{pmatrix} + \begin{pmatrix} [b]^2 \\ [b]^1 \end{pmatrix} \tag{9.3.2}$$

with $[x]^k, [y]^k, [b]^1, [b]^2 \in \mathbb{IR}^n$ and $[z]^k = [x]^k + i[y]^k$, $[b] = [b]^1 + i[b]^2$. Since (9.3.1) is equivalent to

$$\kappa([z]^{k+1}) = \kappa([A][z]^k + [b]) = \kappa([A])\kappa([z]^k) + \kappa([b])$$

and since (9.3.2) is equivalent to

$$\begin{pmatrix} [x]^{k+1} \\ -[y]^{k+1} \end{pmatrix} = \kappa([A]) \begin{pmatrix} [x]^k \\ -[y]^k \end{pmatrix} + \begin{pmatrix} [b]^1 \\ -[b]^2 \end{pmatrix}$$

the iteration (9.3.1) is equivalent to (9.3.2). If $\rho(|[A]|) < 1$, the same holds for $|\kappa([A])|$ by virtue of Theorem 9.1.10 (b). Thus Theorem 5.5.10 can be applied to (9.3.2) and yields convergence to some limit $(([x]^*)^T, ([y]^*)^T)$ which is independent of the starting vector $(([x]^0)^T, ([y]^0)^T)$. It is easy to see that

$$\kappa([z]^*) = \kappa([A])\kappa([z]^*) + \kappa([b]) = \kappa([A][z]^* + [b])$$

holds for $[z]^* = [x]^* + i[y]^*$. This proves the theorem. □

This proof shows how the mapping κ can be applied in order to use known real results when working in $\mathbb{IC}_R$.

Theorem 9.3.1 holds also if the entries in (9.3.1) are from $\mathbb{IC}_C$. In this case the proof follows the lines of that for Theorem 5.5.10. It is left to the reader as Exercise 9.3.1.

As a second application of complex intervals we want to simultaneously enclose the zeros of the polynomial

$$p(z) = \sum_{k=0}^{n} a_k z^k, \quad a_k, z \in \mathbb{C},\ a_n = 1,$$

provided that

$$\text{all zeros } z_i^* \text{ of } p \text{ are simple and} \tag{9.3.3}$$

$$\text{pairwise disjoint initial enclosures } [z]_i \in \mathbb{IC}_R \text{ of } z_i^* \text{ are known.} \tag{9.3.4}$$

The factorization

$$p(z) = \prod_{j=1}^{n} (z - z_j^*) = (z - z_i^*) \prod_{\substack{j=1 \\ j \neq i}}^{n} (z - z_j^*)$$

yields

$$z_i^* = \tilde{z}_i - \frac{p(\tilde{z}_i)}{\prod_{\substack{j=1\\ j\neq i}}^n (\tilde{z}_i - z_j^*)} \in \tilde{z}_i - \frac{p(\tilde{z}_i)}{\prod_{\substack{j=1\\ j\neq i}}^n (\tilde{z}_i - [z]_j)}, \tag{9.3.5}$$

where $\tilde{z}_i$ is any element of $[z]_i$. From (9.3.5) we derive the following two iterative methods in $\mathbb{IC}_R$.

Method 1 (Weierstrass method – total step variant).

$$[z]_i^0 := [z]_i, \quad i = 1, \ldots, n.$$

$$\left.\begin{array}{l} \text{Choose } \tilde{z}_i^k \in [z]_i^k, \\ [z]_i^{k+1} := \left\{ \tilde{z}_i^k - \dfrac{p(\tilde{z}_i^k)}{\prod_{\substack{j=1\\ j\neq i}}^n (\tilde{z}_i^k - [z]_j^k)} \right\} \cap [z]_i^k \end{array}\right\}, \quad \begin{array}{l} i = 1, \ldots, n; \\ k = 0, 1, \ldots . \end{array}$$

Method 2 (Weierstrass method – single step variant).

$$[z]_i^0 := [z]_i, \quad i = 1, \ldots, n.$$

$$\left.\begin{array}{l} \text{Choose } \tilde{z}_i^k \in [z]_i^k, \\ [z]_i^{k+1} := \left\{ \tilde{z}_i^k - \dfrac{p(\tilde{z}_i^k)}{\prod_{j=1}^{i-1} (\tilde{z}_i^k - [z]_j^{k+1}) \prod_{j=i+1}^n (\tilde{z}_i^k - [z]_j^k)} \right\} \cap [z]_i^k \end{array}\right\}, \quad \begin{array}{l} i = 1, \ldots, n; \\ k = 0, 1, \ldots . \end{array}$$

Notice that if

$$0 \notin \prod_{\substack{j=1\\ j\neq i}}^n (\tilde{z}_i^k - [z]_j^k), \quad \text{resp.} \quad 0 \notin \prod_{j=1}^{i-1} (\tilde{z}_i^k - [z]_j^{k+1}) \prod_{j=i+1}^{n} (\tilde{z}_i^k - [z]_j^k),$$

the iterates $[z]_i^k$ are well-defined with $z_i^* \in [z]_i^{k+1} \subseteq [z]_i^k$, $i = 1, \ldots, n$; $k = 0, 1, \ldots$. For further properties in the case $z_i^* \in \mathbb{R}$ we refer to Alefeld, Herzberger [26] and for the case $\mathbb{IC}_C$ to Petković [269].

Exercises

Ex. 9.3.1. Prove Theorem 9.3.1 for $\mathbb{IC}_C$ instead of $\mathbb{IC}_R$.

Notes to Chapter 9

To 9.1: Rectangular complex intervals are considered in Alefeld [5]. Definition 9.1.8 and the two succeeding theorems originate from Arndt [52].

To 9.2: Circular complex intervals are introduced in Henrici [142] and Gargantini, Henrici [112].

To 9.3: The reduction of complex interval problems to real ones using the mapping κ seems to be first presented in Arndt [52]. The two methods for enclosing complex zeros of polynomials can be found in Alefeld, Herzberger [26]. Further methods with complex intervals are considered in Petković [269] and M. S. Petkovic, L. D. Petkovic [270].

Final Remarks

In this book we have studied many topics but had to leave at least as many open. Thus we did not consider enclosures of integrals, of extremal function values, of solutions of integral equations, of initial and boundary value problems for ordinary differential equations, of corresponding problems for partial differential equations. We did not give an overview on software for working with interval arithmetic nor did we raise improper intervals. As a first step towards additional topics we mention the books of Moore [232, 235], Kulisch, Miranker [184], Atanassova, Herzberger [57], Herzberger [144].

For enclosures of integrals we refer to Moore [232, 235], Kelch [162], Adams, Kulisch [2], Petras [271], Moore, Kearfott, Cloud [237].

For optimization problems we cite Bauch et al. [60], Ratschek, Rokne [285], Jansson [155], Kearfott [161], Fiedler et al. [95]. Particular aspects of a verification in optimization are considered in the theses of Ratz [289] and Berner [69].

The linear complementarity problem is treated in Schäfer [325, 328], Alefeld, Chen, Potra [20], Alefeld, Schäfer [47]. Mixed complementarity problems can be found in D. Hammer [128], nonlinear complementarity problems are considered in Wang [359].

Geometric computations with intervals are contained in Ratschek, Rokne [287] and Enger [91].

Integral equations were handled in Kaucher, Miranker [159], Klein [166], Gienger [119].

Ideas for the enclosure of operator equations can be found in Kaucher, Miranker [159] and Moore [236].

For the variety of enclosures in ordinary differential equations we recommend Moore [232], Krückeberg [178], Eijgenraam [90], Adams, Lohner [3], Stetter [347], Nickel [262], Cordes [77], Lohner [192], Corliss [78], Adams, Cordes, Keppler [1], Schulte [334], Plum [273], Neher [251, 252], Rihm [292, 293], Nedialkov, Jackson, Corliss [248], Nedialkov, Jackson [249], Eble [89], and Nagatou [242].

As a starter for the verification and enclosure of solutions of partial differential equation we mention Krückeberg [179], Plum [274], Nakao [246], Alefeld, Mayer [38]. Deeper insight can be obtained by Kaucher, Schulz-Rinne [160], Dobner [84], Nakao [244, 245], Koeber [168], [169], Nagatou [241], Nakao, Yamamoto, Nagatou [247], Nagatou, Hashimoto, Nakao [243], Watanabe et al. [360].

Applications of interval methods in practical computation can be found in T. Maier [197].

DOI 10.1515/9783110499469-011

Appendix

A Proof of the Jordan normal form

The proof follows the lines in Filippov [96].

Proof. First we show the *existence* of a Jordan normal form.

Since A represents a linear mapping ϕ from some n dimensional linear space V into itself, we will base our proof on ϕ instead of A. We will show by induction on the dimension $\dim V = n$ of V that there is a basis $\{b^1, \dots, b^n\}$ of V which either satisfies

$$\phi(b^i) = \lambda b^i \tag{A.1}$$

or

$$\phi(b^i) = \lambda b^i + b^{i-1}. \tag{A.2}$$

Here λ is an eigenvector of ϕ, and b^i from (A.1) is a corresponding eigenvector.

If $n = 1$, then V is spanned by some vector $b = b^1 \neq 0$ which obviously fulfills (A.1).

Assume now that for each linear space of dimension less than n one can find a basis as described above.

Case 1, A is singular: Then $r = \operatorname{rank}(\phi) = \dim \phi(V) < n$. Hence $U = \phi(V)$ together with the restriction $\phi_U : U \to U$, $\phi_U(x) = \phi(x)$ fulfill the induction hypothesis, i.e., there is a basis

$$\{b^1, \dots, b^r\}, \quad r \leq n-1 \tag{A.3}$$

of U which satisfies (A.1) and (A.2). We will extend this basis to a corresponding basis of V which also satisfies (A.1) and (A.2).

To this end, let $Q = \ker(\phi) \cap U$ with $q = \dim Q$, where $\ker(\phi)$ denotes the null space (= kernel) of ϕ. Since A is singular, $\dim \ker(\phi) \neq 0$, while $q = 0$ is possible. Each nonzero element of Q is an eigenvector of ϕ_U with respect to the eigenvalue zero. Therefore, according to the properties of the basis vectors in (A.3), there must be q basis vectors b^i which span Q and which are head vectors of q Jordan chains in U. Let u^i, $i = 1, \dots, q$, be the corresponding end vectors. They coincide with some basis vectors of (A.3), and as elements of U they are images of some vectors $v^i \in V$, $i = 1, \dots, q$. Since these vectors v^i satisfy $\phi(v^i) = 0 \cdot v^i + u^i$ they prolong the q Jordan chains starting in Q.

From $\dim \ker(\phi) = n - r$ there are vectors $w^i \in \ker(\phi) \setminus Q$, $i = 1, \dots s = n - r - q$, which together with the basis vectors for Q form a basis of $\ker \phi$.

If renamed and renumbered appropriately, the vectors

$$b^1, \dots, b^r, \quad v^1, \dots, v^q, \quad w^1, \dots, w^s \tag{A.4}$$

satisfy (A.1) and (A.2). We will show that they are linearly independent, hence they form a basis of V. If the basis vectors are reordered appropriately, the matrix which

DOI 10.1515/9783110499469-012

represents ϕ has already the Jordan normal form J. If S describes the change of basis to that in which ϕ is represented by the given matrix A, we obtain $J = S^{-1}AS$ as required in Theorem 1.7.1.

Let

$$\sum_{i=1}^{r} \alpha_i b^i + \sum_{i=1}^{q} \beta_i v^i + \sum_{i=1}^{s} \gamma_i w^i = 0. \tag{A.5}$$

Then

$$\sum_{i=1}^{r} \alpha_i \phi(b^i) + \sum_{i=1}^{q} \beta_i \phi(v^i) + \sum_{i=1}^{s} \gamma_i \phi(w^i) = \sum_{i=1}^{r} \alpha_i \phi(b^i) + \sum_{i=1}^{q} \beta_i u^i = 0. \tag{A.6}$$

If b^i does not belong to a Jordan chain with respect to zero, then $\phi(b^i) = \lambda b^i$ or $\phi(b^i) = \lambda b^i + b^{i-1}$. In both cases $\phi(b^i)$ is not in the linear subspace spanned by the end vectors $u^1, \ldots, u^q$. If $\phi(b^i)$ belongs to a Jordan chain with respect to zero, then $\phi(b^i) = b^{i-1} \neq u^j, j = 1, \ldots, q$. Therefore, (A.6) implies $\beta_i = 0$, $i = 1, \ldots, q$. Thus the middle sum (A.5) disappears and the remaining two are zero, since the first sum belongs to U while the last one does not unless it is zero. Hence $\alpha_i = 0$, $i = 1, \ldots, r$, and $\gamma_i = 0$, $i = 1, \ldots, s$, which proves the linear independence of the vectors in (A.4).

Case 2, A is regular: Let λ_0 be an eigenvalue of A. Then $A - \lambda_0 I$ is singular and has a Jordan normal form J' by Case 1. With some transformation matrix S we get $J' = S^{-1}(A - \lambda_0 I)S = S^{-1}AS - \lambda_0 I$, hence $S^{-1}AS = J' + \lambda_0 I$ which is a Jordan normal form of A.

Now we prove the *uniqueness* of the Jordan normal form.

From

$$S^{-1}(A - \lambda I)^m S = [S^{-1}(A - \lambda I)S]^m = (J - \lambda I)^m \tag{A.7}$$

we see that $\dim\ker(A - \lambda I)^m = \dim\ker(J - \lambda I)^m$. The representation of $(J - \lambda I)^m$ shows immediately that $\ker(J - \lambda I)^m$ is spanned by eigenvectors which belong to λ and by the corresponding principal vectors of degree $m \geq 2$. Considering $m = 1, 2, \ldots,$ one can see from (A.7) that for two Jordan normal forms J and J', the number of linearly independent eigenvectors with respect to the same eigenvalue coincides and so does the number of the linearly independent principal vectors of degree $m \geq 2$ associated with the same eigenvalue. Therefore, the number and the dimension of the Jordan blocks associated with the same eigenvalue λ are invariants. □

B Two elementary proofs of Brouwer's fixed point theorem

In Section 1.5 we stated Brouwer's fixed point theorem for the unit ball in $\mathbb{R}^n$ as Theorem 1.5.7. We proved this theorem by means of the mapping degree. In this appendix we present two alternative proofs. The first is based on the transformation formula for multidimensional integrals, the second uses the Sperner Lemma.

For our first proof we start with two preliminary lemmata, where

$$\|f\|_\infty = \max_D \|f(x)\|_\infty$$

for continuous functions $f\colon D \to \mathbb{R}^n$, defined on a compact set D.

Lemma B.1. *Let $B(0,1) \subseteq \mathbb{R}^n$ and $f\colon \overline{B}(0,1) \to \overline{B}(0,1)$ be continuous. Then there are functions $f_k\colon \overline{B}(0,1) \to \overline{B}(0,1)$ which are continuously differentiable and satisfy $\lim_{k\to\infty}\|f_k - f\|_\infty = 0$. If each function f_k has a fixed point, then the same holds for f.*

Proof. Weierstrass' approximation theorem 1.4.6 applied to the components of f guarantees functions $\tilde{f}_k\colon \overline{B}(0,1) \to \mathbb{R}^n$ such that $\|f - \tilde{f}_k\|_\infty < \frac{1}{k}$, $k = 1, 2, \ldots$. Define $f_k = \frac{k}{k+1}\tilde{f}_k$. Then

$$\begin{aligned}|f_k(x)| &\le \left|f_k(x) - \frac{k}{k+1}f(x)\right| + \left|\frac{k}{k+1}f(x)\right| \\ &= \frac{k}{k+1}|\tilde{f}_k(x) - f(x)| + \frac{k}{k+1}|f(x)| \le \frac{1}{k+1} + \frac{k}{k+1} = 1,\end{aligned}$$

whence $R_{f_k}(\overline{B}(0,1)) \subseteq \overline{B}(0,1)$. From $\|f - f_k\|_\infty \le \|f - \tilde{f}_k\|_\infty + \|\tilde{f}_k - f_k\|_\infty \le \frac{1}{k} + \frac{1}{k}\|f_k\|_\infty \le \frac{2}{k}$ we obtain the convergence of (f_k) to f.

Let x_k^* be a fixed point of f_k. Since $\overline{B}(0,1)$ is compact there is a convergent subsequence of (x_k^*) which converges to some limit $x^* \in \overline{B}(0,1)$. W.l.o.g. let $\lim_{k\to\infty} x_k^* = x^*$. Then we get

$$\|f(x^*) - x^*\|_\infty \le \|f(x^*) - f(x_k^*)\|_\infty + \|f(x_k^*) - f_k(x_k^*)\|_\infty + \|x_k^* - x^*\|_\infty.$$

The right-hand side tends to zero if k tends to infinity, hence $f(x^*) = x^*$. □

Lemma B.2. *There is no function $f\colon \overline{B}(0,1) \to \partial B(0,1) \subseteq \mathbb{R}^n$ that is continuously differentiable and satisfies $f(x) = x$ for all $x \in \partial B(0,1)$.*

Proof. Assume that there is a continuously differentiable function $f\colon \overline{B}(0,1) \to \partial B(0,1)$ with $f(x) = x$ for all x on the unit sphere $\partial B(0,1)$. Define $g(x) = f(x) - x$ and $f_t(x) = (1-t)x + tf(x) = x + tg(x)$. Then

$$\|f_t(x)\|_2 \le (1-t)\|x\|_2 + t\|f(x)\|_2 \le (1-t) + t = 1$$

for all $x \in \overline{B}(0,1)$, $t \in [0,1]$. Hence $f_t\colon \overline{B}(0,1) \to \overline{B}(0,1)$ is a continuously differentiable function which satisfies $f_t(x) = x$ for all x on the unit sphere since $f(x) = x$,

DOI 10.1515/9783110499469-013

$x \in \partial B(0,1)$, by assumption. Using Theorem 1.6.2 there is a positive constant c such that $\|g(x) - g(y)\|_2 \le c\|x - y\|_2$ holds for all $x, y \in \overline{B}(0,1)$. Choose $t \in [0, 1/(2c)]$ and assume that $f_t(\tilde{x}) = f_t(\tilde{y})$ holds for some $\tilde{x}, \tilde{y} \in \overline{B}(0,1)$ which differ from each other. Then $\tilde{x} - \tilde{y} = -t(g(\tilde{x}) - g(\tilde{y}))$ from which we get the contradiction

$$0 \ne \|\tilde{x} - \tilde{y}\|_2 = t\|g(\tilde{x}) - g(\tilde{y})\|_2 \le tc\|\tilde{x} - \tilde{y}\|_2 \le \frac{1}{2}\|\tilde{x} - \tilde{y}\|_2 .$$

Hence f_t is injective for $t \in [0, 1/(2c)]$. Its Jacobian satisfies $f_t'(x) = I + tg'(x)$, where, by the continuity of g', there is some positive real number t_0 such that $\det f_t'(x) > 0$, $t \in [0, t_0]$. W.l.o.g. choose $t_0 \le 1/(2c)$. Then $R_{f_t}(B(0,1))$ is an open set by the inverse function theorem. We show that $R_{f_t}(B(0,1)) = B(0,1)$ holds for $t \in [0, t_0]$. Otherwise there is some $\tilde{y} \in (\partial R_{f_t}(B(0,1))) \cap B(0,1)$ and a sequence (x^k) in $B(0,1)$ such that $\lim_{k\to\infty} f_t(x^k) = \tilde{y}$. Since $\overline{B}(0,1)$ is compact there is a subsequence of (x^k) which is convergent to some $\tilde{x} \in \overline{B}(0,1)$. W.l.o.g. let $\lim_{k\to\infty} x^k = \tilde{x}$. Then $f_t(\tilde{x}) = \tilde{y}$ whence $\tilde{y} \in f_t(\overline{B}(0,1))$. Since $R_{f_t}(B(0,1))$ is open and $\tilde{y} \in \partial R_{f_t}(B(0,1))$ we must have $\tilde{x} \in \overline{B}(0,1) \setminus B(0,1) = \partial B(0,1)$, whence $\tilde{y} = f_t(\tilde{x}) = \tilde{x} \in \partial B(0,1)$ contradicting $\tilde{y} \in B(0,1)$.

Define the function $\upsilon \colon [0,1] \to \mathbb{R}$ by

$$\upsilon(t) = \int_{B(0,1)} \det f_t'(x)\, dx = \int_{B(0,1)} \det(I + tg'(x))\, dx,$$

which is a polynomial in t. For $t \in [0, t_0]$ we get

$$\begin{aligned}\upsilon(t) &= \int_{B(0,1)} |\det f_t'(x)|\, dx = \int_{R_{f_t}(B(0,1))} 1\, dx \\ &= \operatorname{vol}(R_{f_t}(B(0,1))) = \operatorname{vol}(B(0,1)),\end{aligned}$$

where we used the transformation formula for multidimensional integrals. Here, $\operatorname{vol}(\cdot)$ denotes the n-dimensional volume of the set in brackets. The last equality above shows that the polynomial υ is constant for $t \in [0, t_0]$ and therefore $\upsilon(t) = \operatorname{vol}(B(0,1)) \ne 0$ for *all* $t \in [0,1]$. Since $f_1(x) = f(x) \in \partial B(0,1)$ for all $x \in \overline{B}(0,1)$ we get

$$1 = \|f_1(x)\|_2^2 = f_1(x)^T f_1(x) = \sum_{i=1}^{n} (f_1(x))_i^2,$$

whence

$$0 = \sum_{i=1}^{n} \frac{\partial}{\partial x_j} (f_1(x))_i^2 = \sum_{i=1}^{n} 2(f_1(x))_i \frac{\partial (f_1(x))_i}{\partial x_j}, \quad j = 1, \ldots, n,$$

and $f_1'(x)^T \cdot f_1(x) = 0$. From $\|f_1(x)\|_2 = 1$ we conclude that $f_1'(x)^T$ is singular, hence $\det f_1'(x) = \det f_1'(x)^T = 0$ for all $x \in \overline{B}(0,1)$. This leads to the contradiction $\upsilon(1) = 0$. □

Now we are ready to prove Brouwer's fixed point theorem.

First alternative proof of Brouwer's fixed point theorem

By Lemma B.1 it is sufficient to show that each continuously differentiable function $f\colon \overline{B}(0,1) \to \overline{B}(0,1)$ has a fixed point. To this end assume that there is such a function f which has no fixed point. Then $f(x) - x \neq 0$ for all $x \in \overline{B}(0,1)$ and we can construct a function $h\colon \overline{B}(0,1) \to \partial B(0,1)$ such that $h(x)$ is the intersection of the sphere $\partial B(0,1)$ with the ray from $f(x)$ through x, i.e., $h(x) = f(x) + t(x - f(x))$ with $t > 0$ such that $\|h(x)\|_2 = 1$. This implies $\|f(x)\|_2^2 + 2tf(x)^T(x - f(x)) + t^2\|x - f(x)\|_2^2 = 1$ with the unique positive solution

$$t = \frac{1}{\|x - f(x)\|_2^2}\left(-f(x)^T(x - f(x)) + \sqrt{[f(x)^T(x - f(x))]^2 + \|x - f(x)\|_2^2(1 - \|f(x)\|_2^2)}\right).$$

Using this expression for t we can immediately see that h is continuously differentiable. In addition, $h(x) = x$ for all $x \in \partial B(0,1)$. This contradicts Lemma B.2 and concludes the proof of Brouwer's fixed point theorem. □

Now we prepare our second proof. To this end we start with three definitions. The first recalls the definition of a simplex.

Definition B.3. Let $x^0, \dots, x^n \in \mathbb{R}^n$. The set

$$S = \left\{x \mid x = \sum_{i=0}^{n} \alpha_i x^i,\ \alpha_i \geq 0,\ \sum_{i=0}^{n} \alpha_i = 1\right\}$$

is called a simplex in $\mathbb{R}^n$ with the vertices $x^0, \dots, x^n$. The numbers $\alpha_0, \dots, \alpha_n$ are called barycentric coordinates (with respect to $x^0, \dots, x^n$) of the point $x \in S$. The simplex is *nondegenerate* if the vectors $x^1 - x^0, x^2 - x^0, \dots, x^n - x^0$ are linearly independent. In this case we say that the points $x^0, \dots, x^n$ are in general position and that S has dimension n. □

Notice that for $x \in S$ we have

$$x = x^0 + \sum_{i=1}^{n} \beta_i(x^i - x^0), \quad \beta_i \geq 0, \quad \sum_{i=1}^{n} \beta_i \leq 1.$$

By this and a contradiction one easily proves that for a nondegenerate simplex the barycentric coordinates of a point $x \in S$ are unique.

In $\mathbb{R}^1$ the nondegenerate simplices are line segments, in $\mathbb{R}^2$ they are triangles, and in $\mathbb{R}^3$ they are tetrahedra.

In our next definition we introduce certain subdivisions of a given simplex.

Definition B.4. Let $S \subseteq \mathbb{R}^n$ be a nondegenerate simplex with the vertices $x^0, \dots, x^n$. Then for any $k \in \mathbb{N}$ the points with the barycentric coordinates $k_0/k, \dots, k_n/k$, $k_i \in \mathbb{N}_0$, $\sum_{i=0}^{n} k_i = k$, form the vertices of the k-th barycentric subdivision of S (cf. Figure B.1 for $S \subseteq \mathbb{R}^2$ and $k = 3$). In such a subdivision each nondegenerate simplex of minimal volume is called a cell (in Figure B.1 there are 9 cells, one of which is shaded). □

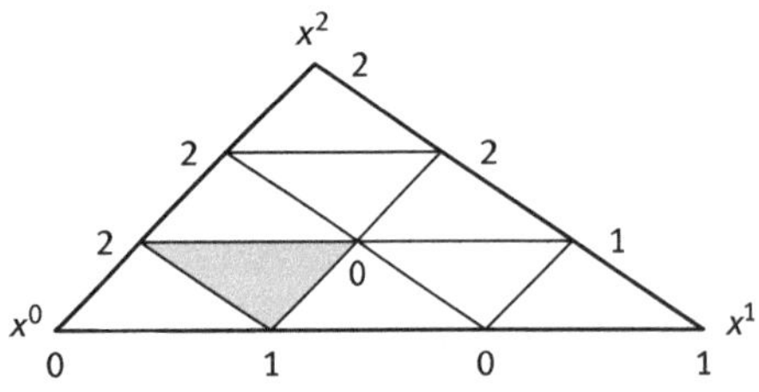

Fig. B.1: Third barycentric subdivision of $S \subseteq \mathbb{R}^2$ with possible values of a Sperner mapping.

In our last definition we consider a particular integer-valued function which is needed later on.

Definition B.5. Let $S \subseteq \mathbb{R}^n$ be a nondegenerate simplex with the vertices $x^0, \ldots, x^n$. A Sperner mapping m is any mapping $m\colon S \to \{0, 1, \ldots, n\}$ with the only restriction $m(x) \neq i$ if $\alpha_i = 0$ for the i-th barycentric coordinate of x. The value $m(x)$ is called the Sperner value. □

Notice that each vertex x^j of a nondegenerate simplex S necessarily has the Sperner value $m(x^j) = j$ since its barycentric coordinates satisfy $\alpha_i = 0$ for $i \neq j$ so that the only possible value which remains for $m(x^j)$ is j.

Now we formulate Sperner's Lemma in which only those cells of a fixed barycentric subdivision of S are counted whose $n + 1$ vertices attain all $n + 1$ possible values of a given Sperner mapping m.

Lemma B.6 (Sperner Lemma; cf. Franklin [100]). *Let $S \subseteq \mathbb{R}^n$ be a nondegenerate simplex and let m be a Sperner mapping. Consider an arbitrary k-th barycentric subdivision of S. Then the number of those cells whose vertices are mapped by m* onto $\{0, 1, \ldots, n\}$ *is odd.*

Proof. We proceed by induction on n. For $n = 1$, the simplex S is a line segment from x^0 to x^1. The vertices v^i, $i = 0, \ldots, k$, of the k-th barycentric subdivision satisfy $v^i = x^0 + (i/k)(x^1 - x^0)$, $i = 0, \ldots, k$. The values of the Sperner mapping m of S are 0 or 1 with $m(x^0) = 0$ and $m(x^1) = 1$ (cf. the remark below Definition B.5). Therefore, the finite sequence $m(v^0), m(v^1), \ldots, m(v^k)$ is a sequence of zeros and ones, which starts with 0 and ends with 1, and its changes from 0 to 1 must be one more than those from 1 to 0. This proves the assertion for $n = 1$ and arbitrary $k \in \mathbb{N}$.

Assume now that the assertion holds for all nondegenerate simplices of dimension less than n and let S be a nondegenerate simplex of dimension n. Consider its k-th barycentric subdivision. Let $F(\ell)$ be the number of cells whose $n + 1$ vertices have the Sperner values $0, \ldots, n - 1$ and the additional value ℓ (which occurs twice if $\ell \neq n$). The order in which these Sperner values appear does not matter. At the moment we do not know whether $F(\ell) \neq 0$ occurs. We are interested in $F(n)$. If we can show that $F(n)$ is odd, we are done.

Each cell of the k-th barycentric subdivision contains simplices of dimension $n-1$ as boundary elements. Such boundary elements are called *faces*. We will compute the number z which counts the occurrence of those faces F_n whose n vertices have the different Sperner values $0, \dots, n-1$. For each cell which contributes to $F(\ell)$ there are exactly two such faces F_n if $\ell \neq n$ and exactly one if $\ell = n$. Hence

$$z = 2 \sum_{\ell=0}^{n-1} F(\ell) + F(n).$$

We will compute z in a second way: If a face F_n does not completely belong to the boundary of S it belongs to two cells of the k-th barycentric subdivision of S. Therefore, it counts twice for z. If it is contained in the boundary of S it lies completely in that unique face S_{n-1} of S whose vertices are $x^0, \dots, x^{n-1}$. This can again be seen from the remark below Definition B.5 and the admissible Sperner values of the vertices of F_n. Therefore, we can express z as

$$z = F_i + F_b$$

where F_i counts the faces F_n which have at least one element in the interior of S and F_b counts the corresponding faces in S_{n-1}. Clearly, F_i is even. By the induction hypotheses F_b is odd, and so are z and $F(n)$. □

Now we prove the following version of Brouwer's fixed point theorem.

Theorem (Brouwer's fixed point theorem for simplices). *If the function f maps a nondegenerate simplex S of $\mathbb{R}^n$ continuously into itself, then it has a fixed point x^*.*

Proof. Let $\alpha_i(x)$, $i = 0, \dots, n$, be the barycentric coordinates of x. First we remark that the fixed point property $x^* = f(x^*)$ is equivalent to

$$\alpha_i(x^*) \geq \alpha_i(f(x^*)), \quad i = 0, \dots, n. \tag{B.1}$$

One direction follows immediately from the uniqueness of the barycentric coordinates for nondegenerate simplices, the converse direction follows from

$$\sum_{i=0}^{n} \alpha_i(x^*) = \sum_{i=0}^{n} \alpha_i(f(x^*)) = 1, \quad \alpha_i(x^*) \geq 0,\ \alpha_i(f(x^*)) \geq 0. \tag{B.2}$$

Assume now that f does not have any fixed point. Then for each $x \in S$ and $y = f(x)$ there is an index i_0 such that $\alpha_{i_0}(x) < \alpha_{i_0}(f(x))$ holds whence by a relation analogous to (B.2) there is an index i_1 such that $\alpha_{i_1}(x) > \alpha_{i_1}(f(x))$. This defines a function $m\colon S \to \{0, 1, \dots, n\}$ by

$$m(x) = \min\{\, j \mid \alpha_j(x) > \alpha_j(f(x)) \,\}.$$

Since $m(x) = i$ implies $\alpha_i(x) > \alpha_i(f(x)) \geq 0$, i.e., $\alpha_i(x) \neq 0$, the function m is a Sperner mapping. According to the Sperner lemma, for each k-th barycentric subdivision of S there is a cell C_k with vertices $x^0(k), \dots, x^n(k)$ which are mapped by m onto $\{0, 1, \dots, n\}$. Eventually, by reordering the vertices $x^i(k)$, $i = 0, \dots, n$, for each fixed k, the 'onto'-property allows that we may assume

$$m(x^i(k)) = i \quad \text{for } i = 0, \dots, n \text{ and } k = 1, 2, \dots. \tag{B.3}$$

Since S is bounded there is a subsequence (k_ℓ) such that the sequence $(x^0(k_\ell))$ has a limit x^*. With increasing ℓ the cells C_{k_ℓ} contract to x^*, i.e., $\lim_{\ell\to\infty} x^i(k_\ell) = x^*$, $i = 0, \dots, n$, hence $\lim_{\ell\to\infty} f(x^i(k_\ell)) = f(x^*)$. By (B.3) and the definition of m we get $\alpha_i(f(x^i(k))) < \alpha_i(x^i(k))$, whence $\alpha_i(f(x^*)) = \lim_{\ell\to\infty} \alpha_i(f(x(k_\ell))) \le \alpha_i(x^*)$ for $i = 0, \dots, n$. By virtue of (B.1) this means $x^* = f(x^*)$, contradicting the assumption. □

In order to deduce Brouwer's fixed point theorem for the closed unit ball $\overline{B}(0, 1)$ from the version which we have just proved we proceed similarly as in Theorem 1.5.8, this time with $D = \overline{B}(0, 1)$ and with any simplex S enclosing $\overline{B}(0, 1)$. Thus S plays the role of $\overline{B}(0, \rho)$ in the proof of Theorem 1.5.8. We leave the details to the reader.

Exercises

Ex. B.1. Let S be a nondegenerate simplex in $\mathbb{R}^n$ with the vertices $x^0, \dots, x^n$. Show that the barycentric coordinates $\alpha_i(x)$, $i = 0, \dots, n$ of $x \in S$ are unique.

Ex. B.2. Let S be a nondegenerate simplex in $\mathbb{R}^n$ with the vertices $x^0, \dots, x^n$. Show that the barycentric coordinates $\alpha_i(x)$, $i = 0, \dots, n$ of $x \in S$ depend continuously on x.
Hint: Use the uniqueness of $\alpha_0(x), \dots, \alpha_n(x)$.

Ex. B.3. Deduce Brouwer's fixed point theorem for the closed unit ball $\overline{B}(0, 1) \subseteq \mathbb{R}^n$ from that for nondegenerate simplices $S \subseteq \mathbb{R}^n$. (Give all details of the proof.)

Notes to Appendix B

Both proofs in Appendix B of Brouwer's fixed point theorem follow the lines of Franklin [100], where even more proofs of this theorem can be found.

C Proof of the Newton–Kantorovich Theorem

We prove here the Newton–Kantorovich Theorem which was stated as Theorem 1.5.10 using the notation and references there without repeating them. The proof follows the lines of Ortega, Rheinboldt [267].

Proof. The numbers $\underline{t}, \overline{t}$ in (1.5.10) are the zeros of the polynomial $p(t) = \frac{1}{2}\beta\gamma t^2 - t + \eta$.

First we consider the case $0 < \alpha \le 1/2$, which implies $\eta > 0$ and $0 < \underline{t} \le \overline{t}$.

Since α is positive, the same holds for $\underline{t}, \overline{t}$. In addition, we have

$$p'(t) = \beta\gamma t - 1 \neq 0 = p'\left(\frac{\underline{t} + \overline{t}}{2}\right), \quad \text{if } t \neq \frac{\underline{t} + \overline{t}}{2}. \tag{C.1}$$

Let t_k be the k-th Newton iterate for p starting with $t_0 = 0$, i.e.,

$$\begin{cases} t_0 = 0, \\ t_{k+1} = t_k - \dfrac{p(t_k)}{p'(t_k)} = t_k - \dfrac{\frac{1}{2}\beta\gamma t_k^2 - t_k + \eta}{\beta\gamma t_k - 1}, \quad k = 0, 1, \dots . \end{cases} \tag{C.2}$$

If $p'(t_k) \neq 0$, then

$$\begin{aligned} t_{k+1} - \underline{t} &= \frac{1}{p'(t_k)}\{t_k p'(t_k) - p(t_k) - \underline{t}p'(t_k)\} \\ &= -\frac{1}{p'(t_k)}\{p(t_k) + p'(t_k)(\underline{t} - t_k) + \tfrac{1}{2}p''(t_k)(\underline{t} - t_k)^2 - \tfrac{1}{2}p''(t_k)(\underline{t} - t_k)^2\} \\ &= \frac{1}{2}\frac{\beta\gamma}{\beta\gamma t_k - 1}(\underline{t} - t_k)^2 \end{aligned} \tag{C.3}$$

holds, where we used the Taylor expansion of $p(\underline{t}) = 0$ at t_k. Since (C.2) implies $p(t_{k-1}) + p'(t_{k-1})(t_k - t_{k-1}) = 0$ we get

$$\begin{aligned} t_{k+1} - t_k &= -\frac{p(t_k)}{p'(t_k)} = -\frac{1}{p'(t_k)}\{p(t_{k-1}) + p'(t_{k-1})(t_k - t_{k-1}) + \tfrac{1}{2}p''(t_{k-1})(t_k - t_{k-1})^2\} \\ &= \frac{1}{2}\frac{\beta\gamma}{1 - \beta\gamma t_k}(t_k - t_{k-1})^2. \end{aligned} \tag{C.4}$$

Assertion 1:

$$t_0 < t_1 < \dots < t_k < \underline{t} \quad \text{for all } k \in \mathbb{N}_0. \tag{C.5}$$

We prove (C.5) by induction. For $k = 0$, we get $0 = t_0 < \underline{t}$. Assume now that $t_k < \underline{t}$ is known. Then $\beta\gamma t_k - 1 < \beta\gamma\underline{t} - 1 = -\sqrt{1 - 2\alpha} \le 0$ by virtue of (iv), whence (C.3), (C.4) imply $t_k < t_{k+1} < \underline{t}$.

Assertion 2: $\lim_{k\to\infty} t_k = \underline{t}$. From Assertion 1 we see that (t_k) is convergent to some real number $\hat{t} \le \underline{t}$. Assume that $\hat{t} < \underline{t}$. Then we have $p'(\hat{t}) = \beta\gamma\hat{t} - 1 \neq 0$ by virtue of (C.1) and with (C.3) we get

$$\hat{t} - \underline{t} = \frac{1}{2}\frac{\beta\gamma}{\beta\gamma\hat{t} - 1}(\underline{t} - \hat{t})^2.$$

DOI 10.1515/9783110499469-014

Dividing by $\underline{t} - \hat{t}$ implies

$$-1 = \frac{1}{2} \frac{\beta\gamma}{\beta\gamma\hat{t} - 1} (\underline{t} - \hat{t}),$$

whence $2 - \beta\gamma\underline{t} = \beta\gamma\hat{t}$. Use $2 - \beta\gamma\underline{t} = 1 + \sqrt{1-2\alpha}$ in order to obtain the contradiction $\hat{t} = \bar{t} \geq \underline{t} > \hat{t}$.

Assertion 3: $f'(x)^{-1}$ exists for all $x \in B(x^0, \underline{t})$, and

$$\|f'(x)^{-1}\| \leq \frac{\beta}{1 - \beta\gamma\|x - x^0\|} \tag{C.6}$$

holds.

We start with

$$\begin{aligned} \|f'(x^0)^{-1}\{f'(x^0) - f'(x)\}\| &\leq \|f'(x^0)^{-1}\| \cdot \|f'(x^0) - f'(x)\| \\ &\leq \beta\gamma\|x^0 - x\| < \beta\gamma\,\underline{t} = 1 - \sqrt{1-2\alpha} \leq 1 \quad \text{for all } x \in B(x^0, \underline{t}). \end{aligned}$$

The Neumann series shows that

$$f'(x) = f'(x^0)\{I - f'(x^0)^{-1}(f'(x^0) - f'(x))\}$$

has an inverse which can be represented as

$$f'(x)^{-1} = \sum_{i=0}^{\infty} \{f'(x^0)^{-1}(f'(x^0) - f'(x))\}^i f'(x^0)^{-1}.$$

This implies

$$\|f'(x)^{-1}\| \leq \sum_{i=0}^{\infty} (\beta\gamma\|x^0 - x\|)^i \cdot \beta = \frac{\beta}{1 - \beta\gamma\|x^0 - x\|},$$

where $1 - \beta\gamma\|x^0 - x\| = -p'(\|x^0 - x\|) \neq 0$ for $\|x^0 - x\| < \underline{t}$ by virtue of (C.1).

Assertion 4: $x^k \in B(x^0, \underline{t})$ and $\|x^k - x^{k-1}\| \leq t_k - t_{k-1}$. In particular, the Newton iterates are well defined.

We prove this assertion by induction. Let $k = 1$. With (iii) we get

$$\begin{aligned} \|x^1 - x^0\| &= \|f'(x^0)^{-1} f(x^0)\| \leq \eta = t_1 - t_0 \\ &< \eta \cdot \frac{2}{1 + \sqrt{1-2\alpha}} = \eta \cdot \frac{2(1 - \sqrt{1-2\alpha})}{2\alpha} = \underline{t}. \end{aligned}$$

Here, the representation $\eta = t_1 - t_0$ follows from (C.2) with $t_0 = 0$, and the second inequality holds because the factor behind η is greater than one.

Now let the assertion be true up to some k. Then by the induction hypothesis we get

$$\|x^k - x^0\| \leq \sum_{i=1}^{k} \|x^i - x^{i-1}\| \leq \sum_{i=1}^{k} (t_i - t_{i-1}) = t_k - t_0 = t_k. \tag{C.7}$$

With $g(x) = x - f'(x)^{-1}f(x)$, $x \in B(x^0, \underline{t})$, and with Assertion 3 we get

$$\begin{aligned}\|x^{k+1} - x^k\| &= \|g(g(x^{k-1})) - g(x^{k-1})\| \\ &= \|g(x^{k-1}) - f'(g(x^{k-1}))^{-1}f(g(x^{k-1})) - g(x^{k-1})\| \\ &\le \frac{\beta}{1 - \beta\gamma\|g(x^{k-1}) - x^0\|} \cdot \|f(g(x^{k-1}))\|.\end{aligned}$$

Here, we used the fact that $g(x^{k-1}) = x^k \in B(x^0, \underline{t})$. Since

$$g(x^{k-1}) - x^{k-1} = -f'(x^{k-1})^{-1}f(x^{k-1})$$

we obtain

$$f(g(x^{k-1})) = f(g(x^{k-1})) - f(x^{k-1}) - f'(x^{k-1})\{g(x^{k-1}) - x^{k-1}\}$$

and

$$\begin{aligned}\|x^{k+1} - x^k\| &\le \frac{\beta}{1 - \beta\gamma\|x^k - x^0\|} \left\| \int_0^1 f'(x^{k-1} + \lambda(g(x^{k-1}) - x^{k-1}))\, d\lambda \cdot \{g(x^{k-1}) - x^{k-1}\} \right. \\ &\qquad \left. - \int_0^1 f'(x^{k-1})\, d\lambda \cdot \{g(x^{k-1}) - x^{k-1}\} \right\| \\ &\le \frac{\beta}{1 - \beta\gamma\|x^k - x^0\|} \int_0^1 \|f'(x^{k-1} + \lambda(g(x^{k-1}) - x^{k-1})) - f'(x^{k-1})\|\, d\lambda \\ &\qquad \cdot \|g(x^{k-1}) - x^{k-1}\| \\ &\le \frac{\beta}{1 - \beta\gamma\|x^k - x^0\|}\, \gamma \int_0^1 \lambda\, d\lambda\, \|g(x^{k-1}) - x^{k-1}\|^2 \\ &= \frac{\beta\gamma}{2(1 - \beta\gamma\|x^k - x^0\|)} \|x^k - x^{k-1}\|^2 \\ &\le \frac{\beta\gamma}{2(1 - \beta\gamma t_k)} (t_k - t_{k-1})^2 = t_{k+1} - t_k.\end{aligned}$$

For the third inequality we used (i), for the last inequality we applied the induction hypothesis and (C.7) combined with $1 - \beta\gamma t_k > 1 - \beta\gamma\underline{t} \ge 0$. The last equality follows from (C.4).

Analogously to (C.7) we can conclude $\|x^{k+1} - x^0\| \le t_{k+1} < \underline{t}$, hence $x^{k+1} \in B(x^0, \underline{t})$.

Assertion 5: $\lim_{k\to\infty} x^k = x^* \in \overline{B}(x^0, \underline{t})$ with $f(x^*) = 0$. Assertion 4 implies

$$\|x^{k+p} - x^k\| \le \sum_{i=k+1}^{k+p} \|x^i - x^{i-1}\| \le \sum_{i=k+1}^{k+p} (t_i - t_{i-1}) = t_{k+p} - t_k.$$

Since (t_k) is convergent, the sequence (x^k) is a Cauchy sequence in $B(x^0, \underline{t})$. Hence $x^* = \lim_{k\to\infty} x^k \in \overline{B}(x^0, \underline{t})$ exists, and by virtue of the continuity of f and f' the Newton equation $-f(x^k) = f'(x^k)(x^{k+1} - x^k)$ tends to $-f(x^*) = 0$ if k tends to infinity.

Assertion 6: The zero x^* of f is unique in $B(x^0, \bar{t}) \cap D$ if $\alpha < 1/2$, and in $\overline{B}(x^0, \underline{t})$ if $\alpha = 1/2$.

Assume that there is another zero y^* of f which lies in $B(x^0, \bar{t})$ if $\alpha < 1/2$ and in $\overline{B}(x^0, \underline{t})$ if $\alpha = 1/2$.

With the Newton iterates x^k define the sequence

$$\begin{cases} y^0 = y^*, \\ y^{k+1} = y^k - f'(x^k)^{-1} f(y^k), \quad k = 0, 1, \ldots . \end{cases}$$

Then $y^k = y^*$, $k = 0, 1, \ldots$.

With t_k from (C.2) define the sequence

$$\begin{cases} s_0 = \|y^* - x^0\| < \bar{t}, \\ s_{k+1} = s_k - \dfrac{p(s_k)}{p'(t_k)}, \quad k = 0, 1, \ldots . \end{cases} \tag{C.8}$$

We conclude the proof by three major steps.

Assertion 6.1:

$$s_{k+1} - t_{k+1} = -\frac{1}{p'(t_k)} \cdot \tfrac{1}{2}\beta\gamma(s_k - t_k)^2 > 0, \text{ if } s_k \neq t_k, \tag{C.9}$$

$$s_{k+1} - \underline{t} = \frac{\beta\gamma}{p'(t_k)} \left(t_k - \tfrac{1}{2}s_k - \tfrac{1}{2}\underline{t}\right)(s_k - \underline{t}). \tag{C.10}$$

The first formula follows from

$$\begin{aligned} s_{k+1} - t_{k+1} &= s_k - t_k - \frac{p(s_k) - p(t_k)}{p'(t_k)} \\ &= \frac{1}{p'(t_k)} \{p'(t_k)(s_k - t_k) + p(t_k) - p(s_k)\} \\ &= \frac{1}{p'(t_k)} \left(-\tfrac{1}{2}p''(t_k)(s_k - t_k)^2\right). \end{aligned}$$

The second formula results from

$$\begin{aligned} s_{k+1} - \underline{t} = s_k - \underline{t} - \frac{p(s_k)}{p'(t_k)} &= \frac{1}{p'(t_k)}\{p'(t_k)(s_k - \underline{t}) - p(s_k)\} \\ &= \frac{1}{p'(t_k)} \Big\{ \underbrace{(p'(\underline{t}) + p''(\underline{t})(t_k - \underline{t}))}_{=p'(t_k)} (s_k - \underline{t}) \\ &\qquad - p(\underline{t}) - p'(\underline{t})(s_k - \underline{t}) - \tfrac{1}{2}p''(\underline{t})(s_k - \underline{t})^2 \Big\} \\ &= \frac{\beta\gamma}{p'(t_k)}(s_k - \underline{t})\left\{ t_k - \underline{t} - \tfrac{1}{2}(s_k - \underline{t}) \right\}. \end{aligned}$$

Assertion 6.2:

$$\lim_{k\to\infty} s_k = \underline{t}. \tag{C.11}$$

If $s_0 = t_0 = 0$, then $s_k = t_k$, $k = 0, 1, \ldots$, by virtue of (C.8), and (C.11) follows from Assertion 2.

If $s_0 = \underline{t}$, then $p(\underline{t}) = 0$ implies $s_k = \underline{t}$, $k = 0, 1, \ldots$, which proves (C.11).

If $t_0 = 0 < s_0 < \underline{t}$, we show by induction that the inequalities $s_0 < s_1 < \cdots < s_k < \underline{t}$ and $t_k < s_k$ hold. For $k = 0$ this is trivial. Assume that they are true up to some integer k. Since $p'(t_k) < 0$, Assertion 6.1 guarantees $s_{k+1} - t_{k+1} > 0$, which proves the last inequality. By the induction hypothesis we can deduce that $p(s_k) > 0$, whence $s_{k+1} > s_k$ by (C.8). From

$$t_k - \tfrac{1}{2}s_k - \tfrac{1}{2}\underline{t} = \tfrac{1}{2}(t_k - \underline{t}) + \tfrac{1}{2}(t_k - s_k) < 0$$

and (C.10) we get $s_{k+1} - \underline{t} < 0$, which prolongs the first chain of inequalities. Combining these inequalities with Assertion 2 proves (C.11).

If $\underline{t} < s_0 < \bar{t}$, which does not occur in the case $\alpha = 1/2$, we show by induction the inequalities $\underline{t} < s_k < \cdots < s_1 < s_0 < \bar{t}$. The case $k = 0$ is again trivial. From $\underline{t} < s_k < \bar{t}$ we can see $p(s_k) < 0$, whence $s_{k+1} < s_k$ as above. With Assertion 6.1 we obtain $t_k - \frac{1}{2}s_k - \frac{1}{2}\underline{t} < t_k - t_k/2 - t_k/2 = 0$, hence (C.10) proves $s_{k+1} - \underline{t} > 0$. Thus the inequalities are proved and show that the sequence (s_k) is convergent to some limit $\hat{s}$. From Assertion 2, (C.1), and (C.8) we see that $p(\hat{s}) = 0$, hence $\hat{s} = \underline{t}$.

Assertion 6.3: $\|y^k - x^k\| \le s_k - t_k$.

Again we proceed by induction. For $k = 0$ the assertion follows from $\|y^0 - x^0\| = \|y^* - x^0\| = s_0 = s_0 - t_0$. If the inequality holds for some integer k, we use Assertion 3 in order to get

$$\begin{aligned}\|y^{k+1} - x^{k+1}\| &= \| - f'(x^k)^{-1}\{-f'(x^k)(y^k - x^k) + f(y^k) - f(x^k)\}\| \\ &\le \frac{\beta}{1 - \beta\gamma\|x^k - x^0\|}\left\| \int_0^1 \{f'(x^k + \lambda(y^k - x^k)) \quad f'(x^k)\}\, d\lambda \right\| \cdot \|y^k - x^k\|.\end{aligned}$$

Analogously to the proof of Assertion 4 and with the induction hypothesis and (C.9) we can overestimate this last expression by

$$\frac{\beta\gamma}{1 - \beta\gamma t_k} \int_0^1 \lambda\, d\lambda\, \|y^k - x^k\|^2 \le \frac{\beta\gamma}{-p'(t_k)} \cdot \frac{1}{2}(s_k - t_k)^2 = s_{k+1} - t_{k+1}.$$

Notice that $s_k = t_k$ implies $s_{k+1} = t_{k+1}$, so that the last equality holds trivially in this special case. From the last two assertions we finally get $\|y^* - x^*\| = \lim_{k\to\infty}\|y^k - x^k\| = \underline{t} - \underline{t} = 0$, which proves the Assertion 6.

It remains to show the error estimate (1.5.11) which we are going to repeat.

Assertion 7: $\|x^k - x^*\| \le \frac{1}{\beta\gamma 2^k}(2\alpha)^{2^k}$, $k = 0, 1, \ldots$.

We proceed in two major steps.

Assertion 7.1: $t_{k+1} - t_k \le \eta \cdot 2^{-k}$.

For $k = 0$ we have equality. If the inequality holds up to some integer k, we use $t_{k+1} = \sum_{i=0}^{k}(t_{i+1} - t_i) \le \sum_{i=0}^{k} \eta \cdot 2^{-i} = 2\eta(1 - 2^{-k-1})$ in order to estimate

$$1 - \beta\gamma t_{k+1} \ge 1 - 2\alpha(1 - 2^{-k-1}) \ge 2^{-k-1}. \tag{C.12}$$

Together with the induction hypothesis and (C.4) we finally get

$$\begin{aligned} t_{k+2} - t_{k+1} &= \frac{\beta\gamma/2}{1 - \beta\gamma t_{k+1}}(t_{k+1} - t_k)^2 \\ &\le \frac{\beta\gamma/2}{2^{-k-1}}\eta^2 2^{-2k} = \eta\alpha \cdot 2^{k-2k} < \eta \cdot 2^{-k-1}. \end{aligned}$$

Assertion 7.2: $\underline{t} - t_k \le \frac{1}{\beta\gamma 2^k}(2\alpha)^{2^k}$.

For $k = 0$ this inequality follows immediately from

$$\underline{t} = \frac{2\alpha}{\beta\gamma(1 + \sqrt{1 - 2\alpha})} \le \frac{2\alpha}{\beta\gamma}.$$

If it holds up to some integer k, we can use (C.3) and (C.12) (with k instead of $k + 1$) to see the estimate

$$\underline{t} - t_{k+1} \le \frac{\beta\gamma/2}{2^{-k}} \frac{1}{(\beta\gamma 2^k)^2}(2\alpha)^{2^k \cdot 2} = \frac{1}{\beta\gamma 2^{k+1}}(2\alpha)^{2^{k+1}}.$$

In order to finish the proof for Assertion 7 we look at the proof for Assertion 5. From there we can conclude $\|x^{k+p} - x^k\| \le t_{k+p} - t_k$, and for the limit $p \to \infty$ we end up with

$$\|x^* - x^k\| \le \underline{t} - t_k \le \frac{1}{\beta\gamma 2^k}(2\alpha)^{2^k}.$$

It remains to consider the case $\alpha = 0 < \min\{\beta, \gamma\}$ for which we use the notation above. Here, $\eta = 0$, $f(x^0) = 0$ whence $t_k = \underline{t} = 0 < \bar{t}$, and $x^k = x^*$ follows for $k = 0, 1, \ldots$. This proves the existence statement of the theorem. The error estimate is trivial now. In order to prove uniqueness we assume $x^* \ne y^*$. With the notation above we notice that (C.9), (C.10) still hold, $0 = \underline{t} < s_0 < \bar{t}$ is trivial, and $s_1 < s_0$ follows from (C.8) as in the proof of Assertion 6.2. From the inequality $\|y^1 - x^1\| \le s_1 - t_1 = s_1$, which can be shown as in the proof of Assertion 6.3, we get the contradiction $s_0 = \|y^0 - x^0\| = \|y^* - x^*\| = \|y^1 - x^1\| \le s_1 < s_0$. □

D Convergence proof of the row cyclic Jacobi method

In this appendix we will prove the convergence of the row cyclic Jacobi method.

Theorem D.1. *Let $A = A^T = A_0 \in \mathbb{R}^{n\times n}$. Then the matrices $A_k = (a_{ij}^{(k)})$, $k = 0, 1, \ldots$, of the row cyclic Jacobi method converge to the Jordan normal form of A, where the order of the diagonal entries depend on A.*

Proof. The proof of this theorem follows the lines of Forsythe, Henrici [99]. As there, we will proceed in several steps.

We will denote the entries of A_k to be zeroed by the k-th Jacobi rotation by $a_{p_k,q_k}^{(k)} = a_{q_k,p_k}^{(k)}$, $p_k < q_k$, so that $a_{p_k,q_k}^{(k+1)} = a_{q_k,p_k}^{(k+1)} = 0$. A measure for the deviation of A_k from its diagonal part is given by the sum

$$S^{(k)} = \sum_{i\neq j} (a_{ij}^{(k)})^2. \tag{D.1}$$

Later we will show $\lim_{k\to\infty} S^{(k)} = 0$. If k is irrelevant for our momentary argumentation, we will drop it, writing A and A' instead of A_k, A_{k+1}, and similarly p, q, S as we already did in the formulae of the Algorithm 7.3.1.

We mention that the multiplication of A with the Jacobi rotation J_{pq} from the right only changes the columns p and q of A, while the multiplication with J_{pq}^T from the left changes the corresponding rows. Therefore, the entries in the columns and rows p, q are called *affected*, the remaining ones *unaffected*. Obviously, the latter satisfy $a'_{ij} = a_{ij}$. From the formulae in Algorithm 7.3.1 and after a simple calculation (cf. Exercise D.1) we get

$$\begin{aligned} (a'_{pj})^2 + (a'_{qj})^2 &= a_{pj}^2 + a_{qj}^2, \\ (a'_{jp})^2 + (a'_{jq})^2 &= a_{jp}^2 + a_{jq}^2 \end{aligned} \tag{D.2}$$

for $j \notin \{p, q\}$. As in Forsythe, Henrici [99] we will call the entries of the pairs (a_{pj}, a_{qj}) and (a_{jp}, a_{jq}), $j \notin \{p, q\}$, *coupled* during the k-th transformation and we speak of *rotated* entries a_{pq}, a_{qp} for those two to be zeroed.

From (D.1), (D.2) we obtain

$$S - S' = 2a_{pq}^2, \tag{D.3}$$

hence $(S^{(k)})$ is monotonically decreasing and trivially bounded from below by zero. Therefore, the sequence $(S^{(k)})$ is convergent, and (D.3) implies

$$\lim_{k\to\infty} a_{p_k,q_k}^{(k)} = 0 \tag{D.4}$$

for the rotated entries.

Our first lemma shows that coupled entries cannot behave independently of each other when being transformed.

DOI 10.1515/9783110499469-015

Lemma D.2. *Let* $\varepsilon > 0$, $j \notin \{p, q\}$,

$$|a_{pj}| < \varepsilon, \quad |a'_{qj}| < \varepsilon. \tag{D.5}$$

Then

$$|a'_{pj}| < c_0\varepsilon, \quad |a_{qj}| < c_0\varepsilon \tag{D.6}$$

holds with the constant $c_0 = 1 + \sqrt{2}$ *which is independent of* k, p_k, q_k.

Proof. First we recall that the rotation angle φ is restricted to $\varphi \in [-\pi/4, \pi/4]$. This bounds c, s in Algorithm 7.3.1 by $c \geq 1/\sqrt{2} \neq 0$, $s \leq 1/\sqrt{2}$. From the formulae of this algorithm we get

$$a'_{pj} = a_{pj} + s(a_{qj} - \tau a_{pj}) = (1 - s\tau)a_{pj} + sa_{qj} = ca_{pj} + sa_{qj} \tag{D.7}$$

$$a'_{qj} = a_{qj} - s(a_{pj} + \tau a_{qj}) = ca_{qj} - sa_{pj}. \tag{D.8}$$

Equation (D.8) implies

$$a_{qj} = (a'_{qj} + sa_{pj})/c. \tag{D.9}$$

Inserting (D.9) in (D.7) results in

$$a'_{pj} = (c^2 a_{pj} + sa'_{qj} + s^2 a_{pj})/c = (a_{pj} + sa'_{qj})/c,$$

which can be estimated by

$$|a'_{pj}| \leq (|a_{pj}| + |a'_{qj}|/\sqrt{2}) \cdot \sqrt{2} = \sqrt{2}|a_{pj}| + |a'_{qj}|.$$

Together with (D.5) this yields the first inequality in (D.6) with $c_0 = 1 + \sqrt{2}$. The second follows similarly from (D.9). □

For the proof of our next lemma we need the submatrix

$$M_{lm}^{(k)} = (a_{ij}^{(k)}) \in \mathbb{R}^{(m-l+1)\times(m-l+1)},$$
$$1 \leq l < m \leq n, \quad l \leq i \leq m, \quad l \leq j \leq m.$$

An index k is said to be *associated* with (l, m) if $(p_k, q_k) = (l, m)$. In this case we write $k = I(l, m)$. The indices $k_1, k_2, \ldots k_r$ are called *cocyclic* if they are contained in one and the same of the intervals $[zN, (z + 1)N - 1]$, where $z = 0, 1, 2, \ldots$ and $N = n(n - 1)/2$ is the number of rotations in a full cycle of the Jacobi method. Similarly to (D.1) we define

$$S_{lm}^{(k)} = \sum_{\substack{l\leq i,j\leq m\\ i\neq j}} (a_{ij}^{(k)})^2.$$

Obviously,

$$S_{l,l+1}^{(k)} = 2(a_{l,l+1}^{(k)})^2 \quad \text{and} \quad S_{1,n}^{(k)} = S^{(k)} \tag{D.10}$$

holds.

Lemma D.3. *Let $\varepsilon > 0$ and k_0 such that $2(a_{p_k,q_k}^{(k)})^2 < \varepsilon$ for $k > k_0$ which is possible by virtue of* (D.4). *Let f, g, h be three cocyclic indices associated with the pairs (l, m), $(l, m+1)$, $(l+1, m+1)$, respectively, where $1 \le l < m < n$. Assume $f > k_0$. Then the inequalities*

$$S_{lm}^{(f)} < \varepsilon, \quad S_{l+1,m+1}^{(h)} < \varepsilon. \tag{D.11}$$

imply

$$S_{l,m+1}^{(g)} < d\varepsilon, \tag{D.12}$$

where $d = d_{m-l} > 0$ depends on $m - l$ and on the constant c_0 of Lemma D.2.

Proof. By virtue of cocyclicity the indices f, g, h satisfy $f < g = f + 1 < h$. We will decompose the submatrix $M_{l,m+1}^{(g)}$ into appropriate blocks and estimate their entries accordingly. Since the entries in $M_{l+1,m-1}^{(f)}$ are unaffected by the rotation f we get

$$S_{l+1,m-1}^{(f)} = S_{l+1,m-1}^{(g)}. \tag{D.13}$$

The entries $a_{lj}^{(f)} = a_{jl}^{(f)}$, $a_{mj}^{(f)} = a_{jm}^{(f)}$ are coupled for $j = l+1, \dots, m-1$ during the f-th Jacobi transformation. In addition, $a_{lm}^{(g)} = a_{lm}^{(f+1)} = 0$ since $a_{lm}^{(f)}$ is a rotated entry for f. Together with (D.2) and (D.13) this implies $S_{lm}^{(f)} - 2(a_{lm}^{(f)})^2 = S_{lm}^{(g)}$, hence

$$S_{lm}^{(g)} \le S_{lm}^{(f)} < \varepsilon \tag{D.14}$$

follows using (D.11). Since $(p_g, q_g) = (l, m+1)$ is a rotated entry for g, our very first assumption guarantees

$$|a_{l,m+1}^{(g)}| < \sqrt{\varepsilon/2}. \tag{D.15}$$

Now we will consider the entries $a_{j,m+1}^{(g)}$, $j = l+1, \dots, m$. After the transformation g these entries are unaffected by the rotations of the entries in position $(l, m+2)$, $\dots, (l, n)$, which proves $(a_{j,m+1}^{(g)})' = a_{j,m+1}^{(g+1)} = a_{j,m+1}^{(t)}$, where t is cocyclic with f, g, h and associated with the pair $(l+1, l+2)$ which follows the pair (l, n) in the row cyclic Jacobi method. We consider the sum

$$s_t = \sum_{j=l+1}^{m} (a_{j,m+1}^{(t)})^2,$$

which is apparently an upper bound for all $((a_{j,m+1}^{(g)})')^2$, $j = l+1, \dots, m$. If $l+2 = m+1$, then $t = h$ and $s_t = s_h = (a_{l+1,m+1}^{(h)})^2 = (a_{m,m+1}^{(h)})^2$. If $l+2 < m+1$, then $a_{l+1,m+1}^{(t)}$, $a_{l+2,m+1}^{(t)}$ are coupled during the t-th Jacobi transformation while $a_{j,m+1}^{(t)}$ is unaffected for $j = l+3, \dots, m$. Therefore, we get

$$\begin{aligned} s_t &= (a_{l+1,m+1}^{(t)})^2 + (a_{l+2,m+1}^{(t)})^2 + \sum_{j=l+3}^{m} (a_{j,m+1}^{(t)})^2 \\ &= (a_{l+1,m+1}^{(t+1)})^2 + (a_{l+2,m+1}^{(t+1)})^2 + \sum_{j=l+3}^{m} (a_{j,m+1}^{(t+1)})^2 = s_{t+1}. \end{aligned}$$

Repeating these ideas leads to $s_t = s_{t+1} = \ldots = s_h$. Since in both cases s_h is part of $S^{(h)}_{l+1,m+1}$ we obtain

$$((a^{(g)}_{j,m+1})')^2 \le s_t = s_h \le \frac{1}{2} S^{(h)}_{l+1,m+1} < \varepsilon/2$$

by means of (D.11). From (D.14) we get $|a^{(g)}_{lj}| < \sqrt{\varepsilon/2}$. Taking into account the symmetry of A_g we apply Lemma D.2 with $(p, q) = (l, m+1)$ and $\sqrt{\varepsilon/2}$ instead of ε. This implies $|a^{(g)}_{j,m+1}| < c_0 \sqrt{\varepsilon/2}$ and

$$\begin{aligned} S^{(g)}_{l,m+1} &= S^{(g)}_{lm} + 2(a^{(g)}_{l,m+1})^2 + 2 \sum_{j=l+1}^{m} (a^{(g)}_{j,m+1})^2 \\ &< \varepsilon + \varepsilon + c_0^2\, \varepsilon\, (m-l) = d\, \varepsilon \end{aligned}$$

with $d = d_{m-l} = 2 + c_0^2\,(m-l)$. □

Lemma D.4. $S^* := \lim_{k\to\infty} S^{(k)} = 0$.

Proof. Let $\delta > 0$ be given and

$$(a^{(k)}_{p_k,q_k})^2 < (d_1 \cdot d_2 \cdots d_{n-2})^{-1} \delta/2$$

for all $k \ge k_0 = I(1,2)$, where we used (D.4) and our previous notation. By virtue of (D.10) this implies

$$S^{(k)}_{l,l+1} < (d_1 \cdot d_2 \cdots d_{n-2})^{-1} \delta$$

for $k = I(l, l+1)$ cocyclic with k_0 and all l such that $1 \le l \le n-1$. This means that (D.11) is fulfilled for *all* $f, g, h \ge k_0$ cocyclic with k_0 provided that $r = m - l = 1$ and $\varepsilon = (d_1 \cdots d_{n-2})^{-1}\delta$. Assume that

$$S^{(k)}_{l,l+r} < (d_r \cdot d_{r+1} \cdots d_{n-2})^{-1} \delta \tag{D.16}$$

holds for some positive integer $r < n-1$ with $k = I(l, l+r)$ cocyclic with k_0 and *all* l such that $1 \le l \le n - r$. Then Lemma D.3 implies

$$S^{(k+1)}_{l,l+r+1} < d_r \cdot (d_r \cdots d_{n-2})^{-1} \delta = (d_{r+1} \cdots d_{n-2})^{-1} \delta$$

which is (D.16) with $r+1$ replacing r. This completes our induction and shows $S^{(k_0+n-1)}_{1,n} < \delta$, where $k_0 + n - 1 = I(1, n)$. Since δ was arbitrary and $S^{(k_0+n-1)}_{1,n} = S^{(k_0+n-1)}$ the limit S^* must be zero. □

Lemma D.5. *For each $k \in \mathbb{N}$ there is a permutation $\pi_k = (\pi_k(i))$ of $1, \ldots, n$ such that*

$$\lim_{k\to\infty} |\lambda_{\pi_k(i)} - a^{(k)}_{ii}| = 0 \tag{D.17}$$

holds.

Proof. Let the eigenvalues λ_i of A, and therefore of A_k, be ordered increasingly, fix k, and denote by $D_k = (d_{ij}^{(k)}) \in \mathbb{R}^{n\times n}$ the diagonal matrix with $d_{ii}^{(k)} = a_{ii}^{(k)}$, $i = 1, \dots, n$. Let $\nu_k = (\nu_k(i))$ be a permutation of $1, \dots, n$ such that the entries $a_{\nu_k(i),\nu_k(i)}^{(k)}$ are ordered increasingly. Then Theorem 1.8.18 with $A = D_k$, $B = A_k$ and the Frobenius norm $\|\cdot\|_F$ implies

$$|\lambda_i - a_{\nu_k(i),\nu_k(i)}^{(k)}| \le \|A_k - D_k\|_F = (S^{(k)})^{1/2}, \quad i = 1, \dots, n. \tag{D.18}$$

If $\pi_k = (\pi_k(i))$ denotes the inverse permutation of ν_k, one can rewrite (D.18) as

$$|\lambda_{\pi_k(i)} - a_{ii}^{(k)}| \le (S^{(k)})^{1/2}, \quad i = 1, \dots, n. \tag{D.19}$$

Now (D.17) is proved by Lemma D.4. □

In order to terminate the proof of Theorem D.1 it remains to show that π_k in Lemma D.5 can be replaced by a permutation π which is independent of k at least for sufficiently large k. This is obviously true if all eigenvalues of A are identical. Otherwise let

$$d = \min_{\lambda_i \ne \lambda_j} |\lambda_i - \lambda_j|, \quad \varepsilon_k = (S^{(k)})^{1/2}, \tag{D.20}$$

and choose k_0 such that $\varepsilon_k < d/3$ holds for all $k \ge k_0$. Then by virtue of (D.19) every diagonal entry $a_{jj}^{(k)}$ of A_k is closest to exactly one number of the set $\Lambda = \{\lambda_i\}$. Denote this eigenvalue by $\lambda(k, j)$. By using the *set* Λ we emphasize that we no longer distinguish between identical eigenvalues. We show that $\lambda(k, j)$ can be chosen independently of k if $k \ge k_0$. To this end we remark that during the k-th Jacobi rotation in the diagonal of A_k only the entries $a_{p_k,p_k}^{(k)}$ and $a_{q_k,q_k}^{(k)}$ are affected. For simplicity we will mostly omit the index k. The formulae of Algorithm 7.3.1, $\varphi \in [-\pi/4, \pi/4]$, and (D.20) imply

$$|a'_{pp} - a_{pp}| = t\,|a_{pq}| \le |a_{pq}| \le \varepsilon_k.$$

From (D.19) and (D.20) we get

$$|\lambda(k, p) - a_{pp}| \le |\lambda_{\pi_k(p)} - a_{pp}| \le \varepsilon_k$$

and similarly

$$|\lambda(k+1, p) - a'_{pp}| \le \varepsilon_{k+1}.$$

Therefore,

$$\begin{aligned} |\lambda(k+1, p) - \lambda(k, p)| &\le |\lambda(k+1, p) - a'_{pp}| + |a'_{pp} - a_{pp}| + |a_{pp} - \lambda(k, p)| \\ &\le \varepsilon_{k+1} + 2\varepsilon_k < d, \end{aligned}$$

which, by virtue of (D.20), proves $\lambda(k+1, p) = \lambda(k, p)$. Similarly, we get $\lambda(k+1, q) = \lambda(k, q)$, so that $\lambda(k+1, i) = \lambda(k, i) =: \lambda(i)$ for $k \ge k_0$ and $i = 1, \dots, n$. Then $\lambda_{\pi_k(i)}$ in (D.17) can be replaced by $\lambda(i)$ whence $\lim_{k\to\infty} |\lambda(i) - a_{ii}^{(k)}| = 0$ follows. With $D = \operatorname{diag}(\lambda(1), \dots, \lambda(n))$ we obtain $\lim_{k\to\infty} A_k = D$, and since the eigenvalues of A_k are those of A including their multiplicity, D must be the Jordan normal form of A. The order of the diagonal entries depends on A because the choice of the rotational entries, Algorithm 7.3.1, and the angle φ there are unique. This finishes the proof of Theorem D.1. □

The convergence of the column cyclic Jacobi method can be proved analogously as was done in Forsythe, Henrici [99]. The speed of convergence is studied in Henrici [141].

Exercises

Ex. D.1. Prove (D.2).

E The CORDIC algorithm

The CORDIC algorithm (Coordinate Rotation Digital Computing) aims to compute elementary functions by a shift-and-add strategy, which means that only shifts of a bit sequence in the arithmetic logic unit (ALU) and additions of binary numbers are required in order to approximate values of selected elementary functions such as $\sin x$, $\cos x$, $\arctan x$, $\sinh x$, $\cosh x$, $\operatorname{artanh}(y/x)$, from which additional ones can be easily derived, for instance $\ln x = 2 \operatorname{artanh} \frac{x-1}{x+1}$, $e^x = \cosh x + \sinh x$. We will explain the algorithm only for computing $\sin x$, $\cos x$, and refer to Muller [240] for modifications in order to obtain other functions from $\mathbb{F}$. We proceed in three steps:

(1) Reduce the argument x by multiples of π such that $\theta := x - k\pi \in [-\frac{\pi}{2}, \frac{\pi}{2}]$, $k \in \mathbb{Z}$, where k is chosen appropriately.
(2) Represent $\theta \in [-\frac{\pi}{2}, \frac{\pi}{2}]$ as $\theta = \sum_{k=0}^{\infty} d_k w_k$, where $d_k = \pm 1$, $w_k = \arctan 2^{-k}$. Here we assume that w_k is taken from a table available for the computer.
(3) Apply the CORDIC algorithm, whose basic form is an iterated application of dilations and rotations.

In view of step (2) we need the following theorem.

Theorem E.1. *Let (w_k) be a monotonously decreasing sequence of positive real numbers for which $s := \sum_{k=0}^{\infty} w_k < \infty$ exists and which satisfies*

$$w_k \le \sum_{i=k+1}^{\infty} w_i, \qquad k \in \mathbb{N}_0. \tag{E.1}$$

Let $t \in [-s, s]$ be fixed and define (t_k) recursively by

$$\begin{cases} t_0 := 0 \\ t_{k+1} := t_k + d_k w_k \end{cases} \qquad \text{with } d_k := \begin{cases} 1, & \text{if } t_k \le t \\ -1, & \text{if } t_k > t \end{cases}. \tag{E.2}$$

Then

$$t = \lim_{k\to\infty} t_k = \sum_{k=0}^{\infty} d_k w_k \tag{E.3}$$

holds.

Proof. By virtue of (E.2) we have $t_k = \sum_{i=0}^{k-1} d_i w_i$, $k \in \mathbb{N}$. We show by induction

$$-\sum_{i=k}^{\infty} w_i \le t - t_k \le \sum_{i=k}^{\infty} w_i \tag{E.4}$$

which implies (E.3).

For $k = 0$ we have $t_0 = 0$, hence $-s \le t - t_0 = t \le s$ follows from our assumption on t.

DOI 10.1515/9783110499469-016

Let (E.4) hold for some k. If $d_k = -1$, then $t_k > t$, $t - t_k < 0$, and

$$t - t_{k+1} = t - (t_k - w_k) < w_k \le \sum_{i=k+1}^{\infty} w_i$$

holds, where the last inequality follows from (E.1). Furthermore,

$$t - (t_k - w_k) \ge -\sum_{i=k}^{\infty} w_i + w_k = -\sum_{i=k+1}^{\infty} w_i$$

by our induction hypothesis. This proves (E.4) in the case $d_k = -1$. The case $d_k = 1$ can be shown analogously. □

Remark E.1.

(a) The choices $w_k = \frac{y}{2^k}$ $(y > 0)$, $w_k = \ln(1 + 2^{-k})$, and $w_k = \arctan 2^{-k}$ fulfill the assumptions of Theorem E.1.
We prove this only for $w_k = \arctan 2^{-k}$ leaving the proof for the other two possibilities to the reader in Exercise E.1.
We start with two properties on $\arctan x$ (principal value):

$$\begin{aligned} &\arctan x \le x && \text{for } x \ge 0, \\ &\arctan(x + y) \le \arctan x + \arctan y && \text{for } x, y \ge 0. \end{aligned}$$

The first inequality can be seen by means of the function $g(x) := \arctan x - x$ which satisfies $g'(x) = -x^2/(1 + x^2) \le 0$ and is therefore monotonously decreasing. This implies $-\infty = \lim_{x\to\infty} g(x) \le g(x) \le g(0) = 0$ for $x \ge 0$.
For the second inequality we use the function

$$h(x) := \arctan(x + y) - \arctan x - \arctan y \quad \text{for fixed } y \ge 0,$$

whose derivative

$$h'(x) = \frac{1}{1 + (x + y)^2} - \frac{1}{1 + x^2}$$

is nonpositive. Hence $h(x) \le h(0) = 0$ holds for $x \ge 0$.
Obviously (w_k) is a monotonously decreasing sequence of positive numbers. Furthermore,

$$s := \sum_{k=0}^{\infty} w_k = \sum_{k=0}^{\infty} \arctan 2^{-k} \le \sum_{k=0}^{\infty} 2^{-k} = 2,$$

in particular, s exists. Finally,

$$w_k = \arctan(2^{-k}) = \arctan\left(\sum_{i=k+1}^{\infty} 2^{-i} \right) \le \sum_{i=k+1}^{\infty} \arctan(2^{-i}) = \sum_{i=k+1}^{\infty} w_i.$$

(b) Based on the error estimate

$$0 \le s - \sum_{i=0}^{k} w_i = \sum_{i=k+1}^{\infty} w_i \le \sum_{i=k+1}^{\infty} 2^{-i} = 2^{-k}$$

the sum $s = \sum_{k=0}^{\infty} w_k$, $w_k = \arctan(2^{-k})$ can be approximated arbitrarily well. One obtains $s = 1.743\,286\,620\,472\,340\,003\,5\ldots.$

CORDIC algorithm – basic version (Volder [357])

Let $x_0, y_0, z_0 \in R$ be given. Iterate according to

$$\left.\begin{aligned} d_k &:= \begin{cases} 1 & \text{if } z_k \ge 0 \\ -1 & \text{otherwise} \end{cases} \\ x_{k+1} &:= x_k - d_k y_k 2^{-k} \\ y_{k+1} &:= y_k + d_k x_k 2^{-k} \\ z_{k+1} &:= z_k - d_k \arctan 2^{-k} \end{aligned}\right\} \quad k = 0, 1, \ldots \tag{E.5}$$

Theorem E.2. *Let (x_k, y_k, z_k) be constructed by the basic version of the CORDIC algorithm. Assume $|z_0| \le s = \sum_{k=0}^{\infty} \arctan 2^{-k}$. Then*

$$\lim_{k\to\infty} \begin{pmatrix} x_k \\ y_k \\ z_k \end{pmatrix} = K \cdot \begin{pmatrix} x_0 \cos z_0 - y_0 \sin z_0 \\ x_0 \sin z_0 + y_0 \cos z_0 \\ 0 \end{pmatrix}$$

holds with

$$K := \prod_{k=0}^{\infty} \sqrt{1 + 2^{-2k}} = 1.646\,760\,258\,121\,065\,648\,366\,051\ldots.$$

The choice

$$x_0 := \frac{1}{K} = 0.607\,252\,935\,008\,881\,256\,169\,4\ldots, \quad y_0 := 0, \quad z_0 := \theta, \quad |\theta| \le s,$$

implies

$$\lim_{k\to\infty} \begin{pmatrix} x_k \\ y_k \\ z_k \end{pmatrix} = \begin{pmatrix} \cos\theta \\ \sin\theta \\ 0 \end{pmatrix}.$$

Proof. Apply Theorem E.1 to $w_k = \arctan 2^{-k} \in (0, \frac{\pi}{4}]$ and $t = z_0$, taking into account Remark E.1. For

$$\begin{cases} t_0 := 0 \\ t_{k+1} := t_k + d_k w_k \end{cases} \quad \text{with} \quad d_k := \begin{cases} 1, & \text{if } t_k \le z_0 \\ -1, & \text{otherwise} \end{cases} \tag{E.6}$$

one gets

$$\lim_{k\to\infty} t_k = \sum_{k=0}^{\infty} d_k w_k = z_0. \tag{E.7}$$

Define

$$\begin{aligned} \begin{pmatrix} x_{k+1} \\ y_{k+1} \end{pmatrix} &:= \frac{1}{\cos w_k} \begin{pmatrix} \cos(d_k w_k) & -\sin(d_k w_k) \\ \sin(d_k w_k) & \cos(d_k w_k) \end{pmatrix} \begin{pmatrix} x_k \\ y_k \end{pmatrix} \\ &= \frac{1}{\cos w_k} \begin{pmatrix} \cos w_k & -d_k \sin w_k \\ d_k \sin w_k & \cos w_k \end{pmatrix} \begin{pmatrix} x_k \\ y_k \end{pmatrix}. \end{aligned}$$

Here the matrix effects a counterclockwise rotation of $(x_k, y_k)^T$ with the angle $d_k w_k$, and the factor $1/\cos w_k > 1$ in front means a dilation since $w_k \in (0, \pi/4]$. From the definition of w_k one gets $\sin w_k = 2^{-k} \cos w_k$, hence

$$\begin{pmatrix} x_{k+1} \\ y_{k+1} \end{pmatrix} = \begin{pmatrix} 1 & -d_k 2^{-k} \\ d_k 2^{-k} & 1 \end{pmatrix} \begin{pmatrix} x_k \\ y_k \end{pmatrix}$$

follows. Since

$$\begin{pmatrix} \cos\alpha & -\sin\alpha \\ \sin\alpha & \cos\alpha \end{pmatrix} \cdot \begin{pmatrix} \cos\beta & -\sin\beta \\ \sin\beta & \cos\beta \end{pmatrix} = \begin{pmatrix} \cos(\alpha+\beta) & -\sin(\alpha+\beta) \\ \sin(\alpha+\beta) & \cos(\alpha+\beta) \end{pmatrix}$$

one finally obtains

$$\lim_{k\to\infty} \begin{pmatrix} x_k \\ y_k \end{pmatrix} = \left(\prod_{k=0}^{\infty} \frac{1}{\cos w_k} \right) \begin{pmatrix} \cos z_0 & -\sin z_0 \\ \sin z_0 & \cos z_0 \end{pmatrix} \begin{pmatrix} x_0 \\ y_0 \end{pmatrix}.$$

Now

$$1 - \cos^2 w_k = \sin^2 w_k = 2^{-2k} \cos^2 w_k,$$

whence

$$\frac{1}{\cos w_k} = \sqrt{1 + 2^{-2k}} \quad \text{and} \quad \prod_{k=0}^{\infty} \frac{1}{\cos w_k} = K$$

follow. Notice that the last infinite product exists since

$$u_m := \ln\left(\prod_{k=0}^{m} \sqrt{1 + 2^{-2k}} \right) = \sum_{k=0}^{m} \frac{1}{2} \ln(1 + 2^{-2k})$$

is monotonously increasing and satisfies

$$u_m \le \frac{1}{2} \sum_{k=0}^{m} 2^{-2k} \le \frac{1}{2} \sum_{k=0}^{\infty} 2^{-2k} = \frac{2}{3}$$

because of $\ln(1 + x) \le x$ for $x \ge 0$; cf. Exercise E.2. Hence (u_m) is convergent to some value u^* and

$$\lim_{m\to\infty} e^{u_m} = e^{u^*} = \prod_{k=0}^{\infty} \sqrt{1 + 2^{-2k}} = K$$

exists. Now let $z_k := z_0 - t_k$, $k = 1, 2, \ldots$. Then

$$\begin{cases} z_0 = z_0 - t_0 & t_0 \text{ given as in (E.6)} \\ z_{k+1} = z_0 - t_{k+1} = z_0 - t_k - d_k w_k = z_k - d_k w_k \end{cases}$$

holds with

$$d_k := \begin{cases} 1 & \text{if } z_k = z_0 - t_k \ge 0 \\ -1 & \text{otherwise} \end{cases}$$

and effects that $z_k = z_0 - \sum_{i=0}^{k} d_i w_i$ converges to $z_0 - z_0 = 0$ because of (E.7). □

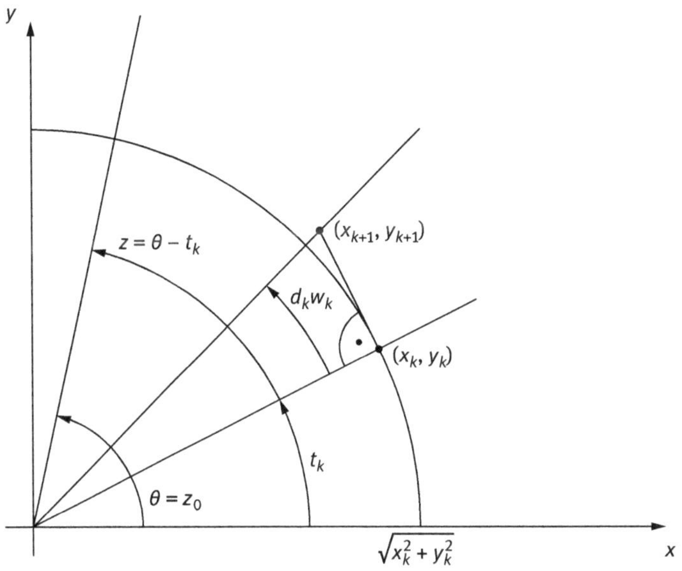

Fig. E.1: An iteration step of the CORDIC algorithm.

Figure E.1 may illustrate the iterate (x_{k+1}, y_{k+1}) if $z_0 = \theta$. Notice that

$$\begin{pmatrix} x_{k+1} - x_k \\ y_{k+1} - y_k \end{pmatrix}^T \begin{pmatrix} x_k \\ y_k \end{pmatrix} = \begin{pmatrix} -d_k 2^{-k} y_k \\ d_k 2^{-k} x_k \end{pmatrix}^T \begin{pmatrix} x_k \\ y_k \end{pmatrix} = 0,$$

i.e., both vectors are orthogonal.

In practice, the basic version of the CORDIC algorithm must be modified in order to be efficient – see Muller [240] for details. It was implemented in the HP35 pocket calculator, and in the Intel processors 8087 up to 80468. Nowadays there are faster algorithms for computing elementary functions; see again Muller [240].

Exercises

Ex. E.1. Prove Remark E.1 (a) for $w_k = y/2^k$ $(y > 0)$ and for $w_k = \ln(1 + 2^{-k})$.

Ex. E.2. Prove $\ln(1 + x) \leq x$ for $x \geq 0$.

F The symmetric solution set – a proof of Theorem 5.2.6

We prove here the missing converse direction '$\Leftarrow$' of Theorem 5.2.6 using the notation of the theorem and its proof '$\Rightarrow$' in Section 5.2. To this end we need some preparations which we will present first.

Definition F.1. Let $p, \tilde{p}, q, q', \tilde{q}, \tilde{q}' \in \{0, 1\}^n$. Then we define

$$\overline{p} = e - p, \quad \overline{q} = e - q, \quad q' = (0, 0, q_3, \ldots, q_n)^T, \quad \tilde{q}' = (0, 0, \tilde{q}_3, \ldots, \tilde{q}_n)^T,$$

$$\begin{aligned}
p_c &= p \mathbin{.*} \tilde{p}, & p_r &= p - p_c, & q_c &= q \mathbin{.*} \tilde{q}, & q_r &= q - q_c,\\
\tilde{p}_c &= \tilde{p} \mathbin{.*} p, & \tilde{p}_r &= \tilde{p} - \tilde{p}_c, & \tilde{q}_c &= \tilde{q} \mathbin{.*} q, & \tilde{q}_r &= \tilde{q} - \tilde{q}_c,\\
& & & & q'_c &= q' \mathbin{.*} \tilde{q}', & q'_r &= q' - q'_c,\\
& & & & \tilde{q}'_c &= \tilde{q}' \mathbin{.*} q', & \tilde{q}'_r &= \tilde{q}' - \tilde{q}'_c,\\
p_C &= p \mathbin{.*} \tilde{q}, & p_R &= p - p_C, & q_C &= q \mathbin{.*} \tilde{p}, & q_R &= q - q_C,\\
\tilde{p}_C &= \tilde{p} \mathbin{.*} q, & \tilde{p}_R &= \tilde{p} - \tilde{p}_C, & \tilde{q}_C &= \tilde{q} \mathbin{.*} p, & \tilde{q}_R &= \tilde{q} - \tilde{q}_C,\\
& & & & q'_C &= q' \mathbin{.*} \tilde{p}, & q'_R &= q' - q'_C,\\
& & & & \tilde{q}'_C &= \tilde{q}' \mathbin{.*} p, & \tilde{q}'_R &= \tilde{q}' - \tilde{q}'_C,
\end{aligned}$$

where $.*$ denotes the Hadamard product as in Section 2.1.

The subscripts remind us of '*common* components', i.e., components that are one for both operands simultaneously, and '*remaining* components'.

Definition F.2. Let $[A] = [A]^T \in \mathbb{IR}^{n\times n}$, $[b] \in \mathbb{IR}^n$, $x \in \mathbb{R}^n$, $p, q \in \{0, 1\}^n$. Then we define

$$\begin{aligned}
\underline{L}_p^q &= x^T D_p(\underline{b} - \overline{A} D_{\overline{q}} x), & \dot{\underline{L}}_p^q &= -x^T D_p \overline{A} D_q x,\\
\overline{L}_p^q &= x^T D_p(\overline{b} - \underline{A} D_{\overline{q}} x), & \dot{\overline{L}}_p^q &= -x^T D_p \underline{A} D_q x.
\end{aligned}$$

The position of the bar in $\underline{L}_p^q, \overline{L}_p^q$ is the same as with $\underline{b}, \overline{b}$. Notice the missing bar at 'q' when defining $\dot{\underline{L}}_p^q, \dot{\overline{L}}_p^q$. If $p_i = q_j = 1$, the entry $\overline{a}_{ij}$ is missing when computing $\underline{L}_p^q$ while it is present when computing $\dot{\underline{L}}_p^q$.

DOI 10.1515/9783110499469-017

Lemma F.3. *With the assumptions and the notations in the Definitions* F.1 *and* F.2 *we get the following properties which (mostly) hold for q and $\overline{L}_p^q$ similarly.*

(a) $p_c = \tilde{p}_c, \quad p + \tilde{p}_r = \tilde{p} + p_r, \quad p_C = \tilde{q}_C, \quad p + \tilde{q}_R = \tilde{q} + p_R$.

(b) *Let $p = 0$, $q = e^{(i)}$, $x \ge 0$, $x_i > 0$. Then the i-th Oettli–Prager inequality in Theorem* 5.2.2 (d) *is equivalent to*

$$\left.\begin{aligned} \underline{L}_0^{e^{(i)}} &= 0 \le \overline{L}_{e^{(i)}}^0 = x_i\left(\overline{b}_i - \sum_{j=1}^n \underline{a}_{ij}x_j\right), \\ \underline{L}_{e^{(i)}}^0 &= x_i\left(\underline{b}_i - \sum_{j=1}^n \overline{a}_{ij}x_j\right) \le 0 = \overline{L}_0^{e^{(i)}}. \end{aligned}\right\} \tag{F.1}$$

(c) $\dot{\underline{L}}_p^q = \dot{\underline{L}}_q^p = \dot{\underline{L}}_{p_c}^q + \dot{\underline{L}}_{p_r}^q$, $(\le \dot{\overline{L}}_p^q$ *if* $x \ge 0)$.

(d) $\underline{L}_p^q = \underline{L}_p^{q+\tilde{q}_r} + \dot{\underline{L}}_{\tilde{q}_r}^p = \underline{L}_{p_c}^q + \underline{L}_{p_r}^q$, $(\le \overline{L}_p^q$ *if* $x \ge 0)$.

(e) $\underline{L}_{p_c}^q + \underline{L}_{\tilde{q}_r}^{\tilde{p}} = \underline{L}_{p_c}^{q+\tilde{q}_r} + \underline{L}_{\tilde{q}_r}^{\tilde{p}_r}$, $\underline{L}_{q_C}^p + \underline{L}_{q_C}^{\tilde{q}} = \underline{L}_{q_C}^{p+\tilde{q}_R} + \underline{L}_{q_C}^{\tilde{q}_C}$.

The proof of Lemma F.3 is based on simple calculations and can be left to the reader.

We are now ready to prove the converse direction '$\Leftarrow$' of Theorem 5.2.6. To this end we remark that the property '$x \in S$' is equivalent to the Oettli–Prager inequality $|\check{A}x - \check{b}| \le \mathrm{rad}([A])|x| + \mathrm{rad}([b])$ according to Theorem 5.2.2.

Proof '$\Leftarrow$' of Theorem 5.2.6. Assume that a given vector x is an element of S and the inequalities (5.2.16) hold for $0 \ne p \prec_{\mathrm{lex}} q$ with $p^Tq = 0$ (which implies $q \ne e$). Then these inequalities are valid for all vectors $p, q \in \{0, 1\}^n$ as we saw in the first part of the proof.

Without loss of generality we assume $x \ge 0$. Otherwise replace $[A]$, $[b]$, x by $[B] = D_x[A]D_x = [B]^T$, $[c] = D_x[b]$, $z = D_x x = |x| \ge 0$ with $D_x = \mathrm{diag}(\sigma_1, \ldots, \sigma_n) \in \mathbb{R}^{n\times n}$, where $\sigma_i = 1$ if $\mathrm{sign}(x_i) = 0$, and $\sigma_i = \mathrm{sign}(x_i)$ otherwise. Then $\mathrm{rad}([B]) = \mathrm{rad}([A])$, $\mathrm{rad}([c]) = \mathrm{rad}([b])$, $|\check{c} - \check{B}z| = |D_x(\check{b} - \check{A}x)| = |D_x r| = |r|$, $z^T(D_p - D_q)(\check{c} - \check{B}z) = x^TD_x(D_p - D_q)D_x r = x^T(D_p - D_q)r$. Therefore, the inequalities (5.2.14), (5.2.16) hold for $[B]$, $[c]$, z if they hold for $[A]$, $[b]$, x, and vice versa.

With $p_iq_i = 0$, $i = 1, \ldots, n$, we get $|D_p - D_q| = D_p + D_q$ in (5.2.16). Using (5.2.22) and the Definitions F.1 and F.2 a simple calculation shows that (5.2.16) is *equivalent* to

$$\underline{L}_p^q \le \overline{L}_q^p, \qquad \underline{L}_q^p \le \overline{L}_p^q \tag{F.2}$$

if $0 \ne p \prec_{\mathrm{lex}} q$ with $p^Tq = 0$. If $p^Tq \ne 0$, then (5.2.16) still *implies* (F.2) because of $|D_p - D_q| \le D_p + D_q$ and (5.2.22).

We now present the ideas for the remaining part of our proof: Successively for each index pair (i, j) with $i < j$ and $x_ix_j > 0$ we will replace the entries $[a]_{ij}$ and $[a]_{ji} = [a]_{ij}$ in $[A]$ simultaneously by some point intervals $[a_{ij}, a_{ij}]$, $[a_{ji}, a_{ji}]$ with $a_{ij} = a_{ji} \in [a]_{ij}$ so that the inequalities (5.2.14), (5.2.16) still hold. At the end of the complete replacement process we will apply Theorem 5.2.3 (a), (d) (= Oettli–Prager criterion) to the resulting final matrix $[A]^{\mathrm{new}} = ([A]^{\mathrm{new}})^T \subseteq [A]$ and to $[b]$ in order to obtain a (possibly unsymmetric) matrix $\tilde{A} \in [A]^{\mathrm{new}}$ and a vector $\tilde{b} \in [b]$ with $\tilde{A}x = \tilde{b}$. From $\tilde{A}$ we can construct at once a symmetric matrix $\tilde{A}^{\mathrm{sym}} \in [A]$ which satisfies

$$\tilde{A}^{\mathrm{sym}}x = \tilde{A}x = \tilde{b} \tag{F.3}$$

and thus implies $x \in S_{\mathrm{sym}}$.

For ease of notation we show how

$$\underline{a}_{12},\ \underline{a}_{21},\ \overline{a}_{12},\ \overline{a}_{21} \tag{F.4}$$

and $[a]_{12} = [a]_{21}$ can be replaced assuming $x_1x_2 > 0$. (If $x_1x_2 = 0$, then $[a]_{12} = [a]_{21}$ remains unchanged for the moment, and another entry is considered.) By virtue of Lemma F.3 (b), we can replace the first two Oettli–Prager inequalities equivalently by those in (F.1) for $i = 1$ and $i = 2$. Therefore, we will enlarge the choice of p and q in Theorem 5.2.6 by the cases $p = 0,\ q = e^{(1)}$ and $p = 0,\ q = e^{(2)}$ without mentioning it further. This extension avoids the study of special subcases. Now we consider only those inequalities (5.2.16) which contain at least one entry of (F.4) explicitly. Taking into account (5.2.15) there are only three cases for this:

Case 1: $q = (1, 0, *)^T,\ p = (0, 0, *)^T$
Case 2: $q = (0, 1, *)^T,\ p = (0, 0, *)^T$
Case 3: $q = (1, 1, *)^T,\ p = (0, 0, *)^T$

Here '$*$' replaces the remaining components of p and q. Notice that for $q = (1, 0, *)^T$, $p = (0, 1, *)^T$ the inequalities (F.2) do not contain one of the entries in (F.4) explicitly. Furthermore, the entries $\underline{a}_{12} = \underline{a}_{21}$ and $\overline{a}_{12} = \overline{a}_{21}$ cannot appear in one and the same of the two inequalities (F.2) because of $p^Tq = 0$.

Our first step consists of isolating the entries (F.4) in (F.2). With the notation of Definition F.1 and $\underline{a}_{12} = \underline{a}_{21}$, $\overline{a}_{12} = \overline{a}_{21}$ we get

$$\underline{a}_{12}x_1x_2 \le \overline{L}_{e^{(1)}}^{e^{(2)}+p} + \overline{L}_{q'}^{p} - \underline{L}_{p}^{q} \qquad \text{Case 1} \tag{F.5}$$

$$\underline{a}_{12}x_1x_2 \le \overline{L}_{e^{(2)}}^{e^{(1)}+p} + \overline{L}_{q'}^{p} - \underline{L}_{p}^{q} \qquad \text{Case 2} \tag{F.6}$$

$$\underline{a}_{12}x_1x_2 \le (\overline{L}_{e^{(1)}}^{e^{(2)}+p} + \overline{L}_{e^{(2)}}^{e^{(1)}+p} + \overline{L}_{q'}^{p} - \underline{L}_{p}^{q})/2 \qquad \text{Case 3} \tag{F.7}$$

and

$$\underline{L}_{e^{(1)}}^{e^{(2)}+p} + \underline{L}_{q'}^{p} - \overline{L}_{p}^{q} \le \overline{a}_{12}x_1x_2 \qquad \text{Case 1} \tag{F.8}$$

$$\underline{L}_{e^{(2)}}^{e^{(1)}+p} + \underline{L}_{q'}^{p} - \overline{L}_{p}^{q} \le \overline{a}_{12}x_1x_2 \qquad \text{Case 2} \tag{F.9}$$

$$(\underline{L}_{e^{(1)}}^{e^{(2)}+p} + \underline{L}_{e^{(2)}}^{e^{(1)}+p} + \underline{L}_{q'}^{p} - \overline{L}_{p}^{q})/2 \le \overline{a}_{12}x_1x_2 \qquad \text{Case 3} \tag{F.10}$$

If we can show (which we will do at the end of the proof) that each left-hand side of (F.8)–(F.10) is less than or equal to each right-hand side of (F.5)–(F.7) (with another admissible choice of p, q), then the same holds for the maximum M_ℓ of all such left-hand sides as compared with the minimum m_r of all such right-hand sides. If $m_r > \overline{a}_{12}x_1x_2$, we redefine it by $\overline{a}_{12}x_1x_2$, knowing by (F.8)–(F.10) that $M_\ell \le \overline{a}_{12}x_1x_2$. Similarly, if

$M_\ell < \underline{a}_{12}x_1x_2$, we redefine it by $\underline{a}_{12}x_1x_2$. Now we choose any number from $[M_\ell, m_r]$. Obviously, it is representable as $a_{12}x_1x_2$ with some number $a_{12} \in [a]_{12}$. The inequalities (F.5)–(F.10) hold with a_{12} in place of $\underline{a}_{12}$, $\overline{a}_{12}$, and so do the inequalities (F.2) if we define $a_{21} = a_{12}$ and replace the entries (F.4) correspondingly. Replacing the entries $[a]_{12}$, $[a]_{21}$ in $[A]$ by $a_{12} = a_{21}$ results in a matrix $[A]'$ for which the assumptions of Theorem 5.2.6 are also satisfied. It forms the starting point of our next replacement. Repeating this process for all entries $[a]_{ij}$, $i<j$, with $x_ix_j > 0$ we finally end up with the matrix $[A]^{\text{new}}$ which we already mentioned at the beginning. It has degenerate symmetric entries $a_{ij} = a_{ji} \in [a]_{ij}$ whenever $x_ix_j > 0$ is true, and it satisfies the inequalities (5.2.14), (5.2.16). Therefore, a matrix $\tilde{A} = (\tilde{a}_{ij}) \in [A]^{\text{new}}$ and a vector $\tilde{b} \in [b]$ exist with $\tilde{A}x = \tilde{b}$. Trivially, $\tilde{a}_{ij} = \tilde{a}_{ji} = a_{ij}$ if $x_ix_j > 0$. If $x_i = 0$ the value of $\tilde{a}_{ji}$ does not matter in the product $\tilde{A}x$. Therefore, we may replace it by $\tilde{a}_{ij}$, a step towards symmetry. Similarly, if $x_j = 0$, we replace $\tilde{a}_{ij}$ by $\tilde{a}_{ji}$, ending up with a symmetric matrix $\tilde{A}^{\text{sym}} \in [A]$ which satisfies (F.3) and finishes the proof.

It remains to show that each left-hand side of (F.8)–(F.10) is less than or equal to each right-hand side of (F.5)–(F.7). To this end we have to combine each right-hand side of (F.5)–(F.7) with each left-hand side of (F.8)–(F.10) which leads to nine combinations.

(1) Case 1 vs. Case 1: Let $\tilde{p}$, $\tilde{q}$ and p, q be chosen according to Case 1, independently of each other. We have to show that

$$\underline{L}_{e^{(1)}}^{e^{(2)}+\tilde{p}} + \underline{L}_{\tilde{q}'}^{\tilde{p}} - \overline{L}_{\tilde{p}}^{\tilde{q}} \le \overline{L}_{e^{(1)}}^{e^{(2)}+p} + \overline{L}_{q'}^{p} - \underline{L}_{p}^{q} \tag{F.11}$$

holds. With the notation of Definition F.1 and with Lemma F.3 we get

$$\begin{aligned}
\underline{L}_{e^{(1)}}^{e^{(2)}+\tilde{p}} + \underline{L}_{\tilde{q}'}^{\tilde{p}} + \underline{L}_{p}^{q} &= \underline{L}_{e^{(1)}}^{e^{(2)}+\tilde{p}} + \underline{L}_{\tilde{q}'}^{\tilde{p}} + \underline{L}_{p_c}^{q} + \underline{L}_{p_r}^{e^{(1)}+q'} \\
&= \underline{L}_{e^{(1)}}^{e^{(2)}+\tilde{p}+p_r} + \underline{L}_{\tilde{q}'}^{\tilde{p}} + \underline{L}_{p_c}^{q} + \underline{L}_{p_r}^{q'} \\
&= \underline{L}_{e^{(1)}}^{e^{(2)}+\tilde{p}+p_r} + \underline{L}_{\tilde{q}'_c}^{\tilde{p}} + (\underline{L}_{\tilde{q}'_r}^{\tilde{p}} + \underline{L}_{p_c}^{q}) + \underline{L}_{p_r}^{q'} \\
&= \underline{L}_{e^{(1)}}^{e^{(2)}+\tilde{p}+p_r} + \underline{L}_{\tilde{q}'_c}^{\tilde{p}} + (\underline{L}_{\tilde{q}'_r}^{\tilde{p}_r} + \underline{L}_{p_c}^{q+\tilde{q}'_r}) + \underline{L}_{p_r}^{q'} \\
&= \underline{L}_{e^{(1)}}^{e^{(2)}+\tilde{p}+p_r} + \underline{L}_{\tilde{q}'_c}^{\tilde{p}+p_r} + \underline{L}_{\tilde{q}'_r}^{\tilde{p}_r} + \underline{L}_{p_c}^{q+\tilde{q}'_r} + \underline{L}_{p_r}^{q'_r}.
\end{aligned}$$

Similarly we obtain

$$\begin{aligned}
\overline{L}_{e^{(1)}}^{e^{(2)}+p} + \overline{L}_{q'}^{p} + \overline{L}_{\tilde{p}}^{\tilde{q}} &= \overline{L}_{e^{(1)}}^{e^{(2)}+p+\tilde{p}_r} + \overline{L}_{q'_c}^{p+\tilde{p}_r} + \overline{L}_{q'_r}^{p_r} + \overline{L}_{\tilde{p}_c}^{\tilde{q}+q'_r} + \overline{L}_{\tilde{p}_r}^{\tilde{q}'_r} \\
&= \overline{L}_{e^{(1)}}^{e^{(2)}+\tilde{p}+p_r} + \overline{L}_{\tilde{q}'_c}^{\tilde{p}+p_r} + \overline{L}_{\tilde{p}_r}^{\tilde{q}'_r} + \overline{L}_{p_c}^{q+\tilde{q}'_r} + \overline{L}_{q'_r}^{p_r}.
\end{aligned}$$

For the last formula we used the equality $q'_r = q_r$, $\tilde{q}'_r = \tilde{q}_r$, which holds by virtue of the particular form of q and $\tilde{q}$ in Case 1.

Comparing both final expressions and using (F.2) (with $(p, q) = (\tilde{p}_r, \tilde{q}'_r)$, and $(p, q) = (p_r, q'_r)$, respectively) and Lemma F.3 (d) proves (F.11).

(2) Case 2 vs. Case 2: is proved similarly.

(3) Case 1 vs. Case 2: Let $\tilde{p}$, $\tilde{q}$ be chosen according to Case 1, and let p, q be chosen according to Case 2. We have to show that

$$\underline{L}_{e^{(1)}}^{e^{(2)}+\tilde{p}} + \underline{L}_{\tilde{q}'}^{\tilde{p}} - \overline{L}_{\tilde{p}}^{\tilde{q}} \le \overline{L}_{e^{(2)}}^{e^{(1)}+p} + \overline{L}_{q'}^{p} - \underline{L}_{p}^{q} \tag{F.12}$$

holds. With $q_R = e^{(2)} + q'_R$, $\tilde{q}_R = e^{(1)} + \tilde{q}'_R$ and $p_C = \tilde{q}_C = \tilde{q}'_C$ we obtain

$$\begin{aligned}
\underline{L}_{e^{(1)}}^{e^{(2)}+\tilde{p}} + \underline{L}_{\tilde{q}'}^{\tilde{p}} + \underline{L}_{p}^{q} &= \underline{L}_{e^{(1)}}^{e^{(2)}+\tilde{p}} + \underline{L}_{\tilde{q}'_C}^{\tilde{p}} + \underline{L}_{\tilde{q}'_R}^{\tilde{p}} + \underline{L}_{p_C}^{q} + \underline{L}_{p_R}^{q} \\
&= (\underline{L}_{e^{(1)}}^{e^{(2)}+\tilde{p}+q'_R} + \dot{\underline{L}}_{e^{(1)}}^{q'_R}) + (\underline{L}_{p_C}^{\tilde{p}} + \underline{L}_{p_C}^{q}) + (\underline{L}_{\tilde{q}'_R}^{\tilde{p}+q'_R} + \dot{\underline{L}}_{\tilde{q}'_R}^{q'_R}) + (\underline{L}_{p_R}^{q+\tilde{p}_R} + \dot{\underline{L}}_{p_R}^{\tilde{p}_R}) \\
&= \underline{L}_{e^{(1)}}^{e^{(2)}+\tilde{p}+q'_R} + (\underline{L}_{p_C}^{\tilde{p}_C} + \underline{L}_{p_C}^{q+\tilde{p}_R}) + (\underline{L}_{\tilde{q}'_R}^{e^{(2)}+\tilde{p}+q'_R} + \dot{\underline{L}}_{e^{(2)}}^{\tilde{q}'_R} + \underline{L}_{p_R}^{q+\tilde{p}_R}) \\
&\quad + \dot{\underline{L}}_{e^{(1)}}^{q'_R} + \dot{\underline{L}}_{\tilde{q}'_R}^{q'_R} + \dot{\underline{L}}_{p_R}^{\tilde{p}_R} \\
&= \underline{L}_{\tilde{q}+p_R}^{q+\tilde{p}_R} + \underline{L}_{p_C}^{\tilde{p}_C} + \dot{\underline{L}}_{e^{(1)}}^{q'_R} + \dot{\underline{L}}_{e^{(2)}}^{\tilde{q}'_R} + \dot{\underline{L}}_{\tilde{q}'_R}^{q'_R} + \dot{\underline{L}}_{p_R}^{\tilde{p}_R}.
\end{aligned}$$

Similarly we obtain

$$\begin{aligned}
\overline{L}_{e^{(2)}}^{e^{(1)}+p} + \overline{L}_{q'}^{p} + \overline{L}_{\tilde{p}}^{\tilde{q}} &= \overline{L}_{q+\tilde{p}_R}^{\tilde{q}+p_R} + \overline{L}_{\tilde{p}_C}^{p_C} + \dot{\overline{L}}_{e^{(2)}}^{\tilde{q}'_R} + \dot{\overline{L}}_{e^{(1)}}^{q'_R} + \dot{\overline{L}}_{q'_R}^{\tilde{q}'_R} + \dot{\overline{L}}_{\tilde{p}_R}^{p_R} \\
&= \overline{L}_{q+\tilde{p}_R}^{\tilde{q}+p_R} + \overline{L}_{\tilde{p}_C}^{p_C} + \dot{\overline{L}}_{e^{(1)}}^{q'_R} + \dot{\overline{L}}_{e^{(2)}}^{\tilde{q}'_R} + \dot{\overline{L}}_{\tilde{q}'_R}^{q'_R} + \dot{\overline{L}}_{p_R}^{\tilde{p}_R},
\end{aligned}$$

which proves (F.12) as above.

(4) Case 3 vs. Case 3: Let $\tilde{p}$, $\tilde{q}$ and p, q be chosen according to Case 3. We have to show that

$$\underline{L}_{e^{(1)}}^{e^{(2)}+\tilde{p}} + \underline{L}_{e^{(2)}}^{e^{(1)}+\tilde{p}} + \underline{L}_{\tilde{q}'}^{\tilde{p}} - \overline{L}_{\tilde{p}}^{\tilde{q}} \le \overline{L}_{e^{(1)}}^{e^{(2)}+p} + \overline{L}_{e^{(2)}}^{e^{(1)}+p} + \overline{L}_{q'}^{p} - \underline{L}_{p}^{q} \tag{F.13}$$

holds. With $\tilde{q}_C = \tilde{q}'_C = p_C$ we obtain

$$\begin{aligned}
&\underline{L}_{e^{(1)}}^{e^{(2)}+\tilde{p}} + \underline{L}_{e^{(2)}}^{e^{(1)}+\tilde{p}} + \underline{L}_{\tilde{q}'}^{\tilde{p}} + \underline{L}_{p}^{q} \\
&\quad = (\underline{L}_{e^{(1)}}^{q+\tilde{p}_R} + \dot{\underline{L}}_{e^{(1)}}^{e^{(1)}+q'_R}) + (\underline{L}_{e^{(2)}}^{q+\tilde{p}_R} + \dot{\underline{L}}_{e^{(2)}}^{e^{(2)}+q'_R}) + (\underline{L}_{\tilde{q}'_C}^{\tilde{p}} + \underline{L}_{\tilde{q}'_R}^{\tilde{p}}) + (\underline{L}_{p_C}^{q} + \underline{L}_{p_R}^{q}) \\
&\quad = \underline{L}_{e^{(1)}+e^{(2)}}^{q+\tilde{p}_R} + (\underline{L}_{p_C}^{q+\tilde{p}_R} + \underline{L}_{p_C}^{\tilde{p}_C}) + (\underline{L}_{\tilde{q}'_R}^{\tilde{p}+q_R} + \dot{\underline{L}}_{\tilde{q}'_R}^{q_R}) + (\underline{L}_{p_R}^{q+\tilde{p}_R} + \dot{\underline{L}}_{p_R}^{\tilde{p}_R}) + \dot{\underline{L}}_{e^{(1)}}^{e^{(1)}+q'_R} + \dot{\underline{L}}_{e^{(2)}}^{e^{(2)}+q'_R} \\
&\quad = \underline{L}_{\tilde{q}+p_R}^{q+\tilde{p}_R} + \underline{L}_{p_C}^{\tilde{p}_C} + \dot{\underline{L}}_{e^{(1)}}^{e^{(1)}+q'_R} + \dot{\underline{L}}_{e^{(2)}}^{e^{(2)}+q'_R} + \dot{\underline{L}}_{\tilde{q}'_R}^{q_R} + \dot{\underline{L}}_{p_R}^{\tilde{p}_R} \\
&\quad = \underline{L}_{\tilde{q}+p_R}^{q+\tilde{p}_R} + \underline{L}_{p_C}^{\tilde{p}_C} + (\dot{\underline{L}}_{e^{(1)}}^{e^{(1)}} + \dot{\underline{L}}_{e^{(2)}}^{e^{(2)}} + \dot{\underline{L}}_{q'_R}^{e^{(1)}+e^{(2)}}) + (\dot{\underline{L}}_{\tilde{q}'_R}^{e^{(1)}+e^{(2)}} + \dot{\underline{L}}_{\tilde{q}'_R}^{q'_R}) + \dot{\underline{L}}_{p_R}^{\tilde{p}_R}.
\end{aligned}$$

Similarly we obtain

$$\begin{aligned}
&\overline{L}_{e^{(1)}}^{e^{(2)}+p} + \overline{L}_{e^{(2)}}^{e^{(1)}+p} + \overline{L}_{q'}^{p} + \overline{L}_{\tilde{p}}^{\tilde{q}} \\
&\quad = \overline{L}_{q+\tilde{p}_R}^{\tilde{q}+p_R} + \overline{L}_{\tilde{p}_C}^{p_C} + (\dot{\overline{L}}_{e^{(1)}}^{e^{(1)}} + \dot{\overline{L}}_{e^{(2)}}^{e^{(2)}} + \dot{\overline{L}}_{\tilde{q}'_R}^{e^{(1)}+e^{(2)}}) + (\dot{\overline{L}}_{q'_R}^{e^{(1)}+e^{(2)}} + \dot{\overline{L}}_{q'_R}^{\tilde{q}'_R}) + \dot{\overline{L}}_{\tilde{p}_R}^{p_R},
\end{aligned}$$

which proves (F.13) as above.

(5) Case 1 vs. Case 3: Let $\tilde{p}$, $\tilde{q}$ be chosen according to Case 1, and let p, q be chosen according to Case 3. We have to show that

$$2\underline{L}_{e^{(1)}}^{e^{(2)}+\tilde{p}} + 2\underline{L}_{\tilde{q}'}^{\tilde{p}} - 2\overline{L}_{\tilde{p}}^{\tilde{q}} \le \overline{L}_{e^{(1)}}^{e^{(2)}+p} + \overline{L}_{e^{(2)}}^{e^{(1)}+p} + \overline{L}_{q'}^{p} - \underline{L}_{p}^{q} \tag{F.14}$$

holds. With $\tilde{q}_r = \tilde{q}_r'$, $q = e^{(1)} + (e^{(2)} + q')$ we obtain

$$\begin{aligned}
2\underline{L}_{e^{(1)}}^{e^{(2)}+\tilde{p}} + 2\underline{L}_{\tilde{q}'}^{\tilde{p}} + \underline{L}_{p}^{q} &= (\underline{L}_{e^{(1)}}^{e^{(2)}+\tilde{p}} + \underline{L}_{\tilde{q}'}^{\tilde{p}} + \underline{L}_{p}^{q}) + (\underline{L}_{e^{(1)}}^{e^{(2)}+\tilde{p}} + \underline{L}_{\tilde{q}'}^{\tilde{p}}) \\
&= (\underline{L}_{e^{(1)}}^{e^{(2)}+\tilde{p}+p_r} + \underline{L}_{\tilde{q}_c'}^{\tilde{p}} + \underline{L}_{\tilde{q}_r'}^{\tilde{p}} + \underline{L}_{p_c}^{q} + \underline{L}_{p_r}^{e^{(2)}+q'}) + (\underline{L}_{e^{(1)}}^{e^{(2)}+\tilde{p}} + \underline{L}_{\tilde{q}'}^{\tilde{p}}) \\
&= (\underline{L}_{e^{(1)}}^{e^{(2)}+\tilde{p}+p_r} + \underline{L}_{\tilde{q}_c'}^{\tilde{p}} + \underline{L}_{\tilde{q}_r'}^{\tilde{p}_r} + \underline{L}_{p_c}^{q+\tilde{q}_r'} + \underline{L}_{p_r}^{e^{(2)}+q'}) + (\underline{L}_{e^{(1)}}^{e^{(2)}+\tilde{p}} + \underline{L}_{\tilde{q}'}^{\tilde{p}}) \\
&= (\underline{L}_{e^{(1)}}^{e^{(2)}+\tilde{p}+p_r} + \underline{L}_{\tilde{q}_c'}^{\tilde{p}+p_r} + \underline{L}_{p_c}^{q+\tilde{q}_r} + \underline{L}_{\tilde{q}_r'}^{\tilde{p}_r}) + (\underline{L}_{e^{(1)}}^{e^{(2)}+\tilde{p}} + \underline{L}_{\tilde{q}'}^{\tilde{p}} + \underline{L}_{p_r}^{e^{(2)}+q_r'}).
\end{aligned}$$

Similarly, with $q_r = e^{(2)} + q_r'$, we get

$$\begin{aligned}
&\overline{L}_{e^{(1)}}^{e^{(2)}+p} + \overline{L}_{e^{(2)}}^{e^{(1)}+p} + \overline{L}_{q'}^{p} + 2\overline{L}_{\tilde{p}}^{\tilde{q}} \\
&\quad = (\overline{L}_{e^{(1)}}^{e^{(2)}+p} + \overline{L}_{q'}^{p} + \overline{L}_{\tilde{p}}^{\tilde{q}}) + (\overline{L}_{e^{(2)}}^{e^{(1)}+p} + \overline{L}_{\tilde{p}}^{\tilde{q}}) \\
&\quad = (\overline{L}_{e^{(1)}}^{e^{(2)}+p+\tilde{p}_r} + \overline{L}_{q_c'}^{p} + \overline{L}_{q_r'}^{p} + \overline{L}_{\tilde{p}_c}^{\tilde{q}} + \overline{L}_{\tilde{p}_r}^{\tilde{q}'}) + (\overline{L}_{e^{(2)}}^{e^{(1)}+p} + \overline{L}_{\tilde{p}}^{\tilde{q}}) \\
&\quad = (\overline{L}_{e^{(1)}}^{e^{(2)}+p+\tilde{p}_r} + \overline{L}_{q_c'}^{p} + \overline{L}_{q_r'}^{p_r} + \overline{L}_{p_c}^{\tilde{q}+q_r'} + \overline{L}_{\tilde{p}_r}^{\tilde{q}'}) + (\overline{L}_{e^{(2)}}^{e^{(1)}+p} + \overline{L}_{\tilde{p}}^{\tilde{q}}) \\
&\quad = (\overline{L}_{e^{(1)}}^{e^{(2)}+p+\tilde{p}_r} + \overline{L}_{q_c'}^{p} + \overline{L}_{p_c}^{q+\tilde{q}_r} + \dot{\overline{L}}_{p_c}^{e^{(2)}} + \overline{L}_{\tilde{p}_r}^{\tilde{q}'}) + (\overline{L}_{e^{(2)}}^{e^{(1)}+p} + \overline{L}_{q_r'}^{p_r} + \overline{L}_{\tilde{p}}^{\tilde{q}}) \\
&\quad = (\overline{L}_{e^{(1)}}^{e^{(2)}+p+\tilde{p}_r} + \overline{L}_{q_c'}^{p+\tilde{p}_r} + \overline{L}_{p_c}^{q+\tilde{q}_r} + \overline{L}_{\tilde{p}_r}^{\tilde{q}_r'}) + (\overline{L}_{e^{(2)}}^{e^{(1)}+p_r} + \overline{L}_{q_r'}^{p_r} + \overline{L}_{\tilde{p}}^{\tilde{q}}) \\
&\quad = (\overline{L}_{e^{(1)}}^{e^{(2)}+\tilde{p}+p_r} + \overline{L}_{\tilde{q}_c'}^{\tilde{p}+p_r} + \overline{L}_{p_c}^{q+\tilde{q}_r} + \overline{L}_{\tilde{p}_r}^{\tilde{q}_r'}) + (\overline{L}_{e^{(2)}}^{e^{(1)}+p_r} + \overline{L}_{q_r'}^{p_r} + \overline{L}_{\tilde{p}}^{\tilde{q}}).
\end{aligned}$$

Comparing the final expressions one sees that (F.14) certainly holds if

$$\underline{L}_{e^{(1)}}^{e^{(2)}+\tilde{p}} + \underline{L}_{\tilde{q}'}^{\tilde{p}} + \underline{L}_{p_r}^{e^{(2)}+q_r'} \le \overline{L}_{e^{(2)}}^{e^{(1)}+p_r} + \overline{L}_{q_r'}^{p_r} + \overline{L}_{\tilde{p}}^{\tilde{q}} \tag{F.15}$$

is true. But this is just Case 1 vs. Case 2 with $\tilde{p}$, $\tilde{q}$ as above and with p, q there being replaced by the present p_r, $e^{(2)} + q_r'$. Therefore, (F.15) holds, and (F.14) is proved.

Case 2 vs. Case 1, Case 3 vs. Case 1, Case 2 vs. Case 3, and Case 3 vs. Case 2 are proved using the same ideas. □

G A short introduction to INTLAB

INTLAB ('Interval laboratory') is a toolbox for MATLAB written completely in MATLAB code by S. M. Rump [320]. It is easy to handle and allows computing with real and complex intervals provided that the computer arithmetic satisfies the IEEE 754 arithmetic standard [151], Bohlender, Ullrich [70]. Basic Linear Algebra Subroutines BLAS 1–3 as described in Lawson et al. [187], Dongarra et al. [86], Dongarra et al. [87] are involved whenever possible in order to speed up the computation. The implementation is based on switching between different rounding modes which is invisible to the user (`setround(-1)` for rounding downwards, `setround(1)` for rounding upwards, `setround(0)` for rounding to nearest). Details on INTLAB and further links can be found at the website http://www.ti3.tuhh.de/rump/ and in Hargreaves [135].

Storage of intervals

The storage of an interval $[a]$ depends on it being real or complex. Real intervals are stored by their endpoints $\underline{a}$, $\overline{a}$, complex intervals by their midpoint $\check{a}$ and their radius $\mathrm{rad}([a]) = r_a$. To this end, the datatype `intval` is introduced which is a structure with five components (cf. Rump [320]): The component `a.complex` is Boolean and true if $[a]$ is complex. So it is the switch for the correct representation. The components `a.inf`, `a.sup` are used for $\underline{a}$, $\overline{a}$ if $[a]$ is real; they are empty for $[a]$ being complex. The components `a.mid`, `a.rad` are used for $\check{a}$, r_a if $[a]$ is complex; they are empty for real intervals $[a]$. This recommends the way to enter an interval. Since a complex interval $[a] = \langle \check{a}, r_a \rangle$ with *real* midpoint is interpreted as *real* interval $\check{a} + r_a[-1, 1]$, the function `cintval(midpoint,radius)` should be used in order to get the right disc in the complex plane. If a complex interval is entered as a rectangle $[\underline{a}, \overline{a}]$, it is stored as the smallest disc enclosing this rectangle using the midpoint $\check{a} = (\underline{a} + \overline{a})/2$. An infsup-output of such a disc $\langle a \rangle$ delivers a rectangle $[a]$ which encloses this disc best possible.

Arithmetic

Independently of the storage of an interval $[a]$, INTLAB provides two interval arithmetics: a fast one working with the midpoint $\check{a}$ and the radius r_a of $[a]$, and a slower sharp one using the lower bound $\underline{a}$ and the upper bound $\overline{a}$. The user can change the arithmetic by the commands

DOI 10.1515/9783110499469-018

```
intvalinit('FastIVMult')
```

and

```
intvalinit('SharpIVMult')
```

which can be done in the initialization file `startintlab.m`, at the beginning of a program, or during an INTLAB session.

With $\boldsymbol{a} = \langle \check{a}, r_a \rangle$, $\boldsymbol{b} = \langle \check{b}, r_b \rangle$ the elementary operations $\oplus$, $\ominus$, $\otimes$, $\oslash$ of the fast arithmetic are defined by

$$\begin{aligned}
\boldsymbol{a} \oplus \boldsymbol{b} &= \langle \check{a} + \check{b}, r_a + r_b \rangle \\
\boldsymbol{a} \ominus \boldsymbol{b} &= \langle \check{a} - \check{b}, r_a + r_b \rangle \\
\boldsymbol{a} \odot \boldsymbol{b} &= \langle \check{a} \cdot \check{b}, |\check{a}| r_b + r_a |\check{b}| + r_a r_b \rangle \\
1 \oslash \boldsymbol{b} &= \langle \check{b}/D, r_b/D \rangle, \quad D = \check{b}^2 - r_b^2,\ 0 \notin \boldsymbol{b} \\
\boldsymbol{a} \oslash \boldsymbol{b} &= \boldsymbol{a} \odot (1 \oslash \boldsymbol{b}), \quad 0 \notin \boldsymbol{b}.
\end{aligned}$$

The operations of the second arithmetic are the same as in Section 2.2. The results coincide in both arithmetics for addition, subtraction and inversion (when computing exactly). They may differ for multiplication as the subsequent example shows.

$$\langle -1, 1 \rangle \odot \langle 1, 1 \rangle = \langle -1, 3 \rangle = [-4, 2] \supset [-4, 0] = [-2, 0] \cdot [0, 2].$$

Therefore, in contrast to the second arithmetic the first one is not always optimal concerning inclusion. Nevertheless, the width of the multiplication is overestimated at most by a factor of 1.5 as was proved in Rump [319]; see also Theorem 9.2.4. This uniform factor is best possible as our example shows. Extensive numerical experiments in Rump [319] indicate, however, that the factor is much smaller in practice, and it does not perpetuate. Moreover, when machine arithmetic is used, the midpoint-radius representation allows tighter intervals than the standard arithmetic – see $\langle 1, 10^{-30} \rangle$ for example when using double precision. Even the multiplication can be better in machine arithmetic, as is illustrated by Rump's example $[1.2, 1.4] \cdot [9.1, 9.3]$ which gives $[10, 14]$ in rounded 2-digit decimal arithmetic while the corresponding operation $\boldsymbol{a} \odot \boldsymbol{b} = \langle 1.3, 0.1 \rangle \odot \langle 9.2, 0.1 \rangle$ yields the better result $\boldsymbol{c} = \langle 12, 1.7 \rangle = [10.3, 13.7]$. Here we used the formulae

$$\begin{aligned}
\check{c} &= \square(\check{a} \cdot \check{b}), \\
r_c &= \triangle(\eta + \varepsilon|\check{c}| + |\check{a}| r_b + r_a |\check{b}| + r_a r_b),
\end{aligned}$$

in which all numbers are interpreted to be machine numbers, $\square(\cdot)$ means rounding to the nearest, $\triangle(\cdot)$ means rounding upwards, η is the smallest representable denormalized positive machine number and ε is the machine precision which is 0.05 in the example. (In IEEE 754 double precision we have $\eta = 2^{-1074}$ and $\varepsilon = 2^{-53}$.)

By virtue of the advantages of the midpoint-radius arithmetic it is not astonishing that this fast arithmetic is recommended whenever there is nothing serious to prevent it. Nevertheless, we frequently will use the sharp arithmetic just for demonstration purposes.

Initialization

The INTLAB package comes with the m-file `startintlab.m` which must be invoked from the INTLAB directory during a MATLAB session by typing in `startintlab` or, shorter, `startup`. This file initializes some basic features for INTLAB – among them the INTLAB path, the arithmetic and the input/output format mostly to be used. It can be changed by the user. For this book we changed the output format `intvalinit('Display_')` into `intvalinit('DisplayInfSup')` in order to see the interval bounds, and we replaced the fast arithmetic initialized by `intvalinit('FastIVMult')` by the slow but exact one using `intvalinit('SharpIVmult')`. We refer to INTLAB Version 9, which was the current one when writing this appendix. The basic features are listed in the file `contents.m`, which is part of the INTLAB subdirectory `intval`. The following copy is an extract of this file.

```
%INTLAB interval toolbox
%
%Interval constructors
%  intval        - Intval constructor
%  cintval       - Complex intval
%  infsup        - Infimum/supremum to interval
%  midrad        - Midpoint/radius to interval
%  gradient      - Gradient to interval
%  hessian       - Hessian to interval
%  slope         - Slope to interval
%  spones        - Sparse interval of ones
%  horzcat       - Horizontal concatenation          [ , ]
%  vertcat       - Vertical concatenation            [ ; ]
%  cat           - Concatenation of arrays
%  subsasgn      - Subscripted assignment A(i,:) = 1
%  subsref       - Subscripted reference r = A(3,4)
%  tocmplx       - Interval to complex
%
%Display of intervals (rigorous) - shortened
%  display       - Command window display of interval
%                  (use user defined default)
%  displaywidth  - Set width of display
%  disp_         - Display with uncertainty
```

```
%  infsup       - Display infimum and supremum
%  midrad       - Display midpoint and radius
%  realimag     - Real and Imaginary part separately
%  plotintval   - Plots real or complex intervals
%
%Interval arithmetic operations - shortened
%  mtimes       - Matrix multiply                      *
%  times        - Elementwise multiply                 .*
%  mldivide     - Backslash, linear system solving     \
%  mrdivide     - Slash or right division              /
%  rdivide      - Elementwise right division           ./
%  mpower       - Matrix power                         ^
%  power        - Elementwise power                    .^
%  intersect    - Intersection
%  hull         - Hull or convex union
%
%Other interval operations - shortened
%  setround     - Set rounding mode
%  getround     - Get rounding mode
%  ctranspose   - Complex conjugate transpose          '
%  transpose    - Transpose                            .'
%  conj         - Conjugate
%  abs          - Interval absolute value, result real interval
%  mag          - Magnitude = absolute value, result double
%  mig          - Mignitude, result double
%  inf          - Infimum
%  inf_         - Infimum (for problems with inf)
%  sup          - Supremum
%  mid          - Midpoint
%  rad          - Radius
%  diam         - Diameter
%  real         - Real part
%  imag         - Imaginary part
%  qdist        - Metric distance
%
%Utility routines - shortened
%  pred         - Predecessor
%  succ         - Successor
%  isnan        - True for Not a Number
%  isreal       - Interval is real
%  isempty      - Interval is empty in Matlab sense, i.e. []
%  emptyintersect  - Check for empty intersection
%                    in mathematical sense
%  iszero       - Interval is zero (componentwise)
%
```

```
%Structural operations - shortened
%  length       - Length
%  size         - Size
%  dim          - Dimension of square matrix
%  reshape      - Reshape
%
%Interval exponential functions - shortened
%  exp          - Exponential
%  log          - Natural logarithm
%  log10        - Logarithm to base 10
%  sqr          - Square
%  sqrt         - Square root
%
%Interval Comparison
%  eq           - Equal                          ==
%  ne           - Not equal                      ~=
%  gt           - Greater than                   >
%  ge           - Greater than or equal          >=
%  lt           - Less than                      <
%  le           - Less than or equal             <=
%  in           - Contained in
%  in0          - Contained in interior
%
%Verification routines and auxiliary - shortened
%  verifylss    - Verified linear system solver including
%                 rectangular and sparse systems
%  lssresidual  - Improved approximation and inclusion
%                 of residual
%  plotlinsol   - Solution of 2x2 and 3x3 interval
%                 linear systems
%  verifyeig    - Verified eigenvalue clusters
%                 and invariant subspaces
%  verifyquad   - Verified quadrature
%  verifynlss   - Verified nonlinear system solver
%  verifynlss2  - Verified nonlinear system solver
%                 for double roots
%  test         - Sample nonlinear test functions
%  inv          - Verified inverse of square matrix
%  typeadj      - Type adjustment
%  typeof       - Type for type adjustment
%
%
% Copyright (c) Siegfried M. Rump,
%               Head of the Institute for Reliable Computing,
%               Hamburg University of Technology
```

We recommend that the user prints out the complete file `contents.m` in order to see what is possible in INTLAB.

Now we want to illustrate some of the preceding commands. But we do not intend to present an exhaustive description of the toolbox nor to go into too much detail. We assume that the reader has some elementary knowledge of MATLAB. For briefness, we will often suppress the MATLAB output although our commands are not terminated by a semicolon, the correct way to work in the background in MATLAB.

Assignment of intervals

(a) Real intervals

```
a = intval(5)        % assignment of the point interval [5,5]
a = intval('3.14_') % assignment of the interval [3.13,3.15]
a = infsup(-3,7)     % assignment of [-3,7]
                     % in infsup representation
a = midrad(2,5)      % assignment of <2,5> = [-3,7]
                     % in midpoint-radius representation
```

The underscore in input and output produces an interval whose lower bound is one unit less than the last digit being shown. Correspondingly, the upper bound is one unit more than the last displayed digit. This results in the interval [3.13, 3.15] above.

(b) Complex intervals

```
a = intval(5 + 2i)     % assignment of the complex
                       % point interval <5+2i,0>
a = midrad(5+0i,1)     % assignment of the real (!)
                       % interval <5,1> = [4,6]
a = cintval(5,1)       % assignment of the complex (!)
                       % interval <5,1>
a = infsup(1+2i,7+10i) % assignment of [a] = [1+2i,7+10i]
                       % in infsup representation stored
                       % as enclosure <4+6i,5>
a = midrad(1+i,2)      % assignment of the complex (!)
                       % interval a = <1+i,2>
                       % in midpoint-radius representation
```

Bounds, midpoint, radius of an interval

(a) Real intervals ($[a] = [-3, 7]$)

```
a.inf       % access to the lower bound -3 of [a]
a.sup       % access to the upper bound  7 of [a]
inf(a)      % access to the lower bound -3 of [a]
sup(a)      % access to the upper bound  7 of [a]
a.mid       % access to the midpoint 2 of [a]
a.rad       % access to the radius   5 of [a]
mid(a)      % access to the midpoint 2 of [a]
rad(a)      % access to the radius   5 of [a]
```

(b) Complex intervals

$[a] = [-3 + i, 5 + 7i]$ stored as $\langle 1 + 4i, 5\rangle$
rectangular output: $[(1 - 5) + (4 - 5)i, (1 + 5) + (4 + 5)i] = [-4 - i, 6 + 9i]$

```
a.inf       % access to the lower bound -4-i
            % of the rectangular output
a.sup       % access to the upper bound  6+9i
            % of the rectangular output
inf(a)      % access to the lower bound -4-i
            % of the rectangular output
sup(a)      % access to the upper bound  6+9i
            % of the rectangular output
a.mid       % access to the midpoint 1+4i of [a]
a.rad       % access to the radius 5 of
            % the enclosure <1+4i,5> of [a]
mid(a)      % access to the midpoint 1+4i of [a]
rad(a)      % access to the radius 5 of
            % the enclosure <1+4i,5> of [a]
```

Auxiliary quantities for $[a] = [-3, 7]$, $[b] = [5, 9]$

```
s = abs(a)        % assignment of the interval [0,7]
                  % = { |x| | x in [a] = [-3,7] }
t = mag(a)        % assignment of |[-3,7]| = 7
u = mig(a)        % assignment of <[-3,7]> = 0
d = diam(a)       % assignment of the diameter 10 of [a]
q = qdist(a,b)    % assignment of the Hausdorff distance q = 8
```

Conversion

In all our examples we assumed that the data are machine numbers. If not, the following phenomena are important to know. To this end recall that the decimal number 0.1 is equal to the periodic dual fraction $0.0\overline{0011}$, hence it is not representable as a floating point number. In double precision, the binary approximation $(1.10011\ldots0011001 + 2^{-52}) \cdot 2^{-4}$ is used with 52 digits after the dual point. This rounded to nearest approximation is apparently greater than the original decimal number 0.1. For internal storage its exponent -4 is transformed to the binary equivalence of $-4 + 1023$, where 1023 is the bias in IEEE double precision.

```
a = intval(0.1)          % 0.1 is not contained in [a]
                         % output: [0.1000,0.1001]
b = intval('0.1')        % 0.1 is     contained in [b]
                         % output: [0.0999,0.1001]
```

In the first case INTLAB first converts 0.1 to the nearest machine number and then assigns this number to the lower and upper bound of $[a]$, i.e., $[a]$ is a point interval as expected. The output is rounded outward. This pretends that $[a]$ is nondegenerate and an enclosing interval for 0.1, which is not the case.

In the second case, `'0.1'` is a string which is converted using INTLAB's conversion routine `str2intval` described in Rump [320]. This results in the nearest nondegenerate machine interval that contains 0.1. So, $[b]$ is nondegenerate in contrast to the usual meaning of `intval( )`. For the diameter of $[a]$ and $[b]$ and the corresponding radii we get the following values:

```
diam(a)            % [a] is a point interval
ans =
    0

rad(a)
ans =
    0

diam(b)            % [b] is a nondegenerate interval
ans =
  1.3878e-17

rad(b)
ans =
  1.3878e-17
```

For the nondegenerate interval $[b]$, radius and diameter coincide. This unexpected result is due to the philosophy of INTLAB to guarantee that the machine interval $[\tilde{a} - \tilde{r}_a, \tilde{a} + \tilde{r}_a]$ contains the exact interval $[a] = [\underline{a}, \overline{a}]$. Since in our example the bounds of $[b]$ differ only in the last bit, the machine number of the midpoint must necessarily be one of these bounds. In fact, it is the upper one as can be seen by the following commands:

```
format hex        % hexadecimal format
0.1               % round to nearest: 3fb999999999999a
                  % exponent: 3fb = 1023 - 4
                  % mantissa: 999999999999a
inf(b)            % lower bound 3fb9999999999999
sup(b)            % upper bound 3fb999999999999a
mid(b)            % midpoint    3fb999999999999a
format short      % default format
```

In order to follow the philosophy, one must obtain the same result for `diam(b)` and `rad(b)`. For thicker intervals this pathology disappears, of course.

Calling up the functions `inf` and `sup` for the intervals $[a]$ and $[b]$ effects another surprise. In all four cases one obtains the answer `ans = 0.1`, although one might expect $\inf(b) = 0.0999$ and $\sup(b) = 0.1001$. Nevertheless, the INTLAB answer 0.1 is correct since the bounds are binary numbers which are rounded to the nearest decimal number for the output, i.e., to 0.1000 in `format short`. Changing the format to `format long e` before typing in `inf(b)` etc. reveals the difference in $[a]$ and $[b]$:

```
format long e             % floating point format with 15 digits

inf(a)
ans =
    1.000000000000000e-001

sup(a)
ans =
    1.000000000000000e-001

inf(b)
ans =
    9.999999999999999e-002

sup(b)
ans =
    1.000000000000000e-001
```

Notice that independently of the format, INTLAB internally computes with MATLAB precision, i.e., with IEEE 754 double precision.

Output format

INTLAB provides several output formats which can be seen by typing in `help format`. Some of them can be established by the following commands.

```
format short        % scaled fixed point format with 5 digits
format short e      % floating point format with 5 digits
format long         % scaled fixed point format with 15 digits
format long e       % floating point format with 15 digits
```

The current format can be changed interactively.

Apart from the format, there are another three ways to influence the output. They are set by the commands

```
intvalinit{'display_'}          % display with uncertainty
                                % (e.g., 3.14_)
intvalinit{'displayInfSup'}     % display infimum/supremum
intvalinit{'displayMidRad'}     % display midpoint/radius
```

The effect of the first one can also be obtained for an interval by the command `disp_(interval)`.

Operations and functions

```
(6.0 + 5 e-1 - 3.5) / (5 * 2.4)     % rational expression
ans =
    0.2500

exp(1) - ( sqr(sin(3.0)) + sqr(cos(3.0)) ) / ...
         sqrt( cosh(1)^2 - sinh(1)^2 ) + 1 + log(1)
                      % arithmetic expression
                      % ellipsis  ...  extends the line
ans =
    2.7183
```

Vectors, matrices

```
u = [1,2,3]                 % row vector
v = u'                      % ' = conjugate transpose vector
                            %   = transpose for real vectors
                            %   = column vector
w = [1;2;3]                 % column vector

A = [[1,2,3];[4,5,6]]       % 2 x 3 matrix
                            % [ , ] : separator in a row
                            % [ ; ] : separator in a column
B = A'                      % '  = conjugate transposed matrix
                            %    = transposed for real matrices
A.'                         % .' = transposed matrix ( = B )
A*B                         % matrix-matrix multiply
C = 7*A                     % scalar * matrix
D = infsup(A,C)             % interval matrix [D] = [A,C]
in(A,D)                     % tests whether A is contained in D
                            % true if result is a 2 x 3 matrix of
                            % ones only
                            % false if this matrix contains at
                            % least one zero
in0(A,D)                    % tests whether A is contained in the
                            % interior of D
                            % true if result is a 2 x 3 matrix of
                            % ones only
                            % false if this matrix contains at
                            % least one zero
isemtpy(intersect(A,D))% tests whether the intersection
                            % of A and D is empty
                            % true if result is a 2 x 3 matrix of
                            % ones only
                            % false if this matrix contains at
                            % least one zero
```

Gradient

INTLAB provides automatic differentiation in forward mode. It is initialized by the command `gx = gradientinit(x)`, where `x` is either a value or a variable to which a value has been assigned. The name `gx` is arbitrary. This creates a structure with the components `gx.x` and `gx.dx`. The first component contains the value x, the second contains the value 1 which is the derivative of x. When plugging `gx` into some function f, say y = f(gx), then the result y also has two components, where y.x contains the value $f(x)$ and y.dx contains the value $f'(x)$ of the corresponding derivative.

```
x = 2                   % assignment of the value 2
gx = gradientinit(x)    % initialization of automatic
                        % differentiation
y = sin(pi*gx)          % y.x  = sin(2*pi)     = 0
                        % y.dx = pi*cos(2*pi)  = pi
                        % (in exact arithmetic)
gradient value y.x =
 -2.4493e-016
gradient derivative(s) y.dx =
    3.1416

x = intval(2)           % point interval [2,2]
gx = gradientinit(x)    % initialization of automatic
                        % differentiation
y = sin(pi*gx)          % y.x  = sin(2*pi)     = 0
                        % y.dx = pi*cos(2*pi)  = pi
                        % (in exact arithmetic)
intval gradient value y.x =                                %
  1.0e-015 *                                               %
[   -0.2450,    -0.2449]                                   %
intval gradient derivative(s) y.dx =                       %
[    3.1415,     3.1416]                                   %
```

The commands AccSum and AccDot

The INTLAB command AccSum abbreviates 'accurate sum', is applied to a real vector, and computes the sum of its components with an error of at most 1 ulp.

The INTLAB command AccDot abbreviates 'accurate dot product' and is applied – among others – to two *real* vectors x, y of the same length. It is called up by AccDot(x',y) and computes the dot product $s = x^T y$ with an error of at most 1 ulp. Apart from x, y, it also allows a third parameter K which regulates the precision of s. This parameter can be a nonnegative integer, the option inf, or the symbol [ ]. If K = 0, the dot product s is rounded to nearest; if K = 1, the result is faithfully rounded; if K > 1, the dot product s is computed with K-fold precision and stored in a cell array of length K as a nonoverlapping sequence. In this case, s can be interpreted as being stored in staggered correction format in the sense of Stetter [346]. It is obvious that AccDot(x',e,K) (e = ones(length(x),1), K > 1) computes the sum of the entries of x with K-fold precision. If K = inf, the result s is stored in a cell array of a sufficiently large length such that s is represented exactly. If K = [ ], the dot product s is enclosed in an interval where the inclusion is best possible if no underflow occurs.

The command `AccDot` works similarly if the vectors x, y are replaced by real matrices $A \in \mathbb{R}^{m\times r}$, $B \in \mathbb{R}^{r\times n}$. This time each cell $s\{i\}$ contains an $m \times n$ matrix. One of the matrices A, B is even allowed to be a cell array; interval matrices are forbidden in INTLAB Version 9. Further interesting information on `AccDot` can be obtained by means of INTLAB's help on this command. In particular, a simple way to get a tight interval enclosure of s if s is computed with K-fold precision is presented there.

Exercises

Ex. G.1. Compute the expressions of Exercise 2.2.2 using INTLAB.

Ex. G.2. Run the following little INTLAB code and explain its output.

```
a = infsup(-3+i,5+7i)
infsup(a.inf,a.sup)
```

Ex. G.3. In Rump [321] the following INTLAB code was presented to invert ill-conditioned matrices:

```
C = inv(A);
for k = 2 : kmax
    CA = AccDot(C,A);
    C = AccDot(inv(CA),C, k);
end
```

Here the MATLAB command `inv(A)` computes an approximation of the inverse of $A \in \mathbb{R}^{n\times n}$ and the integer `kmax` bounds the precision.

(a) Read the MATLAB help on cells and the INTLAB help on `AccDot`.

(b) Test this algorithm for the preconditioned Hilbert matrix $H_n^{\text{prec}} = (h_{ij}) \in \mathbb{R}^{n\times n}$ with the entries $h_{ij} = c_n/(i+j-1)$, $i, j = 1, \ldots, n$, where c_n is the least common multiple of the denominators $1, \ldots, 2n-1$. This factor c_n can be computed, for instance, by a simple loop in MATLAB using the MATLAB command `lcm(·,·)`. It guarantees that the matrix H_n^{prec} has integer entries and is therefore exactly representable on a computer – at least if n is not so large; cf. also Example 7.3.3. According to Gregory, Karney [122] its inverse has the entries

$$\left((H_n^{\text{prec}})^{-1}\right)_{ij} = (-1)^{i+j} \cdot \frac{i+j-1}{c_n} \cdot \binom{n+i-1}{n-j} \times \binom{n+j-1}{n-i} \cdot \binom{i+j-2}{i-1}^2 .$$

For each pair $(n, \texttt{kmax})$ of parameters compute the spectral condition number $\text{cond}(A) = \|A\|_2 \cdot \|A^{-1}\|_2$ with an appropriate command in MATLAB and the error $\|I - C \cdot A\|_\infty$ using INTLAB.
Try $5 \le n \le 20$ and $\texttt{kmax} = 2, \ldots, 6$.

(c) Repeat (b) with Boothroyd–Dekker matrices B_n whose entries are integers defined by

$$(B_n)_{ij} = \frac{n}{i+j-1} \cdot \binom{n+i-1}{i-1} \cdot \binom{n-1}{j-1}, \qquad i, j = 1, \ldots, n.$$

Boothroyd–Dekker matrices are extremely ill-conditioned. Their inverse has the entries $(B_n^{-1})_{ij} = (-1)^{i+j}(B_n)_{ij}$, i.e., B_n^{-1} is the original matrix B_n modified by a checkerboard-like distribution of plus and minus signs.

(d) Experiment as in (b) with various other ill-conditioned matrices as described, for instance, in Gregory, Karney [122], Rump [314], or Zielke [367].

Ex. G.4. Write an INTLAB function `iAccDot(·,·)` which computes an interval dot product $[z]$ of two interval vectors $[x]$, $[y]$ such that the exact interval result $[x]^T[y]$ is contained in $[z]$ but $[z]$ encloses it possibly tighter than by means of the algorithm

```
[z] = 0;
for i = 1:length([x])
    [z] = [z] + [x](i)*[y](i);
end
```

Hint: Use the tight interval enclosure of INTLAB's accurate dot product `AccDot`.

Bibliography

[1] Adams E, Cordes D, Keppler H. Enclosure methods as applied to linear periodic ODEs and matrices. Z. Angew. Math. Mech. 1990; 70(12): 565–578.

[2] Adams E, Kulisch U (eds.). Scientific Computing with Automatic Result Verification. Mathematics in Science and Engineering, Vol. 189. Boston, New York: Academic Press; 1993.

[3] Adams E, Lohner R. Error bounds and sensitivity analysis. In: Stepleman R et al. (eds.). Scientific Computing. IMACS/North-Holland Publishing Company; 1983, pp. 213–222.

[4] Ahlfors LV. Complex Analysis. 2nd edn. New York: McGraw-Hill; 1966.

[5] Alefeld G. Intervallrechnung über den komplexen Zahlen und einige Anwendungen. Thesis, Universität Karlsruhe, Karlsruhe 1968.

[6] Alefeld G. Stets konvergente Verfahren höherer Ordnung zur Berechnung von reellen Nullstellen. Computing. 1974; 13: 55–65.

[7] Alefeld G. Über die Durchführbarkeit des Gaußschen Algorithmus bei Gleichungen mit Intervallen als Koeffizienten. Computing Suppl. 1977; 1: 15–19.

[8] Alefeld G. On the convergence of some interval-arithmetic modifications of Newton's method. SIAM J. Numer. Anal. 1984; 21(2): 363–372.

[9] Alefeld G. The centered form and the mean value form – A necessary condition that they yield the range. Computing. 1984; 33: 165–169.

[10] Alefeld G. Berechenbare Fehlerschranken für ein Eigenpaar unter Einschluß von Rundungsfehlern bei Verwendung des genauen Skalarprodukts. Z. angew. Math. Mech. 1987; 67: 145–152.

[11] Alefeld G. Rigorous error bounds for singular values of a matrix using the precise scalar product. In: Kaucher E, Kulisch U, Ullrich Ch (eds.). Computerarithmetic. Stuttgart: Teubner; 1987, pp. 9–30.

[12] Alefeld G. Über die Konvergenzordnung des Intervall-Newton-Verfahrens. Computing. 1987; 39: 363–369.

[13] Alefeld G. Errorbounds for quadratic systems of nonlinear equations using the precise scalar product. In: Kulisch U, Stetter HJ (eds.). Scientific Computation with Automatic Result Verification. Computing Suppl. 1988; 6: 59–68.

[14] Alefeld G. Berechenbare Fehlerschranken für ein Eigenpaar beim verallgemeinerten Eigenwertproblem. Z. Angew. Math. Mech. 1988; 68: 181–184.

[15] Alefeld G. Über die Divergenzgeschwindigkeit des Intervall-Newton-Verfahrens. Fasculi Mathematici. 1989; 19: 9–13.

[16] Alefeld G. On the approximation of the range of values by interval expressions. Computing. 1990; 44: 273–278.

[17] Alefeld G. Über das Divergenzverhalten des Intervall-Newton-Verfahrens. Computing. 1991; 46: 289–294.

[18] Alefeld G. Inclusion methods for systems of nonlinear equations – The interval Newton method and modifications. In: Herzberger J (ed.). Topics in Validated Computations. Amsterdam: Elsevier; 1994, pp. 7–26.

[19] Alefeld G, Apostolatos N. Praktische Anwendung von Abschätzungsformeln bei Iterationsverfahren. ZAMM. 1968; 48: T46–T49.

[20] Alefeld G, Chen X, Potra F. Numerical validation of solutions of linear complementarity problems. Numer. Math. 1999; 83: 1–23.

[21] Alefeld G, Frommer A, Heindl G, Mayer J. On the existence theorems of Kantorovich, Miranda and Borsuk. Electronic Transactions on Numerical Analysis. 2004; 17: 102–111.

DOI 10.1515/9783110499469-019

[22] Alefeld G, Gienger A, Mayer G. Numerical validation for an inverse matrix eigenvalue problem. Computing. 1994; 53: 311–322.

[23] Alefeld G, Gienger A, Potra F. Efficient numerical validation of solutions of nonlinear systems. SIAM J. Numer. Anal. 1994; 31: 252–260.

[24] Alefeld G, Herzberger J. Über das Newton-Verfahren bei nichtlinearen Gleichungssystemen. ZAMM. 1970; 50: 773–774.

[25] Alefeld G, Herzberger J. Einführung in die Intervallrechnung. Reihe Informatik/12, Mannheim: Bibliographisches Institut; 1974.

[26] Alefeld G, Herzberger J. Introduction to Interval Computations. New York: Academic Press; 1983.

[27] Alefeld G, Hoffmann R, Mayer G. Verification algorithms for generalized singular values. Math. Nachr. 1999; 208: 5–29.

[28] Alefeld G, Kreinovich V, Mayer G. The shape of the symmetric solution set. In: Kearfott RB, Kreinovich V (eds.). Applications of Interval Computations. Boston: Kluwer; 1996, pp. 61–79.

[29] Alefeld G, Kreinovich V, Mayer G. Symmetric linear systems with perturbed input data. In: Alefeld G, Herzberger J (eds.). Numerical Methods and Error Bounds. Berlin: Akademie Verlag; 1996, pp. 16–22.

[30] Alefeld G, Kreinovich V, Mayer G. On the shape of the symmetric, persymmetric, and skew-symmetric solution set. SIAM J. Matrix Anal. Appl. 1997; 18: 693–705.

[31] Alefeld G, Kreinovich V, Mayer G. The shape of the solution set of linear interval equations with dependent coefficients. Math. Nachr. 1998; 192: 23–26.

[32] Alefeld G, Kreinovich V, Mayer G. Modifications of the Oettli–Prager theorem with application to the algebraic eigenvalue problem. In: Alefeld G, Rohn J, Rump SM, Yamamoto T (eds.). Symbolic Algebraic Methods and Verification Methods – Theory and Applications. Wien: Springer; 2001.

[33] Alefeld G, Kreinovich V, Mayer G. On the solution set of particular classes of linear interval systems. J. Comp. Appl. Math. 2003; 152: 1–15.

[34] Alefeld G, Lohner R. On higher order centered forms. Computing. 1985; 35: 177–184.

[35] Alefeld G, Mayer G. The Cholesky method for interval data. Linear Algebra Appl. 1993; 194: 161–182.

[36] Alefeld G, Mayer G. A computer aided existence and uniqueness proof for an inverse matrix eigenvalue problem. Interval Computations. 1994; 1994/1: 4–27.

[37] Alefeld G, Mayer G. On the symmetric and unsymmetric solution set of interval systems. SIAM J. Matrix Anal. Appl. 1995; 16: 1223–1240.

[38] Alefeld G, Mayer G. Interval analysis: Theory and applications. J. Comp. Appl. Math. 2000; 121: 421–464, Special Issue: Wuytack L, Wimp J (eds.). Numerical Analysis in the 20th Century, Vol. I. Approximation Theory.

[39] Alefeld G, Mayer G. On singular interval systems. In: Alt R, Frommer A, Kearfott RB, Luther W (eds.). Numerical Software with Result Verification (Platforms, Algorithms, Applications in Engineering, Physics and Economics). Lecture Notes in Computer Science. Berlin: Springer; 2004, pp. 191–197.

[40] Alefeld G, Mayer G. Enclosing solutions of singular interval systems iteratively. Reliable Computing. 2005; 11: 165–190.

[41] Alefeld G, Mayer G. New criteria for the feasibility of the Cholesky method for interval data. SIAM J. Matrix Anal. Appl. 2008; 30(4): 1392–1405.

[42] Alefeld G, Platzöder L. A quadratically convergent Krawczyk-like algorithm. SIAM J. Numer. Anal. 1983; 20: 211–219.

[43] Alefeld G, Potra F. A new class of interval methods with higher order of convergence. Computing. 1989; 42: 69–80.

[44] Alefeld G, Potra F. Some efficient methods for enclosing simple zeros of nonlinear equations. BIT. 1992; 22: 334–344.
[45] Alefeld G, Potra F, Shen Z. On the existence theorems of Kantorovich, Moore and Miranda. Comput. Suppl. 2001; 15: 21–28.
[46] Alefeld G, Potra F, Shi Y. On enclosing simple roots of nonlinear equations. Math. Comp. 1993; 61(204): 733–744.
[47] Alefeld G, Schäfer U. Iterative methods for linear complementarity problems with interval data. Computing. 2003; 70: 235–259.
[48] Alefeld G, Spreuer H. Iterative improvement of componentwise errorbounds for invariant subspaces belonging to a double or nearly double eigenvalue. Computing. 1986; 36: 321–334.
[49] Anderson E, Bai Z, Bischof C, Blackford S, Demmel J, Dongarra J, Du Croz J, Greenbaum A, Hammarling S, McKenney A, Sorensen D. LAPACK User's Guide. 3rd edn. Philadelphia: SIAM; 1999.
[50] ANSI/IEEE 754–1985: IEEE Standard for Binary Floating-point Arithmetic. New York; 1985.
[51] Apostolatos N, Kulisch U. Grundzüge einer Intervallrechnung für Matrizen und einige Anwendungen. Elektron. Rechenanl. 1968; 10: 73–83.
[52] Arndt H-R. Zur Lösungseinschließung bei linearen Gleichungssystemen mit ungenauen Eingangsdaten, Thesis, Universität Rostock, 2005.
[53] Arndt H-R. On interval systems $[x] = [A][x] + [b]$ and the powers of interval matrices in complex arithmetics. Reliable Comput. 2007; 13: 245–259.
[54] Arndt H-R, Mayer G. On the semi-convergence of interval matrices. Linear Algebra Appl. 2004; 393: 15–37.
[55] Arndt H-R, Mayer G. On the solutions of the interval system $[x] = [A][x] + [b]$. Reliable Computing. 2005; 11: 87–103.
[56] Arndt H-R, Mayer G. New criteria for the semi-convergence of interval matrices. SIAM J. Matrix Anal. Appl. 2005; 27(3): 689–711.
[57] Atanassova L, Herzberger J (eds.). Computer Arithmetic and Enclosure Methods. Amsterdam: North-Holland; 1992.
[58] Auzinger W, Stetter HJ. Accurate arithmetic results for decimal data and non-decimal computers. Computing. 1985; 35: 141–151.
[59] Barth W, Nuding E. Optimale Lösung von Intervallgleichungssystemen. Computing. 1974; 12: 117–125.
[60] Bauch H, Jahn K-U, Oelschlägel D, Süsse H, Wiebigke V. Intervallmathematik. Theorie und Anwendungen. BSB BG. Leipzig: Teubner; 1987.
[61] Baur W, Strassen V. The complexity of partial derivatives. Theoretical Computer Science. 1983; 22: 317–330.
[62] Beeck H. Über Struktur und Abschätzungen der Lösungsmenge von linearen Gleichungssystemen mit Intervallkoeffizienten. Computing. 1972; 10: 231–244.
[63] Beeck H. Zur scharfen Außenabschätzung der Lösungsmenge bei linearen Intervallgleichungssystemen. Z. Angew. Math. Mech. 1974; 54: T208–T209.
[64] Beeck H. Zur Problematik der Hüllenbestimmung von Intervallgleichungssystemen. In: Nickel K (ed.). Interval Mathematics, Lecture Notes in Computer Science 29. Berlin: Springer; 1975. pp. 150–159.
[65] Behnke H. Inclusion of eigenvalues of general eigenvalue problems for matrices. In: Kulisch U, Stetter HJ (eds.). Scientific Computation with Automatic Result Verification. Computing Suppl. 1988; 6: 69–78.
[66] Behnke H. Die Bestimmung von Eigenwertschranken mit Hilfe von Variationsmethoden und Intervallarithmetik. Thesis, Institut für Mathematik, Technische Universität Clausthal, 1989.

[67] Behnke H. The determination of guaranteed bounds to eigenvalues with the use of variational methods II. In: Ullrich C (ed.). Computer Arithmetic and Self-Validating Numerical Methods. Boston, New York: Academic Press; 1990, pp. 155–170.
[68] Berman A, Plemmons RJ. Nonnegative Matrices in the Mathematical Sciences. Classics in Applied Mathematics 9. Philadelphia: SIAM: 1994.
[69] Berner S. Ein paralleles Verfahren zur verifizierten globalen Optimierung. Thesis, Universität Wupppertal. Aachen: Shaker Verlag; 1995.
[70] Bohlender G, Ullrich C. Standards zur Computerarithmetik. In: Herzberger J (ed.). Wissenschaftliches Rechnen. Eine Einführung in das Scientific Computing. Berlin: Akademie Verlag; 1995, pp. 9–52.
[71] Borsuk K. Drei Sätze über die n-dimensionale Sphäre. Fund. Math. 1933; 20: 177–190.
[72] Bronstein IN, Semendjajew KA, Musiol G, Mühlig H. Taschenbuch der Mathematik. Thun, Frankfurt am Main: Verlag Harri Deutsch; 1993.
[73] Brouwer LEJ. Über Abbildung von Mannigfaltigkeiten. Math. Ann. 1912; 71: 97–115.
[74] Bunch JR, Kaufman L, Parlett BN. Decomposition of a symmetric matrix. Numer. Math. 1976; 27: 95–109.
[75] Caprani O, Madsen K. Iterative methods for interval inclusion of fixed points. BIT. 1978; 18: 42–51.
[76] Chaitin-Chatelin F. Lectures on Finite Precision Computations. Philadelphia: SIAM; 1996.
[77] Cordes D. Verifizierter Stabilitätsnachweis für Lösungen von Systemen periodischer Differentialgleichungen auf dem Rechner mit Anwendungen. Thesis, Universität Karlsruhe, Karlsruhe, 1987.
[78] Corliss GF. Survey of interval algorithms for ordinary differential equations. Appl. Math. Comput. 1989; 31: 112–120.
[79] Corliss GF. Automatic differentiation bibliography. In: Griewank A, Corliss G (eds.). Automatic Differentiation of Algorithms: Theory, Implementation, and Applications. Philadelphia: SIAM; 1991, pp. 331–353.
[80] Cornelius H, Lohner R. Computing the range of values of real functions with accuracy higher than second order. Computing. 1984; 33: 331–347.
[81] Deif AS. The interval eigenvalue problem. Z. Angew. Math. Mech. 1991; 71: 61–64.
[82] Deif AS. Singular values of an interval matrix. Linear Algebra Appl. 1991; 151: 125–133.
[83] Deimling K. Nonlinear Functional Analysis. Berlin: Springer; 1985.
[84] Dobner H-J. Einschließungsalgorithmen für hyperbolische Differentialgleichungen. Thesis, Universität Karlsruhe, Karlsruhe, 1986.
[85] Dongarra JJ, Moler CB, Wilkinson JH. Improving the accuracy of computed eigenvalues and eigenvectors. SIAM J. Numer. Anal. 1983; 20: 23–45.
[86] Dongarra JJ, Du Croz J, Hammarling S, Hanson RJ. An extended set of FORTRAN basic linear algebra subroutines. ACM Trans. Math. Soft. 1988; 14: 1–17.
[87] Dongarra JJ, Du Croz J, Duf, IS, Hammarling S. A set of level 3 basic linear algebra subprograms. ACM Trans. Math. Soft. 1990; 16: 1–17.
[88] Dwyer PS. Linear Computations. New York: Wiley; 1951.
[89] Eble I. Über Taylor-Modelle. Thesis, Universität Karlsruhe, Karlsruhe, 2007.
[90] Eijgenraam P. The solution of initial value problems using interval arithmetic. Formulation and analysis of an algorithm. Thesis, Mathematisch Centrum, Amsterdam, 1981.
[91] Enger W. Interval Ray Tracing – ein Divide-and-Conquer Verfahren für photorealistische Computergrafik. Thesis, Universität Freiburg, Freiburg, 1990.
[92] Fan K. Topological proof for certain theorems on matrices with non-negative elements. Monatsh. Math. 1958; 62: 219–237.
[93] Farkas J. Theorie der einfachen Ungleichungen. J. Reine und Angew. Math. 1902; 124: 1–27.

[94] Fernando KV, Pont MW. Computing accurate eigenvalues of a Hermitian matrix. In: Ullrich C, Wolff von Gudenberg J (eds.). Accurate Numerical Algorithms. A Collection of Research Papers. Research Reports ESPRIT, Project 1072, DIAMOND, Vol. 1. Berlin: Springer; 1989, pp. 104–115.
[95] Fiedler M, Nedoma J, Ramík J, Rohn J, Zimmermann K. Linear Optimization Problems with Inexact Data. New York: Springer; 2006.
[96] Filippov AF. A short proof of the theorem on reduction of a matrix to Jordan normal form. Moscow Univ. Bull. 1971; 26(2): 18–19.
[97] Fischer HC. Schnelle automatische Differentiation, Einschließungsmethoden und Anwendungen. Thesis, University of Karlsruhe, Karlsruhe, 1990.
[98] Fischer HC. Automatic Differentiation and Applications. In: Adams E, Kulisch U. (eds.): Scientific Computing with Automatic Result Verification. Mathematics in Science and Engineering, Vol. 189. Boston, New York: Academic Press; 1993.
[99] Forsythe GE, Henrici P. The cyclic Jacobi method for computing the principal values of a complex matrix. Trans. Amer. Math. Soc. 1960; 94: 1–23.
[100] Franklin JN. Methods of Mathematical Economics. Linear and Nonlinear Programming, Fixed-Point Theorems. Classics in Applied Mathematics 37. Philadelphia: SIAM; 2002.
[101] Friedland S. Inverse eigenvalue problems. Linear Algebra Appl. 1977; 17: 15–51.
[102] Friedland S, Nocedal J, Overton ML. The formulation and analysis of numerical methods for inverse eigenvalue problems. SIAM J. Numer. Anal. 1987; 24: 634–667.
[103] Frommer A. A feasibility result for interval Gaussian elimination relying on graph structure. In: Alefeld G, Rohn J, Rump S, Yamamoto T. (eds.). Symbolic Algebraic Methods and Verification Methods. Wien: Springer; 2001, pp. 79–86.
[104] Frommer A. Proving conjectures by use of interval arithmetic. In: Kulisch U, Lohner R, Facius A (eds.). Perspectives of Enclosure Methods. Wien: Springer; 2001, pp. 1–13.
[105] Frommer A, Hoxha F, Lang B. Proving the existence of zeros using the topological degree and interval arithmetic. J. Comp. Appl. Math. 2007; 199: 397–402.
[106] Frommer A, Lang B. On preconditioners for the Borsuk existence test. Proc. Applied Math. Mech. 2004; 4: 638–639.
[107] Frommer A, Lang, B. Existence tests for solutions of nonlinear equations using Borsuk's theorem. SIAM J. Numer. Anal. 2005; 43: 1348–1361.
[108] Frommer A, Lang B, Schnurr M. A comparison of the Moore and Miranda existence tests. Computing. 2004; 72: 349–354.
[109] Frommer A, Mayer G. A new criterion to guarantee the feasibility of the interval Gaussian algorithm. SIAM J. Matrix Anal. Appl. 1993; 14: 408–419.
[110] Frommer A, Mayer G. Linear systems with Ω-diagonally dominant matrices and related ones. Linear Algebra Appl. 1993; 186: 165–181.
[111] Gantmacher FR. Matrizentheorie. Berlin, New York: Springer; 1986.
[112] Gargantini J, Henrici P. Circular arithmetic and the determination of polynomial zeros. Numer. Math. 1972; 18: 305–320.
[113] Garey ME, Johnson DS. Computers and Intractability. A Guide to the Theory of NP-Completeness. San Francisco: Freeman; 1979.
[114] Garloff J. Block methods for the solution of linear interval equations. SIAM J. Matrix Anal. Appl. 1990; 11(1): 89–106.
[115] Garlof, J. Interval Gaussian elimination with pivot tightening. SIAM J. Matrix Anal. Appl. 2009; 30: 1761–1772.
[116] Garlof, J. Pivot tightening for the interval Cholesky method. PAMM, Proc. Appl. Math. Mech. 2010; 10: 549–550.

[117] Garloff J. Pivot tightening for direct methods for solving symmetric positive definite systems of linear interval equations. Computing, Online First. October 29, 2011.
[118] George A, Liu JW. Computer Solution of Large Sparse Positive Definite Systems. Prentice Hall Series in Computational Mathematics. Englewood Cliffs, NJ: Prentice Hall; 1981.
[119] Gienger A. Zur Lösungsverifikation bei Fredholmschen Integralgleichungen zweiter Art. Thesis, Universität Karlsruhe, Karlsruhe, 1997.
[120] Goldstejn A. Comparison of the Hansen–Sengupta and the Frommer–Lang–Schnurr existence tests. Computing. 2007; 79: 53–60.
[121] Golub GH, van Loan CF. Matrix Computations. Baltimore, Maryland: The Johns Hopkins University Press; 1983.
[122] Gregory RT, Karney DL. A Collection of Matrices for Testing Computational Algorithms. New York: Wiley; 1969.
[123] Griewank A. On automatic differentiation. Mathematical Programming. 1989.
[124] Grüner K. Solving the complex algebraic eigenvalue problem with verified high accuracy. In: Ullrich C, Wolff von Gudenberg J (eds.). Accurate Numerical Algorithms. A Collection of Research Papers. Research Reports ESPRIT, Project 1072, DIAMOND, Vol. 1. Berlin: Springer; 1989, pp. 87–103.
[125] Hämmerlin G, Hoffmann K-H. Numerische Mathematik. Springer Lehrbuch, 4. Aufl. Berlin: Springer; 1994.
[126] Hales T. A proof of the Kepler conjecture. Annals of Mathematics. 2005; 162: 1063–1183.
[127] Hales T, Ferguson SP. A formulation of the Kepler conjecture. Discrete and Computational Geometry. 2006; 36: 21–69.
[128] Hammer D. Methoden zur Lösungsverifikation und Einschließungsverfahren bei gemischten Komplementaritätsproblemen. Thesis, Karlsruher Institut für Technologie (KIT), Karlsruhe, 2012.
[129] Hammer R, Hocks M, Kulisch U, Ratz D. Numerical Toolbox for Verified Computing I. Basic Numerical Problems. Springer Series in Computational Mathematics 21. Berlin: Springer; 1993.
[130] Hammer R, Hocks M, Kulisch U, Ratz D. C++ Toolbox for Verified Computing. Berlin: Springer; 1995.
[131] Hansen ER. Bounding the solution of linear interval equations. SIAM J. Numer. Anal. 1992; 29: 1493–1503.
[132] Hansen E, Sengupta S. Bounding solutions of systems of equations using interval analysis. BIT. 1981; 21: 203–211.
[133] Hansen E, Walster GW. Global Optimization Using Interval Analysis. 2nd edn., revised and expanded. New York, Basel: Marcel Dekker; 2004.
[134] Hardy GH. Sur les zéros de la fonction $\zeta(s)$ de Riemann. C. R. Acad. Sci. Paris. 1914; 158: 1012–1014.
[135] Hargreaves GI. Interval Analysis in MATLAB. Numerical Analysis Report No. 416. Manchester Centre for Computational Mathematics, Department of Mathematics, University of Manchester; 2002.
[136] Hartfiel DJ. Concerning the solution set of $Ax = b$ where $P \le A \le Q$ and $p \le b \le q$. Numer. Math. 1980; 35: 355–359.
[137] Hass J, Schlafly R. Double bubbles minimize. Ann. Math. 2000; 151(2): 459–515.
[138] Hebgen M. Eine scaling-invariante Pivotsuche für Intervallmatrizen. Computing. 1974; 12: 99–106.
[139] Heindl G. Computable hypernorm bounds for the error of an approximate solution to a system of nonlinear equations. In: Atanassova L, Herzberger J (eds.). Computer Arithmetic and Enclosure Methods. Amsterdam: North-Holland; 1992, pp. 451–461.

[140] Heindl G. Some inclusion results based on a generalized version of the Oettli–Prager theorem. Z. angew. Math. Mech. 1996; 76: S3, 263–266.

[141] Henrici P. On the speed of convergence of cyclic and quasi-cyclic Jacobi methods for computing eigenvalues of Hermitian matrices. SIAM J. Appl. Math. 1958; 6: 144–162.

[142] Henrici P. Circular arithmetic and the determination of polynomial zeros. Springer Lecture Notes in Mathematics 228. Berlin, New York: Springer; 1971, pp. 86–92.

[143] Hertling P. A lower bound for range enclosure in interval arithmetic. Theoretical Computer Science. 2002; 279: 83–95.

[144] Herzberger J. Topics in Validated Computations. Amsterdam: Elsevier; 1994.

[145] Heuser H. Lehrbuch der Analysis. Teil 2. Stuttgart: Teubner; 1981.

[146] Heuser H. Funktionalanalysis. Theorie und Anwendung. 4. Aufl. Stuttgart: Teubner; 2006.

[147] Higham JN. Accuracy and Stability of Numerical Algorithms. Philadelphia: SIAM; 1996.

[148] Hille E. Analytic Function Theory. Vol. I. 2nd edn. New York: Chelsea; 1959.

[149] Hladík M. Description of symmetric and skew-symmetric solution set. SIAM J. Matrix Anal. Appl. 2008; 30 (2): 509–521.

[150] Horn RA, Johnson CR. Matrix Analysis. Cambridge: Cambridge University Press; 1994.

[151] IEEE Task P754. ANSI/IEEE 754–1985, Standard for Binary Floating-Point Arithmetic. New York: IEEE; 1985.

[152] Isträţescu VI. Fixed Point Theory. Mathematics and its Applications, Vol. 7. Dordecht, Holland: D. Reidel Publishing Co.; 1981.

[153] Jansson C. Rigorous sensitivity analysis for real symmetric matrices with uncertain data. In: Kaucher E, Markov SM, Mayer G (eds.). Computer Arithmetic, Scientific Computation and Mathematical Modelling. Basel: Baltzer; 1991, pp. 293– 316.

[154] Jansson C. Interval linear systems with symmetric matrices, skew-symmetric matrices and dependencies in the right hand side. Computing. 1991; 46: 265–274.

[155] Jansson C. On self-validating methods for optimization problems. In: Herzberger J (ed.). Topics in Validated Computations. Amsterdam: Elsevier; 1994, pp. 381–438.

[156] Jansson C. Calculation of exact bounds for the solution set of linear interval systems. Linear Algebra Appl. 1997; 251: 321–340.

[157] Jordan DW, Smith P. Nonlinear Differential Equations (Second Edition). Oxford: Clarendon Press; 1987.

[158] Kantorovich L. On Newton's method for functional equations. (Russian) Dokl. Akad. Nauk SSSR. 1940, 59. 1237–1240.

[159] Kaucher EW, Miranker WL. Self-Validating Numerics for Function Space Problems. Computation with Guarantee for Differential and Integral Equations. Orlando, New York: Academic Press; 1984.

[160] Kaucher E, Schulz-Rinne C. Aspects of self-validating numerics in Banach spaces. In: Ullrich C (ed.). Computer Arithmetic and Self-Validating Numerical Methods. Boston, New York: Academic Press; 1990, pp. 269–299.

[161] Kearfott RB. Rigorous Global Search: Continuous Problems. Nonconvex Optimization and Its Applications, Vol. 13. Dordrecht: Kluwer; 1996.

[162] Kelch R. Ein adaptives Verfahren zur Numerischen Quadratur mit automatischer Ergebnisverifikation. Thesis, Universität Karlsruhe, Karlsruhe, 1989.

[163] Kioustelidis JB. Algorithmic error estimation for approximate solutions on nonlinear systems of equations. Computing. 1978; 19: 313–320.

[164] Klatte R, Kulisch U, Neaga M, Ratz D, Ullrich C. PASCAL-XSC, Sprachbeschreibung mit Beispielen. Berlin: Springer; 1991.

[165] Klatte R, Kulisch U, Wiethof, A, Lawo C, Rauch M. C-XSC, A C++ Class Library for Extended Scientific Computing. Berlin: Springer; 1993.

[166] Klein W. Zur Einschließung der Lösung von linearen und nichtlinearen Fredholmschen Integralgleichungssystemen zweiter Art. Thesis, Universität Karlsruhe, Karlsruhe, 1990.
[167] Koeber M. Private communication, Karlsruhe, 1993.
[168] Koeber M. Lösungseinschließung bei Anfangswertproblemen für quasilineare hyperbolische Differentialgleichungen. Thesis, Universität Karlsruhe, Karlsruhe, 1997.
[169] Koeber M. Inclusion of solutions of Cauchy problems for quasilinear hyperbolic equations. In: Fey M, Jeltsch R. Hyperbolic Problems: Theory, Numerics, Applications, Vol. II. International Series of Numerical Mathematics, Vol. 130. Basel: Birkhäuser: 1999, pp. 569–578.
[170] König S. On the inflation parameter used in self-validating methods. In: Ullrich C (ed.). Contributions to Computer Arithmetic and Self-Validating Numerical Methods. IMACS Ann. Comput. Appl. Math. 7. Basel: Baltzer, IMACS; 1990, pp. 127–132.
[171] Krawczyk R. Iterative Verbesserung von Schranken für Eigenwerte und Eigenvektoren reeller Matrizen. Z. Angew. Math. Mech. 1968; 48: T80–T83.
[172] Krawczyk R. Newton-Algorithmen zur Bestimmung von Nullstellen mit Fehlerschranken. Computing. 1969; 4: 187–201.
[173] Krawczyk R. Fehlerabschätzung reeller Eigenwerte und Eigenvektoren von Matrizen. Computing. 1969; 4: 281–293.
[174] Krawczyk R. Verbesserung von Schranken für Eigenwerte und Eigenvektoren von Matrizen. Computing. 1970; 5: 200–206.
[175] Kreinovich V, Lakeyev AV, Rohn J, Kahl PT. Computational Complexity and Feasibility of Data Processing and Interval Computations. Dordrecht: Springer; 1998.
[176] Kreß N. Über Fehlerabschätzungen für Eigenwerte und Eigenvektoren reellsymmetrischer Matrizen. Thesis, Universität Karlsruhe, Karlsruhe 1973. (Interner Bericht Nr. 73/4, Institut für Praktische Mathematik, Universität Karlsruhe)
[177] Krier N. Komplexe Kreisarithmetik. Thesis, Universität Karlsruhe, Karlsruhe 1973. (Interner Bericht Nr. 73/2, Institut für Praktische Mathematik, Universität Karlsruhe)
[178] Krückeberg F. Ordinary differential equations. In: Hansen E (ed.). Topics in Interval Analysis. Oxford: Oxford University Press; 1969, pp. 91–97.
[179] Krückeberg F. Partial differential equations. In: Hansen E (ed.). Topics in Interval Analysis. Oxford: Oxford University Press; 1969, pp. 98–101.
[180] Kulisch U. Grundzüge der Intervallrechnung. Überblicke Mathematik 2. Mannheim: Bibliographisches Institut; 1969.
[181] Kulisch U. Grundlagen des numerischen Rechnens. Reihe Informatik/19. Mannheim: Bibliographisches Institut; 1976.
[182] Kulisch U. Computer Arithmetic and Validity. Theory, Implementation, Applications. Studies in Mathematics 33. Berlin, New York: De Gruyter; 2008.
[183] Kulisch UW, Miranker WL. Computer Arithmetic in Theory and Practice. New York: Academic Press; 1981.
[184] Kulisch UW, Miranker WL (eds.). A New Approach to Scientific Computing. New York: Academic Press; 1983.
[185] Kuttler J. A fourth order finite-difference approximation for the fixed membrane eigenproblem. Math. Comput. 1971; 25: 237–256.
[186] Lakeyev AV, Noskov SI. On the set of solutions of a linear equation with intervally defined operator and right-hand side. In Russian. Siberian Mathematical Journal. 1994; 35: 1074–1084.
[187] Lawson CL, Hanson RJ, Kincaid D, Krogh FT. Basic linear algebra subprograms for Fortran usage. ACM Trans. Math. Soft. 1979; 5: 308–323.
[188] Lehmann NJ. Beiträge zur Lösung linearer Eigenwertprobleme I. Z. Angew. Math. Mech. 1949; 29: 341–356.

[189] Lehmann NJ. Beiträge zur Lösung linearer Eigenwertprobleme II. Z. Angew. Math. Mech. 1949; 30: 1–16.
[190] Ljapin ES. Semigroups. Revised edition. Providence: Amer. Math. Soc.; 1968.
[191] Leontief WW. The Structure of the American Economy, 1919–1939: An Empirical Application of Equilibrium Analysis. 2nd edn. New York: Oxford University Press; 1951.
[192] Lohner R. Einschließung der Lösung gewöhnlicher Anfangs- und Randwertaufgaben und Anwendungen. Thesis, Universität Karlsruhe, Karlsruhe, 1988.
[193] Lohner R. Enclosing all eigenvalues of symmetric matrices. In: Ullrich C, Wolff von Gudenberg J (eds.). Accurate Numerical Algorithms. A Collection of Research Papers. Research Reports ESPRIT, Project 1072, DIAMOND, Vol. 1. Berlin: Springer; 1989, pp. 87–103.
[194] Lohner RL. Interval arithmetic in staggered correction format. In: Adams E, Kulisch U (eds.). Scientific Computing with Automatic Result Verification. New York: Academic Press; 1993, pp. 301–321.
[195] Love ER. The electrostatic field of two equal circular conducting disks. Quart. J. Mech. Appl. Math. 2 1949; 428–451.
[196] van de Lune J, te Riele HJJ, Winter DT. On the zeros of the Riemann zeta function in the critical strip. IV. Math. Comp. 1986; 46: 667–681.
[197] Maier T. Intervall-Input-Output-Rechnung. Königstein/Ts: Hain; 1985.
[198] Mayer G. On the convergence of powers of interval matrices. Linear Algebra Appl. 1984; 58: 201–216.
[199] Mayer G. On the convergence of powers of interval matrices (2). Numer. Math. 1985; 46: 69–83.
[200] Mayer G. On the convergence of the Neumann series in interval analysis. Linear Algebra Appl. 1985; 65: 63–70.
[201] Mayer G. On a theorem of Stein–Rosenberg type in interval analysis. Numer. Math. 1986; 50: 17–26.
[202] Mayer G. On the asymptotic convergence factor of the total step method in interval computation. Linear Algebra Appl. 1987; 85: 153–164.
[203] Mayer G. Comparison theorems for iterative methods based on strong splittings. SIAM J. Numer. Anal. 1987; 24: 215–227.
[204] Mayer G. Enclosing the solutions of systems of linear equations by interval iterative processes. In: Kulisch U, Stetter HJ (eds.). Scientific Computation with Automatic Result Verification. Computing Suppl. 1988; 6: 47–58.
[205] Mayer G. Old and new aspects of the interval Gaussian algorithm. In: Kaucher E, Markov SM, Mayer G (eds.). Computer Arithmetic, Scientific Computation and Mathematical Modelling. Basel: Baltzer; 1991, pp. 329–349.
[206] Mayer G. Some remarks on two interval-arithmetic modifications of the Newton method. Computing. 1992; 48: 125–128.
[207] Mayer G. Enclosures for eigenvalues and eigenvectors. In: Atanassova L, Herzberger J (eds.). Computer Arithmetic and Enclosure Methods. Amsterdam: North-Holland; 1992, pp. 49–67.
[208] Mayer G. Result verification for eigenvectors and eigenvalues. In: Herzberger J (ed.). Topics in Validated Computations. Amsterdam: Elsevier; 1994, pp. 209–276.
[209] Mayer, G. A unified approach to enclosure methods for eigenpairs. Z. Angew. Math. Mech. 1994; 74: 115–128.
[210] Mayer G. Ergebnisverifikation beim algebraischen Eigenwertproblem. In: Chatterji SD, Fuchssteiner B, Kulisch U, Liedl R (eds.). Jahrbuch Überblicke Mathematik 1994. Braunschweig: Vieweg Verlag; 1994, pp. 111–130.
[211] Mayer G. Epsilon-inflation in verification algorithms. J. Comp. Appl. Math. 1995; 60: 147–169.
[212] Mayer G. Epsilon-inflation with contractive interval functions. Appl. Math. 1998; 43: 241–254.

[213] Mayer G. A new way to describe the symmetric solution set S_{sym} of linear interval systems. In: Alefeld G, Chen X (eds.). Topics in Numerical Analysis With Special Emphasis on Nonlinear Problems. Computing Suppl. 2001; 15: 151–163.
[214] Mayer G. A contribution to the feasibility of the interval Gaussian algorithm. Reliable Computing. 2006; 12(2): 79–98.
[215] Mayer G. On the interval Gaussian algorithm. In: Luther W, Otten W (eds.). IEEE-Proceedings of SCAN 2006, 12th GAMM – IMACS International Symposium on Scientific Computing, Computer Arithmetic and Validated Numerics. Duisburg, Germany, September 26–29, 2006. Washington, DC: IEEE Computer Society; ISBN 0-7695-2821-X 2007; CD.
[216] Mayer G. A survey on properties and algorithms for the symmetric solution set. Preprint 12/2 from the series 'Preprints aus dem Institut für Mathematik', Universität Rostock, Rostock, 2012, 58 pages.
[217] Mayer G. An Oettli–Prager like theorem for the symmetric solution set and for related solution sets. SIAM J. Matrix Anal. Appl. 2012; 33(3): 979–999.
[218] Mayer G. On an expression for the midpoint and the radius of the product of two intervals. Reliable Computing. 2012; 16: 210–224.
[219] Mayer G. Three short descriptions of the symmetric solution set and of the skew-symmetric solution set. Linear Algebra Appl. 2015; 475: 73–79.
[220] Mayer G, Pieper L. A necessary and sufficient criterion to guarantee the feasibility of the interval Gaussian algorithm for a class of matrices. Appl. Math. 1993; 38: 205–220.
[221] Mayer G, Rohn J. On the applicability of the interval Gaussian algorithm. Reliable Computing. 1998; 4(3): 205–222.
[222] Mayer G, Warnke I. On the shape of the fixed points of $[f]([x]) = [A][x] + [b]$. In: Alefeld G, Rohn J, Rump SM, Yamamoto T (eds.). Symbolic Algebraic Methods and Verification Methods – Theory and Applications. Wien: Springer; 2001, pp. 11–20.
[223] Mayer G, Warnke I. On the limit of the total step method in interval analysis. In: Kulisch U, Lohner R, Facius A (eds.). Perspectives on Enclosure Methods. Wien: Springer; 2001, pp. 157–172.
[224] Mayer G, Warnke I. On the fixed points of the interval function $[f]([x]) = [A][x] + [b]$. Linear Algebra Appl. 2003; 363: 201–216.
[225] Mayer J. An approach to overcome division by zero in the interval Gaussian algorithm. Reliable Computing. 2002; 8(3) 229–237.
[226] Mayer O. Über die in der Intervallrechnung auftretenden Räume und einige Anwendungen. Thesis, Universität Karlsruhe, Karlsruhe, 1968.
[227] Mayer O. Algebraische und metrische Strukturen in der Intervallrechnung und einige Anwendungen. Computing. 1970; 5: 144–162.
[228] Miranda C. Un' osservazione su un Teorema di Brouwer. Boletino Unione Mat. Ital. Ser. II 1940; 3: 5–7.
[229] Miyajima S, Ogita T, Rump SM, Oishi S. Fast verification for all eigenpairs in symmetric positive definite generalized eigenvalue problems. Reliable Computing. 2010; 14: 24 –45.
[230] Moore RE. Automatic error analysis in digital computation. Technical Report LMSD-48421. Lockheed Missiles and Space Division. Sunnyvale, CA; 1959.
[231] Moore RE. Interval arithmetic and automatic error analysis in digital computing. Thesis, Applied Mathematics and Statistics Laboratories, Report 25. Stanford University, 1962.
[232] Moore RE. Interval Analysis. Englewood Cliffs, NJ: Prentice-Hall; 1966.
[233] Moore RE. A test for existence of solutions to nonlinear systems. SIAM J. Numer. Anal. 1977; 14: 611–615.
[234] Moore RE. A computational test for convergence of iterative methods for nonlinear systems. SIAM J. Numer. Anal. 1978; 15(6): 1194–1196.

[235] Moore RE. Methods and Applications of Interval Analysis. SIAM Studies 2. Philadelphia: SIAM; 1979.
[236] Moore RE. Computational Functional Analysis. Chichester: Ellis Horwood Ltd; 1985.
[237] Moore RE, Kearfott RB, Cloud MJ. Introduction to Interval Analysis. Philadelphia: SIAM; 2009.
[238] Moore RE, Kioustelidis JB. A simple test for accuracy of approximate solutions to nonlinear (or linear) systems. SIAM J. Numer. Anal. 1980; 17: 521–529.
[239] Moore RE, Qi L. A successive interval test for nonlinear systems. SIAM J. Numer. Anal. 1982; 19(4): 845–850.
[240] Muller J-M. Elementary Functions. Algorithms and Implementation. 2nd edn. Basel: Birkhäuser; 2006.
[241] Nagatou K. A numerical method to verify the elliptic eigenvalue problems including a uniqueness property. Computing. 1999; 63: 109–130.
[242] Nagatou K. Validated computations for fundamental solutions of linear ordinary differential operators. In: Bandle C, Gilányi A, Losonczi L, Páles Z, Plum M (eds.). Inequalities and Applications. International Series of Numerical Mathematics. 2008; 157: 43–50.
[243] Nagatou K, Hashimoto K, Nakao MT. Numerical verification of stationary solutions for Navier-Stokes problems. J. Comp. Appl. Math. 2007; 199: 445–451.
[244] Nakao MT. Solving nonlinear parabolic problems with result verification: Part I. J. Comp. Appl. Math. 1991; 38: 323–334.
[245] Nakao MT. A numerical verification method for the existence of weak solutions for nonlinear boundary value problems. J. Math. Anal. Appl. 1992; 164: 489–507.
[246] Nakao MT. State of the art for numerical computation with guaranteed accuracy. Math. Japanese 1998; 48: 323–338.
[247] Nakao MT, Yamamoto N, Nagatou K. Numerical verifications for eigenvalues of second-order elliptic operators. Japan J. Ind. Appl. Math. 1999; 16(3): 307–320.
[248] Nedialkov NS, Jackson KR, Corliss GF. Validated solutions of initial value problems for ordinary differential equations. Appl. Math. Comp. 1999; 105(1) 21–68.
[249] Nedialkov NS, Jackson KR. A new perspective on the wrapping effect in interval methods for initial value problems for ordinary differential equations. In: Kulisch U, Lohner R, Facius A. (eds.). Perspectives on Enclosure Methods. Wien: Springer-Verlag; 2001, pp. 219–264.
[250] Neher M. Private communication, Karlsruhe, 1993.
[251] Neher M. Ein Einschließungsverfahren für das inverse Dirichletproblem. Thesis, Universität Karlsruhe, Karlsruhe, 1993.
[252] Neher M. An enclosure method for the solution of linear ODEs with polynomial coefficients. Numer. Funct. Anal. Optim. 1999; 20: 779–803.
[253] Neumaier A. Eigenwertabschätzungen für Matrizen. J. Reine und Angew. Math. 1984; 354: 206–212.
[254] Neumaier A. New techniques for the analysis of linear interval equations. Linear Algebra Appl. 1984; 58: 273–325.
[255] Neumaier A. Further results on linear interval equations. Linear Algebra Appl. 1987; 87: 155–179.
[256] Neumaier A. The enclosure of solutions of parameter-dependent systems of equations. In: Moore RE (ed.). Reliability in Computing. The Role of Interval Methods in Scientific Computing. Perspectives in Computing 19. Boston: Academic Press; 1988, pp. 269–286.
[257] Neumaier A. Interval Methods for Systems of Equations. Cambridge: Cambridge University Press; 1990.
[258] Neumaier A. The wrapping effect, ellipsoid arithmetic, stability and confidence regions. In: Albrecht R, Alefeld G, Stetter HJ (eds.). Validation Numerics. Theory and Applications. Computing Suppl. 9, Wien: Springer; 1993, pp. 175–190.

[259] Neumaier A, Shen Z. The Krawczyk operator and Kantorovich's theorem. J. Math. Anal. Appl. 1990; 149: 437–443.
[260] Nickel K. On the Newton method in interval analysis. MRC Technical Summary Report #1136. University of Wisconsin, Madison WI, 1971.
[261] Nickel K. Eine Überschätzung des Wertebereichs einer Funktion in der Intervallrechnung mit Anwendung auf lineare Gleichungssysteme. Computing. 1977; 18: 15–36.
[262] Nickel K. Using interval methods for the numerical solution of ODE's. Z. Angew. Math. Mech. 1986; 66: 513–523.
[263] Ning S, Kearfott RB. A comparison of some methods for solving linear interval equations. SIAM J. Numer. Anal. 1997; 34: 1289–1305.
[264] Oettli W, Prager W. Compatibility of approximate solution of linear equations with given error bounds for coefficients and right-hand sides. Numer. Math. 1964; 6: 405–409.
[265] Oishi S, Rump SM. Fast verification of solutions of matrix equations. Numer. Math. 2002; 90: 755–773.
[266] Ortega JM. Numerical Analysis: A Second Course. Classics in Applied Mathematics 3. Philadelphia: SIAM; 1990.
[267] Ortega JM, Rheinboldt WC. Iterative Solution of Nonlinear Equations in Several Variables. New York: Academic Press; 1970.
[268] Pang C-T, Lur Y-Y, GuuS-M. A new proof of Mayer's theorem. Linear Algebra Appl. 2002; 350: 273–278.
[269] Petković M. Iterative Methods for Simultaneous Inclusion of Polynomial Zeros. Springer Lecture Notes in Mathematics, Vol. 1387. Berlin: Springer; 1989.
[270] Petković MS, Petković LD. Complex Interval Arithmetic and Its Applications. Wiley-VCH Mathematical Research, Vol. 105. Berlin: Wiley-VCH; 1998.
[271] Petras K. Principles of verified numerical integration. J. Comput. Appl. Math. 2007; 199(2): 317–328.
[272] Pissanetzky S. Sparse Matrix Technology. London: Academic Press; 1984.
[273] Plum M. Computer-assisted existence proofs for two-point boundary value problems. Computing. 1991; 46: 19–34.
[274] Plum M. Inclusion methods for elliptic boundary value problems. In: Herzberger J (ed.). Topics in Validated Computations. Amsterdam: Elsevier; 1994, pp. 323–379.
[275] Popova ED. Multiplication distributivity or proper and improper intervals. Reliable Computing. 2001; 7: 129–140.
[276] Popova ED. Quality of the solution sets of parameter-dependent interval linear systems. Z. Angew. Math. Mech. 2002; 82: 1–5.
[277] Qi L. A generalization of the Krawczyk–Moore algorithm. In: Nickel K (ed.). Interval Mathematics 1980. New York: Academic Press; 1980, pp. 481–488.
[278] Qi L. A note on the Moore test for nonlinear systems. SIAM J. Numer. Anal. 1982; 19(4): 851–857.
[279] Rall LB. A comparison of the existence theorems of Kantorovich and Moore. SIAM J. Numer. Anal. 1980; 17: 148–161.
[280] Rall LB. Automatic Differentiation. Springer Lecture Notes in Computer Science 120. Berlin, New York: Springer; 1981.
[281] Ratschek H. Die binären Systeme der Intervallarithmetik. Computing. 1970; 6: 295–308.
[282] Ratschek H. Die Subdistributivität in der Intervallarithmetik. Z. Angew. Math. Mech. 1971; 51: 189–192.
[283] Ratschek H. Teilbarkeitskriterien der Intervallarithmetik. J. Reine und Angew. Math. 1972; 252: 128–138.

[284] Ratschek H, Rokne J. Computer Methods for the Range of Functions. Chichester: Ellis Horwood Ltd; 1984.
[285] Ratschek H, Rokne J. New Computer Methods for Global Optimization. Chichester: Ellis Horwood Ltd; 1988.
[286] Ratschek H, Rokne JG. Formulas for the width of interval products. Reliable Comput. 1995; 1(1): 9–14.
[287] Ratschek H, Rokne JG. Geometric Computations with Intervals and New Robust Methods. Applications in Computer Graphics, GIS and Computational Geometry. Chichester: Horwood Publishing: 2003.
[288] Ratschek H, Sauer W. Linear interval equations. Computing. 1982; 28: 105–115.
[289] Ratz D. Automatische Ergebnisverifikation bei globalen Optimierungsproblemen. Thesis, Universität Karlsruhe, Karlsruhe 1992.
[290] Reichmann K. Ein hinreichendes Kriterium für die Durchführbarkeit des Intervall-Gauß-Algorithmus bei Intervall-Hessenberg-Matrizen ohne Pivotsuche. Z. Angew. Math. Mech. 1979; 59: 373–379.
[291] Reichmann K. Abbruch beim Intervall-Gauss-Algorithmus. Computing. 1979; 22: 355–361.
[292] Rihm R. Über Einschließungsverfahren für gewöhnliche Anfangswertprobleme und ihre Anwendung auf Differentialgleichungen mit unstetiger rechter Seite. Thesis, Universität Karlsruhe, Karlsruhe, 1993.
[293] Rihm R. Interval methods for initial value problems in ODEs. In: Herzberger J (ed.). Topics in Validated Computations. Amsterdam: Elsevier; 1994, pp. 173–207.
[294] Ris FN. Interval Analysis and Applications to Linear Algebra, Thesis, Oxford University, 1972.
[295] Ris FN. Tools for the analysis of interval arithmetic. In: Nickel K (ed.). Interval Mathematics. Lecture Notes in Computer Science, Vol. 29. Berlin: Springer; 1975, pp. 75–98.
[296] Roesler F. Riemann's hypothesis as an eigenvalue problem. Linear Algebra Appl. 1986; 81: 153–198.
[297] Rohn J.Input-Output planning with inexact data. Freiburger Intervall-Berichte. 1978; 78/9: 1–16.
[298] Rohn J. Interval linear systems. Freiburger Intervall-Berichte.1984; 84/7: 33–58.
[299] Rohn J. Inner solutions of linear interval systems. In: Nickel K (ed.). Interval Mathematics 1985. Lecture Notes in Computer Science 212. Berlin: Springer; 1986, pp. 157–158.
[300] Rohn J. On nonconvexity of the solution set of a system of linear interval equations. BIT. 1989; 30: 161–165.
[301] Rohn J. Systems of linear interval equations. Linear Algebra Appl. 1989; 126: 39–78.
[302] Rohn J. Cheap and tight bounds: The recent result by E. Hansen can be made more efficient. Interval Computations. 1993; 4: 13–21.
[303] Rohn J. Enclosing solutions of linear interval equations is NP-hard. Computing. 1994; 53: 365–368.
[304] Rohn J. On overestimation produced by the interval Gaussian algorithm. Reliable Computing. 1997; 3(4): 363–368.
[305] Rohn J. Personal Communication. Berlin, 2001.
[306] Rohn J. A Method for Handling Dependent Data in Interval Linear Systems. Technical Report No. 911. Institute of Computer Science, Academy of Sciences of the Czech Republic, July 7, 2004, 1–7.
[307] Rohn J. An algorithm for computing the hull of the solution set of interval linear equations. Linear Algebra Appl. 2011; 435: 193–201.
[308] Rohn J. Personal Communication via Email, October 3, 2014.
[309] Rohn J, Kreinovich V. Computing exact componentwise bounds on solutions of linear systems with interval data is NP-hard. SIAM J. Matrix Anal. Appl. 1995; 16: 415–420.

[310] Rokne J, Lancaster P. Complex interval arithmetic. Comm. ACM. 1971; 14: 111–112.
[311] Rump SM. Kleine Fehlerschranken bei Matrixproblemen. Thesis, Universität Karlsruhe, 1980.
[312] Rump SM. Solving algebraic problems with high accuracy. In: Kulisch UW, Miranker WL (eds.). A New Approach to Scientific Computation. New York: Academic Press; 1983, pp. 51–120.
[313] Rump SM. Rigorous sensitivity analysis for systems of linear and nonlinear equations. Math. Comp. 1990; 54: 721–736.
[314] Rump SM. A Class of arbitrarily ill-conditioned floating-point matrices. SIAM J. Matrix Anal. Appl. 1991; 12(4): 645–653.
[315] Rump SM. On the solution of interval linear systems. Computing. 1992; 47: 337–353.
[316] Rump SM. Inclusion of the solution of large linear systems with positive definite symmetric M-matrix. In: Atanassova L, Herzberger J (eds.). Computer Arithmetic and Enclosure Methods. Amsterdam: North-Holland; 1992, pp. 339–347.
[317] Rump SM. Validated solution of large linear systems. In: Albrecht R, Alefeld G, Stetter HJ (eds.). Validation Numerics. Theory and Applications. Computing Suppl. 9. Wien: Springer; 1993, pp. 191–212.
[318] Rump SM. Verification methods for dense and sparse systems of equations. In: Herzberger J (ed.). Topics in Validated Computations. Amsterdam: Elsevier; 1994, pp. 63–135.
[319] Rump SM. Fast and parallel interval arithmetic. BIT. 1999; 39(3): 534–554.
[320] Rump SM. INTLAB – INTerval LABoratory. In: Csendes T (ed.). Developments in Reliable Computing. Dordrecht: Kluwer; 1999, pp. 77–104.
[321] Rump SM. Inversion of extremely ill-conditioned matrices using faithfully rounded dot product. In: Research Society of Nonlinear Theory and its Applications (ed.). Proceedings of the 2007 International Symposium on Nonlinear Theory and Applications (NOLTA'07), Vancouver, Canada, September 16–19. Japan: IEICE; 2007, pp. 168–171.
[322] Rump SM. Verification methods: Rigorous results using floating-point arithmetic. Acta Numerica. 2010; 19: 287–449.
[323] Rump SM, Ogita T, Oishi S. Accurate floating point summation I: Faithful rounding. SIAM J. Sci. Comput. 2008; 31(1): 189–224.
[324] Rump SM, Ogita T, Oishi S. Accurate floating point summation II: Sign, K-fold faithful and rounding to nearest. SIAM J. Sci. Comput. 2008; 31(2): 1269–1302.
[325] Schäfer U. Das lineare Komplementaritätsproblem mit Intervalleinträgen. Thesis, Universität Karlsruhe, Karlsruhe, 1999.
[326] Schäfer U. Two ways to extend the Cholesky decomposition to block matrices with interval entries. Reliable Computing. 2002; 8(1): 1–20.
[327] Schäfer U. Aspects for a block version of the interval Cholesky algorithm. J. Comput. Appl. Math. 2003; 152 (1): 481–491.
[328] Schäfer U. Das lineare Komplementaritätsproblem – Eine Einführung. Berlin: Springer; 2008.
[329] Schäfer U, Schnurr, M. A comparison of simple tests for accuracy of approximate solutions to nonlinear systems with uncertain data. Preprint Nr. 04/22, Universität Karlsruhe, 2004.
[330] Schnurr M. On the proofs of some statements concerning the theorems of Kantorovich, Moore, Miranda. Reliable Computing. 2005; 11: 77–85.
[331] Schnurr M. Steigungen höherer Ordnung zur verifizierten globalen Optimierung. Thesis, Universität Karlsruhe, Universitätsverlag Karlsruhe, Karlsruhe, 2007.
[332] Schrijver A. Theory of Linear and Integer Programming. New York: Wiley; 1986.
[333] Schröder J. Das Iterationsverfahren bei allgemeinerem Abstandsbegriff. Math. Zeitschr. 1956; 66: 111–116.
[334] Schulte U. Einschließungsverfahren zur Bewertung von Getriebeschwingungsmodellen. Reihe 11: Schwingungstechnik, Nr. 146. Düsseldorf: VDI-Verlag; 1991.

[335] Schwandt H. Schnelle fast global konvergente Verfahren für die Fünf-Punkt-Diskretisierung mit Dirichletschen Randbedingungen auf Rechteckgebieten. Thesis, Fachbereich Mathematik, TU Berlin, Berlin 1981.
[336] Sharaya IA. On unbounded tolerable solution sets. Reliable Computing. 2005; 11: 425–432.
[337] Shary SP. On controlled solution set of interval algebraic systems. Interval Computations. 1992; 6: 66–75.
[338] Shary SP. Solving the linear interval tolerance problem. Math. Comput. Simulation. 1995; 39: 53–85.
[339] Shary SP. Algebraic approach to the interval linear static identification, tolerance, control problems, or one more application of Kaucher arithmetic. Reliable Computing. 1996; 2(1): 3–33.
[340] Shary SP. Controllable solution set to interval static systems. Appl. Math. Comput. 1997; 86: 185–196.
[341] Shary SP. Algebraic approach in the "Outer Problem" for interval linear equations. Reliable Computing. 1997; 3(2): 103–135.
[342] Shary SP. A new technique in systems analysis under interval uncertainty and ambiguity. Reliable Computing. 2002; 8(5): 321–418.
[343] Shen Z, Neumaier A. A note on Moore's interval test for zeros of nonlinear systems. Computing. 1988; 40: 85–90.
[344] Shen Z, Wolfe MA. A note on the comparison of the Kantorovich and Moore theorems. Nonlinear Analysis, Theory, Methods and Appl. 1990; 15: 229–232.
[345] Spaniol O. Die Distributivität in der Intervallarithmetik. Computing. 1970; 5: 6–16.
[346] Stetter, HJ. Sequential defect correction for high-accuracy floating-point algorithms. Lecture Notes in Mathematics 1066. Berlin: Springer; 1984, pp. 186–202.
[347] Stetter HJ. Validated solution of initial value problems for ODE. In: Ullrich C (ed.). Computer Arithmetic and Self-Validating Numerical Methods. Boston, New York: Academic Press; 1990, pp. 171–187.
[348] Stoer J, Bulirsch R. Introduction to Numerical Analysis. 2nd. edn. New York: Springer; 1993.
[349] Strubecker K. Einführung in die Höhere Mathematik. Band I: Grundlagen. 2nd edn. München, Wien: R. Oldenbourg Verlag; 1966.
[350] Suchowitzki SI, Andejewa LI. Lineare und konvexe Programmierung. München, Wien: R. Oldenbourg Verlag; 1969.
[351] Sunaga T. Theory of an interval algebra and its application to numerical analysis. RAAG Memoirs. 1958; 2: 29–46.
[352] Tucker WA. Rigorous ODE solver and Smale's 14th problem. Found. Comp. Math. 2002; 2(1): 53–117.
[353] Tucker W. Validated Numerics. Princeton and Oxford: Princeton University Press; 2011.
[354] Unger H. Nichtlineare Behandlung von Eigenwertaufgaben. Z. Angew. Math. Mech. 1950; 30 (8/9): 281–282.
[355] van Leeuwen J. Handbook of Theoretical Computer Science. Vol. A, Algorithms and Complexity. Amsterdam: Elsevier; 1998.
[356] Varga RS. Matrix Iterative Analysis. Englewood Cliffs, NJ: Prentice-Hall; 1962.
[357] Volder J. The CORDIC computing technique. IRE Transactions on Electronic Computers EC-8. 1959; (3): 330–334. Reprinted in Swartzlander EE. Computer Arithmetic, Vol. 1. Los Alamitos, CA: IEEE Computer Society Press Tutorial; 1990.
[358] Vrahatis MN. A short proof and a generalization of Miranda's existence theorem. Proc. Amer. Math. Soc. 1989; 107(3): 701–703.

[359] Wang Z. Semilocal convergence of Newton's method for finite-dimensionale variational inequalities and nonlinear complementarity problems. Thesis, Karlsruher Institut für Technologie (KIT), Karlsruhe, 2005.
[360] Watanabe Y, Nagatou K, Plum M, Nakao MT. Verified computations of eigenvalue exclosures for eigenvalue problems in Hilbert spaces. SIAM J. Numer. Anal. 2014; 52(2): 975–992.
[361] Wilkinson JH. Rounding Errors in Algebraic Processes. Notes on Applied Science 32. London: National Physical Laboratory; 1963.
[362] Wilkinson JH. The Algebraic Eigenvalue Problem. Oxford: Clarendon; 1988.
[363] Wilkinson JH, Reinsch C. Handbook for Automatic Computation. Volume II, Linear Algebra. Berlin: Springer; 1971.
[364] Wongwises P. Experimentelle Untersuchungen zur numerischen Auflösung von linearen Gleichungssystemen mit Fehlererfassung. Thesis, Int. Ber. des Inst. für Prakt. Math. 75/1, Universität Karlsruhe, Karlsruhe, 1975.
[365] Yakubovich VA, Starzhinskii, VM. Linear Differential Equations with Periodic Coefficients. Vol. 1 + 2. New York: J. Wiley & Sons; 1975.
[366] Young RC. The algebra of many-valued quantities. Math. Anal. 1931; 104: 260–290.
[367] Zielke G. Testmatrizen mit maximaler Konditionszahl. Computing. 1974; 13: 33–54.

Symbol Index

conv S	convex hull of a set S of real vectors	181
$\emptyset$	empty set	3
$\square S$	interval hull of a bounded set S	76
∂S	boundary of a set S in a metric space	3
$(S, \|\cdot\|)$	normed linear space with linear space S and norm $\|\cdot\|$	7
(S, m)	metric space with set S and metric $m(\cdot,\cdot)$	2
$B(x, r)$	open ball in a metric space with center x and radius r	3
$\overline{B}(x, r)$	closed ball in a metric space with center x and radius r	3
$\overline{S}$	closure of a set S in a metric space	3
S	solution set of an interval linear system $[A]x = [b]$	160
S_{control}	control solution set	163
S_{per}	persymmetric solution set	175
S_{skew}	skew-symmetric solution set	175
S_{sym}	symmetric solution set	164
S_{tol}	tolerance solution set	162
inf S	infimum of an (ordered) set S	76
sup S	supremum of an (ordered) set S	76
$:=$	definition	2
$=:$	definition	2
$\diamond\!\!\!\circ$	one of the binary machine interval operations $+, -, \cdot, /$ with outward rounding	105
$\lvert\cdot\rvert$	absolute value of a number, vector, matrix, interval, interval vector, or interval matrix	9
$\circ$	one of the binary interval operations $+, -, \cdot, /$	77
δ_{ij}	Kronecker symbol	2
$\div$	integral division	99
$.*$	Hadamard product or, equivalently, entrywise product of two matrices	110
$\|\cdot\|$	norm	7
$\|\cdot\|_1$	l_1 norm of a vector or column sum norm of a matrix	8
$\|\cdot\|_2$	Euclidean norm of a vector or spectral norm of a matrix	8
$\|\cdot\|_F$	Frobenius norm	10
$\|\cdot\|_S$	maximum norm scaled with the matrix S^{-1} or corresponding operator norm	10
$\|\cdot\|_\infty$	maximum norm of a vector or row sum norm of a matrix	8
$\|\cdot\|_p$	l_p norm	8
$\|\cdot\|_z$	scaled maximum norm $\|\cdot\|_{D_z}$ of a vector or corresponding operator norm of a matrix	10
$\underline{\cup}$	convex union	76
$o(\cdot)$, $O(\cdot)$	Landau symbols	2
$\mathbb{C}$	set of complex numbers	1
$\mathbb{E}$	set of programmable scalar mathematical expressions	125
$\mathbb{F}$	set of particular elementary functions	97
$\mathbb{F}_b$	set of binary operations $+ - \cdot, /$	125
$\mathbb{F}_u$	set of unary operations $+, -$	125
$\mathbb{I}(S)$	set of all compact intervals, resp. interval vectors, resp. interval matrices, contained in the set S	75

$\mathbb{IC}_C$	set of circular complex intervals $\langle \check{z}, r_z \rangle$	423
$\mathbb{IC}_R$	set of rectangular complex intervals $[a] + i[b]$	418
$\mathbb{IC}_R^n$	set of rectangular complex interval vectors	420
$\mathbb{IC}_R^{m\times m}$	set of rectangular complex $m \times m$ matrices	420
$\mathbb{IR}$	set of all compact intervals $[a]$ contained in the set $\mathbb{R}$	75
$\mathbb{IR}^{m\times n}$	set of interval matrices with m rows and n columns	109
$\mathbb{IR}^n$	set of interval vectors with n components	109
$\mathbb{IR}_B$	set of all intervals $[a]$ with $\underline{a}, \overline{a} \in \mathbb{R}_B$	104
$\mathbb{IR}_M$	set of all machine intervals	104
$\mathbb{K}$	set $\mathbb{R}$ of reals or set $\mathbb{C}$ of complex numbers	1
$\mathbb{N}$	set of positive integers	1
$\mathbb{N}_0$	set of nonnegative integers	1
$\mathbb{N}_M$	particular finite subset of $\mathbb{N}_0$	100
$\mathbb{R}$	set of real numbers	1
$\mathbb{R}^+$	set of positive reals	1
$\mathbb{R}_0^+$	set of nonnegative reals	1
$\mathbb{R}_B$	particular subset of $\mathbb{R}$ containing all normalized floating point numbers	101
$\mathbb{R}_M$	set of floating point numbers	100
$\mathbb{R}^{m\times n}$	set of real $m \times n$ matrices	1
$\mathbb{R}^n$	set of vectors with n real components	1
$\mathbb{Z}$	set of integers	1
$\mathbb{Z}_M$	set of machine integers	99
$\beta([a],[b])$	Neumaier's β-function of two intervals $[a]$, $[b]$	97
$\chi([a])$	Ratschek's χ-function of an interval $[a]$	82
$d([a])$	diameter or, equivalently, width of the interval $[a]$	75
$q([a],[b])$	distance of two intervals $[a]$, $[b]$	93
r_a	short form of the radius of the interval $[a]$	75
r_a^+	real number $\max\{\lvert\check{a}\rvert,\ r_a\}$	89
r_a^-	real number $\min\{\lvert\check{a}\rvert,\ r_a\}$	89
$[a] < [b]$	comparison between two intervals $[a]$, $[b]$	76
$[a] > [b]$	comparison between two intervals $[a]$, $[b]$	76
$[a] \ge [b]$	comparison between two intervals $[a]$, $[b]$	76
$[a] \le [b]$	comparison between two intervals $[a]$, $[b]$	76
$[a]$	compact interval $[\underline{a}, \overline{a}]$	75
$[a]^2$	square of the interval $[a]$	230
$[a]^{-1}$	interval division $1/[a]$	77
$[a]^{1/2}$	square root of the interval $[a]$	230
$\lvert[a]\rvert$	magnitude or, equivalently, absolute value of the interval $[a]$	88
$\check{a}$	short form of the midpoint of the interval $[a]$	75
$\mathrm{int}([a])$	interior of the interval $[a]$	75
$\langle[a]\rangle$	mignitude of the interval $[a]$	88
$\mathrm{mid}([a])$	midpoint of the interval $[a]$	75
$\mathrm{rad}([a])$	radius of the interval $[a]$	75
$\underline{a}$, $\overline{a}$	lower, resp. upper bound of the interval $[a] = [\underline{a}, \overline{a}]$	75
a_l^u	with a finite decimal number a and l, $u \in \mathbb{N}_0$; interval $[a] = [\underline{a}, \overline{a}]$ with $\underline{a} = al$, $\overline{a} = au$	75
$[x]$	interval vector with components $[x]_i$	109
$[x]^C$	interval vector resulting from the interval Cholesky method	231
$[x]^G$	interval vector resulting from the interval Gaussian algorithm	193

$[x]^H$	interval hull of the solution set S	181
$q([A],[B])$	distance of two interval matrices $[A]$, $[B]$	109
r_A	short form of the radius of the interval matrix $[A]$	109
r_A^+	matrix $(r_{a_{ij}}^+)$	110
r_A^-	matrix $(r_{a_{ij}}^-)$	110
$[A] \le [B]$	comparison of two interval matrices $[A]$, $[B]$	110
$[A]$	interval matrix with bounds $\underline{A}, \overline{A}$ and entries $[a]_{ij} = [\underline{a}_{ij}, \overline{a}_{ij}]$	109
$[A]^T$	transposed of the interval matrix $[A]$	110
$[A]^k$	k-th power of the $n \times n$ interval matrix $[A]$ with $[A]^k = [A]^{k-1} \cdot [A]$	115
$[A]^{(k)}$	interval matrices occurring in the Gaussian algorithm	193
$[A]^{-1}$	interval hull of the set of all inverses $\tilde{A}^{-1}$ with $\tilde{A} \in [A]$	112
$\lvert [A]^C \rvert$	particular nonnegative matrix associated with the interval Cholesky method applied to the interval matrix $[A]$	231
$\check{A}$	short form of the midpoint of the interval matrix $[A]$	109
$\langle [A] \rangle$	comparison matrix or, equivalently, Ostrowski matrix of an $n \times n$ interval matrix $[A]$	110
$\mathrm{mid}([A])$	midpoint of the interval matrix $[A]$	109
$\mathrm{rad}([A])$	radius of the interval matrix $[A]$	109
$\lvert [A] \rvert$	absolute value of the interval matrix $[A]$	109
$\lvert [A]^G \rvert$	particular real matrix associated with the interval Gaussian algorithm applied to the interval matrix $[A]$	194
${}^k[A]$	k-th power of the $n \times n$ interval matrix $[A]$ with $[A]^k = [A] \cdot [A]^{k-1}$	115
$([M],[N])$	splitting of the interval matrix $[A]$; $[A] = [M] - [N]$	243
$(f)_k$	k-th Taylor coefficient of the function f	407
(x^k)	sequence of vectors	2
(x_k)	sequence (of numbers, e.g.)	2
$[b]^{(k)}$	interval vectors occurring in the Gaussian algorithm	193
$[b]_C^i$	centered Borsuk intervals	337
$[b]_F^i$	face-centered Borsuk intervals	337
$[b]_M^i$	midpoint based Borsuk intervals	337
$[b]_N^i$	naive Borsuk intervals	337
$[b]_S^i$	scalar product based Borsuk intervals	336
$[f]$	interval function	125
$[f]_M([x])$	mean value form	129
$[m]_C^{i,\pm}$	centered Miranda intervals	327
$[m]_F^{i,\pm}$	face-centered Miranda intervals	327
$[m]_N^{i,\pm}$	naive Miranda intervals	327
$\deg(f, D, y)$	mapping degree of a function f at y with respect to the set D	21
$\deg(i) = \deg_{\overline{G}([A])}(i)$	degree of the node i in the undirected graph $\overline{G}([A])$	211
$\det A$	determinant of a square matrix A	1
$\mathrm{diag}(z_1, \ldots, z_n)$	diagonal matrix with diagonal entries $z_1, \ldots, z_n$	1
$\mathrm{ICh}([A],[b])$	interval vector resulting from the interval Cholesky method; same as $[x]^C$	231
$\mathrm{IGA}([A])$	particular interval matrix associated with the interval Gaussian algorithm	194
$\mathrm{IGA}([A],[b])$	interval vector resulting from the interval Gaussian elimination process applied to $[A]$ and $[b]$; same as $[x]^G$	193

$\operatorname{ind}(A)$	index of inertia of a symmetric matrix A	39
$\langle \check{z}, r_z \rangle$	circular complex interval with midpoint $\check{z}$ and radius r_z	423
$\langle A \rangle$	comparison matrix of a matrix A; Ostrowski matrix	69
$\mathfrak{I}$	any interval iterative method	243
$\overline{G}(A) = (N, E)$	undirected graph of a matrix A with the set N of nodes and the set E of edges	211
$\rho(A)$	spectral radius of a square matrix A	10
$\operatorname{supp}(f)$	support of a function f	20
$Z^{n\times n}$	set of $n \times n$ matrices with non-positive off-diagonal entries	65
$A = (a_{ij})$	matrix with entries a_{ij}	1
$A([x], \tilde{x}, \tilde{C})$	Alefeld operator	309
A^0	zeroth power of a square matrix A; identity matrix	1
A^H	conjugate transpose of a matrix A	1
A^k	k-th power of a square matrix A	1
A^T	transpose of a matrix A	1
$A^{1/2}$	square root of a symmetric positive definite matrix A	44
A_k	k-th leading principal submatrix of the matrix A	198
$A_{*,j}$	j-th column of a matrix A	1
$A_{i,*}$	i-th row of a matrix A	1
b_{nk}	k-th Bernstein polynomial of degree n	14
$C(\mathfrak{I}, [x]^*)$	set of all output sequences of the interval iterative method $\mathfrak{I}$ which converge to $[x]^*$	243
$C(D)$	set of all continuous functions f defined on D	2
$C(D, Y)$	set of all continuous functions $f\colon D \to Y$	2
$C^0(D)$	set of all continuous functions f defined on D	2
$C^0(D, Y)$	set of all continuous functions $f\colon D \to Y$	2
$C^k(D)$	set of all continuous functions f defined on D with continuous partial derivatives up to the order k	2
$C^k(D, Y)$	set of all continuous functions $f\colon D \to Y$ with continuous partial derivatives up to the order k	2
D_z	diagonal matrix $\operatorname{diag}(z_1, \dots, z_n)$ associated with a vector z	1
e	vector whose components all are one	1
$e^{(i)}$	i-th column of the identity matrix I; i-th coordinate unit vector	1
$f([x])$	interval arithmetic evaluation of f at an interval vector $[x]$	125
$f^{-1}(T)$	subset of the domain D of f with $f(x) \in T$	2
$fl(r)$	floating point number associated with $r \in \mathbb{R}_B$	101
$fl_c(r)$	rounding of r by chopping	102
$fl_n(r)$	rounding of r to nearest	102
$fl_\diamond([a])$	particular outward rounding of the interval $[a] \in \mathbb{IR}_B$	104
$fl_\downarrow$	downward directed rounding	101
$fl_\triangledown(r)$	particular downward directed rounding of r	102
$fl_\vartriangle(r)$	particular upward directed rounding of r	102
$fl_\uparrow$	upward directed rounding	101
$G(A) = (N, E)$	directed graph of a matrix A with the set N of nodes and the set E of edges	34
$G_k = (N^k, E^k)$	k-th elimination graph	211
$H([x], \tilde{x}, C)$	Hansen–Sengupta operator	320
$H^{\mathrm{mod}}([x], \tilde{x}, C)$	modified Hansen–Sengupta operator	324
H_n	Hilbert matrix	369

H_n^{prec}	preconditioned Hilbert matrix	369
I	identity matrix	1
I_n	$n \times n$ identity matrix	1
$J(x, y)$	matrix of the multidimensional mean value theorem	28
J_{pq}	Jacobi (or Givens) rotation matrix	363
$K([x], \tilde{x}, C)$	Krawczyk operator	304
$K_s([x], \tilde{x}, [Y], C)$	slope based Krawczyk operator	314
$L([x]^0)$	set of all functions which are Lipschitz continuous in $\mathbb{I}([x]^0)$	131
L^C	lower triangular matrix of the Cholesky decomposition	45
l_f	Lipschitz constant	131
$M([x], \tilde{x}, C)$	Moore–Qi operator	310
m_i	i-th moment of a symmetric matrix A and a vector x	46
$N([x], \tilde{x})$	multidimensional interval Newton operator	298
O	zero matrix or orthant in the cartesian coordinate system of $\mathbb{R}^n$	1
$O_R(\mathfrak{I})$	R-order of the method $\mathfrak{I}$	244
$q(\cdot, \cdot)$	Hausdorff metric, Hausdorff distance	7
$R_f(D)$	range of a function f with domain D	2
$R_p([x]^k)$	R_p-factor of the sequence $([x]^k)$	243
$R_p(\mathfrak{I})$	R_p-factor of the method $\mathfrak{I}$	243
R_x	Rayleigh quotient of a matrix A with respect to a vector x	46
$s(x, z)$	slope between x and z	128
$T_x(\alpha)$	Temple quotient of a matrix A with respect to a vector x and a real number α	46
W_{2n+1}^+	Wilkinson matrix	370
$x = (x_i)$	vector with components x_i	1
x^+	nonnegative vector $(\max\{x_i, 0\})$	179
x^-	nonnegative vector $(\max\{-x_i, 0\})$	179
x^D	particular vector	180
z''	two's complement of a machine integer z	99
z'	one's complement of a machine integer z	99
trace(A)	trace of the matrix A	37
bvp	boundary value problem	318
NaN	not a number	100
ulp	unit of the last place	104

Author Index

Adams, E. 433, 483, 487, 491
Ahlfors, L. V. 5, 417, 425, 483
Albrecht, R. 493, 496
Alefeld, G. ix, 27, 107, 123, 157, 166, 176, 199, 238, 287, 293, 302, 308, 315, 318, 325, 344, 359, 373, 375, 377, 381, 394, 397, 401, 404, 405, 424, 425, 431, 433, 483–485, 487, 492, 493, 496
Alt, R. 484
Andejewa, L. I. 497
Anderson, E. 102, 485
Apostolatos, N. 123, 157, 483, 485
Arndt, H.-R. 123, 265, 288, 420, 485
Atanassova, L. 433, 485, 491, 496
Auzinger, W. 105, 485
Awdejewa, L. I. 287

Bai, Z. 485
Banach, S. 4
Bandle, C. 493
Barth, W. 123, 187, 485
Bauch, H. 433, 485
Baur, W. 415, 485
Beeck, H. 187, 485
Behnke, H. ix, 345, 382, 384, 391, 405, 485, 486
Berman, A. 73, 486
Berner, S. 433, 486
Bernstein, S. N. 14
Birkhoff, G. D. 19
Bischof, C. 485
Blackford, S. 485
Bogolyubov, N. 46
Bohlender, G. 100, 107, 469, 486
Boothroyd, J. 482
Borel, E. 5
Borsuk, K. ix, 21, 72, 335, 341, 486
Bronstein, I. N. 428, 486
Brouwer, L. E. J. vii, 22, 72, 439, 486
Bulirsch, R. 19, 72, 286, 497
Bunch, J. R. 40, 382, 486

Caprani, O. 151, 157, 486
Chaitin-Chatelin, F. 103, 486
Chatterji, S. D. 491
Chen, X. 433, 483, 492
Cholesky, A.-L. 45
Cloud, M. J. 433, 493
Cordes, D. 115, 123, 433, 483, 486
Corliss, G. F. 416, 433, 486, 493
Cornelius, H. 135, 139, 157, 486
Courant, R. 47
Cramer, G. 38, 162, 182
Csendes, T. 496

Deif, A. S. 405, 486
Deimling, K. 72, 486
Dekker, T. J. 482
Demmel, J. 485
Dobner, H.-J. 433, 486
Dongarra, J. J. 373, 469, 485, 486
Du Croz, J. 485, 486
Duff, I. S. 486
Dwyer, P. S. 106, 486

Eble, I. 433, 486
Eijgenraam, P. 433, 486
Enger, W. 433, 486

Facius, A. 487, 492, 493
Fan, K. 486
Farkas, J. 177, 181, 182, 287, 486
Ferguson, S. P. viii, 488
Fernando, K. V. 405, 487
Fey, M. 490
Fiedler, M. 160, 287, 433, 487
Filippov, A. F. ix, 437, 487
Fischer, E. 47
Fischer, H. C. 415, 416, 487
Floquet, G. 123
Forsythe, G. E. x, 365, 371, 451, 456, 487
Fourier, J. 164
Franklin, J. N. 72, 442, 444, 487
Fredholm, E. I. 259
Friedland, S. 401, 403, 487
Frobenius, F. G. 10, 12, 58, 62, 343
Frommer, A. viii, 27, 287, 327, 333, 334, 344, 483, 484, 487
Fuchssteiner, B. 491

Gantmacher, F. R. 214, 487
Garey, M. E. 487
Gargantini, J. 432, 487
Garloff, J. 252, 287, 487, 488
Gauss, C. F. 191, 239, 241, 248

George, A. 488
Gerling, C. L. 239
Gershgorin, S. A. ix, 48, 343, 345, 363
Gienger, A. 315, 318, 401, 405, 433, 484, 488
Gilányi, A. 493
Givens, J. W. 55
Goldstejn, A. 324, 488
Golub, G. H. x, 73, 488
Gram, J. P. 366
Greenbaum, A. 485
Gregory, R. T. 360, 369, 391, 482, 488
Griewank, A. 415, 416, 486, 488
Grüner, K. 405, 488
Guu, S.-M. 123, 494

Hadamard, J. 110, 262, 463
Hales, T. viii, 488
Hammarling, S. 485, 486
Hammer, D. 433, 488
Hammer, R. 416, 488
Hämmerlin, G. 72, 488
Hansen, E. ix, 157, 287, 310, 320, 325, 340, 344, 488, 490
Hanson, R. J. 486, 490
Hardy, G. H. 488
Hargreaves, G. I. 469, 488
Hartfiel, D. J. 161, 182, 488
Hashimoto, K. 433, 493
Hass, J. viii, 488
Hausdorff, F. 7, 93, 139
Hebgen, M. 229, 288, 488
Heindl, G. 27, 287, 344, 483, 488, 489
Heine, E. 5
Helmholtz, H. 415
Henrici, P. x, 365, 371, 432, 451, 456, 487, 489
Hermite, C. 18
Hertling, P. 489
Herzberger, J. x, 107, 157, 160, 288, 303, 344, 404, 424, 425, 432, 433, 484–486, 489, 491, 494–496
Hessenberg, K. 208, 239, 287
Heuser, H. x, 4, 72, 489
Higham, J. N. 103, 489
Hilbert, D. 369, 481
Hille, E. 489
Hladík, M. x, 166, 175, 287, 489
Hocks, M. 416, 488
Hoffmann, K.-H. 72, 488
Hoffmann, R. 405, 484
Hooke, R. 398
Horn, R. A. 72, 489
Householder, A. S. 43
Hoxha, F. 333, 334, 344, 487

Istrăţescu, V. I. 489

Jackson, K. R. 433, 493
Jacobi, C. G. J. ix, 240, 241, 248, 345, 363, 371, 451
Jahn, K.-U. 485
Jansson, C. 162, 176, 283, 287, 288, 433, 489
Jeltsch, R. 490
Johnson, C. R. 72, 489
Johnson, D. S. 487
Jordan, D. W. 489
Jordan, M. E. C. ix, 29, 437

Kahl, P. T. 490
Kantorovich, L. x, 26, 72, 307, 315, 329, 445, 489
Karney, D. L. 360, 369, 391, 482, 488
Kaucher, E. 433, 489, 491
Kaufman, L. 40, 382, 486
Kearfott, R. B. 287, 433, 489, 493, 494
Kearfott, R B. 484
Kelch, R. 433, 489
Keppler, H. 433, 483
Kincaid, D. 490
Kioustelidis, J. B. 325, 326, 344, 489, 493
Klatte, R. 152, 366, 377, 489
Klein, W. 433, 490
Koeber, M. 72, 141, 433, 490
König, S. 157, 490
Krawczyk, R. ix, 240, 278, 303, 307, 314, 320, 325, 338, 344, 345, 349, 405, 490
Kreinovich, V. 160, 166, 176, 287, 405, 484, 490, 495
Kreß, N. 369, 405
Kreß, O. 490
Krier, N. 426, 490
Krogh, F. T. 490
Kronecker, L. 2
Krückeberg, F. 433, 490
Krylov, N. 46
Krückeberg, F. 433
Kulisch, U. x, 105, 107, 123, 366, 416, 433, 483, 485, 487–493, 496
Kuttler, J. 121, 490

Lakeyev, A. V. 163, 490
Lancaster, P. 496
Landau, E. 2
Lang, B. ix, 327, 333, 334, 344, 487
Lawo, C. 366, 489
Lawson, C. L. 469, 490
Lehmann, N. J. ix, 50, 73, 490, 491
Leontief, W. 69, 159, 162, 286, 491
Leray, J. 22
Liedl, R. 491
Lipschitz, R. 6, 29, 131, 142
Liu, J. W. 488
Ljapin, E. S. 82, 491
Lohner, R. ix, 105, 115, 135, 139, 152, 157, 363, 370, 405, 416, 433, 483, 484, 486, 487, 491–493
Lorenz, E. viii
Losonczi, L. 493
Love, E. R. 259, 491
Lur, Y.-Y. 123, 494
Luther, W. 484, 492

Madsen, K. 151, 157, 486
Maier, T. 433, 491
Markov, S. M. 489, 491
Mayer, G. 46, 123, 141, 157, 166, 175–177, 238, 265, 287, 288, 345, 360, 401, 404, 405, 433, 484, 485, 487, 489, 491, 492
Mayer, J. 27, 344, 483, 492
Mayer, O. 157, 244, 492
Mc Kenney, A. 485
Miranda, C. ix, 24, 72, 325, 338, 340, 492
Miranker, W. L. 105, 107, 433, 490, 496
Miyajima, S. 405, 492
Möbius, A. F. 425
Moler, C. B. 373, 486
Moore, E. H. 286
Moore, R. E. 106, 123, 127, 310, 325, 326, 344, 433, 492, 493
Motzkin, T. 164
Mühlig, H. 486
Muller, J.-M. 461, 493
Musiol, G. 486

Nagatou, K. 433, 493, 498
Nakao, M. T. 433, 493, 498
Neaga, M. 366, 489
Nedialkov, N. S. 433, 493
Nedoma, J. 487
Neher, M. 72, 433, 493
Neumaier 195
Neumaier, A. x, 46, 72, 97, 107, 123, 157, 164, 177, 194, 196, 214, 220, 228, 239, 278, 280, 287, 330, 342, 344, 416, 493, 494, 497
Neumann, C. 31
Newton, I. vii, 26, 72, 289, 293, 315, 329, 398, 445
Nickel, K. 157, 344, 433, 485, 494, 495
Ning, S. 287, 494
Nocedal, J. 403, 487
Noskov, S. I. 163, 490
Nuding, E. 123, 187, 485

Oelschlägel, D. 485
Oettli, W. 161, 464, 494
Ogita, T. 405, 492, 496
Oishi, S. 288, 405, 492, 494, 496
Ortega, J. M. x, 20, 55, 72, 494
Ostrowski, A. 69, 110
Otten, W. 492
Overton, M. L. 403, 487

Páles, Z. 493
Pang, C.-T. 123, 494
Parlett, B. N. 40, 382, 486
Pascal, B. 286
Penrose, R. 286
Perron, O. 58
Petković, L. D. 432, 494
Petković, M. S. 432, 494
Petras, K. 433, 494
Pieper, L. 287, 492
Pissanetzky, S. 36, 494
Platzöder, L. 344, 484
Plemmons, R. J. 73, 486
Plum, M. 433, 493, 494, 498
Poincaré, H. 47
Pont, M. W. 405, 487
Popova, E. D. 106, 287, 494
Potra, F. 315, 318, 344, 433, 483–485
Prager, W. 161, 464, 494

Qi, L. 310, 325, 493, 494

Rall, L. B. 307, 329, 331, 344, 416, 494
Ramík, J. 487
Ratschek, H. 82, 84, 106, 107, 157, 433, 494, 495

Ratz, D. 366, 416, 433, 488, 489, 495
Rauch, M. 366, 489
Rayleigh, J. W. S. 46
Reichmann, K. 219, 220, 235, 287, 495
Reinsch, C. 381, 498
Rheinboldt, W. C. x, 20, 72, 494
Richardson, L. 241, 248, 261
Rihm, R. 115, 433, 495
Ris, F. N. 107, 495
Ritz, W. 47
Roesler, F. 361, 495
Rohn, J. 160, 176, 179, 183, 228, 286, 287, 487, 490, 492, 495
Rokne, J. 107, 157, 433, 495, 496
Rump, S. M. x, 105, 107, 151, 157, 228, 288, 363, 404, 405, 424, 426, 469, 470, 482, 487, 492, 494, 496

Sauer, W. 495
Schäfer, U. 288, 344, 433, 485, 496
Schauder, J. 22
Schlafly, R. viii, 488
Schmidt, E. 366
Schnurr, M. 344, 487, 496
Schrijver, A. 164, 287, 496
Schröder, J. 496
Schulte, U. 433, 496
Schulz-Rinne, C. 433, 489
Schur, I. 32, 55
Schwandt, H. 287, 298, 497
Seidel, L. 240, 241, 248
Semendjajew, K. A. 486
Sengupta, S. ix, 310, 320, 325, 340, 344, 488
Sharaya, I. A. 287, 497
Shary, S. P. 164, 287, 497
Shen, Z. 330, 342, 344, 485, 494, 497
Shi, Y. 344, 485
Smale, S. viii
Smith, P. 489
Sorensen, D. 485
Spaniol, O. 106, 497
Sperner, E. 439, 442
Spreuer, H. ix, 373, 375, 377, 485
Starzhinskii, V. M. 123, 498
Stetter, H. J. 105, 433, 480, 485, 491, 493, 496, 497
Stieltjes, T. J. 65, 235
Stoer, J. 19, 72, 286, 497
Strassen, V. 415, 485
Strubecker, K. 428, 497
Sturm, J. C. F. 43
Suchowitzki, S. I. 287, 497
Sunaga, T. 106, 123, 344, 497
Süsse, H. 485

te Riele, H. J. J. 491
Temple, G. 46
Toeplitz, O. 164
Tucker, W. A. viii, 416, 497

Ullrich, C. 100, 107, 366, 469, 486–491, 497
Unger, H. 404, 497

van de Lune, J. 491
van Leeuwen, J. 497
van Loan, C. F. x, 73, 488
Varga, R. S. x, 73, 497
Volder, J. 104, 459, 497
Vrahatis, M. N. 72, 497

Walster, G. W. 488
Wang, Z. 433, 498
Warnke, I. 288, 492
Watanabe, Y. 433, 498
Weierstrass, K. 15, 431
Weinstein, D. H. 46
Weyl, H. 47
Wiebigke, V. 485
Wielandt, H. 46
Wiethoff, A. 366, 489
Wilkinson, J. H. ix, 54, 73, 345, 363, 370, 373, 381, 486, 498
Winter D. T. 491
Wolfe, M. A. 344, 497
Wolff von Gudenberg, J. 487, 488, 491
Wongwises, P. 228, 288, 498

Yakubovich, V. A. 123, 498
Yamamoto, N. 433, 493
Yamamoto, T. 487, 492
Young, R. C. 106, 498

Zielke, G. 482, 498
Zimmermann, K. 487

Subject Index

B-orthogonal 49
B-orthonormal 49, 385, 388
B-orthonormal basis 49
H-matrix 69
– properties 122
M-matrix 65, 200
– properties 122
M-splitting 243, 252
P-contraction ix, 140
– local 144
R-factor of a
– method 243
– sequence 243
R-order 243
β-function 97, 215
χ-function 82
– properties 82
ε-inflation ix, 151
l_1 norm 8
l_p norm 8

a-posteriori error estimate 4
a-priori error estimate 4
abs 388
absolute norm 9
absolute value 8, 9, 88, 109
– circular complex interval 428
– rectangular complex interval 420
absolute value of a matrix 56
AccDot 359, 363, 366–368, 480
AccSum 367, 480
adder 103
addition 103
affected entries 451
Alefeld method 309
Alefeld operator 309
Alefeld test 309
algorithm of Bunch/Kaufman/Parlett 40
– modified 382, 385
antisymmetric rounding 101
arithmetic
– circular 423
– optimal circular enclosing 424
– rectangular 418
arrowhead matrix 213
associated index 452
associativity 81, 111
asymptotic convergence factor 244
automatic differentiation ix, 407
– backward mode ix
– forward mode ix
AWA 115
AWA software package 152

backward mode 407, 413
backward substitution 193
ball 3
Banach space 8
Banach's fixed point theorem ix, 4
barycentric coordinates 441
barycentric subdivision 441
basis
– orthonormal 39
Behnke method 382
Bernstein polynomial 14
bias 100
binary number
– conversion 99
– finite 99
– representation 99
Birkhoff interpolation 19
bisection 384
bisection method 342
Bogolyubov 46
Bolzano's theorem 26
Boothroyd–Dekker matrix 482
Borsuk test 338
Borsuk's theorem ix, 21, 26, 72
boundary 3
boundary value problem 318
bounded sequence 4
bounded set 3
Brouwer's fixed point theorem ix, 22, 72, 439
Bunch/Kaufman/Parlett algorithm 40, 73, 384
– modified 382, 385

C-XSC 366
CAGD 14
Cauchy sequence 3
cell 366, 368, 441
cell array 105
center 3, 129
centered form viii, 128
centro-symmetric 177

centro-symmetric solution set 177
centroskew-symmetric 177
centroskew-symmetric solution set 177
characteristic polynomial 29, 36
Cholesky decomposition 45
Cholesky method ix
circular arithmetic 423
– optimal enclosing 424
circular complex interval 423
closed ball 3
closed set 3
closure of a set 3
cluster of eigenvalues 345, 346, 367, 369, 392
cocyclic indices 452
column cyclic Jacobi method 456
commutativity 81, 111
compact set 5
comparison matrix 69, 110
comparison theorem 59
compatible 11
complement 99
complementarity problem 433
complete metric space 4, 94
complex interval ix
– circular 423
– rectangular 418
condition number 482
congruence transformation 40
connected graph 211
continuity
– eigenvalues 37
– eigenvectors 37
– zeros of polynomials 13
continuous 5
– function 2
– interval function 93
contraction 4
contraction constant 4
control solution set 163
convergent interval sequence 93
convergent matrix 31, 115
convergent sequence 3
conversion error 102
convex hull 181
convex union 76
CORDIC algorithm x, 104, 457
coupled entries 451
Cramer's rule 38, 162
cycle 34, 211
cyclic matrix 62
cyclic normal form 62

decimal number 99
definiteness
– metric 2
– norm 7
degenerate interval 75
degenerate interval matrix 110
degree integral 21
denormalized floating point number 100
dependency of variables 128
dependent variables 128
determinant 1
diagonal matrix 121
diameter 75
– circular complex interval 428
– rectangular complex interval 419
differential equation 433
differentiation
– automatic 407
– backward mode 413
– numerical 407
– symbolic 407
dimension 109
direct method 191
direct splitting 247
directed graph of a matrix 34
discrete metric 3
distance 93, 109
– circular complex interval 428
– rectangular complex interval 419
distributivity 84
divergence 293, 302
division 104
domain 2
double 100
double bubble conjecture viii
double eigenvalues 371
downward directed rounding 101
downward rounding 102

edge 34
eigenpair 30, 345
– normalized 345
eigenvalue 30, 36
– algebraic multiplicity 30
– cluster 384
– double 345, 371
– geometric multiplicity 30

– multiple 38, 345, 369, 382
– multiplicity 30
– nearly double 371
– simple 38, 345, 391
eigenvalue problem ix, 345
– generalized ix, 379, 382, 405
– interval 345, 405
– inverse 398, 405
– non-algebraic 346
eigenvector 30
– left 34
elementary divisor 29
elementary function 97, 104, 457
elementary interval function viii, 98
elimination graph 211
empty
– product 2
– set 3
– sum 2
Euclidean norm 8, 12
exponent 100
expression 125

face 443
factor sequence 125
factorable function 125
Farkas lemma 177, 181
finite subcovering 5
fixed point 20
fixed point iteration
– general 315
fixed point theorem
– Banach ix
– Brouwer ix
floating point number 100
– denormalized 100
– normalized 100
format double 100
forward mode 407
forward substitution 193
Fourier–Motzkin elimination 164
Frobenius companion matrix 343
Frobenius norm 10, 12, 455

Gauss 239
Gaussian algorithm 191
Gauss–Seidel method 240, 241
– componentwise intersection 241
Gauss–Seidel-like Krawczyk method 310
general fixed point iteration 315
general position 441
generalized eigenvalue problem ix, 49, 346, 379, 382, 405
generalized eigenvector 30
generalized singular values 405
generator 372
geometric mean 318, 359
Gerling 239
Gershgorin disc 48
Gershgorin interval 368, 369, 392
– isolated 369
Gershgorin test 388
Gershgorin theorem ix, 48, 368, 388
Givens rotation 363
Gram–Schmidt orthogonalization 366
graph 211
– connected 211
– undirected 211
graph of a matrix 34

Hadamard product 110, 463
Hansen–Sengupta method ix, 320, 325
Hansen–Sengupta operator 320
Hansen–Sengupta test 321
Hausdorff distance
– properties 95
Hausdorff metric 93
Heine–Borel property 5
Helmholtz energy function 415
Hermite interpolation 18, 136
Hermitian matrix 12, 405
Hessenberg matrix 208, 239
hidden bit 100
Hilbert matrix viii, 369, 481
– preconditioned 369, 481
homogeneity 7

IEEE standard 754–1985 100
improper interval 433
inclusion isotony 79, 98, 126
inclusion monotony 80, 98, 111, 126
inclusion property 79, 98, 111, 126
index of inertia 39
inertia 39
– Sylvester's law 39
infimum 76
inner enclosure 281
input–output model 69, 159, 162
integral 433
integral division 99

integral equation 433
interior 75
interval 2, 75
– degenerate 75
– non-degenerate 75
interval H-matrix viii, 121
interval M-matrix viii, 121
interval arithmetic viii, 77
– properties 80
interval arithmetic evaluation viii, 125
interval Cholesky method 229
– feasibility 235
interval eigenvalue problem 345, 405
interval function viii, 75
interval Gaussian algorithm ix, 193, 388
– feasibility 199
– pivoting 228
interval hull ix, 76, 177
interval linear system ix, 160
interval matrix viii, 109
– degenerate 110
– irreducible 115
– non-degenerate 110
– reducible 115
– regular 112
– singular 112
– strongly regular 112
interval Newton method ix
– multidimensional case 298
– one-dimensional case 290
interval vector viii, 109
INTLAB x, 105, 283, 319, 359, 363, 366–368, 370, 388, 392, 401, 427, 469
inverse eigenvalue problem ix, 398, 405
inverse function theorem 440
inverse positive matrix 71, 121, 187
inverse stable matrix viii, 122, 186
irreducible interval matrix 115
irreducible matrix 35
irreducibly diagonally dominant 67, 200
iteration
– residual 280
iterative method
– convergent 243

Jacobi cycle 367
Jacobi method ix, 240, 241, 364, 451
– column cyclic 456
– row cyclic x, 451
Jacobi method for eigenpairs 363, 371
– row cyclic 371
– row cyclic variant 364
– threshhold variant 365
Jacobi rotation 363, 367
Jordan block 29, 31
Jordan chain 30
Jordan measurable set 20
Jordan normal form ix, 29, 37

Kepler conjecture viii
Krawczyk iteration
– slope based 315, 320
Krawczyk method ix, 240, 303, 304
– Gauss–Seidel-like 310
– residual form 278
Krawczyk operator 304
– slope based 314
Krawczyk residual iteration 308
Krawczyk test 304
Krawczyk-like method for eigenpairs 349
Krawczyk-method
– interval linear systems 278
Kronecker symbol 2
Krylov 46
Kuttler's theorem 121

Landau symbols 2
leading principal submatrix 1
left eigenvector 34
left singular vector 33, 393
Lehmann's theorem 50, 73
length 211
Leontief 69, 159, 162
Leray–Schauder's fixed point theorem 22, 26
lexicographic ordering 166
linear complementary problem 179
linear elementary divisor 29
Lipschitz continuous 6, 29, 131
list 295
localization theorems 46
locally quadratically convergent 289
Lohner method ix, 363, 370, 405
loop 34
Love integral equation 259
LU-decomposition 193

machine epsilon 102
machine integer 99
machine interval 104

machine interval arithmetic viii, 99
machine number
– integer 99
– maximal integer 99
machine precision 102, 317
magnitude 88
– properties 89
mantissa 100
MAPLE 106, 407
mapping degree ix, 20
– properties 21
mapping degree test 333
MATLAB x, 105, 106, 283, 338, 359, 363, 366, 368, 385, 392, 401, 427, 469
matrix
– absolute value 56
– Boothroyd–Dekker 482
– conjugate transposed 1
– convergent 31, 115
– cyclic 62
– diagonal 1
– directed graph 34
– graph 34
– Hermitian 1
– Hilbert viii, 369, 481
– identity 1
– inverse stable viii
– nonnegative 56
– orthogonal 1
– oscillatory 213
– partial ordering 56
– positive 56
– power 1
– primitive 62
– semi-convergent 31, 115
– signature 1
– skew-symmetric 1
– strongly regular 112
– symmetric 1, 345
– totally nonnegative 213
– totally positive 213
– trace 37
– transposed 1
– unitarian 1
– Wilkinson 370
– zero 1
matrix norm 10
– monotone 109
maximum norm 8
max–min principle 47
mean value form viii, 129
mean value theorem 28
metric space 2
midpoint 3, 75, 109
– properties 90
– rectangular complex interval 419
mignitude 88, 110
– properties 89
minimum degree 211
minor 1
min–max principle 47
Miranda test 327
modified algorithm of Bunch/Kaufman/Parlett 382, 385
modified Hansen–Sengupta operator 325
modified Hansen–Sengupta test 325
modified relative width 359, 361
moments 46
monotone matrix norm 109
monotone norm 9
monotone rounding 101
Moore test 306, 315
Moore–Kioustelidis test 326
Moore–Qi operator 310
Moore–Qi test 311
multiple eigenvalues 38, 345, 369, 382
multiple precision 105

nearly double eigenvalues 371
Neumann series 31
neutral element 81, 111
Newton method ix
– modified 294, 296
– multidimensional case 297
– one-dimensional case 289
– quadratic divergence 293, 302
– simplified 303
Newton operator 297
Newton test 299
Newton-like method 303
Newton–Kantorovich test 329
Newton–Kantorovich theorem 26, 72
node 34
non-algebraic eigenvalue problem 346
non-degenerate interval 75
non-degenerate interval matrix 110
nonlinear equation ix
nonnegative matrix viii, 56

norm 7
– Frobenius 455
norm equivalence 8
normal form of matrices 29
normalized eigenpair 345
normalized floating point number 100
normed linear space 7
notation 1
numerical differentiation 407

Oettli–Prager inequality 161, 464
Oettli–Prager theorem 161
– generalized 287
one's complement 99
open ball 3
open set 3
operator equation 433
operator norm 10
optimal outward rounding 105
optimization 433
order of approximation 130
orthogonal matrix 12
orthogonalization 366
orthonormal basis 39
oscillatory matrix 213
Ostrowski matrix 69, 110
outer enclosure 281
outward rounding 105
– optimal 105
overflow 101

partial differential equation 433
partial ordering 76
partial ordering of a matrix 56
Pascal matrix 286
PASCAL-XSC x, 366
path 34, 211
– length 211
permutation matrix 34
Perron vector 58
perskew-symmetric 177
perskew-symmetric solution set 177
persymmetric 175
persymmetric solution set 175, 177
pivot element 193
pivoting 193
point interval 75
point matrix 110
polynomial 12
positive definite 44, 382
positive matrix 56
power of a matrix 1
precision
– multiple 105
preconditioned Hilbert matrix 369, 481
prime counting function $\pi(x)$ 362
primitive matrix 62
principal minor 1
principal submatrix 1
principal vector 30

quadratic divergence 293, 302
quadratic system ix, 346
quotient theorem 59

radius 3, 75, 109
– inequalities 91
– particular cases 92
– properties 90
– rectangular complex interval 419
range 2, 97
Rayleigh quotient 46
rectangular arithmetic 418
rectangular complex interval 418
rectangular complex interval matrix 420
rectangular complex interval vector 420
reducible interval matrix 115
reducible matrix 35
reducible normal form 36
reduction rule 81
regular
– strongly 112, 199
regular interval matrix 112
– strongly 112
regular splitting 71, 243
Reichmann example 219
reorthogonalization 366
residual iteration 280
residual Krawczyk iteration 278
Richardson iteration 241, 261
Riemann hypothesis 361
Riemann's zeta function 362
right singular vector 33, 393
root convergence factor 243
rotated entries 451
rounding 101
– antisymmmetric 101
– by chopping 102, 103
– downward 102
– downward directed 101

– error 102
– monotone 101
– optimal outward 105
– outward 105
– to nearest 102, 103
– upward 102
– upward directed 101
row cyclic Jacobi method x, 451

Schur complement 195
Schur decomposition 32
Schur normal form 32, 37, 55
semi-convergent matrix 31, 115
sequentially compact 5
shift-and-add strategy 457
sign accord algorithm 180
– modification 188, 264
signature matrix 1
signed exponent 100
similarity transformation 40
simple eigenvalue 30, 38, 345, 391
simplex 441
– dimension 441
– nondegenerate 441
simplified Newton method 303
singular *M*-matrix 69
singular interval matrix 112
singular linear systems 193
singular value ix, 33, 393
– generalized 405
singular value decomposition 33, 393
singular vector 33
– left 393
– right 393
skew-symmetric 175
skew-symmetric solution set ix, 175, 177
slope 129
slope based Krawczyk iteration 315, 320
slope based Krawczyk operator 314
slope matrix 315
Smale's 14th problem viii
small of second order 282
solution set ix, 160
– characterizations 160
spectral condition number 482
spectral norm 12
spectral radius 10, 39
Sperner lemma 442
Sperner mapping 442
Sperner value 442
splitting 243
– *M*- 243
– direct 247
– regular 243
– triangular 243
square root of a matrix 44
stability
– initial value problem 114
stable 115
– asymptotically 115
stack 295
staggered correction format 105, 366, 480
Stieltjes matrix 65, 67
strict Miranda test 340
strictly diagonally dominant 67, 200
strong Alefeld test 309
strong Hansen–Sengupta test 321
strong Krawczyk test 304
strong Moore test 315
strong Moore–Qi test 311
strongly regular 112, 199
subdistributivity 81, 82, 111
– complex rectangular arithmetic 419
submultiplicativity 10
subtraction 103
support of a function 20
supremum 76
Sylvester's law of inertia 39, 385
symbolic differentiation 407
symmetric Gauss–Seidel method 256
symmetric interval matrix 110
symmetric matrix 345
symmetric solution set ix, x, 164
symmetry
– matrix 1
– metric 2

Taylor coefficient ix
Temple 46
Temple quotient 46
test
– Alefeld 309
– Borsuk 338
– Gershgorin 388
– Hansen–Sengupta 321
– Krawczyk 304
– mapping degree 333
– Miranda 327

– modified Hansen–Sengupta 325
– Moore 315
– Moore–Kioustelidis 326
– Moore–Qi 311
– Newton–Kantorovich 329
– strict Miranda test 340
theorem
– Borsuk ix
– Gershgorin 368, 388
– Miranda ix, 24, 26, 72
– Newton–Kantorovich x
– Perron and Frobenius 58
– Weyl 47
tolerance solution set 162
topology 3
– metric 3
totally nonnegative matrix 213
totally positive matrix 213
transpose 110
tree 211
triangular decomposition 195
triangular inequality
– metric 2
– norm 7
triangular matrix 121, 242
triangular splitting 243, 248
two's complement 99

ulp 104, 480
unaffected entries 451
underflow 101
undirected graph 211
undirected path 211
uniformly continuous 6
unit of the last place 104, 480
unit rounding 103
unit roundoff 103
upward directed rounding 101
upward rounding 102

Weierstrass method
– single step variant 431
– total step variant 431
Weierstrass' approximation theorem 15
Weinstein 46
width 75
– modified relative 359, 361
– rectangular complex interval 419
Wielandt 46
Wilkinson matrix 370
Wilkinson theorem 54, 363

zero 20
zero divisor free 81
zero-symmetric 75
zero-symmetric interval
– properties 82
zero-symmetric interval matrix 110, 120
zeta function 362

De Gruyter Studies in Mathematics

Volume 64
Dorina Mitrea, Irina Mitrea, Marius Mitrea, Michael Taylor
The Hodge-Laplacian. Boundary Value Problems on Riemannian Manifolds, 2016
ISBN 978-3-11-048266-9, e-ISBN 978-3-11-048438-0, Set-ISBN 978-3-11-048439-7

Volume 63
Evguenii A. Rakhmanov
Orthogonal Polynomials
ISBN 978-3-11-031385-7, e-ISBN 978-3-11-031386-4

Volume 62
Ulrich Krause
Positive Dynamical Systems in Discrete Time, 2015
ISBN 978-3-11-036975-5, e-ISBN 978-3-11-036569-6

Volume 61
Francesco Altomare, Mirella Cappelletti Montano, Vita Leonessa, Ioan Rasa
Markov Operators, Positive Semigroups and Approximation Processes, 2015
ISBN 978-3-11-037274-8, e-ISBN 978-3-11-036697-6

Volume 60
Vladimir A. Mikhailets, Alexandr A. Murach, Peter V. Malyshev
Hörmander Spaces, Interpolation, and Elliptic Problems, 2014
ISBN 978-3-11-029685-3, e-ISBN 978-3-11-029689-1

Volume 59
Jan de Vries
Topological Dynamical Systems, 2014
ISBN 978-3-11-034073-0, e-ISBN 978-3-11-034240-6, Set-ISBN 978-3-11-034241-3

Volume 58
Lubomir Banas, Zdzislaw Brzezniak, Mikhail Neklyudov, Andreas Prohl
Stochastic Ferromagnetism: Analysis and Numerics, 2014
ISBN 978-3-11-030699-6, e-ISBN 978-3-11-030710-8, Set-ISBN 978-3-11-030711-5

Bei Fragen zur Produktsicherheit wenden Sie sich bitte an:
If you have any questions regarding product safety,
please contact:

Walter de Gruyter GmbH
Genthiner Straße 13
10785 Berlin
productsafety@degruyterbrill.com